CAMBRIDGE LIBRARY COLLECTION

Books of enduring scholarly value

Mathematical Sciences

From its pre-historic roots in simple counting to the algorithms powering modern desktop computers, from the genius of Archimedes to the genius of Einstein, advances in mathematical understanding and numerical techniques have been directly responsible for creating the modern world as we know it. This series will provide a library of the most influential publications and writers on mathematics in its broadest sense. As such, it will show not only the deep roots from which modern science and technology have grown, but also the astonishing breadth of application of mathematical techniques in the humanities and social sciences, and in everyday life.

A Treatise on Universal Algebra

Alfred North Whitehead (1861-1947) was equally celebrated as a mathematician, a philosopher and a physicist. He collaborated with his former student Bertrand Russell on the first edition of Principia Mathematica (published in three volumes between 1910 and 1913), and after several years teaching and writing on physics and philosophy of science at University College London and Imperial College, was invited to Harvard to teach philosophy and the theory of education. A Treatise on Universal Algebra was published in 1898, and was intended to be the first of two volumes, though the second (which was to cover quaternions, matrices and the general theory of linear algebras) was never published. This book discusses the general principles of the subject and covers the topics of the algebra of symbolic logic and of Grassmann's calculus of extension.

Cambridge University Press has long been a pioneer in the reissuing of out-of-print titles from its own backlist, producing digital reprints of books that are still sought after by scholars and students but could not be reprinted economically using traditional technology. The Cambridge Library Collection extends this activity to a wider range of books which are still of importance to researchers and professionals, either for the source material they contain, or as landmarks in the history of their academic discipline.

Drawing from the world-renowned collections in the Cambridge University Library, and guided by the advice of experts in each subject area, Cambridge University Press is using state-of-the-art scanning machines in its own Printing House to capture the content of each book selected for inclusion. The files are processed to give a consistently clear, crisp image, and the books finished to the high quality standard for which the Press is recognised around the world. The latest print-on-demand technology ensures that the books will remain available indefinitely, and that orders for single or multiple copies can quickly be supplied.

The Cambridge Library Collection will bring back to life books of enduring scholarly value across a wide range of disciplines in the humanities and social sciences and in science and technology.

A Treatise on Universal Algebra

With Applications

ALFRED NORTH WHITEHEAD

CAMBRIDGE
UNIVERSITY PRESS

CAMBRIDGE UNIVERSITY PRESS

Cambridge New York Melbourne Madrid Cape Town Singapore São Paolo Delhi

Published in the United States of America by Cambridge University Press, New York

www.cambridge.org
Information on this title: www.cambridge.org/9781108001687

This edition first published 1898
This digitally printed version 2009

ISBN 978-1-108-00168-7

A TREATISE

ON

UNIVERSAL ALGEBRA.

A TREATISE

ON

UNIVERSAL ALGEBRA

WITH APPLICATIONS.

BY

ALFRED NORTH WHITEHEAD, M.A.

FELLOW AND LECTURER OF TRINITY COLLEGE, CAMBRIDGE.

VOLUME I.

CAMBRIDGE:
AT THE UNIVERSITY PRESS.
1898

PREFACE.

IT is the purpose of this work to present a thorough investigation of the various systems of Symbolic Reasoning allied to ordinary Algebra. The chief examples of such systems are Hamilton's Quaternions, Grassmann's Calculus of Extension, and Boole's Symbolic Logic. Such algebras have an intrinsic value for separate detailed study; also they are worthy of a comparative study, for the sake of the light thereby thrown on the general theory of symbolic reasoning, and on algebraic symbolism in particular.

The comparative study necessarily presupposes some previous separate study, comparison being impossible without knowledge. Accordingly after the general principles of the whole subject have been discussed in Book I. of this volume, the remaining books of the volume are devoted to the separate study of the Algebra of Symbolic Logic, and of Grassmann's Calculus of Extension, and of the ideas involved in them. The idea of a generalized conception of space has been made prominent, in the belief that the properties and operations involved in it can be made to form a uniform method of interpretation of the various algebras.

Thus it is hoped in this work to exhibit the algebras both as systems of symbolism, and also as engines for the investigation of the possibilities of thought and reasoning connected with the abstract general idea of space. A natural mode of comparison between the algebras is thus at once provided by the unity of the subject-matters of their interpretation. The detailed comparison of their symbolic structures has been adjourned to the second volume, in which it is intended to deal with Quaternions, Matrices, and the general theory of Linear Algebras. This comparative anatomy of the subject was originated by B. Peirce's paper on Linear Associative Algebra*, and has been carried forward by more recent investigations in Germany.

* First read before the National Academy of Sciences in Washington, 1871, and republished in the *American Journal of Mathematics*, vol. iv., 1881.

The general name to be given to the subject has caused me much thought: that finally adopted, Universal Algebra, has been used somewhat in this signification by Sylvester in a paper, *Lectures on the Principles of Universal Algebra*, published in the *American Journal of Mathematics*, vol. vi., 1884. This paper however, apart from the suggestiveness of its title, deals explicitly only with matrices.

Universal Algebra has been looked on with some suspicion by many mathematicians, as being without intrinsic mathematical interest and as being comparatively useless as an engine of investigation. Indeed in this respect Symbolic Logic has been peculiarly unfortunate; for it has been disowned by many logicians on the plea that its interest is mathematical, and by many mathematicians on the plea that its interest is logical. Into the quarrels of logicians I shall not be rash enough to enter. Also the nature of the interest which any individual mathematician may feel in some branch of his subject is not a matter capable of abstract argumentation. But it may be shown, I think, that Universal Algebra has the same claim to be a serious subject of mathematical study as any other branch of mathematics. In order to substantiate this claim for the importance of Universal Algebra, it is necessary to dwell shortly upon the fundamental nature of Mathematics.

Mathematics in its widest signification is the development of all types of formal, necessary, deductive reasoning.

The reasoning is formal in the sense that the meaning of propositions forms no part of the investigation. The sole concern of mathematics is the inference of proposition from proposition. The justification of the rules of inference in any branch of mathematics is not properly part of mathematics: it is the business of experience or of philosophy. The business of mathematics is simply to follow the rule. In this sense all mathematical reasoning is necessary, namely, it has followed the rule.

Mathematical reasoning is deductive in the sense that it is based upon definitions which, as far as the validity of the reasoning is concerned (apart from any existential import), need only the test of self-consistency. Thus no external verification of definitions is required in mathematics, as long as it is considered merely as mathematics. The subject-matter is not necessarily first presented to the mind by definitions: but no idea, which has not been completely defined as far as concerns its relations to other ideas involved in the subject-matter, can be admitted into the reasoning. Mathematical definitions are always to be construed as limitations as well as definitions;

namely, the properties of the thing defined are to be considered for the purposes of the argument as being merely those involved in the definitions.

Mathematical definitions either possess an existential import or are conventional. A mathematical definition with an existential import is the result of an act of pure abstraction. Such definitions are the starting points of applied mathematical sciences; and in so far as they are given this existential import, they require for verification more than the mere test of self-consistency.

Hence a branch of applied mathematics, in so far as it is applied, is not merely deductive, unless in some sense the definitions are held to be guaranteed a priori as being true in addition to being self-consistent.

A conventional mathematical definition has no existential import. It sets before the mind by an act of imagination a set of things with fully defined self-consistent types of relation. In order that a mathematical science of any importance may be founded upon conventional definitions, the entities created by them must have properties which bear some affinity to the properties of existing things. Thus the distinction between a mathematical definition with an existential import and a conventional definition is not always very obvious from the form in which they are stated. Though it is possible to make a definition in form unmistakably either conventional or existential, there is often no gain in so doing. In such a case the definitions and resulting propositions can be construed either as referring to a world of ideas created by convention, or as referring exactly or approximately to the world of existing things. The existential import of a mathematical definition attaches to it, if at all, quâ mixed mathematics; quâ pure mathematics, mathematical definitions must be conventional*.

Historically, mathematics has, till recently, been confined to the theories of Number, of Quantity (strictly so-called), and of the Space of common experience. The limitation was practically justified: for no other large systems of deductive reasoning were in existence, which satisfied our definition of mathematics. The introduction of the complex quantity of ordinary algebra, an entity which is evidently based upon conventional definitions, gave rise to the wider mathematical science of to-day. The realization of wider conceptions has been retarded by the habit of mathematicians, eminently useful and indeed necessary for its own purposes, of extending all names to apply to new ideas as they arise. Thus the name

* Cf. Grassmann, *Ausdehnungslehre von* 1844, Einleitung.

of quantity was transferred from the quantity, strictly so called, to the generalized entity of ordinary algebra, created by conventional definition, which only includes quantity (in the strict sense) as a special case.

Ordinary algebra in its modern developments is studied as being a large body of propositions, inter-related by deductive reasoning, and based upon conventional definitions which are generalizations of fundamental conceptions. Thus a science is gradually being created, which by reason of its fundamental character has relation to almost every event, phenomenal or intellectual, which can occur. But these reasons for the study of ordinary Algebra apply to the study of Universal Algebra; provided that the newly invented algebras can be shown either to exemplify in their symbolism, or to represent in their interpretation interesting generalizations of important systems of ideas, and to be useful engines of investigation. Such algebras are mathematical sciences, which are not essentially concerned with number or quantity; and this bold extension beyond the traditional domain of pure quantity forms their peculiar interest. The ideal of mathematics should be to erect a calculus to facilitate reasoning in connection with every province of thought, or of external experience, in which the succession of thoughts, or of events can be definitely ascertained and precisely stated. So that all serious thought which is not philosophy, or inductive reasoning, or imaginative literature, shall be mathematics developed by means of a calculus.

It is the object of the present work to exhibit the new algebras, in their detail, as being useful engines for the deduction of propositions; and in their several subordination to dominant ideas, as being representative symbolisms of fundamental conceptions. In conformity with this latter object I have not hesitated to compress, or even to omit, developments and applications which are not allied to the dominant interpretation of any algebra. Thus unity of idea, rather than completeness, is the ideal of this book. I am convinced that the comparative neglect of this subject during the last forty years is partially due to the lack of unity of idea in its presentation.

The neglect of the subject is also, I think, partially due to another defect in its presentation, which (for the want of a better word) I will call the lack of independence with which it has been conceived. I will proceed to explain my meaning.

Every method of research creates its own applications: thus Analytical Geometry is a different science from Synthetic Geometry, and both these sciences are different from modern Projective Geometry. Many propositions are identical in all three sciences, and the general subject-matter, Space, is the same throughout. But it would be a serious mistake in the development of one of the three merely to take a list of the propositions as they occur in the others, and to endeavour to prove them by the methods of the one in hand. Some propositions could only be proved with great difficulty, some could hardly even be stated in the technical language, or symbolism, of the special branch. The same applies to the applications of the algebras in this book. Thus Grassmann's Algebra, the Calculus of Extension, is applied to Descriptive Geometry, Line Geometry, and Metrical Geometry, both non-Euclidean and Euclidean. But these sciences, as here developed, are not the same sciences as developed by other methods, though they apply to the same general subject-matter. Their combination here forms one new and distinct science, as distinct from the other sciences, whose general subject-matters they deal with, as is Analytical Geometry from Pure Geometry. This distinction, or independence, of the application of any new algebra appears to me to have been insufficiently realized, with the result that the developments of the new Algebras have been cramped.

In the use of symbolism I have endeavoured to be very conservative. Strange symbols are apt to be rather an encumbrance than an aid to thought: accordingly I have not ventured to disturb any well-established notation. On the other hand I have not hesitated to introduce fresh symbols when they were required in order to express new ideas.

This volume is divided into seven books. In Book I. the general principles of the whole subject are considered. Book II. is devoted to the Algebra of Symbolic Logic; the results of this book are not required in any of the succeeding books of this volume. Book III. is devoted to the general principles of addition and to the theory of a Positional manifold, which is a generalized conception of Space of any number of dimensions without the introduction of the idea of distance. The comprehension of this book is essential in reading the succeeding books. Book IV. is devoted to the principles of the Calculus of Extension. Book V. applies the Calculus of Extension to the theory of forces in a Positional manifold of three dimensions. Book VI. applies the Calculus of Extension to Non-Euclidean Geometry, considered, after Cayley, as being the most general theory of distance in a Positional manifold; the comprehension of this book is not necessary in reading the succeeding book. Book VII. applies the Calculus of Extension to ordinary Euclidean Space of three dimensions.

It would have been impossible within reasonable limits of time to have made an exhaustive study of the many subjects, logical and mathematical, on which this volume touches; and, though the writing of this volume has been continued amidst other avocations since the year 1890, I cannot pretend to have done so. In the subject of pure Logic I am chiefly indebted to Mill, Jevons, Lotze, and Bradley; and in regard to Symbolic Logic to Boole, Schröder and Venn. Also I have not been able in the footnotes to this volume adequately to recognize my obligations to De Morgan's writings, both logical and mathematical. The subject-matter of this volume is not concerned with Quaternions; accordingly it is the more necessary to mention in this preface that Hamilton must be regarded as a founder of the science of Universal Algebra. He and De Morgan (cf. note, p. 131) were the first to express quite clearly the general possibilities of algebraic symbolism.

The greatness of my obligations in this volume to Grassmann will be understood by those who have mastered his two *Ausdehnungslehres*. The technical development of the subject is inspired chiefly by his work of 1862, but the underlying ideas follow the work of 1844. At the same time I have tried to extend his Calculus of Extension both in its technique and in its ideas. But this work does not profess to be a complete interpretation of Grassmann's investigations, and there is much valuable matter in his *Ausdehnungslehres* which it has not fallen within my province to touch upon. Other obligations, as far as I am aware of them, are mentioned as they occur. But the book is the product of a long preparatory period of thought and miscellaneous reading, and it was only gradually that the subject in its full extent shaped itself in my mind; since then the various parts of this volume have been systematically deduced, according to the methods appropriate to them here, with hardly any aid from other works. This procedure was necessary, if any unity of idea was to be preserved, owing to the bewildering variety of methods and points of view adopted by writers on the various subjects of this volume. Accordingly there is a possibility of some oversights, which I should very much regret, in the attribution of ideas and methods to their sources. I should like in this connection to mention the names of Arthur Buchheim and of Homersham Cox as the mathematicians whose writings have chiefly aided me in the development of the Calculus of Extension (cf. notes, pp. 248, 346, 370, and 575). In the development of Non-Euclidean Geometry the ideas of Cayley, Klein, and Clifford have been chiefly followed; and in the development of the theory of Systems of Forces I am indebted to Sir R. S. Ball, and to Lindemann.

I have added unsystematically notes to a few theorems or methods, stating that they are, as far as I know, now enunciated for the first time. These notes are unsystematic in the double sense that I have not made a systematic search in the large literatures of the many branches of mathematics with which this book has to do, and that I have not added notes to every theorem or method which happens to be new to me.

My warmest thanks for their aid in the final revision of this volume are due to Mr Arthur Berry, Fellow of King's College, to Mr W. E. Johnson, of King's College, and Lecturer to the University in Moral Science, to Prof. Forsyth, Sadlerian Professor to the University, who read the first three books in manuscript, and to the Hon. B. Russell, Fellow of Trinity College, who has read many of the proofs, especially in the parts connected with Non-Euclidean Geometry.

Mr Johnson not only read the proofs of the first three books, and made many important suggestions and corrections, but also generously placed at my disposal some work of his own on Symbolic Logic, which will be found duly incorporated with acknowledgements.

Mr Berry throughout the printing of this volume has spared himself no trouble in aiding me with criticisms and suggestions. He undertook the extremely laborious task of correcting all the proofs in detail. Every page has been improved either substantially or in expression owing to his suggestions. I cannot express too strongly my obligations to him both for his general and detailed criticism.

The high efficiency of the University Press in all that concerns mathematical printing, and the courtesy which I have received from its Officials, also deserve grateful acknowledgements.

CAMBRIDGE,

December, 1897.

CONTENTS.

The following Books and Chapters are not essential for the comprehension of the subsequent parts of this volume: Book II, Chapter V of Book IV, Book VI.

BOOK I.

PRINCIPLES OF ALGEBRAIC SYMBOLISM.

CHAPTER I.

ON THE NATURE OF A CALCULUS.

CHAPTER II.

MANIFOLDS.

CHAPTER III.

Principles of Universal Algebra.

BOOK II.

THE ALGEBRA OF SYMBOLIC LOGIC.

CHAPTER I.

The Algebra of Symbolic Logic.

CHAPTER II.

The Algebra of Symbolic Logic (*continued*).

CHAPTER III.

Existential Expressions.

CHAPTER IV.

Application to Logic.

CHAPTER V.

Propositional Interpretation.

BOOK III.

POSITIONAL MANIFOLDS.

CHAPTER I.

FUNDAMENTAL PROPOSITIONS.

CHAPTER II.

STRAIGHT LINES AND PLANES.

CHAPTER III.

QUADRICS.

CHAPTER IV.

INTENSITY.

BOOK IV.

CALCULUS OF EXTENSION.

CHAPTER I.

COMBINATORIAL MULTIPLICATION.

CHAPTER II.

REGRESSIVE MULTIPLICATION.

CHAPTER III.

SUPPLEMENTS.

CHAPTER IV.

DESCRIPTIVE GEOMETRY.

CHAPTER V.

DESCRIPTIVE GEOMETRY OF CONICS AND CUBICS.

CHAPTER VI.

MATRICES.

BOOK V.

EXTENSIVE MANIFOLDS OF THREE DIMENSIONS.

CHAPTER I.

SYSTEMS OF FORCES.

CHAPTER II.

Groups of Systems of Forces.

CHAPTER III.

Invariants of Groups.

CHAPTER IV.

Matrices and Forces.

BOOK VI.

THEORY OF METRICS.

CHAPTER I.

Theory of Distance.

CHAPTER II.

Elliptic Geometry.

CHAPTER III.

Extensive Manifolds and Elliptic Geometry.

CHAPTER IV.

Hyperbolic Geometry.

CHAPTER V.

Hyperbolic Geometry (*continued*).

CHAPTER VI.

KINEMATICS IN THREE DIMENSIONS.

CHAPTER VII.

CURVES AND SURFACES.

CHAPTER VIII.

Transition to Parabolic Geometry.

BOOK VII.

APPLICATION OF THE CALCULUS OF EXTENSION TO GEOMETRY.

CHAPTER I.

Vectors.

CHAPTER II.

VECTORS (*continued*).

CHAPTER III.

CURVES AND SURFACES.

CHAPTER IV.

Pure Vector Formulæ.

BOOK I.

PRINCIPLES OF ALGEBRAIC SYMBOLISM.

CHAPTER I.

On the nature of a Calculus.

1. Signs. Words, spoken or written, and the symbols of Mathematics are alike signs. Signs have been analysed* into (α) suggestive signs, (β) expressive signs, (γ) substitutive signs.

A suggestive sign is the most rudimentary possible, and need not be dwelt upon here. An obvious example of one is a knot tied in a handkerchief to remind the owner of some duty to be performed.

In the use of expressive signs the attention is not fixed on the sign itself but on what it expresses; that is to say, it is fixed on the meaning conveyed by the sign. Ordinary language consists of groups of expressive signs, its primary object being to draw attention to the meaning of the words employed. Language, no doubt, in its secondary uses has some of the characteristics of a system of substitutive signs. It remedies the inability of the imagination to bring readily before the mind the whole extent of complex ideas by associating these ideas with familiar sounds or marks; and it is not always necessary for the attention to dwell on the complete meaning while using these symbols. But with all this allowance it remains true that language when challenged by criticism refers us to the meaning and not to the natural or conventional properties of its symbols for an explanation of its processes.

A substitutive sign is such that in thought it takes the place of that for which it is substituted. A counter in a game may be such a sign: at the end of the game the counters lost or won may be interpreted in the form of money, but till then it may be convenient for attention to be concentrated on the counters and not on their signification. The signs of a Mathematical Calculus are substitutive signs.

The difference between words and substitutive signs has been stated thus, 'a word is an instrument for thinking about the meaning which it

* Cf. Stout, 'Thought and Language,' *Mind*, April, 1891, repeated in the same author's *Analytic Psychology*, (1896), ch. x. § 1: cf. also a more obscure analysis to the same effect by C. S. Peirce, *Proc. of the American Academy of Arts and Sciences*, 1867, Vol. vii. p. 294.

expresses; a substitute sign is a means of not thinking about the meaning which it symbolizes*.' The use of substitutive signs in reasoning is to economize thought.

2. Definition of a Calculus. In order that reasoning may be conducted by means of substitutive signs, it is necessary that rules be given for the manipulation of the signs. The rules should be such that the final state of the signs after a series of operations according to rule denotes, when the signs are interpreted in terms of the things for which they are substituted, a proposition true for the things represented by the signs.

The art of the manipulation of substitutive signs according to fixed rules, and of the deduction therefrom of true propositions is a Calculus.

We may therefore define a sign used in a Calculus as 'an arbitrary mark, having a fixed interpretation, and susceptible of combination with other signs in subjection to fixed laws dependent upon their mutual interpretation†.'

The interpretation of any sign used in a series of operations must be fixed in the sense of being the same throughout, but in a certain sense it may be ambiguous. For instance in ordinary Algebra a letter x may be used in a series of operations, and x may be defined to be any algebraical quantity, without further specification of the special quantity chosen. Such a sign denotes any one of an assigned class with certain unambiguously defined characteristics. In the same series of operations the sign must always denote the same member of the class; but as far as any explicit definitions are concerned any member will do.

When once the rules for the manipulation of the signs of a calculus are known, the art of their practical manipulation can be studied apart from any attention to the meaning to be assigned to the signs. It is obvious that we can take any marks we like and manipulate them according to any rules we choose to assign. It is also equally obvious that in general such occupations must be frivolous. They possess a serious scientific value when there is a similarity of type of the signs and of the rules of manipulation to those of some calculus in which the marks used are substitutive signs for things and relations of things. The comparative study of the various forms produced by variation of rules throws light on the principles of the calculus. Furthermore the knowledge thus gained gives facility in the invention of some significant calculus designed to facilitate reasoning with respect to some given subject.

It enters therefore into the definition of a calculus properly so called that the marks used in it are substitutive signs. But when a set of marks and the rules for their arrangements and rearrangements are analogous to

* Cf. Stout, 'Thought and Language,' *Mind*, April, 1891.

† Boole, *Laws of Thought*, Ch. II.

those of a significant calculus so that the study of the allowable forms of their arrangements throws light on that of the calculus,—or when the marks and their rules of arrangement are such as appear likely to receive an interpretation as substitutive signs or to facilitate the invention of a true calculus, then the art of arranging such marks may be called—by an extension of the term—an uninterpreted calculus. The study of such a calculus is of scientific value. The marks used in it will be called signs or symbols as are those of a true calculus, thus tacitly suggesting that there is some unknown interpretation which could be given to the calculus.

3. Equivalence. It is necessary to note the form in which propositions occur in a calculus. Such a form may well be highly artificial from some points of view, and may yet state the propositions in a convenient form for the eliciting of deductions. Furthermore it is not necessary to assert that the form is a general form into which all judgments can be put by the aid of some torture. It is sufficient to observe that it is a form of wide application.

In a calculus of the type here considered propositions take the form of assertions of equivalence. One thing or fact, which may be complex and involve an inter-related group of things or a succession of facts, is asserted to be equivalent in some sense or other to another thing or fact.

Accordingly the sign $=$ is taken to denote that the signs or groups of signs on either side of it are equivalent, and therefore symbolize things which are so far equivalent. When two groups of symbols are connected by this sign, it is to be understood that one group may be substituted for the other group whenever either occurs in the calculus under conditions for which the assertion of equivalence holds good.

The idea of equivalence requires some explanation. Two things are equivalent when for some purpose they can be used indifferently. Thus the equivalence of distinct things implies a certain defined purpose in view, a certain limitation of thought or of action. Then within this limited field no distinction of property exists between the two things.

As an instance of the limitation of the field of equivalence consider an ordinary algebraical equation, $f(x, y) = 0$. Then in finding $\frac{dy}{dx}$ by the formula, $\frac{dy}{dx} = -\frac{\partial f}{\partial x} \bigg/ \frac{\partial f}{\partial y}$, we may not substitute 0 for f on the right-hand side of the last equation, though the equivalence of the two symbols has been asserted in the first equation, the reason being that the limitations under which $f = 0$ has been asserted are violated when f undergoes partial differentiation.

The idea of equivalence must be carefully distinguished from that of mere identity*. No investigations which proceed by the aid of propositions merely asserting identities such as A is A, can ever result in anything but barren identities†. Equivalence on the other hand implies non-identity as its general case. Identity may be conceived as a special limiting case of equivalence. For instance in arithmetic we write, $2+3=3+2$. This means that, in so far as the total number of objects mentioned, $2+3$ and $3+2$ come to the same number, namely 5. But $2+3$ and $3+2$ are not identical; the order of the symbols is different in the two combinations, and this difference of order directs different processes of thought. The importance of the equation arises from its assertion that these different processes of thought are identical as far as the total number of things thought of is concerned.

From this arithmetical point of view it is tempting to define equivalent things as being merely different ways of thinking of the same thing as it exists in the external world. Thus there is a certain aggregate, say of 5 things, which is thought of in different ways, as $2+3$ and as $3+2$. A sufficient objection to this definition is that the man who shall succeed in stating intelligibly the distinction between himself and the rest of the world will have solved the central problem of philosophy. As there is no universally accepted solution of this problem, it is obviously undesirable to assume this distinction as the basis of mathematical reasoning.

Thus from another point of view all things which for any purpose can be conceived as equivalent form the extension (in the logical sense) of some universal conception. And conversely the collection of objects which together form the extension of some universal conception can for some purpose be treated as equivalent. So $b=b'$ can be interpreted as symbolizing the fact that the two individual things b and b' are two individual cases of the same general conception B‡. For instance if b stand for $2+3$ and b' for $3+2$, both b and b' are individual instances of the general conception of a group of five things.

The sign $=$ as used in a calculus must be discriminated from the logical copula 'is.' Two things b and b' are connected in a calculus by the sign $=$, so that $b=b'$, when both b and b' possess the attribute B. But we may not translate this into the standard logical form, b is b'. On the contrary, we say, b is B, and b' is B; and we may not translate these standard forms of formal logic into the symbolic form, $b=B$, $b'=B$; at least we may not do so, if the sign $=$ is to have the meaning which is assigned to it in a calculus.

It is to be observed that the proposition asserted by the equation, $b=b'$, consists of two elements; which for the sake of distinctness we will name, and will call respectively the 'truism' and the 'paradox.' The truism is the partial identity of both b and b', their common B-ness. The paradox is the

* Cf. Lotze, *Logic*, Bk. I. Ch. II. Art. 64.

† Cf. Bradley, *Principles of Logic*, Bk. I. Ch. V.

‡ *Ibid.* Bk. II. Pt. I. Ch. IV. Art. 3 (β).

distinction between b and b', so that b is one thing and b' is another thing: and these things, as being different, must have in some relation diverse properties. In assertions of equivalence as contained in a calculus the truism is passed over with the slightest possible attention, the main stress being laid on the paradox. Thus in the equation $2+3=3+2$, the fact that both sides represent a common five-ness of number is not even mentioned explicitly. The sole direct statement is that the two *different* things $3+2$ and $2+3$ are in point of number equivalent.

The reason for this unequal distribution of attention is easy to understand. In order to discover new propositions asserting equivalence it is requisite to discover easy marks or tests of equivalent things. These tests are discovered by a careful discussion of the truism, of the common B-ness of b and b'. But when once such tests have been elaborated, we may drop all thought of the essential nature of the attribute B, and simply apply the superficial test to b and b' in order to verify $b=b'$. Thus in order to verify that thirty-seven times fifty-six is equal to fifty-six times thirty-seven, we may use the entirely superficial test applicable to this case that the same factors are mentioned as multiplied, though in different order.

This discussion leads us at once to comprehend the essence of a calculus of substitutive signs. The signs are by convention to be considered equivalent when certain conditions hold. And these conditions when interpreted imply the fulfilment of the tests of equivalence.

Thus in the discussion of the laws of a calculus stress is laid on the truism, in the development of the consequences on the paradox.

4. Operations. Judgments of equivalence can be founded on direct perception, as when it is judged by direct perception that two different pieces of stuff match in colour. But the judgment may be founded on a knowledge of the respective derivations of the things judged to be equivalent from other things respectively either identical or equivalent. It is this process of derivation which is the special province of a calculus. The derivation of a thing p from things $a, b, c, \dots$, can also be conceived as an operation on the things $a, b, c, \dots$, which produces the thing p. The idea of derivation includes that of a series of phenomenal occurrences. Thus two pieces of stuff may be judged to match in colour because they were dyed in the same dipping, or were cut from the same piece of stuff. But the idea is more general than that of phenomenal sequence of events: it includes purely logical activities of the mind, as when it is judged that an aggregate of five things has been presented to the mind by two aggregates of three things and of two things respectively. Another example of derivation is that of two propositions a and b which are both derived by strict deductive reasoning from the same propositions c, d, and e. The two propositions are either both proved or both unproved according as c, d, and e are granted or disputed. Thus a and b are so far equivalent. In other words a and b may be considered as the equivalent results of two operations on c, d and e.

The words operation, combination, derivation, and synthesis will be used to express the same general idea, of which each word suggests a somewhat specialized form. This general idea may be defined thus: A thing a will be said to result from an operation on other things, c, d, e, etc., when a is presented to the mind as the result of the presentations of c, d and e, etc. under certain conditions; and these conditions are phenomenal events or mental activities which it is convenient to separate in idea into a group by themselves and to consider as defining the nature of the operation which is performed on c, d, e, etc.

Furthermore the fact that c, d, e, etc. are capable of undergoing a certain operation involving them all will be considered as constituting a relation between c, d, e, etc.

Also the fact that c is capable of undergoing an operation of a certain general kind will be considered as a property of c. Any additional specialization of the kind of operation or of the nature of the result will be considered as a mode of that property.

5. Substitutive Schemes. Let a, a', etc., b, b', etc., ...z, z', etc., denote any set of objects considered in relation to some common property which is symbolized by the use of the italic alphabet of letters. The common property may not be possessed in the same mode by different members of the set. Their equivalence, or identity in relation to this property, is symbolized by a literal identity. Thus the fact that the things a and m' are both symbolized by letters from the italic alphabet is here a sign that the things have some property in common, and the fact that the letters a and m' are different letters is a sign that the two things possess this common property in different modes. On the other hand the two things a and a' possess the common property in the same mode, and as far as this property is concerned they are equivalent. Let the sign = express equivalence in relation to this property, then $a = a'$, and $m = m'$.

Let a set of things such as that described above, considered in relation to their possession of a common property in equivalent or in non-equivalent modes be called a *scheme* of things; and let the common property of which the possession by any object marks that object as belonging to the scheme be called the *Determining Property of the Scheme*. Thus objects belonging to the same scheme are equivalent if they possess the determining property in the same mode.

Now relations must exist between non-equivalent things of the scheme which depend on the differences between the modes in which they possess the determining property of the scheme. In consequence of these relations from things a, b, c, etc. of the scheme another thing m of the scheme can be derived by certain operations. The equivalence, $m = m'$, will exist between m and m', if m and m' are derived from other things of the scheme by operations which only differ in certain assigned modes. The modes in which processes of derivation of equivalent things m and m' from other things of the scheme can differ without destroying the equivalence of m and m' will be called the *Characteristics* of the scheme.

Now it may happen that two schemes of things—with of course different determining properties—have the same characteristics. Also it may be possible to establish an unambiguous correspondence between the things of the two schemes, so that if a, a', b, etc., belong to one scheme and α, α', β, etc., belong to the other, then a corresponds to α, a' to α', b to β and so on. The essential rule of the correspondence is that if in one scheme two things, say a and a', are equivalent, then in the other scheme their corresponding things α and α' are equivalent. Accordingly to any process of derivation in the italic alphabet by which m is derived from a, b, etc. there must correspond a process of derivation in the Greek alphabet by which μ is derived from α, β, etc.

In such a case instead of reasoning with respect to the properties of one scheme in order to deduce equivalences, we may substitute the other scheme, or conversely; and then transpose at the end of the argument. This device of reasoning, which is almost universal in mathematics, we will call the method of substitutive schemes, or more briefly, the method of substitution.

These substituted things belonging to another scheme are nothing else than substitutive signs. For in the use of substituted schemes we cease to think of the original scheme. The rule of reasoning is to confine thought to those properties, previously determined, which are shared in common with the original scheme, and to interpret the results from one set of things into the other at the end of the argument.

An instance of this process of reasoning by substitution is to be found in the theory of quantity. Quantities are measured by their ratio to an arbitrarily assumed quantity of the same kind, called the unit. Any set of quantities of one kind can be represented by a corresponding set of quantities of any other kind merely in so far as their numerical ratios to their unit are concerned. For the representative set have only to bear the same ratios to their unit as do the original set to their unit.

6. Conventional Schemes. The use of a calculus of substitutive signs in reasoning can now be explained.

Besides using substitutive schemes with naturally suitable properties, we may by convention assign to arbitrary marks laws of equivalence which are identical with the laws of equivalence of the originals about which we desire to reason. The set of marks may then be considered as a scheme of things with properties assigned by convention. The determining property of the scheme is that the marks are of certain assigned sorts arranged in certain types of sequence. The characteristics of the scheme are the conventional laws by which certain arrangements of the marks in sequence on paper are to be taken as equivalent. As long as the marks are treated as mutually determined by their conventional properties, reasoning concerning the marks will hold good concerning the originals for which the marks are substitutive signs. For instance in the employment of the marks x, y, $+$, the equation, $x+y=y+x$, asserts that a certain union on paper of x and y possesses the conventional quality that the order of x and y is indifferent. Therefore any union of two things with a result independent of any precedence of one thing before the other possesses so far properties identical with those of the union above set down between x and y. Not only can the reasoning be transferred from the originals to the substitutive signs, but the imaginative thought itself can in a large measure be avoided. For whereas combinations of the original things are possible only in thought and by an act of the imagination, the combinations of the conventional substitutive signs of a calculus are physically made on paper. The mind has simply to attend to the rules for transformation and to use its experience and imagination to suggest likely methods of procedure. The rest is merely physical actual interchange of the signs instead of thought about the originals.

A calculus avoids the necessity of inference and replaces it by an external demonstration, where inference and external demonstration are to be taken in the senses assigned to them by F. H. Bradley*. In this connexion a demonstration is to be defined as a process of combining a complex of facts, the data, into a whole so that some new fact is evident. Inference is an ideal combination or construction within the mind of the reasoner which results in the intuitive evidence of a new fact or relation between the data. But in the use of a calculus this process of combination is externally performed by the combination of the concrete symbols, with the result of a new fact respecting the symbols which arises for sensuous perception†. When this new fact is treated as a symbol carrying a meaning, it is found to mean the fact which would have been intuitively evident in the process of inference.

7. Uninterpretable Forms. The logical difficulty‡ involved in the use of a calculus only partially interpretable can now be explained. The

* Cf. Bradley, *Principles of Logic*, Bk II. Pt I. Ch. III.

† Cf. C. S. Peirce, *Amer. Journ. of Math.* Vol. VII. p. 182: 'As for algebra, the very idea of the art is that it presents formulæ which can be manipulated, and that by observing the effects of such manipulation we find properties not otherwise to be discovered.'

‡ Cf. Boole, *Laws of Thought*, Ch. V. § 4.

discussion of this great problem in its application to the special case of $(-1)^{\frac{1}{2}}$ engaged the attention of the leading mathematicians of the first half of this century, and led to the development on the one hand of the Theory of Functions of a Complex Variable, and on the other hand of the science here called Universal Algebra.

The difficulty is this: the symbol $(-1)^{\frac{1}{2}}$ is absolutely without meaning when it is endeavoured to interpret it as a number; but algebraic transformations which involve the use of complex quantities of the form $a+bi$, where a and b are numbers and i stands for the above symbol, yield propositions which do relate purely to number. As a matter of fact the propositions thus discovered were found to be true propositions. The method therefore was trusted, before any explanation was forthcoming why algebraic reasoning which had no intelligible interpretation in arithmetic should give true arithmetical results.

The difficulty was solved by observing that Algebra does not depend on Arithmetic for the validity of its laws of transformation. If there were such a dependence, it is obvious that as soon as algebraic expressions are arithmetically unintelligible all laws respecting them must lose their validity. But the laws of Algebra, though suggested by Arithmetic, do not depend on it. They depend entirely on the convention by which it is stated that certain modes of grouping the symbols are to be considered as identical. This assigns certain properties to the marks which form the symbols of Algebra. The laws regulating the manipulation of the algebraic symbols are identical with those of Arithmetic. It follows that no algebraic theorem can ever contradict any result which could be arrived at by Arithmetic; for the reasoning in both cases merely applies the same general laws to different classes of things. If an algebraic theorem is interpretable in Arithmetic, the corresponding arithmetical theorem is therefore true. In short when once Algebra is conceived as an independent science dealing with the relations of certain marks conditioned by the observance of certain conventional laws, the difficulty vanishes. If the laws be identical, the theorems of the one science can only give results conditioned by the laws which also hold good for the other science; and therefore these results, when interpretable, are true.

It will be observed that the explanation of the legitimacy of the use of a partially interpretable calculus does not depend upon the fact that in another field of thought the calculus is entirely interpretable. The discovery of an interpretation undoubtedly gave the clue by means of which the true solution was arrived at. For the fact that the processes of the calculus were interpretable in a science so independent of Arithmetic as is Geometry at once showed that the laws of the calculus might have been defined in reference to geometrical processes. But it was a paradox to assert that a science like Algebra, which had been studied for centuries without reference to Geometry, was after all dependent upon Geometry for its first principles. The step to the true explanation was then easily taken.

But the importance of the assistance given to the study of Algebra by the discovery of a complete interpretation of its processes cannot be over-estimated. It is natural to think of the substitutive set of things as assisting the study of the properties of the originals. Especially is this the case with a calculus of which the interest almost entirely depends upon its relation to the originals. But it must be remembered that conversely the originals give immense aid to the study of the substitutive things or symbols.

The whole of Mathematics consists in the organization of a series of aids to the imagination in the process of reasoning; and for this purpose device is piled upon device. No sooner has a substitutive scheme been devised to assist in the investigation of any originals, than the imagination begins to use the originals to assist in the investigation of the substitutive scheme. In some connexions it would be better to abandon the conception of originals studied by the aid of substitutive schemes, and to conceive of two sets of inter-related things studied together, each scheme exemplifying the operation of the same general laws. The discovery therefore of the geometrical representation of the algebraical complex quantity, though unessential to the logic of Algebra, has been quite essential to the modern developments of the science.

CHAPTER II.

MANIFOLDS.

8. MANIFOLDS. The idea of a manifold was first explicitly stated by Riemann*; Grassmann† had still earlier defined and investigated a particular kind of manifold.

Consider any number of things possessing any common property. That property may be possessed by different things in different modes: let each separate mode in which the property is possessed be called an element. The aggregate of all such elements is called the manifold of the property.

Any object which is specified as possessing a property in a given mode corresponds to an element in the manifold of that property. The element may be spoken of as representing the object or the object as representing the element according to convenience. All such objects may be conceived as equivalent in that they represent the same element of the manifold.

Various relations can be stated between one mode of a property and another mode; in other words, relations exist between two objects, whatever other properties they may possess, which possess this property in any two assigned modes. The relations will define how the objects necessarily differ in that they possess this property differently: they define the distinction between two sorts of the same property. These relations will be called relations between the various elements of the manifold of the property; and the axioms from which can be logically deduced the whole aggregate of such relations for all the elements of a given manifold are called the characteristics of the manifold.

The idea of empty space referred to coordinate axes is an example of a manifold. Each point of space represents a special mode of the common property of spatiality. The fundamental properties of space expressed in terms of these coordinates, i.e. all geometrical axioms, form the characteristics of this manifold.

* *Ueber die Hypothesen, welche der Geometrie zu Grunde liegen*, *Gesammelte Mathematische Werke*; a translation of this paper is to be found in Clifford's *Collected Mathematical Papers*.

† *Ausdehnungslehre von* 1844.

It is the logical deductions from the characteristics of a manifold which are investigated by means of a calculus. The manifolds of separate properties may have the same characteristics. In such a case all theorems which are proved for one manifold can be directly translated so as to apply to the other. This is only another mode of stating the ideas explained in Chapter I. §§ 3, 4, 5.

The relation of a manifold of elements to a scheme of things (cf. § 5), is that of the abstract to the concrete. Consider as explained in § 5 the scheme of things represented by a, a' etc., b, b' etc., ...z, z' etc. Then these concrete things are not elements of a manifold. But to such a scheme a manifold always corresponds, and conversely to a manifold a scheme of things corresponds. The abstract property of a common A-ness which makes the equivalence of a, a', etc., in the scheme is an element of the manifold which corresponds to this scheme. Thus the relation of a thing in a scheme to the corresponding element of the corresponding manifold is that of a subject of which the element can be predicated. If A be the element corresponding to a, a' etc., then a is A, and a' is A. Thus if we write $2+3=5$ at length, the assertion is seen to be

$$(1+1)+(1+1+1)=1+1+1+1+1;$$

this asserts that two methods of grouping the marks of the type 1 are equivalent as far as the common five-ness of the sum on each side.

The manifold corresponding to a scheme is the manifold of the determining property of the scheme. The characteristics of the manifold correspond to the characteristics of the scheme.

9. SECONDARY PROPERTIES OF ELEMENTS. In order to state the characteristics of a manifold it may be necessary to ascribe to objects corresponding to the elements the capability of possessing other properties in addition to that definite property in special modes which the elements represent. Thus for the purpose of expressing the relation of an element A of a manifold to the elements B and C it may be necessary to conceive an object corresponding to A which is either a_1 or a_2, or a_3, where the suffix denotes the possession of some other property, in addition to the A-ness of A, in some special mode which is here symbolized by the suffix chosen. Such a property of an object corresponding to A, which is necessary to define the relation of A to other elements of the manifold, is called a Secondary Property of the element A.

Brevity is gained by considering each element of the manifold, such as A, as containing within itself a whole manifold of its secondary properties. Thus with the above notation A stands for any one of A_1, A_2, A_3 etc., where the suffix denotes the special mode of the secondary property. Hence the object a_1, mentioned above, corresponds to A_1, and a_2 to A_2, and so on.

And the statement of the relation between two elements of the original manifold, such as A and B, requires the mention of a special A, say A_2 and of a special B, say B_4.

For example consider the manifold of musical notes conceived as representing every note so far as it differs in pitch and quality from every other note. Thus each element is a note of given pitch and given quality. The attribute of loudness is not an attribute which this manifold represents: but it is a secondary property of the elements. For consider a tone A and two of its overtones B and C, and consider the relations of A, B, C to a note P which is of the same pitch as A and which only involves the overtones B and C. Then P can be described as the pitch and quality of the sound produced by the simultaneous existence of concrete instances of A, B and C with certain relative loudnesses. Hence the relation of P to A, B, C requires the mention of the loudness of each element in order to express it. Thus if A_2, B_3, C_4 denote A, B, C with the required ratio of their loudnesses, P might be expressed as the combination of A_2, B_3, C_4.

The sole secondary property with which in this work we shall be concerned is that of intensity. Thus in some manifolds each element is to be conceived as the seat of a possible intensity of any arbitrarily assumed value, and this intensity is a secondary property necessary to express the various relations of the elements.

10. DEFINITIONS. To partition a manifold is to make a selection of elements possessing a common characteristic: thus if the manifold be a plane, a selection may be made of points at an equal distance from a given point. The selected points then form a circle. The selected elements of a partitioned manifold form another manifold, which may be called a submanifold in reference to the original manifold.

Again the common attribute C, which is shared by the selected elements of the original manifold A, may also be shared by elements of another manifold B. For instance in the above illustration other points in other planes may be at the same distance from the given point. We thus arrive at the conception of the manifold of the attribute C which has common elements with the manifolds A and B. This conception undoubtedly implies that the three manifolds A, B and C have an organic connection, and are in fact parts of a manifold which embraces them all three.

A manifold will be called the complete manifold in reference to its possible submanifolds; and the complete manifold will be said to contain its submanifolds. The submanifolds will be said to be incident in the complete manifold.

One submanifold may be incident in more than one manifold. It will then be called a common submanifold of the two manifolds. Manifolds will be said to intersect in their common submanifolds.

11. SPECIAL MANIFOLDS. A few definitions of special manifolds will both elucidate the general explanation of a manifold given above and will serve to introduce the special manifolds of which the properties are discussed in this work.

A manifold may be called self-constituted when only the properties which the elements represent are used to define the relations between elements; that is, when there are no secondary properties.

A manifold may be called extrinsically constituted when secondary properties have to be used to define these relations.

The manifold of integral numbers is self-constituted, since all relations of such numbers can be defined in terms of them.

A uniform manifold is a manifold in which each element bears the same relation as any other element to the manifold considered as a whole.

If such a manifold be a submanifold of a complete manifold, it is not necessary that each element of the uniform submanifold bear the same relation to the complete manifold as any other element of that submanifold.

Space, the points being elements, forms a uniform manifold. Again the perimeter of a circle, the points being elements, forms a uniform manifold. The area of a circle does not form a uniform manifold.

A simple serial manifold is a manifold such that the elements can be arranged in one series. The meaning of this property is that some determinate process of deriving the elements in order one from the other exists (as in the case of the successive integral numbers), and that starting from some initial element all the other elements of the manifold are derived in a fixed order by the successive application of this process. Since the process is determinate for a simple serial manifold, there is no ambiguity as to the order of succession of elements. The elements of such a manifold are not necessarily numerable. A test of a simple serial manifold is that, given any three elements of the manifold it may be possible to conceive their mutual relations in such a fashion that one of them can be said to lie between the other two. If a simple serial manifold be uniform it follows that any element can be chosen as the initial element.

A manifold may be called a complex serial manifold when all its elements belong to one or more submanifolds which are simple serial manifolds, but when it is not itself a simple serial manifold. A surface is such a manifold, while a line is a simple serial manifold.

Two manifolds have a one to one correspondence* between their elements if to every element of either manifold one and only one element of the other manifold corresponds, so that the corresponding elements bear a certain defined relation to each other.

* The subject of the correspondence between the elements of manifolds has been investigated by G. Cantor, in a series of memoirs entitled, 'Ueber unendliche, lineare Punktmannichfaltigkeiten,' *Math. Annalen*, Bd. 15, 17, 20, 21, 23, and *Borchardt's Journal*, Bd. 77, 84.

A quantitively defined manifold is such that each element is specified by a definite number of measurable entities of which the measures for any element are the algebraic quantities ξ, η, ζ, etc., so that the manifold has a one to one correspondence with the aggregate of sets of simultaneous values of these variables.

A quantitively defined manifold is a manifold of an algebraic function when each element represents in some way the value of an algebraic quantity w for a set of simultaneous values of ξ, η, ζ, etc., where w is a function of ξ, η, ζ, etc., in the sense that it can be constructed by definite algebraic operations on ξ, η, ζ, etc., regarded as irresoluble magnitudes, real or imaginary*.

A quantitively defined manifold in which the elements are defined by a single quantity ξ is a simple serial manifold as far as real values of ξ are concerned. For the elements can be conceived as successively generated in the order in which they occur as ξ varies from $-\infty$ to $+\infty$.

If an element of the manifold corresponds to each value of ξ as it varies continuously through all its values, then the manifold may be called continuous. If some values of ξ have no elements of the manifold corresponding to them, then the manifold may be called discontinuous.

A quantitively defined manifold depending on more than one quantity is a complex serial manifold. For if the quantities defining it ξ, η, ζ, etc. be put equal to arbitrary functions of any quantity τ, so that $\xi = f_1(\tau)$, $\eta = f_2(\tau)$, etc., then a submanifold is formed which is a quantitively defined manifold depending on the single quantity τ. This submanifold is therefore a simple serial manifold. But by properly choosing the arbitrary functions such a submanifold may be made to contain any element of the complete manifold. Hence the complete manifold is a complex serial manifold.

The quantitively defined manifold is continuous if an element corresponds to every set of values of the variables.

A quantitively defined manifold which requires for its definition the absolute values (as distinct from the ratios) of ν variables is said to be of ν dimensions.

A continuous quantitively defined manifold of ν dimensions may also be called a ν-fold extended continuous manifold†.

* Cf. Forsyth, *Theory of Functions*, Ch. I. §§ 6, 7.

† Cf. Riemann, loc. cit. section I. § 2.

CHAPTER III.

PRINCIPLES OF UNIVERSAL ALGEBRA.

12. INTRODUCTORY. Universal Algebra is the name applied to that calculus which symbolizes general operations, defined later, which are called Addition and Multiplication. There are certain general definitions which hold for any process of addition and others which hold for any process of multiplication. These are the general principles of any branch of Universal Algebra. These principles, which are few in number, will be considered in the present chapter. But beyond these general definitions there are other special definitions which define special kinds of addition or of multiplication. The development and comparison of these special kinds of addition or of multiplication form special branches of Universal Algebra. Each such branch will be called a special algebraic calculus, or more shortly, a special algebra, and the more important branches will be given distinguishing names. Ordinary algebra will, when there is no risk of confusion, be called simply algebra: but when confusion may arise, the term ordinary will be prefixed.

13. EQUIVALENCE. It has been explained in § 3 that the idea of equivalence requires special definition for any subject-matter to which it is applied. The definitions of the processes of addition and multiplication do carry with them this required definition of equivalence as it occurs in the field of Universal Algebra. One general definition holds both for addition and multiplication, and thus through the whole field of Universal Algebra. This definition may be framed thus: In any algebraic calculus only one recognized type of equivalence exists.

The meaning of this definition is that if two symbols a and a' be equivalent in that sense which is explicitly recognized in some algebraic calculus by the use of the symbol =, then either a or a' may be used indifferently in any series of operations of addition or multiplication of the type defined in that calculus.

This definition is so far from being obvious or necessary for any symbolic calculus, that it actually excludes from the scope of Universal Algebra the

Differential Calculus, excepting limited parts of it. For if $f(x, y)$ be a function of two independent variables x and y, and the equivalence $f(x, y)=0$, be asserted, then $\frac{\partial}{\partial x}f(x, y)$ and $\frac{\partial}{\partial y}f(x, y)$ are not necessarily zero, whereas $\frac{\partial}{\partial x}0$ and $\frac{\partial}{\partial y}0$ are necessarily zero. Hence the symbols $f(x, y)$ and 0 which are recognized by the sign of equality as equivalent according to one type of equivalence are not equivalent when submitted to some operations which occur in the calculus.

14. Principles of Addition. The properties of the general operation termed addition will now be gradually defined by successive specifications.

Consider a group of things, concrete or abstract, material things or merely ideas of relations between other things. Let the individuals of this group be denoted by letters $a, b \ldots z$. Let any two of the group of things be capable of a synthesis which results in some third thing.

Let this synthesis be of such a nature that all the properties which are attributed to any one of the original group of things can also be attributed to this result of the synthesis. Accordingly the resultant thing belongs to the original group.

Let the idea of *order* between the two things be attributable to their synthesis. Thus if a and b be the two things of which the synthesis is being discussed, orders as between a first or b first can be attributed to this synthesis. Also let only *two* possible alternative orders as between a and b be material, so as to be taken into explicit consideration when judging that things are or are not equivalent.

Let the result of the synthesis be unambiguous, in the sense that all possible results of a special synthesis in so far as the process is varied by the variation of non-apparent details are to be equivalent. It is to be noted in this connection that the properties of the synthesis which are explicitly mentioned cannot be considered as necessarily defining its nature unambiguously. The present assumption therefore amounts to the statement that the same words (or symbols) are always to mean the same thing, at least in every way which can affect equivalence.

This process of forming a synthesis between two things, such as a and b, and then of considering a and b, thus united, as a third resultant thing, may be symbolized by $a \frown b$*. Here the order is symbolized by the order in which a and b are mentioned; accordingly $a \frown b$ and $b \frown a$ symbolize two different things. Then by definition the only question of order as between a and b which can arise in this synthesis is adequately symbolized. Also $a \frown b$ whenever it occurs must always mean the same thing, or at least stand for some one of a set of equivalent things.

* Cf. Grassmann, *Ausdehnungslehre von* 1844, Preface.

Further $a \frown b$ is by assumption a thing capable of the same synthesis with any other of the things $a, b, \ldots x$. Accordingly we may write

$$p \frown (a \frown b) \quad \text{and} \quad (a \frown b) \frown p$$

to represent the two possible syntheses of the type involving p and $a \frown b$. The bracket is to have the usual meaning that the synthesis within the bracket is to be performed first and the resultant thing then to be combined as the symbols indicate.

According to the convention adopted here the symbol $a \frown b$ is to be read from left to right in the following manner: a is to be considered as given first, and b as joined on to it according to the manner prescribed by the symbol $\frown$. Thus $(a \frown b) \frown p$ means that the result of $a \frown b$ is first obtained and then p is united to it. But $a \frown b$ is obtained by taking a and joining b on to it. Thus the total process may equally well be defined by $a \frown b \frown p$. Hence, since both its right-hand and left-hand sides have been defined to have the same meaning, we obtain the equation

$$a \frown b \frown p = (a \frown b) \frown p.$$

Definition. Let any one of the symbols, either a single letter or a complex of letters, which denotes one of the group of things capable of this synthesis be called a term. Let the symbol $\frown$ be called the sign of the operation of this synthesis.

It will be noticed that this synthesis has essentially been defined as a synthesis between two terms, and that when three terms such as a, b, p, are indicated as subjects of the synthesis a sequence or time-order of the operations is also unambiguously defined. Thus in the syntheses $(a \frown b) \frown p$ there are two separate ideas of *order* symbolized; namely, the determined but unspecified idea of order of synthesis as between the two terms which is involved by hypothesis in the act of synthesis, and further the sequence of the two successive acts of synthesis, and this time-order involves the sequence in which the various terms mentioned are involved in the process. Thus $a \frown b \frown p$ and $p \frown (a \frown b)$ both involve that the synthesis $a \frown b$ is to be first performed and then the synthesis of $a \frown b$ and p according to the special order of synthesis indicated.

In the case of three successive acts of synthesis an ambiguity may arise. Consider the operations indicated in the symbols

$$a \frown b \frown c \frown d, \quad c \frown (a \frown b) \frown d.$$

No ambiguity exists in these two expressions; each of them definitely indicates that the synthesis $a \frown b$ is to be made first, then a synthesis with c, and then a synthesis of this result with d. Similarly each of the two expressions $d \frown (a \frown b \frown c)$, and $d \frown \{c \frown (a \frown b)\}$ indicates unambiguously the same sequence of operations, though in the final synthesis of d with the result of the previous syntheses the alternative order of synthesis is adopted to that adopted in the two previous examples.

But consider the expressions

$$(a \frown b) \frown (c \frown d) \quad \text{and} \quad (c \frown d) \frown (a \frown b).$$

Here the two syntheses $a \frown b$ and $c \frown d$ are directed to be made and then the resulting terms to be combined together. Accordingly there is an ambiguity as to the sequence in which these syntheses $a \frown b$, $c \frown d$ are to be performed. It has been defined however that $a \frown b$ and $c \frown d$ are always to be unambiguous and mean the same thing. This definition means that the synthesis $\frown$ depends on no previous history and no varying part of the environment. Accordingly $a \frown b$ is independent of $c \frown d$ and these operations may take place in any sequence of time.

The preceding definitions can be connected with the idea of a manifold. All equivalent things must represent the same element of the manifold. The synthesis $a \frown b$ is a definite unambiguous union which by hypothesis it is always possible to construct with any two things representing any two elements of the manifold. This synthesis, when constructed and represented by its result, represents some third element of the manifold. It is also often convenient to express this fact by saying that $a \frown b$ represents a relation between two elements of the manifold by which a third element of the manifold is generated; or that the term $a \frown b$ represents an element of the manifold. An element may be named after a term which represents it: thus the element x is the element represented by the term x. The same element might also be named after any term equivalent to x.

It is obvious that any synthesis of the two terms a and b may be conceived as an operation performed on one of them with the help of the other. Accordingly it is a mere change of language without any alteration of real meaning, if we sometimes consider $a \frown b$ as representing an operation performed on b or on a.

15. Addition. Conceive now that this synthesis which has been defined above is such that it follows the *Commutative* and *Associative* Laws.

The Commutative Law asserts that

$$a \frown b = b \frown a.$$

Hence the two possible orders of synthesis produce equivalent results.

It is to be carefully noticed that it would be erroneous to state the commutative law in the form that, order is not involved in the synthesis $a \frown b$. For if order is not predicable of the synthesis, then the equation, $a \frown b = b \frown a$, must be a proposition which makes no assertion at all. Accordingly it is essential to the importance of the commutative law that order should be involved in the synthesis, but that it should be indifferent as far as equivalence is concerned.

The Associative Law is symbolized by

$$a \frown b \frown c = a \frown (b \frown c);$$

where $a \frown b \frown c$ is defined in § 14.

The two laws combined give the property that the element of the manifold identified by three given terms in successive synthesis is independent of the order in which the three terms are chosen for the operation, and also of the internal order of each synthesis.

Let a synthesis with the above properties be termed *addition*; and let the manifold of the corresponding type be called an *algebraic* manifold; and let a scheme of things representing an algebraic manifold be called an algebraic scheme. Let addition be denoted by the sign +. Accordingly it is to be understood that the symbol $a+b$ represents a synthesis in which the above assumptions are satisfied.

The properties of this operation will not be found to vary seriously in the different algebras. The great distinction between these properties turns on the meaning assigned to the addition of a term to itself. Ordinary algebra and most special algebras distinguish between a and $a+a$. But the algebra of Symbolic Logic identifies a and $a+a$. The consequences of these assumptions will be discussed subsequently.

16. Principles of Subtraction. Let a and b be terms representing any two given elements of an algebraic manifold. Let us propose the problem, to find an element x of the manifold such that

$$x+b=a.$$

There may be no *general* solution to this problem, where a and b are connected by no special conditions. Also when there is one solution, there may be more than one solution. It is for instance easy to see that in an algebra which identifies a and $a+a$, there will be at least two solutions if there be one. For if x be one answer, then $x+b=x+b+b=a$. Hence $x+b$ is another answer.

If there be a solution of the above equation, let it be written in the form, $a \smile b$. Then it is assumed that $a \smile b$ represents an element of the manifold, though it may be ambiguous in its signification.

The definition of $a \smile b$ is

$$a \smile b+b=a \qquad \text{.............................(1).}$$

If c be another element of the manifold let us assume that $(a \smile b) \smile c$ symbolizes the solution of a double problem which has as its solution or solutions one or more elements of the manifold.

Then
$$\begin{aligned} a \smile b \smile c+(b+c) &= a \smile b \smile c+(c+b) \\ &= a \smile b \smile c+c+b \\ &= a \smile b+b=a. \end{aligned}$$

It follows that the problem proposed by the symbol $a \smile (b+c)$ has one or more solutions, and that the solutions to the problem $a \smile b \smile c$ are included in them.

Conversely suppose that the problem $a \smile (b+c)$ is solved by one or more elements of the manifold.

Then by hypothesis $a \smile (b+c) + (b+c) = a$; and hence

$$\{a \smile (b+c)\} + c + b = a \smile (b+c) + (b+c) = a.$$

But if $d+c+b=a$, then $d+c$ is one value of $a \smile b$ and d is one value of $a \smile b \smile c$.

Accordingly $a \smile b \smile c$ is a problem which by hypothesis must have one or more solutions, and the solutions to $a \smile (b+c)$ are included in them.

Hence since the solutions of each are included in those of the other, the two problems must have the same solutions. Therefore whatever particular meaning (in the choice of ambiguities) we assign to one may also be assigned to the other. We may therefore write

$$a \smile (b+c) = a \smile b \smile c \quad\quad\quad (2).$$

Again we have

$$a \smile (b+c) = a \smile (c+b).$$

Hence from equation (2),

$$a \smile b \smile c = a \smile c \smile b \quad\quad\quad (3).$$

It may be noted as a consequence of equations (2) and (3), that if $a \smile (b+c)$ admit of solutions, then also both $a \smile b$ and $a \smile c$ admit of solutions.

Hence if $a \smile b$ and $b \smile c$ admit of solutions; then $a \smile b = a \smile (b \smile c + c)$; and it follows from the above note that $a \smile (b \smile c)$ admits of a solution.

Also in this case

$$a \smile b + c = a \smile (b \smile c + c) + c = a \smile (b \smile c) \smile c + c,$$

from equation (2).

Hence $$a \smile b + c = a \smile (b \smile c) \quad\quad\quad (4).$$

We cannot prove that $a \smile b + c = a + c \smile b$, and that $a + (b \smile c) = a + b \smile c$, without making the assumption that $a \smile b$, if it exists, is unambiguous.

Summing up: for three terms a, b and c there are four equivalent forms symbolized by

$$(a \smile b) \smile c = (a \smile c) \smile b = a \smile (b+c) = a \smile (c+b):$$

also there are three sets of forms, the forms in each set being equivalent but not so forms taken from different sets, namely

$$(a \smile b) + c = a \smile (b \smile c) = c + (a \smile b) \quad\quad\quad (\alpha),$$

$$(c \smile b) + a = c \smile (b \smile a) = a + (c \smile b) \quad\quad\quad (\beta),$$

$$(a+c) \smile b = (c+a) \smile b \quad\quad\quad (\gamma).$$

Subtraction. Let us now make the further assumption that the reverse analytical process is unambiguous, that is to say that only one element of the manifold is represented by a symbol of the type $a \smile b$. Let us replace in this case the sign $\smile$ by $-$, and call the process subtraction.

Now at least one of the solutions of $a + b \smile b$ is a. Hence in subtraction the solution of $a + b - b$ is a, or symbolically $a + b - b = a$. But by definition, $a - b + b = a$.

Hence,
$$a + b - b = a - b + b = a \quad\text{.........................(5).}$$

We may note that the definition, $a - b + b = a$, assumes that the question $a - b$ has an answer. But equation (5) proves that a manifold may always without any logical contradiction be assumed to exist in which the subtractive question $a-b$ has an answer independently of any condition between a and b. For from the definition, $a - b + b$, where $a - b$ is assumed to have an answer, can then be transformed into the equivalent form $a + b - b$, which is a question capable of an answer without any condition between a and b. But it may happen that in special interpretations of an algebra $a - b$, though unambiguous, has no solution unless a and b satisfy certain conditions. The remarks of § 7 apply here.

Again
$$\begin{aligned} a + b - c &= a + (b - c + c) - c \\ &= a + (b - c) + c - c \\ &= a + (b - c) \quad\text{..............................(6).} \end{aligned}$$

17. THE NULL ELEMENT. On the assumption that to any question of the type $a - b$ can be assigned an answer, some meaning must be assigned to the term $a - a$.

Now if c be any other term,
$$c + a - a = c = c + b - b.$$

Hence it may be assumed that
$$a - a = b - b.$$

Thus we may put
$$a - a = 0 \quad\text{..................................(7);}$$
where 0 represents an element of the manifold independent of a. Let the element 0 be called the null element. The fundamental property of the null element is that the addition of this element and any other element a of the manifold yields the same element a. It would be wrong to think of 0 as necessarily symbolizing mere nonentity. For in that case, since there can be no differences in nonentities, its equivalent forms $a - a$ and $b - b$ must be not only equivalent, but absolutely identical; whereas they are palpably different. Let any term, such as $a - a$, which represents the null element be called a null term.

The fundamental property of 0 is,
$$a + 0 = a \quad\text{..................................(8).}$$

Other properties of 0 which can be derived from this by the help of the previous equations are,

$$0+0=0;$$

and $$a-0=a-(b-b)=a-b+b=a.$$

Again forms such as $0-a$ may have a meaning and be represented by definite elements of the manifold.

The fundamental properties of $0-a$ are symbolized by

$$b+(0-a)=b+0-a=b-a,$$

and $$b-(0-a)=b-0+a=b+a.$$

Since in combination with any other element the null element 0 disappears, the symbolism may be rendered more convenient by writing $-a$ for $0-a$. Thus $-a$ is to symbolize the element $0-a$.

18. Steps. We notice that, since $a=0+a$, we may in a similar way consider a or $+a$ as a degenerate form of $0+a$. From this point of view every element of the manifold is defined by reference to its relation with the null element. This relation with the null element may be called the *step* which leads from the null element to the other element. And by fastening the attention rather on the method of reaching the final element than on the element itself when reached, we may call the symbol $+a$ the symbol of the step by which the element a of the manifold is reached.

This idea may be extended to other elements besides the null element. For we may write $b=a+(b-a)$; and $b-a$ may be conceived as the *step* from a to b. The word step has been used* to imply among other things a quantity; but as *defined* here there is no necessary implication of quantity. The step $+a$ is simply the process by which any term p is transformed into the term $p+a$. The two steps $+a$ and $-a$ may be conceived as exactly opposed in the sense that their successive application starting from any term p leads back to that term, thus $p+a-a=p$. In relation to $+a$, the step $-a$ will be called a negative step; and in relation to $-a$, the step $+a$ will be called a positive step. The fundamental properties of steps are (1) that they can be taken in any order, which is the commutative law, and (2) that any number of successive steps may be replaced by one definite resultant step, which is the associative law.

The introduction of the symbols $+a$ and $-a$ involves the equations

$$\left.\begin{aligned} +(+a)&=+(0+a)=0+a=+a=a,\\ -(+a)&=-(0+a)=-0-a=-a,\\ +(-a)&=+(0-a)=0-a=-a,\\ -(-a)&=-(0-a)=-0+a=+a=a. \end{aligned}\right\}\quad\ldots\ldots\ldots\ldots(9).$$

19. Multiplication. A new mode of synthesis, multiplication, is now to be introduced which does not, like addition, necessarily concern terms of a

* Cf. Clifford, *Elements of Dynamic*.

single algebraic scheme (cf. § 15), nor does it necessarily reproduce as its result a member of one of the algebraic schemes to which the terms synthesized belong. Again, the commutative and associated laws do not necessarily hold for multiplication: but a new law, the distributive law, which defines the relation of multiplication to addition holds. Any mode of synthesis for which this relation to addition holds is here called multiplication. The result of multiplication like that of addition is unambiguous.

Consider two algebraic manifolds; call them the manifolds A and B. Let a, a', a'' etc., be terms denoting the various elements of A, and let b, b', b'' etc., denote the various elements of B. Assume that a mode of synthesis is possible between any two terms, one from each manifold. Let this synthesis result in some third thing, which is the definite unambiguous product under all circumstances of this special synthesis between those two elements.

Also let the idea of order between the two things be attributable to their union in this synthesis. Thus if a and b be the two terms of which the synthesis is being discussed, an order as between a first or b first can be attributed to this synthesis. Also let only two possible alternative orders as between a and b exist.

Let this mode of synthesis be, for the moment, expressed by the sign $\eqsim$. Thus between two terms a, b from the respective manifolds can be generated the two things $a \eqsim b$ and $b \eqsim a$.

All the things thus generated may be represented by the elements of a third manifold, call it C. Also let the symbols $a \eqsim b$ and $b \eqsim a$ conceived as representing such things be called terms. Now assume that the manifold C is an algebraic manifold, according to the definition given above (§ 15). Then its corresponding terms are capable of addition. And we may write $(a \eqsim b) + (b' \eqsim a'') +$ etc.; forming thereby another term representing an element of the manifold C.

The definition of the algebraic nature of C does not exclude the possibility that elements of C exist which cannot be formed by this synthesis of two elements from A and B respectively. For $(a \eqsim b) + (b'' \eqsim a')$ is by definition an element of C; but it will appear that this element cannot in general be formed by a single synthesis of either of the types $a^{(p)} \eqsim b^{(q)}$ or $b^{(q)} \eqsim a^{(p)}$.

Again $a + a' + a'' +$ etc., represents an element of the manifold A, and $b + b' + b'' +$ etc., represents an element of the manifold B. Hence there are elements of the manifold C represented by terms of the form

$$(a + a' + a'' + \text{etc.}) \eqsim (b + b' + b'' + \ldots),$$

and

$$(b + b' + b'' + \text{etc.}) \eqsim (a + a' + a'' + \ldots).$$

Now let this synthesis be termed *Multiplication*, when such expressions as the above follow the distributive law as defined by equations (10) below.

For multiplication let the synthesis be denoted by $\times$ or by mere juxtaposition. Then the definition of multiplication yields the following symbolic statements

$$\left.\begin{aligned} a(b+b') &= ab+ab', \\ (a+a')b &= ab+a'b, \\ b(a+a') &= ba+ba', \\ (b+b')a &= ba+b'a. \end{aligned}\right\} \quad \text{......................(10).}$$

It will be noticed that the general definition of multiplication does not involve the associative or the commutative law.

20. Orders of Algebraic Manifolds. Consider a single algebraic manifold A, such that its elements can be multiplied together. Call such a manifold a self-multiplicative manifold of the first order. Now the products of the elements, namely aa, aa', $a'a$, etc., by hypothesis form another algebraic manifold; call it B. Then B will be defined to be a manifold of the second order.

Now let the elements of A and B be capable of multiplication, thus forming another algebraic manifold C. Let C be defined to be a manifold of the third order. Also in the same way the elements of A and C form by multiplication an algebraic manifold, D, of the fourth order; and so on.

Further let the elements of any two of these manifolds be capable of multiplication, and each manifold be self-multiplicative.

Let the following law hold, which we may call the associative law for manifolds. The elements formed by multiplying elements of the manifold of the mth order with elements of the manifold of the nth order belong to the manifold of the $(m+n)$th order.

Thus the complete manifold of the mth order is formed by the multiplication of the elements of any two manifolds, of which the sum of the orders forms m, and also by the elements deduced by the addition of elements thus formed.

For instance aa', $aa'a''$, $aa'a''a'''$, represent elements of the manifolds of the second, third, and fourth orders respectively; also aa represents an element of the manifold of the second order. Also $a''(aa')$ is an element of the manifold of the third order: and $(aa')(a''a''')$ is an element of the manifold of the fourth order; and $aa'(aa'a''a''')$ is an element of the manifold of the sixth order; and so on.

Such a system of manifolds will be called a complete algebraic system.

In special algebras it will be found that the manifold of some order, say the mth, is identical with the manifold of the first order. Then the manifold of the $m+1$th order is identical with that of the second order, and so on.

Such an algebra will be said to be of the $m-1$th species. In an algebra of the first species only the manifold of the first order can occur. Such an algebra is called linear. The Calculus of Extension, which is a special algebra invented by Grassmann, can be of any species.

It will save symbols, where no confusion results, to use dots instead of brackets. Thus $a''(aa')$ is written $a''.aa'$, and $(aa')(a''a''')$ is written $aa'.a''a'''$, and so on. A dot will be conceived as standing for two opposed bracket signs, thus)(, the other ends of the two brackets being either other dots or the end or beginning of the row of letters. Thus $ab.cd$ stands for $(ab)(cd)$, and is not $(ab)cd$, unless in the special algebra considered, the two expressions happen to be identical; also $ab.cde.fg$ stands for $(ab)(cde)(fg)$. It will be noticed that in these examples each dot has been replaced by two opposed bracket signs. An ingenious use of dots has been proposed by Mr W. E. Johnson which entirely obviates the necessity for the use of brackets. Thus $a\{b(cd)\}$ is written $a..b.cd$, and $a[b\{c(de)\}]$ is written $a...b..c.de$. The principle of the method is that those multiplications indicated by the fewest dots are the first performed. Thus $a\{b(cd)\}(ef)$ is written $a..b.cd..ef$, and $a\{b(cd)\}ef$ is written $a..b.cd..e...f$, where in the case of equal numbers of dots the left-hand multiplication is first performed.

21. THE NULL ELEMENT. Returning to the original general conception of two algebraic manifolds A and B of which the elements can be multiplied together, and thus form a third algebraic manifold C: let 0_1 be the null element of A, 0_2 the null element of B, and 0_3 the null element of C.

Then if a and b represent any two elements of the manifolds A and B respectively, we have

$$a + 0_1 = a, \text{ and } b + 0_2 = b.$$

Hence $$(a + 0_1)b = ab = ab + 0_1 b.$$

Accordingly, $$0_1 b = 0_3.$$

Similarly, $$b0_1 = 0_3 = a0_2 = 0_2 a.$$

No confusion can arise if we use the same symbol 0 for the null elements of each of the three manifolds.

Accordingly, $$0a = a0 = 0b = b0 = 0 \qquad (11).$$

It will be observed that a null element has not as yet been defined for the algebraic manifold in general; but only for those which allow of the process of subtraction, as defined in § 16. Thus manifolds for which the relation $a + a = a$ holds are excluded from the definition.

In order to include these manifolds let now the null element be defined as that single definite element, if it exist, of the manifold for which the equation

$$a + 0 = a,$$

holds, where a is *any* element of the manifold.

It will be noted that for the definite element a the same property may hold for a as well as for 0; since in some algebras $a+a=a$. But 0 is defined to be the single element which retains this property with all elements. Then in the case of multiplication equations (11) hold.

22. Classification of Special Algebras. The succeeding books of this work will be devoted to the discussion and comparison of the leading special algebras. It remains now to explain the plan on which this investigation will be conducted.

It follows from a consideration of the ideas expounded in Chapter I. that it is desirable to conduct the investigation of a calculus strictly in connection with its interpretations, and that without some such interpretation, however general, no great progress is likely to be made. Therefore each special algebra will, as far as possible, be interpreted concurrently with its investigation. The interpretation chosen, where many are available, will be that which is at once most simple and most general; but the remaining applications will also be mentioned with more or less fulness according as they aid in the development of the calculus. It must be remembered, however, in explanation of certain obvious gaps that the investigation is primarily for the sake of the algebra and not of the interpretation.

No investigation of ordinary algebra will be attempted. This calculus stands by itself in the fundamental importance of the theory of quantity which forms its interpretation. Its formulæ will of course be assumed throughout when required.

In the classification of the special algebras the two genera of addition form the first ground for distinction.

For the purpose of our immediate discussion it will be convenient to call the two genera of algebras thus formed the non-numerical genus and the numerical genus.

In the non-numerical genus investigated in Book II. the two symbols a and $a+a$, where a represents any element of the algebraic manifold, are equivalent, thus $a=a+a$. This definition leads to the simplest and most rudimentary type of algebraic symbolism. No symbols representing number or quantity are required in it. The interpretation of such an algebra may be expected therefore to lead to an equally simple and fundamental science. It will be found that the only species of this genus which at present has been developed is the Algebra of Symbolic Logic, though there seems no reason why other algebras of this genus should not be developed to receive interpretations in fields of science where strict demonstrative reasoning without relation to number and quantity is required. The Algebra of Symbolic Logic is the simplest possible species of its genus and has accordingly the simplest interpretation in the field of deductive logic. It is however always desirable while developing the symbolism of a calculus to reduce the interpretation to the utmost simplicity consistent with complete generality.

Accordingly in discussing the main theory of this algebra the difficulties peculiar to Symbolic Logic will be avoided by adopting the equally general interpretation which considers merely the intersection or non-intersection of regions of space. This interpretation will be developed concurrently with the algebra. After the main theory of the algebra has been developed, the more abstract interpretation of Symbolic Logic will be introduced.

In the numerical genus the two symbols a and $a+a$ are not equivalent. The symbol $a+a$ is shortened into $2a$; and by generalization of this process a symbol of the form ξa is created, where ξ is an ordinary algebraical quantity, real or imaginary. Hence the general type of addition for this genus is symbolized by $\xi a + \eta b + \zeta c +$ etc., where a, b, c, etc. are elements of the algebraic manifold, and ξ, η, ζ, etc. are any ordinary algebraic quantities (such quantities being always symbolized by Greek letters, cf. Book III. Chapter I. below). There are many species of algebra with important interpretations belonging to this genus; and an important general theory, that of Linear Associative Algebras, connecting and comparing an indefinitely large group of algebras belonging to this genus.

The special manifolds, which respectively form the interpretation of all the special algebras of this genus, have all common properties in that they all admit of a process symbolized by addition of the numerical type. Any manifold with these properties will be called a 'Positional Manifold.' It is therefore necessary in developing the complete theory of Universal Algebra to enter into an investigation of the general properties of a positional manifold, that is, of the properties of the general type of numerical addition. It will be found that the idea of a positional manifold will be made more simple and concrete without any loss of generality by identifying it with the general idea of space of any arbitrarily assigned number of dimensions, but excluding all metrical spatial ideas. In the discussion of the general properties of numerical addition this therefore will be the interpretation adopted as being at once the most simple and the most general. All the properties thus deduced must necessarily hold for any special algebra of the genus, though the scale of the relative importance of different properties may vary in different algebras. Positional manifolds are investigated in Book III.

Multiplication in algebras of the numerical genus of course follows all the general laws investigated in this chapter. There is also one other general law which holds throughout this genus. The product of ξa and ηb (ξ and η being numbers) is defined to be equivalent to the product of $\xi\eta$ (ordinary multiplication) into the product of a and b. Thus in symbols

$$\xi a \,.\, \eta b = \xi\eta ab, \quad \eta b \,.\, \xi a = \xi\eta ba:$$

where the juxtaposition of ξ and η always means that they are to be multiplied according to the ordinary law of multiplication for numbers.

If this law be combined with equation 10 of § 19, the following general equation must hold: let $e_1, e_2, \ldots e_\nu$ be elements of the manifold, and let Greek letters denote numbers (*i.e.* ordinary algebraic quantities, real or imaginary), then

$$(\alpha_1 e_1 + \alpha_2 e_2 + \ldots + \alpha_\nu e_\nu)(\beta_1 e_1 + \beta_2 e_2 + \ldots + \beta_\nu e_\nu)$$
$$= \alpha_1\beta_1 e_1 e_1 + \alpha_1\beta_2 e_1 e_2 + \alpha_2\beta_1 e_2 e_1 + \ldots + \alpha_\nu\beta_\nu e_\nu e_\nu.$$

It follows that in the numerical genus of algebras the successive derived manifolds are also positional manifolds, as well as the manifold of the first order.

In the classification of the special algebras of this genus the nature of the process of multiplication as it exists in each special algebra is the guide.

The first division must be made between those algebras which involve a complete algebraical system of more than one manifold and those which involve only one manifold, that is, between algebras of an order higher than the first and between linear algebras (cf. § 20). It is indeed possible to consider all algebras as linear. But this simplification, though it has very high authority, is, according to the theory expounded in this work, fallacious. For it involves treating elements for which addition has no meaning as elements of one manifold; for instance in the Calculus of Extension it involves treating a point element and a linear element as elements of one manifold capable of addition, though such addition is necessarily meaningless.

The only known algebra of a species higher than the first is Grassmann's Calculus of Extension; that is to say, this is the only algebra for which this objection to its simplification into a linear algebra holds good. The Calculus of Extension will accordingly be investigated first among the special algebras of the numerical genus. It can be of any species (cf. § 20). The general type of manifold of the first algebraic order in which the algebra finds its interpretation will be called an Extensive Manifold. Thus an extensive manifold is also a positional manifold.

In Book IV. the fundamental definitions and formulæ of the Calculus of Extension will be stated and proved. The calculus will also be applied in this book to an investigation of simple properties of extensive manifolds which, though deduced by the aid of this calculus, belong equally to the more general type of positional manifolds. One type of formulæ of the algebra will thus receive investigation. Other types of formulæ of the same algebra are developed in Books V., VI. and VII., each type being developed in conjunction with its peculiar interpretation. The series of interpretations will form, as they ought to do, a connected investigation of the general theory of spatial ideas of which the foundation has been laid in the discussion of positional manifolds in Book III.

This spatial interpretation, which also applies to the algebra of Symbolic Logic, will in some form or other apply to every special algebra, in so far as interpretation is possible. This fact is interesting and deserves investigation.

The result of it is that a treatise on Universal Algebra is also to some extent a treatise on certain generalized ideas of space.

In order to complete this subsidiary investigation an appendix on a mode of arrangement of the axioms of geometry is given at the end of this volume.

The second volume of this work will deal with Linear Algebras. In addition to the general theory of their classification and comparison, the special algebras of quaternions and matrices will need detailed development.

NOTE. The discussions of this chapter are largely based on the 'Uebersicht der allgemeinen Formenlehre' which forms the introductory chapter to Grassmann's *Ausdehnungslehre von* 1844.

Other discussions of the same subject are to be found in Hamilton's *Lectures on Quaternions*, Preface; in Hankel's *Vorlesungen über Complexe Zahlen* (1867); and in De Morgan's *Trigonometry and Double Algebra*, also in a series of four papers by De Morgan, '*On the Foundation of Algebra*,' Transactions of the Cambridge Philosophical Society, vols. VII. and VIII., (1839, 1841, 1843, 1844).

BOOK II.

THE ALGEBRA OF SYMBOLIC LOGIC.

CHAPTER I.

The Algebra of Symbolic Logic.

23. Formal Laws. The Algebra of Symbolic Logic* is the only known member of the non-numerical genus of Universal Algebra (cf. Bk. I., Ch. III., § 22).

It will be convenient to collect the formal laws which define this special algebra before considering the interpretations which can be assigned to the symbols. The algebra is a linear algebra (cf. § 20), so that all the terms used belong to the same algebraic scheme and are capable of addition.

Let a, b, c, etc. be terms representing elements of the algebraic manifold of this algebra. Then the following symbolic laws hold.

(1) The general laws of addition (cf. Bk. I. Ch. III., §§ 14, 15):

$$a + b = b + a,$$
$$a + b + c = (a + b) + c = a + (b + c).$$

(2) The special law of addition (cf. § 22):

$$a + a = a.$$

(3) The definition of the null element (cf. § 21):

$$a + 0 = a.$$

(4) The general laws of multiplication (cf. § 19):

$$c(a + b) = ca + cb,$$
$$(a + b)c = ac + bc.$$

(5) The special laws of multiplication:

$$ab = ba,$$
$$abc = ab . c = a . bc,$$
$$aa = a.$$

* This algebra in all essential particulars was invented and perfected by Boole, cf. his work entitled, *An Investigation of the Laws of Thought*, London, 1854.

(6) The law of 'absorption':

$$a + ab = a.$$

This law includes the special law (2) of addition.

(7) The definition of the 'Universe.' This is a special element of the manifold, which will be always denoted in future by i, with the following property:

$$ai = a.$$

(8) Supplementary elements. An element b will be called supplementary to an element a if both $a + b = i$, and $ab = 0$. It will be proved that only one element supplementary to a given element can exist; and it will be assumed that one such element always does exist. If a denote the given element, $\bar{a}$ will denote the supplementary element. Then $\bar{a}$ will be called the supplement of a. The supplement of an expression in a bracket, such as $(a + b)$, will be denoted by $\overline{}(a + b)$.

The theorem that any element a has only one supplement follows from the succeeding fundamental proposition which develops a method of proof of the equivalence of two terms.

PROPOSITION I. If the equations $xy = xz$, and $x + y = x + z$, hold simultaneously, then $y = z$.

Multiply the second equation by $\bar{x}$, where $\bar{x}$ is one of the supplements of x which by hypothesis exists.

Then $$\bar{x}(x + y) = \bar{x}(x + z).$$

Hence by (4) $$\bar{x}x + \bar{x}y = \bar{x}x + \bar{x}z,$$

hence by (8) and (3) $$\bar{x}y = \bar{x}z.$$

Add this to the first equation, then by (4)

$$(x + \bar{x})y = (x + \bar{x})z,$$

hence by (8) $$iy = iz,$$

hence by (7) $$y = z.$$

COROLLARY I. There is only one supplement of any element x. For if possible let $\bar{x}$ and x' be two supplements of x.

Then $$x\bar{x} = 0 = xx', \text{ and } x + \bar{x} = i = x + x'.$$

Hence by the proposition, $$\bar{x} = x'.$$

COROLLARY II. If $x = y$, then $\bar{x} = \bar{y}$.

COROLLARY III. $$\bar{i} = 0, \text{ and } \bar{0} = i.$$

COROLLARY IV. $$\bar{\bar{x}} = x;$$

where $\bar{\bar{x}}$ means the supplement of the supplement of x. The proofs of these corollaries can be left to the reader.

PROPOSITION II. $(x+y)(x+z)=x+yz.$

For $(x+y)(x+z)=xx+xy+xz+yz=x+x(y+z)+yz$
$=x+yz,$ by (6).

PROPOSITION III. $x0=0=0x$, and $x+i=i=i+x$.

The first is proved in § 21. The proof of the second follows at once from (6) and (7).

24. RECIPROCITY BETWEEN ADDITION AND MULTIPLICATION. A reciprocity between addition and multiplication obtains throughout this algebra; so that corresponding to every proposition respecting the addition and multiplication of terms there is another proposition respecting the multiplication and addition of terms. The discovery of this reciprocity was first made by C. S. Peirce*; and later independently by Schröder†.

The mutual relations between addition and multiplication will be more easily understood if we employ the sign $\times$ to represent multiplication. The definitions and fundamental propositions of this calculus can now be arranged thus.

The Commutative Laws are (cf. (1) and (5))

$$\left.\begin{aligned} x+y&=y+x,\\ x\times y&=y\times x.\end{aligned}\right\} \quad \ldots\ldots\ldots\ldots(A).$$

The Distributive Laws are (cf. (4) and Prop. II.)

$$\left.\begin{aligned} x\times(y+z)&=(x\times y)+(x\times z),\\ x+(y\times z)&=(x+y)\times(x+z).\end{aligned}\right\} \quad \ldots\ldots\ldots\ldots(B).$$

The Associative Laws are (cf. (1) and (5))

$$\left.\begin{aligned} x+(y+z)&=x+y+z,\\ x\times(y\times z)&=x\times y\times z.\end{aligned}\right\} \quad \ldots\ldots\ldots\ldots(C).$$

The Laws of Absorption are (cf. (6))

$$\left.\begin{aligned} x+(x\times y)&=x=x+x,\\ x\times(x+y)&=x=x\times x.\end{aligned}\right\} \quad \ldots\ldots\ldots\ldots(D).$$

The properties of the Null element and of the Universe are (cf. (3), (7), and Prop. III.),

$$\left.\begin{aligned} x+i&=i,\\ x\times 0&=0.\end{aligned}\right\} \quad \ldots\ldots\ldots\ldots(E),$$

$$\left.\begin{aligned} x+0&=x,\\ x\times i&=x.\end{aligned}\right\} \quad \ldots\ldots\ldots\ldots(F).$$

The definition of the supplement of a term gives (cf. (8) and Prop. I.)

$$\left.\begin{aligned} x+\bar{x}&=i,\\ x\times\bar{x}&=0.\end{aligned}\right\} \quad \ldots\ldots\ldots\ldots(G).$$

* *Proc. of the American Academy of Arts and Sciences*, 1867.

† *Der Operationskreis des Logikkalküls*, 1877.

There can therefore be no distinction in properties between addition and multiplication. All propositions in this calculus are necessarily divisible into pairs of reciprocal propositions; and given one proposition the reciprocal proposition can be immediately deduced from it by interchanging the signs $+$ and $\times$, and the terms i and 0. An independent proof can of course always be found: it will in general be left to the reader.

Also any interpretation of which the calculus admits can always be inverted so that the interpretation of addition is assigned to multiplication, and that of multiplication to addition, also that of i to 0 and that of 0 to i.

25. Interpretation. It is desirable before developing the algebraic formulæ to possess a simple and general form of interpretation (cf. § 7 and § 22).

Let the elements of this algebraic manifold be regions in space, each region not being necessarily a continuous portion of space. Let any term symbolize the mental act of determining and apprehending the region which it represents. Terms are equivalent when they place the same region before the mind for apprehension.

Let the operation of addition be conceived as the act of apprehending in the mind the complete region which comprises and is formed by all the regions represented by the terms added. Thus in addition the symbols represent firstly the act of the mind in apprehending the component regions represented by the added terms and then its act in apprehending the complete region. This last act of apprehension determines the region which the resultant term represents. This interpretation of terms and of addition both satisfies and requires the formal laws (1) and (2) of § 23. For the complete region does not depend on the order of apprehension of the component regions; nor does it depend on the formation of subsidiary complete regions out of a selection of the added terms. Hence the commutative and associative laws of addition are required. The law, $a + a = a$, is satisfied since a region is in no sense reduplicated by being placed before the mind repeatedly for apprehension. The complete region represented by $a + a$ remains the region represented by a. This is called the Law of Unity by Jevons (cf. *Pure Logic*, ch. VI).

The null element must be interpreted as denoting the non-existence of a region. Thus if a term represent the null element, it symbolizes that the mind after apprehending the component regions (if there be such) symbolized by the term, further apprehends that the region placed by the term before the mind for apprehension does not exist. It may be noted that the addition of terms which are not null cannot result in a null term. A null term can however arise in the multiplication of terms which are not null.

Let the multiplication of terms result in a term which represents the entire region common to the terms multiplied. Thus xyz represents the entire region which is at once incident in the regions x and y and z. Hence the term xy symbolizes the mental acts first of apprehending the regions symbolized by x and y, and then of apprehending the region which is their complete intersection. This final act of apprehension determines the region which xy represents.

This interpretation of multiplication both satisfies and requires the distributive law, numbered (4) in § 23, and the commutative and associative laws marked (5) in § 23. The law, $aa = a$, which also occurs in (5) of § 23 is satisfied: for the region which is the complete intersection of the region a with itself is again the region a. This is called the Law of Simplicity by Jevons (cf. *loc. cit.*).

The Law of Absorption (cf. (6) § 23) is also required and satisfied. For the complete region both formed by and comprising the regions a and ab is the region a, and the final act of apprehension symbolized by $a + ab$ is that of the region a. Hence

$$a + ab = a.$$

This interpretation also requires that if $x + y = x$, then $y = xy$. And this proposition can be shown to follow from the formal laws (cf. § 26, Prop. VIII.).

The element called the Universe (cf. § 23 (7)), must be identified with all space; or if discourse is limited to an assigned portion of space which is to comprise all the regions mentioned, then the Universe is to be that complete region of space.

The term supplementary (cf. § 23 (8)) to any term a represents that region which includes all the Universe with the exception of the region a. The two regions together make up the Universe; but they do not overlap, so that their region of intersection is non-existent.

It follows that the supplement of the Universe is a non-existent region, and that the supplement of a non-existent region is the Universe (cf. Prop. I. Cor. 3).

26. ELEMENTARY PROPOSITIONS. The following propositions of which the interpretation is obvious can be deduced from the formal laws and from the propositions already stated.

PROPOSITION IV. If $x + y = 0$, then $x = 0$, $y = 0$.

For multiplying by x, $\quad x(x + y) = 0$.

But $\quad x(x + y) = x$, by (6) § 23.

Hence $x = 0$. Similarly, $y = 0$.

The reciprocal theorem is, if $xy = i$, then $x = i$, $y = i$.

PROPOSITION V. $x + y = x + y\bar{x}$, and $xy = x(y + \bar{x})$.

For $\quad x + y = x + y(x + \bar{x}) = x + yx + y\bar{x} = x + y\bar{x}$.

The second part is the reciprocal proposition to the first part.

PROPOSITION VI. $\quad {}^-(xy) = \bar{x} + \bar{y},$

and $\quad {}^-(x + y) = \bar{x}\bar{y}.$

For by Prop. V., $\quad \bar{x} + \bar{y} = \bar{x} + \bar{\bar{x}}\bar{y} = \bar{x} + x\bar{y}.$

Hence $\quad xy + (\bar{x} + \bar{y}) = xy + x\bar{y} + \bar{x} = x(y + \bar{y}) + \bar{x}$

$$= x + \bar{x} = i.$$

Also $\quad xy(\bar{x} + \bar{y}) = x\bar{x}y + xy\bar{y} = 0.$

Hence by § 23 (8) $\quad \bar{x} + \bar{y} = {}^-(xy).$

The second part is the reciprocal of the first part. Also it can be deduced from the first part thus:

$${}^-(\bar{x}\bar{y}) = \bar{\bar{x}} + \bar{\bar{y}} = x + y.$$

Taking the supplements of both sides

$${}^-(x + y) = {}^=(\bar{x}\bar{y}) = \bar{x}\bar{y}.$$

COROLLARY. The supplement of any complex expression is found by replacing each component term by its supplement and by interchanging + and × throughout.

PROPOSITION VII. If $xy = xz$, then

$$x\bar{y} = x\bar{z}, \text{ and } \bar{x} + y = \bar{x} + z.$$

For taking the supplement of both sides of the given equation, by Prop. VI.,

$$\bar{x} + \bar{y} = \bar{x} + \bar{z}.$$

Multiplying by x, $\quad x\bar{y} = x\bar{z}.$

Again taking the supplement of this equation, then

$$\bar{x} + y = \bar{x} + z.$$

The reciprocal proposition is, if $x + y = x + z$, then $x + \bar{y} = x + \bar{z}$, and $\bar{x}y = \bar{x}z$.

PROPOSITION VIII. The following equations are equivalent, so that from any one the remainder can be derived:

$$y = xy, \; x + y = x, \; \bar{x}y = 0, \; x + \bar{y} = i.$$

Firstly: assume $\quad y = xy.$

Then $\quad x + y = x + xy = x.$

And $\quad \bar{x}y = \bar{x}xy = 0.$

And $\quad x + \bar{y} = x + {}^-(xy) = x + \bar{x} + \bar{y} = i.$

Secondly: assume $\quad x + y = x.$

Then $\quad xy = (x + y)y = y.$

Hence the other two equations can be derived as in the first case.

Thirdly: assume $\quad \bar{x}y = 0.$

Then $$y = (x + \bar{x})\, y = xy + \bar{x}y = xy.$$

Hence the other two equations can be derived.

Fourthly: assume $$x + \bar{y} = i.$$

Then taking the supplements of both sides

$$\overline{\ \ }(x + \bar{y}) = \bar{x}y = 0.$$

Hence by the third case the other equations are true.

Corollary. By taking the supplements of the first and second equations two other forms equivalent to the preceding can be derived, namely

$$\bar{y} = \bar{x} + \bar{y},\ \ \bar{x}\bar{y} = \bar{x}.$$

Proposition IX. If $x = xyz$, then $x = xy = xz$, and if $x = x + y + z$, then $x = x + y = x + z$.

For $xy = xy\,(z + \bar{z}) = xyz + xy\bar{z} = x + xy\bar{z} = x$, from (6) § 23.

The second part of the proposition is the reciprocal theorem to the first part.

Corollary. A similar proof shews that if $z = z\,(xu + yv)$, then $z = z\,(x + y)$; and that if $z = z + (x + u)(y + v)$, then $z = z + xy$.

27. Classification. The expression $x + y + z + \ldots$, which we can denote by u for the sake of brevity, is formed by the addition of the regions x, y, etc. Now these regions may be overlapping regions: we require to express u as a sum of regions which have no common part. To this problem there exists the reciprocal problem, given that u stands for the product $xyz\ldots$, to express u as a product of regions such that the sum of any two completes the universe. These problems may be enunciated and proved symbolically as follows.

Proposition X. To express $u\,(= x + y + z + \ldots)$, in the form

$$X + Y + Z + \ldots;$$

where X, Y, Z have the property that for any two of them, Y and Z say, the condition $YZ = 0$, holds.

Also to express $u\,(= xyz\ldots)$ in the form $XYZ\ldots$; where X, Y, Z have the property that for any two of them, Y and Z say, the condition $Y + Z = i$, holds.

Now from Prop. IV., if $x\,(y + z) = 0$, then $xy = 0$, $xz = 0$. Hence for the first part of the proposition the conditions that X, Y, Z, etc. must satisfy can be expressed thus,

$$X\,(Y + Z + T + \ldots) = 0,\ \ Y\,(Z + T + \ldots) = 0,\ \ Z\,(T + \ldots) = 0, \text{ etc.}$$

Now by Prop. V.,
$$u = x + y + z + \ldots$$
$$= x + \bar{x}\,(y + z + \ldots);$$

and $$y+z+t+\ldots = y+\bar{y}(z+t+\ldots):$$
and $$z+t+\ldots = z+\bar{z}(t+\ldots).$$

Proceeding in this way, we find

$$u = x+\bar{x}y+\bar{x}\bar{y}z+\bar{x}\bar{y}\bar{z}t+\ldots.$$

Hence we may write

$$X = x,\quad Y = \bar{x}y,\quad Z = \bar{x}\bar{y}z, \text{ etc.}$$

It is obvious that there is more than one solution of the problem.

Again for the second part of the proposition, consider

$$u = \bar{x}+\bar{y}+\bar{z}+\ldots.$$

By the first part of the proposition,

$$\bar{u} = x+x\bar{y}+xy\bar{z}+xyz\bar{t}+\ldots.$$

Here any two terms, Y and Z, satisfy the condition $YZ=0$.

Taking the supplements of these equations,

$$\begin{aligned} u &= {}^{-}(\bar{x}+\bar{y}+\bar{z}+\ldots) \\ &= xyz\ldots \\ &= {}^{-}(\bar{x}+x\bar{y}+xy\bar{z}+\ldots) \\ &= x(\bar{x}+y)(\bar{x}+\bar{y}+z)(\bar{x}+\bar{y}+\bar{z}+t)\ldots. \end{aligned}$$

Hence we may write $XYZ\ldots$ for $xyz\ldots$, where $X=x$, $Y=\bar{x}+y$, $Z=\bar{x}+\bar{y}+z$, etc. and any two of X, Y, Z, etc., for instance Y and Z, satisfy the condition $Y+Z=i$.

It is obvious that there is more than one solution of this problem.

These problems are of some importance in the logical applications of the algebra.

28. Incident Regions. The symbolic study of regions incident (cf. § 10) in other regions has some analogies to the theory of inequalities in ordinary algebra. These relationships also partly possess the properties of algebraic equations. Two mixed symbols have therefore been adopted to express them, namely $\mathrel{=\!\!\!<}$ and $\mathrel{>\!\!\!=}$ (cf. Schröder, *Algebra der Logik*). Then, $y \mathrel{=\!\!\!<} x$, expresses that y is incident in x; and $x \mathrel{>\!\!\!=} y$ expresses that x contains y. Expressions of this kind will be called, borrowing a term from Logic, subsumptions. Then a subsumption has analogous properties to an inequality. The Theory of Symbolic Logic has been deduced by C. S. Peirce from the type of relation symbolized by $\mathrel{>\!\!\!=}$, cf. *American Journal of Mathematics*, Vols. III and VII (1880, 1885). His investigations are incorporated in Schröder's *Algebra der Logik*.

In order to deduce the properties of a subsumption as far as possible purely symbolically by the methods of this algebra, it is necessary to start from a proposition connecting subsumptions with equations. Such an initial proposition must be established by considering the meaning of a subsumption.

Proposition XI. If $y ⋹ x$, then $y = xy$; and conversely.

For if y be incident in x, then y and xy denote the same region.

The converse of this proposition is also obvious.

It is obvious that any one of the equations proved in § 26, Prop. VIII., to be equivalent to $y = xy$ is equivalent to $y ⋹ x$. In fact the subsumption $y ⋹ x$ may be considered as the general expression for that relation between x and y which is implied by any one of the equations of Prop. VIII. It follows that an equation is a particular case of a subsumption.

Corollary. $xz ⋹ x ⋹ x + z$.

Proposition XII. If $x ⋺ y$, and $y ⋺ z$;

then $$x ⋺ z.$$

For by Prop. XI. and by § 26, Prop. IX.

$$z = zy = zxy = zx.$$

Hence $$x ⋺ z.$$

Proposition XIII. If $x ⋺ y$, and $y ⋺ x$; then $x = y$.

For since $x ⋺ y$, then $y = xy$.

And since $y ⋺ x$, then $y = x + y$.

Hence $y = xy = x(x + y) = x$.

Proposition XIV. If $x ⋺ y$, and $u ⋺ v$;

then $ux ⋺ vy$, and $u + x ⋺ v + y$.

For $y = yx$, and $v = vu$; hence $vy = yxvu = vy \,.\, xu$.

Therefore $$ux ⋺ vy.$$

Also $x = x + y$, and $u = u + v$;

hence $$x + u = x + y + u + v = (x + u) + (y + v).$$

Therefore $$x + u ⋺ y + v.$$

Corollary. If $x ⋺ y$, and $u = v$; then $ux ⋺ vy$, and $u + x ⋺ v + y$.

For $v = vu$, and $u = u + v$; hence the proof can proceed exactly as in the proposition.

The proofs of the following propositions may be left to the reader.

Proposition XV. If $x ⋺ y$, then $\bar{y} ⋺ \bar{x}$.

Proposition XVI. If $z ⋹ xy$, then $z ⋹ x$, $z ⋹ y$, $z ⋹ x + y$.

Proposition XVII. If $z ⋹ xy$, then $\overline{xy} ⋹ \bar{z}$, $\bar{x} + \bar{y} ⋹ \bar{z}$.

Proposition XVIII. If $z ⋺ x + y$, then $z ⋺ x$, $z ⋺ y$, $z ⋺ xy$.

Proposition XIX. If $z ⋺ x + y$, then $\overline{xy} ⋺ \bar{z}$, $\bar{x} + \bar{y} ⋺ \bar{z}$.

Proposition XX. If $xz ⋹ y$, and $x ⋹ y + z$, then $x ⋹ y$.

PROPOSITION XXI. If $z \notin xu + yv$, then $z \notin x + y$.

The importance of Prop. XXI. demands that its proof be given.

By Prop. IX. Cor., $z = z(xu + yv) = z(x + y)$.

Therefore $$z \notin x + y.$$

COROLLARY. If $z = xu + yv$, then $z \notin x + y$;
that is, $$xu + yv \notin x + y.$$

Prop.* XXII. If $z \not\ni (x + u)(y + v)$, then $z \not\ni xy$.

COROLLARY. $(x + u)(y + v) \not\ni xy$.

* This proposition, which I had overlooked, was pointed out by Mr W. E. Johnson.

CHAPTER II.

The Algebra of Symbolic Logic (*continued*).

29. Development. (1) The expression for any region whatsoever may be written in the form $ax + b\bar{x}$; where x represents any region.

For let z be any region. Now $x + \bar{x} = i$.

Hence
$$z \mathrel{\text{⋹}} x + \bar{x}$$
$$= zx + z\bar{x}.$$

Now let $a = zx + u\bar{x}$, and $b = z\bar{x} + vx$, where u and v are restricted by no conditions.

Then
$$ax + b\bar{x} = (zx + u\bar{x})\,x + (z\bar{x} + vx)\,\bar{x} = zx + z\bar{x} = z.$$

Hence by properly choosing a and b, $ax + b\bar{x}$ can be made to represent any region z without imposing any condition on x.

Again the expression for any region can be written in the form
$$(a + x)\,(b + \bar{x}),$$
where x represents any other region. For by multiplication
$$(a + x)\,(b + \bar{x}) = ab + a\bar{x} + bx = (a + ab)\,\bar{x} + (b + ab)\,x = a\bar{x} + bx.$$
This last expression has just been proved to represent the most general region as far as its relation to the term x is concerned.

(2) Binomial expressions of the form $ax + b\bar{x}$ have many important properties which must be studied. It is well to notice at once the following transformations:
$$ax + b\bar{x} = (b + x)\,(a + \bar{x}):$$
$$\overline{(ax + b\bar{x})} = (\bar{a} + \bar{x})\,(\bar{b} + x)$$
$$= \bar{a}x + \bar{b}\bar{x};$$
$$(ax + b\bar{x})\,(cx + d\bar{x}) = acx + bd\bar{x};$$
$$ax + b\bar{x} + c = (a + c)\,x + (b + c)\,\bar{x}:$$
$$ax + b\bar{x} = ax + b\bar{x} + ab.$$

(3) Let $f(x)$ stand for any complex expression formed according to the processes of this algebra by successive multiplications and additions of x and $\bar{x}$ and other terms denoting other regions. Then $f(x)$ denotes some region with a specified relation to x. But by (1) of this article $f(x)$ can also be written in the form $ax + b\bar{x}$. Furthermore a and b can be regarded as specified by multiplications and additions of the other terms involved in the formation of $f(x)$ without mention of x. For if a be a complex expression, it must be expressible, by a continual use of the distributive law, as a sum of products of which each product either involves x or $\bar{x}$ or neither. Since a only appears when multiplied by x, any of these products involving $\bar{x}$ as a factor can be rejected, since $x\bar{x} = 0$; also any of these products involving x as a factor can be written with the omission of x, since $xx = x$. Hence a can be written in a form not containing x or $\bar{x}$. Similarly b can be written in such a form.

(4) Boole has shown how to deduce immediately from $f(x)$ appropriate forms for a and b. For write $f(x) = ax + b\bar{x}$. Let i be substituted for x, then

$$f(i) = ai + b\bar{i} = ai + 0 = a.$$

Again let 0 be substituted for x, then

$$f(0) = a0 + b\bar{0} = 0 + bi = b.$$

Hence $$f(x) = f(i)\,x + f(0)\,\bar{x}.$$

For complicated expressions the rule expressed by this equation shortens the process of simplification. This process is called by Boole the development of $f(x)$ with respect to x.

The reciprocity between multiplication and addition gives the reciprocal rule

$$f(x) = \{f(0) + x\}\,\{f(i) + \bar{x}\}.$$

(5) The expressions $f(i)$ and $f(0)$ may involve other letters y, z, etc. They may be developed in respect to these letters also.

Consider for example the expression $f(x, y, z)$ involving three letters.

$$\begin{aligned}
f(x, y, z) &= f(i, y, z)\,x + f(0, y, z)\,\bar{x},\\
f(i, y, z) &= f(i, i, z)\,y + f(i, 0, z)\,\bar{y},\\
f(i, i, z) &= f(i, i, i)\,z + f(i, i, 0)\,\bar{z},\\
f(i, 0, z) &= f(i, 0, i)\,z + f(i, 0, 0)\,\bar{z},\\
f(0, y, z) &= f(0, i, z)\,y + f(0, 0, z)\,\bar{y},\\
f(0, i, z) &= f(0, i, i)\,z + f(0, i, 0)\,\bar{z},\\
f(0, 0, z) &= f(0, 0, i)\,z + f(0, 0, 0)\,\bar{z}.
\end{aligned}$$

Hence by substitution

$$\begin{aligned}
f(x, y, z) = {} & f(i, i, i)\,xyz + f(0, i, i)\,\bar{x}yz + f(i, 0, i)\,x\bar{y}z + f(i, i, 0)\,xy\bar{z}\\
& + f(i, 0, 0)\,x\bar{y}\bar{z} + f(0, i, 0)\,\bar{x}y\bar{z} + f(0, 0, i)\,\bar{x}\bar{y}z + f(0, 0, 0)\,\bar{x}\bar{y}\bar{z}.
\end{aligned}$$

The reciprocal formula, owing to the brackets necessary, becomes too complicated to be written down here.

Let any term in the above developed expression for $f(x, y, z)$, say $f(0, i, 0)\,\bar{x}y\bar{z}$, be called a constituent term of the type $\bar{x}y\bar{z}$ in the development.

(6) The rule for the supplement of a binomial expression given in subsection (1) of this article, namely $^{-}(ax + b\bar{x}) = \bar{a}x + \bar{b}\bar{x}$, can be extended to an expression developed with respect to any number of terms $x, y, z, \ldots$. The extended rule is that if

$$f(x, y, z, \ldots) = axyz \ldots + \ldots + g\bar{x}\bar{y}\bar{z} \ldots,$$

then

$$^{-}f(x, y, z, \ldots) = \bar{a}xyz \ldots + \ldots + \bar{g}\bar{x}\bar{y}\bar{z} \ldots.$$

In applying this proposition any absent constituent term must be replaced with 0 as its coefficient and any constituent term with the form $xyz\ldots$ must be written $ixyz\ldots$ so that i is its coefficient.

For assume that the rule is true for n terms $x, y, z\ldots$ and let t be an $(n + 1)$th term.

Then developing with respect to the n terms $x, y, z, \ldots$

$$f(x, y, z, \ldots t) = Axyz \ldots + \ldots + G\bar{x}\bar{y}\bar{z} \ldots,$$

where the products such as $xyz \ldots, \ldots, \bar{x}\bar{y}\bar{z} \ldots$ do not involve t, and

$$A = at + a'\bar{t}, \ldots, G = gt + g'\bar{t}.$$

Then the letters $a, a', \ldots, g, g'$ are the coefficients of constituent terms of the expression as developed with respect to the $n + 1$ terms $x, y, z, \ldots t$.

Hence by the assumption

$$^{-}f(x, y, z, \ldots t) = \bar{A}xyz \ldots + \ldots + \bar{G}\bar{x}\bar{y}\bar{z} \ldots.$$

But by the rule already proved for one term,

$$\bar{A} = \bar{a}t + \bar{a}'\bar{t}, \ldots, \bar{G} = \bar{g}t + \bar{g}'\bar{t}.$$

Hence the rule holds for $(n + 1)$th terms. But the rule has been proved for one term. Thus it is true always.

30. ELIMINATION. (1) The object of elimination may be stated thus: Given an equation or a subsumption involving certain terms among others, to find what equations or subsumptions can be deduced not involving those terms.

The leading propositions in elimination are Propositions XXI. and XXII. of the last chapter, namely that, if $z \nleftarrow xu + yv$, then $z \nleftarrow x + y$; and if $z \nrightarrow (x + u)(y + v)$, then $z \nrightarrow xy$; and their Corollaries that, $xu + yv \nleftarrow x + y$, and, $(x + u)(y + v) \nrightarrow xy$.

(2) To prove that if $ax + b\bar{x} = c$, then $a + b \nrightarrow c \nrightarrow ab$.

Eliminating x and $\bar{x}$ by the above-mentioned proposition from the equation

$$c = ax + b\bar{x}, \quad c \nleftarrow a + b.$$

Also taking the supplementary equation,

$$\bar{c} = \bar{a}x + \bar{b}\bar{x}.$$

Hence from above

$$\bar{c} \Leftarrow \bar{a} + \bar{b}.$$

Taking the supplementary subsumption,

$$ab = {}^{-}(\bar{a} + \bar{b})$$
$$\Leftarrow c.$$

Therefore finally

$$a + b \Rightarrow c \Rightarrow ab.$$

The second part can also be proved* by taking the reciprocal equation, $c = (a + x)(b + \bar{x})$, and by using Prop. XXII. Corollary.

The same subsumptions, written in the supplementary form, are

$$\bar{a} + \bar{b} \Rightarrow \bar{c} \Rightarrow \bar{a}\bar{b}.$$

(3) By Prop. XI. each of these subsumptions is equivalent to an equation, which by Prop. VIII. can be put into many equivalent forms.

Thus $a + b \Rightarrow c$, can be written

$$c = c(a + b);$$

and this is equivalent to

$$\bar{a}\bar{b}c = 0.$$

And $c \Rightarrow ab$, can be written

$$ab = abc;$$

and this is equivalent to $\quad abc̄ = 0.$

(4) Conversely, if

$$\left.\begin{aligned} c &\Leftarrow a + b, \\ &\Rightarrow ab, \end{aligned}\right\}$$

then it has to be shown that we can write $ax + b\bar{x} = c$; where we have to determine the conditions that x must fulfil. This problem amounts to proving that the equation $ax + b\bar{x} = c$ has a solution, when the requisite conditions between a, b, c are fulfilled. The solution of the problem is given in the next article (cf. § 31 (9)).

The equation $ax + b\bar{x} = c$ includes a number of subsidiary equations.

For instance $ax = cx$; thence $\bar{a} + \bar{x} = \bar{c} + \bar{x}$, and thence $\bar{a}x = \bar{c}x$. Similarly $b\bar{x} = c\bar{x}$, and $\bar{b}\bar{x} = \bar{c}\bar{x}$. The solution of the given equation will satisfy identically all these subsidiary equations.

(5) *Particular Cases.* There are two important particular cases of this equation, when $c = i$, and when $c = 0$.

Firstly, $c = i$. Then $\quad ax + b\bar{x} = i.$

* Pointed out to me by Mr W. E. Johnson.

Hence $$a + b \ni i.$$

But the only possible case of this subsumption is

$$a + b = i.$$

Also $ab \in i$, which is necessarily true.

Therefore finally, $a + b = i$, is the sole deduction independent of x.

Secondly, $c = 0$. Then $$ax + b\bar{x} = 0.$$

Hence $a + b \ni 0$, which is necessarily true.

Also $ab \in 0$. But the only possible case of this subsumption is $ab = 0$.

Therefore finally, $ab = 0$, is the only deduction independent of x. If the equation be written $f(x) = 0$, the result of the elimination becomes

$$f(i) f(0) = 0.$$

These particular cases include each other. For if

$$ax + b\bar{x} = i,$$

then $$-(ax + b\bar{x}) = 0,$$

that is $$\bar{a}x + b\bar{x} = 0.$$

And $a + b = i$ is equivalent to $\bar{a}\bar{b} = 0$.

(6) *General Equation.* The general form $\phi(x) = \psi(x)$, where $\phi(x)$ and $\psi(x)$ are defined in the same way as $f(x)$ in § 29 (3), can be reduced to these cases. For this equation is equivalent to

$$\phi(x)\bar{\psi}(x) + \bar{\phi}(x)\psi(x) = 0.$$

This is easily proved by noticing that the derived equation implies

$$\phi(x)\bar{\psi}(x) = 0, \quad \bar{\phi}(x)\psi(x) = 0,$$

that is $$\phi(x) \in \psi(x), \quad \psi(x) \in \phi(x),$$

that is $$\phi(x) = \psi(x).$$

Hence the equation $\phi(x) = \psi(x)$ can be written by § 29 (4) in the form

$$\{\phi(i)\bar{\psi}(i) + \bar{\phi}(i)\psi(i)\}x + \{\phi(0)\bar{\psi}(0) + \bar{\phi}(0)\psi(0)\}\bar{x} = 0.$$

Hence the result of eliminating x from the general equation is

$$\{\phi(i)\bar{\psi}(i) + \bar{\phi}(i)\psi(i)\}\{\phi(0)\bar{\psi}(0) + \bar{\phi}(0)\psi(0)\} = 0.$$

This equation includes the four equations

$$\phi(i)\phi(0)\bar{\psi}(i)\bar{\psi}(0) = 0, \quad \phi(i)\bar{\phi}(0)\bar{\psi}(i)\psi(0) = 0,$$

$$\bar{\phi}(i)\phi(0)\psi(i)\bar{\psi}(0) = 0, \quad \bar{\phi}(i)\bar{\phi}(0)\psi(i)\psi(0) = 0.$$

The reduction of the general equation to the form with the right-hand side null is however often very cumbrous. It is best to take as the standard form

$$ax + b\bar{x} = cx + d\bar{x} \quad \text{..............................(1).}$$

This form reduces to the form, $ax + b\bar{x} = c$, when $c = d$. For

$$cx + d\bar{x} = c(x + \bar{x}) = c.$$

The equation is equivalent to the two simultaneous equations

$$ax = cx, \quad b\bar{x} = d\bar{x};$$

as may be seen by multiplying the given equation respectively by x and $\bar{x}$.

Let the equation $ax = cx$ be called the positive constituent equation, and the equation $b\bar{x} = d\bar{x}$ be called the negative constituent equation of equation (1).

Taking the supplements

$$\bar{a} + \bar{x} = \bar{c} + \bar{x}, \quad \bar{b} + x = \bar{d} + x.$$

Hence multiplying by x and $\bar{x}$ respectively

$$\bar{a}x = \bar{c}x, \quad \text{and} \quad \bar{b}\bar{x} = \bar{d}\bar{x}.$$

So equation (1) can also be written

$$\bar{a}x + b\bar{x} = \bar{c}x + d\bar{x};$$

and the two supplementary forms give

$$\bar{a}x + \bar{b}\bar{x} = \bar{c}x + \bar{d}\bar{x},$$

$$ax + \bar{b}x = cx + \bar{d}x.$$

The elimination of x can also be conducted thus.

Put each side of equation (1) equal to z.

Then
$$ax + b\bar{x} = z,$$

$$cx + d\bar{x} = z.$$

Hence the following subsumptions hold,

$$a + b \ni z \ni ab;$$

$$c + d \ni z \ni cd.$$

Therefore
$$a + b \ni cd,$$

$$c + d \ni ab.$$

Also similarly from the form, $\bar{a}x + b\bar{x} = \bar{c}x + d\bar{x}$, we find the subsumptions

$$\bar{a} + b \ni \bar{c}d,$$

$$\bar{c} + d \ni \bar{a}b.$$

These four subsumptions contain (cf. § 31 (9), below) the complete result of eliminating x from the given equation (1). The two supplementary forms give the same subsumptions, only in their supplementary forms, but involving no fresh information.

From these four subsumptions it follows that,

$$ab\bar{c}\bar{d} = \bar{a}\bar{b}cd = a\bar{b}\bar{c}d = \bar{a}bc\bar{d} = 0.$$

These are obviously the four equations found by the other method, only written in a different notation.

From these equations the original subsumptions can be deduced. For $ab\bar{c}\bar{d}=0$ can be written

$$ab \mathrel{\not\in} {}^-(\bar{c}\bar{d}),$$

and therefore $$ab \mathrel{\not\in} c+d.$$

Similarly for the other subsumptions.

Also it can easily be seen that the four subsumptions can be replaced by the more symmetrical subsumptions, which can be expressed thus,

(The sum of any two coefficients, one from each constituent equation) $\mathrel{\not\ni}$ (The product of the other two).

(7) *Discriminants.* All these conditions and (as it will be shown) the solution of the equation can be expressed compendiously by means of certain functions of the coefficients which will be called the Discriminants of the equation.

The discriminant of $ax=cx$ is $ac+\bar{a}\bar{c}$. Let it be denoted by A.

The discriminant of $b\bar{x}=d\bar{x}$ is $bd+\bar{b}\bar{d}$. Let it be denoted by B.

Then A and B will respectively be called the positive and negative discriminants of the equation

$$ax+b\bar{x}=cx+d\bar{x} \quad \text{..........................(1).}$$

Now $$\bar{A}={}^-(ac+\bar{a}\bar{c})=\bar{a}c+a\bar{c},$$

and $$\bar{B}={}^-(bd+\bar{b}\bar{d})=\bar{b}d+b\bar{d}.$$

Therefore, remembering that $y+z=0$ involves $y=0$, $z=0$, it follows that all the conditions between the coefficients a, b, c, d can be expressed in the form

$$\bar{A}\bar{B}=0.$$

This equation will be called the resultant* of the equation

$$ax+b\bar{x}=cx+d\bar{x}.$$

It can be put into the following forms,

$$\bar{A} \mathrel{\not\in} B,\ \bar{B} \mathrel{\not\in} A,\ \bar{A}B=\bar{A},\ \bar{B}A=\bar{B},\ \bar{A}+B=B,\ \bar{B}+A=A.$$

It is shown below in § 31 (9) that the resultant includes every equation between the coefficients and not containing x which can be deduced from equation (1).

The equation $ax+b\bar{x}=cx+d\bar{x}$ when written with its right-hand side null takes the form

$$\bar{A}x+\bar{B}\bar{x}=0.$$

(8) Again let there be n simultaneous equations $a_1x=c_1x$, $a_2x=c_2x$, ..., $a_nx=c_nx$, and let A_1, A_2, ... A_n be the discriminants of the successive equations respectively; then their product $A_1A_2 \ldots A_n$ is called the resultant

* Cf. Schröder, *Algebra der Logik*, § 21.

discriminant of the n equations. It will be denoted by $\Pi(A_r)$, or more shortly A. Similarly let there be n simultaneous equations $b_1\bar{x} = d_1\bar{x}$ $b_2\bar{x} = d_2\bar{x}, \ldots b_n\bar{x} = d_n\bar{x}$; and let $B_1, B_2 \ldots B_n$ be the discriminants. Then $B_1B_2 \ldots B_n$ is the resultant discriminant of the n equations. It will be called $\Pi(B_r)$ or B.

The n equations
$$a_1x + b_1\bar{x} = c_1x + d_1\bar{x},$$
$$a_2x + b_2\bar{x} = c_2x + d_2\bar{x},$$
$$\ldots\ldots\ldots\ldots\ldots\ldots\ldots\ldots$$
$$a_nx + b_n\bar{x} = c_nx + d_n\bar{x},$$
involve the $2n$ equations just mentioned and conversely.

The functions A and B are called the positive and negative resultant discriminants of these equations.

Now
$$\bar{A} = \bar{A}_1 + \bar{A}_2 + \ldots + \bar{A}_n,$$
$$\bar{B} = \bar{B}_1 + \bar{B}_2 + \ldots + \bar{B}_n.$$

Hence
$$\bar{A}\bar{B} = \Sigma\, \bar{A}_r\bar{B}_s.$$

Now any equation $a_rx = c_rx$ may be joined with any equation $b_s\bar{x} = d_s\bar{x}$ to form the equation $a_rx + b_s\bar{x} = c_rx + d_s\bar{x}$. Hence all the relations between the coefficients are included in all the equations of the type,
$$\bar{A}_r\bar{B}_s = 0.$$

But these equations are all expressed by the equation
$$\bar{A}\bar{B} = 0.$$

This equation may therefore conveniently be called the resultant of the n equations.

This is the complete solution of the problem of the elimination of a single letter which satisfies any number of equations.

The single equation
$$\bar{A}x + \bar{B}\bar{x} = 0,$$
is equivalent to the n given equations.

It must be carefully noticed that in this algebra the distinctions of procedure, which exist in ordinary algebra according to the number of equations given, do not exist. For here one equation can always be found which is equivalent to a set of equations, and conversely a set of equations can be found which are equivalent to one equation.

(9) *More than one Unknown.* The general equation involving two unknowns, x and y, is of the type
$$axy + bx\bar{y} + c\bar{x}y + d\bar{x}\bar{y} = exy + fx\bar{y} + g\bar{x}y + h\bar{x}\bar{y}.$$

This equation is equivalent to the separate constituent equations, $axy = exy$, $bx\bar{y} = fx\bar{y}$, etc. Let a constituent equation involving x (as distinct from $\bar{x}$) be called a constituent positive with respect to x, and let a constituent equation involving $\bar{x}$ (as distinct from x) be called a constituent negative with respect to x. Thus, $axy = exy$, is positive with respect both to x and y; $bx\bar{y} = fx\bar{y}$, is positive with respect to x, negative with respect to y, and so on.

Let A, B, C, D stand for the discriminants of these constituents. Thus $A = ae + \bar{a}\bar{e}$, $B = bf + \bar{b}\bar{f}$, $C = cg + \bar{c}\bar{g}$, $D = dh + \bar{d}\bar{h}$. Then the discriminant A is called the discriminant positive with respect to x and y, B is the discriminant positive with respect to x and negative with respect to y, and so on.

The equation can be written in the form

$$(ay + b\bar{y})\,x + (cy + d\bar{y})\,\bar{x} = (ey + f\bar{y})\,x + (gy + h\bar{y})\,\bar{x}.$$

If we regard x as the only unknown, the positive discriminant is

$$(ay + b\bar{y})(ey + f\bar{y}) + (\bar{a}y + \bar{b}\bar{y})(\bar{e}y + \bar{f}\bar{y}),$$

that is

$$Ay + B\bar{y}.$$

The negative discriminant is $Cy + D\bar{y}$.

The resultant is $\quad (\bar{A}y + \bar{B}\bar{y})(\bar{C}y + \bar{D}\bar{y}) = 0;$

that is

$$\bar{A}\bar{C}y + \bar{B}\bar{D}\bar{y} = 0.$$

This is the equation satisfied by y when x is eliminated. It will be noticed that A and C are the discriminants of the given equation positive with respect to y, and B and D are the discriminants negative with respect to y.

Similarly the equation satisfied by x when y is eliminated is

$$\bar{A}\bar{B}x + \bar{C}\bar{D}\bar{x} = 0.$$

The resultant of either of these two equations is

$$\bar{A}\bar{B}\bar{C}\bar{D} = 0.$$

This is therefore the resultant of the original equation.

The original equation when written with its right-hand side null takes the form

$$\bar{A}xy + \bar{B}x\bar{y} + \bar{C}\bar{x}y + \bar{D}\bar{x}\bar{y} = 0 \text{ ....(1)}.$$

Again suppose there are n simultaneous equations of the above type the coefficients of which are distinguished by suffixes 1, 2, ... n.

Then it may be shown just as in the case of a single unknown x, that all equations of the type, $\bar{A}_p\bar{B}_q\bar{C}_r\bar{D}_s = 0$, hold.

Hence if A stand for $A_1A_2 \dots A_n$, and B for $B_1B_2 \dots B_n$, C for $C_1C_2 \dots C_n$, and D for $D_1D_2 \dots D_n$, the resultant of the equations is $\bar{A}\bar{B}\bar{C}\bar{D} = 0$.

The n equations can be replaced by a single equation of the same form as (1) above.

Also the equation satisfied by x, after eliminating y only is

$$\bar{A}\bar{B}x + \bar{C}\bar{D}\bar{x} = 0,$$

where A and B are the positive discriminants with respect to x, and C and D are the negative discriminants. The equation satisfied by y is

$$\bar{A}\bar{C}y + \bar{B}\bar{D}\bar{y} = 0,$$

where a similar remark holds.

(10) This formula can be extended by induction to equations involving any number of unknowns. For the sake of conciseness of statement we will only give the extension from two unknowns to three unknowns, though the reasoning is perfectly general.

The general equation for three unknowns can be written in the form

$$(az + a'\bar{z})\,xy + (bz + b'\bar{z})\,x\bar{y} + (cz + c'\bar{z})\,\bar{x}y + (dz + d'\bar{z})\,\bar{x}\bar{y}$$
$$= (ez + e'\bar{z})\,xy + (fz + f'\bar{z})\,x\bar{y} + (gz + g'\bar{z})\,\bar{x}y + (hz + h'\bar{z})\,\bar{x}\bar{y}.$$

Then, if $A = ae + \bar{a}\bar{e}$, $A' = a'e' + \bar{a}'\bar{e}'$, and so on, A is the discriminant positive with respect to x, y, and z, and A' is the discriminant positive with respect to x and y, but negative with respect to z; and so on.

If x and y be regarded as the only unknowns, then the two discriminants positive with respect to x are

$$(az + a'\bar{z})(ez + e'\bar{z}) + (\bar{a}z + \bar{a}'\bar{z})(\bar{e}z + \bar{e}'\bar{z}),$$

and

$$(bz + b'\bar{z})(fz + f'\bar{z}) + (\bar{b}z + \bar{b}'\bar{z})(\bar{f}z + \bar{f}'\bar{z}),$$

that is,

$$Az + A'\bar{z}, \text{ and } Bz + B'\bar{z}.$$

Similarly the two discriminants negative with respect to x are

$$Cz + C'\bar{z}, \text{ and } Dz + D'\bar{z}.$$

Hence the equation for x after eliminating y is

$$(\bar{A}z + \bar{A}'\bar{z})(\bar{B}z + \bar{B}'\bar{z})\,x + (\bar{C}z + \bar{C}'\bar{z})(\bar{D}z + \bar{D}'\bar{z})\,\bar{x} = 0,$$

that is

$$(\bar{A}\bar{B}z + \bar{A}'\bar{B}'\bar{z})\,x + (\bar{C}\bar{D}z + \bar{C}'\bar{D}'\bar{z})\,\bar{x} = 0.$$

The result of eliminating z from this equation is

$$\bar{A}\bar{A}'\bar{B}\bar{B}'x + \bar{C}\bar{C}'\bar{D}\bar{D}'\bar{x} = 0.$$

Hence the equation for x after eliminating the other unknowns is of the form, $Px + Q\bar{x} = 0$, where P is the product of the supplements of the discriminants positive with respect to x, and Q is the product of the supplements of the discriminants negative with respect to x.

The resultant of the whole equation is

$$\bar{A}\bar{A}'\bar{B}\bar{B}'\bar{C}\bar{C}'\bar{D}\bar{D}' = 0,$$

that is the product of the supplements of the discriminants is zero.

The given equation when written with its right-hand side null takes the form

$$\bar{A}xyz + \bar{A}'xy\bar{z} + \bar{B}x\bar{y}z + \bar{B}'x\bar{y}\bar{z} + \bar{C}\bar{x}yz + \bar{C}'\bar{x}y\bar{z} + \bar{D}\bar{x}\bar{y}z + \bar{D}'\bar{x}\bar{y}\bar{z} = 0.$$

The same formulæ hold for any number of equations with any number of variables, if resultant discriminants are substituted for the discriminants of a single equation.

(11) It is often convenient to notice that if

$$\phi(x, y, z, \ldots) = \psi(x, y, z, \ldots),$$

be an equation involving any number of variables, then any discriminant is of the form

$$\phi\begin{pmatrix} i, & i, & i, & \ldots \\ 0, & 0, & 0, & \ldots \end{pmatrix} \psi\begin{pmatrix} i, & i, & i, & \ldots \\ 0, & 0, & 0, & \ldots \end{pmatrix} + \bar{\phi}\begin{pmatrix} i, & i, & i, & \ldots \\ 0, & 0, & 0, & \ldots \end{pmatrix} \bar{\psi}\begin{pmatrix} i, & i, & i, & \ldots \\ 0, & 0, & 0, & \ldots \end{pmatrix},$$

where i is substituted for each of the unknowns with respect to which the discriminant is positive and 0 is substituted for each of the unknowns with respect to which the discriminant is negative.

(12) The formula for the elimination of some of the unknowns, say, $u, v, w, \ldots$, from an equation involving any number of unknowns, $x, y, z, \ldots u, v, w, \ldots$, can easily be given. For example, consider only four unknowns, x, y, z, t, and let it be desired to eliminate z and t from this equation, so that a resultant involving only x and y is left. Let any discriminant of the equation be written $D\begin{pmatrix} i, & i, & i, & i \\ 0, & 0, & 0, & 0 \end{pmatrix}$, where either i or 0 is to be written according to the rule of subsection (11). The equation can be written

$$\{\bar{D}(i, i, i, i)\,xyz + \bar{D}(i, i, 0, i)\,xy\bar{z} + \bar{D}(i, 0, i, i)\,x\bar{y}z + \bar{D}(0, i, i, i)\,\bar{x}yz$$
$$+ \bar{D}(i, 0, 0, i)\,x\bar{y}\bar{z} + \bar{D}(0, i, 0, i)\,\bar{x}y\bar{z} + D(0, 0, i, i)\,\bar{x}\bar{y}z + \bar{D}(0, 0, 0, i)\,\bar{x}\bar{y}\bar{z}\}\,t$$
$$+ \{\bar{D}(i, i, i, 0)\,xyz + \bar{D}(i, i, 0, 0)\,xy\bar{z} + \ldots\}\,\bar{t} = 0.$$

Hence eliminating t, the resultant is

$$\bar{D}(i, i, i, i)\,\bar{D}(i, i, i, 0)\,xyz + \bar{D}(i, i, 0, i)\,\bar{D}(i, i, 0, 0)\,xy\bar{z}$$
$$+ \bar{D}(i, 0, i, i)\,\bar{D}(i, 0, i, 0)\,x\bar{y}z + \ldots + \bar{D}(0, 0, 0, i)\,\bar{D}(0, 0, 0, 0)\,\bar{x}\bar{y}\bar{z} = 0.$$

Again eliminating z by the same method, the resultant is

$$\bar{D}(i, i, i, i)\,\bar{D}(i, i, i, 0)\,\bar{D}(i, i, 0, i)\,\bar{D}(i, i, 0, 0)\,xy$$
$$+ \bar{D}(i, 0, i, i)\,\bar{D}(i, 0, i, 0)\,\bar{D}(i, 0, 0, i)\,\bar{D}(i, 0, 0, 0)\,x\bar{y}$$
$$+ \bar{D}(0, i, i, i)\,\bar{D}(0, i, i, 0)\,\bar{D}(0, i, 0, i)\,\bar{D}(0, i, 0, 0)\,\bar{x}y$$
$$+ \bar{D}(0, 0, i, i)\,\bar{D}(0, 0, i, 0)\,\bar{D}(0, 0, 0, i)\,\bar{D}(0, 0, 0, 0)\,\bar{x}\bar{y} = 0.$$

It is evident from the mode of deduction that the same type of formula holds for any number of unknowns.

31. SOLUTION OF EQUATIONS WITH ONE UNKNOWN. (1) The solutions of equations will be found to be of the form of sums of definite regions together with sums of undetermined portions of other definite regions; for example to be of the form $a + v_1 b + v_2 c$, where a, b, c are defined regions and v_1, v_2 are entirely arbitrary, including i or 0.

Now it is to be remarked that $u(b + c)$, where u is arbitrary, is as general as $v_1b + v_2c$. For writing $u = v_1b + v_2(\bar{b} + bc)$, which is allowable since u is entirely arbitrary, then

$$u(b+c) = \{v_1b + v_2(\bar{b} + bc)\}\{b + c\}$$
$$= v_1b + v_2(bc + \bar{b}c)$$
$$= v_1b + v_2c.$$

Hence it will always be sufficient to use the form $u(b+c)$, unless v_1 and v_2 are connected by some condition in which case $v_1b + v_2c$ may be less general than $u(b+c)$.

(2) $$ax = cx.$$

Then by § 30, (7) $$\bar{A}x = 0.$$

Hence by § 26, Prop. VIII. $$x = Ax.$$

But instead of x on the right-hand side of this last equation, $(x + v\bar{A})$ may be substituted, where v is subject to no restriction. But the only restriction to which x is subjected by this equation is that it must be incident in A. Hence $x + v\bar{A}$ is perfectly arbitrary.

Thus finally $$x = vA;$$
where v is arbitrary.

(3) $$b\bar{x} = d\bar{x}.$$

From subsection (2); $$\bar{x} = \bar{u}B.$$

Hence $$x = u + \bar{B}.$$

(4) $$ax + b\bar{x} = cx + d\bar{x}; \text{ where } \bar{A}\bar{B} = 0.$$

From the equation $ax = cx$, it follows that $x = uA$;

and from $b\bar{x} = d\bar{x}$, that $x = v + \bar{B}$.

Hence $$uA = v + \bar{B}.$$

Therefore $$v\bar{A} = 0.$$

Hence $$v = wA.$$

Finally, $$x = \bar{B} + wA;$$
where w is arbitrary.

This solution can be put into a more symmetrical form, remembering that
$$\bar{B} + A = A.$$

For $$x = \bar{B}(w + \bar{w}) + wA = \bar{B}\bar{w} + w(A + \bar{B}) = wA + \bar{w}\bar{B}.$$

Hence the solution can be written

$$\left.\begin{aligned} x &= \bar{B} + wA, \\ \bar{x} &= \bar{A} + \bar{w}B. \end{aligned}\right\}$$

Or
$$\left.\begin{array}{l} x = wA + \bar{w}\bar{B}, \\ \bar{x} = w\bar{A} + \bar{w}B. \end{array}\right\}$$

The first form of solution has the advantage of showing at a glance the terms definitely given and those only given with an undetermined factor.

(5) To sum up the preceding results in another form: the condition that the equations $ax = cx$, $b\bar{x} = d\bar{x}$ may be treated as simultaneous is
$$\bar{A}\bar{B} = 0.$$

The solution which satisfies both equations is
$$x = \bar{B} + uA.$$

The solution which satisfies the first and not necessarily the second is
$$x = uA.$$

The solution which satisfies the second and not necessarily the first is
$$\bar{x} = \bar{u}B, \text{ that is } x = u + \bar{B}.$$

In all these cases u is quite undetermined and subject to no limitation.

(6) The case $ax + b\bar{x} = c$, is deduced from the preceding by putting $d = c$.

Then
$$A = ac + \bar{a}\bar{c}, \quad B = bc + \bar{b}\bar{c}.$$

The solutions retain the same form as in the general case.

The relations between a, b, c are all included in the two subsumptions
$$a + b \mathrel{\ni} c \mathrel{\ni} ab.$$

The case $ax + b\bar{x} = 0$ is found by putting $c = d = 0$.

The equation can be written
$$ax + b\bar{x} = 0x + 0\bar{x}.$$

The positive discriminant is $a0 + \bar{a}\bar{0}$, that is $\bar{a}$, the negative is $\bar{b}$. The resultant is $ab = 0$. The solution is
$$x = b + u\bar{a}.$$

(7) The solution for n simultaneous equations can be found with equal ease.

Let x satisfy the n equations
$$a_1x + b_1\bar{x} = c_1x + d_1\bar{x},$$
$$a_2x + b_2\bar{x} = c_2x + d_2\bar{x},$$
$$\dots\dots\dots\dots\dots\dots\dots$$
$$a_nx + b_n\bar{x} = c_nx + d_n\bar{x}.$$

Then x satisfies the two groups of n equations each, namely
$$a_1x = c_1x, \quad a_2x = c_2x \dots a_nx = c_nx;$$
and
$$b_1\bar{x} = d_1\bar{x}, \quad b_2\bar{x} = d_2\bar{x} \dots b_n\bar{x} = d_n\bar{x}.$$

From the first group

$$x = u_1A_1 = u_2A_2 = \dots = u_nA_n.$$

Hence $x = uA$; where u is not conditioned.

Similarly from the second group

$$\bar{x} = \bar{v}_1B_1 = \bar{v}_2B_2 = \dots = \bar{v}_nB_n.$$

Hence $\bar{x} = \bar{v}B$, $x = v + \bar{B}$; where v is not conditioned.

Therefore $$uA = v + \bar{B}.$$

Hence $$v\bar{A} = 0, \text{ that is } v = wA.$$

So finally the solution of the n equations is

$$\left.\begin{aligned} x &= \bar{B} + wA = wA + \bar{w}\bar{B}, \\ \bar{x} &= \bar{A} + \bar{w}B = w\bar{A} + \bar{w}B \end{aligned}\right\}.$$

The group $a_1x = c_1x$, $a_2x = c_2x$, etc. can always be treated as simultaneous, and so can the group of typical form $b_r\bar{x} = d_r\bar{x}$.

The condition that the two groups can be treated as simultaneous is

$$\bar{A}\bar{B} = 0.$$

(8) It has been proved that the solution $\bar{B} + uA$ satisfies the equation, $\bar{A}x + \bar{B}\bar{x} = 0$, without imposing any restriction on u. It has now to be proved that any solution of the equation can be represented by $\bar{B} + uA$, when u has some definite value assigned to it.

For if some solution cannot be written in this form, it must be capable of being expressed in the form $m\bar{B} + wA + n\bar{A}B$.

But $\bar{A}x = 0$, and $\bar{A}\bar{B} = 0$, hence, by substituting for x its assumed form, $n\bar{A}B = 0$. Thus the last term can be omitted.

Again, $\bar{B}\bar{x} = 0$; and $\bar{A}\bar{B} = 0$, hence $\bar{B}(\bar{m} + B)(\bar{w} + \bar{A}) = 0$; that is $\bar{m}\bar{w}\bar{B} = 0$. Hence $\bar{m} = p(w + B)$, and therefore $m = \bar{p} + \bar{w}\bar{B}$.

Therefore the solution becomes

$$\begin{aligned} x &= m\bar{B} + wA = (\bar{p} + \bar{w}\bar{B})\bar{B} + w(A + \bar{B}), \\ &= \bar{p}\bar{B} + \bar{B} + wA = \bar{B} + wA, \end{aligned}$$

Thus the original form contains all the solutions.

(9) To prove that the resultant $\bar{A}\bar{B} = 0$, includes all the equations to be found by eliminating x from

$$\bar{A}x + \bar{B}\bar{x} = 0.$$

For $x = \bar{B} + wA$ satisfies the equation on the assumption that $\bar{A}\bar{B} = 0$, and without any other condition.

Hence $\bar{A}\bar{B}$ is the complete resultant.

It easily follows that for more than one unknown the resultants found in § 30 are the complete resultants.

(10) Subsumptions of the general type

$$ax + b\bar{x} \ni cx + d\bar{x}$$

can be treated as particular cases of equations.

For the subsumption is equivalent to the equation

$$\begin{aligned} cx + d\bar{x} &= (cx + d\bar{x})(ax + b\bar{x}) \\ &= acx + bd\bar{x}. \end{aligned}$$

Hence

$$\begin{aligned} A &= ac + \bar{c}^{\,-}(ac) = ac + \bar{c} = a + \bar{c}, \\ B &= bd + \bar{d}^{\,-}(bd) = bd + \bar{d} = b + \bar{d}, \\ \bar{A} &= c(\bar{a} + \bar{c}) = \bar{a}c, \\ \bar{B} &= d(\bar{b} + \bar{d}) = \bar{b}d. \end{aligned}$$

Therefore the resultant $\bar{A}\bar{B} = 0$ is equivalent to $\bar{a}\bar{b}cd = 0$. This is the only relation between the coefficients to be found by eliminating x.

The given subsumption is equivalent to the two subsumptions

$$ax \ni cx, \quad b\bar{x} \ni d\bar{x};$$

that is, to the two equations

$$cx = acx, \quad d\bar{x} = bd\bar{x}.$$

The solution of $\quad ax \ni cx$ is $x = uA = u(a + \bar{c})$.

The solution of $\quad b\bar{x} \ni d\bar{x}$ is $x = u + \bar{B} = u + \bar{b}d$.

The solution of

$$ax + b\bar{x} \ni cx + d\bar{x}$$

is

$$\begin{aligned} x &= \bar{B} + uA = \bar{b}d + u(a + \bar{c}) \\ &= \bar{u}\bar{B} + uA = u(a + \bar{c}) + \bar{u}\bar{b}d. \end{aligned}$$

The case of n subsumptions of the general type with any number of unknowns can be treated in exactly the same way as a special type of equation.

32. ON LIMITING AND UNLIMITING EQUATIONS. (1) An equation $\phi(x, y, z, \ldots t) = \psi(x, y, z, \ldots t)$ involving the n unknowns $x, y, z, \ldots t$ is called unlimiting with respect to any of its unknowns (x say), if any arbitrarily assigned value of x can be substituted in it and the equation can be satisfied by solving for the remaining unknowns $y, z, \ldots t$; otherwise the equation is called limiting with respect to x. The equation is unlimiting with respect to a set of its variables $x, y, z, \ldots$, if the above property is true for each one of the unknowns of the set. The equation is unlimiting with respect to all its unknowns, if the above is true for each one of its unknowns. Such an equation is called an unlimiting equation.

The equation is unlimiting with respect to a set of its unknowns *simultaneously*, if arbitrary values of each of the set of unknowns can be *simultaneously* substituted in the equation.

It is obvious that an equation cannot be unlimiting with respect to all its unknowns simultaneously, unless it be an identity.

(2) The condition that any equation is unlimiting with respect to an unknown x is found from § 30 (10). For let P be the product of the supplements of the discriminants positive with respect to x and Q be the product of the supplements of the discriminants negative with respect to x. Then the equation limiting the arbitrary choice of x is, $Px + Q\bar{x} = 0$. Hence if the given equation be unlimiting with respect to x, the equation just found must be an identity. Hence $P = 0$, $Q = 0$.

(3) The condition that the equation be unlimiting with respect to a set of its unknowns is that the corresponding condition hold for each variable.

(4) The condition that the equation is unlimiting with respect to a set $x, y, z, \dots$ of its unknowns simultaneously is that the equation found after eliminating the remaining unknowns $t, u, v, \dots$ should be an identity. The conditions are found by reference to § 30 (12) to be that each product of supplements of all the discriminants of the same denomination (positive or negative) with respect to each unknown of the set, but not necessarily of the same denomination for different unknowns of the set, vanishes.

(5) Every equation can be transformed into an unlimiting equation. For let the equation involve the unknowns $x, y, z, \dots t$: and let the resultant of the elimination of all the unknowns except x be, $Px + Q\bar{x} = 0$.

Then $x = Q + u\bar{P}$, and if u be assigned any value without restriction, then x will assume a suitable value which may be substituted in the equation previous to solving for the other unknowns. Thus if all the equations of the type $Px + Q\bar{x} = 0$, be solved, and the original equation be transformed by substitution of, $x = Q + u\bar{P}$, $y = S + v\bar{R}$, etc., then the new equation between $u, v, \dots$ is unlimiting.

(6) The field of an unknown which appears in an equation is the collection of values, any one of which can be assigned to the unknown consistently with the solution of the equation. If the equation be unlimiting with respect to an unknown, the field of that unknown is said to be unlimited; otherwise the field is said to be limited.

Let the unknown be x, and with the notation of subsection (5), let the resultant after eliminating the other unknowns be $Px + Q\bar{x} = 0$. Then $x = Q + u\bar{P}$. Hence the field of x is the collection of values found by substituting all possible values for u, including i and 0. Thus every member of the field of x contains Q; and $\bar{P}$ contains every member of the field, since $PQ = 0$. The field of x will be said to have the minimum extension Q and the maximum extension $\bar{P}$.

33. ON THE FIELDS OF EXPRESSIONS. (1) *Definition.* The 'field of the expression $\phi(x, y, z, \dots t)$' will be used to denote the collection of values which the expression $\phi(x, y, z, \ldots t)$ can be made to assume by different choices of the unknowns $x, y, z, \ldots t$. If $\phi(x, y, z, \ldots t)$ can be made to assume any assigned value by a proper choice of $x, y, z, \ldots t$, then the field of $\phi(x, y, z, \ldots t)$ will be said to be unlimited. But if $\phi(x, y, z, \ldots t)$ cannot by any choice of $x, y, z, \ldots t$, be made to assume some values, then the field of $\phi(x, y, z, \ldots t)$ will be said to be limited.

(2) To prove that

$$axyz \ldots t + bxyz \ldots \bar{t} + \ldots k\bar{x}\bar{y}\bar{z} \ldots \bar{t},$$

is capable of assuming the value $a + b + \ldots + k$. This problem is the same as proving that the equation

$$axyz \ldots t + bxyz \ldots \bar{t} + \ldots + k\bar{x}\bar{y}\bar{z} \ldots \bar{t} = a + b + \ldots + k,$$

is always possible.

The discriminants (cf. § 30 (11)) are

$$A = a + \bar{a}\bar{b} \ldots \bar{k},\quad B = b + \bar{a}\bar{b} \ldots \bar{k},\quad \ldots K = k + \bar{a}\bar{b} \ldots \bar{k}.$$

Hence

$$\bar{A} = \bar{a}(b + c + \ldots k),\ \bar{B} = \bar{b}(a + c + \ldots + k),\ \ldots \bar{K} = \bar{k}(a + b + c + \ldots).$$

Hence the resultant $\bar{A}\bar{B} \ldots \bar{K} = 0$ is satisfied identically.

It is obvious that each member of the field of the expression must be incident in the region $a + b + c + \ldots + k$: for $a + b + c + \ldots + k$ is the value assumed by the expression when i is substituted for each product $xyz \ldots t$, $xyz \ldots \bar{t}, \ldots \bar{x}\bar{y}\bar{z} \ldots \bar{t}$. But this value certainly contains each member of the field.

(3) To prove that any member of the field of

$$axyz \ldots t + bxyz \ldots \bar{t} + \ldots + k\bar{x}\bar{y}\bar{z} \ldots \bar{t}$$

contains the region $abc \ldots k$.

For let $\phi(x, y, z, \ldots t)$, stand for the given expression. Then the region containing any member of the field of $\bar{\phi}(x, y, z, \ldots t)$ by the previous subsection is $\bar{a} + \bar{b} + \bar{c} + \ldots + \bar{k}$. Hence the region contained by any member of the field of $\phi(x, y, z, \ldots t)$ is $abc \ldots k$. Hence combining the results of the previous and present subsections

$$a + b + c + \ldots + k \mathrel{\Rrightarrow} \phi(x, y, z, \ldots t) \mathrel{\Rrightarrow} abc \ldots k.$$

The field of $\phi(x, y, z \ldots t)$ will be said to be contained between the maximum extension $a + b + \ldots + k$, and the minimum extension $ab \ldots k$.

(4) The most general form of p, where

$$a + b + c + \ldots + k \mathrel{\Rrightarrow} p \mathrel{\Rrightarrow} abc \ldots k,$$

is

$$p = abc \ldots k + u(a + b + c + \ldots + k).$$

In order to prove that the fields of

$$\phi(x, y, z, \ldots t) \text{ and } abc \ldots k + u(a + b + c + \ldots k),$$

are identical, it is necessary to prove that the equation

$$\phi(x, y, z, \ldots t) = abc \ldots k + u(a + b + c + \ldots + k),$$

is unlimiting as regards u.

The equation can be written

$$axyz \ldots t + bxyz \ldots \bar{t} + \ldots + k\bar{x}\bar{y}\bar{z} \ldots \bar{t} = abc \ldots k\bar{u} + (a + b + c + \ldots k)u.$$

The discriminants positive with respect to u are (cf. § 30 (11))

$$a(a + b + c \ldots + k) + \bar{a}.\bar{a}\bar{b}\bar{c} \ldots \bar{k}, \text{ that is, } a + \bar{a}\bar{b}\bar{c} \ldots \bar{k},$$

and $$b + \bar{a}\bar{b}\bar{c} \ldots \bar{k},\ c + \bar{a}\bar{b}\bar{c} \ldots \bar{k},\ \ldots k + \bar{a}\bar{b}\bar{c} \ldots \bar{k}.$$

Their supplements are

$$\bar{a}(b + c + \ldots + k),\ \bar{b}(a + c + \ldots + k),\ \bar{c}(a + b + \ldots + k),\ \ldots \bar{k}(a + b + c + \ldots).$$

Hence the product of the supplements is identically zero.

Similarly the discriminants negative with respect to u are

$$abc \ldots k + \bar{a}(\bar{a} + \bar{b} + \ldots + \bar{k}),\ abc \ldots k + \bar{b}(\bar{a} + \bar{b} + \ldots + \bar{k}),$$

and so on. Their supplements are $a(\bar{b} + \bar{c} + \ldots + \bar{k})$, and so on. The product of the supplements is identically zero. Hence (cf. § 32 (2)) the equation is unlimiting with respect to u.

Thus* the fields of $\phi(x, y, z, \ldots t)$ and of $abc \ldots k + u(a + b + c + \ldots + k)$ are identical and therefore without imposing any restriction on u we may write

$$\phi(x, y, z, \ldots t) = abc \ldots k + u(a + b + c + \ldots + k).$$

(5) The conditions that the field of $\phi(x, y, z, \ldots t)$ may be unlimited are obviously $abc \ldots k = 0$, $a + b + c + \ldots + k = i$.

The two conditions may also be written

$$abc \ldots k = 0 = \bar{a}\bar{b}\bar{c} \ldots \bar{k}.$$

(6) Consider the two expressions

$$axyz \ldots t + bxyz \ldots \bar{t} + \ldots + k\bar{x}\bar{y}\bar{z} \ldots \bar{t},$$

and $$a_1uvw \ldots p + b_1uvw \ldots \bar{p} + \ldots h_1\bar{u}\bar{v}\bar{w} \ldots \bar{p},$$

not necessarily involving the same number of unknowns. Call them $\phi(x, y, z \ldots t)$ and $\psi(u, v, w \ldots p)$. The conditions that the field of $\phi(x, y, z \ldots t)$ may contain the field of $\psi(u, v, w \ldots p)$, i.e. that all the values which ψ may assume shall be among those which ϕ may assume, are $abc \ldots k \mathrel{\not\leftarrow} a_1b_1c_1 \ldots h_1$, and $a + b + c \ldots + k \mathrel{\not\rightarrow} a_1 + b_1 + c_1 + \ldots + h_1$.

The two conditions may also be written

$$abc \ldots k \mathrel{\not\leftarrow} a_1b_1c_1 \ldots h_1,$$
$$\bar{a}\bar{b}\bar{c} \ldots \bar{k} \mathrel{\not\leftarrow} \bar{a}_1\bar{b}_1\bar{c}_1 \ldots \bar{h}_1.$$

* Cf. Schröder, *Algebra der Logik*, Lecture 10, § 19, where this theorem is deduced by another proof.

(7) The conditions that the fields of $\phi(x, y, z \dots t)$ and $\psi(u, v, w \dots p)$ may be identical are obviously

$$abc \dots k = a_1 b_1 c_1 \dots h_1,$$
$$\bar{a}\bar{b}\bar{c} \dots \bar{k} = \bar{a}_1 \bar{b}_1 \bar{c}_1 \dots \bar{h}_1.$$

(8) To find the field of $f(x, y, z \dots t)$, when the unknowns are conditioned by any number of equations of the general type

$$\phi_r(x, y, z \dots t) = \psi_r(x, y, z \dots t).$$

Write $p = f(x, y, z \dots t)$; and eliminate $x, y, z \dots t$ from this equation and the equations of condition. Let the discriminant of the typical equation of condition positive with respect to all the variables be A_r, let the discriminant positive with respect to all except t be B_r, and so on, till all the discriminants are expressed. Then the resultant discriminants (cf. § 30 (8) and (9)) of these equations are $A = \Pi(A_r)$, $B = \Pi(B_r)$, etc.

Also let $f(x, y, z \dots t)$ be developed with respect to all its unknowns, so that we may write

$$p = axyz \dots t + bxyz \dots \bar{t} + \dots.$$

The discriminants of this equation are $pa + \bar{p}\bar{a}$, $pb + \bar{p}\bar{b}$, etc. Hence the resultant after eliminating $x, y, z \dots t$ is

$$^-\{(pa + \bar{p}\bar{a}) A\} \; ^-\{(pb + \bar{p}\bar{b}) B\} \dots = 0,$$

that is, $\quad \{p(\bar{a} + \bar{A}) + \bar{p}(a + \bar{A})\} \{p(\bar{b} + \bar{B}) + \bar{p}(b + \bar{B})\} \dots = 0.$

Hence $\quad p(\bar{a} + \bar{A})(\bar{b} + \bar{B}) \dots + \bar{p}(a + \bar{A})(b + \bar{B}) \dots = 0.$

Thus (cf. § 32 (6)) the field of p is comprised between

$$(a + \bar{A})(b + \bar{B}) \dots \text{ and } aA + bB + \dots.$$

But apart from the conditioning equations the field of p is comprised between $abc \dots$ and $a + b + c + \dots$. Thus the effect of the equations in limiting the field of p is exhibited.

The problem of this subsection is Boole's general problem of this algebra, which is stated by him as follows (cf. *Laws of Thought*, Chapter IX. § 8): 'Given any equation connecting the symbols $x, y, \dots w, z, \dots$, required to determine the logical expression of any class expressed in any way by the symbols $x, y \dots$ in terms of the remaining symbols, w, z, etc.' His mode of solution is in essence followed here, $w, z, \dots$ being replaced by the coefficients and discriminants. Boole however did not notice the distinction between expressions with limited and unlimited fields, so that he does not point out that the problem may also have a solution where no equation of condition is given.

A particular case of this general problem is as follows:

Given n equations of the type $a_r x + b_r \bar{x} = c_r x + d_r \bar{x}$,

to determine z, where z is given by $z = ex + f\bar{x}$.

Let the discriminants of the n equations be A and B, those of the equation which defines z are $ez + \bar{e}\bar{z}$, $\quad fz + \bar{f}\bar{z}$.

Hence the resultant is $^-(eAz + \bar{e}A\bar{z})^-(fBz + \bar{f}B\bar{z}) = 0$,

that is $(\bar{e}\bar{f} + f\bar{A} + \bar{e}\bar{B})z + (ef + f\bar{A} + e\bar{B})\bar{z} = 0$; where $\bar{A}\bar{B} = 0$.

Hence $$z = (ef + f\bar{A} + e\bar{B}) + u(eA + fB).$$

Another mode of solution, useful later, of this particular case is as follows:

The solution for x of the equations is $x = \bar{B} + vA$, $\bar{x} = \bar{A} + \bar{v}B$.

Substitute this value of x in the expression for z.

Then $$z = e\bar{B} + f\bar{A} + veA + \bar{v}fB = (eA + f\bar{A})v + (fB + e\bar{B})\bar{v}.$$

It is easy to verify by the use of subsection (7) that this solution is equivalent to the previous solution.

(9) An example of the general problem of subsection (8), which leads to important results later (cf. § 36 (2) and (3)), is as follows.

Given the equation $\bar{A}xy + \bar{B}x\bar{y} + \bar{C}\bar{x}y + \bar{D}\bar{x}\bar{y} = 0$, to determine xy, $x\bar{y}$, $\bar{x}y$, $\bar{x}\bar{y}$, $xy + \bar{x}\bar{y}$, $x\bar{y} + \bar{x}y$.

Put $z = xy$, then by comparison with subsection (8) $a = i$, $b = 0 = c = d$. Hence $(a + \bar{A})(b + \bar{B})(c + \bar{C})(d + \bar{D})$ becomes $\bar{B}\bar{C}\bar{D}$, and $aA + bB + cC + dD$ becomes A.

Thus, remembering that $\bar{A}\bar{B}\bar{C}\bar{D} = 0$,

$$xy = \bar{B}\bar{C}\bar{D} + uA = A(\bar{B}\bar{C}\bar{D} + u).$$

Similarly
$$x\bar{y} = \bar{A}\bar{C}\bar{D} + uB = B(\bar{A}\bar{C}\bar{D} + u),$$
$$\bar{x}y = \bar{A}\bar{B}\bar{D} + uC = C(\bar{A}\bar{B}\bar{D} + u),$$
$$\bar{x}\bar{y} = \bar{A}\bar{B}\bar{C} + uD = D(\bar{A}\bar{B}\bar{C} + u).$$

Also
$$xy + \bar{x}\bar{y} = \bar{B}\bar{C} + u(A + D) = (A + D)\{\bar{B}\bar{C} + u\},$$
$$x\bar{y} + \bar{x}y = \bar{A}\bar{D} + u(B + C) = (B + C)\{\bar{A}\bar{D} + u\}.$$

It is to be noticed that the arbitrary term u of one equation is not identical with the arbitrary term u of any other equation. But relations between the various u's must exist, since $xy + x\bar{y} + \bar{x}y + \bar{x}\bar{y} = i$.

(10) It is possible that the dependence of the value of an expression $f(x, y, z \ldots t)$ on the value of any one of the unknowns may be only apparent. For instance if $f(x)$ stand for $x + \bar{x}$, then $f(x)$ is always i for all values of x.

It is required to find the condition that, when the values of $y, z, \ldots t$ are given, the value of $f(x, y, z \ldots t)$ is also given.

For let $f(x, y, z \ldots t) = xf_1 + \bar{x}f_2$, where f_1 and f_2 are functions of $y, z \ldots t$ only. Then on the right-hand side either i or 0 may by hypothesis be put for x without altering the value of the function.

Hence $$f_1 = f(x, y \ldots t) = f_2.$$

Thus $f_1 = f_2$ is the requisite condition.

Let $f(x, y, z \dots t)$ be written in the form

$$x(ayz \dots t + byz \dots \bar{t} + \dots) + \bar{x}(a'yz \dots t + b'yz \dots \bar{t} + \dots),$$

then the required condition is $a = a'$, $b = b'$, etc.

34. Solution of Equations with more than one Unknown. (1) Any equation involving n unknowns, $x, y, z \dots r, s, t$ can always be transformed into an equation simultaneously unlimiting with respect to a set of any number of its unknowns, say with respect to $x, y, z \dots$. For let P_1 be the product of the supplements of the discriminants positive with respect to x, and Q_1 the product of the supplements of those negative with respect to x. Then (cf. § 30 (11)) the resultant after the elimination of all unknowns except x is,

$$P_1 x + Q_1 \bar{x} = 0.$$

Hence we may write, $x = Q_1 + \bar{P}_1 x_1 = Q_1 \bar{x}_1 + \bar{P}_1 x_1$, where x_1 is perfectly arbitrary. Substitute this value of x in the given equation, then the transformed equation is unlimiting with respect to its new unknown x_1.

Again, in the original equation treat x as known, and eliminate all the other unknowns except y.

Then the resultant is an equation of the form

$$(Rx + S\bar{x})\,y + (Tx + U\bar{x})\,\bar{y} = 0,$$

where R, S, T, U can easily be expressed in terms of the products of the supplements of discriminants of the original equation. The discriminants in each product are to be selected according to the following scheme (cf. § 30 (12)):

	R,	S,	T,	U
x	+,	−,	+,	−
y	+,	+,	−,	−

Now substitute for x in terms of x_1, and the resultant becomes

$$P_2 y + Q_2 \bar{y} = 0,$$

where P_2 and Q_2 are functions of x_1.

Solving, $$y = Q_2 + \bar{P}_2 y_2 = Q_2 \bar{y}_2 + \bar{P}_2 y_2;$$

where y_2 is an arbitrary unknown.

If this value for y be substituted in the transformed equation, then an equation between $x_1, y_2, z \dots r, s, t$ is found which is unlimiting with respect to x_1 and y_2 simultaneously.

Similarly in the original equation treat x, y as known, and eliminate all the remaining unknowns except z: a resultant equation is found of the type

$$(V_1 xy + V_2 x\bar{y} + V_3 \bar{x}y + V_4 \bar{x}\bar{y})\,z + (W_1 xy + W_2 x\bar{y} + W_3 \bar{x}y + W_4 \bar{x}\bar{y})\,\bar{z} = 0;$$

where the V's and W's are products of the supplements of discriminants selected according to an extension of the above scheme. Now substitute for x and y in terms of x_1 and y_2, and there results an equation of the type

$$P_3 z + Q_3 \bar{z} = 0;$$

where P_3 and Q_3 contain x_1, y_2.

Solving, $$z = Q_3 + \bar{P}_3 z_3 = Q_3 \bar{z}_3 + \bar{P}_3 z_3,$$

where z_3 is an arbitrary unknown. Then substituting for z, the transformed equation involving $x_1, y_2, z_3 \ldots r, s, t$ is unlimiting with regard to x_1, y_2, z_3 simultaneously.

Thus by successive substitutions, proceeding according to this rule, any set of the unknowns can be replaced by a corresponding set with respect to which the transformed equation is simultaneously unlimiting.

(2) If this process has been carried on so as to include the $n-1$ unknowns $x, y, z \ldots s$, then the remaining unknown t is conditioned by the equation $P_n t + Q_n \bar{t} = 0$; where P_n and Q_n involve $x_1, y_2 \ldots s_{n-1}$ which are unlimited simultaneously.

Solving for t, $$t = Q_n + \bar{P}_n t_n = Q_n \bar{t}_n + \bar{P}_n t_n;$$

where t_n is an arbitrary unknown.

Thus the general equation is solved by the following system of values,

$$x = Q_1 + \bar{P}_1 x_1,\ y = Q_2 + \bar{P}_2 y_2, \ldots s = Q_{n-1} + \bar{P}_{n-1} s_{n-1},\ t = Q_n + \bar{P}_n t_n;$$

where $x_1, y_2 \ldots t_n$ are arbitrary unknowns.

(3) The generality of the solution, namely the fact that the field of the solution for any variable is identical with the field of that variable as implicitly defined by the original equation, is proved by noting that each step of the process of solution is either a process of forming the resultant of an equation or of solving an equation for one unknown. But since the resultant thus formed is known to be the complete resultant (cf. § 31 (9)), and the solution of the equation for one unknown is known to be the complete solution (cf. § 31 (8)), it follows that the solutions found are the general solutions.

It follows from this method of solution that the general solution of the general equation involving n unknowns requires n arbitrary unknowns.

(4) Consider, as an example*, the general equation involving two unknowns,

$$axy + bx\bar{y} + c\bar{x}y + d\bar{x}\bar{y} = exy + fx\bar{y} + g\bar{x}y + h\bar{x}\bar{y}.$$

Let A, B, C, D be its discriminants.

Then $$\bar{x} = C\bar{D} + {}^{-}(\bar{A}\bar{B})\, x_1 = {}^{-}(\bar{A}\bar{B})\, x_1 + \bar{C}\bar{D}\bar{x}_1,$$

$$\bar{x} = \bar{A}\,\bar{B} x_1 + {}^{-}(\bar{C}\bar{D})\, \bar{x}_1.$$

* Cf. Schröder, *Algebra der Logik*, § 22.

Also $$(\bar{A}x + \bar{C}\bar{x})\, y + (\bar{B}x + \bar{D}\bar{x})\, \bar{y} = 0.$$

Hence

$$y = \bar{B}x + \bar{D}\bar{x} + (Ax + C\bar{x})\, y_2 = (Ax + C\bar{x})\, y_2 + (\bar{B}x + \bar{D}\bar{x})\, \bar{y}_2$$
$$= \{(A + \bar{A}\bar{B}C)\, x_1 + (C + A\bar{C}\bar{D})\, \bar{x}_1\}\, y_2 + \{(A\bar{B} + \bar{A}\bar{B}\bar{D})\, x_1 + (C\bar{D} + \bar{B}\bar{C}\bar{D})\, \bar{x}_1\}\, \bar{y}_2.$$

As a verification it may be noticed that the field of y as thus expressed is contained between $A + C$ and $\bar{B}\bar{D}$. This is easily seen to be true, remembering that $\bar{A}\bar{B}\bar{C}\bar{D} = 0$.

(5) The equation involving two unknowns may be more symmetrically solved by substituting (cf. § 32 (5))

$$x = \bar{C}\bar{D} + {}^{-}(\bar{A}\bar{B})\, u = \bar{C}\bar{D}\bar{u} + {}^{-}(\bar{A}\bar{B})\, u,$$
$$y = \bar{B}\bar{D} + {}^{-}(\bar{A}\bar{C})\, v = \bar{B}\bar{D}\bar{v} + {}^{-}(\bar{A}\bar{C})\, v.$$

Then u and v are connected by the equation*,

$$\bar{A}BCuv + A\bar{B}Du\bar{v} + A\bar{C}D\bar{u}v + BC\bar{D}\bar{u}\bar{v} = 0.$$

This is an unlimiting equation: thus either u or v may be assumed arbitrarily and the other found by solving the equation.

Thus $$v = A\bar{B}Du + BC\bar{D}\bar{u} + {}^{-}(\bar{A}BCu + A\bar{C}D\bar{u})\, p,$$
or $$u = A\bar{C}Dv + BC\bar{D}\bar{v} + {}^{-}(\bar{A}BCv + A\bar{B}D\bar{v})\, q;$$
where p and q are arbitrary.

35. Symmetrical Solution of Equations with two unknowns.

(1) Schröder† has given a general symmetrical solution of the general equation involving two unknowns in a form involving three arbitrary unknowns.

The following method of solution includes his results but in a more general form.

(2) Consider any unlimiting equation involving two unknowns. Let A, B, C, D be its four discriminants. Then the equation can be written in the form

$$\bar{A}xy + \bar{B}x\bar{y} + \bar{C}\bar{x}y + \bar{D}\bar{x}\bar{y} = 0 \quad \text{.....................}(\alpha).$$

Now put $$x = a_1 uv + b_1 u\bar{v} + c_1 \bar{u}v + d_1 \bar{u}\bar{v} \quad \text{.......................}(\beta),$$
$$y = a_2 uv + b_2 u\bar{v} + c_2 \bar{u}v + d_2 \bar{u}\bar{v} \quad \text{.......................}(\gamma).$$

Since the equation (α) is unlimiting (cf. § 32 (2)),

$$\bar{A}\bar{B} = \bar{A}\bar{C} = \bar{B}\bar{D} = \bar{C}\bar{D} = 0.$$

* This equation was pointed out to me by Mr W. E. Johnson and formed the starting-point for my investigations into limiting and unlimiting equations and into expressions with limited and unlimited fields. As far as I am aware these ideas have not previously been developed, nor have the general symmetrical solutions for equations involving three or more unknowns been previously given, cf. §§ 35—37.

† *Algebra der Logik*, Lecture XII. § 24.

Also since the fields both of x and y are unlimited, then (cf. § 33 (5))

$$a_1b_1c_1d_1 = 0 = \bar{a}_1\bar{b}_1\bar{c}_1\bar{d}_1 = a_2b_2c_2d_2 = \bar{a}_2\bar{b}_2\bar{c}_2\bar{d}_2.$$

Substitute for x and y from (β) and (γ) in (α), and write $\phi(p, q)$ for the expression

$$\bar{A}pq + \bar{B}p\bar{q} + \bar{C}\bar{p}q + \bar{D}\bar{p}\bar{q}.$$

Then the equation between u and v is found to be

$$\phi(a_1, a_2)\,uv + \phi(b_1, b_2)\,u\bar{v} + \phi(c_1, c_2)\,\bar{u}v + \phi(d_1, d_2)\,\bar{u}\bar{v} = 0 ......(\delta).$$

Equation (δ) is the result of a general transformation from unknowns x and y to unknowns u and v.

(3) If the forms (β) and (γ) satisfy equation (α) identically for any two simultaneous values of u and v, then

$$\phi(a_1, a_2) = 0 = \phi(b_1, b_2) = \phi(c_1, c_2) = \phi(d_1, d_2).$$

Thus if the pairs (a_1, a_2), (b_1, b_2), (c_1, c_2), (d_1, d_2) be any pairs of simultaneous particular solutions of the original equation, then (β) and (γ) are also solutions.

(4) Assuming that $(a_1, a_2) \ldots (d_1, d_2)$ are pairs of simultaneous particular solutions of (α), it remains to discover the condition that the expressions (β) and (γ) for x and y give the general form of the solution.

This condition is discovered by noting that the solution is general, if when x has any arbitrarily assigned value, the field of y as defined by equation (α) is the same as the field of y as defined by (γ) when u and v are conditioned by equation (β).

Now equation (α) can be written

$$(\bar{A}x + \bar{C}\bar{x})\,y + (\bar{B}x + \bar{D}\bar{x})\,\bar{y} = 0.$$

Hence the field of y as defined by (α) is contained between the maximum extension (cf. § 32 (6)) $Ax + C\bar{x}$ and the minimum extension $\bar{B}x + \bar{D}\bar{x}$.

Now let A_x, B_x, C_x, D_x be the discriminants of (β) considered as an equation between u and v. Then

$$A_x = a_1x + \bar{a}_1\bar{x}, \quad B_x = b_1x + \bar{b}_1\bar{x}, \quad C_x = c_1x + \bar{c}_1\bar{x}, \quad D_x = d_1x + \bar{d}_1\bar{x}.$$

But by § 33 (8) the field of y as defined by (γ), where u and v are conditioned by (β) is contained between the maximum extension

$$a_2A_x + b_2B_x + c_2C_x + d_2D_x,$$

and the minimum extension

$$(a_2 + \bar{A}_x)(b_2 + \bar{B}_x)(c_2 + \bar{C}_x)(d_2 + \bar{D}_x):$$

that is, between the maximum extension

$$(a_1a_2 + b_1b_2 + c_1c_2 + d_1d_2)\,x + (\bar{a}_1a_2 + \bar{b}_1b_2 + \bar{c}_1c_2 + \bar{d}_1d_2)\,\bar{x},$$

and the minimum extension

$$(\bar{a}_1 + a_2)(\bar{b}_1 + b_2)(\bar{c}_1 + c_2)(\bar{d}_1 + d_2)\,x + (a_1 + a_2)(b_1 + b_2)(c_1 + c_2)(d_1 + d_2)\,\bar{x}.$$

If the field of y be the same according to both definitions, then

$$a_1a_2 + b_1b_2 + c_1c_2 + d_1d_2 = A \quad\text{.................}(\epsilon),$$

$$\bar{a}_1a_2 + \bar{b}_1b_2 + \bar{c}_1c_2 + \bar{d}_1d_2 = C \quad\text{.................}(\zeta),$$

$$(\bar{a}_1 + a_2)(\bar{b}_1 + b_2)(\bar{c}_1 + c_2)(\bar{d}_1 + d_2) = \bar{B} \quad\text{.................}(\eta),$$

$$(a_1 + a_2)(b_1 + b_2)(c_1 + c_2)(d_1 + d_2) = \bar{D} \quad\text{.................}(\theta).$$

These equations can be rewritten in the form

$$a_1a_2 + b_1b_2 + c_1c_2 + d_1d_2 = A \quad\text{.................}(\epsilon_1),$$

$$\bar{a}_1a_2 + \bar{b}_1b_2 + \bar{c}_1c_2 + \bar{d}_1d_2 = C \quad\text{.................}(\zeta_1),$$

$$a_1\bar{a}_2 + b_1\bar{b}_2 + c_1\bar{c}_2 + d_1\bar{d}_2 = B \quad\text{.................}(\eta_1),$$

$$\bar{a}_1\bar{a}_2 + \bar{b}_1\bar{b}_2 + \bar{c}_1\bar{c}_2 + \bar{d}_1\bar{d}_2 = D \quad\text{.................}(\theta_1).$$

It follows from their symmetry that if y be given, the field of x as defined by (β) and conditioned by (γ) is the same as the field of x as defined by (α).

By adding ϵ_1 and η_1, $\quad a_1 + b_1 + c_1 + d_1 = A + B.$

Hence $\quad \bar{a}_1\bar{b}_1\bar{c}_1\bar{d}_1 = \bar{A}\bar{B}.$

By adding (ζ_1) and (θ_1),

$$\bar{a}_1 + \bar{b}_1 + \bar{c}_1 + \bar{d}_1 = C + D.$$

Hence $\quad a_1b_1c_1d_1 = \bar{C}\bar{D}.$

Similarly, $\quad \bar{a}_2\bar{b}_2\bar{c}_2\bar{d}_2 = \bar{A}\bar{C},\ a_2b_2c_2d_2 = \bar{B}\bar{D}.$

Thus if the conditions between A, B, C, D of subsection (2) are fulfilled, then the conditions between a_1, b_1, c_1, d_1 and a_2, b_2, c_2, d_2 of subsection (2) are also fulfilled.

Hence finally if (a_1, a_2), (b_1, b_2), (c_1, c_2), (d_1, d_2) be any pairs of simultaneous solutions of (α) which satisfy equations (ϵ_1), (ζ_1), (η_1), (θ_1), then the expressions (β) and (γ) for x and y form the general solution of equation (α).

(5) Now take one pair of coefficients, say a_1 and a_2, to be any pair of particular simultaneous solutions of the equations

$$\bar{A}xy + \bar{B}x\bar{y} + \bar{C}\bar{x}y + \bar{D}\bar{x}\bar{y} = 0 \quad\text{........................}(\kappa),$$

and

$$xy = A \quad\text{..................................}(\lambda).$$

These two equations can be treated as simultaneous; for the discriminants of (λ) are A, $\bar{A}$, $\bar{A}$, $\bar{A}$. Hence the complete resultant of the two equations is

$$\bar{A}(\bar{B} + A)(\bar{C} + A)(\bar{D} + A) = 0,$$

that is

$$\bar{A}\bar{B}\bar{C}\bar{D} = 0;$$

and this equation is satisfied by hypothesis. Thus (κ) and (λ) can be combined into the single equation

$$\bar{A}xy + (\bar{B} + A)x\bar{y} + (\bar{C} + A)\bar{x}y + (\bar{D} + A)\bar{x}\bar{y} = 0;$$

that is, since $\bar{A}\bar{B}=\bar{A}\bar{C}=0$,

$$\bar{A}xy+Ax\bar{y}+A\bar{x}y+(\bar{D}+A)\bar{x}\bar{y}=0.$$

Any solution of this equation gives $xy=A$, $x\bar{y}\not\leqslant B$, $\bar{x}y\not\leqslant C$, $\bar{x}\bar{y}\not\leqslant D$; and hence any solution is consistent with equations (ϵ_1), (ζ_1), (η_1), (θ_1).

This equation is a limiting equation. By § 34 (5) it can be transformed into an unlimiting equation.

Put $$x=A+k,\ y=A+l.$$
Then the equation becomes

$$\bar{A}kl+\bar{A}\bar{D}\bar{k}\bar{l}=0.$$

Let another pair of the coefficients, say b_1 and b_2, be chosen to be any particular solutions of the equations

$$\bar{A}xy+\bar{B}x\bar{y}+\bar{C}\bar{x}y+\bar{D}\bar{x}\bar{y}=0;$$
$$x\bar{y}=B.$$

These equations can be treated as simultaneous; and are equivalent to the single equation

$$Bxy+\bar{B}x\bar{y}+(\bar{C}+B)\bar{x}y+B\bar{x}\bar{y}=0.$$

Any solutions of this equation give $xy\not\leqslant A$, $x\bar{y}=B$, $\bar{x}y\not\leqslant C$, $\bar{x}\bar{y}\not\leqslant D$.

To transform into an unlimiting equation, put $x=B+m$, $\bar{y}=B+n$. Then the equation becomes

$$\bar{B}mn+\bar{B}\bar{C}\bar{m}\bar{n}=0.$$

Let another pair of the coefficients, say c_1 and c_2, be chosen to be any particular solutions of the equations

$$\bar{A}xy+\bar{B}x\bar{y}+\bar{C}\bar{x}y+\bar{D}\bar{x}\bar{y}=0,$$
$$\bar{x}y=C.$$

These equations can be treated as simultaneous; and are equivalent to the single equation

$$Cxy+(\bar{B}+C)x\bar{y}+\bar{C}\bar{x}y+C\bar{x}\bar{y}=0.$$

Any solutions of this equation give $xy\not\leqslant A$, $x\bar{y}\not\leqslant B$, $\bar{x}y=C$, $\bar{x}\bar{y}\not\leqslant D$.

To transform into an unlimiting equation, put $\bar{x}=C+p$, $y=C+q$. Then the equation becomes

$$\bar{B}\bar{C}\bar{p}\bar{q}+\bar{C}pq=0.$$

Let the last pair of coefficients, namely d_1 and d_2, be chosen to be any particular solutions of the equations

$$\bar{A}xy+\bar{B}x\bar{y}+\bar{C}\bar{x}y+\bar{D}\bar{x}\bar{y}=0,$$
$$\bar{x}\bar{y}=D.$$

These equations can be treated as simultaneous; and are equivalent to the single equation

$$(\bar{A}+D)xy+Dx\bar{y}+D\bar{x}y+\bar{D}\bar{x}\bar{y}=0.$$

Any solutions of this equation give

$$xy \nleftarrow A, \quad x\bar{y} \nleftarrow B, \quad \bar{x}y \nleftarrow C, \quad \bar{x}\bar{y} = D.$$

To transform into an unlimiting equation, put $\bar{x} = D + r$, $\bar{y} = D + s$, Then the equation becomes

$$\bar{A}\bar{D}\bar{r}\bar{s} + \bar{D}rs = 0.$$

If the coefficients $a_1, a_2 \ldots d_2$, have these values, then the equations (ϵ), (ζ), (η), (θ) are necessarily satisfied.

Hence finally we have the result that the most general solution of the unlimiting equation

$$\bar{A}xy + \bar{B}x\bar{y} + \bar{C}\bar{x}y + \bar{D}\bar{x}\bar{y} = 0,$$

can be written

$$x = (A + k)\,uv + (B + m)\,u\bar{v} + \bar{C}\bar{p}\bar{u}v + \bar{D}\bar{r}\bar{u}\bar{v},$$
$$y = (A + l)\,uv + \bar{B}\bar{n}u\bar{v} + (C + q)\,\bar{u}v + \bar{D}\bar{s}\bar{u}\bar{v};$$

where u, v are arbitrary unknowns, and k and l, m and n, p and q, r and s, are any particular pairs of simultaneous solutions of

$$\left.\begin{aligned} \bar{A}kl + \bar{A}\bar{D}\bar{k}\bar{l} &= 0,\\ \bar{B}mn + \bar{B}\bar{C}\bar{m}\bar{n} &= 0,\\ \bar{C}pq + \bar{B}\bar{C}\bar{p}\bar{q} &= 0,\\ \bar{D}rs + \bar{A}\bar{D}\bar{r}\bar{s} &= 0. \end{aligned}\right\}$$

Let these equations be called the auxiliary equations.

The auxiliary equations can also be written,

$$\bar{A}\phi(k, l) = 0, \quad \bar{B}\phi(m, \bar{n}) = 0, \quad \bar{C}\phi(\bar{p}, q) = 0, \quad \bar{D}\phi(\bar{r}, \bar{s}) = 0.$$

(6) As an example, we may determine k, l, m, n, p, q, r, s so that the general solution has a kind of skew symmetry; namely so that x has the same relation to A as $\bar{y}$ has to D.

Thus put $k = 0$, $l = \bar{A}$; $m = B$, $\bar{n} = C$; $q = C$, $\bar{p} = B$; $s = 0$, $r = \bar{D}$. These satisfy the auxiliary equations. Hence the general solution can be written, remembering that $\bar{B}C = \bar{B}$, $\bar{C}B = \bar{C}$,

$$x = Auv + Bu\bar{v} + \bar{C}\bar{u}v, \qquad \bar{x} = \bar{A}uv + \bar{B}u\bar{v} + C\bar{u}v + \bar{u}\bar{v},$$
$$y = uv + \bar{B}u\bar{v} + C\bar{u}v + \bar{D}\bar{u}\bar{v}, \quad \bar{y} = Bu\bar{v} + \bar{C}\bar{u}v + D\bar{u}\bar{v}.$$

Again, put $k = i$, $l = 0$; $m = 0$, $n = i$; $p = 0$, $q = i$; $r = i$, $s = 0$. The solution takes the skew symmetrical form

$$x = uv + Bu\bar{v} + \bar{C}\bar{u}v, \quad \bar{x} = \bar{B}u\bar{v} + C\bar{u}v + \bar{u}\bar{v};$$
$$y = Auv + \bar{u}v + \bar{D}\bar{u}\bar{v}, \quad \bar{y} = \bar{A}uv + u\bar{v} + D\bar{u}\bar{v}.$$

As another example, notice that the auxiliary equations are satisfied by $k = w$, $l = \bar{w}$, $m = w$, $n = \bar{w}$, $p = \bar{w}$, $q = w$, $r = w$, $s = \bar{w}$.

Hence the general solution can be written

$$x = (A + w)\,uv + (B + w)\,u\bar{v} + \bar{C}w\bar{u}v + \bar{D}\bar{w}\bar{u}\bar{v},$$
$$y = (A + \bar{w})\,uv + \bar{B}wu\bar{v} + (C + w)\,\bar{u}v + \bar{D}w\bar{u}\bar{v};$$

where u, v and w are unrestricted, and any special value can be given to w without limiting the generality of the solution.

(7) The general symmetrical solution of the limiting equation can now be given. Let $\bar{A}xy + \bar{B}x\bar{y} + \bar{C}\bar{x}y + \bar{D}\bar{x}\bar{y} = 0$ be the given equation.

By § 34 (5), put $x = \bar{C}\bar{D} + (A + B)\,X$, $y = \bar{B}\bar{D} + (A + C)\,Y$; where X and Y are conditioned by

$$\bar{A}BCXY + A\bar{B}DX\bar{Y} + A\bar{C}D\bar{X}Y + BC\bar{D}\bar{X}\bar{Y} = 0.$$

The general symmetrical solution for X and Y is therefore by (5) of this section,

$$X = (A + \bar{B} + \bar{C} + k)\,uv + (\bar{A} + B + \bar{D} + m)\,u\bar{v} + A\bar{C}D\bar{p}\bar{u}v + BC\bar{D}\bar{r}\bar{u}\bar{v},$$
$$Y = (A + \bar{B} + \bar{C} + l)\,uv + A\bar{B}D\bar{n}u\bar{v} + (\bar{A} + C + \bar{D} + q)\,\bar{u}v + BC\bar{D}\bar{s}\bar{u}\bar{v};$$

where k, l; m, n; p, q; r, s are any simultaneous particular solutions of the auxiliary equations

$$\left.\begin{aligned} \bar{A}BCkl + \bar{A}BC\bar{D}\bar{k}\bar{l} &= 0,\\ A\bar{B}Dmn + A\bar{B}\bar{C}D\bar{m}\bar{n} &= 0,\\ A\bar{C}Dpq + A\bar{B}\bar{C}\bar{D}\bar{p}\bar{q} &= 0,\\ BC\bar{D}rs + \bar{A}BC\bar{D}\bar{r}\bar{s} &= 0.\end{aligned}\right\}$$

(8) As a particular example, adapt the first solution of subsection (6) of this section. Then a general solution of the equation is

$$x = \bar{C}\bar{D} + (A + B\bar{C})\,uv + (B + A\bar{D})\,u\bar{v} + A\bar{C}D\,\bar{u}v,$$
$$y = \bar{B}\bar{D} + (A + C)\,uv + A\bar{B}Du\bar{v} + (C + A\bar{D})\,\bar{u}v + BC\bar{D}\bar{u}\bar{v}.$$

(9) If a number of equations of the type,

$$\psi_1(x, y) = \chi_1(x, y), \quad \psi_2(x, y) = \chi_2(x, y), \text{ etc.},$$

be given, then (assuming that they satisfy the condition for their possibility) their solution can be found by substituting their resultant discriminants (cf. § 30, (8), (9)) for the discriminants of the single equation which has been considered in the previous subsections of this article.

(10) The symmetrical solution of an equation with two unknowns has been obtained in terms of two arbitrary unknowns, and of one or more unknowns to which any arbitrary particular values can be assigned without loss of the generality of the solution. It was proved in § 34 (3) that no solution with less than two unknowns could be general. It is of importance in the following articles to obtain the general symmetrical

solution with more than two arbitrary unknowns. For instance take three unknowns, u, v, w (though the reasoning will apply equally well to any number). Let the given unlimiting equation be

$$\bar{A}xy + \bar{B}x\bar{y} + \bar{C}\bar{x}y + \bar{D}\bar{x}\bar{y} = 0 \text{ ....}(\alpha).$$

Put

$$\left.\begin{aligned} x &= a_1\, uvw + b_1\, u\bar{v}w + c_1\, \bar{u}vw + d_1\, \bar{u}\bar{v}w \\ &\quad + a_1' uv\bar{w} + b_1' u\bar{v}\bar{w} + c_1' uv\bar{w} + d_1' \bar{u}\bar{v}\bar{w}, \\ y &= a_2\, uvw + \ldots \ldots + d_2\, \bar{u}\bar{v}w \\ &\quad + a_2' uv\bar{w} + \ldots\ldots + d_2' \bar{u}\bar{v}\bar{w}. \end{aligned}\right\} \text{ ....}(\beta).$$

Consider x as a known, then the maximum extension of the field of y as defined by (α) is $Ax + C\bar{x}$, and its minimum extension is $\bar{B}x + \bar{D}\bar{x}$.

Also the maximum extension of the field of y as defined by (β) is $\Sigma a_1 a_2 \,.\, x + \Sigma \bar{a}_1 a_2 \,.\, \bar{x}$, and its minimum extension is $\Pi(\bar{a}_1 + a_2)\, x + \Pi(a_1 + a_2) \,.\, \bar{x}$.

Hence, if (β) is the general solution of (α), the following four conditions must hold

$$\Sigma a_1 a_2 = A, \quad \Sigma a_1 \bar{a}_2 = B, \quad \Sigma \bar{a}_1 a_2 = C, \quad \Sigma \bar{a}_1 \bar{a}_2 = D.$$

Also a_1, a_2; $b_1, b_2 \ldots d_1', d_2'$; must be pairs of simultaneous solutions of the given equation (α).

36. Johnson's Method. (1) The following interesting method of solving symmetrically equations, limiting or unlimiting, involving any number of unknowns is due to Mr W. E. Johnson.

(2) *Lemma.* To divide $a + b$ into two mutually exclusive parts x and y, such that $x \mathrel{\not\leftarrow} a$ and $y \mathrel{\not\leftarrow} b$.

The required conditions are

$$x + y = a + b, \quad \bar{a}x = 0, \quad \bar{b}y = 0, \quad xy = 0.$$

These can be written $xy + \bar{a}x\bar{y} + \bar{b}\bar{x}y + (a + b)\,\bar{x}\bar{y} = 0$.

Hence by § 34 (5), $\left.\begin{aligned} x &= a\bar{b} + au = a(\bar{b} + u), \\ y &= \bar{a}b + bv = b(\bar{a} + v); \end{aligned}\right\}$ ....(1),

where $$ab\,(uv + \bar{u}\bar{v}) = 0 \text{ ....}(2).$$

Solving (2) for v in terms of u, by § 31 (5), $v = ab\bar{u} + (\bar{a} + \bar{b} + \bar{u})\, w$.

Substituting for v in (1) and simplifying,

$$x = a(\bar{b} + u), \quad y = b(\bar{a} + \bar{u}).$$

(3) Let the equation, limiting or unlimiting, be

$$\bar{A}xy + \bar{B}x\bar{y} + \bar{C}\bar{x}y + \bar{D}\bar{x}\bar{y} = 0 \text{.. ....}(3).$$

The resultant of elimination can be written $A + B + C + D = i$.

Also $xy + \bar{x}\bar{y}$ and $x\bar{y} + \bar{x}y$ are mutually exclusive, their sum

$$= i = A + B + C + D,$$

and obviously from the given equation $xy + \bar{x}\bar{y} \mathrel{\not\leftarrow} A + D$, $x\bar{y} + \bar{x}y \mathrel{\not\leftarrow} B + C$.

Hence by the lemma

$$xy + \bar{x}\bar{y} = (A + D)(\bar{B}\bar{C} + u), \quad x\bar{y} + \bar{x}y = (B + C)(\bar{A}\bar{D} + \bar{u}) \ldots\ldots(4).$$

The course of the proof has obviously secured that u does not have to satisfy some further condition in order that equation (4) may express the full knowledge concerning $xy + \bar{x}\bar{y}$ and $x\bar{y} + \bar{x}y$, which can be extracted from equation (3).

Also, as an alternative proof of this point, § 33 (9) secures that equation (4) represents the complete solution for these expressions.

Again, by equations (4) $xy + \bar{x}\bar{y} \mathrel{\ne\!\!<} \bar{B}\bar{C} + u$, hence $xy \mathrel{\ne\!\!<} \bar{B}\bar{C} + u$.

Also by equation (3), $xy \mathrel{\ne\!\!<} A$. Hence by § 28, Prop. XIV., $xy \mathrel{\ne\!\!<} A(\bar{B}\bar{C} + u)$.

Similarly $\bar{x}\bar{y} \mathrel{\ne\!\!<} D(\bar{B}\bar{C} + u)$.

Therefore by the lemma and equations (4) and simplifying

$$\left.\begin{aligned} xy &= A(\bar{B}\bar{C} + u)(\bar{D} + p), \\ \bar{x}\bar{y} &= D(\bar{B}\bar{C} + u)(\bar{A} + \bar{p}). \end{aligned}\right\} \ldots\ldots\ldots\ldots\ldots\ldots(5).$$

Also, as before, it follows that equations (5) are a complete expression of the information respecting xy and $\bar{x}\bar{y}$ to be extracted from equation (3).

Similarly

$$\left.\begin{aligned} x\bar{y} &= B(\bar{A}\bar{D} + \bar{u})(\bar{C} + q), \\ \bar{x}y &= C(\bar{A}\bar{D} + \bar{u})(\bar{B} + \bar{q}). \end{aligned}\right\} \ldots\ldots\ldots\ldots\ldots\ldots(6).$$

Adding appropriate equations out of (5) and (6),

$$\left.\begin{aligned} x &= A(\bar{B}\bar{C} + u)(\bar{D} + p) + B(\bar{A}\bar{D} + \bar{u})(\bar{C} + q), \\ y &= A(\bar{B}\bar{C} + u)(\bar{D} + p) + C(\bar{A}\bar{D} + \bar{u})(\bar{B} + \bar{q}). \end{aligned}\right\}$$

This symmetrical solution with three arbitraries is the symmetrical solution first obtained by Schröder (cf. loc. cit.).

(4) A simplified form of this expression has also been given by Johnson.

For $A(\bar{B}\bar{C} + u)(\bar{D} + p) = A(\bar{B}\bar{C}\bar{u} + u)(\bar{D} + p)$,

and $B(\bar{A}\bar{D} + \bar{u})(\bar{C} + q) = B(\bar{A}\bar{D}u + \bar{u})(\bar{C} + q)$.

Hence

$$\begin{aligned} x &= u\{A\bar{D} + Ap + B\bar{A}\bar{D}(\bar{C} + q)\} + \bar{u}\{A\bar{B}\bar{C}(\bar{D} + p) + B\bar{C} + Bq\} \\ &= u\{A\bar{D} + Ap + B\bar{D}(\bar{C} + q)\} + \bar{u}\{A\bar{C}(\bar{D} + p) + B\bar{C} + Bq\} \\ &= A(\bar{C} + u)(\bar{D} + p) + B(\bar{D} + \bar{u})(\bar{C} + q). \end{aligned}$$

Similarly $y = A(\bar{B} + u)(\bar{D} + p) + C(\bar{D} + \bar{u})(\bar{B} + \bar{q})$.

(5) This method of solution can be applied to equations involving any number of unknowns. The proof is the same as for two unknowns, and the headings of the argument will now be stated for three unknowns.

Consider the equation

$$\bar{A}xyz + \bar{B}x\bar{y}z + \bar{C}\bar{x}yz + \bar{D}\bar{x}\bar{y}z + \bar{A}'xy\bar{z} + \bar{B}'x\bar{y}\bar{z} + \bar{C}'\bar{x}y\bar{z} + \bar{D}'\bar{x}\bar{y}\bar{z} = 0 \ldots(1).$$

The resultant is $A + B + C + D + A' + B' + C' + D' = i$.

Also from (1) $xyz + x\overline{yz} + \bar{x}y\bar{z} + \overline{xy}z \nleftarrow A + D + B' + C'$,

$$\bar{x}yz + x\bar{y}z + xy\bar{z} + \overline{xyz} \nleftarrow B + C + A' + D'.$$

Hence by the lemma, cf. subsection (2)

$$\left.\begin{aligned} xyz + x\overline{yz} + \bar{x}y\bar{z} + \overline{xy}z &= (A + D + B' + C')(\bar{B}\bar{C}\bar{A}'\bar{D}' + s),\\ \bar{x}yz + x\bar{y}z + xy\bar{z} + \overline{xyz} &= (B + C + A' + D')(\bar{A}\bar{D}\bar{B}'\bar{C}' + \bar{s}).\end{aligned}\right\} \quad \ldots\ldots(2).$$

Again, from (2) and (1),

$$x\overline{yz} + \bar{x}y\bar{z} \nleftarrow (B' + C')(\bar{B}\bar{C}\bar{A}'\bar{D}' + s),$$
$$xyz + \overline{xy}z \nleftarrow (A + D)(\bar{B}\bar{C}\bar{A}'\bar{D}' + s).$$

Hence from the lemma, cf. subsection (2), and simplifying,

$$\left.\begin{aligned} x\overline{yz} + \bar{x}y\bar{z} &= (B' + C')(\bar{B}\bar{C}\bar{A}'\bar{D}' + s)(\bar{A}\bar{D} + m),\\ xyz + \overline{xy}z &= (A + D)(\bar{B}\bar{C}\bar{A}'\bar{D}' + s)(\bar{B}'\bar{C}' + \bar{m}).\end{aligned}\right\} \quad \ldots\ldots\ldots\ldots(3).$$

Similarly

$$\left.\begin{aligned} x\bar{y}z + \bar{x}yz &= (B + C)(\bar{A}\bar{D}\bar{B}'\bar{C}' + \bar{s})(\bar{A}'\bar{D}' + n),\\ xy\bar{z} + \overline{xyz} &= (A' + D')(\bar{A}\bar{D}\bar{B}'\bar{C}' + \bar{s})(\bar{B}\bar{C} + \bar{n}).\end{aligned}\right\} \quad \ldots\ldots\ldots\ldots(4).$$

Again, from equations (3) and (1),

$$xyz \nleftarrow A(\bar{B}\bar{C}\bar{A}'\bar{D}' + s)(\bar{B}'\bar{C}' + \bar{m}), \quad \overline{xy}z \nleftarrow D(\bar{B}\bar{C}\bar{A}'\bar{D}' + s)(\bar{B}'\bar{C}' + \bar{m}).$$

Hence by similar reasoning to that above

$$xyz = A(\bar{B}\bar{C}\bar{A}'\bar{D}' + s)(\bar{B}'\bar{C}' + \bar{m})(\bar{D} + q), \quad \overline{xy}z = D(\bar{B}\bar{C}\bar{A}'\bar{D}' + s)(\bar{B}'\bar{C}' + \bar{m})(\bar{A} + \bar{q}).$$

Similarly,

$$x\overline{yz} = B'(\bar{A}'\bar{B}\bar{C}\bar{D}' + s)(\bar{A}\bar{D} + m)(\bar{C}' + p), \quad \bar{x}y\bar{z} = C'(\bar{A}'\bar{B}\bar{C}\bar{D}' + s)(\bar{A}\bar{D} + m)(\bar{B}' + \bar{p}),$$

$$\bar{x}yz = C(\bar{A}\bar{B}'\bar{C}'\bar{D} + \bar{s})(\bar{A}'\bar{D}' + n)(\bar{B} + t), \quad x\bar{y}z = B(\bar{A}\bar{B}'\bar{C}'\bar{D} + \bar{s})(\bar{A}'\bar{D}' + n)(\bar{C} + \bar{t}),$$

$$xy\bar{z} = A'(\bar{A}\bar{B}'\bar{C}'\bar{D} + \bar{s})(\bar{B}\bar{C} + \bar{n})(\bar{D}' + l), \quad \overline{xyz} = D'(\bar{A}\bar{B}'\bar{C}'\bar{D} + \bar{s})(\bar{B}\bar{C} + \bar{n})(\bar{A}' + \bar{l}).$$

By adding the appropriate equations we determine x, y, z.

This method is applicable to an equation involving n unknowns, and in this case the solution will involve $2^n - 1$ arbitraries.

37. Symmetrical Solution of Equations with three unknowns.

(1) Consider the unlimiting equation

$$\begin{aligned} &\bar{A}xyz + \bar{B}x\overline{yz} + C\bar{x}yz + \bar{D}\overline{xy}z \\ &+ A'xy\bar{z} + \bar{B}'x\overline{yz} + \bar{C}'\bar{x}y\bar{z} + \bar{D}'\overline{xyz} = 0 \end{aligned} \quad \ldots\ldots\ldots\ldots\ldots(\alpha).$$

The conditions that the equation is unlimiting are (cf. § 32 (3)),

$$\bar{A}\bar{B}\bar{C}\bar{D} = 0 = \bar{A}'\bar{B}'\bar{C}'\bar{D}' = \bar{A}\bar{A}'\bar{C}\bar{C}' = \bar{B}\bar{B}'\bar{D}\bar{D}' = \bar{A}\bar{A}'\bar{B}\bar{B}' = \bar{C}\bar{C}'\bar{D}\bar{D}'.$$

Let the left-hand side of (α) be written $\phi(x, y, z)$ for brevity.

Let the general solution of (α) be

$$\left.\begin{aligned} x &= a_1 uvw + b_1 u\bar{v}w + c_1 \bar{u}vw + d_1 \bar{u}\bar{v}w \\ &\quad + a_1' uv\bar{w} + b_1' u\bar{v}\bar{w} + c_1' \bar{u}v\bar{w} + d_1' \bar{u}\bar{v}\bar{w}, \\ y &= a_2 uvw + b_2 u\bar{v}w + c_2 \bar{u}vw + d_2 \bar{u}\bar{v}w \\ &\quad + a_2' uv\bar{w} + b_2' u\bar{v}\bar{w} + c_2' \bar{u}v\bar{w} + d_2' \bar{u}\bar{v}\bar{w}, \\ z &= a_3 uvw + b_3 u\bar{v}w + c_3 \bar{u}vw + d_3 \bar{u}\bar{v}w \\ &\quad + a_3' uv\bar{w} + b_3' \bar{u}\bar{v}w + c_3' \bar{u}v\bar{w} + d_3' \bar{u}\bar{v}\bar{w}; \end{aligned}\right\} \quad \ldots\ldots\ldots\ldots\ldots (\beta)$$

where u, v, w are arbitrary unknowns.

By substituting for x, y, z from equations (β) in equation (α) the conditions that (β) should be some solution of (α) without restricting u, v, w are found to be,

$$\phi(a_1, a_2, a_3) = 0 = \phi(b_1, b_2, b_3) = \phi(c_1, c_2, c_3) = \phi(d_1, d_2, d_3)$$
$$= \phi(a_1', a_2', a_3') = \phi(b_1', b_2', b_3) = \phi(c_1', c_2', c_3') = \phi(d_1', d_2', d_3').$$

Thus the corresponding triplets of coefficients must be solutions of the given equation.

(2) It remains to find the conditions that (β) may represent the general solution of (α). Eliminate z from (α), the resultant is

$$\bar{A}\bar{A}'\,xy + \bar{B}\bar{B}'x\bar{y} + \bar{C}\bar{C}'\bar{x}y + \bar{D}\bar{D}'\overline{xy} = 0.$$

By § 35 (10), the conditions that the first two equations of (β) should be the general solution of this equation are

$$\Sigma a_1 a_2 = A + A', \quad \Sigma a_1\bar{a}_2 = B + B', \quad \Sigma \bar{a}_1 a_2 = C + C', \quad \Sigma \bar{a}_1\bar{a}_2 = D + D'.$$

Similarly eliminating y from equation (α), the resultant is

$$\bar{A}\bar{B}xz + \bar{A}'\bar{B}'x\bar{z} + \bar{C}\bar{D}\bar{x}z + \bar{C}'\bar{D}'\overline{xz} = 0.$$

The conditions that the first and third of equations (β) should form the general solution of this equation are

$$\Sigma a_1 a_3 = A + B, \quad \Sigma a_1\bar{a}_3 = A' + B', \quad \Sigma \bar{a}_1 a_3 = C + D, \quad \Sigma \bar{a}_1\bar{a}_3 = C' + D'.$$

Lastly, eliminating x from equation (α), the resultant is

$$\bar{A}\bar{C}yz + A'\bar{C}'y\bar{z} + \bar{B}\bar{D}\bar{y}z + \bar{B}'\bar{D}'\overline{yz} = 0.$$

The conditions that the second and third of equations (β) should form the general solution of this equation are

$$\Sigma a_2 a_3 = A + C, \quad \Sigma a_2\bar{a}_3 = A' + C', \quad \Sigma \bar{a}_2 a_3 = B + D, \quad \Sigma \bar{a}_2\bar{a}_3 = B' + D'.$$

(3) Again, if y and z be conceived as given, the field of x as defined by equation (α) is contained between the maximum extension

$$Ayz + A'y\bar{z} + B\bar{y}z + B'\overline{yz},$$

and the minimum extension

$$\bar{C}yz + \bar{C}'y\bar{z} + \bar{D}\bar{y}z + \bar{D}'y\bar{z}.$$

But (cf. § 33 (8)) the field of x as defined by equations (β) is contained between the maximum extension $\Sigma a_1(a_2 y + \bar{a}_2\bar{y})(a_3 z + \bar{a}_3\bar{z})$, and the minimum extension $\Pi \{a_1 + \bar{a}_2 y + a_2\bar{y} + \bar{a}_3 z + a_3\bar{z})$; that is, between the maximum extension

$$\Sigma a_1 a_2 a_3 \,.\, yz + \Sigma a_1 a_2 \bar{a}_3 \,.\, y\bar{z} + \Sigma a_1 \bar{a}_2 a_3 \,.\, \bar{y}z + \Sigma a_1 \bar{a}_2 \bar{a}_3 \,.\, \bar{y}\bar{z}$$

and the minimum extension

$$\Pi(a_1 + \bar{a}_2 + \bar{a}_3) \,.\, yz + \Pi(a_1 + \bar{a}_2 + a_3) \,.\, y\bar{z}$$
$$+ \Pi(a_1 + a_2 + \bar{a}_3) \,.\, \bar{y}z + \Pi(a_1 + a_2 + a_3) \,.\, \bar{y}\bar{z}.$$

Hence since the extensions as defined by (α) and (β) must be identical, we find by comparison

$$\Sigma a_1 a_2 a_3 = A, \quad \Sigma a_1 \bar{a}_2 a_3 = B, \quad \Sigma \bar{a}_1 a_2 a_3 = C, \quad \Sigma \bar{a}_1 \bar{a}_2 a_3 = D,$$
$$\Sigma a_1 a_2 \bar{a}_3 = A', \quad \Sigma a_1 \bar{a}_2 \bar{a}_3 = B', \quad \Sigma \bar{a}_1 a_2 \bar{a}_3 = C' \quad \Sigma \bar{a}_1 \bar{a}_2 \bar{a}_3 = D'.$$

The symmetry of these equations shows that, if z and x be conceived as given, the field of y as defined by (α) is the same as that defined by (β), and that if x and y be conceived as given, the same is true for z.

By adding the appropriate pairs of this set of equations it can be seen at once that these eight conditions include the twelve conditions of subsection (2).

Hence finally equations (β) form the general solution of equation (α), if the triplets a_1, a_2, a_3; b_1, b_2, b_3; ...; d_1', d_2', d_3'; are any simultaneous sets of solutions of the given equation which satisfy the eight conditions above.

(4) Now following the method of § 35 (5), let a_1, a_2, a_3 be any particular simultaneous solutions of the equations

$$\phi(x, y, z) = 0,$$

and

$$xyz = A.$$

These equations can be treated as simultaneous and are equivalent to the single equation

$$\bar{A} xyz + (\bar{B} + A) x\bar{y}z + (\bar{C} + A)\bar{x}yz + (\bar{D} + A)\bar{x}\bar{y}z$$
$$+ (\bar{A}' + A) xy\bar{z} + (\bar{B}' + A) x\bar{y}\bar{z} + (\bar{C}' + A)\bar{x}y\bar{z} + (\bar{D}' + A)\bar{x}\bar{y}\bar{z} = 0.$$

This equation is in general a limiting equation. It can be transformed into an unlimiting equation by writing (cf. § 32 (5))

$$x = (\bar{C} + A)(\bar{C}' + A)(\bar{D} + A)(\bar{D}' + A) + {}^{-}(\bar{A}\,\bar{A}'\bar{B}\bar{B}')\,p_1,$$
$$y = (\bar{B} + A)(\bar{B}' + A)(\bar{D} + A)(\bar{D}' + A) + {}^{-}(\bar{A}\,\bar{A}'\bar{B}\bar{B}')\,p_2,$$
$$z = (\bar{A}' + A)(\bar{B}' + A)(\bar{C}' + A)(\bar{D}' + A) + {}^{-}(\bar{A}\,\bar{B}\bar{C}\bar{D})\,p_3.$$

The conditions in subsection (1) that the original equation may be unlimiting reduce these formulæ of transformation to

$$x = A + p_1, \quad y = A + p_2, \quad z = A + p_3.$$

Then p_1, p_2, p_3 satisfy the unlimiting equation

$$\bar{A}\, p_1 p_2 p_3 + \bar{A}\,\bar{B} p_1 \bar{p}_2 p_3 + \bar{A}\,\bar{C}\bar{p}_1 p_2 p_3 + \bar{A}\,\bar{D}\bar{p}_1\bar{p}_2 p_3$$
$$+ \bar{A}\,\bar{A}' p_1 p_2 \bar{p}_3 + \bar{A}\,\bar{B}' p_1 \bar{p}_2 \bar{p}_3 + \bar{A}\,\bar{C}' \bar{p}_1 p_2 \bar{p}_3 + \bar{A}\,\bar{D}' \bar{p}_1 \bar{p}_2 \bar{p}_3 = 0 \ldots\ldots(1).$$

Similarly put $b_1 = B + q_1$, $\bar{b}_2 = B + \bar{q}_2$, $b_3 = B + q_3$, where q_1, q_2, q_3 satisfy the unlimiting equation

$$\bar{A}\bar{B}q_1q_2q_3 + \bar{B}q_1\bar{q}_2q_3 + \bar{B}\bar{C}\bar{q}_1q_2q_3 + \bar{B}\bar{D}\bar{q}_1\bar{q}_2q_3$$
$$+ \bar{B}\bar{A}'q_1q_2\bar{q}_3 + \bar{B}\bar{B}'q_1\bar{q}_2\bar{q}_3 + \bar{B}\bar{C}'\bar{q}_1q_2\bar{q}_3 + \bar{B}\bar{D}'\bar{q}_1\bar{q}_2\bar{q}_3 = 0 \text{......}(2).$$

Similarly put $\bar{c}_1 = C + \bar{r}_1$, $c_2 = C + r_2$, $c_3 = C + r_3$, where r_1, r_2, r_3 satisfy the unlimiting equation

$$\bar{A}\bar{C}r_1r_2r_3 + \bar{B}\bar{C}r_1\bar{r}_2r_3 + \bar{C}\bar{r}_1r_2r_3 + \bar{C}\bar{D}\bar{r}_1\bar{r}_2r_3$$
$$+ \bar{C}\bar{A}'r_1r_2\bar{r}_3 + \bar{C}\bar{B}'r_1\bar{r}_2\bar{r}_3 + \bar{C}\bar{C}'\bar{r}_1r_2\bar{r}_3 + \bar{C}\bar{D}'\bar{r}_1\bar{r}_2\bar{r}_3 = 0 \text{......}(3).$$

And so on for the remaining triplets of coefficients, putting

$$\begin{array}{lll} \bar{d}_1 = D + \bar{s}_1, & \bar{d}_2 = D + \bar{s}_2, & d_3 = D + s_3, \\ a_1' = A' + p_1', & a_2' = A' + p_2', & \bar{a}_3' = A' + \bar{p}_3', \\ b_1' = B' + q_1', & \bar{b}_2' = B' + \bar{q}_2', & \bar{b}_3' = B' + \bar{q}_3', \\ \bar{c}_1' = C' + \bar{r}_1', & c_2' = C' + r_2', & \bar{c}_3' = C' + \bar{r}_3', \\ \bar{d}_1' = D' + \bar{s}_1' & \bar{d}_2' = D' + \bar{s}_2', & \bar{d}_3' = D' + \bar{s}_3'. \end{array}$$

And the sets s_1, s_2, s_3; p_1', p_2', p_3'; ...; s_1', s_2', s_3'; satisfy unlimiting equations formed according to the same law as (1), (2), (3). These other equations will be numbered (4), (5), (6), (7), (8).

Let the equations (1)...(8) be called the auxiliary equations. When the coefficients a_1, a_2, a_3; b_1, b_2, b_3; ...; d_1', d_2', d_3' have the values here assigned, the eight equations of condition of subsection (3) are identically satisfied.

(5) Hence the general solution of the equation

$$\bar{A}\,xyz + \bar{B}\,x\bar{y}z + \bar{C}\,\bar{x}yz + \bar{D}\,\bar{x}\bar{y}z$$
$$+ \bar{A}'xy\bar{z} + \bar{B}'x\bar{y}\bar{z} + \bar{C}'\bar{x}y\bar{z} + \bar{D}'\bar{x}\bar{y}\bar{z} = 0,$$

is given by

$$x = (A + p_1)\,uvw + (B + q_1)\,u\bar{v}w + \bar{C}r_1\bar{u}vw + \bar{D}s_1\bar{u}\bar{v}w$$
$$+ (A' + p_1')\,uv\bar{w} + (B' + q_1')\,u\bar{v}\bar{w} + \bar{C}'r_1'\bar{u}v\bar{w} + \bar{D}'s_1'\bar{u}\bar{v}\bar{w},$$
$$y = (A + p_2)\,uvw + \bar{B}q_2u\bar{v}w + (C + r_2)\,\bar{u}vw + \bar{D}s_2\bar{u}\bar{v}w$$
$$+ (A' + p_2')\,uv\bar{w} + \bar{B}'q_2'u\bar{v}\bar{w} + (C' + r_2')\,\bar{u}v\bar{w} + \bar{D}'s_2'\bar{u}\bar{v}\bar{w},$$
$$z = (A + p_3)\,uvw + (B + q_3)\,u\bar{v}w + (C + r_3)\,\bar{u}vw + (D + s_3)\,\bar{u}\bar{v}w$$
$$\bar{A}'p_3'uv\bar{w} + \bar{B}'q_3'u\bar{v}\bar{w} + \bar{C}'r_3'\bar{u}v\bar{w} + \bar{D}'s_3'\bar{u}\bar{v}\bar{w};$$

where p_1, p_2, p_3; q_1, q_2, q_3; ...; s_1', s_2', s_3'; respectively are any sets of particular solutions of the auxiliary equations (1), (2)...(8). These equations can be written,

$$\bar{A}\phi\,(p_1, p_2, p_3) = 0 \text{ ....}(1),$$
$$\bar{B}\phi\,(q_1, q_2, q_3) = 0 \text{ ....}(2),$$
$$\bar{C}\phi\,(r_1, r_2, r_3) = 0 \text{ ....}(3),$$
$$\bar{D}\phi\,(s_1, s_2, s_3) = 0 \text{ ....}(4),$$

$$\bar{A}'\phi(p_1', p_2', p_3') = 0 \quad \text{..........................(5)},$$
$$\bar{B}'\phi(q_1', q_2', q_3') = 0 \quad \text{..........................(6)},$$
$$\bar{C}'\phi(r_1', r_2', r_3') = 0 \quad \text{..........................(7)},$$
$$\bar{D}'\phi(s_1', s_2', s_3') = 0 \quad \text{..........................(8)},$$

where $\phi(x, y, z)$ stands for the left-hand side of the given equation.

It will be observed that we may put

$$p_1 = q_1 = \ldots = s_1' = t_1, \quad p_2 = q_2 = \ldots = s_2' = t_2, \quad p_3 = q_3 = \ldots = s_3' = t_3,$$

where t_1, t_2, t_3 form any particular solution of the given equation,

$$\phi(x, y, z) = 0.$$

(6) The general solution of a limiting equation involving three unknowns is found, first by transforming it into an unlimiting equation according to § 32 (5), and then by applying the solution of subsection (5) of the present section.

(7) The method of reasoning of the present section and the result are both perfectly general. Thus the general equation involving three unknowns can be solved with a redundant unknown by the method of § 35 (10). Then by the method of the present section the equation involving four unknowns can be solved in a general symmetrical form. And the auxiliary equations will take the same form: and so on for any number of unknowns.

(8) As an example consider the equations

$$yz = a, \quad zx = b, \quad xy = c.$$

These equations can be combined into a single equation with its right-hand side zero by finding their resultant discriminants (cf. § 30 (9)).

The discriminants, cf. § 30 (11), of the first equation positive with respect to x and y are a and $\bar{a}$.

The discriminants of the first equation positive with respect to x and negative with respect to y are $\bar{a}$ and $\bar{a}$.

The discriminants negative with respect to x of the first equation are the same as those positive with respect to x.

Hence the following scheme holds for the discriminants:

Constituent......	xyz	$xy\bar{z}$	$x\bar{y}z$	$x\overline{yz}$	$\bar{x}yz$	$\bar{x}y\bar{z}$	$\overline{xy}z$	$\overline{xyz}$
1st Equation ...	a	$\bar{a}$	$\bar{a}$	$\bar{a}$	a	$\bar{a}$	$\bar{a}$	$\bar{a}$
2nd Equation...	b	$\bar{b}$	b	$\bar{b}$	$\bar{b}$	$\bar{b}$	$\bar{b}$	$\bar{b}$
3rd Equation...	c	c	$\bar{c}$	$\bar{c}$	$\bar{c}$	$\bar{c}$	$\bar{c}$	$\bar{c}$
Resultant Discriminants	abc	$\bar{a}\bar{b}c$	$\bar{a}b\bar{c}$	$\bar{a}\bar{b}\bar{c}$	$a\bar{b}\bar{c}$	$\bar{a}\bar{b}\bar{c}$	$\bar{a}\bar{b}\bar{c}$	$\bar{a}\bar{b}\bar{c}$

The resultant is

$$(\bar{a}+\bar{b}+\bar{c})(a+b+\bar{c})(a+\bar{b}+c)(a+b+c)(\bar{a}+b+c)=0;$$

that is $$\bar{a}bc+a\bar{b}c+ab\bar{c}=0.$$

This equation can be solved for c in terms of a and b.

The positive discriminant is $^{-}(a\bar{b}+\bar{a}b)$, that is $ab+\bar{a}\bar{b}$, the negative one is $\bar{a}+\bar{b}$.

The resultant is $ab(a\bar{b}+\bar{a}b)=0$, which is identically true.

The solution is $$c=ab+u(ab+\bar{a}\bar{b})=ab+u\bar{a}\bar{b}.$$

Hence the solution for x is found from

$$x=(\bar{a}+b+c)(a+b+c)+u(abc+\bar{a}\bar{b}c+\bar{a}b\bar{c}+\bar{a}\bar{b}\bar{c})=b+c+u\bar{a}\bar{b}\bar{c}$$
$$=b+c+u\bar{a}.$$

Similarly $$y=c+a+v\bar{b},\quad z=a+b+w\bar{c};$$
where u, v, w satisfy three unlimiting equations, found by substituting for x, y, z in the original equations.

Thus substituting in $yz=a$, we obtain

$$a+bc+a\bar{b}v+a\bar{c}w+\bar{b}\bar{c}vw=a;$$

that is $$a+bc+\bar{b}\bar{c}vw=a, \text{ that is } \bar{a}bc+\bar{a}\bar{b}\bar{c}vw=0.$$

But $\bar{a}bc=0$. Hence the equation becomes $\bar{a}\bar{b}\bar{c}vw=0$.

Similarly $$\bar{a}\bar{b}\bar{c}wu=0=\bar{a}\bar{b}\bar{c}uv.$$

Thus u, v, w satisfy the equation $\bar{a}\bar{b}\bar{c}(uv+vw+wu)=0$.

Comparing this with the typical form (α) given in § 37,

$$\bar{A}=\bar{a}\bar{b}\bar{c}=\bar{B}=\bar{C}=\bar{A}',\quad \bar{D}=0=\bar{B}'=\bar{C}'=\bar{D}'.$$

Also a particular solution of the given equation is $u=v=w=0$. Hence from subsection (5)

$$u=(a+b+c)(UVW+U\bar{V}W+UV\bar{W})+U\bar{V}\bar{W},$$
$$v=(a+b+c)(UVW+\bar{U}VW+UV\bar{W})+\bar{U}V\bar{W},$$
$$w=(a+b+c)(UVW+\bar{U}VW+U\bar{V}W)+\bar{U}\bar{V}W.$$

Hence the general solutions for x, y, z are

$$x=b+c+u\bar{a}=b+c+\bar{a}U\bar{V}\bar{W},$$
$$y=c+a+v\bar{b}=c+a+\bar{b}\bar{U}V\bar{W},$$
$$z=a+b+w\bar{c}=a+b+\bar{c}\bar{U}\bar{V}W;$$

where U, V, W are arbitrary unknowns.

38. Subtraction and Division. (1) The Analytical (or reverse) processes, which may be called subtraction and division, have now to be discussed.

Let the expressions $a - b$ and $a \div b$ satisfy the following general conditions:

I. That they denote regions, as do all other expressions of the algebra; so that they can be replaced by single letters which have all the properties of other letters of the algebra.

II. That they satisfy respectively the following equations:

$$(a - b) + b = a; \; (a \div b) \times b = a.$$

(2) Let x stand for $a - b$. Then x is given by the equation

$$x + b = a.$$

The positive discriminant is $ai + \bar{a}\bar{i}$, that is a, the negative is $ab + \bar{a}\bar{b}$.

The resultant is $\quad \bar{a}(\bar{a}b + a\bar{b}) = 0$, that is $\bar{a}b = 0$;

hence

$$b \notin a.$$

The solution is $\quad x = \bar{a}b + a\bar{b} + ua = a\bar{b} + ua.$

Hence for the symbol $a - b$ to satisfy the required conditions it is necessary that $b \notin a$. Furthermore negative terms in combination with positive terms do not obey the associative law. For by definition,

$$b + (a - b) = (a - b) + b = a.$$

Also since $b \notin a$, and therefore $b + a = a$, it follows that

$$(b + a) - b = a - b = a\bar{b} + ua.$$

Therefore $b + (a - b)$ is not equal to $(b + a) - b$.

This difficulty may be evaded for groups of terms by supposing that all the positive terms are added together first and reduced to the mutually exclusive form of § 27, Prop. X. Such groups of terms must evidently be kept strictly within brackets. It is to be further noticed that the result of subtraction is indeterminate.

(3) Again, for division, let $x = a \div b$.

Then

$$bx = a.$$

The positive discriminant is $ab + \bar{a}\bar{b}$, the negative is $a0 + \bar{a}\bar{0}$, that is $\bar{a}$.

The resultant is $\quad a(\bar{a}b + a\bar{b}) = 0$; that is $a\bar{b} = 0$.

Hence

$$a \notin b.$$

The solution is $\quad x = a + v(ab + \bar{a}\bar{b}) = a + v\bar{a}\bar{b} = a + v\bar{b}.$

Factors with the symbol of division prefixed are not associative with those with the symbol of multiplication prefixed (or supposed).

For $b(a \div b) = (a \div b)b = a$, by definition.

Also since $ab = a$,

$$(ba) \div b = a \div b = a + v\bar{b}.$$

This difficulty can be evaded by suitable assumptions just as in the case of subtraction. The result of division is indeterminate.

(4) Owing to these difficulties with the associative law the processes of subtraction and division are not of much importance in this algebra.

All results which might depend on them can be obtained otherwise*. They are useful at times since thereby the introduction of a fresh symbol may be avoided. Thus instead of introducing x, defined by $x + b = a$, we may write $(a - b)$, never however omitting the brackets.

Similarly we may write $(a \div b)$ instead of x, defined by $bx = a$.

But great care must be taken even in the limited use of these symbols not to be led away by fallacious analogies.

For $(a - b) = a\bar{b} + ua$; with the condition $b \nleftarrow a$.

But $\{(a + c) - (b + c)\} = (a + c)^{-}(b + c) + u(a + c)$

$$= a\bar{b}\bar{c} + u(a + c).$$

These two symbols are not identical unless

$$a\bar{b}\bar{c} = a\bar{b}, \text{ and } a + c = a.$$

From the first condition $c \nleftarrow \bar{a} + b$,

that is $ac \nleftarrow ab$

$\nleftarrow b$; since $ab = b$.

From the second condition $ac = c$.

Hence from the two conditions $c \nleftarrow b$.

Again, $(a \div b) = a + v\bar{b}$; with the condition $a \nleftarrow b$.

But $\{(ac) \div (bc)\} = ac + v\,^{-}(bc) = ac + v(\bar{b} + \bar{c})$.

These two symbols are therefore not identical unless

$$ac = a, \text{ and } \bar{b} + \bar{c} = \bar{a}\bar{b}.$$

From the second condition $bc = a + b = b$, hence $b \nleftarrow c$. This includes the first condition which can be written $a \nleftarrow c$. But $a \nleftarrow b$. Hence the final condition is $b \nleftarrow c$.

(5) It can be proved that

$$^{-}(a - b) = (\bar{a} \div \bar{b}), \quad ^{-}(a \div b) = (\bar{a} - \bar{b}).$$

For both $(a - b)$ and $(\bar{a} \div \bar{b})$ involve the same condition, namely $b \nleftarrow a$, or as it may be written $\bar{a} \nleftarrow \bar{b}$.

Again, $a - b = a\bar{b} + va$.

Therefore $^{-}(a - b) = (\bar{a} + \bar{v})(\bar{a} + b) = \bar{a} + \bar{v}b$.

But $(\bar{a} \div \bar{b}) = \bar{a} + ub$.

Therefore the two forms are identical in meaning.

Similarly $^{-}(a \div b) = (\bar{a} - \bar{b})$.

* First pointed out by Schröder, *Der Operationkreis des Logikkalküls*, Leipsic, 1877.

CHAPTER III.

EXISTENTIAL EXPRESSIONS.

39. EXISTENTIAL EXPRESSIONS. (1) Results which are important in view of the logical application of the algebra are obtained by modifying the symbolism so as to express information as to whether the regions denoted by certain of the terms either are known to be existent (i.e. the terms are then not null), or are known not to include the whole of space (i.e. the terms are then not equal to the universe). If this information is expressed the terms, besides representing regions, give also the additional information, that they are not 0, or are not i. When this additional existential information is being given let the symbol $\equiv$ be used instead of the symbol $=$; and let the use of $\equiv$ be taken to mean that, in addition to the regions respectively represented by the combinations of symbols on either side of it being the same, the existential information on the right-hand side can be derived from that on the left-hand side.

The symbolism wanted is one which will adapt itself to the various transformations through which expressions may be passed. If all regions were denoted by single letters, it would be possible simply to write capital letters for regions known to exist, thus X instead of x, and then the information required, namely that X exists, would be preserved through all transformations. Thus $\bar{X}$ at once tells us that X exists and that $\bar{X}$ does not embrace all the universe i. But this notation of capitals is not sufficiently flexible. For instance it is not possible to express by it that the region ab exists: this requires that a exists, that b exists, and in addition that they overlap, and this last piece of information is not conveyed by AB.

The merit of the symbolism now to be developed is that the new symbols go through exactly the same transformations as the old symbols, and thus two sorts of information, viz. the denotation of regions and the implication of their existence, are thrown into various equivalent forms by the same process of transformation.

(2) Any term x can be written in the form xi. Now when the fact has to be expressed that x is not null, let i be modified into j; so that xj expresses that x exists, the j being added after the symbol on which it operates.

Furthermore any term x can be written in the form $x+0$. Now when the fact has to be expressed that x does not exhaust the whole region of discourse, that is to say is not i, let the 0 be modified into ω. Then $x+\omega$ expresses that x is not equivalent to i.

Let any combination of symbols involving j or ω be called an existential expression.

Thus j may be looked on as an affirmative symbol, giving assurance of reality, and ω as a limitative symbol restraining from undue extension. They have no meaning apart from the terms to which they are indissolubly attached, the attachment being indicated by brackets when necessary, i.e. by (xj) and by $(x+\omega)$.

It is to be noted that xj or $x+\omega$ can be read off as assertions: thus xj states that x is not 0, $x+\omega$ that x is not i.

(3) The symbol $xy.j$ will be taken to mean that j operates on xy, so that xy exists. Thus $xy.j$ implies xj and yj; but the converse does not hold.

The mode of attachment of j to the term on which it operates has some analogy to multiplication as it obtains in this algebra. Thus

$$xj.yj.j \equiv xy.j,$$

though $xj.yj$ is not equivalent to $xy.j$ as far as its existential information is concerned.

Again, if $x, y, z, u, \ldots$ represent any number of regions, then

$$(xyzu\ldots)j \equiv (xj.yj.zj\ldots)j;$$

but the final j cannot be omitted, if the existential information is to be the same on both sides.

(4) The distributive laws have now to be examined as regards the multiplication and addition of existential expressions.

Consider in the first place the expression $(x+y)j$.

Now if $x=0$ and $y=0$, then $x+y=0$. Hence $(x+y)j$ implies either xj or yj or both. Thus we may adapt the symbolism so as to write

$$(x+y)j \equiv xj_1 + yj_1;$$

where the suffixes of the j's weaken the meaning to this extent, that one of the j's with this suffix is to hold good as to its existential information but not necessarily both. We define therefore

$$xj_1 + yj_1 + zj_1 + \ldots$$

to mean that one of the terms at least is not 0.

The other formal properties (cf. subsection (3)) of j evidently hold good, retaining always this weakened meaning.

The only point requiring notice is that $xj_1j \equiv xj$; for j_1 has the same meaning as j in a weakened hypothetical form.

Further $$(xj+y)j \equiv xjj_1 + yj_1 \equiv xj+y;$$

for the j_1 can be omitted, since it is known that x exists.

In using the multiplication of existential symbols the dots (or brackets) must be carefully attended to. For instance

$$(x+y) \, . \, zj \equiv x \, . \, zj + y \, . \, zj.$$

But $$(x+y) z \, . \, j \equiv (xz + yz) j \equiv xz \, . \, j_1 + xz \, . \, j_1.$$

In the first expression $(x+y) \, . \, zj$, the j simply asserts that z exists; in the second expression it asserts that $(x+y) z$ exists.

Again, $$xy + z = (x+z)(y+z).$$

Also $xy \, . \, j + z$ implies $(x+z)(y+z) \, . \, j$.

But though $xyj + z \equiv (x+z)(y+z) \, . \, j$; the left-hand side gives more definite information than the right-hand side. For

$$(x+z)(y+z) \, . \, j \equiv xy \, . \, j_1 + zj_1.$$

Also $xy \, . \, j$ implies xj, yj.

Hence $$xy \, . \, j + z \equiv (xj + z)(yj + z) j \equiv xj \, . \, yj \, . \, j_1 + zj_1.$$

But still the right-hand side does not give as much information as the left-hand side; for $xj \, . \, yj \, . \, j_1$ is not equivalent to $xy \, . \, j$.

Hence the distributive power of addition in reference to multiplication to some extent has been lost. It cannot be employed in this instance without some loss of existential information.

(5) The symbol $(x+y+\omega)$ will be taken to mean that ω operates on $x+y$, and therefore that $x+y$ is not i. Thus $x+y+\omega$ implies $x+\omega$ and $y+\omega$; but the converse does not hold. The mode of attachment of ω to the term on which it operates has some analogy to addition as it obtains in this algebra. Thus

$$(x+\omega) + (y+\omega) + \omega \equiv x + y + \omega,$$

though $(x+\omega)+(y+\omega)$ is not equivalent to $(x+y+\omega)$ as far as its existential information is concerned.

Again, if x, y, z, u, ... denote any number of regions, then

$$(x+y+z+u+ \ldots + \omega) \equiv \{(x+\omega) + (y+\omega) + (z+\omega) + \ldots + \omega\};$$

but the final ω cannot be omitted if the existential information is to be the same on both sides.

The distributive law of addition in relation to multiplication (cf. § 24, equation B) does not hold completely.

Consider the expression $xy + \omega$. Now xy can only be i, if both x and y are equivalent to i.

Hence $xy + \omega$ implies either $x+\omega$ or $y+\omega$ or both. Thus we may adapt the symbolism so as to write

$$xy + \omega \equiv (x + \omega_1)(y + \omega_1);$$

where the suffixes of the ω's weaken the meaning to this extent, that one of the ω's is to hold good but not necessarily both. We define therefore $(x+\omega_1)(y+\omega_1)(z+\omega_1)\ldots$ to mean that one at least of the terms $x, y, z, \ldots$ is not i.

It is obvious that $(x+\omega_1+\omega)(y+\omega_1)(z+\omega_1)\ldots \equiv (x+\omega)\,yz\ldots$, since $x+\omega_1+\omega$ ensures definitely that x is not i.

For example,

$$(x+\omega)\,y+\omega \equiv (x+\omega+\omega_1)(y+\omega_1) \equiv (x+\omega)\,y.$$

Let the symbols such as j_1 or ω_1 be called weak symbols in contrast to j or ω which are strong symbols. Then a strong symbol absorbs a weak symbol of the same name (j or ω) when they both operate on the same term, and destroys all the companion weak symbols. Thus

$$xj_1j+yj_1 \equiv xj+y,\quad (x+\omega_1+\omega)(y+\omega_1) \equiv (x+\omega)\,y.$$

(6) The chief use of this notation arises from its adaptation to the ordinary transformations owing to the following consideration. If x exists, then $\bar{x}$ cannot be i; and conversely if $\bar{x}$ be not i, then x exists.

Hence $\quad {}^{-}(xj) \equiv \bar{x}+\omega$, and ${}^{-}(\bar{x}+\omega) \equiv xj.$

But by analogy to § 26, Prop. VI.

$${}^{-}(xj) \equiv \bar{x}+\bar{j},\text{ and } {}^{-}(\bar{x}+\omega) \equiv x\bar{\omega}.$$

Hence we may write $\bar{j}=\omega$, and $\bar{\omega}=j$, corresponding to $\bar{i}=0$, and $\bar{0}=i$.

Thus the original existential information can be retained through any transformations of the algebra.

40. Umbral Letters. (1) This existential notation can be extended. Let the letters of the Greek alphabet be taken to correspond to the letters of the Roman alphabet, so that α corresponds to a, β to b, and so on.

Let $x\alpha$ mean that the regions x and a overlap; in other words $x\alpha$ implies $xa\,.\,j$, but the symbol $x\alpha$ in itself denotes only the region x; it only *implies* this extra information. Also let $x+\bar{\alpha}$, while denoting only the region x, imply that x does not include all the region a; in other words $x+\bar{\alpha}$ implies $\bar{x}a\,.\,j$, that is, it implies $x+\bar{a}+\omega$. Thus $x\alpha$ implies aj and xj and $xa\,.\,j$; while $x+\bar{\alpha}$ implies aj, $\bar{x}j$ and $\bar{x}a\,.\,j$. Also $x\alpha$ does not necessarily exclude $x\bar{\alpha}$, and $x+\bar{\alpha}$ does not necessarily exclude $x+\alpha$.

(2) Now if x includes some of α, it follows that $\bar{x}$ cannot include all α. Hence if $x\alpha$, then $\bar{x}+\bar{\alpha}$. This can be expressed by the equation

$${}^{-}(x\alpha) \equiv \bar{x}+\bar{\alpha}.$$

Thus for instance, $\quad {}^{-}(x\alpha)\,.\,y \equiv (\bar{x}+\bar{\alpha})\,y.$

Also it follows that $\quad {}^{=}(x\alpha) \equiv {}^{-}(\bar{x}+\bar{\alpha});$

and hence $\quad {}^{-}(\bar{x}+\bar{\alpha}) \equiv x\alpha.$

But by analogy to § 26, Prop. VI.

$${}^{-}(\bar{x}+\bar{\alpha}) = \bar{\bar{x}}\bar{\bar{\alpha}} = x\bar{\bar{\alpha}}.$$

Hence we may write $\bar{\bar{\alpha}} = \alpha$; though as a matter of fact the Greek letters have no meaning apart from the Roman letters to which they assign properties, and therefore should not be written alone.

(3) Let these Greek letters be called shadows or umbral letters; and let the Roman letters denoting regions be called regional letters.

Then the umbral letters essentially refer to some regional letters or groups of letters and are never to be separated from them. Thus $a(b+\bar{\gamma})$ cannot be transformed into $ab + a\bar{\gamma}$; the symbol $(b+\bar{\gamma})$ is essentially one whole, and the bracket can never be broken. Similarly $a \,.\, b\gamma$ cannot be transformed into $ab \,.\, \gamma$; since $b\gamma$ is one indivisible symbol.

But with this limitation—that brackets connecting regional and umbral letters are never to be broken—it will be found that the umbral letters follow all the laws of transformation of regional letters.

(4) In accordance with our previous definitions it may be noted that $x(\alpha+\beta)$ implies $x(a+b) \,.\, j$, and $(x+\alpha+\beta)$ implies that x does not include all $^-(a+b)$.

Also $x\alpha\beta$ implies $xab \,.\, j$, and $(x+\alpha\beta)$ implies that x does not include all $^-(ab)$, that is, all $(\bar{a}+\bar{b})$.

It is further to be remarked that $x(\alpha+\beta)$ is not identical in meaning with $x\alpha + x\beta$. For $x(\alpha+\beta)$ implies $x(a+b) \,.\, j$, that is either $xa \,.\, j$ or $xb \,.\, j$ or both, while $x\alpha + x\beta$ implies both $xa \,.\, j$ and $xb \,.\, j$.

Now $x\alpha\beta$ implies $xab \,.\, j$, that is both $xa \,.\, j$ and $xb \,.\, j$ as well as $xab \,.\, j$. Hence $x\alpha\beta$ implies all that $x\alpha + x\beta$ implies and more, and $x\alpha + x\beta$ implies all that $x(\alpha+\beta)$ implies and more; while all three expressions represent the same region, namely x.

(5) The shadows follow among themselves all the symbolic laws of ordinary letters.

For
$$x(\alpha+\beta) \equiv x(\beta+\alpha), \quad x+(\alpha+\beta) \equiv x+(\beta+\alpha),$$
$$x\alpha\beta \equiv x\beta\alpha, \quad x+\alpha\beta \equiv x+\beta\alpha,$$
$$x(\alpha+\alpha) \equiv x\alpha, \quad x+(\alpha+\alpha) \equiv x+\alpha,$$
$$x\alpha(\beta+\gamma) \equiv x(\alpha\beta+\alpha\gamma), \quad x+\alpha(\beta+\gamma) \equiv x+(\alpha\beta+\alpha\gamma),$$
$$x\,^-(\alpha\beta) \equiv x(\bar{\alpha}+\bar{\beta}), \quad x+\,^-(\alpha\beta) \equiv x+(\bar{\alpha}+\bar{\beta}),$$
$$x\,^-(\alpha+\beta) \equiv x\bar{\alpha}\bar{\beta}, \quad x+\,^-(\alpha+\beta) \equiv x+\bar{\alpha}\bar{\beta}.$$

Apart from this detailed consideration it is obvious that the same laws must hold; for the shadows also represent regions, though these shadowed regions are only mentioned in the equations for the sake of indicating properties of other regions in reference to them.

It should also be noticed that since $x\alpha\beta$ implies $xab \,.\, j$, it also implies $ab \,.\, j$.

Other transformations are

$$-\{x(\alpha+\beta)\} \equiv \bar{x} + {}^{-}(\alpha+\beta) \equiv \bar{x} + \alpha\bar{\beta},$$
$$-\{x+(\alpha+\beta)\} \equiv \bar{x}\,{}^{-}(\alpha+\beta) \equiv \bar{x}\bar{\alpha}\bar{\beta},$$
$$-\{x\alpha\beta\} \equiv \bar{x} + {}^{-}(\alpha\beta) \equiv \bar{x} + (\bar{\alpha}+\beta),$$
$$-\{x+\alpha\beta\} = \bar{x}\,{}^{-}(\alpha\beta) \equiv \bar{x}(\bar{\alpha}+\bar{\beta}),$$
$$-\{x\,{}^{-}(\alpha\beta)\} \equiv \bar{x} + \alpha\beta,$$
$$-\{x + {}^{-}(\alpha\beta)\} = \bar{x}\alpha\beta,$$
$$-\{x\,{}^{-}(\alpha+\beta)\} = \bar{x} + (\alpha+\beta).$$

It is to be noted that with the symbol $x(\alpha+\beta)$, we may not transform to $x\alpha + x\beta$, and thence infer $x\alpha$ and $x\beta$; the true transformation is

$$x(\alpha+\beta) \equiv x\alpha_1 + x\beta_1,$$

where α_1 and β_1 are weak forms of α and β.

Similarly we may not transform $x+(\alpha+\beta)$ into $(x+\alpha)+(x+\beta)$ and thence infer $x+\alpha$ and $x+\beta$.

(6) Each complex umbral symbol should be treated as one whole as far as symbolic transformations are concerned. Thus the laws of unity and simplicity (cf. § 25) have to be partially suspended. For instance $x\alpha + x\beta$ denotes only the region x, but for the purposes of the existential shadow letters $x\alpha$ and $x\beta$ must be treated as distinct symbols. Similarly $x\alpha \,.\, x\beta$ denotes only the region x, but it does not mean the same as $x\alpha\beta$; for $x\alpha \,.\, x\beta$ denotes the region x and implies $xa\,.\,j$ and $xb\,.\,j$, whereas $x\alpha\beta$ denotes the region x and implies $xab\,.\,j$. The second implication includes the first, but not the first the second. Hence for the purposes of multiplication $x\alpha$ and $x\beta$ must be treated as different symbols. The suspension of these laws of unity and simplicity causes no confusion, for the symbols are only to be treated as different symbols (although denoting the same region) when they are so obviously to the eye; thus $x\alpha$ and $x\beta$ are obviously different symbols.

(7) When the same regional letter is combined with various umbral letters, the same result is obtained whether the expressions are added or multiplied*.

Thus

$$x\alpha + x\beta \equiv x\alpha \,.\, x\beta,$$
$$x\alpha + \beta \equiv (x+\beta)\,\alpha.$$

(8) This notation enables existential expressions to be transformed.

Thus if ξ corresponds to x, η to y, and ζ to z,

$$xy\,.\,j \equiv x\eta \,.\, y\xi.$$

Hence
$$xy\,.\,j + z \equiv (x\eta + z)(y\xi + z);$$

and in this case the connotation is exactly the same on both sides. Hence the distributive power of addition in reference to multiplication has now

* This remark is due to Mr W. E. Johnson.

been retained. It may be noticed that the right-hand side might have been written $(x\eta + z)(y + z)$ without alteration of connotation; for $x\eta$ implies xj, yj, $xy . j$, and the ξ affixed to y implies no more.

Again, $$x + y + \omega \equiv (x + \eta) + (y + \xi),$$
where $(x + \eta)$ implies that x does not include all $\bar{y}$ and $y + \xi$ implies that y does not include all $\bar{x}$.

Thus $$(x + y + \omega) z \equiv (x + \eta) z + (y + \xi) z,$$
the connotation of both sides is the same. Thus the distributing power of multiplication in reference to addition has now been retained.

It is to be noticed that symbols like $x + \eta$ and $x\eta$ are to be treated as indivisible wholes.

Again as examples consider the transformations
$$^{-}(xy . j) \equiv {}^{-}(x\eta . y\xi) \equiv (\bar{x} + \bar{\eta}) + (\bar{y} + \bar{\xi});$$
and
$$^{-}(x + y + \omega) = {}^{-}\{(x + \eta) + (y + \xi)\} = \bar{x}\bar{\eta} . \bar{y}\bar{\xi}$$
$$= \bar{x}\bar{y} . j.$$

41. ELIMINATION. (1) It is in general possible to eliminate $x, y, z, \dots$ from existential expressions of the forms
$$f(x, y, z, \dots t) j \text{ and } f(x, y, z, \dots t) + \omega.$$

Consider first the form $f(x, y, z, \dots t) j$.

Let $f(x, y, z, \dots t)$ be developed and take the form
$$axyz \dots t + bxyz \dots \bar{t} + \dots + g\bar{x}\bar{y}\bar{z} \dots \bar{t}.$$
By § 33 (2) the maximum extension of the field of this expression is
$$a + b + \dots + g.$$
Hence if $f(x, y, z, \dots t) j$, the maximum extension cannot be null. Thus
$$(a + b + \dots + g) j$$
is the resultant expression when $x, y, z, \dots t$ have been eliminated.

(2) Consider the form, $f(x, y, z, \dots t) + \omega$.

This is equivalent to $\bar{f}(x, y, z, \dots t) . j$.

If $f(x, y, z, \dots t)$ be developed as in (1), then the existential expression becomes
$$(\bar{a}xyz \dots t + \bar{b}xyz \dots \bar{t} + \dots + \bar{g}\bar{x}\bar{y}\bar{z} \dots \bar{t}) j.$$

Hence by (1) $$(\bar{a} + \bar{b} + \dots + \bar{g}) j,$$
that is $$ab \dots g + \omega.$$

This result might also have been deduced by noticing that $ab \dots g$ is the minimum extension of the field of $f(x, y, z, \dots t)$; and therefore is necessarily not i, if $f(x, y, z, \dots t)$ is not i.

(3) As particular cases of the above two subsections, note that

$$(ax + b\bar{x}).j \text{ yields } (a + b)j,$$
$$(au + bv).j \text{ yields } (a + b)j,$$
$$(ax + b\bar{x}) + \omega \text{ yields } ab + \omega.$$

Also note that $(au + bv) + \omega$ yields no information respecting a and b; for when the formula of (2) is applied to its developed form the resultant becomes $0 + \omega$, which is an identity.

(4) To eliminate $x, y, z, \dots t$ from $f(x, y, z, \dots t)j$ and from n equations involving them.

Let $f(x, y, z, \dots t)$ be developed as in (1), and let the corresponding resultant discriminants of the equations be $A, B, C, \dots G$.

Then the maximum extension of the field of $f(x, y, z, \dots t)$ as conditioned by the equations is $aA + bB + \dots + gG$.

Now $f(x, y, z \dots t)j$ requires that the maximum extension shall not be null. Hence the complete existential expression* to be found by elimination is

$$(aA + bB + \dots + gG)j.$$

Let this be called the existential resultant.

The resultant found by elimination of $x, y, z, \dots t$ from the equations is

$$\bar{A}\bar{B}\dots\bar{G} = 0.$$

The existential resultant and the resultant of the equations contain the complete information to be obtained from the given premises after the elimination of $x, y, \dots t$.

(5) An allied problem to that of the previous subsection is to find the condition that the existential expression may not condition $x, y, z, \dots t$ any further than they are already conditioned by the equations.

The minimum extension of the field of $f(x, y, z, \dots t)$ as conditioned by the equations is by § 33 (8),

$$(a + \bar{A})(b + \bar{B})\dots(g + \bar{G}).$$

Hence if $$(a + \bar{A})(b + \bar{B})\dots(g + \bar{G})j,$$
then $f(x, y, z, \dots t)j$, for all values of $x, y, z, \dots t$; and thus $f(x, y, z, \dots t)j$ does not condition $x, y, z, \dots t$.

The condition can also be written

$$(\bar{a}A + \bar{b}B + \dots \bar{g}G + \omega).$$

(6) A special case of (5) arises when there are no equations; the existential expression does not condition the unknowns, if

$$abc \dots g . j.$$

* This expression found by another method was pointed out to me by Mr W. E. Johnson.

(7) If the existential expression be $f(x, y, z, \ldots t) + \omega$, then by reasoning similar to that in subsections (4) and (5) the existential resultant is

$$(a+\bar{A})(b+\bar{B}) \ldots (g+\bar{G}) + \omega.$$

The condition that the unknowns are not conditioned by the existential expression is

$$(aA + bB + \ldots + gG + \omega).$$

These conditions may respectively be written

$$(\bar{a}A + \bar{b}B + \ldots + \bar{g}G)\, j.$$

and

$$(\bar{a} + \bar{A})(\bar{b} + \bar{B}) \ldots (\bar{g} + \bar{G})\, j.$$

42. Solutions of Existential Expressions with one unknown. (1) *Solution of* $ax \,.\, j$. The form of solution for x can be written in two alternative forms by using symbols for undetermined regions: thus

$$x \equiv wa \,.\, j + u\bar{a} \equiv p\alpha.$$

The first form states explicitly that x is some (not none) undetermined part of the region a together with some (or none) of $\bar{a}$. The second form states the same solution more concisely but perhaps less in detail: it states that x may be any region p, so long as p is assumed to include some (not none) of the region α. There is no reason in future to write p for the undetermined region denoted by x. Thus we shall say that the solution of $ax \,.\, j$ is $x \equiv x\alpha$.

(2) *Solution of* $b\bar{x} \,.\, j$. From the preceding proposition

$$\bar{x} \equiv \overline{w}b \,.\, j + \bar{u} \equiv \bar{x}\beta.$$

Hence

$$x \equiv {}^{-}(\overline{w}b \,.\, j + \bar{u}) \equiv {}^{-}(\bar{x}\beta)$$

$$\equiv u(w + \bar{b} + \omega) \equiv x + \bar{\beta}.$$

The form $u(w + \bar{b} + \omega)$ states that x must be some (or none) of a region which is composed of all $\bar{b}$ and of any other region, except that the total region must not comprise all the Universe. The form $x + \bar{\beta}$ states that x may be any region so long as it does not comprise all b.

(3) *Solution of any number of expressions* $ax \,.\, j$, $a'x \,.\, j$, $\ldots$ $a^n x \,.\, j$.

The required solution is obviously

$$x \equiv x\alpha + x\alpha' + \ldots + x\alpha^n$$

$$\equiv \Sigma x\alpha \text{ (say)}.$$

It may be noticed that $x(\alpha + \alpha' + \ldots + \alpha^n)$ is not the required solution, since it is only equivalent to the weakened form $x\alpha_1 + x\alpha_1' + \ldots + x\alpha_1^n$; also that $x\alpha\alpha' \ldots \alpha^n$ implies $aa' \ldots a^n \,.\, j$ and $xaa' \ldots a^n \,.\, j$, which is more than is given by the equations.

By § 40 (7) the solution can also be written

$$x \equiv x\alpha \,.\, x\alpha' \ldots x\alpha^n \equiv \Pi(x\alpha) \text{ (say)}.$$

(4) *Solution of any number of expressions of the types*

$$b\bar{x}\,.\,j,\quad b'\bar{x}\,.\,j,\ \dots\ b^n\bar{x}\,.\,j.$$

The required solution is

$$x \equiv (x+\bar{\beta}) + (x+\bar{\beta}') + \dots + (x+\bar{\beta}^n)$$
$$\equiv \Sigma\,(x+\bar{\beta})\ \text{(say)}.$$

(5) *Solution of any number of expressions of the types*

$$ax\,.\,j,\quad a'x\,.\,j,\ \dots\ a^nx\,.\,j,\quad b\bar{x}\,.\,j,\quad b'\bar{x}\,.\,j,\dots\ b^m\bar{x}\,.\,j.$$

The solution is obviously

$$x \equiv \sum_{\alpha}\sum_{\beta}\,(x+\bar{\beta})\,\alpha.$$

If there are only two such expressions, namely $ax\,.\,j$ and $b\bar{x}\,.\,j$, the solution becomes

$$x \equiv (x+\bar{\beta})\,\alpha.$$

(6) *Solution of* $(ax+b\bar{x})\,.\,j.$

Now $(ax+b\bar{x})\,j \equiv ax\,.\,j_1 + b\bar{x}\,.\,j_1.$

By subsection (5) $ax\,.\,j_1$ and $b\bar{x}\,.\,j_1$ imply

$$x \equiv (x+\bar{\beta}_1)\,\alpha_1;$$

where α_1 and β_1 are alternative weakened forms of shadows.

But this expression does not necessarily imply any restriction on x. For $ax+b\bar{x}$ can only vanish if $ab=0$.

Hence $(ax+b\bar{x})j$ either implies $ab\,.\,j$ and x entirely unconditioned, or $ab=0$ and

$$x \equiv (x+\bar{\beta}_1)\,\alpha_1.$$

(7) *Solution of* $ax+\omega$. Now $ax+\omega$ implies $^-(ax+\omega)$, that is $(\bar{a}+\bar{x})\,j$. But $(\bar{a}+\bar{x})j \equiv (\bar{a}x+\bar{x})\,j$. This implies either $\bar{a}j$ and x entirely unconditioned, or $\bar{a}=0$, that is $a=i$, and $x+\omega$.

(8) *Solution of* $b\bar{x}+\omega$.

Now $b\bar{x}+\omega$ implies $^-(b\bar{x}+\omega)$, that is $(\bar{b}+x)\,j$.

But $(\bar{b}+x)j \equiv (\bar{b}\bar{x}+x)\,j$. This implies either $\bar{b}j$ and x entirely unconditioned, or $\bar{b}=0$, that is $b=i$, and xj.

(9) *Solution of* $ax+b\bar{x}+\omega$.

Now $ax+b\bar{x}+\omega$ implies $^-(ax+b\bar{x}+\omega)$.

But $$^-(ax+b\bar{x}+\omega) \equiv (\bar{a}x+\bar{b}\bar{x})\,j.$$

Hence either $\bar{a}\bar{b}\,.\,j$ (that is, $a+b+\omega$) and x is entirely unconditioned, or $\bar{a}\bar{b}=0$ and

$$x=(x+\beta_1)\,\alpha_1;$$

where α_1 and β_1 are weak forms.

43. Existential Expressions with two unknowns. (1) The general form of the existential expression involving two unknowns, x and y, is

$$(axy + bx\bar{y} + c\bar{x}y + d\bar{x}\bar{y})\,j.$$

Let $f(x, y)$ stand for the expression $axy + bx\bar{y} + c\bar{x}y + d\bar{x}\bar{y}$.

If $abcd\,.\,j$, the above existential expression does not condition x and y in any way (cf. § 41 (6)).

But if $abcd = 0$, then $f(x, y)$ vanishes (cf. § 34 (5)), if

$$x = cd + u(\bar{a} + \bar{b}), \qquad y = bd + v(\bar{a} + \bar{c}) \;\ldots\ldots\ldots\ldots\ldots\ldots(1);$$

where u and v satisfy the unlimiting equation

$$a\bar{b}\bar{c}uv + \bar{a}b\bar{d}u\bar{v} + \bar{a}c\bar{d}\bar{u}v + \bar{b}\bar{c}d\bar{u}\bar{v} = 0.$$

Thus if $f(x, y)$ is to vanish the minimum extension of the field of x is cd, its maximum extension is $\bar{a} + \bar{b}$, the minimum extension of the field of y is bd, its maximum extension is $\bar{a} + \bar{c}$.

Accordingly, $f(x, y)\,j$ and $abcd = 0$, yield three cases:

(α) x lies outside its above-mentioned field, and y is unrestricted:

(β) y lies outside its above-mentioned field, and x is unrestricted:

(γ) both x and y lie within their respective fields, but do not occupy *simultaneous positions* within their fields. That is to say, x and y can both be expressed by equations (1), but $(a\bar{b}\bar{c}uv + \bar{a}b\bar{d}u\bar{v} + \bar{a}c\bar{d}\bar{u}v + \bar{b}\bar{c}d\bar{u}\bar{v})\,j$.

If $f(x, y) = 0$ be an unlimiting equation for x and y, then cases (α) and (β) necessarily cannot be realized; and the existential expression in case (γ) becomes $f(u, v)\,j$, where u and v are written instead of x and y.

Case (α) is symbolized by $x \equiv \{x + {}^{-}(\chi\delta)_1\}\,(\alpha\beta)_1$, where χ is the umbral letter of c and the suffixes denote alternative weak forms. This existential expression for x implies that either x does not include all cd, or x does include some region not $(\bar{a} + \bar{b})$.

Case (β) is symbolized by $y \equiv \{y + {}^{-}(\beta\delta)_1\}\,(\alpha\chi)_1$.

Case (γ) requires that the problem of the next subsection be first considered.

(2) To solve for x and y from the expression, $f(x, y)\,j$; where $f(x, y) = 0$ is an unlimiting equation.

No expression for x or for y can be given, which taken by itself will satisfy $f(x, y)\,j$: for since the equation, $f(x, y) = 0$, is unlimiting any value of x or of y is consistent with its satisfaction. Thus to secure the satisfaction of $f(x, y)\,j$, either x or y must be assumed to have been assigned and then the suitable expression for the other (i.e. y or x) can be given. Thus write

$$f(x, y)\,j \equiv \{(ay + b\bar{y})\,x + (cy + d\bar{y})\,\bar{x}\}\,j.$$

Then by § 42 (6), if y be conceived as given,

$$x \equiv \{x + (\bar{\chi}\eta + \bar{\delta}\bar{\eta})_1\}\,(\alpha\eta + \beta\bar{\eta})_1.$$

Similarly if x be conceived as given

$$y \equiv \{y + (\bar{\beta}\xi + \bar{\delta}\bar{\xi})_1\}(\alpha\xi + \chi\bar{\xi})_1.$$

Both these expressions for x and y hold concurrently, and either of them expresses the full solution of the problem.

(3) Returning to the general problem of the solution of

$$(axy + bx\bar{y} + c\bar{x}y + d\bar{x}\bar{y})\,j,$$

where $abcd = 0$; the different cases can be symbolized thus:

$$(\alpha)\quad x \equiv \{x + {}^{-}(\chi\delta)_1\}(\alpha\beta)_1,$$

$$(\beta)\quad y \equiv \{y + {}^{-}(\beta\delta)_1\}(\alpha\chi)_1,$$

$$(\gamma)\quad \left.\begin{array}{l} x \equiv \{x + (\bar{\chi}\eta + \bar{\delta}\bar{\eta})_1\}(\alpha\eta + \beta\bar{\eta})_1, \\ y \equiv \{y + (\bar{\beta}\xi + \bar{\delta}\bar{\xi})_1\}(\alpha\xi + \chi\bar{\xi})_1 \end{array}\right\},$$

or (the second line of (γ) above), where x and y have the forms assigned in equations (1) of subsection (1).

44. Equations and Existential Expressions with one unknown. (1) Let there be n equations of the type $a_r x + b_r \bar{x} = c_r x + d_r \bar{x}$;

and an existential expression of the type $ex\,.\,j$.

Let A and B be the resultant discriminants of the n equations. Then the total amount of information to be got from the equations alone is (cf. § 30),

$$\bar{A}\bar{B} = 0, \text{ and } x = \bar{B} + uA.$$

The full information to be obtained by eliminating x is (cf. § 41 (4)),

$$\bar{A}\bar{B} = 0,\quad eA\,.\,j.$$

In considering the effect of the existential proposition on the solution for x two cases arise. For $x = \bar{B} + uA$, where u is conditioned by

$$e(\bar{B} + uA)\,.\,j.$$

Hence either (1) $e\bar{B}\,.\,j$, $x \equiv \bar{B}\epsilon + uA$, in which case u is entirely unconditioned (cf. § 41 (5)); or (2) $e\bar{B} = 0$, and $ueA\,.\,j$.

If the coefficients such as e, a_r, b_r, etc. be supposed to be known, then any result not conditioning u may be supposed to give no fresh information. Thus in case (1), where $e\bar{B}\,.\,j$, this result must be supposed to have been previously known, and therefore the existential expression $ex\,.\,j$ adds nothing to the equations. But in the case (2), $ueA\,.\,j$ gives $u = ue\alpha$, where α is the umbral letter of A. Hence the solution for x is

$$x \equiv \bar{B} + uA\,.\,e\alpha.$$

Here the existential expression $ex\,.\,j$ has partially conditioned u, and thus has given fresh information.

(2) Let there be n equations of the type $a_r x + b_r \bar{x} = c_r x + d_r \bar{x}$, and an existential expression of the type $e\bar{x} \,.\, j$.

The resultant of the equations is $\bar{A}\bar{B} = 0$, and their solution is

$$x = \bar{B} + uA.$$

Hence $$\bar{x} = \bar{A} + \bar{u}B.$$

Hence $$e(\bar{A} + \bar{u}B) \,.\, j.$$

The resultants $\bar{A}\bar{B} = 0$, $eB \,.\, j$ contain the full information to be found by eliminating x (cf. § 41 (4)).

The solution for x falls into two cases; either (1) $e\bar{A} \,.\, j$, and u is not conditioned (cf. § 41 (5)); or (2) $e\bar{A} = 0$, and $\bar{u}eB \,.\, j$.

If the coefficients be assumed to be known apart from these given equations, then the solution in case (1) must be taken to mean that the existential expression adds nothing to the determination of x beyond the information already contained in the equations. But in case (2) u is partially determined; for from $\bar{u}eB \,.\, j$, we deduce $\bar{u} \equiv \bar{u}\epsilon\beta$, where β is the umbral letter of B.

Hence $$u \equiv (u + \bar{\epsilon} + \bar{\beta}).$$

Therefore if $e\bar{A} = 0$,

$$x \equiv B + (u + \bar{\epsilon} + \bar{\beta})\, A.$$

In this case the existential expression has given fresh information.

(3) Let there be n equations of the type $a_r x + b_r \bar{x} = c_r x + d_r \bar{x}$, and an existential expression of the type $(ex + g\bar{x})j$.

The resultant of the equations is $\bar{A}\bar{B} = 0$, and their solution is

$$x = \bar{B} + uA, \quad \bar{x} = \bar{A} + \bar{u}B.$$

Hence $$\{e\bar{B} + g\bar{A} + eAu + gB\bar{u}\}\, j.$$

The resultants $\bar{A}\bar{B} = 0$, $(eA + gB)j$ contain the full information to be found by eliminating x (cf. § 41 (4)).

The solution for x falls into two cases, according as the existential expression $\{e\bar{B} + g\bar{A} + eAu + gB\bar{u}\}\, j$ does not or does condition u.

Case (1). If $(e\bar{B} + g\bar{A} + egAB)\, j$, then the above existential expression does not condition u at all (cf. § 41 (5)).

Hence if the coefficients are assumed to be known apart from the information of the given equations and existential expression, then the existential expression must be considered as included in the equations.

Case (2). If $(e\bar{B} + g\bar{A} + egAB) = 0$, then the existential expression for u reduces to $(eAu + gB\bar{u})\, j$, where $egAB = 0$.

Hence (cf. § 42 (5)) the solution for u is

$$u \equiv \{u + {}^{-}(\gamma\beta)_1\}\, (\epsilon\alpha)_1,$$

where the suffix 1 to the brackets of the umbral letters implies that they are alternative weak forms.

Hence the solution for x is in this case

$$x \equiv \bar{B} + A\,\{u + {}^{-}(\gamma\beta)_1\}\,.\,(\epsilon\alpha)_1.$$

In this case the given existential expression is to be considered as giving fresh information.

45. Boole's General Problem. (1) This problem (cf. § 33 (8)) can be adapted to the case when existential expressions are given, as in the following special case.

Let there be given n equations of the type $a_r x + b_r \bar{x} = c_r x + d_r \bar{x}$, and an existential expression of the type $gx\,.\,j$; it is required to determine z, where z is given by

$$z = ex + f\bar{x}.$$

By § 33 (8), $$z = e\bar{B} + f\bar{A} + ueA + \bar{u}fB,$$
where $$x = \bar{B} + uA.$$

Hence from above either (1) $g\bar{B}\,.\,j$, and u is unconditioned by the existential expression, or (2) $g\bar{B} = 0$, $gAu\,.\,j$. In the second case $u \equiv u\gamma\alpha$.

Hence if $g\bar{B}\,.\,j$ the existential expression adds nothing to the solution, assuming that the coefficients are already known; if $g\bar{B} = 0$, then

$$z \equiv e\bar{B} + f\bar{A} + eA\,.\,u\gamma\alpha + fB\,(\bar{u} + \bar{\gamma} + \bar{\alpha}).$$

It is to be noticed that even in the second case the existential expression gives no positive information as to z, and that it only suggests a possibility. For the solution asserts that u contains some of gA, but eA need not overlap that part of gA contained in u. Similarly the umbral letters in the expression $fB\,(\bar{u} + \bar{\gamma} + \bar{\alpha})$ give no definite information as to the nature of the term.

(2) If the existential expression in this problem be of the type $g\bar{x}\,.\,j$, then if $g\bar{A}\,.\,j$, it is included in the equations. But if $g\bar{A} = 0$, the solution for z is

$$z \equiv e\bar{B} + f\bar{A} + eA\,(u + \bar{\beta} + \bar{\gamma}) + fB\,.\,\bar{u}\beta\gamma.$$

Similar remarks apply to this solution as apply to that of the previous form of the problem.

(3) If the existential expression be of the type $gz\,.\,j$ or $g\bar{z}\,.\,j$, then more definite information can be extracted. Take the first case, namely $gz\,.\,j$, as an example.

The solution for z from the equations is (cf. § 33 (8))

$$z = (ef + e\bar{B} + f\bar{A}) + u\,(eA + fB),$$
where $$ef + e\bar{B} + f\bar{A} \nleftarrow eA + fB.$$

The existential expression requires the condition

$$g(eA + fB) . j.$$

If the coefficients are assumed to be well-known, then if

$$g(ef + e\bar{B} + f\bar{A}) . j,$$

no information is added by the existential expression. But if

$$g(ef + e\bar{B} + f\bar{A}) = 0,$$

then $$z \equiv (ef + e\bar{B} + f\bar{A}) + (eA + fB)\, u\, (\gamma\epsilon\alpha + \gamma\phi\beta),$$

where ϕ is the umbral letter of f.

The solution for $g\bar{z} . j$ is similar in type.

46. Equations and Existential Propositions with many unknowns. (1) A more complicated series of problems is arrived at by considering the set of n equations involving two unknowns of the type

$$a_r xy + b_r x\bar{y} + c_r \bar{x}y + d_r \bar{x}\bar{y} = a_r' xy + b_r' x\bar{y} + c_r' \bar{x}y + d_r' \bar{x}\bar{y} \ \ldots\ldots \ (1);$$

combined with the existential expression of the type

$$(exy + fx\bar{y} + g\bar{x}y + l\bar{x}\bar{y}) . j \ \ldots\ldots\ldots\ldots \ (2).$$

The various discriminants of the typical equation are

$$A_r = a_r a_r' + \bar{a}_r \bar{a}_r', \ B_r = b_r b_r' + \bar{b}_r \bar{b}_r', \ C_r = c_r c_r' + \bar{c}_r \bar{c}_r', \ D_r = d_r d_r' + \bar{d}_r \bar{d}_r'.$$

Also the resultant discriminants are

$$A = \Pi(A_r), \ B = \Pi(B_r), \ C = \Pi(C_r), \ D = \Pi(D_r).$$

Then from § 30 (9) the resultant of the equations is $\bar{A}\bar{B}\bar{C}\bar{D} = 0$, and from § 41 (4) the existential resultant is $(eA + fB + gC + lD) . j$.

If $$(e + \bar{A})(f + \bar{B})(g + \bar{C})(l + \bar{D}) . j,$$

then by § 41 (5) the existential expression (2) adds nothing to the equations (1) as regards the determination of x, assuming that the coefficients are well-known.

Assume that $$(e + \bar{A})(f + \bar{B})(g + \bar{C})(l + \bar{D}) = 0.$$

The solutions of the equations for x and y can be written

$$x = (A + B)\, u + \bar{C}\bar{D}\bar{u}, \quad y = (A + C)\, v + \bar{B}\bar{D}\bar{v},$$

where u and v satisfy

$$\bar{A}BCuv + A\bar{B}Duv + A\bar{C}D\bar{u}v + BC\bar{D}\bar{u}\bar{v} = 0 \ \ldots\ldots\ldots\ldots \ (3).$$

Substituting in (2) for x and y,

$$[\{e(A + BC) + f\bar{A}B\bar{C} + g\bar{A}\bar{B}C + l\bar{A}\bar{B}\bar{C}\}\, uv$$
$$+ \{eA\bar{B}\bar{D} + f(B + AD) + g\bar{A}\bar{B}\bar{D} + l\bar{A}\bar{B}D\}\, u\bar{v}$$
$$+ \{eA\bar{C}\bar{D} + f\bar{A}\bar{C}\bar{D} + g(C + AD) + l\bar{A}\bar{C}D\}\, \bar{u}v$$
$$+ \{e\bar{B}\bar{C}\bar{D} + fB\bar{C}\bar{D} + g\bar{B}C\bar{D} + l(D + BC)\}\, \bar{u}\bar{v}] . j \ \ldots\ldots\ldots \ (4).$$

The equation (3) is unlimiting and the problem now becomes that of the next subsection.

(2) Given an unlimiting equation (5) and an existential expression

$$(exy + fx\bar{y} + g\bar{x}y + l\bar{x}\bar{y}) \,.\, j \quad \text{..........} \quad \text{..............(6)}$$

to find the solution for x and y.

Let A, B, C, D be the discriminants of the equation (5). Then, as before, the condition that x and y are conditioned by (6) is

$$(e + \bar{A})(f + \bar{B})(g + \bar{C})(l + \bar{D}) = 0.$$

Since the equation (5) is unlimiting, this equation can be written

$$efgl + fgl\bar{A} + gle\bar{B} + lef\bar{C} + efg\bar{D} = 0.$$

Let a symmetrical solution of the equation (5) according to the method of § 35 be

$$x = a_1uv + b_1u\bar{v} + c_1\bar{u}v + d_1\bar{u}\bar{v},$$

$$y = a_2uv + b_2u\bar{v} + c_2\bar{u}v + d_2\bar{u}\bar{v}.$$

Let the expression (6) be written $f(x, y) \,.\, j$ for brevity.

Then substituting in (6) for x and y, as in § 33 (2), the expression becomes

$$\{f(a_1, a_2)\, uv + f(b_1, b_2)\, u\bar{v} + f(c_1, c_2)\, \bar{u}v + f(d_1, d_2)\, \bar{u}\bar{v}\} \,.\, j.$$

But this expression has been solved in § 43.

NOTE.—In this discussion of Existential Expressions valuable hints have been taken from the admirable paper, 'On the Algebra of Logic,' by Miss Christine Ladd (Mrs Franklin) in the book entitled *Studies in Deductive Logic, by Members of the Johns Hopkins University* But Mrs Franklin's calculus does not conform to the algebraic type considered in this book; and the discussion of Existential Expressions given here will, it is believed, be found to have been developed on lines essentially different to the discussion in that paper.

CHAPTER IV.

Application to Logic.

47. Propositions. (1) It remains to notice the application of this algebra to Formal Logic, conceived as the Art of Deductive Reasoning. It seems obvious that a calculus—beyond its suggestiveness—can add nothing to the theory of Reasoning. For the use of a calculus is after all nothing but a way of avoiding reasoning by the help of the manipulation of symbols.

(2) The four traditional forms of proposition of Deductive Logic are

All a is b ....(A),

No a is b ....(E),

Some a is b ....(I),

Some a is not b ....(O).

Proposition A can be conceived as stating that the region of a's is included within that of b's, the regions of space being correlated to classes of things. It is unnecessary to enquire here whether this is a satisfactory mode of stating the proposition for the purpose of explaining the theory of judgment: it is sufficient that it is a mode of expressing what the proposition expresses.

(3) Accordingly in the notation of the Algebra of Symbolic Logic proposition A can be represented by

$$a \mathrel{<\!\!\!\!\mid} b \quad \text{..(A),}$$

where a symbolizes the class of things each a, and b the class of things each b.

By § 26, Prop. VIII, and § 28, this proposition can be put into many equivalent symbolic forms, namely $a = ab$, $b = a + b$.

Also into other forms involving i, $\bar{a}$ and $\bar{b}$; namely,

$$\bar{b} \mathrel{<\!\!\!\!\mid} \bar{a}, \quad a\bar{b} = 0, \quad \bar{a} = \bar{a} + \bar{b}, \quad \bar{a} + b = i, \quad \bar{b} = \bar{a}\bar{b}.$$

Also into other forms involving the mention of an undetermined class u; namely

$$a = ub, \quad b = a + u, \quad \bar{a} = \bar{b} + u.$$

(4) According to this interpretation i must symbolize that limited class of things which is the special subject of discourse on any particular occasion. Such a class was called by De Morgan, the Universe of Discourse. Hence the name, Universe, which has been adopted for it here.

(5) Proposition E can be construed as denying that the regions of a's and b's overlap. Its symbolic expression is therefore

$$ab = 0 \quad \text{..................................(E).}$$

This can be converted into the alternative forms

$$a \notin \bar{b}, \quad b \notin \bar{a}, \quad a = a\bar{b}, \quad b = b\bar{a}, \quad \bar{a} + \bar{b} = i, \quad \bar{a} = \bar{a} + b, \quad \bar{b} = \bar{b} + a.$$

Thus, allowing the introduction of i, there are eight equivalent symbolic forms of the universal negative proposition, as well as eight forms of the universal affirmative. But if the introduction of i be not allowed, there is but one form, namely, $ab = 0$; remembering that the supplement of a term by its definition [cf. § 23 (8)] implies i.

On the other hand if the introduction of an undefined class symbol (u) be allowed, then four other forms appear, namely,

$$a = u\bar{b}, \quad \bar{a} = \bar{u} + b, \quad b = u\bar{a}, \quad \bar{b} = \bar{u} + a.$$

(6) Proposition I can be construed as affirming that the regions of the a's and b's overlap. Hence it affirms that the region ab exists. This is symbolically asserted by

$$ab \,.\, j \text{..................................(I).}$$

Equivalent forms are (cf. § 40) $a\beta \,.\, b\alpha$; $\bar{a} + \bar{b} + \omega$; $(\bar{a} + \bar{\beta}) + (\bar{b} + \bar{a})$.

Also if the introduction of undefined class symbols be allowed, then other equivalent forms are,

$$a \equiv wb \,.\, j + u; \quad b \equiv wa \,.\, j + u; \quad \bar{a} \equiv \bar{u}\,(\bar{w} + \bar{b} + \omega); \quad \bar{b} \equiv \bar{u}\,(\bar{w} + \bar{a} + \omega).$$

(7) Proposition O affirms that the regions of a's and $\bar{b}$'s overlap. This is symbolically expressed by

$$a\bar{b} \,.\, j \text{..................................(O).}$$

Equivalent forms are $a\bar{\beta} \,.\, \bar{b}\alpha$; $\bar{a} + b + \omega$; $(\bar{a} + \beta) + (b + \alpha)$.

Also using undefined class symbols,

$$a = w\bar{b} \,.\, j + u; \quad b = u\,(w + \bar{a} + \omega); \quad \bar{a} = \bar{u}\,(\bar{w} + b + \omega); \quad \bar{b} = \bar{w}a \,.\, j + \bar{u}.$$

48. Exclusion of Nugatory Forms. (1) It is sometimes necessary to symbolize propositions of the type A, so as to exclude nugatory forms; for instance when it is desired to infer symbolically a particular proposition from two universals.

(2) In order to avoid the form of nugatoriness which would arise from $a = 0$, in $a \notin b$, we can write

$$aj \notin bj \text{..................................(1),}$$

or

$$aj \equiv ab \,.\, j \text{..................................(2).}$$

The series of other forms can be deduced by mere symbolical reasoning from this form. Thus $b = b + ab$; also bj, $ab \,.\, j$, and $ab \,.\, j \equiv aj$; hence

$$bj \equiv bj + aj \qquad \text{(3)}.$$

Again, by taking the supplement of bj, we deduce $\bar{b} + \omega$. Multiplying (2) by $(\bar{b} + \omega)$, we find

$$aj \,.\, (\bar{b} + \omega) = 0 \qquad \text{(4)}.$$

By taking supplements of (1),

$$\bar{b} + \omega \notin \bar{a} + \omega \qquad \text{(5)}.$$

By taking supplements of (2)

$$\bar{a} + \omega \equiv \bar{a} + \bar{b} + \omega \qquad \text{(6)}.$$

By taking supplements of (3)

$$\bar{b} + \omega \equiv (\bar{b} + \omega)(\bar{a} + \omega) \qquad \text{(7)}.$$

By taking supplements of (4)

$$(\bar{a} + \omega) + bj \equiv i \qquad \text{(8)}.$$

Thus the eight forms of the proposition (A) (excluding those with undetermined class symbols) have been symbolized so as to exclude the nugatory form which arises when $a = 0$.

(3) Another nugatory form arises when $b = i$, this form can be excluded by the forms

$$a + \omega \notin b + \omega, \text{ or } (a + \omega) \equiv (a + \omega)(b + \omega).$$

By comparing these forms with equations (5) and (7) in subsection (2) it is easy to write down the remaining six forms.

It is also possible to combine the symbolism of both cases and thus to exclude both forms of nugatoriness, viz. $a = 0$, or $b = i$. But it is rarely that reasoning requires both forms to be excluded simultaneously, so there is no gain in the additional complication of the symbolism.

49. SYLLOGISM. (1) The various figures of the traditional syllogisms are as follows, where a is the minor term, b the middle term and c the major term:

First Figure.

A, All b is c,	E, No b is c,	A, All b is c,	E, No b is c,
A, All a is b,	A, All a is b,	I, Some a is b,	I, Some a is b,
therefore	therefore	therefore	therefore
A, All a is c.	E, No a is c.	I, Some a is c.	O, Some a is not c.

Second Figure.

E, No c is b,	A, All c is b,	E, No c is b,	A, All c is b,
A, All a is b,	E, No a is b,	I, Some a is b,	O, Some a is not b,
therefore	therefore	therefore	therefore
E, No a is c.	E, No a is c.	O, Some a is not c.	O, Some a is not c.

Third Figure.

A, All b is c,
A, All b is a,
therefore
I, Some a is c.

E, No b is c,
A, All b is a,
therefore
O, Some a is not c.

I, Some b is c,
A, All b is a,
therefore
I, Some a is c.

A, All b is c,
I, Some b is a,
therefore
I, Some a is c.

O, Some b is not c,
A, All b is a,
therefore
O, Some a is not c.

E, No b is c,
I, Some b is a,
therefore
O, Some a is not c.

Fourth Figure.

A, All c is b,
A, All b is a,
therefore
I, Some a is c.

A, All c is b,
E, No b is a,
therefore
O, Some a is not c.

I, Some c is b,
A, All b is a,
therefore
I, Some a is c.

E, No c is b,
A, All b is a,
therefore
O, Some a is not c.

E, No c is b,
I, Some b is a,
therefore
O, Some a is not c.

(2) The first mood of the first figure can be symbolized thus:

$$b \nleftarrow c,\quad a \nleftarrow b,\ \text{therefore}\ a \nleftarrow c:$$

or thus: $b = bc,\ a = ab,\ \text{therefore}\ a = ab = abc = ac$:

or thus: $b\bar{c} = 0,\ a\bar{b} = 0,\ \text{therefore}\ ac = a(b + \bar{b})\,\bar{c} = a\,.\,b\bar{c} + a\bar{b}\,.\,\bar{c} = 0$:

or thus: $\bar{c} \nleftarrow \bar{b},\ \bar{b} \nleftarrow \bar{a},\ \text{therefore}\ \bar{c} \nleftarrow \bar{a}$:

or thus: $c = b + c,\ b = a + b,\ \text{therefore}\ c = b + c = a + b + c = a + c$:

or thus: $\bar{b} = \bar{b} + \bar{c},\ \bar{a} = \bar{a} + \bar{b},\ \text{therefore}\ \bar{a} = \bar{a} + \bar{b} = \bar{a} + \bar{b} + \bar{c} = \bar{a} + \bar{c}$:

or thus: $\bar{b} + c = i,\ \bar{a} + b = i,\ \text{therefore}\ \bar{a} + c = \bar{a} + b\bar{b} + c$
$$= (\bar{a} + \bar{b} + c)(\bar{a} + b + c) = i:$$

or thus: $\bar{c} = \bar{b}\bar{c},\ \bar{b} = \bar{a}\bar{b},\ \text{therefore}\ \bar{c} = \bar{b}\bar{c} = \bar{a}\bar{b}\bar{c} = \overline{ac}.$

One half of these forms can be deduced from the other half by taking supplements.

In each case the two premises, which are each of the type A, have been written down in the same form. By combining two different methods of exhibiting symbolically propositions of the type A many other methods of conducting the reasoning symbolically can be deduced. It is unnecessary to state them here.

(3) The second mood of the first figure can be symbolized thus:

$$bc = 0,\ a = ab,\ \text{therefore}\ ac = abc = 0:$$

or thus: $b \nleftarrow \bar{c},\ a \nleftarrow b,\ \text{therefore}\ a \nleftarrow \bar{c}$:

or thus: $c \nleftarrow \bar{b},\ \bar{b} \nleftarrow \bar{a},\ \text{therefore}\ c \nleftarrow \bar{a}$:

or thus: $b = b\bar{c},\ a = ab,\ \text{therefore}\ a = ab\bar{c} = a\bar{c}$:

or thus: $c = c\bar{b},\ \bar{b} = \bar{a}\bar{b},\ \text{therefore}\ c = c\bar{b} = c\bar{a}\bar{b} = c\bar{a}$:

or thus: $\bar{b}+\bar{c}=i$, $a\bar{b}=0$, therefore $a=a(\bar{b}+\bar{c})=a\bar{b}+a\bar{c}=a\bar{c}$:

or thus: $\bar{b}=\bar{b}+c$, $a\bar{b}=0$, therefore $a(\bar{b}+c)=ac=a\bar{b}=0$:

or thus: $\bar{c}=\bar{c}+b$, $a=ab$, therefore $a\bar{c}=a(\bar{c}+b)=a\bar{c}+a=a$.

Eight forms have been given here but many others could be added by combining otherwise the modes of symbolizing propositions of the type A and E.

50. Symbolic Equivalents of Syllogisms. (1) It is better however at once to generalize the point of view of this symbolic discussion of the syllogism. It is evident that each syllogism is simply a problem of elimination of the middle term, and the symbolic discussions can be treated as special cases of the general methods already developed. Also the symbolic equivalence of all the forms of a proposition makes it indifferent which special form of a proposition is chosen as typical.

(2) Consider the first mood of the first figure: the term b is to be eliminated from $b=bc$, $a=ab$.

The positive discriminant of $b=bc$, is c, the negative discriminant is i.

The equation, $a=ab$, can be written $ab+a\bar{b}=ab$. The positive discriminant is i; its negative discriminant is $\bar{a}$.

Hence all the information to be found by eliminating b is

$$^-(ci)\times{}^-(i\bar{a})=0;$$

that is

$$a\bar{c}=0.$$

(3) Consider the second mood: the term b is to be eliminated from $bc=0$, $a=ab$.

The discriminants of the first equation are $\bar{c}$ and i; and of the second equation are i and $\bar{a}$.

Hence the elimination of b gives

$$^-(\bar{c}i)\times{}^-(i\bar{a})=0;$$

that is

$$ac=0.$$

It is obvious that the first and second moods of the second figure are symbolically the same problem as this mood.

(4) The third mood of the first figure is symbolically stated thus:

$$b=bc,\ ab\,.j.$$

Hence eliminating b by § 41 (4), the existential resultant is $ac\,.j$.

This is symbolically the same problem as the third and fourth moods of the third figure, and the third of the fourth figure.

(5) The fourth mood of the first figure can be symbolized thus:

$$bc=0,\ ab\,.j.$$

Hence eliminating b by § 41 (4), the existential resultant is $a\bar{c}\,.j$.

This is symbolically the same problem as the third mood of the second figure, the sixth of the third figure, and the fifth of the fourth figure.

(6) The only mood in the second figure not already discussed is the fourth; it can be symbolically stated thus: $c\bar{b}=0$, $a\bar{b}\,.\,j$.

Hence eliminating b by § 41 (4), the existential resultant is $a\bar{c}\,.\,j$.

(7) In the first mood of the third figure a particular proposition is inferred from two universal premises. It is necessary therefore in order to symbolize this mood that universal propositions as symbolically expressed be put on the same level as particular propositions in regard to the exclusion of nugatory forms. The syllogism can be symbolized thus,

$$bj \equiv bc\,.\,j,\ \ bj \equiv ba\,.\,j,$$

hence $bj = bc\,.\,j = bac\,.\,j$, hence $ac\,.\,j$.

(8) It is immediately evident that the premises assume more than is necessary to prove the conclusion, thus $b = bc$, instead of $bj = bc\,.\,j$, and $ab\,.\,j$, instead of $bj \equiv ab\,.\,j$, would have been sufficient. This is not a syllogism with what is technically known as a weakened conclusion, since no stronger conclusion of this type could have been drawn. It might be called a syllogism with over-strong premises. The syllogism of the same type with its premises not over-strong is the third of the first figure. Hence the symbolic treatment of that mood would serve for this one.

(9) The second mood of the third figure can be symbolized thus,

$$bc = 0,\ \ bj \equiv ab\,.\,j,\ \text{now}\ ab\,.\,j \equiv ab\,(c+\bar{c})\,.\,j \equiv ab\bar{c}\,.\,j,\ \text{hence}\ a\bar{c}\,.\,j.$$

This is obviously a syllogism with over-strong premises, since $bc = 0$, $ab\,.\,j$, would have been sufficient for the conclusion. The syllogism of the same type with sufficient premises is the fourth of the first figure.

(10) The fifth mood of the third figure can be symbolized thus: $b\bar{c}\,.\,j$, $\bar{a}b = 0$.

Hence eliminating b by § 41 (4), the existential resultant is $a\bar{c}\,.\,j$.

(11) The first mood of the fourth figure is symbolized thus,

$$cj \equiv bc\,.\,j,\ \ bj \equiv ab\,.\,j,\ \text{hence}\ bc\,.\,j \equiv abc\,.\,j,\ \text{hence}\ ac\,.\,j.$$

This is a syllogism with over-strong premises, the corresponding syllogism with sufficient premises is the third of the first figure.

(12) The second mood of the fourth figure is symbolized thus,

$$cj \equiv bc\,.\,j,\ \ ab = 0,\ \text{therefore}\ bc\,.\,j \equiv bc\,(a+\bar{a})\,.\,j \equiv bc\bar{a}\,.\,j,\ \text{hence}\ c\bar{a}\,.\,j.$$

This is a syllogism with over-strong premises; the corresponding syllogism with sufficient premises is the fourth of the first figure.

(13) The fourth mood of the fourth figure is symbolized thus,

$$bc = 0,\ \ bj \equiv ab\,.\,j,\ \text{therefore}\ ab\,.\,j \equiv ab\,(c+\bar{c})\,.\,j \equiv ab\bar{c}\,.\,j,\ \text{hence}\ a\bar{c}\,.\,j.$$

This is a syllogism with over-strong premises; the corresponding syllogism with sufficient premises is the fourth of the first figure.

(14) Since the conclusion of any syllogism can be obtained from the premises by the purely symbolic methods of this algebra, it follows that the conclusion of any train of reasoning, valid according to the formal canons of the traditional Deductive Logic, can also be obtained from the premises by the use of the algebra, using purely symbolic transformations.

51. GENERALIZATION OF LOGIC. (1) This discussion of the various moods of Syllogism suggests[1] that the processes of elimination and solution as applied to a system of equations and existential expressions developed in the preceding chapters of this Book can be construed as being a generalization of the processes of syllogism and conversion of common Logic.

It will be seen by reference to § 47 that a universal proposition is symbolized in the form of an equation, and a particular proposition in the form of an existential expression. Hence the most general form of equation may be conceived as a complex universal proposition, and a set of equations as a set of universal propositions. Also the most general form of an existential expression is the most general form of a particular proposition, and a set of such expressions is a set of particular propositions.

(2) The most general form of a system, entirely of universal propositions and involving one element to be determined, is given in Chapter II, §§ 29, 30. It is

$$a_1x + b_1\bar{x} = c_1x + d_1\bar{x},$$
$$\cdots\cdots\cdots\cdots\cdots\cdots$$
$$a_rx + b_r\bar{x} = c_rx + d_r\bar{x},$$
$$\cdots\cdots\cdots\cdots\cdots\cdots$$
$$a_nx + b_n\bar{x} = c_nx + d_n\bar{x}.$$

Here x is supposed to be the class to be further determined, and the other symbols all refer to well-known classes.

Then the information wanted is found by forming n functions of the type, $A_r = a_rc_r + \bar{a}_r\bar{c}_r$, and n of the type, $B_r = b_rd_r + \bar{b}_r\bar{d}_r$, and by forming the products $A = A_1A_2 \ldots A_n$, $B = B_1B_2 \ldots B_n$. Then $x = \bar{B} + uA$; with the condition that $\bar{A}\bar{B} = 0$, which is probably well-known.

(3) The essential part of this process is the formation of the two regions A and B out of the well-defined regions involved in the system of propositions. This composition of the two discriminants is a process of rearranging our original knowledge so as to express in a convenient form the fresh information conveyed in the system. Formally it is a mere picking out of certain regions defined by the inter-relations of the known regions which are the coefficients of the equations: but the process in practice may result in a real addition to knowledge of the true definition of x. For instance rationality and animality may have been the characteristics of two regions among the

[1] Cf. Boole, *Laws of Thought*, chapter IX. § 8, chapter XV.

coefficients in the system; but in A and B the common part of the regions may only occur: then it is at once known that x only involves the ideas of rationality and animality in so far as it involves those of humanity—a very real addition to knowledge, though formally it is only a question of better arrangement as compared to the original system.

(4) The undefined nature of the information given by particular propositions makes it usually desirable not to deal with such propositions in a mass, but to sort them one by one, comparing their information with that derived from the known system of universal propositions.

Thus let the above system of universal propositions be known, and also the proposition of the type I, viz. $ex \,.\, j$.

Then from § 41 the full information to be found by eliminating x is, $\bar{A}\bar{B} = 0$, $eA \,.\, j$; and the solution for x is, either

$$(1)\ e\bar{B} \,.\, j,\ x \equiv \bar{B}\epsilon + uA, \text{ or } (2)\ e\bar{B} = 0,\ x \equiv \bar{B} + uA \,.\, \epsilon\alpha.$$

Now propositions including a common term x are in general accumulated in science or elsewhere just because information concerning x is required. Also the propositions will as far as possible connect x with thoroughly well-known terms. If we conceived this process as performed with ideal success, then the coefficients of x and $\bar{x}$ in the above equations and existential expression must be conceived as completely known, and no information concerning their relations will be fresh. Hence in case (1), when $e\bar{B} \,.\, j$, the particular proposition ($ex \,.\, j$) is included in the universal propositions; but in case (2) the particular proposition has added fresh information.

But this sharp division between things known and things unknown is not always present in reasoning. In such a case the universals and the particular perform a double function, they both define more accurately the properties of things already fairly well-known, and determine the things x which are comparatively unknown.

The discussion of this typical case may serve to exemplify the logical interpretation of the problems of the previous chapters.

CHAPTER V.

Propositional Interpretation.

52. Propositional Interpretation. (1) There is another possible mode of interpreting the Algebra of Symbolic Logic which forms another application of the calculus to Logic.

Let any letter of the calculus represent a proposition or complex of propositions. The propositions represented are to be either simple categorical propositions, or complexes of such propositions of one or other of two types. One type is the complex proposition which asserts two or more simple propositions to be conjointly true; such a proposition asserts the truth of all its simple components, and the proposer is prepared to maintain any one of them. The verbal form by which such propositions are coupled together is a mere accident: the essential point to be noticed is that the complex proposition is conceived as the product of a set of simple propositions, marked off from all other propositions, and set before the mind by some device, linguistic or otherwise, in such fashion that each single proposition of the set is stated as valid. Hence if one single proposition of the set be disproved, the complex proposition is disproved. Let such a complex of propositions be called a conjunctive complex.

(2) The other type of complex proposition is that which asserts that one at least out of a group of simple propositions, somehow set before the mind, is true. Here again the linguistic device is immaterial, the essential point is that the group of propositions is set before the mind with the understood assertion that one at least is true. Let such a type of complex of propositions be called a disjunctive complex.

(3) Furthermore we may escape the difficult (and perhaps unanswerable or even unmeaning) question of deciding what propositions are to be regarded as simple propositions. The simplicity which is here asserted of certain propositions, is, so to speak, a simplicity *de facto* and not *de jure*. All that is meant is that a simple proposition is one which as a matter of fact for the purpose in hand is regarded, and is capable of being regarded, as a simple

assertion of a fact, which fact may be indefinitely complex and capable of further analysis.

Thus a conjunctive or a disjunctive complex may each of them be regarded as a simple proposition by directing attention to the single element of assertion which binds together the different component propositions of a complex of either type.

(4) To sum up: all propositions symbolized, actually or potentially, by single letters can be regarded as simple propositions: and the only analysis of simple propositions is to be their analysis either into conjunctive or disjunctive complexes of simple propositions. Also a simple proposition is a proposition which can be regarded as containing a single element of categorical assertion.

53. Equivalent Propositions. Two propositions, x and y, will be said to be equivalent, the equivalence being expressed by $x=y$, when they are equivalent in validity. By this is meant that any motives (of those motives which are taken account of in the particular discourse) to assent, which on presentation to the mind induce assent to x, also necessarily induce assent to y and conversely.

54. Symbolic Representation of Complexes. (1) Let the disjunctive complex formed out of the component propositions $a, b, c\ldots$ be symbolized by $(a+b+c\ldots)$. This symbolism is allowable since the disjunctive complex has the properties of addition: for (1) the result of the synthesis of the propositions is a definite unique thing of the same type as the thing synthesized, namely another proposition: (2) the order which is conceivable in the mental arrangement of the propositions is immaterial as far as the equivalence of the resulting proposition is concerned: (3) the components of a disjunctive complex may be associated in any way into disjunctive complexes; so that the associative law holds.

(2) Let the conjunctive complex formed out of the component propositions $a, b, c\ldots$ be symbolized by $abc\ldots$. This symbolism by the sign of multiplication is allowable: (1) since the result of the synthesis of a number of component propositions into a conjunctive complex is definite and unique, being in fact another proposition which can be regarded as a simple proposition; (2) since the conjunctive complex formed out of the proposition a and the complex $b+c$ is the same proposition as the disjunctive complex formed by ab and ac; in other words $a(b+c)=ab+ac$.

55. Identification with the Algebra of Symbolic Logic. (1) It now remains to identify the addition and multiplication of propositions, as here defined, with the operations of the Algebra of Symbolic Logic.

The disjunctive complex $x+x$ is the same as the simple proposition x.

For $x+x$ means either the proposition x or the proposition x, and this is nothing else than the proposition x. Hence $x+x=x$.

(2) The conjunctive complex obeys the associative law: for to assert a and b and c conjointly is the same as asserting b and c conjointly and asserting a conjointly with this complex assertion. Hence $abc=a\,.\,bc$.

(3) The conjunctive complex also obviously obeys the commutative law: thus

$$abc=acb=bac.$$

(4) The conjunctive complex formed of a and a is the same as the simple proposition a; hence $aa=a$.

(5) *The null-element* of the manifold of the Algebra corresponds to the absolute rejection of all motives for assent to a proposition, and further to the consequent rejection of the validity of the proposition. Hence $x=0$, comes to mean the rejection of x from any process of reason, or from any act of assertion. In so far as they are thus rejected all such propositions are equivalent. Thus if $x=0$, $y=0$, then $x=y=0$. Furthermore if $b=0$, the proposition $a+b$ is equivalent to the proposition a alone; for the motives of validity of b being absolutely rejected, those for the validity of a alone remain.

Hence if $b=0$, $$a+b=a.$$

Again, if $b=0$, then $ab=0$; for ab means that a and b are asserted conjointly, and if the motives for b be rejected, then the motives for the complex proposition are rejected.

The class of propositions to be thus absolutely rejected is best discussed later, after the discussion of the other special element.

(6) *The Universe.* The other special element of the manifold is that which has been called the Universe. Those propositions, or that class of perhaps an indefinite number of propositions, will be severally considered as equivalent to the Universe when their validity has acquired some special absoluteness of assent, either conventionally (for the sake of argument), or naturally.

This class of propositions may be fixed by sheer convention: certain propositions may be arbitrarily enumerated and to them may be assigned the absolute validity which is typified by the element called the Universe. Or some natural characteristic may be assigned as the discriminating mark of propositions which are equivalent to the universe. For instance, propositions which while reasoning on a given subject matter are implied in reasoning without rising to explicit consciousness or needing explicit statement at any stage of the argument might be equated to the Universe.

The laws of thought as stated in Logic are such propositions. Again in a discussion between two billiard markers on a game of billiards the proposition, that two of the balls were white and the third red, might be of this character. For billiard markers such a proposition rises to the level of a law of thought.

Again, in legal arguments before an inferior court the judgments of the Supreme Court of Judicature might be considered as propositions each equivalent to the Universe.

In this interpretation the name of the Universe as applied to this element is unfortunate: the Truism would be a better name for it. Let all propositions equivalent to the Universe be termed self-evident.

(7) The properties assigned to the Universe (i) in relation to any proposition x are (cf. § 23 (6) and (7))

$$x + i = i,$$

$$xi = x.$$

The validity of any proposition equivalent to the Universe being taken as absolute, the validity of the disjunctive complex formed of this proposition and some other proposition x cannot be anything else but the absolute validity of the Universe. Hence the equation $x + i = i$ is valid for the present interpretation.

Again, in the conjunctive complex formed of any proposition and a proposition equivalent to the Universe, the validity of the second proposition being unquestioned, the validity of the whole is regulated by that of the first proposition. Hence the equation $xi = x$ is also valid.

(8) This conception of a class of propositions either conventionally or naturally of absolute validity gives rise for symbolic purposes in this chapter to an extension of the traditional idea of the conversion of propositions. If the Universe be narrowed down to the Laws of Thought, then all the propositions which can be derived from any given proposition x taken in connection with the propositions of the Universe are those propositions which arise in the traditional theory of the conversion of propositions. Hence if we extend the Universe of self-evident propositions either by some natural or conventional definition, we may extend the conception of conversion to include any proposition which can be derived from a given proposition x taken in connection with the assigned propositions equivalent to the Universe.

Thus if i be any proposition equivalent to the Universe, xi will be considered to be simply the proposition x in another form.

(9) The supplementary proposition, $\bar{x}$, of the proposition x is defined by the properties,

$$x\bar{x} = 0, \quad x + \bar{x} = i.$$

Whatever the propositions of the Universe may be, even if they are reduced to the minimum of the Laws of Thought, the logical contradictory of x satisfies these conditions and therefore is a form of the supplementary proposition. But by the aid of the propositions of the Universe there are other more special forms into which the contradictory can be 'converted.'

Any such form, equivalent to the contradictory, is with equal right called the supplement of x. Thus to the billiard markers cited above the supplement to the proposition, the ball is red, is the proposition, the ball is white; for one of the two must be true and they cannot both be true.

(10) It is now possible clearly to define the class, necessarily of indefinite number, of propositions which are to be equated to the null element. This equation must not rest merely on the empirical negative fact of the apparent absence of motives for assent; but on the positive fact of inconsistency with the propositions which are equated to the Universe. If the Universe be reduced to the Laws of Thought, then all propositions equated to null are self-contradictory. With a more extended Universe, all propositions equated to null are those which contradict the fundamental assumptions of our reasoning. Let all propositions equated to the null-element be called self-condemned.

(11) The hypothetical relation between two propositions x and y, namely, If y be true then x is true, implies that the motives for assent to y are included among those for assent to x. Hence the relation can be expressed by $y \ni x$, or by any of the equivalent equational forms of § 26, Prop. VIII. And y may be said to be incident in x.

We have now examined all the fundamental principles of the Algebra of Symbolic Logic and shown that our present symbolism for propositions agrees with and interprets them all. Hence the development of this symbolism is simply the development of the Algebra which has been already carried out.

56. EXISTENTIAL EXPRESSIONS. The symbol $x.j$ denotes the proposition x and implies that it is not self-condemned. The symbol $x + \omega$ denotes the proposition x and implies that it is not self-evident. Hence, $-(xj) = \bar{x} + \omega$, implies that the supplement of a proposition not self-condemned is itself not self-evident.

Umbral letters. The symbol $x\eta$ denotes the proposition x and implies that xy is not self-condemned: the symbol $x + \bar{\eta}$ implies that $x + \bar{y}$ is not self-evident (cf. § 40 (1)). The whole use of umbral letters therefore receives its interpretation.

57. SYMBOLISM OF THE TRADITIONAL PROPOSITIONS. (1) This system of interpretation, which in its main ideas is a modification of that due to Boole[1], has perhaps the best right to be called a system of Symbolic Logic. It assumes the existence of an unquestioned sphere of knowledge, and traces generally the consequences which can be deduced from any categorical proposition or set of categorical propositions taken in connection with this sphere of knowledge. The former mode of interpretation, by class inclusion

[1] Cf. *Laws of Thought*, chap. XI.

and exclusion, only applied to propositions of the subsumption type: the present mode applies to any categorical proposition, that is to any proposition depending on a single element of assertion. Further it can symbolize any relation in which two such propositions can stand to each other, namely, (1) the disjunctive relation, in either of the two forms, namely, either when the propositions can be both true or when only one can be true (i.e. by the forms $x+y$ and $x+y\bar{x}$); (2) the conjunctive relation; (3) the hypothetical relation (i.e. by the equation $y=xy$).

(2) A defect of the method at first sight is that it seemingly cannot exhibit the process of thought in a syllogism.

Thus if x and y be the two premises, and z be the conclusion, then z is true if xy be true: hence $xy=xy \,.\, z$, or $xy \mathrel{\not\prec} z$ are two of the forms in which an argument from two propositions to a third can be exhibited. But this symbolism only exhibits the fact that z has been concluded from xy, and in no way traces the course of thought.

(3) The defect is remedied by McColl (*Proc. London Math. Soc.*, Vols. IX., X., XI., XIII.), by means of the device of analysing a proposition of one of the traditional types, A, E, I, O, into a relation between other propositions—thus instead of, All A is B, consider the propositions, It is A, It is B; then, All A is B, is the same thing as saying that the proposition, It is A, is equivalent in validity to the conjunctive complex, It is A and It is B. Hence if one proposition is a, and the other b, the original proposition is symbolized by $a=ab$. In other words, the hypothetical relation mentioned in § 55 (11) holds between the propositions a and b.

This analysis is certainly possible; and it is not necessary for the symbolism that it should be put forward as a fundamental analysis, but merely as possible. It requires however some careful explanation in order to understand the possible relations and transformations of such propositions as, It is A.

58. Primitive Predication. (1) Let a proposition of the type, It is A, be called a primitive predication. In such a proposition the subject is not defined in the proposition itself; it is supposed to be known, either by direct intuition, or as the result of previous discourse. In the latter case the proposition must not be considered as an analytical deduction from previous propositions defining the 'it.' The previous discourse is simply a means of bringing the subject before the mind: and when the subject is so brought before the mind, the proposition is a fresh synthetic proposition. A primitive predication necessarily implies the existence of the subject. The proposition may be in error; but without a subject, instead of a proposition there is a mere exclamation.

(2) If the predicate be a possible predicate, either because it is not self-contradictory, or further because its possibility is not inconsistent with the

rest of knowledge, primitive predication can only be tested as to its truth or falsehood by an act of intuition. For a primitive predication is essentially a singular act having relation to a definite intuition; and it is only knowledge based on definite intuitions having concrete relations with this intuition which can confirm or invalidate it.

The propositions taken as equivalent to the Universe in the present symbolism must be propositions deducible from propositions relating universal ideas or be such propositions themselves. Hence if x stand for a proposition which is a primitive predication, then x can only be self-condemned if the predicate be self-contradictory or inconsistent with the propositions equivalent to the Universe.

Also x can only itself be equivalent to the universe, if there be the convention that during the given process of inference the ultimate subject of every proposition shall have certain assigned attributes. Then an act of primitive predication attributing one of these attributes to a subject is equivalent to the Universe, that is, is self-evident.

(3) If x be a primitive predication, $\bar{x}$ is not a primitive predication; it may be called a primitive negation. Thus if x stands for, It is man, then $\bar{x}$ stands for, It is not man; that is to say, the subject may have any possible attribute except that of man. If x be self-condemned, then $\bar{x}$ states that the subject may have any possible attribute; thus $\bar{x}=i$, since it is an obvious presupposition of all thought that a subject undefined except by the fact of an act of intuition may have any possible attribute.

If $x=i$, then $\bar{x}$ is a denial that the subject referred to has a certain attribute, which by hypothesis all subjects under consideration do possess; hence $\bar{x}$ is self-condemned: that is, $\bar{x}=0$.

(4) A primitive negation does not necessarily occur merely as the denial of a primitive predication. The relations of the two types of proposition may be inverted. The fundamental proposition may be the denial that a certain predicate is attributable to the subjects within a certain field of thought. If this proposition, which relates universal ideas, be included among propositions which are self-evident, then any primitive denial which denies the certain predicate is also self-evident; and its supplement, which is a primitive predication, is self-condemned.

59. EXISTENTIAL SYMBOLS AND PRIMITIVE PREDICATION. (1) If x stand for a primitive predication, then xj implies that the predicate is a possible predicate of a subject in so far as the self-evident propositions regulate our knowledge of possibility. Now xj implies $\bar{x}+\omega$; this last expression implies that the denial of the primitive predication cannot be deduced to be true for all possible subjects of predication by means of the self-evident propositions. This deduction is an obvious consequence of xj.

(2) Also $\bar{x}j$ implies that the denial of the primitive predication is consistent with the self-evident propositions as far as some possible subjects of predication are concerned. Now $\bar{x}j$ implies $x + \omega$, and this implies that the primitive predication is not self-evident for all possible subjects of predication.

(3) If x, y, z, etc., all stand for separate primitive predications, then in any complex, either conjunctive or disjunctive, which comprises two or more of these propositions, the propositions are to be understood to refer to the same subject. Otherwise, since the propositions are singular acts, the propositions can have no relation to each other. Thus xy, i.e. x with y, stands for the combined assertions, It is X and it is Y, or in other words, It is both X and Y. Also $x + y$ stands for, it is either X or Y or both. Similarly primitive denials occurring together in a complex must both refer to the same subject; so also must primitive predications and primitive denials occurring together in a complex.

(4) The symbol $x\eta$ stands for the proposition, It is X, and also implies the consistency with the self-evident propositions of the proposition, It is Y, as applied to the same subject as x. The umbral letter η affixed to x is in fact a reminder that xy is consistent with the self-evident propositions for some possible subjects of predication.

60. Propositions. (1) It is now possible to symbolize the traditional forms of logical proposition.

Proposition A. All X is Y, takes the form, if x then y, where x and y are the primitive predications, It is X, It is Y. Hence the proposition takes the symbolic forms

$$x \mathrel{\not\!\!+} y,\ x = xy, \text{ or any symbolically equivalent form.}$$

(2) Proposition E. No X is Y, takes the form, If x then $\bar{y}$. Hence the proposition takes the symbolic forms

$$x \mathrel{\not\!\!+} \bar{y},\ x = x\bar{y}, \text{ or any symbolically equivalent form.}$$

(3) Proposition I. Some X is Y, takes the form that the conjunctive complex xy is not self-condemned; if the denial of all predicates or combinations of predicates, which do not actually occur in subjects belonging to the field of thought considered, be included among the self-evident propositions. Hence the proposition can be put in the symbolic form, $xy\,.\,j$, or in any symbolically equivalent form.

It must be carefully noticed that it is the connotation of $xy\,.\,j$ which expresses the Proposition I and not the conjunctive complex xy, which stands for, It is X and Y. Thus the supplement of $xy\,.\,j$, namely, $^{-}(xy\,.\,j)$, or $\bar{x} + \bar{y} + \omega$, does not express the contradictory of the Proposition I, but the contradictory of the conjunctive complex xy. On the contrary the connotation of $\bar{x} + \bar{y} + \omega$ still expresses the same Proposition I.

(4) Proposition O. Some X is not Y, takes the form that the conjunctive complex $x\bar{y}$ is not self-condemned; where the same hypothesis as to the self-evident propositions is made as in the case of Proposition I. The symbolic form is therefore, $x\bar{y} \, . \, j$, or any equivalent symbolic form.

(5) The universal Propositions A, E as symbolized above give no existential import to their subjects. But the symbolism as there explained has the further serious defect that there is no symbolic mode of giving warning of the nugatoriness of the propositions when the subject is non-existent. But this can be easily remedied by including among the self-evident propositions the denial of any predicates which do not appear in an existent subject in the field of thought. This is the same supposition as had to be made in order to symbolize I and O. Hence in the proposition $x = xy$, if there be no X's, then $x = 0$.

Also if it be desired to exclude this nugatory case, then the proposition can be written

$$xj \equiv xy \, . \, j.$$

(6) It has now been proved that the present form of interpretation includes that of the preceding chapter as a particular case. Thus all the results of the previous chapter take their place as particular cases of the interpretations of this present chapter.

Historical Note. The Algebra of Symbolic Logic, viewed as a distinct algebra, is due to Boole, whose 'Laws of Thought' was published in 1854. Boole does not seem in this work to fully realize that he had discovered a system of symbols distinct from that of ordinary algebra. In fact the idea of 'extraordinary algebras' was only then in process of formation and he himself in this work was one of its creators. Hamilton's Lectures on Quaternions were only published in 1853 (though his first paper on Quaternions was published in the Philosophical Magazine, 1844), and Grassmann's *Ausdehnungslehre* of 1844 was then unknown. The task of giving thorough consistency to Boole's ideas and notation, with the slightest possible change, was performed by Venn in his 'Symbolic Logic,' (1st Ed. 1881, 2nd Ed. 1894). The non-exclusive symbolism for addition (i.e. $x + y$ instead of $x + y\bar{x}$) was introduced by Jevons in his 'Pure Logic,' 1864, and by C. S. Peirce in the *Proceedings of the American Academy of Arts and Sciences*, Vol. VII, 1867. Peirce continued his investigations in the *American Journal of Mathematics*, Vols. III. and VII. The later articles also contain the symbolism for a subsumption, and many further symbolic investigations of logical ideas, especially in the Logic of Relatives, which it does not enter into the plan of this treatise to describe. These investigations of Peirce form the most important contribution to the subject of Symbolic Logic since Boole's work.

Peirce (*loc. cit.* 1867) and Schröder in his important pamphlet, *Operationskreis des Logikkalküls*, 1877, shewed that the use of numerals, retained by Boole, was unnecessary, and also exhibited the reciprocity between multiplication and addition; Schröder (*loc. cit.*) also shewed that the operations of subtraction and division might be dispensed with. Schröder has since written a very complete treatise on the subject, 'Vorlesungen über die Algebra der Logik,' Teubner, Leipsic, Vol. I, 1890, Vol. II, 1891, Vol. III, 1895; Vol. III. deals with the Logic of Relatives.

A small book entitled 'Studies in Deductive Logic,' Boston 1883, has in it suggestive papers, especially one by Miss Ladd (Mrs Franklin) 'On the Algebra of Logic,' and one by Dr Mitchell 'On a new Algebra of Logic.'

A most important investigation on the underlying principles and assumptions which belong equally to the ordinary Formal Logic, to Symbolic Logic, and to the Logic of Relatives is given by Mr W. E. Johnson in three articles, 'The Logical Calculus,' in *Mind*, Vol. I, New Series, 1892. His symbolism is not in general that of the Algebraic type dealt with in this work.

The propositional interpretation in a different form to that given in this work was given by Boole in his book: modifications of it have been given by Venn (Symbolic Logic), Peirce (*loc. cit.*), H. McColl in the *Proceedings of the London Mathematical Society*, Vols. IX, X, XI, XIII, 'On the Calculus of Equivalent Statements.' The latter also introduces some changes in notation and some applications to the limits of definite integrals, which are interesting to mathematicians.

A large part of Boole's 'Laws of Thought' is devoted to the application of his method to the Theory of Probability.

Both Venn and Schröder give careful bibliographies in their works. These two works, Johnson's articles in *Mind*, and of course Boole's 'Laws of Thought,' should be the first consulted by students desirous of entering further into the subject.

There is a hostile criticism of the utility of the whole subject from a logical point of view in Lotze's Logic.

BOOK III.

POSITIONAL MANIFOLDS.

CHAPTER I.

Fundamental Propositions.

61. Introductory. (1) In all algebras of the numerical genus (cf. § 22) any element of the algebraic manifold of the first order can be expressed in the form $\alpha_1 e_1 + \alpha_2 e_2 + \dots + \alpha_\nu e_\nu$, where $e_1, e_2, \dots e_\nu$ are ν elements of this manifold and $\alpha_1, \alpha_2, \dots \alpha_\nu$ are numbers, where number here means a quantity of ordinary algebra, real or imaginary. It will be convenient in future invariably to use ordinary Roman or italic letters to represent the symbols following the laws of the special algebra considered: thus also each group of such letters is a symbol following the laws of the special algebra. Such letters or such group of letters may be called extraordinaries* to indicate that in their mutual relations they do not follow the laws of ordinary algebra. Greek letters will be strictly confined to representing numbers, and will in their mutual relations therefore follow all the laws of ordinary algebra.

(2) The properties of a positional manifold will be easily identified with the descriptive properties of Space of any number of dimensions, to the exclusion of all metrical properties. It will be convenient therefore, without effecting any formal identification, to use spatial language in investigating the properties of positional manifolds.

A positional manifold will be seen to be a quantitively defined manifold, and therefore also a complex serial manifold (cf. § 11).

(3) The fundamental properties which must belong, in some form or other, to any positional manifold must now be discussed. The investigation of §§ 62 63 will be conducted according to the same principles as that of §§ 14—18, which will be presupposed throughout. The present investigation is an amplification of those articles, stress being laid on the special properties of algebraic manifolds of the numerical genus.

62. Intensity. (1) Each thing denoted by an extraordinary, representing an element of a positional manifold, involves a quantity special to it, to be called its intensity. The special characteristic of intensity is that in general the thing is absent when the intensity is zero, and is never absent

* This name was used by Cayley.

unless the intensity is zero. There is, however, an exceptional case discussed in Chapter IV. of this book.

(2) Two things alike in all respects, except that they possess intensities of different magnitudes, will be called things of the same kind. They represent the same element of the positional manifold, the intensity being in fact a secondary property of the elements of the manifold (cf. § 9).

(3) Let any arbitrary intensity of a thing representing a certain element be chosen as the unit intensity, then the numerical measure of the intensity of another thing representing the same element is the ratio of its intensity to the unit intensity. Let the letter e denote the thing at unit intensity, then a thing of the same kind at intensity α, where α is some number, will be denoted by αe or by $e\alpha$, which will be treated as equivalent symbols.

(4) Let the intensity of a thing which is absent be denoted by 0. Then by the definition of intensity,

$$0e = e0 = 0.$$

(5) Further, two things representing the same element at intensities α and β are to be conceived as capable of a synthesis so as to form one thing representing the same element at intensity $\alpha + \beta$. This synthesis is unambiguous and unique, and such as can be symbolized by the laws of addition. Hence

$$\alpha e + \beta e = (\alpha + \beta) e = (\beta + \alpha) e = \beta e + \alpha e.$$

The equation

$$(\alpha + \beta) e = \alpha e + \beta e,$$

involves the formal distributive law of multiplication (cf. § 19). Accordingly in the symbol αe, we may conceive α and e, as multiplied together.

(6) Conversely a thing of intensity $\alpha + \beta$ is to be conceived as analysable into the two things representing the same element at intensities α and β. Then it is to be supposed that one of the things at intensity β can be removed, and only the thing at intensity α left. This process can be conceived as and symbolized by subtraction. Its result is unambiguous and unique. Hence

$$(\alpha + \beta) e - \beta e = \alpha e.$$

(7) If corresponding to any thing αe there can be conceived another thing, such that a synthesis of addition of the two annihilates both, then this second thing may be conceived as representing the same element as the first but of negative intensity $-\alpha$ [cf. § 89 in limitation of this statement].

Thus $$\alpha e + (-\alpha e) = \alpha e - \alpha e = 0e = 0.$$

Complex intensities of the form $\alpha + i\beta$ can also be admitted (i being $\sqrt{-1}$). It was explained in § 7 that the logical admissibility of their use was altogether independent of the power of interpreting them.

(8) Thus finally, if α, β, γ, δ be any numbers, real or complex, and e an extraordinary, we have

$$\alpha\delta e + \beta\delta e - \gamma\delta e = (\alpha\delta + \beta\delta - \gamma\delta)\, e = \delta\,(\alpha + \beta - \gamma)\, e = \delta\,(\alpha e + \beta e - \gamma e);$$

also
$$0e = 0.$$

All the general laws of addition and subtraction (cf. §§ 14—18) can be easily seen to be compatible with the definitions and explanations given above.

(9) It must be remembered that other quantities may be involved in a thing αe besides its own intensity. But such quantities are to be conceived as defining the quality, or character, of the thing, in other words, the element of the manifold which the thing represents; as for instance its pitch defines in part the character of a wrench. If any of these quantities alter, the thing alters and either it ceases to be capable of representation by any multiple of e, or e can represent more than one element [cf. § 89 (2)].

63. Things representing different elements. (1) Let e_1, $e_2 \dots e_\nu$ denote ν things representing different elements each at unit intensity. Let things at any intensities of these kinds be capable of a synthesis giving a resultant thing; and let the laws of this synthesis be capable of being symbolized by addition.

Let a be the resultant of $\alpha_1 e_1$, $\alpha_2 e_2$, ... $\alpha_\nu e_\nu$;

then
$$a = \alpha_1 e_1 + \alpha_2 e_2 + \alpha_3 e_3 + \dots + \alpha_\nu e_\nu.$$

(2) By these principles and by the previous definitions of the present chapter,

$$\begin{aligned} 2a = a + a &= (\alpha_1 e_1 + \alpha_2 e_2 + \dots + \alpha_\nu e_\nu) + (\alpha_1 e_1 + \dots + \alpha_\nu e_\nu) \\ &= (\alpha_1 + \alpha_1)\, e_1 + (\alpha_2 + \alpha_2)\, e_2 + \dots + (\alpha_\nu + \alpha_\nu)\, e_\nu \\ &= 2\alpha_1 e_1 + 2\alpha_2 e_2 + \dots + 2\alpha_\nu e_\nu. \end{aligned}$$

Similarly if β be any real positive number, integral or fractional,

$$\beta a = \beta\alpha_1 e_1 + \beta\alpha_2 e_2 + \dots + \beta\alpha_\nu e_\nu.$$

Let this law be extended by definition to the case of negative and complex numbers.

Hence for all values of β

$$\beta a = \beta\,(\alpha_1 e_1 + \alpha_2 e_2 + \dots \alpha_\nu e_\nu) = \beta\alpha_1 e_1 + \beta\alpha_2 e_2 + \dots + \beta\alpha_\nu e_\nu.$$

Then
$$0a = 0e_1 + 0e_2 + \dots + 0e_\nu = 0.$$

(3) The resultant of an addition is a thing possessing a character (in that it represents a definite element) and intensity of its own. The character is completely defined (cf. Prop. II. following) by the ratios

$$\alpha_1 : \alpha_2 : \alpha_3 \dots : \alpha_\nu.$$

Hence the intensities are secondary properties of elements according to the definition of § 9.

The comparison of the intensities of things representing different elements may be possible. The whole question of such comparison will be discussed later in chapter IV. of this book. But it is only in special developments that the comparison of intensities assumes any importance: the more general formulæ do not assume any definite law of comparison.

(4) *Definition of Independent Units.* Let $e_1, e_2 \ldots e_\nu$ be defined to be such that no one of them can be expressed as the sum of the rest at any intensities. Symbolically this definition states that no one of these letters, e_1, say, can be expressed in the form $\alpha_2 e_2 + \alpha_3 e_3 + \ldots + \alpha_\nu e_\nu$.

Then $e_1, e_2 \ldots e_\nu$ are said to be mutually independent. If $e_1, e_2 \ldots e_\nu$ are all respectively at unit intensity, then they are said to be independent units. Any one of them is said to be independent of the rest.

64. Fundamental Propositions. (1) A group of propositions* can now be proved; they will be numbered because of their importance and fundamental character.

Prop. I. If $e_1, e_2 \ldots e_\nu$ be ν independent extraordinaries, then the equation,

$$\alpha_1 e_1 + \alpha_2 e_2 + \ldots + \alpha_\nu e_\nu = 0,$$

involves the n simultaneous equations, $\alpha_1 = 0,\ \alpha_2 = 0 \ldots \alpha_\nu = 0$.

Suppose firstly that all the coefficients are zero except one, α, say, then $\alpha_1 e_1 = 0$. And by definition this involves $\alpha_1 = 0$.

Again assume that a number of the coefficients, including α_1, are not zero. Then we can write

$$e_1 = -\frac{\alpha_2}{\alpha_1} e_2 - \frac{\alpha_3}{\alpha_1} e_3 - \ldots - \frac{\alpha_\nu}{\alpha_1} e_\nu .$$

But this is contrary to the supposition that $e_1, e_2 \ldots e_\nu$ are independent. Hence finally all the coefficients must separately vanish.

Prop. II. If the two sums $\alpha_1 e_1 + \alpha_2 e_2 + \ldots + \alpha_\nu e_\nu$ and $\beta_1 e_1 + \beta_2 e_2 + \ldots + \beta_\nu e_\nu$ are multiples of the same extraordinary, where $e_1, e_2 \ldots e_\nu$ are independent extraordinaries, then $\alpha_1/\beta_1 = \alpha_2/\beta_2 = \ldots = \alpha_\nu/\beta_\nu$.

For by hypothesis $\beta_1 e_1 + \beta_2 e_2 + \ldots + \beta_\nu e_\nu = \gamma(\alpha_1 e_1 + \alpha_2 e_2 + \ldots + \alpha_\nu e_\nu)$.

Hence $\quad (\beta_1 - \gamma\alpha_1) e_1 + (\beta_2 - \gamma\alpha_2) e_2 + \ldots + (\beta_\nu - \gamma\alpha_\nu) e_\nu = 0.$

Therefore by Proposition I, $\beta_1 - \gamma\alpha_1 = 0,\ \beta_2 - \gamma\alpha_2 = 0 \ldots \beta_\nu - \gamma\alpha_\nu = 0$.

Hence $\quad \beta_1/\alpha_1 = \beta_2/\alpha_2 = \ldots = \beta_\nu/\alpha_\nu = \gamma.$

It follows [cf. § 62 (2)], as has been explained in § 63 (3), that the ratios of the coefficients of a sum define the character of the resultant, that is to say, the element represented by the resultant. Only it must be remembered that the extraordinaries have to be independent.

(2) These propositions make a few definitions and recapitulations desirable. If two terms a and b both represent the same element, but at different intensities, then a and b will be said to be congruent to each other. The

* Cf. Grassmann, *Ausdehnungslehre* of 1862; also De Morgan, *Transactions of the Cambridge Philosophical Society*, 1844.

fact that the extraordinaries a and b are congruent will be expressed by $a \equiv b$. This relation implies an equation of the form $a = \lambda b$, where λ is some number. The symbol $\equiv$ will also be used to imply that an equation, concerned solely with the quantities of ordinary algebra, is an identity.

(3) The extraordinary $\alpha_1 e_1 + \alpha_2 e_2 + \ldots + \alpha_\nu e_\nu$ will be said to be dependent on the extraordinaries $e_1, e_2 \ldots e_\nu$; and the element represented by

$$\alpha_1 e_1 + \alpha_2 e_2 + \ldots + \alpha_\nu e_\nu$$

will be said to be dependent on the elements represented by $e_1, e_2 \ldots e_\nu$,

An expression of the form $\alpha_1 e_1 + \alpha_2 e_2 + \ldots + \alpha_\nu e_\nu$ is often written $\Sigma \alpha e$.

(4) Let the ν given independent extraordinaries be called the original defining extraordinaries, or the original defining units, if they are known to be at unit intensity. They define a positional manifold of $\nu - 1$ dimensions (cf. § 11). Any element of the type $\Sigma \alpha e$ belongs to this manifold. This complete positional manifold, found by giving all values (real or complex) to $\alpha_1, \alpha_2, \ldots \alpha_\nu$, will be called the complete region. Any ρ of these ν defining extraordinaries define a positional manifold of $\rho - 1$ dimensions. It is incident in the complete region, and will be called a subregion of the complete region.

A region or subregion defined by $e_1, e_2 \ldots e_\rho$ will be called the region or subregion $(e_1, e_2 \ldots e_\rho)$.

(5) As far as has been shown up to the present, the ν defining units represent elements which appear to have a certain special function and preeminence in the complete region. It will be proved in the succeeding propositions that this is not really the case, but that any two elements are on an equality like two points in space.

(6) If letters $a, b, c \ldots$ denoting elements of the region be not mutually independent, then at least one equation of the form,

$$\alpha a + \beta b + \gamma c + \ldots = 0,$$

exists between them, where $\alpha, \beta, \gamma \ldots$ are not all zero.

Let such equations be called the addition relations between the mutually dependent letters.

(7) Prop. III. An unlimited number of groups of ν independent extraordinaries can be found in a region of $\nu - 1$ dimensions.

Let the region be $(e_1, e_2 \ldots e_\nu)$.

It is possible in an unlimited number of ways to find ν^2 numbers, real or complex, $\alpha_1, \alpha_2 \ldots \alpha_\nu$, $\beta_1, \beta_2 \ldots \beta_\nu \ldots \kappa_1, \kappa_2 \ldots \kappa_\nu$, such that the determinant

$$\begin{vmatrix} \alpha_1, & \alpha_2 & \ldots \alpha_\nu \\ \beta_1, & \beta_2 & \ldots \beta_\nu \\ \ldots & \ldots & \ldots \\ \kappa_1, & \kappa_2 & \ldots \kappa_\nu \end{vmatrix}$$

is not zero.

Let
$$a = \alpha_1 e_1 + \alpha_2 e_2 + \dots + \alpha_\nu e_\nu,$$
$$b = \beta_1 e_1 + \beta_2 e_2 + \dots + \beta_\nu e_\nu;$$
$$\dots\dots\dots\dots\dots\dots\dots\dots\dots$$
$$k = \kappa_1 e_1 + \kappa_2 e_2 + \dots + \kappa_\nu e_\nu.$$

Now let $\xi, \eta \dots \chi$ be numbers such that (if possible)
$$\xi a + \eta b + \dots + \chi k = 0.$$

Then substituting for $a, b \dots k$, we find
$$(\xi\alpha_1 + \eta\beta_1 + \dots + \chi\kappa_1) e_1 + (\xi\alpha_2 + \eta\beta_2 + \dots + \chi\kappa_2) e_2 + \dots$$
$$+ (\xi\alpha_\nu + \eta\beta_\nu + \dots + \chi\kappa_\nu) e_\nu = 0.$$

Hence by Proposition I, the ν coefficients are separately zero.

But since the determinant written above is not zero, these ν equations involve $\xi = 0, \eta = 0 \dots \chi = 0$. Hence the ν letters $a, b \dots k$ are mutually independent.

Prop. IV. No group containing more than ν independent letters can be found in a region of $\nu - 1$ dimensions.

Let the region be defined by $e_1, e_2 \dots e_\nu$, and let $a_1, a_2 \dots a_\nu$ be ν independent letters in the region. Then by solving for $e_1, e_2 \dots e_\nu$ in terms of $a_1, a_2 \dots a_\nu$, we can write
$$e_1 = \alpha_{11} a_1 + \alpha_{12} a_2 + \dots + \alpha_{1\nu} a_\nu,$$
$$e_2 = \alpha_{21} a_1 + \alpha_{22} a_2 + \dots + \alpha_{2\nu} a_\nu,$$
$$\dots\dots\dots\dots\dots\dots\dots\dots\dots$$
$$e_\nu = \alpha_{\nu 1} a_1 + \alpha_{\nu 2} a_2 + \dots + \alpha_{\nu\nu} a_\nu.$$

Now any other letter b in the region is of the form
$$b = \beta_1 e_1 + \beta_2 e_2 + \dots + \beta_\nu e_\nu;$$
hence substituting for $e_1, e_2 \dots e_\nu$ in terms of $a_1, a_2 \dots a_\nu$,
$$b = \gamma_1 a_1 + \gamma_2 a_2 + \dots + \gamma_\nu a_\nu.$$
Thus b cannot be independent of $a_1, a_2 \dots a_\nu$.

Prop. V. If $a_1, a_2 \dots a_\rho$ be ρ independent extraordinaries in a region of $\nu - 1$ dimensions, where ν is greater than ρ, then another extraordinary can be found in an infinite number of ways which is independent of the ρ independent extraordinaries.

Let the complete region be defined by $e_1, e_2, \dots e_\nu$. Then the expressions for $a_1, a_2 \dots a_\rho$ in terms of the units must involve at least ρ of the defining extraordinaries with non-vanishing coefficients in such a way that they cannot be simultaneously eliminated. For if not, then the defining extraordinaries involved define a region containing $a_1, a_2, \dots a_\rho$ and of less than $\rho - 1$ dimensions. But since $a_1, a_2, \dots a_\rho$ are independent, by Prop. IV. this is impossible.

Let the extraordinaries $e_1, e_2, \ldots e_\rho$ at least be involved in the expressions for $a_1, a_2, \ldots a_\rho$. Then by solving for $e_1, e_2 \ldots e_\rho$, we have

$$e_1 = \alpha_{11} a_1 + \alpha_{12} a_2 + \ldots + \alpha_{1\rho} a_\rho + \alpha_{1,\rho+1} e_{\rho+1} + \ldots + \alpha_{1\nu} e_\nu,$$

$$\ldots\ldots\ldots\ldots\ldots\ldots\ldots\ldots\ldots\ldots\ldots\ldots$$

$$e_\rho = \alpha_{\rho 1} a_1 + \alpha_{\rho 2} a_2 + \ldots + \alpha_{\rho\rho} a_\rho + \alpha_{\rho,\rho+1} e_{\rho+1} + \ldots + \alpha_{\rho\nu} e_\nu.$$

Let b be any other letter, defined by

$$b = \xi_1 e_1 + \xi_2 e_2 + \ldots + \xi_\nu e_\nu.$$

Substituting for $e_1, e_2 \ldots e_\rho$,

$$b = \eta_1 a_1 + \ldots + \eta_\rho a_\rho + \zeta_{\rho+1} e_{\rho+1} + \ldots + \zeta_\nu e_\nu;$$

where any one of the ζ's, say ζ_ν, is of the form

$$\alpha_{1\nu}\xi_1 + \alpha_{2\nu}\xi_2 + \ldots + \alpha_{\rho\nu}\xi_\rho + \xi_\nu.$$

Thus there are ν undetermined numbers, $\xi_1, \xi_2 \ldots \xi_\nu$, and $\nu - \rho$ coefficients of $e_{\rho+1}, e_{\rho+2} \ldots e_\nu$ (viz. $\zeta_{\rho+1}, \ldots \zeta_\nu$). Hence it is possible in an infinite number of ways to determine the numbers so that all these coefficients do not simultaneously vanish. And in such a case b is independent of the group $a_1 \ldots a_\rho$.

COROLLARY. By continually adding another independent letter to a group of independent letters, it is obvious that any group of independent letters can be completed so as to contain a number of letters one more than the number of dimensions of the region.

(8) By the aid of these propositions it can be seen that any group of ν independent extraordinaries can be taken as defining the complete region.

The original units have only the advantage that their unit intensities are known (cf. chapter IV. following).

Any group of ν independent elements which are being used to define a complete region will be called coordinate elements of the region, or more shortly coordinates of the region.

65. SUBREGIONS. (1) The definition of a subregion in § 64 (4) can be extended. A region defined by *any* ρ independent letters lying in a region of $\nu - 1$ dimensions, where ρ is less than ν, is called a subregion of the original region. The original region is the complete region, and the subregion is incident in the complete region.

(2) A region of no dimensions consists of a single element. It is analogous to a point in space.

A subregion of one dimension, defined by a, b, is in its real part the collection of elements found from $\xi a + \eta b$, where ξ/η is given all real values. Hence it contains a singly infinite number of elements, which will be called real elements when a and b are considered to be real elements. It is analogous to a straight line. And like a straight line it is given an imaginary extension by the inclusion of all elements found by giving ξ/η all imaginary values.

A subregion of two dimensions, defined by a, b, c, is the collection of elements found from $\xi a + \eta b + \zeta c$, where ξ/ζ, η/ζ are given all values. Hence its real part contains a doubly infinite number of elements. It is analogous to a plane of ordinary space.

Similarly a subregion of three dimensions contains in its real part a trebly infinite number of elements, and is analogous to space of three dimensions; and so on.

(3) A subregion is called a co-ordinate region when, being itself of $\rho - 1$ dimensions, ρ of the co-ordinate (or reference) elements of the complete region have been taken in it.

For example, if $e_1, e_2 \ldots e_\nu$ be the co-ordinate elements, then the region defined by $e_1, e_2 \ldots e_\rho$ is a co-ordinate region of $\rho - 1$ dimensions.

(4) The complete region being of $\nu - 1$ dimensions, to every co-ordinate subregion of $\rho - 1$ dimensions there corresponds another co-ordinate subregion of $(\nu - \rho - 1)$ dimensions, so that the two do not overlap, and the co-ordinate elements of the two together define the complete region. Such co-ordinate regions will be called supplementary, and one will be said to be supplementary to the other.

(5) If two co-ordinate subregions of $\rho - 1$ and $\sigma - 1$ dimensions do not overlap, then there is in the complete region a remaining co-ordinate subregion of $(\nu - \rho - \sigma - 1)$ dimensions belonging to neither; where of course $\rho + \sigma$ is less than ν.

If the co-ordinate regions do overlap and have a common subregion of $\tau - 1$ dimensions, then the remaining co-ordinate subregion is of $(\nu + \tau - \rho - \sigma - 1)$ dimensions; also the common subregion of $\tau - 1$ dimensions must be a co-ordinate region.

If $\rho + \sigma$ is greater than ν, then the subregions must overlap and have in common a subregion of at least $(\rho + \sigma - \nu - 1)$ dimensions.

(6) Let two regions of $\rho - 1$ and $\sigma - 1$ dimensions overlap and have in common a subregion of $\tau - 1$ dimensions. Let the subregion be defined by the terms $a_1, a_2 \ldots a_\tau$. Then the region of $\rho - 1$ dimensions can be defined by the ρ terms

$$a_1, a_2 \ldots a_\tau, b_{\tau+1} \ldots b_\rho;$$

and the region of $\sigma - 1$ dimensions by the σ terms

$$a_1, a_2 \ldots a_\tau, c_{\tau+1} \ldots c_\sigma;$$

where the terms $b_{\tau+1} \ldots b_\rho$ are independent of the terms $c_{\tau+1} \ldots c_\sigma$. For if the b's and the c's do not together form a group of independent letters then another common letter $a_{\tau+1}$ can be found independent of the other a's and can be added to them to define a common subregion of τ dimensions.

The region defined by the letters

$$a_1, a_2 \ldots a_\tau, b_{\tau+1} \ldots b_\rho, c_{\rho+1} \ldots c_\sigma$$

is called the containing region of the two overlapping regions; and is of $(\rho + \sigma - \tau - 1)$ dimensions.

It is the region of fewest dimensions which contains both regions as subregions, whether the regions do or do not overlap.

(7) Every complete region and every subregion can be conceived of as a continuous whole. For any element of a subregion can be represented by

$$x = \xi_1 a_1 + \xi_2 a_2 + \ldots + \xi_\rho a_\rho;$$

and by a gradual modification of the values of the coefficients x can be gradually altered so as to represent any element of the region. Hence x can be conceived as representing a gradually altering element which successively coincides with all the elements of the region. The region can always therefore be conceived as continuous.

(8) Also the subregions must not be conceived as bounding each other. Each subregion has no limits, and may be called therefore unlimited. For any region is an aggregation of elements, and no one of these elements is more at the boundary or more in the midst of the region than any other element. Overlapping regions are not in any sense bounded by their common subregion. For any subregion of a region may be common to another region also. Regions, therefore, are like unlimited lines or surfaces, either stretching in all directions to infinity or returning into themselves so as to be closed; two infinite planes cutting each other in a line are not bounded by this line, which is a subregion common to both.

(9) Consider the one-dimensional subregion defined by the two elements a_1 and a_2. Any extraordinary x which represents an element belonging to the subregion is of the form $\xi_1 a_1 + \xi_2 a_2$. As ξ_2/ξ_1 takes all real positive values from 0 to $+\infty$, x may be conceived as representing a variable element travelling through a continuous series of elements arranged in order, starting from a_1 and ending at a_2. Again, as ξ_2/ξ_1 takes all real negative values from $-\infty$ to 0, x may be conceived as travelling through another continuous series of elements arranged in order, starting from a_2 and ending in a_1.

It is for the purposes of this book the simplest and most convenient supposition to conceive a one-dimensional subregion defined by two elements as formed by a continuous series of elements arranged in order, and such that by starting from any one element a_1 and proceeding through the continuous series in order a variable element finally returns to a_1.

This supposition might be replaced* in investigations in which the object was to illustrate the Theory of Functions by another one. The element x starts from a_1 and passes through the series of elements given by ξ_2/ξ_1 positive and varying from 0 to ∞, and thus reaches a_2; then as ξ_2/ξ_1 varies from $-\infty$ to 0, x passes through another series of elements and finally reaches an element which in our symbolism is a_1. But a_1, *as thus arrived at,* may be conceived to be a different element from the element a_1 from which x started.

* Cf. Klein, *Nicht-Euklidische Geometrie*, Vorlesungen, 1889—1890.

This conception has no natural symbolism in the investigations of this work, and therefore will not be adopted; but in other modes of investigation it is imperative that it be kept in view. Call this second a_1 the a_1 of the first arrival, and denote it by ${}_1a_1$. Similarly we might find an a_1 of the second arrival and denote it by ${}_2a_1$, and so on. Finally the analysis might suggest the identification of the a_1 of the ρth arrival with the original a_1. Thus $a_1 = {}_\rho a_1$. It is sufficient in this treatise simply to have noticed these possibilities.

66. Loci. (1) A locus is a more general conception than a subregion; it is an aggregation of a number (in general infinite) of elements determined according to some law. Thus if x denote the element $\xi_1 e_1 + \xi_2 e_2 + \ldots \xi_\nu e_\nu$, in the region defined by $e_1, e_2 \ldots e_\nu$, then the equation $\phi(\xi_1, \xi_2 \ldots \xi_\nu) = 0$, where ϕ is a homogeneous function, limits the arbitrary nature of the ratios $\xi_1 : \xi_2 \ldots : \xi_\nu$. The corresponding values of x form therefore a special aggregation of the elements out of the whole number in the region. But these elements, except in the special case of flat loci (cf. subsection (5) of this section), do not form themselves a subregion, according to the definition of a subregion given in this book; they are parts of many subregions.

(2) A locus may be defined by more than one equation: thus the equations $\phi_1(\xi_1 \ldots \xi_\nu) = 0$, $\phi_2(\xi_1 \ldots \xi_\nu) = 0$, ..., $\phi_\rho(\xi_1 \ldots \xi_\nu) = 0$, where the left-hand sides are all homogeneous, define a locus when treated as simultaneous. If there be $\nu - 1$ independent equations, they determine a finite definite number of elements; and more than $\nu - 1$ equations cannot in general be simultaneously satisfied.

(3) A locus defined by ρ simultaneous equations will be said to be of $\nu - \rho - 1$ dimensions when the case is excluded in which the satisfaction of some of the equations secures that of others of the equations. In a region of $\nu - 1$ dimensions there cannot be a locus of more than $\nu - 2$ dimensions, and a locus containing an infinite number of elements must be of at least one dimension. Hence such a locus cannot be defined by more than $\nu - 2$ equations. In a locus of one dimension the number of elements is singly infinite: in a locus of two dimensions it is doubly infinite, and so on.

(4) Let $\rho + \sigma$ equations define a locus of $\nu - \rho - \sigma - 1$ dimensions. These equations may be split into two groups of ρ equations and of σ equations respectively. The group of ρ equations defines a locus of $\nu - \rho - 1$ dimensions, and the group of σ equations defines a locus of $\nu - \sigma - 1$ dimensions. The original locus is contained in both these loci. Hence the locus of $\nu - \rho - \sigma - 1$ dimensions may be conceived as formed by the intersection of two loci of $\nu - \rho - 1$ dimensions and of $\nu - \sigma - 1$ dimensions respectively. Similarly these intersecting loci can be split up into the intersections of other loci of higher dimensions. So finally the locus of $\nu - \rho - 1$ dimensions

may be conceived as the intersection of ρ loci of $\nu-2$ dimensions; each of these loci being given by one of the simultaneous equations.

(5) The locus corresponding to an equation of the first degree, namely,

$$\alpha_1\xi_1 + \alpha_2\xi_2 + \ldots + \alpha_\nu\xi_\nu = 0,$$

is also a subregion of $\nu-2$ dimensions, as well as being a locus of the same number of dimensions.

For if $x_1, x_2 \ldots x_\nu$ be ν elements in the locus, given by

$$\begin{aligned} x_1 &= \xi_{11}e_1 + \xi_{12}e_2 + \ldots \xi_{1\nu}e_\nu, \\ x_2 &= \xi_{21}e_1 + \xi_{22}e_2 + \ldots \xi_{2\nu}e_\nu, \\ &\ldots\ldots\ldots\ldots\ldots\ldots\ldots\ldots \\ x_\nu &= \xi_{\nu1}e_1 + \xi_{\nu2}e_2 + \ldots \xi_{\nu\nu}e_\nu; \end{aligned}$$

then the ν equations of the type,

$$\alpha_1\xi_{\rho1} + \alpha_2\xi_{\rho2} + \ldots + \alpha_\nu\xi_{\rho\nu} = 0,$$

involve the vanishing of the determinant

$$\begin{vmatrix} \xi_{11} & \xi_{12} & \ldots & \xi_{1\nu} \\ \xi_{21} & \xi_{22} & \ldots & \xi_{2\nu} \\ \ldots & \ldots & \ldots & \ldots \\ \xi_{\nu1} & \xi_{\nu2} & \ldots & \xi_{\nu\nu} \end{vmatrix}.$$

Hence an addition relation of the form,

$$\lambda_1x_1 + \lambda_2x_2 + \ldots + \lambda_\nu x_\nu = 0;$$

exists between the elements. The ν elements are therefore not independent. But $\nu-1$ independent points can be determined. Again, since the equation of the locus is linear, if x_1 and x_2 be two elements in it, then $\lambda x_1 + \mu x_2$ also lies in the locus. Hence the whole region of the $\nu-1$ independent elements is contained in the locus, and *vice versa*; the region and locus coinciding. A locus defined by ρ simultaneous independent linear equations can in the same way be proved to contain groups of $\nu-\rho$ independent elements. It therefore in like manner can be proved to be a subregion of $\nu-\rho-1$ dimensions. Let such a locus be called 'flat.' Then a flat locus is a subregion.

(6) A locus defined by $\rho+\sigma$ equations, of which ρ are linear, can therefore be treated as a locus of $\nu-\rho-\sigma-1$ dimensions in a region of $\nu-\rho-1$ dimensions.

(7) There is a great distinction between the region defined by the combined elements which define subregions and the flat locus determined by the simultaneous satisfaction of the equations of other flat loci. Consider for instance, in a complete region of two dimensions, the two subregions defined by e_1, e_2 and e_3, e_4 respectively. The region defined by the four elements e_1, e_2, e_3, e_4 includes not only the elements of the subregion e_1e_2, of the form $\xi_1e_1 + \xi_2e_2$, and of the subregion e_3e_4, of the form $\xi_3e_3 + \xi_4e_4$; but also it includes all elements of the form

$$\xi_1e_1 + \xi_2e_2 + \xi_3e_3 + \xi_4e_4,$$

which includes elements not lying in the subregions. But the equations of the loci taken together indicate a locus which is the region (in this case a single element) common to the two regions e_1e_2, e_3e_4.

67. SURFACE LOCI AND CURVE LOCI. (1) Let a locus which is of one dimension less than a region or subregion containing it be called a surface locus in this region. Let this region, which contains the surface locus and is of one dimension more than the surface locus, be called the containing region. In other words [cf. § 66 (6)] a surface locus can be defined by $\rho+1$ equations, of which only one at most is non-linear and ρ (defining the containing region) are linear. A surface locus defined by an equation of the μth degree will be called a surface locus of the μth degree. For example, in this nomenclature we may say that a surface locus of the first degree is flat.

(2) In reference to a complete region of $\nu-1$ dimensions a flat surface locus of $\nu-2$ dimensions will be called a plane. A flat locus of $\nu-3$ dimensions will be called a subplane in a complete region of $\nu-1$ dimensions. Subplanes are planes of planes.

(3) A subregion is either contained in any surface locus of the complete region or intersects it in a locus which is another surface locus contained in the subregion.

Let the complete region be of $\nu-1$ dimensions and the subregion of $\nu-\rho-1$ dimensions. Then the subregion is a flat locus defined by ρ equations. The intersection of the subregion and the surface locus is therefore defined by $\rho+1$ equations and is therefore of $\nu-\rho-2$ dimensions. Also it lies in the subregion which is of $\nu-\rho-1$ dimensions. Hence it is a surface locus; unless the satisfaction of the ρ equations of the flat locus also secures the satisfaction of the equation of the surface locus. In this case the subregion is contained in the original surface locus.

(4) A locus of one dimension either intersects a surface locus in a definite number of elements or lies completely in the surface locus. For if the complete region be of $\nu-1$ dimensions, the locus of one dimension is determined by $\nu-2$ equations; and these together with the equation of the surface locus give $\nu-1$ equations which, if independent, determine a definite number of elements. If the equation of the surface locus does not form an additional independent equation, it must be satisfied when the equations of the one dimensional locus are satisfied; that is to say the one dimensional locus lies in the surface locus.

A locus of one dimension will be called a curvilinear locus. A flat curvilinear locus is a region of one dimension, and will be called a straight line.

(5) A locus of $\tau-1$ dimensions which cannot be contained in a region of τ dimensions (i.e. which is not a surface locus) will be called a curve locus. Thus if the locus be determined by $\rho+\sigma$ equations, of which ρ are linear and σ non-linear, then $\sigma>1$; and the locus is of $\nu-\rho-\sigma-1$ dimensions.

(6) A curve locus formed by the intersection of ρ surface loci, each in the same containing region, and such that it cannot be contained in a region of fewer dimensions than this common containing region, will be called a curve locus of the $(\rho-1)$th order of tortuosity contained in this region. Thus the order of tortuosity of a surface locus is zero.

Each of the surface loci, which form the curve locus by their intersection, will be called a containing locus.

(7) In general the intersection of a curve locus of any order of tortuosity with a subregion is another curve locus contained in that subregion and of the same order of tortuosity. For the curve locus may be conceived as defined by ρ equations of the first degree which define the containing region of $\nu-\rho-1$ dimensions; and by σ equations

$$\phi_1(\xi_1 \dots \xi_\nu)=0, \dots, \phi_\sigma(\xi_1 \dots \xi_\nu)=0,$$

each of a degree higher than the first degree, which define the tortuosity. Now the subregion which intersects the curve locus is defined by τ equations of the first degree in addition to the equations defining the containing region. These $\rho+\tau$ equations of the first degree and the σ equations $\phi_1=0, \dots \phi_\sigma=0$, in general define a curve locus of the $(\sigma-1)$th order of tortuosity in the subregion.

(8) A plane curve in geometry is a surface locus; a plane curve of the first order of tortuosity consists of a finite number of points.

In three dimensions an ordinary surface is a surface locus, an ordinary tortuous curve is a curve locus of the first order of tortuosity; and a finite number of points not in one plane form a curve locus of the second order of tortuosity.

There are of course exceptional cases in relation to the tortuosity of curve loci when the equations are not all independent. It is not necessary now to enter into them.

NOTE. The analytical part of this chapter follows closely the methods of Grassmann's *Ausdehnungslehre von* 1862, chapter I. §§ 1—36. This theory of Grassmann is a generalization of Möbius' Der Barycentrischer Calcul (1827), in which the addition of points is defined and considered. Hamilton's Quaternions also involve the same theory of the addition of extraordinaries (the number of independent extraordinaries being however limited to four). This theory is considered in his 'Lectures on Quaternions' (1853), Lecture I, and in his 'Elements of Quaternions,' Part I. The idea in Hamilton's works was a generalization of the composition of velocities according to the parallelogram law. Hamilton's first paper on Quaternions was published in the *Philosophical Magazine* (1844); De Morgan in his last paper, 'On the Foundation of Algebra' (loc. cit. 1844) writes of it, 'To this paper I am indebted for the idea of inventing a distinct system of unit-symbols, and investigating or assigning relations which define their mode of action on each other.'

Some simple ideas which arise in the study of Descriptive Geometry of any number of dimensions have been discussed in §§ 66, 67 as far as they will be wanted in this treatise. On this subject Cayley's 'Memoir on Abstract Geometry,' *Phil. Trans.* Vol. CLX. 1870 (and *Collected Mathematical Papers*, Vol. VI. No. 413), should be studied. It enters into the subject with a complete generality of treatment which is not necessary here.

CHAPTER II.

STRAIGHT LINES AND PLANES.

68. INTRODUCTORY. The theorems of Projective Geometry extended to any number of dimensions can be deduced as necessary consequences of the definitions of a positional manifold. Grassmann's 'Calculus of Extension' (to be investigated in Book IV.) forms a powerful instrument for such an investigation; the properties also can to some extent be deduced by the methods of ordinary co-ordinate Geometry. Only such theorems will be now investigated which are either useful subsequently in this treatise or exemplify in their proof the method of the addition of extraordinaries.

69. ANHARMONIC RATIO. (1) Any point p on the straight line aa' can be written in the form $\xi a + \xi' a'$, where the position of p is defined by the ratio ξ/ξ'. If p_1 be another point, $\xi_1 a + \xi_1' a'$, on the same line, then the ratio $\xi\xi_1'/\xi'\xi_1$ is called the *anharmonic ratio* of the range (aa', pp_1). It is to be carefully noticed that the anharmonic ratio of a range of four collinear elements is here defined apart from the introduction of any idea of distance. It is also independent of the intensities at which a and a' happen to represent their elements. For it is obviously unaltered if a, a' are replaced by αa, $\alpha' a'$, α and α' being any arbitrary quantities.

(2) If the anharmonic ratio of (aa', pp_1) be -1, the range is said to be harmonic; and p and p_1 can then be written respectively in the forms

$$\xi a + \xi' a' \text{ and } \xi a - \xi' a'.$$

(3) Let p_1, p_2, p_3, p_4 be any four points, $\xi_1 a + \xi_1' a'$, etc. Then

$$a = (\xi_2' p_1 - \xi_1' p_2)/(\xi_1 \xi_2');$$

where $(\xi_1\xi_2')$ stands for the determinant $\xi_1\xi_2' - \xi_2\xi_1'$.

Similarly
$$a' = (\xi_1 p_2 - \xi_2 p_1)/(\xi_1\xi_2').$$

Hence
$$p_3 = \xi_3 a + \xi_3' a' = \{(\xi_3\xi_2')\, p_1 - (\xi_3\xi_1')\, p_2\}/(\xi_1\xi_2');$$

and
$$p_4 = \{(\xi_4\xi_2')\, p_1 - (\xi_4\xi_1')\, p_2\}/(\xi_1\xi_2').$$

Hence the anharmonic ratio of the range $(p_1 p_2,\ p_3 p_4)$ is

$$(\xi_3\xi_2')\,(\xi_4\xi_1')/(\xi_3\xi_1')\,(\xi_4\xi_2').$$

70. Homographic Ranges. (1) Let

$$(b_1 b_2 p_1 p_2 p_3 \ldots) \text{ and } (c_1 c_2 q_1 q_2 q_3 \ldots)$$

be two ranges of corresponding points such that the anharmonic ratio of the four points $(b_1 b_2,\ p_\rho p_{\rho+1})$, and that of the corresponding points $(c_1 c_2,\ q_\rho q_{\rho+1})$ are equal, where ρ is any one of the suffixes 1, 2, 3, etc.

(2) It can now be proved that the anharmonic ratio of any four points $(p_\lambda p_\mu,\ p_\rho p_\sigma)$ of the first range is equal to that of the corresponding points $(q_\lambda q_\mu,\ q_\rho q_\sigma)$ of the second range.

For let $$p_\rho = \xi_\rho b_1 + \xi_\rho' b_2,\quad q_\rho = \eta_\rho c_1 + \eta_\rho' c_2.$$

Then by definition $$\xi_\rho \xi'_{\rho+1}/\xi_{\rho+1}\xi_\rho' = \eta_\rho \eta'_{\rho+1}/\eta_{\rho+1}\eta_\rho'.$$

Now replace ρ in turn by $\rho+1,\ \rho+2, \ldots \sigma-1$, and multiply together corresponding sides of the equations, so obtained. Finally we deduce

$$\xi_\rho \xi_\sigma'/\xi_\sigma \xi_\rho' = \eta_\rho \eta_\sigma'/\eta_\sigma \eta_\rho';$$

and hence by subtracting 1 from both sides,

$$(\xi_\rho \xi_\sigma')/\xi_\sigma \xi_\rho' = (\eta_\rho \eta_\sigma')/\eta_\sigma \eta_\rho'.$$

It follows that the anharmonic ratio of any four points $(p_\lambda p_\mu,\ p_\rho p_\sigma)$ of the first range is equal to that of the four corresponding points $(q_\lambda q_\mu,\ q_\rho q_\sigma)$ of the second range.

Such corresponding ranges are called homographic.

71. Linear Transformations. (1) Let e_1 and e_2 be any two points, and let

$$a_1 = \alpha_1 e_1 + \alpha_1' e_2,\quad a_2 = \alpha_2 e_1 + \alpha_2' e_2,\quad a_3 = \alpha_3 e_1 + \alpha_3' e_2,\quad p = \xi e_1 + \xi' e_2,$$

be three given points and any fourth point on one range of points; also let

$$b_1 = \beta_1 e_1 + \beta_1' e_2,\quad b_2 = \beta_2 e_1 + \beta_2' e_2,\quad b_3 = \beta_3 e_1 + \beta_3' e_2,\quad q = \eta e_1 + \eta' e_2,$$

be the corresponding points on a second range homographic to the first range.

Then $$(\alpha_3 \alpha_1')(\xi \alpha_2')/(\alpha_3 \alpha_2')(\xi \alpha_1') = (\beta_3 \beta_1')(\eta \beta_2')/(\beta_3 \beta_2')(\eta \beta_1').$$

Therefore ξ/ξ' and η/η' are connected by a relation of the form

$$\lambda \xi \eta + \mu \xi \eta' + \mu' \xi' \eta + \lambda' \xi' \eta' = 0 \ \ldots\ldots\ldots\ldots\ldots\ldots\ldots\text{(A)},$$

where λ, μ, μ', λ' are constants depending on the arbitrarily chosen points

$$a_1,\ a_2,\ a_3,\ b_1,\ b_2,\ b_3.$$

This equation can also be written in the form

$$\frac{\alpha_{11}\xi + \alpha_{12}\xi'}{\eta} = \frac{\alpha_{21}\xi + \alpha_{22}\xi'}{\eta'} = \rho \ \ldots\ldots\ldots\ldots\ldots\ldots\ldots\text{(A')};$$

where α_{11}, α_{12}, α_{21}, α_{22} are constants which determine the nature of the transformation, and ρ must be chosen so that the point

$$q = \eta e_1 + \eta' e_2$$

may have the desired intensity.

Such transformations as those represented algebraically by equations (A) or (A′) are called linear transformations. Only real transformations will be considered, namely those for which the coefficients $\alpha_{11}, \alpha_{12}, \alpha_{21}, \alpha_{22}$ of equation (A′) are real.

(2) There are in general two points which correspond to themselves on the two ranges. For by substituting ξ, ξ' for η, η' in the above equations and eliminating we find

$$(\alpha_{11} - \rho)(\alpha_{22} - \rho) - \alpha_{12}\alpha_{21} = 0 \text{..........................(B)},$$

an equation which determines two values of ρ real and unequal or real and equal or imaginary; and each value of ρ determines $\xi : \xi'$ and $\eta : \eta'$ uniquely.

(3) Let ρ_1 and ρ_2 be the two roots of this quadratic, and first let them be assumed to be unequal.

Then by substituting ρ_1 in one of the equations (A′) a self-corresponding point d_1 is determined by

$$\xi/\xi' = \alpha_{12}/(\rho_1 - \alpha_{11}) = (\rho_1 - \alpha_{22})/\alpha_{21}.$$

Similarly a self-corresponding point d_2 is determined by

$$\xi/\xi' = \alpha_{12}/(\rho_2 - \alpha_{11}) = (\rho_2 - \alpha_{22})/\alpha_{21}.$$

Let these self-corresponding points be the reference points, so that any point is determined by $\xi d_1 + \xi' d_2$.

Then the equations defining the transformation take the form

$$\frac{\rho_1 \xi}{\eta} = \frac{\rho_2 \xi'}{\eta'} = \rho \text{ ....(C)}.$$

By putting ν for ρ_1/ρ_2, this equation can be written

$$\eta/\eta' = \nu \xi/\xi' \text{ ....(C′)}.$$

(4) Linear transformations fall into three main classes, according as the roots are (1) real and unequal, (2) imaginary, (3) equal.

In transformations of the first class the two points d_1 and d_2 are real. Then ν is real, and is positive when any point in the first range and its corresponding point on the second range both lie between d_1 and d_2.

(5) In transformations of the second class the two points d_1 and d_2 are imaginary. Then ν is complex, and it can be proved that mod. $\nu = 1$, assuming that real points are transformed into real points.

For ρ_1 and ρ_2 are conjugate complexes, and can be written

$$\sigma e^{i\delta} \text{ and } \sigma e^{-i\delta}.$$

Accordingly $$\nu = \rho_1/\rho_2 = e^{2i\delta}.$$

Hence mod. $\nu = 1$, and $\log \nu = 2i\delta$; where δ is real.

(6) The linear transformations of the first class are called hyperbolic; and those of the second class elliptic.

(7) The third special class of linear transformations exists in the case when the roots of the quadratic are equal, that is to say when the two points a_1 and a_2 coincide. Linear transformations of this class are called parabolic.

The condition for this case is that in equation (A), modified by substituting ξ and ξ' for η and η' respectively, the following relation holds between the coefficients,

$$4\lambda\lambda' = (\mu + \mu')^2.$$

Let u be the double self-corresponding point and e any other point, and let these points replace e_1 and e_2 in subsection (1) above. Then the modified equation (A), regarded as a quadratic in $\xi : \xi'$, must have two roots infinite: hence

$$\lambda = 0, \quad \mu + \mu' = 0.$$

Therefore if a point

$$p\ (= \xi u + \xi' e)$$

be transformed into $\quad q\ (= \eta u + \eta' e)$,

equation (A) takes the form

$$\mu\,(\xi\eta' - \xi'\eta) + \lambda'\xi'\eta' = 0;$$

that is $\quad \xi/\xi' = \eta/\eta' + \text{constant}$ ....(D).

(8) By a linear transformation a series of points $p_1, p_2, p_3 \ldots$ can be determined with the property that the range $(p_1, p_2, p_3 \ldots)$ is homographic with the range $(p_2, p_3, p_4 \ldots)$.

Firstly, let the linear transformation be elliptic or hyperbolic and let the co-ordinate points e_1, e_2 be the pair of self-corresponding points of the two ranges.

Let $\quad p_1 = \xi e_1 + \xi' e_2.$

Then if ν be any arbitrarily chosen constant, the points

$$p_2 = \nu\xi e_1 + \xi' e_2, \quad p_3 = \nu^2\xi e_1 + \xi' e_2, \ldots p_\rho = \nu^{\rho-1}\xi e_1 + \xi' e_2$$

satisfy the required condition.

(9) Secondly, let the linear transformations be parabolic. Let u be the double self-corresponding point, and let e be another arbitrarily chosen reference point.

Let $p_1 = \xi u + e$, and let δ be the arbitrarily chosen constant of the transformation.

Then by equation (D) the other points of the range are successively given by

$$p_2 = (\xi + \delta)\,u + e, \quad p_3 = (\xi + 2\delta)\,u + e, \ldots$$

$$p_\rho = (\xi + \overline{\rho - 1}\,\delta)\,u + e.$$

These results will be found to be of importance in the discussion in Book VI., chapter I., on the Cayley-Klein theory of distance.

72. Elementary Properties. (1) Let the ν independent elements $e_1, e_2, \ldots e_\nu$ define the complete region of $\nu - 1$ dimensions. Then the ρ elements $e_1, e_2, \ldots e_\rho$ $(\rho < \nu)$ define a subregion of $\rho - 1$ dimensions. Any point in this region can be written in the form

$$\xi_1 e_1 + \xi_2 e_2 + \ldots + \xi_\rho e_\rho.$$

Thus any point on the straight line defined by e_1, e_2 can be written

$$\xi_1 e_1 + \xi_2 e_2;$$

and any point on the subregion of two dimensions (or ordinary geometrical plane) defined by e_1, e_2, e_3 can be written

$$\xi_1 e_1 + \xi_2 e_2 + \xi_3 e_3.$$

(2) Any $\rho + 1$ points $x_1; x_2 \ldots x_{\rho+1}$ in the subregion $e_1, e_2 \ldots e_\rho$ can be connected by at least one equation of the form

$$\xi_1 x_1 + \xi_2 x_2 + \ldots + \xi_{\rho+1} x_{\rho+1} = 0.$$

Let such an equation be called the addition relation between the dependent points $x_1, x_2, \ldots x_{\rho+1}$.

Thus any three points x_1, x_2, x_3 on a straight line satisfy an equation of the form

$$\xi_1 x_1 + \xi_2 x_2 + \xi_3 x_3 = 0;$$

and similarly for any four points on a two-dimensional subregion.

(3) If $e_1, e_2 \ldots e_\nu$ be the independent reference units of the complete region, and any point be written in the form $\Sigma \xi e$, then the quantities

$$\xi_1, \xi_2, \ldots \xi_\nu$$

are called the co-ordinates of the point.

The locus denoted by

$$\lambda_1 \xi_1 + \lambda_2 \xi_2 + \ldots + \lambda_\nu \xi_\nu = 0$$

is a plane (i.e. a subregion of $\nu - 2$ dimensions).

The intersection of the ρ planes $(\rho < \nu)$

$$\begin{gathered}\lambda_{11}\xi_1 + \lambda_{21}\xi_2 + \ldots + \lambda_{\nu 1}\xi_\nu = 0,\\ \lambda_{12}\xi_1 + \lambda_{22}\xi_2 + \ldots + \lambda_{\nu 2}\xi_\nu = 0,\\ \ldots\ldots\ldots\ldots\ldots\ldots\ldots\ldots\\ \lambda_{1\rho}\xi_1 + \lambda_{2\rho}\xi_2 + \ldots + \lambda_{\nu\rho}\xi_\nu = 0,\end{gathered}$$

is a subregion of $\nu - \rho - 1$ dimensions.

(4) The intersection of $\nu - 1$ such planes is a single point which can be written in the form

$$\begin{vmatrix} e_1, & e_2, & \ldots e_\nu, \\ \lambda_{11}, & \lambda_{21}, & \ldots \lambda_{\nu 1}, \\ \lambda_{12}, & \lambda_{22}, & \ldots \lambda_{\nu 2}, \\ \ldots & \ldots & \ldots \\ \lambda_{1, \nu-1}, & \lambda_{2, \nu-1}, & \ldots \lambda_{\nu, \nu-1} \end{vmatrix}$$

For instance, in the region $e_1e_2e_3$ of two dimensions, the two straight lines

$$\lambda_{11}\xi_1 + \lambda_{21}\xi_2 + \lambda_{31}\xi_3 = 0,$$
$$\lambda_{12}\xi_1 + \lambda_{22}\xi_2 + \lambda_{32}\xi_3 = 0,$$

intersect in the point

$$(\lambda_{21}\lambda_{32} - \lambda_{31}\lambda_{22})\, e_1 + (\lambda_{31}\lambda_{12} - \lambda_{11}\lambda_{32})\, e_2 + (\lambda_{11}\lambda_{22} - \lambda_{21}\lambda_{12})\, e_3.$$

Returning to the general case of $\nu - 1$ planes, it is obvious that their point of intersection lies on the plane

$$\alpha_1\xi_1 + \alpha_2\xi_2 + \ldots + \alpha_\nu\xi_\nu = 0,$$

if the determinant

$$\begin{vmatrix} \alpha_1, & \alpha_2, & \ldots \alpha_\nu, \\ \lambda_{11}, & \lambda_{21}, & \ldots \lambda_{\nu 1}, \\ \ldots & \ldots & \ldots \\ \lambda_{1,\nu-1}, & \lambda_{2,\nu-1}, & \ldots \lambda_{\nu,\nu-1} \end{vmatrix}$$

vanishes.

(5) To prove that it is in general impossible in a complete region of $\nu - 1$ dimensions to draw a straight line from any given point to intersect two non-intersecting subregions of $\rho - 1$ and $\sigma - 1$ dimensions respectively, where ρ and σ are arbitrarily assigned; and that, when it is possible, only one such straight line can be drawn. Since the subregions are non-intersecting [cf. § 65 (5)]

$$\rho + \sigma \leqq \nu.$$

Of the reference elements let ρ be chosen in the subregion of $\rho - 1$ dimensions, namely $j_1, j_2, \ldots j_\rho$, and let σ be chosen in the subregion of $\sigma - 1$ dimensions, namely $k_1, k_2, \ldots k_\sigma$, and $\nu - \rho - \sigma$ must be chosen in neither region, namely $e_1, e_2, \ldots e_{\nu-\rho-\sigma}$.

Let the given point be $p = \Sigma\alpha j + \Sigma\beta k + \Sigma\gamma e$.

Let $\Sigma\xi j$ be any point in the subregion $j_1 \ldots j_\rho$. Then $p + \lambda\Sigma\xi j$ can be made to be any point on the line joining p and $\Sigma\xi j$ by properly choosing λ. But if this line intersect the subregion $k_1 \ldots k_\sigma$, then for some value of λ, say λ_1, $p + \lambda_1\Sigma\xi j$ depends on $k_1 \ldots k_\sigma$ only.

Hence either $\gamma_1 = \gamma_2 = \ldots = \gamma_{\nu-\rho-\sigma} = 0$, in which case p cannot be any arbitrarily assigned point; or $\rho + \sigma = \nu$, and there are no reference points of the type

$$e_1, e_2, \ldots e_{\nu-\rho-\sigma}.$$

Hence we find the condition $\rho + \sigma = \nu$.

Again $$p + \lambda_1\Sigma\xi j = \Sigma\beta k.$$

Hence also $\xi_1 = \alpha_1$, $\xi_2 = \alpha_2$, ... $\xi_\nu = \alpha_\nu$, and $\lambda_1 = -1$.

Thus the line through p intersecting the two regions intersects them in $\Sigma\alpha j$ and $\Sigma\beta k$ respectively. Accordingly there is only one such line through any given point p.

73. Reference-Figures. (1) The figure formed in a region of $\nu-1$ dimensions by constructing the straight lines connecting every pair of ν independent elements is the analogue of the triangle in plane geometry and of the tetrahedron in space of three dimensions, the sides of the triangle and edges of the tetrahedron being supposed to be produced indefinitely. Let such a figure be called a reference-figure, because its corner points can be taken as reference points to define the region. Let the straight lines be called the edges of the figure.

(2) Let $e_1, e_2, \dots e_\nu$ be the corners of such a figure, and let them also be taken as reference points. Consider the points in which any plane, such as

$$\xi_1/\alpha_1 + \xi_2/\alpha_2 + \dots + \xi_\nu/\alpha_\nu = 0,$$

cuts the edges. The point in which the edge e_1e_2 is cut is found by putting

$$\xi_3 = 0 = \xi_4 = \dots = \xi_\nu.$$

Hence the point is $\alpha_1e_1 - \alpha_2e_2$. Similarly the point in which the edge $e_\rho e_\sigma$ is cut is $\alpha_\rho e_\rho - \alpha_\sigma e_\sigma$.

(3) Consider the points of the typical form $\alpha_\rho e_\rho + \alpha_\sigma e_\sigma$. Then the range

$$e_\rho,\ e_\sigma,\ \alpha_\rho e_\rho - \alpha_\sigma e_\sigma,\ \alpha_\rho e_\rho + \alpha_\sigma e_\sigma$$

is harmonic. Also any point on the plane defined by the point $\alpha_1e_1 + \alpha_2e_2$ and the remaining corners of the reference-figure, namely by $e_3, e_4, \dots e_\nu$, is

$$\alpha_1e_1 + \alpha_2e_2 + \xi_3e_3 + \dots + \xi_\nu e_\nu.$$

Hence all such planes pass through the point

$$\alpha_1e_1 + \alpha_2e_2 + \alpha_3e_3 + \dots + \alpha_\nu e_\nu.$$

And conversely the planes through this point and all the corners but two cut the edge defined by the remaining two corners in the points

$$\alpha_1e_1 + \alpha_2e_2, \dots \alpha_\rho e_\rho + \alpha_\sigma e_\sigma.$$

The harmonic conjugates of these points with respect to the corresponding corners are of the typical form $\alpha_\rho e_\rho - \alpha_\sigma e_\sigma$, and these points are coplanar and lie on

$$\xi_1/\alpha_1 + \xi_2/\alpha_2 + \dots + \xi_\nu/\alpha_\nu = 0.$$

(4) The point $\Sigma\alpha e$ may be called the pole of the plane $\Sigma\xi/\alpha = 0$ with respect to the given reference-figure, and the plane may be called the polar of the point.

These properties are easily seen to be generalizations of the familiar properties of triangles (cf. Lachlan, *Modern Pure Geometry*, § 110).

(5) Let points be assumed one on each edge, of the typical form

$$\alpha_{\rho\sigma}e_\rho + \alpha_{\sigma\rho}e_\sigma.$$

It is required to find the condition that they should be coplanar.

Consider the $\nu-1$ edges joining the corner e_1 to the remaining corners. There are $\nu-1$ assumed points of the typical form

$$\alpha_{1\sigma}e_1 + \alpha_{\sigma 1}e_\sigma$$

on such edges, and these points define a plane. It remains, therefore, to determine the condition that any other point of the form

$$\alpha_{\lambda\mu} e_\lambda + \alpha_{\mu\lambda} e_\mu$$

(where neither λ nor μ is unity) lies on this plane. It must be possible to choose $\xi_2, \xi_3, \ldots \xi_\nu$ and η so as to fulfil the condition

$$\sum_{\sigma=2}^{\nu} \xi_\sigma (\alpha_{1\sigma} e_1 + \alpha_{\sigma 1} e_\sigma) + \eta (\alpha_{\lambda\mu} e_\lambda + \alpha_{\mu\lambda} e_\mu) = 0.$$

This requires that the coefficients of $e_1, e_2, \ldots e_\nu$ should be separately zero. Hence if σ be not equal to λ or μ, we find $\xi_\sigma = 0$; and also the three equations

$$\xi_\lambda \alpha_{1\lambda} + \xi_\mu \alpha_{1\mu} = 0,\ \xi_\lambda \alpha_{\lambda 1} + \eta \alpha_{\lambda\mu} = 0,\ \xi_\mu \alpha_{\mu 1} + \eta \alpha_{\mu\lambda} = 0.$$

Hence $$\alpha_{1\lambda} \alpha_{\lambda\mu} \alpha_{\mu 1} = - \alpha_{\lambda 1} \alpha_{1\mu} \alpha_{\mu\lambda}.$$

But this is the condition that three points on the edges joining e_1, e_λ, e_μ should be collinear. Hence since e_1, e_λ, e_μ are any three of the corners of the given reference-figure, the necessary and sufficient condition is that the assumed points lying on the edges which join any three corners in pairs should be collinear.

(6) It follows from (3) of this section that the condition for the concurrence of the planes joining each assumed point of the form

$$\alpha_{\rho\sigma} e_\rho + \alpha_{\sigma\rho} e_\sigma$$

with the corners, not lying on the edge on which the point itself lies, is for the three points on the edges joining in pairs $e_1 e_\lambda e_\mu$

$$\alpha_{1\lambda} \alpha_{\lambda\mu} \alpha_{\mu 1} = \alpha_{\lambda 1} \alpha_{1\mu} \alpha_{\mu\lambda}.$$

Hence if any three edges be taken forming a triangle with the corners as vertices, the three lines joining each assumed point with the opposite vertex are concurrent.

74. PERSPECTIVE. (1) The perspective properties of triangles can be generalized for reference-figures in regions of $\nu - 1$ dimensions*.

Let $e_1 e_2 \ldots e_\nu$ and $e_1' e_2' \ldots e_\nu'$ be two reference-figures, and let the ν lines

$$e_1 e_1',\ e_2 e_2',\ \ldots e_\nu e_\nu'$$

be concurrent and meet in the point g. Then it is required to show that the corresponding edges are concurrent in points which are coplanar.

Since g is in $e_1 e_1', e_2 e_2', \ldots e_\nu e_\nu'$, it follows that

$$\lambda_1 e_1 + \lambda_1' e_1' = \lambda_2 e_2 + \lambda_2' e_2' = \ldots = \lambda_\nu e_\nu + \lambda_\nu' e_\nu' = g.$$

Hence $\lambda_1 e_1 - \lambda_2 e_2 = \lambda_2' e_2' - \lambda_1' e_1'$, with similar equations.

But $\lambda_1 e_1 - \lambda_2 e_2$ is on the edge $e_1 e_2$, and $\lambda_1' e_1' - \lambda_2' e_2'$ is on the edge $e_1' e_2'$. But these are the same point. Hence the edges $e_1 e_2$ and $e_1' e_2'$ are concurrent in this point.

* The theorems of subsections (1) to (4) of this article, proved otherwise, were first given by Veronese, cf. "Behandlung der projectivischen Verhältnisse der Räume von verschiedenen Dimensionen durch das Princip des Projicirens und Schneidens," *Math. Annalen*, Bd. 19 (1882).

But taking $e_1 e_2 \dots e_\nu$ as reference points, it has been proved that the points of the typical form $\lambda_1 e_1 - \lambda_2 e_2$ are coplanar and lie on the plane, $\Sigma\xi/\lambda = 0$. Hence the theorem is proved.

(2) Conversely, if the corresponding edges, such as $e_1 e_2$ and $e_1' e_2'$, intersect in coplanar points, then the lines $e_1 e_1'$, $e_2 e_2'$, ... $e_\nu e_\nu'$ are concurrent.

For let two edges such as $e_1 e_2$, $e_1' e_2'$, intersect in the point d_{12}. Let the plane, on which the points such as d_{12} lie, have for its equation referred to $e_1 e_2 \dots e_\nu$,

$$\xi_1/\lambda_1 + \xi_2/\lambda_2 + \dots + \xi_\nu/\lambda_\nu = 0,$$

and referred to $e_1' e_2' \dots e_\nu'$ let it have for its equation

$$\xi_1'/\lambda_1' + \xi_2'/\lambda_2' + \dots + \xi_\nu'/\lambda_\nu' = 0.$$

Then any such point d_{12} can be written

$$\lambda_1 e_1 - \lambda_2 e_2 \text{ or } \lambda_1' e_1' - \lambda_2' e_2'.$$

But it has not yet been proved that these alternative forms can be assumed to be at the same intensity. Now consider any corresponding triangles with corners such as $e_1 e_2 e_3$ and $e_1' e_2' e_3'$. Write

$$d_{12} = \lambda_1 e_1 - \lambda_2 e_2 = \kappa_{12}(\lambda_1' e_1' - \lambda_2' e_2'),$$
$$d_{23} = \lambda_2 e_2 - \lambda_3 e_3 = \kappa_{23}(\lambda_2' e_2' - \lambda_3' e_3'),$$
$$d_{31} = \lambda_3 e_3 - \lambda_1 e_1 = \kappa_{31}(\lambda_3' e_3' - \lambda_1' e_1').$$

Hence $$d_{12} + d_{23} + d_{31} = 0, \quad d_{12}/\kappa_{12} + d_{23}/\kappa_{23} + d_{31}/\kappa_{31} = 0.$$

But if these relations are independent, the three points d_{12}, d_{23}, d_{31} must coincide, which is not true. Hence $\kappa_{12} = \kappa_{23} = \kappa_{31}$; and by altering the intensity of all the points $e_1' e_2' \dots e_\nu'$ in the same ratio, each factor such as $\kappa_{\rho\sigma}$ can be made equal to -1.

Hence $$d_{12} = \lambda_1 e_1 - \lambda_2 e_2 = \lambda_2' e_2' - \lambda_1' e_1' \quad \text{.......................(A)}.$$

Thus $$\lambda_1 e_1 + \lambda_1' e_1' = \lambda_2 e_2 + \lambda_2' e_2' = \dots = \lambda_\nu e_\nu + \lambda_\nu' e_\nu' = g \quad \text{...........(B)}.$$

Hence the point g is the point of concurrence of

$$e_1 e_1',\ e_2 e_2',\ \dots\ e_\nu e_\nu'.$$

(3) Let the point g be called the centre of perspective and the plane of the points, $d_{\rho\sigma}$, the axal plane of perspective of the two reference-figures.

The equation of the axal plane referred to $e_1 e_2 \dots e_\nu$ is with the previous notation $\Sigma\xi/\lambda = 0$: its equation referred to $e_1' e_2' \dots e_\nu'$ is $\Sigma\xi'/\lambda' = 0$. Let g, the centre of perspective be expressed in the form

$$g = \lambda_1 \alpha_1 e_1 + \lambda_2 \alpha_2 e_2 + \dots + \lambda_\nu \alpha_\nu e_\nu.$$

Then by eliminating e_1, $e_2 \dots e_\nu$ by means of equations (B) above,

$$\lambda_1' \alpha_1 e_1' + \lambda_2' \alpha_2 e_2' + \dots + \lambda_\nu' \alpha_\nu e_\nu' = (\alpha_1 + \alpha_2 + \dots + \alpha_\nu - 1) g.$$

Hence g, though of different intensities, can be expressed in the two forms

$$\Sigma\lambda\alpha e \text{ and } \Sigma\lambda'\alpha e'.$$

Since $\alpha_1, \alpha_2 \dots \alpha_\nu$ can be assumed in independence of $\lambda_1, \lambda_2, \dots \lambda_\nu$, it follows that, given one reference-figure, it is possible to find another reference-figure in perspective with it having any assigned centre of perspective and axal plane of perspective.

(4) Suppose that the corresponding edges of three reference-figures

$$e_1 e_2 \dots e_\nu,\ e_1' e_2' \dots e_\nu',\ e_1'' e_2'' \dots e_\nu''$$

intersect in coplanar points, so that each triad of corresponding edges is concurrent; and let g, g', g'' be the three corresponding centres of perspective.

Consider the three edges $e_\rho e_\sigma$, $e_\rho' e_\sigma'$, $e_\rho'' e_\sigma''$. Then we may assume that

$$\lambda_\rho e_\rho - \lambda_\sigma e_\sigma = \lambda_\rho' e_\rho' - \lambda_\sigma' e_\sigma' = \lambda_\rho'' e_\rho'' - \lambda_\sigma'' e_\sigma'' = d_{\rho\sigma};$$

and hence that

$$g = \lambda_\rho' e_\rho' - \lambda_\rho'' e_\rho'',\ g' = \lambda_\rho'' e_\rho'' - \lambda_\rho e_\rho,\ g'' = \lambda_\rho e_\rho - \lambda_\rho' e_\rho'.$$

Hence $$g + g' + g'' = 0.$$

Hence the three centres of perspective are collinear.

(5) Let there be ν reference-figures such that each pair is in perspective, all pairs having the same centre of perspective g. It is required to show that all the axal planes of perspective are concurrent.

Let the reference-figures be

$$e_{11} e_{12} e_{13} \dots e_{1\nu},\ e_{21} e_{22} \dots e_{2\nu},\ \dots,\ e_{\nu 1} e_{\nu 2} \dots e_{\nu\nu}.$$

Consider the $\nu - 1$ pairs of figures formed by taking the first reference-figure successively with each of the remainder. Let g be the given centre of perspective, and let the equation of the axal plane of perspective of the pair comprising the first and the ρth figure be, referred to the first figure,

$$\xi_1/{}_1\lambda_{1\rho} + \xi_2/{}_2\lambda_{1\rho} + \dots + \xi_\nu/{}_\nu\lambda_{1\rho} = 0 \quad \dots\dots\dots\dots(1),$$

and referred to the ρth figure,

$$\xi_1/{}_1\lambda_{\rho 1} + \xi_2/{}_2\lambda_{\rho 1} + \dots + \xi_\nu/{}_\nu\lambda_{\rho 1} = 0 \quad \dots\dots\dots\dots(2).$$

Hence two typical sets of equations are

$$\left.\begin{aligned} {}_1\lambda_{1\rho} e_{11} + {}_1\lambda_{\rho 1} e_{\rho 1} &= {}_2\lambda_{1\rho} e_{12} + {}_2\lambda_{\rho 1} e_{\rho 2} = \dots = {}_\nu\lambda_{1\rho} e_{1\nu} + {}_\nu\lambda_{\rho 1} e_{\rho\nu} \equiv g = \kappa_{1\rho} g \\ {}_1\lambda_{1\sigma} e_{11} + {}_1\lambda_{\sigma 1} e_{\sigma 1} &= {}_2\lambda_{1\sigma} e_{12} + {}_2\lambda_{\sigma 1} e_{\sigma 2} = \dots = {}_\nu\lambda_{1\sigma} e_{1\nu} + {}_\nu\lambda_{\sigma 1} e_{\sigma\nu} \equiv g = \kappa_{1\sigma} g \end{aligned}\right\} \dots(3).$$

From equation (1) the point (p) of concurrence of the $\nu - 1$ planes of this type is, when referred to the first figure, given by

$$p \equiv \begin{vmatrix} e_{11}, & e_{12}, & \dots & e_{1\nu}, \\ 1/{}_1\lambda_{12}, & 1/{}_2\lambda_{12}, & \dots & 1/{}_\nu\lambda_{12}, \\ \dots & \dots & \dots & \dots \\ 1/{}_1\lambda_{1\nu}, & 1/{}_2\lambda_{1\nu}, & \dots & 1/{}_\nu\lambda_{1\nu} \end{vmatrix}.$$

But by the first of the set of equations (3),

$$e_{11} = \kappa_{1\rho} g/{}_1\lambda_{1\rho} - {}_1\lambda_{\rho 1} e_{\rho 1}/{}_1\lambda_{1\rho},\ \dots\ e_{1\nu} = \kappa_{1\rho} g/{}_\nu\lambda_{1\rho} - {}_\nu\lambda_{\rho 1} e_{\rho\nu}/{}_\nu\lambda_{1\rho}.$$

Hence substituting in the expression for p, and noticing that the coefficient of g vanishes, we obtain

$$p \equiv \begin{vmatrix} {}_1\lambda_{\rho 1} e_{\rho 1}/{}_1\lambda_{1\rho}, & {}_2\lambda_{\rho 1} e_{\rho 2}/{}_2\lambda_{1\rho}, & \ldots, & {}_\nu\lambda_{\rho 1} e_{\rho \nu}/{}_\nu\lambda_{1\rho}, \\ 1/{}_1\lambda_{12}, & 1/{}_2\lambda_{12}, & \ldots, & 1/{}_\nu\lambda_{12}, \\ \cdots & \cdots & \cdots & \cdots \\ 1/{}_1\lambda_{1\nu}, & 1/{}_2\lambda_{1\nu}, & \ldots, & 1/{}_\nu\lambda_{1\nu} \end{vmatrix}.$$

This is the point of concurrence of the $\nu - 1$ planes, referred to the ρth figure.

Now by eliminating $e_{11}, e_{12}, \ldots e_{1\nu}$ from equations (3), we obtain

$$\frac{{}_1\lambda_{\rho 1}\, {}_1\lambda_{1\sigma} e_{\rho 1} - {}_1\lambda_{\sigma 1}\, {}_1\lambda_{1\rho} e_{\sigma 1}}{\kappa_{1\rho}\, {}_1\lambda_{1\sigma} - \kappa_{1\sigma}\, {}_1\lambda_{1\rho}} = \frac{{}_2\lambda_{\rho 1}\, {}_2\lambda_{1\sigma} e_{\rho 2} - {}_2\lambda_{\sigma 1}\, {}_2\lambda_{1\rho} e_{\sigma 2}}{\kappa_{1\rho}\, {}_2\lambda_{1\sigma} - \kappa_{1\sigma}\, {}_2\lambda_{1\rho}} = \ldots$$

$$= \frac{{}_\nu\lambda_{\rho 1}\, {}_\nu\lambda_{1\sigma} e_{\rho \nu} - {}_\nu\lambda_{\sigma 1}\, {}_\nu\lambda_{1\rho} e_{\sigma \nu}}{\kappa_{1\rho}\, {}_\nu\lambda_{1\sigma} - \kappa_{1\sigma}\, {}_\nu\lambda_{1\rho}} = g.$$

Hence the equation of the axal plane of perspective of the ρth and σth figures is, referred to the ρth figure,

$$\xi_1 \left(\frac{\kappa_{1\rho}}{{}_1\lambda_{\rho 1}} - \frac{\kappa_{1\sigma}\, {}_1\lambda_{1\rho}}{{}_1\lambda_{\rho 1}\, {}_1\lambda_{1\sigma}} \right) + \xi_2 \left(\frac{\kappa_{1\rho}}{{}_2\lambda_{\rho 1}} - \frac{\kappa_{1\sigma}\, {}_2\lambda_{1\rho}}{{}_2\lambda_{\rho 1}\, {}_2\lambda_{1\sigma}} \right) + \ldots$$

$$+ \xi_\nu \left(\frac{\kappa_{1\rho}}{{}_\nu\lambda_{\rho 1}} - \frac{\kappa_{1\sigma}\, {}_\nu\lambda_{1\rho}}{{}_\nu\lambda_{\rho 1}\, {}_\nu\lambda_{1\sigma}} \right) = 0.$$

Now by § 72 (4) the point p lies on this plane, if the determinant formed by substituting the coefficients of $\xi_1, \xi_2, \ldots \xi_\nu$ in this equation for

$$e_{\rho 1}, e_{\rho 2}, \ldots e_{\rho \nu}$$

in the determinant, which is the expression for p, vanishes. The determinant so formed can be expressed as the sum of two determinants, one with $\kappa_{1\rho}$ as a factor, the other with $\kappa_{1\sigma}$ as a factor. The determinant with $\kappa_{1\rho}$ as a factor vanishes because it has two rows of the form

$$1/{}_1\lambda_{1\rho}, 1/{}_2\lambda_{1\rho}, \ldots 1/{}_\nu\lambda_{1\rho}.$$

The determinant with $\kappa_{1\sigma}$ as a factor vanishes because it has two rows of the form

$$1/{}_1\lambda_{1\sigma}, 1/{}_2\lambda_{1\sigma}, \ldots 1/{}_\nu\lambda_{1\sigma}.$$

Hence all the axal planes are concurrent in the same point.

The particular case of this theorem for triangles in two dimensions is well-known.

75. QUADRANGLES. (1) As a simple example of this type of reasoning, let us investigate the properties of a quadrangle in a two-dimensional region.

Any four points a, b, c, d are connected by the addition relation

$$\alpha a + \beta b + \gamma c + \delta d = 0.$$

Hence $\alpha a + \beta b$ and $\gamma c + \delta d$ represent the same point, namely the point of intersection of the lines ab and cd.

(2) Consider the six lines joining these four points. Let the three pairs which do not intersect in a, b, c, d, intersect in e, f, g. Then

$$e = \gamma c + \alpha a = -(\beta b + \delta d),$$
$$f = \alpha a + \beta b = -(\gamma c + \delta d),$$
$$g = \beta b + \gamma c = -(\delta d + \alpha a).$$

Hence
$$f - g = \alpha a - \gamma c;$$
and
$$f + g = \beta b - \delta d.$$

From the form of these expressions it follows that $f - g$ is the point where fg intersects ac.

Also it follows that e and $f - g$ are harmonic conjugates with respect to a and c.

Similarly $f + g$ is the point where fg intersects bd; and $f + g$ and e are harmonic conjugates with respect to b and d.

Furthermore $f - g$ and $f + g$ are harmonic conjugates with respect to f and g.

The points $g \pm e$, and $e \pm f$, have similar properties.

Thus the harmonic properties of a complete quadrilateral are immediately obvious.

(3) Again the six points $f \pm g$, $g \pm e$, $e \pm f$ lie by threes on four straight lines. For identically

$$(f - g) + (g - e) + (e - f) = 0,$$
$$(f - g) + (g + e) - (e + f) = 0,$$
$$(f + g) - (g - e) - (e + f) = 0,$$
$$(f + g) - (g + e) + (e - f) = 0.$$

In the accompanying figure h and k stand for $f \mp g$ respectively, l and m for $e \mp f$ respectively, n and p for $g \pm e$ respectively.

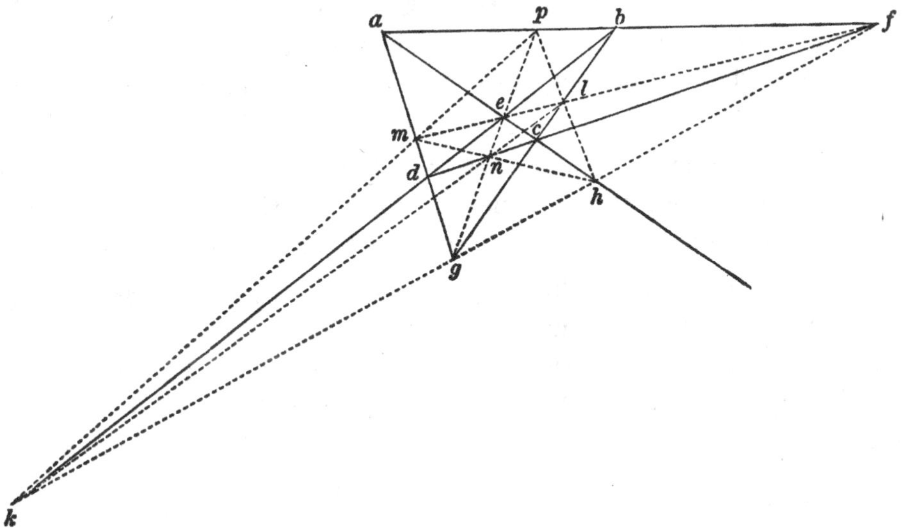

CHAPTER III.

QUADRICS.

76. INTRODUCTORY. (1) Let a surface locus of the second degree be called a quadric surface. Let a curve locus which can be defined as the intersection of ρ ($\rho < \nu$) quadric surfaces be called a quadriquadric curve locus. If it is impossible to define the locus as the intersection of $\rho - \sigma$ quadric surfaces and of σ plane subregions ($\sigma < \rho$), then the quadriquadric curve locus is said to be the $(\rho - 1)$th order of tortuosity [cf. § 67 (6)].

(2) Let the ν reference elements be $e_1, e_2, \dots e_\nu$, and let any point x be defined by $\xi_1 e_1 + \dots + \xi_\nu e_\nu$, which is shortened into $\Sigma\xi e$.

Let the quadric form $\alpha_{11}\xi_1^2 + 2\alpha_{12}\xi_1\xi_2 + \dots$ be written $(\alpha \between x)^2$. Then $(\alpha \between x)^2 = 0$ is the equation of a quadric surface.

Let the lineo-linear form $\alpha_{11}\xi_1\eta_1 + \alpha_{12}(\xi_1\eta_2 + \xi_2\eta_1) + \dots$ be written $(\alpha \between x \between y)$; where $x \equiv \Sigma\xi e$, and $y \equiv \Sigma\eta e$.

77. ELEMENTARY PROPERTIES. (1) If the element z be of the form $\lambda x + \mu y$, then

$$(\alpha \between z)^2 = \lambda^2 (\alpha \between x)^2 + 2\lambda\mu (\alpha \between x \between y) + \mu^2 (\alpha \between y)^2 \dots\dots\dots\dots\dots(A).$$

If more than two elements of a subregion of one dimension lie on a quadric, the whole subregion lies on it. This follows evidently from equation (A).

(2) If a quadric contain one plane subregion of the same dimensions as itself, it must consist of two plane loci taken together. For if there is one linear factor of a quadric form $(\alpha \between x)^2$ the remaining factor must be linear.

(3) A subregion of any dimensions either intersects a quadric surface in a quadric surface locus contained in that subregion as its containing region or itself lies entirely in the quadric. For if the subregion be of $\rho - 1$ dimensions, it may be chosen as a co-ordinate region containing the ρ reference elements $e_1, e_2, \dots e_\rho$. Hence any element in the region has the $\nu - \rho$ co-ordinates

$$\xi_{\rho+1}, \xi_{\rho+2}, \dots \xi_\nu$$

respectively zero. Thus the intersection of $(\alpha \between x)^2 = 0$ with the subregion is found by putting these co-ordinates zero in the quadric equation. Thus either the equation is left as a quadric equation between the remaining ρ co-ordinates; or the left-hand side vanishes identically.

(4) It follows as a corollary from the two previous subsections that a subregion, which intersects a quadric in one subregion of dimensions lower by one than itself, intersects it also in another such subregion; and that these two flat loci together form the entire intersection of the subregion with the quadric.

78. Poles and Polars. (1) The equation, $(\alpha \between x \between x') = 0$, may be conceived as defining the locus of one of the two elements x or x', when the other is fixed.

If x' be fixed, the locus will be called the polar of x' with respect to the quadric surface, $(\alpha \between x)^2 = 0$. The polar of an element is obviously a plane.

The element x' will be called the pole of the plane $(\alpha \between x \between x') = 0$.

(2) The ordinary theorems respecting poles and polars obviously hold.

If x be on the polar of x', then x' lies in the polar of x. For in either case the condition is $(\alpha \between x \between x') = 0$. Two elements for which this condition holds will be called reciprocally polar with respect to the quadric.

If a pole x' lie in its polar $(\alpha \between x \between x') = 0$, then

$$(\alpha \between x' \between x') \equiv (\alpha \between x')^2 = 0.$$

Hence the element x' lies on the quadric. Thus all elements on the quadric may be conceived as reciprocally polar to themselves: they may be called self-polar.

The polars of all elements lying in a plane must pass through the polar of the plane.

(3) By means of these theorems on poles and polars with respect to any assumed quadric a correspondence is established between the elements of a region of $\nu - 1$ dimensions and the subregions of $\nu - 2$ dimensions.

Corresponding to an element there is its polar subregion: corresponding to elements lying in a plane of $\nu - 2$ dimensions there are polars all containing the pole of this plane: corresponding to elements lying in a subregion of $\nu - \rho - 1$ dimensions there are polars all containing a common subregion of $\rho - 1$ dimensions.

(4) Again, consider the elements in which the linear subregion through two reciprocally polar elements x and x' intersects the quadric. Let $\lambda x + \mu x'$ be one of these elements, then from equation (A),

$$\lambda^2 (\alpha \between x)^2 + \mu^2 (\alpha \between x')^2 = 0.$$

The two points of intersection must accordingly be of the form $\lambda x \pm \mu x'$. It follows that two reciprocally polar elements and the two elements in which the straight line containing them intersects the quadric together form a harmonic range.

(5) When the element x' is on the quadric, its polar, viz. $(\alpha \between x \between x') = 0$, will be called a tangential polar of the quadric. Let x lie on the polar of any point x' on the quadric and let $\lambda x + \mu x'$ be on the quadric. Then substituting in the equation of the quadric, the equation to determine λ/μ becomes $\lambda^2 (\alpha \between x)^2 = 0$.

Now in general $(\alpha \between x)^2$ is not zero. Hence λ is zero and both roots of the quadratic are zero. Thus all straight lines drawn through an element x' on the quadric and lying in the polar of x' intersect the quadric in two coincident elements at x'.

(6) Let any plane be represented by the equation

$$\lambda_1 \xi_1 + \lambda_2 \xi_2 + \ldots + \lambda_\nu \xi_\nu = 0.$$

The condition that this plane should be a tangential polar of the quadric is obviously

$$\begin{vmatrix} \alpha_{11}, & \alpha_{12}, & \ldots & \alpha_{1\nu}, & \lambda_1 \\ \alpha_{12}, & \alpha_{22}, & \ldots & \alpha_{2\nu}, & \lambda_2 \\ \cdots & \cdots & \cdots & \cdots & \cdots \\ \alpha_{1\nu}, & \alpha_{2\nu}, & \ldots & \alpha_{\nu\nu}, & \lambda_\nu \\ \lambda_1, & \lambda_2, & \ldots & \lambda_\nu, & 0 \end{vmatrix} = 0.$$

This condition can be written in the form

$$\mathfrak{a}_{11} \lambda_1^2 + 2\mathfrak{a}_{12} \lambda_1 \lambda_2 + \ldots = 0.$$

Let Δ stand for the determinant

$$\begin{vmatrix} \alpha_{11}, & \alpha_{12}, & \ldots & \alpha_{1\nu} \\ \alpha_{12}, & \alpha_{22}, & \ldots & \alpha_{2\nu} \\ \cdots & \cdots & \cdots & \cdots \\ \alpha_{1\nu}, & \alpha_{2\nu}, & \ldots & \alpha_{\nu\nu} \end{vmatrix};$$

then $$\mathfrak{a}_{11} = \frac{d\Delta}{d\alpha_{11}}, \quad 2\mathfrak{a}_{12} = \frac{d\Delta}{d\alpha_{12}}, \text{ etc.}$$

Now let the plane be denoted by L, then the condition that this plane may be a tangential polar of the quadric may be written by analogy

$$(\mathfrak{a} \between L)^2 = 0.$$

Hence corresponding to the condition, $(\alpha \between x)^2 = 0$, that the element x lies on the quadric there is the condition, $(\mathfrak{a} \between L)^2 = 0$, that the plane L is a tangential polar of the quadric.

(7) It will be found on developing the theory of multiplication of Grassmann's Calculus of Extension (cf. Book IV. ch. I.) that, analogously to the notation by which an element can be written $\xi_1 e_1 + \xi_2 e_2 + \ldots + \xi_\nu e_\nu$ where

$e_1, e_2, \dots e_\nu$ are extraordinaries denoting reference elements, the plane L can be written in the form $\lambda_1 E_1 + \lambda_2 E_2 + \dots + \lambda_\nu E_\nu$ where $E_1, E_2 \dots E_\nu$ are extraordinaries denoting reference planes. Thus the theory of duality will receive a full expression later [cf. § 110 (4) and § 123] and need not be pursued now, except to state the fundamental properties.

(8) The equation, $(\alpha \between x)^2 = 0$, will be called the point-equation of the quadric, and the equation, $(\alpha \between L)^2 = 0$, will be called the plane-equation of the quadric. The two equations will be called reciprocal to each other.

(9) It is possible in general to find sets of ν independent elements reciprocally polar to each other.

For let e_1 be any point not on the quadric. Its polar plane is of $\nu - 2$ dimensions and does not contain e_1. The intersection of this plane with the quadric is another quadric of $\nu - 3$ dimensions contained in it. Take any point e_2 in this polar plane not on the quadric. Again take any point e_3 on the intersection of the polar planes of e_1 and e_2; then e_4 on the intersection of the polar planes of e_1, e_2, e_3; and so on. Thus ultimately ν independent points are found all reciprocally polar to each other.

If such points be taken as reference elements, the equation of the quadric becomes

$$\alpha_1 \xi_1^2 + \alpha_2 \xi_2^2 + \dots + \alpha_\nu \xi_\nu^2 = 0.$$

If the elements lie on the quadric it will be proved in the next article that $\nu/2$ or $(\nu - 1)/2$, according as ν is even or odd, independent elements can be found reciprocally polar to each other.

79. Generating Regions. (1) A quadric surface contains within it an infinite number of flat loci, or subregions, real or imaginary, according to the nature of the quadric. Let such contained regions be called generating regions. If the complete region be of 2μ or of $2\mu - 1$ dimensions, the subregions, real or imaginary, contained within any quadric surface will be proved to be of $\mu - 1$ dimensions*.

If b_1 be any point on the quadric, it lies on its polar $(\alpha \between b_1 \between x) = 0$. Let b_2 be another element on the quadric lying in the polar of b_1. Then $(\alpha \between b_1 \between b_2) = 0$, and each point lies in the polar of the other. Hence any element $\lambda_1 b_1 + \lambda_2 b_2$ lies in both polars and on the quadric.

But the polars of b_1 and b_2 intersect in a subregion of $\nu - 3$ dimensions, where ν is put for $2\mu + 1$ or 2μ as the case may be.

Take a third point b_3 on the intersection of this subregion with the quadric. Then the three points b_1, b_2, b_3 are reciprocally polar, and any point of the form $\lambda_1 b_1 + \lambda_2 b_2 + \lambda_3 b_3$ lies in the intersection of the three polars and on the quadric.

* This theorem is due to Veronese, cf. *loc. cit.*

Proceed in this way till ρ points $b_1, b_2, \dots b_\rho$ are determined such that each lies on the polars of all the others and on the quadric, and therefore on its own polar. But the ρ polars, if $b_1, b_2, \dots b_\rho$ be independent, intersect in a region of $\nu - \rho - 1$ dimensions, which contains the ρ independent points.

Hence $$\nu - \rho \geqq \rho.$$

Hence the greatest value of ρ is the greatest integer in $\frac{1}{2}\nu$.

If $\nu = 2\mu$, or $2\mu + 1$, then $\rho = \mu$; there are therefore μ independent points and these define a subregion of $\mu - 1$ dimensions contained in the quadric.

This proposition is a generalization of the proposition that generating lines, real or imaginary, can be drawn through every point of a conicoid.

(2) If ν be even, then each generating region of a quadric is defined by $\frac{1}{2}\nu$ independent points. Hence by § 72 (5) from any point one straight line can be drawn intersecting two non-intersecting generating regions.

If the point from which the line be drawn be on the quadric and do not lie in either of the generating regions, the line meets the quadric in three points, and therefore lies wholly on the quadric.

Hence from any point on a quadric one line and only one line can be drawn meeting any two non-intersecting generating regions and thus lying wholly in the quadric.

(3) If ν be odd, then each generating region is defined by $\frac{1}{2}(\nu - 1)$ independent elements. Hence from § 72 (5) it is in general impossible to draw a line from any point, on or off the quadric, intersecting two non-intersecting generating regions.

80. Conjugate Co-ordinates. (1) Let the ν co-ordinate elements be a reciprocally polar set. Let the equation of the quadric be

$$\xi_1^2/\beta_1 + \xi_2^2/\beta_2 + \dots + \xi_\nu^2/\beta_\nu = 0.$$

The elements

$$\beta_1^{\frac{1}{2}} e_1 \pm (-\beta_2)^{\frac{1}{2}} e_2$$

are on the quadric. They can be assumed to be any two points on the quadric, not in the same generating region, since e_1 can be any point not on the quadric and e_2 any point on the polar of e_1.

(2) Firstly, let $\nu = 2\mu$. The set of μ elements,

$$\beta_1^{\frac{1}{2}} e_1 + (-\beta_2)^{\frac{1}{2}} e_2,$$
$$\beta_3^{\frac{1}{2}} e_3 + (-\beta_4)^{\frac{1}{2}} e_4, \dots$$
$$\beta^{\frac{1}{2}}_{\mu-1} e_{\mu-1} + (-\beta_\mu)^{\frac{1}{2}} e_\mu,$$

are all on the quadric and reciprocally polar to each other. Hence they define a generating region on the quadric.

Similarly the set,

$$\beta_1^{\frac{1}{2}} e_1 - (-\beta_2)^{\frac{1}{2}} e_2, \text{ etc.}$$

define another generating region on the quadric. Also the μ elements of the first set are independent of the μ elements of the second set. Therefore the two generating regions do not overlap at all.

(3) Let the elements of the first set be named in order $j_1, j_2, \dots j_\mu$, and of the second set $k_1, k_2, \dots k_\mu$. Then any element x of the form

$$\Sigma \xi j \equiv \xi_1 j_1 + \xi_2 j_2 + \dots + \xi_\mu j_\mu$$

lies in the generating region $j_1 j_2 \dots j_\mu$, and any element y of the form $\Sigma \eta k$ lies in the generating region $k_1 k_2 \dots k_\mu$.

(4) Again, j_1 is reciprocally polar to all the k's except k_1. Hence

$$(j_1, k_2, k_3, \dots k_\mu)$$

is a generating region, and

$$(j_1, j_2, k_3, k_4, \dots k_\mu)$$

is another, and so on.

Accordingly given one generating region $(j_1, j_2, \dots j_\mu)$ including a given element, other generating regions including that element can be found which either overlap the given region in that element only, or in regions of

$$1, 2 \dots \mu - 2$$

dimensions respectively. Also regions can be found which do not overlap the given region at all.

(5) The 2μ elements,

$$j_1, j_2, \dots j_\mu, \; k_1, k_2, \dots k_\mu$$

can be taken as co-ordinate elements. Let them be called a system of conjugate co-ordinate elements. The properties of such a system are that they are all on the quadric, and that any pair of elements, with the exception of pairs having the same suffix, are reciprocally polar, namely j_1 not with k_1, j_2 not with k_2, and so on.

Let j_1 and k_1, j_2 and k_2, etc. be called conjugate pairs. It can easily be seen by the method of subsections (1) and (2) of this section that in any two non-intersecting generating regions of a quadric a conjugate set of elements can be found, so that $j_1, j_2, \dots j_\mu$ are in one region, and $k_1, k_2, \dots k_\mu$ in the other.

Let any element be written in the form

$$\xi_1 j_1 + \xi_2 j_2 + \dots + \xi_\mu e_\mu + \eta_1 k_1 + \eta_2 k_2 + \dots + \eta_\mu k_\mu.$$

Then, from the definitions of the conjugate elements in subsections (2) and (3) above, the equation of the quadric takes the form

$$\xi_1 \eta_1 + \xi_2 \eta_2 + \dots + \xi_\rho \eta_\rho + \dots + \xi_\mu \eta_\mu = 0.$$

(6) The polar of the element j_1 is $\eta_1 = 0$, that is to say, is the region defined by the elements

$$j_1, j_2, \ldots j_\mu, k_2, k_3, \ldots k_\mu.$$

The intersection of the polar of j_1 with the quadric is

$$\xi_2\eta_2 + \ldots + \xi_\rho\eta_\rho + \ldots + \xi_\mu\eta_\mu = 0,$$

and the $2\mu - 1$ co-ordinate elements

$$j_1, j_2, \ldots j_\mu, k_2, k_3, \ldots k_\mu,$$

define its containing region.

This quadric is contained in a region of $2\mu - 2$ dimensions. In such a region quadrics in general have generating regions of $\mu - 2$ dimensions. But in this quadric all regions of the type

$$(j_1, j_2 \ldots j_r, k_{r+1} \ldots k_\mu)$$

are generating regions, being of $\mu - 1$ dimensions.

(7) The coefficient of j_1 in the expression defining an element does not appear in the equation of the quadric. Hence all one dimensional regions defined by j_1 and any point on the quadric lie entirely in the quadric. Such a surface will be called a conical quadric; the point with the property of j_1 will be called its vertex.

(8) Accordingly the intersection of the polar of any element of a quadric in a region of $2\mu - 1$ dimensions with the quadric is a conical quadric of which the given element is the vertex. Thus in three dimensions, the intersection of a tangent plane with a quadric is two straight lines, that is to say a conical quadric in two dimensions.

(9) Secondly, let $\nu = 2\mu + 1$. Then the system of 2μ conjugate elements

$$j_1, j_2, j_\mu, k_1, \ldots k_\mu,$$

can be found by the same process as in the first case; but do not define the complete region. In forming $j_1, \ldots j_\nu, k_1, \ldots k_\nu$ from the elements $e_1 \ldots e_{2\mu+1}$ the last element $e_{2\mu+1}$ was left over. This element, which will be called simply e, leaving out the suffix, is reciprocally polar to all the other elements $j_1 \ldots k_\mu$, but does not lie on the quadric.

(10) Let any element in the region be denoted by

$$\xi_1 j_1 + \ldots + \xi_\mu j_\mu + \eta_1 k_1 + \ldots + \eta_\mu k_\mu + \zeta e.$$

Then the equation of the quadric becomes

$$\xi_1\eta_1 + \xi_2\eta_2 + \ldots + \xi_\mu\eta_\mu + \zeta^2 = 0.$$

(11) The polar of the element j_1 is given by the equation $\eta_1 = 0$.

The intersection of the polar and the quadric is another quadric

$$\xi_2\eta_2 + \xi_3\eta_3 + \ldots + \xi_\mu\eta_\mu + \zeta^2 = 0;$$

which lies in the region $j_1, j_2, \ldots j_\mu, k_2, \ldots k_\nu, e$. All these co-ordinate elements are reciprocally polar and all, except e, lie on the quadric.

This quadric is contained in a region of $2\mu-1$ dimensions, and its generating regions passing through j_1 are of $\mu-1$ dimensions, not more than the number of dimensions of the generating regions of any quadric in this region.

(12) Since the coefficient ξ_1 of j_1 does not appear in the equation of the quadric, if any point x be on the quadric then the region (j_1, x) lies entirely in the quadric. Hence the quadric is a conical quadric.

So finally we find the general proposition that the intersection of the polar of an element on a quadric with the quadric is a conical quadric with its vertex at the element.

(13) The reduction of the equation of a quadric contained in a region of 2μ dimensions to the form,

$$\xi_1\eta_1 + \ldots + \xi_\mu\eta_\mu + \zeta^2 = 0,$$

is a generalization of the reduction of the equation of a conic section to the form, $LM + R^2 = 0$ (cf. Salmon's *Conic Sections*).

Applying to space of four dimensions the above proposition on the intersection of polars with quadrics, we see that if our flat three-dimensional space be any intersecting region, it intersects the quadric in some conicoid. But if the space be the polar of some element of the quadric, it intersects the quadric in a cone with its vertex at the element on the quadric which is the pole of the space.

A quadric in five-dimensional space has two-dimensional flat spaces as generating regions.

(14) The co-ordinates $j_1 \ldots j_\mu$, $k_1 \ldots k_\mu$, e of a complete region of 2μ dimensions, giving the equation of some quadric in the form

$$\xi_1\eta_1 + \ldots + \xi_\mu\eta_\mu + \zeta^2 = 0,$$

are such that e is the pole of the region $j_1 \ldots k_\mu$. Now e may be any point in the region. Hence the polar of any point (i.e. any plane) can be defined by two not overlapping generating regions of the quadric, viz. $j_1 \ldots j_\mu$ and $k_1 \ldots k_\mu$, which are also generating regions of the quadric formed by the intersection of the polar with the original quadric. This includes the case of space of two dimensions.

If, however, the complete region be of $2\mu-1$ dimensions, the polar of any point intersects the quadric in another quadric which only contains generating regions of $\mu-2$ dimensions; any two such regions cannot serve to define the polar which is of $2\mu-2$ dimensions. This includes the case of space of three dimensions.

81. QUADRIQUADRIC CURVE LOCI. (1) Consider the general case of the curve locus formed by the intersection of the ρ quadric surfaces,

$$(\alpha_1 \between x)^2 = 0,\ (\alpha_2 \between x)^2 = 0, \ldots (\alpha_\rho \between x)^2 = 0.$$

Let b_1 be any point on the locus. Then the polar planes of b_1 are

$$(\alpha_1 \between b_1 \between x) = 0,\ (\alpha_2 \between b_1 \between x) = 0, \dots (\alpha_\rho \between b_1 \between x) = 0.$$

The intersection of the ρ polar planes forms a subregion of $\nu - \rho - 1$ dimensions; where the complete region is of $\nu - 1$ dimensions.

The intersection of this region with the curve locus is another quadriquadric curve locus of the same order of tortuosity, namely $\rho - 1$ [cf. § 67 (7)].

Now find another point b_2 in this second quadriquadric curve locus, then all the ρ polars of b_2 contain b_1. Also the 2ρ polars of $b_1 b_2$ form by their intersection a region of $\nu - 2\rho - 1$ dimensions, and this region intersects the quadriquadric curve locus in another quadriquadric curve locus of the same order of tortuosity, namely $\rho - 1$.

Also it can easily be seen, as in § 79 (1), that the region (b_1, b_2) lies entirely in this last curve locus.

Continuing in this way and taking σ points $b_1, b_2, \dots b_\sigma$, successively, each in the quadriquadric curve locus lying in the region of the intersection of the polars of all the preceding points, we find a subregion defined by $b_1, b_2, \dots b_\sigma$, lying entirely in the original quadriquadric curve locus. Also it must lie in the region of dimensions $\nu - \sigma\rho - 1$ formed by the intersection of the polars. Hence we must have

$$\sigma \leqq \nu - \sigma\rho;$$

that is

$$\sigma \leqq \nu/(\rho + 1).$$

Now let $I(\lambda)$ denote the greatest integer in the number λ. Then we have proved that it is always possible to proceed as above till

$$\sigma = I\{\nu/(\rho + 1)\}.$$

Hence a quadriquadric curve locus, apart from any special relation between the intersecting quadric surfaces, of tortuosity $\rho - 1$, in a complete region of $\nu - 1$ dimensions contains subregions (real or imaginary) defined by

$$I\{\nu/(\rho + 1)\} \text{ elements,}$$

that is to say, of

$$I\{(\nu - \rho - 1)/(\rho + 1)\} \text{ dimensions*.}$$

(2) Hence the least dimensions of a complete region such that a curve locus, of order of tortuosity $\rho - 1$, apart from special conditions must contain a region of one dimension is $2\rho + 1$.

For example, space of five dimensions is of the lowest dimensions for which it is the case that the intersection of two quadric surfaces (a curve locus of order of tortuosity 1) must contain straight lines.

* This generalization of Veronese's Theorem, cf. § 79 1), has not been stated before, as far as I am aware.

Also space of eight dimensions is of the lowest dimensions for which it is the case that a quadriquadric curve locus, of order of tortuosity 1, must contain subregions of two dimensions.

82. Closed Quadrics. (1) A quadric will be called a closed quadric if points not on the quadric exist such that any straight line drawn through one of them must cut the quadric in real points. Such points will be said to be within the quadric: other points not on the quadric which do not possess this property will be said to be without the quadric.

(2) Let a straight line be drawn through any point p cutting the quadric

$$(\alpha \between x)^2 = 0,$$

in two points y_1 and y_2 real or imaginary; and let x be any other real point on this line. Also let

$$y_1 = \lambda_1 p + \mu_1 x, \quad y_2 = \lambda_2 p + \mu_2 x:$$

then λ_1/μ_1 and λ_2/μ_2 are the roots of the equation

$$\lambda^2 (\alpha \between p)^2 + 2\lambda\mu (\alpha \between p \between x) + \mu^2 (\alpha \between x)^2 = 0 \text{.................}(1).$$

The roots of this equation are real or imaginary according as

$$(\alpha \between x)^2 (\alpha \between p)^2 - \{(\alpha \between p \between x)\}^2$$

is negative or positive.

(3) Now choose as the co-ordinate elements a reciprocally polar system with respect to the quadric, and let p be one point of this system.

Let the system be $p, e_2, e_3, \ldots e_\nu$, and let

$$x = \xi p + \xi_2 e_2 + \ldots + \xi_\nu e_\nu.$$

Then $(\alpha \between x)^2$ takes the form

$$\alpha\xi^2 + \alpha_2 \xi_2^2 + \ldots + \alpha_\nu \xi_\nu^2.$$

Hence the roots of equation (1) are real or imaginary according as

$$\alpha (\alpha_2 \xi_2^2 + \alpha_3 \xi_3^2 + \ldots + \alpha_\nu \xi_\nu^2)$$

is negative or positive.

(4) If all lines through p meet the quadric in imaginary points, then the expression is positive for all values of $\xi_2, \xi_3, \ldots \xi_\nu$. Hence $\alpha\alpha_2, \alpha\alpha_3, \ldots \alpha\alpha_\nu$ must be all positive; and therefore $\alpha, \alpha_2, \ldots \alpha_\nu$ must all be of the same sign. The equation of the quadric takes the form,

$$\kappa^2\xi^2 + \kappa_2^2 \xi_2^2 + \ldots + \kappa_\nu^2 \xi_\nu^2 = 0;$$

and the quadric is therefore entirely imaginary.

(5) Again, if all lines drawn through p meet the quadric in real points, then $\alpha\alpha_2, \alpha\alpha_3, \ldots \alpha\alpha_\nu$ must all be negative.

Hence $a_2, a_3, \ldots a_\nu$ are of one sign and α is of the other. The equation of the quadric can therefore be written in the form

$$\kappa_2^2 \xi_2^2 + \kappa_3^2 \xi_3^2 + \ldots + \kappa_\nu^2 \xi_\nu^2 - \kappa^2 \xi^2 = 0 \quad \ldots\ldots\ldots\ldots\ldots(2);$$

where the co-ordinate point p is within the quadric and the remaining $\nu - 1$ co-ordinate points can easily be proved to be without the quadric on the polar of p.

(6) It also follows that the polar of a point inside the quadric does not intersect the quadric in real points. For the polar of p is $\xi = 0$, and its points of intersection with the quadric lie on the imaginary quadric of $\nu - 2$ dimensions given by

$$\kappa_2^2 \xi_2^2 + \ldots + \kappa_\nu^2 \xi_\nu^2 = 0.$$

It has been proved by Sylvester that if a quadric expression referred to one set of reciprocal co-ordinate elements has ρ positive terms and $\nu - \rho$ negative terms, then when referred to any other set of reciprocal elements it still has ρ positive terms and $\nu - \rho$ negative terms (or vice versa).

Hence if the given quadric of equation (2) be referred to any other reciprocal set of co-ordinates, it still takes the form of (2) as far as the signs of its terms are concerned. Thus if the quadric considered be a closed quadric, one element of a reciprocal set of elements is within the quadric and the remaining elements are without the quadric.

(7) The polar of a point without a closed quadric necessarily cuts the quadric and contains points within the quadric. For considering the quadric of equation (2) of subsection (5), the polar of e_2 is, $\xi_2 = 0$. Its intersection with (2) is the quadric

$$\kappa_3^2 \xi_3^2 + \ldots + \kappa_\nu^2 \xi_\nu^2 - \kappa^2 \xi^2 = 0,$$

lying in the plane $\xi_2 = 0$, that is in the region of $p, e_3, \ldots e_\nu$. Now if e_2 be first chosen, p may be any point in this region and within this quadric, which is a real closed quadric.

Hence the polar of any point without a closed quadric necessarily cuts the quadric in real points and contains points within the quadric.

(8) It may be noted that no real generating regions exist on closed quadrics.

(9) Again, choosing any reference points whatsoever, $(a \,\rangle\!\langle\, x)^2$ and $(a \,\rangle\!\langle\, y)^2$ are of the same sign if both x and y be inside the closed surface, or if both be outside the surface, but are of opposite signs if one be inside the surface and one be outside.

For let x' and y' be two points respectively on the polars of x and y. Then

$$\lambda x + \lambda' x' \quad \text{and} \quad \mu y + \mu' y'$$

are two points on the lines xx' and yy'.

The points where these lines cut the surface are given by

$$\lambda^2 (\alpha \between x)^2 + \lambda'^2 (\alpha \between x')^2 = 0,$$

and

$$\mu^2 (\alpha \between y)^2 + \mu'^2 (\alpha \between y')^2 = 0.$$

Firstly, let x and y be both within the surface. Then their polars intersect in a region of $\nu - 3$ dimensions without the surface. Let x' and y' both denote the same point z in this region. Then since the roots of the quadratics for λ/λ' and μ/μ' are both real, $(\alpha \between z)^2$ has opposite signs to both

$$(\alpha \between x)^2 \text{ and } (\alpha \between y)^2.$$

Hence $(\alpha \between x)^2$ and $(\alpha \between y)^2$ have the same sign.

(10) Secondly let x and y be both without the quadric. Then their polars both cut the quadric, hence x' and y' may both be chosen within the quadric. Hence

$$(\alpha \between x')^2 \text{ and } (\alpha \between y')^2$$

have both the same sign. Also the straight lines xx' and yy' both cut the quadric in real points, since x' and y' lie within it. Hence $(\alpha \between x)^2$ has the opposite sign to $(\alpha \between x')^2$ and $(\alpha \between y)^2$ has the opposite sign to $(\alpha \between y')^2$. Accordingly $(\alpha \between x)^2$ and $(\alpha \between y)^2$ have the same sign.

(11) Thirdly, let x be within the quadric and y without the quadric. Then any point x' on the polar of x lies without the quadric. Also

$$(\alpha \between x)^2 \text{ and } (\alpha \between x')^2$$

have opposite signs, and $(\alpha \between x')^2$ and $(\alpha \between y)^2$ have the same sign because both lie without the quadric. Hence $(\alpha \between x)^2$ and $(\alpha \between y)^2$ have opposite signs.

83. Conical Quadric Surfaces. (1) To find the condition that $S \equiv (\alpha \between x)^2 = 0$, should be a conical quadric.

Let b be the vertex and x any point on the surface. Then $\lambda x + \mu b$ lies on the surface for all values of λ and μ. Hence $(\alpha \between b \between x) = 0$; where x is any point on the surface. Therefore, if $b = \Sigma \beta e$, there are ν equations of the type,

$$\alpha_{1\rho}\beta_1 + \alpha_{2\rho}\beta_2 + \dots + \alpha_{\rho\rho}\beta_\rho + \dots + \alpha_{\nu\rho}\beta_\nu = 0,$$

found by putting ρ equal to 1, 2, ... ν, in turn. It follows that the equation $(\alpha \between b \between x) = 0$ holds for all positions of x.

And eliminating the β's, the required condition is found to be,

$$\Delta \equiv \begin{vmatrix} \alpha_{11}, & \alpha_{12}, & \dots & \alpha_{1\nu} \\ \alpha_{12}, & \alpha_{22}, & \dots & \alpha_{2\nu} \\ \dots & \dots & \dots & \dots \\ \alpha_{1\nu}, & \alpha_{2\nu}, & \dots & \alpha_{\nu\nu} \end{vmatrix} = 0.$$

(2) Also the vertex b is the point

$$\begin{vmatrix} e_1, & e_2, & \dots & e_\nu \\ \alpha_{12}, & \alpha_{22}, & \dots & \alpha_{2\nu} \\ \dots & \dots & \dots & \dots \\ \alpha_{1\nu}, & \alpha_{2\nu}, & \dots & \alpha_{\nu\nu} \end{vmatrix}.$$

(3) Again, consider the quadriquadric curve locus of the first order of tortuosity, defined by

$$(\alpha \between x)^2 = 0, \text{ and } (\alpha' \between x)^2 = 0.$$

Any quadric surface intersecting both surfaces in this curve locus is

$$(\alpha \between x)^2 + \lambda (\alpha' \between x)^2 = 0.$$

This surface is a conical quadric if

$$\begin{vmatrix} \alpha_{11} + \lambda\alpha_{11}', & \alpha_{12} + \lambda\alpha_{12}', & \dots & \alpha_{1\nu} + \lambda\alpha_{1\nu}' \\ \alpha_{12} + \lambda\alpha_{12}', & \alpha_{22} + \lambda\alpha_{22}', & \dots & \alpha_{2\nu} + \lambda\alpha_{2\nu}' \\ \dots & \dots & \dots & \dots \\ \alpha_{1\nu} + \lambda\alpha_{1\nu}', & \alpha_{2\nu} + \lambda\alpha_{2\nu}', & \dots & \alpha_{\nu\nu} + \lambda\alpha_{\nu\nu}' \end{vmatrix} = 0.$$

Hence in general ν conical quadric surfaces can be drawn intersecting two quadrics in their common curve locus.

(4) Let b be the vertex of one of these conical quadrics. Take $\nu - 1$ independent points in the quadriquadric curve locus, so as to make with b an independent set of elements. Join b by a straight line with any one of these points; the straight line cuts the quadrics again in another common point. Hence by the harmonic property proved in § 78 (4) it cuts the two polars of b with respect to the two quadrics in a common point. Hence these polars have $\nu - 1$ independent common points. Hence they are identical.

Hence the equations $(\alpha \between b \between x) = 0$, and $(\alpha' \between b \between x) = 0$, are identical.

(5) Let the reference points e_1 and e_2 be the vertices of two such conical quadrics. Then the equations

$$(\alpha \between e_1 \between x) = 0, \text{ and } (\alpha' \between e_1 \between x) = 0,$$

are identical: that is

$$\left.\begin{aligned} \alpha_{11}\xi_1 + \alpha_{12}\xi_2 + \alpha_{13}\xi_3 + \dots + \alpha_{1\nu}\xi_\nu = 0, \\ \alpha_{11}'\xi_1 + \alpha_{12}'\xi_2 + \alpha_{13}'\xi_3 + \dots + \alpha_{1\nu}'\xi_\nu = 0 \end{aligned}\right\} \quad \dots\dots (1)$$

are identical.

Similarly the equations

$$\left.\begin{aligned} \alpha_{12}\xi_1 + \alpha_{22}\xi_2 + \alpha_{23}\xi_3 + \dots + \alpha_{2\nu}\xi_\nu = 0, \\ \alpha_{12}'\xi_1 + \alpha_{22}'\xi_2 + \alpha_{23}'\xi_3 + \dots + \alpha_{2\nu}'\xi_\nu = 0 \end{aligned}\right\} \quad \dots\dots (2)$$

are identical.

From equations (1) it follows that

$$\frac{\alpha_{11}}{\alpha_{11}'} = \frac{\alpha_{12}}{\alpha_{12}'} = \frac{\alpha_{13}}{\alpha_{13}'} = \ldots = \frac{\alpha_{1\nu}}{\alpha_{1\nu}'} \quad\ldots\ldots\ldots\ldots(3),$$

from equations (2) it follows that

$$\frac{\alpha_{12}}{\alpha_{12}'} = \frac{\alpha_{22}}{\alpha_{22}'} = \frac{\alpha_{23}}{\alpha_{23}'} = \ldots = \frac{\alpha_{2\nu}}{\alpha_{2\nu}'} \quad\ldots\ldots\ldots\ldots(4).$$

Hence either $\alpha_{12} = 0 = \alpha_{12}'$; or else if p be any element $\lambda e_1 + \mu e_2$ on the line $e_1 e_2$, then the two polars

$$(\alpha \between p \between x) = 0, \text{ and } (\alpha' \between p \between x) = 0,$$

are identical. Excluding the second alternative, which is obviously a special case, it follows that any two of the ν vertices lie each on the polar of the other with respect to either quadric. Thus the ν vertices form ν independent elements and can be taken as reference elements.

(6) It follows that in general any two quadrics have one common system of polar reciprocal elements, and that these elements are the ν vertices of the ν conical quadrics which can be drawn through the intersection of the two given quadrics.

(7) Let this system of polar reciprocal elements be taken as co-ordinate elements. The equations of the quadrics become

$$\gamma_{11}\xi_1^2 + \gamma_{22}\xi_2^2 + \ldots + \gamma_{\nu\nu}\xi_\nu^2 = 0,$$

and

$$\gamma_{11}'\xi_1^2 + \gamma_{22}'\xi_2^2 + \ldots + \gamma_{\nu\nu}'\xi_\nu^2 = 0.$$

And the ratios γ_{11}/γ_{11}', γ_{22}/γ_{22}', etc. are the roots, with their signs changed, given by the above equation [cf. subsection (4)] of the νth degree determining λ.

(8) One means of making the properties of conical quadrics more evident is to take the vertex as one of the co-ordinate elements of the complete region. Let e_1 be the vertex of the quadric. Then if x be any element on the quadric, by hypothesis $\theta e_1 + x$ is on the quadric, θ being arbitrary. Hence if $x = \Sigma\xi e$, the element

$$(\theta + \xi_1) e_1 + \xi_2 e_2 + \ldots + \xi_\nu e_\nu$$

is also on the quadric. It follows that ξ_1 cannot occur at all in the equation of the quadric. Accordingly the expression $(a \between x)^2$ reduces to

$$a_{22}\xi_2^2 + 2a_{23}\xi_2\xi_3 + \ldots.$$

84. Reciprocal Equations and Conical Quadrics. (1) When the quadric $(a \between x)^2 = 0$ is conical, the reciprocal equation of the quadric, namely, $(\mathbf{a} \between L)^2 = 0$, has peculiar properties.

It has been proved that if b be the vertex, then

$$(a \between x \between b) = 0,$$

whatever element x may be. Hence the polar of any element x passes through the vertex b.

Let L be the plane $\lambda_1\xi_1 + \lambda_2\xi_2 + \ldots + \lambda_\nu\xi_\nu = 0$.

Then it is proved in Salmon's *Higher Algebra*, Lesson V. and elsewhere that, when $\Delta = 0$,

$$(\alpha \mathbin{\S} L)^2 \equiv \frac{1}{\alpha_{11}} (\alpha_{11}\lambda_1 + \alpha_{12}\lambda_2 + \ldots + \alpha_{1\nu}\lambda_\nu)^2.$$

But we may write by § 83 (3)

$$b = \alpha_{11}e_1 + \alpha_{12}e_2 + \ldots + \alpha_{1\nu}e_\nu.$$

Hence, $(\alpha \mathbin{\S} L)^2 = 0$, reduces to, $\Sigma\beta\lambda = 0$; that is to the condition that L pass through the vertex. But this is the property of all polars and not merely of tangential polars. Thus in this particular case of conical quadrics the reciprocal equation to the ordinary point-equation, deduced as in the general case, merely defines the vertex of the quadric.

(2) In order to find the nature of the condition which L must satisfy in order to be a tangential polar of the conical quadric, suppose that e_1 has been chosen to be the vertex of the quadric.

Then $$(\alpha \mathbin{\S} x)^2 \equiv \alpha_{22}\xi_2^2 + 2\alpha_{23}\xi_2\xi_3 + \ldots = 0$$
is the equation of the quadric.

And $$(\alpha \mathbin{\S} x \mathbin{\S} x') \equiv \xi_2(\alpha_{22}\xi_2' + \alpha_{23}\xi_3' + \ldots) + \xi_3(\alpha_{23}\xi_2' + \alpha_{33}\xi_3' + \ldots) + \text{etc.}$$

Hence, as before, the conditions that L, which is the locus defined by

$$\lambda_1\xi_1 + \lambda_2\xi_2 + \ldots + \lambda_\nu\xi_\nu = 0,$$

should touch the quadric are $\lambda_1 = 0$, and

$$\alpha_{22}'\lambda_2^2 + 2\alpha_{23}'\lambda_2\lambda_3 + \ldots = 0;$$

where α_{22}', α_{23}', ... are the minors of α_{22}, α_{23}, ... in the determinant

$$\begin{vmatrix} \alpha_{22}, & \alpha_{23}, & \ldots & \alpha_{2\nu} \\ \alpha_{23}, & \alpha_{33}, & \ldots & \alpha_{3\nu} \\ \cdots & \cdots & \cdots & \cdots \\ \alpha_{2\nu}, & \alpha_{3\nu}, & \ldots & \alpha_{\nu\nu} \end{vmatrix}.$$

Let this determinant be called Δ'.

The first condition, $\lambda_1 = 0$, is the condition given by the ordinary reciprocal equation, namely that the polar should pass through the vertex.

Let the second condition be called the conical reciprocal equation.

(3) Now if we transform to any co-ordinate elements whatever, so that any point b is the vertex, these equations become, $\Sigma\beta\lambda = 0$, which is the condition that b should be the vertex; and $(\alpha \mathbin{\S} L)^2 = 0$, where the coefficients α_{11}, α_{12}, ... satisfy the condition

$$\begin{vmatrix} \alpha_{11}, & \alpha_{12}, & \ldots & \alpha_{1\nu}, \\ \alpha_{12}, & \alpha_{22}, & \ldots & \alpha_{2\nu}, \\ \cdots & \cdots & \cdots & \cdots \\ \alpha_{1\nu}, & \alpha_{2\nu}, & \ldots & \alpha_{\nu\nu} \end{vmatrix} = 0.$$

In accordance with the notation explained § 78 (6) this determinant will be called Δ_1.

(4) Suppose now that we are simply given the equation,

$$(\alpha \between L)^2 = 0.$$

We have to determine what it is to be conceived as denoting when the above determinant vanishes.

If we had the two equations

$$(\alpha \between L)^2 = 0, \text{ and } \Sigma\beta\lambda = 0,$$

a conical quadric of vertex b would be determined. Hence it is possible to conceive $(\alpha \between L)^2 = 0$ as denoting a conical quadric with an undetermined vertex. This, however, is not satisfactory.

Let the reciprocal point-equation be formed. It follows from the previous investigation that this equation is

$$({}_1\alpha_{11}\xi_1 + {}_1\alpha_{12}\xi_2 + \ldots)^2 = 0;$$

where ${}_1\alpha_{11}$, ${}_1\alpha_{12}$, ... are the minors of α_{11}, α_{12}, α_{13}, ... in the determinant Δ_1. This locus is two coincident planes forming a quadric.

If we choose $\nu - 1$ co-ordinate elements in this region and any co-ordinate element e_1 outside it, then $(\alpha \between L)^2 = 0$ takes the form

$$\gamma_{22}\lambda_2^2 + 2\gamma_{23}\lambda_2\lambda_3 + \ldots = 0.$$

Hence it is best to consider, $(\alpha \between L)^2 = 0$, as denoting in the reciprocal point-form the region,

$${}_1\alpha_{11}\xi_1 + {}_1\alpha_{12}\xi_2 + \ldots = 0,$$

taken twice over, and as denoting in the original plane-form a quadric surface of $\nu - 3$ dimensions lying entirely within this region.

(5) If any vertex b be assumed and all the one dimensional regions joining b to elements of this quadric of $\nu - 3$ dimensions be drawn, then a conical quadric of $\nu - 2$ dimensions is obtained.

The vertex b should not be in the region,

$${}_1\alpha_{11}\xi_1 + {}_1\alpha_{12}\xi_2 + \text{etc.} = 0,$$

which contains the quadric surface of $\nu - 3$ dimensions, if a true conical quadric is to be obtained. Hence we may call the region,

$${}_1\alpha_{11}\xi_1 + {}_1\alpha_{12}\xi_2 + \ldots = 0,$$

the 'non-vertical region'. Also call the quadric of $\nu - 3$ dimensions lying in it the 'contained quadric'.

(6) Accordingly, summing up, given the equation, $(\alpha \between L)^2 = 0$, where $\Delta_1 = 0$, we derive the reciprocal equation

$${}_1\alpha_{11}\xi_1 + {}_1\alpha_{12}\xi_2 + \text{etc.} = 0.$$

These two equations taken together represent a non-vertical region and a contained quadric. This may be considered as the degenerate form of a quadric defined by either of its two reciprocal equations.

Accordingly the equation, $(\alpha \between L)^2 = 0$ (when $\Delta_1 = 0$), gives the condition to be satisfied by the co-ordinates of all regions of $\nu - 2$ dimensions whose intersection with the non-vertical region is a tangential polar of the contained quadric.

(7) Let us consider as a special case of the above investigations Geometry of two dimensions. Let e_1, e_2, e_3 be the co-ordinate points, and E_1, E_2, E_3 the corresponding symbols denoting the straight lines e_2e_3, e_3e_1, e_1e_2.

Then any point can be denoted by

$$x = \xi_1 e_1 + \xi_2 e_2 + \xi_3 e_3,$$

and any straight line L by the equation

$$\lambda_1 \xi_1 + \lambda_2 \xi_2 + \lambda_3 \xi_3 = 0.$$

Also the equation,

$$\lambda_1 \xi_1 + \lambda_2 \xi_2 + \lambda_3 \xi_3 = 0,$$

denotes either that x lies on L or that L passes through x.

Any quadric $(a \between x)^2 = 0$ is a conic; the determinant Δ is

$$a_{11} a_{22} a_{33} + 2a_{12} a_{23} a_{31} - a_{11} a_{23}^2 - a_{22} a_{31}^2 - a_{33} a_{12}^2.$$

A conical quadric, for which $\Delta = 0$, is two straight lines. The reciprocal equation is the tangential equation.

Conversely given the tangential equation $(\alpha \between L)^2 = 0$, the point-equation is formed from it by the same law. Also if $\Delta_1 = 0$, then

$$(\alpha \between L)^2 = 0$$

splits up into two factors.

In this case let

$$(\alpha \between L)^2 \equiv (\varpi_1 \lambda_1 + \varpi_2 \lambda_2 + \varpi_3 \lambda_3)(\rho_1 \lambda_1 + \rho_2 \lambda_2 + \rho_3 \lambda_3).$$

Let p be the point

$$\varpi_1 e_1 + \varpi_2 e_2 + \varpi_3 e_3,$$

and let r be the point

$$\rho_1 e_1 + \rho_2 e_2 + \rho_3 e_3.$$

Then $(\alpha \between L)^2 = 0$ is the condition to be satisfied by all lines which pass through either p or r.

Also $\quad \alpha_{11} = \varpi_1 \rho_1$, etc.; and $\alpha_{23} = \frac{1}{2}(\rho_2 \varpi_3 + \rho_3 \varpi_2)$, etc.

Hence $\quad {}_1\alpha_{11} = \alpha_{22}\alpha_{33} - \alpha_{23}^2 = -\frac{1}{4}(\rho_2 \varpi_3 - \rho_3 \varpi_2)^2,$

and $\quad {}_1\alpha_{12} = \alpha_{23}\alpha_{31} - \alpha_{33}\alpha_{12} = -\frac{1}{4}(\rho_2 \varpi_3 - \rho_3 \varpi_2)(\rho_3 \varpi_1 - \rho_1 \varpi_3),$

and $\quad {}_1\alpha_{13} = -\frac{1}{4}(\rho_2 \varpi_3 - \rho_3 \varpi_2)(\rho_1 \varpi_2 - \rho_2 \varpi_1).$

Accordingly the non-vertical region is the straight line

$$(\rho_2\varpi_3 - \rho_3\varpi_2)\,\xi_1 + (\rho_3\varpi_1 - \rho_1\varpi_3)\,\xi_2 + (\rho_1\varpi_2 - \rho_2\varpi_1)\,\xi_3 = 0.$$

This is the straight line joining the points p and q. Thus the non-vertical region is a straight line, and the contained quadric is two points in it.

This agrees with Cayley's statements in his 'Sixth Memoir on Quantics*,' respecting conics in two dimensions.

NOTE. For further information in regard to what is known of the projective geometry of many dimensions, cf. Veronese's treatise, *Fondamenti di geometria* (Padova, 1891), translated into German under the title, *Grundzüge der Geometrie von mehreren Dimensionen und mehreren Arten gradliniger Einheiten in elementarer Form entwickelt* (Leipzig, 1894).

* Cf. *Phil. Trans.* 1859 and *Collected Mathematical Papers*, vol. II., no. 158.

CHAPTER IV.

INTENSITY.

85. DEFINING EQUATION OF INTENSITY. (1) Let a complete region of $\nu - 1$ dimensions be defined by the units $e_1, e_2, \ldots e_\nu$. Then by hypothesis the intensity of the element represented by $\alpha_\rho e_\rho$ is α_ρ, since the intensity of e_ρ is by definition unity. But no principle has as yet been laid down whereby the intensity of a derived element a can be determined; where

$$a = \alpha_1 e_1 + \alpha_2 e_2 + \ldots + \alpha_\nu e_\nu = \Sigma \alpha e, \text{ say.}$$

(2) Let α be the intensity, then we assume that α is some function of

$$\alpha_1, \alpha_2, \ldots \alpha_\nu.$$

Hence we can write

$$\alpha = f(\alpha_1, \alpha_2, \ldots \alpha_\nu).$$

Now the intensity of μa is $\mu\alpha$; therefore we have the condition

$$f(\mu\alpha_1, \mu\alpha_2, \ldots \mu\alpha_\nu) = \mu f(\alpha_1, \alpha_2, \ldots \alpha_\nu).$$

Accordingly $f(\alpha_1, \alpha_2, \ldots \alpha_\nu)$ must be a homogeneous function of the first degree.

(3) If a be at unit intensity, then the coefficients must satisfy the equation

$$f(\alpha_1, \alpha_2, \ldots \alpha_\nu) = 1 \text{..............................(A).}$$

This equation will be called 'the defining equation'; since it defines the unit intensities of elements of the region. It will be noticed that the equation does not in any way limit the ratios

$$\alpha_2/\alpha_1, \alpha_3/\alpha_1 \ldots \alpha_\nu/\alpha_1$$

which determine the character, or position, of the element represented by a. Furthermore this equation essentially refers to the ν special units $e_1, e_2, \ldots e_\nu$ which have been chosen as defining (or co-ordinate) units of the region.

(4) Let $\phi_\lambda(\alpha_1, \ldots \alpha_\nu)$ denote a rational integral homogeneous function of the λth degree, and $\phi_\mu(\alpha_1, \ldots \alpha_\nu)$ a rational integral homogeneous function of the μth degree. Then the most general algebraic form of $f(\alpha_1, \ldots \alpha_\nu)$ is

$$\Sigma \left\{\frac{\phi_\lambda(\alpha_1 \ldots \alpha_\nu)}{\phi_\mu(\alpha_1 \ldots \alpha_\nu)}\right\}^{\frac{1}{\lambda-\mu}}.$$

If there be only one term in this expression, then equation (A) can be written in the form

$$\phi_\lambda(\alpha_1, \alpha_2, \dots \alpha_\nu) = \phi_\mu(\alpha_1, \alpha_2, \dots \alpha_\nu) \dots\dots\dots\dots\dots\dots (A').$$

(5) Let $c_1 c_2 \dots c_\nu$ be any other group of independent letters which can be chosen as co-ordinates of the region. Let

$$a = \gamma_1 c_1 + \gamma_2 c_2 + \dots + \gamma_\nu c_\nu = \Sigma\gamma c.$$

Then the γ's are homogeneous linear functions of the α's.

Hence the defining equation (A′) with reference to any other co-ordinate elements becomes

$$\psi_\lambda(\gamma_1, \gamma_2, \dots \gamma_\nu) = \psi_\mu(\gamma_1, \gamma_2, \dots \gamma_\nu);$$

where ψ_λ and ψ_μ are homogeneous functions of the λth and μth degrees respectively.

(6) Again, when $\alpha_2, \alpha_3, \dots \alpha_\nu$ all simultaneously vanish, the intensity is unity when $\alpha_1 = 1$. But in this case equation (A′) reduces to an equation of the form

$$\xi_1 \alpha_1^{\lambda} = \eta_1 \alpha_1^{\mu}.$$

Therefore for this to be satisfied by $\alpha_1 = 1$, we must have $\xi_1 = \eta_1$. So the coefficients of the highest powers of $\alpha_1, \alpha_2, \dots \alpha_\nu$ on the two sides of the equation are respectively equal.

(7) In the subsequent work, unless otherwise stated, we will assume the defining equation to be of the form

$$\{\phi_\mu(\alpha_1, \alpha_2, \dots \alpha_\nu)\}^{\frac{1}{\mu}} = 1;$$

where $\phi_\mu(\alpha_1, \alpha_2, \dots \alpha_\nu)$ is a rational integral homogeneous function of the μth degree of the form

$$\alpha_1^{\mu} + \alpha_2^{\mu} + \dots + \alpha_\nu^{\mu} + \Sigma\rho\alpha_1^{\sigma_1}\alpha_2^{\sigma_2}\dots\alpha_\nu^{\sigma_\nu},$$

ρ being an arbitrary coefficient and

$$\sigma_1 + \sigma_2 + \dots + \sigma_\nu = \mu.$$

Raising each side of the defining equation to the μth power it becomes,

$$\phi_\mu(\alpha_1, \alpha_2, \dots \alpha_\nu) = 1.$$

Coefficients which satisfy the defining equation will be called the co-ordinates of an element. They define the element at unit intensity. The co-ordinates of an element must be distinguished from the co-ordinate elements of a region, which have been defined before [cf. § 64 (8)].

86. Locus of Zero Intensity. (1) It is obvious that there is one locus of $\nu - 2$ dimensions with exceptional properties in regard to the intensities of its elements. For the equation

$$\phi_\mu(\alpha_1, \dots \alpha_\nu) = 0,$$

is a relation between the ratios $\alpha_2/\alpha_1, \alpha_3/\alpha_1, \dots \alpha_\nu/\alpha_1$; and it, therefore, determines a locus which is such that all the elements of it are necessarily at zero intensity, according to this mode of defining the intensity, and yet do not themselves vanish, since the coefficients of the co-ordinate extraordinaries do not vanish separately. Therefore in relation to the given definition of unit intensity, elements of this locus are all at zero intensity.

(2) Some other law of intensity is necessarily required in the locus of zero intensity, at least in idea as a possibility in order to prevent the introduction of fallacious reasoning. For if two terms a and a' represent the same element at the same intensity, then $a = a'$, and the coefficients of the co-ordinate elements in a and a' are respectively equal in pairs. But if the element be in the locus of zero intensity a and ρa are both zero intensity according to the old definition. Hence from the above argument $a = \rho a$, and therefore $(\rho - 1) a = 0$; and since $\rho - 1$ is not zero, $a = 0$, which is untrue. Therefore in the locus of zero intensity in order to preserve generality of expression some other definition of intensity is to be substituted, at least in idea if not actually formulated. An analogy to this property of points in the locus of zero intensity is found in the fact that two zero forces at infinity are not therefore identical in effects, and that for such forces another definition of intensity is substituted, namely the moment of the force about any point, or in other words the moment of the couple.

(3) If the properties of the region with respect to the intensity are to be assumed to be continuous, at any point of the locus of zero intensity one or more of the co-ordinates must be infinite.

For the equation

$$\phi_\mu (\alpha_1, \dots \alpha_\nu) = 1,$$

viewed as an equation connecting the absolute magnitudes of the co-ordinates, can only be satisfied simultaneously with

$$\phi_\mu (\alpha_1, \dots \alpha_\nu) = 0,$$

viewed as an equation connecting the ratios of the co-ordinates if one or more of the α's are infinite. In this case we can write the first equation in the form

$$\phi_\mu \left(\frac{\alpha_1}{\alpha_\rho}, \dots \frac{\alpha_\nu}{\alpha_\rho}\right) = \frac{1}{\alpha_\rho}.$$

Then if α_ρ becomes infinite, the equation,

$$\phi_\mu \left(\frac{\alpha_1}{\alpha_\rho}, \dots \frac{\alpha_\nu}{\alpha_\rho}\right) = 0,$$

between the ratios of the co-ordinates is simultaneously satisfied.

87. Plane Locus of Zero Intensity. (1) There are two special cases of great importance, one when the locus of zero intensity is plane, the other when it is a quadric.

Considering the case of a plane locus, by a proper choice of the unit intensities of the co-ordinate elements of the complete region the equation of the locus can be written in the form

$$\xi_1 + \xi_2 + \dots + \xi_\nu = 0.$$

(2) Let $e_1, e_2, \dots e_\nu$ be these co-ordinate elements, and let $a_1, a_2, \dots a_\nu$ be another set of independent elements at unit intensity to be used as a new set of co-ordinate elements.

Let
$$\begin{aligned} a_1 &= \alpha_{11} e_1 + \alpha_{12} e_2 + \dots + \alpha_{1\nu} e_\nu, \\ a_2 &= \alpha_{21} e_1 + \alpha_{22} e_2 + \dots + \alpha_{2\nu} e_\nu, \\ &\dots\dots\dots\dots\dots\dots\dots\dots \\ a_\nu &= \alpha_{\nu 1} e_1 + \alpha_{\nu 2} e_2 + \dots + \alpha_{\nu\nu} e_\nu. \end{aligned}$$

Then by hypothesis $\alpha_{11} + \alpha_{12} + \dots + \alpha_{1\nu} = 1,$
with $\nu - 1$ other equations of the same type.

Let any element x at unit intensity be given by $\Sigma\xi e$ and also by $\Sigma\eta a$. Then
$$\xi_1 + \xi_2 + \dots + \xi_\nu = 1.$$
But by comparison $\xi_1 = \alpha_{11}\eta_1 + \alpha_{21}\eta_2 + \dots + \alpha_{\nu 1}\eta_\nu,$
with $\nu - 1$ other equations of the same type.

Hence substituting for the ξ's in the defining equation, we get

$$(\alpha_{11} + \alpha_{12} + \dots + \alpha_{1\nu})\,\eta_1 + (\alpha_{21} + \alpha_{22} + \dots + \alpha_{2\nu})\,\eta_2 + \dots = 1.$$

Therefore using the defining equations for the α's, there results

$$\eta_1 + \eta_2 + \dots + \eta_\nu = 1,$$

as the defining equation for the new co-ordinates. It follows that if this special type of defining equation of the first degree hold for one set of co-ordinate elements it holds for all sets of co-ordinate elements.

(3) Any one-dimensional region meets the locus of zero intensity in one element only, unless it lies wholly in the locus. Let a and b be two elements at unit intensity defining a one-dimensional region. Then by subsection (2) the intensity of any element $\xi a + \eta b$ in the region ab is $\xi + \eta$. Hence $b - a$ is of zero intensity. Let $b - a = u$; then u is the only element in ab at zero intensity according to the original law of intensity, but possessing a finite intensity according to some new definition.

(4) If $\rho - 1$ of the co-ordinate elements, where $\rho \leqq \nu$, be assumed in the region of zero intensity and the remaining $\nu - \rho + 1$ outside that region, then the defining equation takes a peculiar form. For let $u_1, u_2, \dots u_{\rho-1}$ be the co-ordinate elements in the region of zero intensity, and $e_\rho, e_{\rho+1}, \dots e_\nu$ the remaining co-ordinate elements. Then any element can be expressed in the form

$$\Sigma\lambda u + \Sigma\xi e.$$

Now any element of the form $\Sigma\lambda u$ lies in the region of zero intensity. Hence the defining equation must take the form

$$\xi_\rho + \xi_{\rho+1} + \dots + \xi_\nu = 1.$$

If $\rho = \nu$, then the co-ordinate elements $u_1, \dots u_{\nu-1}$ completely define the region of zero intensity.

Let e denote the remaining co-ordinate element. Any element can be written $\Sigma\lambda u + \xi e$. The defining equation becomes, $\xi = 1$.

88. Quadric Locus of Zero Intensity. (1) Let the intensity of the point x be $+\{(\alpha \between x)^2\}^{\frac{1}{2}}$.

Then the locus of zero intensity is the quadric surface $(\alpha \between x)^2 = 0$.

(2) Let us assume this quadric to be closed, or imaginary with real coefficients [cf. § 82 (4)]. If x lie within this quadric and y lie without it (the quadric being real and closed), then by § 82 (11)

$$(\alpha \between x)^2 \text{ and } (\alpha \between y)^2$$

are of opposite sign. Suppose for example that $(\alpha \between x)^2$ is positive for elements within the quadric. And let

$$(\alpha \between x)^2 = \mu^2 \text{ and } (\alpha \between y)^2 = -\nu^2;$$

where μ and ν are by hypothesis real, since the co-ordinates of x and y are real. Then the intensities of x and y as denoted by the symbols x and y are μ and $\sqrt{(-\nu^2)}$ respectively.

(3) The symbols which denote these points at unit intensity are x/μ, and $y/\sqrt{(-\nu^2)}$. Hence although the element y is defined by real ratios, its co-ordinates at unit intensity are imaginaries of the form $i\eta_1, i\eta_2, \dots$, where $\eta_1, \eta_2 \dots$ are real.

Such elements will be called '*intensively imaginary elements.*' If the element be defined by real co-ordinates, its intensity is imaginary. Those elements such that real co-ordinates define a real intensity will be called '*intensively real elements.*'

(4) If intensively real elements lie without the quadric of zero intensity, then intensively imaginary elements lie within it, and conversely. It is to be noted that both sets of elements are real in the sense that the ratios of their co-ordinates are real.

(5) If the quadric of zero intensity be imaginary, then all real elements are intensively real.

89. Antipodal elements and opposite intensities. (1) Since

$$(\alpha \between x)^2 = (\alpha \between -x)^2,$$

the intensities of x and $-x$ are both positive and equal, when the locus of zero intensity is a quadric. An exception, therefore, arises to the law that if $\Sigma\xi e$ and $\Sigma\xi' e$ denote the same element at the same intensity, then

$$\xi_1 = \xi_1', \ \xi_2 = \xi_2', \text{ etc.}$$

Let the generality of this law be saved by considering the intensities denoted by x and $-x$, though numerically the same, to differ by another quality which we will call oppositeness.

(2) Another method of evading this exception to the general law is to regard x and $-x$ as two different elements at the same intensity. This is really a special case of the supposition alluded to in § 65 (9). Let x and $-x$ be called *antipodal* elements.

In this method the quality of oppositeness has been assigned to the intrinsic nature of the element denoted, whereas in the first method it was assigned to the intensity. When the quadric locus of zero intensity is real and closed, the first method is most convenient; when it is imaginary, either method can be chosen.

(3) Antipodal elements have special properties.

If any locus include an element, it also includes its antipodal element.

If two one-dimensional regions intersect, they also intersect in the antipodal element. Hence two one-dimensional regions, if they intersect, intersect in two antipodal points.

A one-dimensional region meets a quadric in four points, real or imaginary, namely in two pairs of antipodal points.

(4) The sign of congruence, namely $\equiv$ [cf. § 64 (2)], connects symbols representing antipodal points as well as symbols representing the same point.

90. THE INTERCEPT BETWEEN TWO ELEMENTS. (1) The one-dimensional region which includes e_1 and e_2 may be conceived as divided by the elements $e_1 e_2$ into two or more intercepts. For the element $\xi_1 e_1 + \xi_2 e_2$ may be conceived as traversing one real portion of the region from e_1 to e_2, if it takes all positions expressed by the continuous variation of ξ_2/ξ_1 from 0 to $+\infty$. Similarly it may travel from e_1 to e_2 through the remaining real portion of the region by assuming all the positions expressed by the continuous variation of ξ_2/ξ_1 from 0 to $-\infty$.

A one-dimensional region may, therefore, be considered as unbounded and as returning into itself [cf. § 65 (9)].

(2) Assume that the expression for the intensity is linear and of the form $\Sigma\xi$. Then the locus of zero intensity cuts the region $e_1 e_2$ at the element defined by $\xi_2/\xi_1 = -1$.

Hence as ξ_2/ξ_1 varies continuously from 0 to $+\infty$, the element x does not pass through the locus of zero intensity, and its intensity cannot change sign, if ξ_2 and ξ_1 do not change sign.

Let this portion of the region be called the intercept between e_1 and e_2. Also let the other portion of the region be called external to the portion limited by e_1 and e_2, which is the intercept.

(3) An element on the intercept between e_1 and e_2 will be said to *lie between* e_1 and e_2.

Also the external portion of the region is divided into two parts by the element of zero intensity, $e_2 - e_1$. Let the continuous portion bounded by e_1 and $e_2 - e_1$ and not including e_2 be called the portion *beyond* e_1, and let the portion bounded by e_2 and $e_2 - e_1$ and not including e_1 be called the portion beyond e_2.

(4) Assume the intensity to be $\{(\alpha \between x)^2\}^{\frac{1}{2}}$.

Let the locus of zero intensity be the real closed quadric, $(\alpha \between x)^2 = 0$.

Firstly, assume that $\pm x$ denote the same element at opposite intensities.

Let the two elements e_1 and e_2 both belong to the intensively real part of the region. Then x moves from e_1 to e_2, as ξ_2/ξ_1 varies from 0 to ∞ or from 0 to $-\infty$. Now [cf. § 82 (9)] since $(\alpha \between e_1)^2$ and $(\alpha \between e_2)^2$ are both of the same sign, as x moves from e_1 to e_2 by either route it must either cut the surface of zero intensity twice or not at all. Call the latter route the intercept between e_1 and e_2. The intercept only contains intensively real elements.

(5) If the quadric $(\alpha \between x)^2 = 0$ be imaginary, then the one-dimensional region e_1e_2 does not cut it at all in real points. Hence there is no fundamental distinction between the two routes from e_1 to e_2, and both of them may be called intercepts between e_1 and e_2. Also all real elements are intensively real.

Hence a one-dimensional region is to be conceived as a closed region, such that two elements e_1e_2 divide it into two parts.

(6) Secondly, assume that $\pm x$ denote two antipodal elements.

Assume that the quadric $(\alpha \between x)^2 = 0$ is entirely imaginary. The two routes from e_1 to e_2 are discriminated by the fact that the one contains both antipodal points $-e_1$ and $-e_2$, and the other contains neither. Let the latter portion of the region be called the intercept, and the former portion the antipodal intercept.

The case when $(\alpha \between x)^2 = 0$ is real is of no practical importance, and need not be discussed.

NOTE. Grassmann does not consider the general question of the comparison of intensities. In the *Ausdehnungslehre von* 1844, 2nd Part, Chapter I., §§ 94—100, he assumes in effect a linear defining equation without considering any other possibility. In the *Ausdehnungslehre von* 1862 no general discussion of the subject is given; but in Chapter V., 'Applications to Geometry,' a linear defining equation for points is in effect assumed, and a quadric defining equation for vectors—assumptions which are obvious and necessary in Euclidean Geometry. It should also be mentioned that the general idea of a defining equation, different for different manifolds, and the idea of a locus of zero intensity do not occur in either of these works. Also v. Helmholtz in his *Handbuch der Physiologischen Optik*, § 20, pp. 327 to 330 (2nd Edition) apparently assumes that only a linear defining equation is possible.

BOOK IV.

CALCULUS OF EXTENSION.

CHAPTER I.

Combinatorial Multiplication.

91. Introductory. (1) The preceding book has developed the general theory of addition for algebras of the numerical genus (cf. § 22). The first special algebra to be discussed is Grassmann's Calculus of Extension*. This algebra requires for its interpretation a complete algebraic system of manifolds (cf. § 20). The manifold of the first order is a positional manifold of $\nu - 1$ dimensions, where ν is any assigned integer; the successive manifolds of the second and higher orders are also positional manifolds (cf. § 22); the manifold of the νth order reduces to a single element; the manifold of the $(\nu+1)$th order is identical with that of the first order. Hence (cf. § 20), when the manifold of the first order is of $\nu - 1$ dimensions, the algebra is of the νth species.

(2) It follows from the general equation for multiplication of algebras of the numerical genus given in § 22, that if two points $a\,(=\Sigma\alpha e)$ and $b\,(=\Sigma\beta e)$ be multiplied together, where $e_1, e_2 \dots e_\nu$ are any ν reference points, then

$$ab = \Sigma\alpha e\,.\,\Sigma\beta e = \Sigma\Sigma\,(\alpha_\rho\beta_\sigma e_\rho e_\sigma);$$

where $\begin{cases}\rho\\ \sigma\end{cases} = 1, 2 \dots \nu$, and $\alpha_\rho\beta_\sigma$ are multiplied together according to the rules of ordinary algebra.

(3) Thus the products two together of the reference elements $e_1, e_2 \dots e_\nu$ yield ν^2 new elements of the form (e_1e_1), (e_2e_2), (e_1e_2), (e_2e_1), etc. These ν^2 elements (which may not all be independent) are conceived as defining a fresh positional manifold of $\nu^2 - 1$ dimensions at most, and ab is an element of this manifold. This is the most general conception possible of a relation between any two elements of a positional manifold which may be symbolized by a multiplication.

(4) No necessary connection exists between the symbols (e_1e_1), (e_1e_2), (e_2e_1), (e_2e_2), etc.: they may therefore, as far as the logic of the formal symbolism is concerned, be conceived as given independent reference elements

* Cf. *Die Ausdehnungslehre von* 1844, and *Die Ausdehnungslehre von* 1862, both by H. Grassmann.

of a new positional manifold. But on the other hand we are equally at liberty to assume that some addition equations exist between these ν^2 products, whereby the number of them, which can be assumed as forming a complete set of independent elements, is reduced. These products of elements are then interpreted as symbolizing relations between the elements of the manifold of the first order which form the factors; and thus the manifolds of orders higher than the first represent properties of the manifold of the first order which it possesses in addition to its properties as a positional manifold. Let any addition equations which exist between products of the reference elements $e_1, e_2 \dots e_\nu$ be called 'equations of condition' of that type of multiplication which is under consideration.

92. Invariant Equations of Condition. (1) The equations of condition will be called invariant, when the same equations of condition hold whatever set of ν independent reference elements be chosen in the manifold of the first order*.

(2) For products of two elements of the first order, there are only two types of multiplication with invariant equations of condition, namely that type for which the equations of condition are of the form

$$(e_\rho e_\sigma) + (e_\sigma e_\rho) = 0, \quad (e_\rho e_\rho) = 0 \text{ ....(1);}$$

and that type for which the equations of condition are of the form

$$(e_\rho e_\sigma) = (e_\sigma e_\rho) \text{ ....(2).}$$

For assume an equation of condition of the most general form possible, namely

$$\alpha_{11}(e_1e_1) + \alpha_{12}(e_1e_2) + \alpha_{21}(e_2e_1) + \dots = 0 \text{ ....(3).}$$

Then if $x_1, x_2 \dots x_\nu$ be any ν independent elements, this equation (3) is to persist unchanged when $x_1, x_2 \dots x_\nu$ are respectively substituted for $e_1, e_2 \dots e_\nu$.

Thus in equation (3) change e_1 into ξe_1, where ξ is any arbitrary number, not unity. Subtract equation (3) from this modified form, and divide by $\xi - 1$, which by hypothesis is not zero. Then

$$(\xi + 1)\,\alpha_{11}(e_1e_1) + \sum_\rho \{\alpha_{1\rho}(e_1e_\rho) + \alpha_{\rho 1}(e_\rho e_1)\} = 0.$$

Hence since ξ is arbitrary,

$$\alpha_{11}(e_1e_1) = 0, \; \sum_{\rho=2}^{\nu} \{\alpha_{1\rho}(e_1e_\rho) + \alpha_{\rho 1}(e_\rho e_1)\} = 0 \text{ ....(4).}$$

Therefore by hypothesis these forms are to be invariant equations of condition. Hence the second of equations (4) must still hold when ξe_2 is substituted

* The type of multiplication is then called by Grassmann (cf. *Ausdehnungslehre von* 1862, § 50) 'linear.' But this nomenclature clashes with the generally accepted meaning of a 'linear algebra' as defined by B. Peirce in his paper on Linear Associative Algebra, *American Journal of Mathematics*, vol. IV. (1881). The theorem of subsection (2) is due to Grassmann, cf. *loc. cit.*

for e_2, ξ being any number not unity. Thus, as before, by subtraction and division by $\xi - 1$,

$$\alpha_{12}(e_1e_2) + \alpha_{21}(e_2e_1) = 0.$$

Since this equation is invariant, it must hold when e_1 and e_2 are interchanged, thus by subtraction,

$$(\alpha_{12} - \alpha_{21})\{(e_1e_2) - (e_2e_1)\} = 0.$$

(3) Firstly assume, $\alpha_{12} = \alpha_{21}$. Then, if e_1 and e_2 are any two of the reference elements,

$$(e_1e_2) + (e_2e_1) = 0 \qquad \text{.................................(5).}$$

Now since this equation must be invariant, put $e_2 + \xi e_1$ for e_2, where ξ is any number not unity; then by subtraction we find the typical form

$$(e_1e_1) = 0 \qquad \text{..................................(6),}$$

and this satisfies the first of equations (4).

It is evident and is formally proved in § 93 (3) that equations of condition, of which equations (5) and (6) are typical forms, actually are invariant.

(4) Secondly assume, $(e_1e_2) = (e_2e_1)$, as the typical form of equation of condition. Then it is immediately evident that $(xy) = (yx)$, where x stands for $\Sigma\xi e$ and y for $\Sigma\eta e$. Thus this form of equation of condition is invariant. Also substituting x instead of e_1 in the first of equations (4), which is invariant, it is obvious that $\alpha_{11} = 0$. Hence this equation introduces no further equation of condition.

Thus there are only two types of multiplication of two elements of the manifold of the first order which have invariant equations of condition.

93. Principles of Combinatorial Multiplication. (1) Let the multiplication be called 'combinatorial' when the following relations hold:

$$\left.\begin{array}{c}(e_\rho e_\sigma) + (e_\sigma e_\rho) = 0 \\ (e_1e_1) = (e_2e_2) = \ldots = (e_\nu e_\nu) = 0\end{array}\right\} \text{.........................(1),}$$

$$(e_1e_2 \ldots e_\rho)(e_{\rho+1}e_{\rho+2} \ldots e_\sigma) = (e_1e_2 \ldots e_\rho e_{\rho+1} \ldots e_\sigma) \text{..............(2).}$$

(2) The second of equations (1) follows from the first, if the first equation be understood to hold in the case when $\rho = \sigma$. For then $(e_\rho e_\sigma) = (e_{\rho\rho})$, and therefore $(e_\rho e_\sigma) + (e_\sigma e_\rho) = 2(e_\rho e_\rho) = 0$.

(3) Equations (1) and (2) as they stand apply to one given set of independent elements, $e_1, e_2 \ldots e_\nu$. Now if $a = \Sigma\alpha e$ and $b = \Sigma\beta e$, then the product ab takes the form $\Sigma(\alpha_\rho\beta_\sigma - \alpha_\sigma\beta_\rho)(e_\rho e_\sigma)$; and the number of independent reference elements of the type $e_\rho e_\sigma$ in the new manifold of the second order created by the products of the reference elements of the first order is $\frac{1}{2}\nu(\nu - 1)$. Similarly the product ba becomes $\Sigma(\alpha_\sigma\beta_\rho - \alpha_\rho\beta_\sigma)(e_\rho e_\sigma)$.

Thus for any two elements a and b of the manifold of the first order equations of the same type as equation (1) hold, namely

$$(ab) + (ba) = 0, \quad (aa) = (bb) = 0.$$

(4) Equation (2) expresses what is known as the associative law of multiplication. It has been defined to hold for products of the ν independent elements $e_1, e_2 \dots e_\nu$. It follows from this law and from equation (1) that

$$e_1e_2 \dots e_\rho e_{\rho+1} \dots e_\sigma = (e_1e_2 \dots e_{\rho-1})(e_\rho e_{\rho+1})(e_{\rho+2} \dots e_\sigma)$$
$$= -(e_1e_2 \dots e_{\rho-1})(e_{\rho+1}e_\rho)(e_{\rho+2} \dots e_\sigma)$$
$$= -(e_1e_2 \dots e_{\rho-1}e_{\rho+1}e_\rho e_{\rho+2} \dots e_\sigma).$$

Accordingly any two adjacent factors may be interchanged, if the sign of the whole product be changed.

By a continually repeated interchange of adjacent factors any two factors can be interchanged if the sign of the whole product be changed.

Again, if the same element appear twice among the factors of a product, the product is null. For by interchanges of factors the product can be written in the form $(e_1e_1 . e_2e_3 \dots)$, where e_1 is the repeated factor. But $(e_1e_1) = 0$. Hence by § 21 $(e_1e_1e_2e_3 \dots) = 0$. It follows from this that in a region of $\nu - 1$ dimensions products of more than ν factors following this combinatorial law are necessarily null, for one factor at least must be repeated.

(5) It remains to extend the associative law and the deductions from it in the previous subsection to products of any elements. In the left-hand side of equation (2) let any element, say e_ρ, be replaced by $e_\rho' = e_\rho + \theta e_\lambda$, where θ is any arbitrary number.

Then by the distributive law of multiplication (cf. § 19),

$$(e_1e_2 \dots e_\rho')(e_{\rho+1}e_{\rho+2} \dots e_\sigma) = (e_1e_2 \dots e_\rho)(e_{\rho+1}e_{\rho+2} \dots e_\sigma) + \theta(e_1e_2 \dots e_\lambda)(e_{\rho+1}e_{\rho+2} \dots e_\sigma)$$
$$= (e_1e_2 \dots e_\rho e_{\rho+1}e_{\rho+2} \dots e_\sigma) + \theta(e_1e_2 \dots e_\lambda e_{\rho+1}e_{\rho+2} \dots e_\sigma)$$
$$= (-1)^{\rho-1}(e_\rho . e_1e_2 \dots e_{\rho+1}e_{\rho+2} \dots e_\sigma) + (-1)^{\rho-1}\theta(e_\lambda . e_1e_2 \dots e_{\rho+1}e_{\rho+2} \dots e_\sigma)$$
$$= (-1)^{\rho-1}(e_\rho' . e_1e_2 \dots e_{\rho+1}e_{\rho+2} \dots e_\sigma).$$

Also by a similar proof

$$e_1e_2 \dots e_\rho' e_{\rho+1}e_{\rho+2} \dots e_\sigma = (-1)^{\rho-1}(e_\rho' . e_1e_2 \dots e_{\rho+1}e_{\rho+2} \dots e_\sigma).$$

Hence $$(e_1e_2 \dots e_\rho')(e_{\rho+1}e_{\rho+2} \dots e_\sigma) = e_1e_2 \dots e_\rho' e_{\rho+1}e_{\rho+2} \dots e_\sigma.$$

Hence the associative law holds when e_ρ has been modified into e_ρ'; and by successive modifications of this type $e_1, e_2 \dots e_\nu$ can be modified into $a_1, a_2 \dots a_\nu$, where $a_1, a_2 \dots a_\nu$ are any independent elements.

(6) The only deduction in subsection (4) requiring further proof to extend it to any product is the last, that in a region of $\nu - 1$ dimensions products of more that ν factors are necessarily null. This theorem is a special instance of the more general theorem, that products of elements which are not independent are necessarily null. For let $a_1, a_2 \dots a_\rho$ be independent, but let $a_{\rho+1} = \alpha_1 a_1 + \dots + \alpha_\rho a_\rho$.

Then $$(a_1a_2 \dots a_\rho a_{\rho+1}) = \alpha_1(a_1a_2 \dots a_\rho a_1) + \dots + \alpha_\rho(a_1a_2 \dots a_\rho a_\rho).$$

Thus every product on the right-hand side has a repeated factor and is therefore null.

(7) Conversely, let it be assumed that a product formed by any number of reference elements is not null, when no reference element is repeated as a factor.

94. Derived Manifolds. (1) There are $\nu!/(\nu-\rho)!\,\rho!$ combinations of the ν independent elements $e_1, e_2 \dots e_\nu$ taken ρ together ($\rho < \nu$). Let the result of multiplying the ρ elements of any one combination together in any arbitrary succession so as to form a product of the ρth order be called a 'multiplicative combination' of the ρth order of the elements $e_1, e_2 \dots e_\nu$.

There are obviously $\nu!/(\nu-\rho)!\,\rho!$ such multiplicative combinations of the ρth order.

(2) It is easy to prove formally that these multiplicative combinations are independent elements of the derived manifold of the ρth order (cf. §§ 20 and 22).

For let $E_1, E_2, \dots E_\sigma, \dots$ be these multiplicative combinations. Assume that

$$\alpha_1 E_1 + \alpha_2 E_2 + \dots + \alpha_\sigma E_\sigma + \dots = 0.$$

Then if E_1 denote the multiplicative combination $(e_1 e_2 \dots e_\rho)$, the $\nu-\rho$ elements $e_{\rho+1}, e_{\rho+2} \dots e_\nu$ do not occur in E_1, and one at least of these elements must occur in each of the other multiplicative combinations.

Now multiply the assumed equation successively by $e_{\rho+1}, e_{\rho+2} \dots e_\nu$, then by § 93 (4) all the terms become null, except the first term.

Accordingly $$\alpha_1 (e_1 e_2 \dots e_\nu) = 0.$$

But $(e_1 e_2 \dots e_\nu)$ is not zero by § 93 (7). So $\alpha_1 = 0$. Similarly $\alpha_2 = 0$, $\alpha_3 = 0$, and so on.

It follows that the sum of different multiplicative combinations cannot itself be a multiplicative combination of the same set of reference elements.

(3) The complete derived manifold of the ρth order is the positional manifold defined by the $\nu!/(\nu-\rho)!\,\rho!$ independent multiplicative combinations of the ρth order formed out of the ν reference elements of the first order. Thus the manifold of the second order is defined by reference elements of the type $(e_\rho e_\sigma)$, of which there are $\frac{1}{2}\nu(\nu-1)$; the manifold of the $(\nu-1)$th order is defined by reference elements of the type $(e_1 e_2 \dots e_{\nu-1})$, of which there are ν; the manifold of the νth order reduces to the single element $(e_1 e_2 \dots e_\nu)$.

(4) The product of any number of independent elements of the first order is never null, no factor being repeated.

For let $a_1, a_2, \dots a_\sigma$, ($\sigma < \nu$) be σ independent elements of the first order. Then by § 64, Prop. v. Corollary, $\nu-\sigma$ other elements $a_{\sigma+1}, \dots a_\nu$ can be added to these elements, so as to form an independent system of ν elements.

Let ν equations hold of the typical form

$$a_\rho = \alpha_{1\rho} e_1 + \alpha_{2\rho} e_2 + \dots \alpha_{\nu\rho} e_\nu .$$

Then by § 93,

$$(a_1 a_2 \dots a_\sigma)(a_{\sigma+1} a_{\sigma+2} \dots a_\nu) = (a_1 a_2 \dots a_\sigma a_{\sigma+1} \dots a_\nu) = \Delta (e_1 e_2 \dots e_\nu);$$

where Δ is the determinant $\Sigma \pm \alpha_{11} \alpha_{22} \dots \alpha_{\nu\nu}$.

Now Δ is not null, since all the elements $a_1, a_2, \dots a_\nu$ are independent, and $(e_1 e_2 \dots e_\nu)$ is not zero by § 93 (7). Hence $(a_1 a_2 \dots a_\sigma)$ cannot be null (cf. § 21).

95. Extensive Magnitudes. (1) Consider a product of μ elements, where $\mu < \nu$: let all these elements, namely $a_1, a_2 \dots a_\mu$, be assumed to be independent. Then they define a subregion of $\mu - 1$ dimensions, which we will call the subregion A_μ. Let $a_1', a_2' \dots a_\mu'$ be μ other independent elements lying in the same subregion A_μ. Then μ equations of the following type must be satisfied, namely

$$\begin{aligned} a_1' &= \lambda_{11} a_1 + \lambda_{12} a_2 + \dots + \lambda_{1\mu} a_\mu, \\ a_2' &= \lambda_{21} a_1 + \lambda_{22} a_2 + \dots + \lambda_{2\mu} a_\mu, \\ &\dots\dots\dots\dots\dots\dots\dots\dots \\ a_\mu' &= \lambda_{\mu 1} a_1 + \lambda_{\mu 2} a_2 + \dots + \lambda_{\mu\mu} a_\mu. \end{aligned}$$

Hence by multiplication we find, remembering the law of interchange of factors,

$$a_1' a_2' \dots a_\mu' = \Delta (a_1 a_2 \dots a_\mu);$$

where Δ stands for the determinant

$$\begin{vmatrix} \lambda_{11}, & \lambda_{12}, & \dots & \lambda_{1\mu} \\ \lambda_{21}, & \lambda_{22}, & \dots & \lambda_{2\mu} \\ \dots & \dots & \dots & \dots \\ \lambda_{\mu 1}, & \lambda_{\mu 2}, & \dots & \lambda_{\mu\mu} \end{vmatrix}.$$

If the elements $a_1', a_2' \dots a_\mu'$ be not independent, then $(a_1' a_2' \dots a_\mu') = 0$, and $\Delta = 0$; and hence in this case also

$$(a_1' a_2' \dots a_\mu') = \Delta (a_1 a_2 \dots a_\mu).$$

Thus if $a_1', a_2' \dots a_\mu'$ and $a_1, a_2 \dots a_\mu$ be respectively two sets of independent elements defining the same subregion, then [cf. § 64 (2)]

$$(a_1' a_2' \dots a_\mu') \equiv (a_1 a_2 \dots a_\mu).$$

(2) Conversely, if $(a_1' a_2' \dots a_\mu') \equiv (a_1 a_2 \dots a_\mu)$, where neither product is zero, then $a_1', a_2' \dots a_\mu'$ and $a_1, a_2 \dots a_\mu$ define the same region: or in words, two congruent products respectively define by their factors of the first order the same subregion of the manifold of the first order.

For we may write $(a_1' a_2' \dots a_\mu') = \lambda (a_1 a_2 \dots a_\mu)$. Multiply both sides by a_1, then $(a_1' a_2' \dots a_\mu' a_1) = 0$. Hence by § 93 (6) and § 94 (4) a_1 lies in the region $(a_1', a_2' \dots a_\mu')$. Similarly a_2 lies in the same region, and so on. Thus the two regions are identical.

(3) A product of μ elements of the first order represents an element of the derived manifold of the μth order (cf. § 20) at a given intensity; two congruent but not equivalent products represent the same element but at different intensities. Now an element of the manifold of the μth order, which is represented by a product, may by means of subsections (1) and (2) be identified with the subregion of the manifold of the first order defined by that product. Thus the product is to be conceived as representing the subregion at a given intensity. Then we shall, consistently with this conception, use the symbol for a product, such as A_μ (where $A_\mu = a_1 a_2 \ldots a_\mu$), also as the name of the subregion represented by the product.

(4) This symbolism and its interpretations can have no application unless a subregion is more than a mere aggregate of its contained elements. It is essentially assumed that a subregion can be treated as a whole and that it possesses certain properties which are symbolized by the relations between the elements of the derived manifold of the appropriate order. Thus a subregion of the manifold of the first order, conceived as an element of a positional manifold of a higher order, is the seat of an intensity and the term which symbolizes it always symbolizes it as at a definite intensity.

(5) A positional manifold whose subregions possess this property will be called an *extensive manifold.*

Let a product of ρ ($\rho < \nu$) elements of the first order (points) be called a regional element of the ρth order, and also a simple extensive magnitude of the ρth order.

Let regional elements of the first order be also called points, as was done in Book III.: let regional elements of the second order be also called linear elements or forces: let regional elements of the $(\nu - 1)$th order be called planar elements.

Also it will be convenient to understand 'regions' to mean regions of the manifold of the first order, unless it is explained otherwise.

(6) Elements of the extensive manifold of the first order (i.e. points) will be denoted exclusively by small Roman letters. Elements of the derived manifolds, when denoted by single letters, will be denoted exclusively by capital Roman letters.

96. Simple and Compound Extensive Magnitudes. (1) There is one difficulty in this theory of derived manifolds which must be carefully noted. For example let the original manifold be of three dimensions defined by reference elements e_1, e_2, e_3, e_4. The reference linear elements of the manifold of the second order are $(e_1e_2), (e_3e_4), (e_1e_3), (e_4e_2), (e_1e_4), (e_2e_3)$.

Then any element P of the positional manifold defined by these six elements is expressed by

$$P = \pi_{12}(e_1e_2) + \pi_{34}(e_3e_4) + \pi_{31}(e_3e_1) + \pi_{24}(e_2e_4) + \pi_{14}(e_1e_4) + \pi_{23}(e_2e_3).$$

But if an element of this manifold of the second order represent the product of two elements $\Sigma\xi e$ and $\Sigma\eta e$ of the original manifold, it can be expressed as

$$(\xi_1\eta_2)(e_1e_2) + (\xi_3\eta_4)(e_3e_4) + (\xi_1\eta_3)(e_1e_3) + (\xi_4\eta_2)(e_4e_2) + (\xi_1\eta_4)(e_1e_4) + (\xi_2\eta_3)(e_2e_3);$$

where $(\xi_\rho\eta_\sigma)$ stands for $\xi_\rho\eta_\sigma - \xi_\sigma\eta_\rho$.

But the following identity holds

$$(\xi_1\eta_2)(\xi_3\eta_4) + (\xi_1\eta_3)(\xi_4\eta_2) + (\xi_1\eta_4)(\xi_2\eta_3) = 0.$$

Accordingly P does not represent a product of elements of the original manifold unless

$$\pi_{12}\pi_{34} + \pi_{13}\pi_{42} + \pi_{14}\pi_{23} = 0.$$

Thus only the elements lying on a quadric surface locus in the positional manifold of five dimensions (which is the manifold of the second order) represent products of elements of the original manifold.

(2) Let a derived manifold of the ρth order be understood to denote the complete positional manifold which is defined by the $\nu!/\rho!\,(\nu-\rho)!$ independent reference elements. Let those elements of this derived manifold which can be represented as products of elements of the original manifold be called 'simple': let the other elements be called 'compound.' Let the term regional element [cf. § 95 (5)] be restricted to simple extensive magnitudes; and let compound elements be termed compound extensive magnitudes or a system of regional elements. The latter term is used since every compound element can be represented as a sum of simple elements. Thus an extensive magnitude of the ρth order is an element of the derived manifold of the ρth order, and may be either simple or compound.

(3) The associative law of multiplication identifies the product of two simple elements (E_ρ and E_σ) of derived manifolds of the ρth and σth orders ($\rho + \sigma < \nu$) with the simple element of the derived manifold of the $(\rho+\sigma)$th order formed by multiplying in any succession the elements of the original manifold which are the factors of E_ρ and E_σ.

Thus the product of any two elements, simple or compound, respectively belonging to manifolds of the ρth and σth orders yield an element, simple or compound, of the manifold of the $(\rho+\sigma)$th order. But the product of two compound elements may be simple.

In the case of simple elements, E_ρ and E_σ, the subregions E_ρ and E_σ of the original manifold may be said to be multiplied together.

97. FUNDAMENTAL PROPOSITIONS. *Prop.* I. If S_ρ be an element (simple or compound) of the derived manifold of the ρth order, and if $(aS_\rho)=0$, where a is a point [cf. § 95 (6)], then S_ρ can be written in the form $(aS_{\rho-1})$; where $S_{\rho-1}$ is an element of the derived manifold of the $(\rho-1)$th order.

For the reference elements of the original manifold may be assumed to be ν independent elements a, b, c.... Let A_1, A_2... B_1, B_2... be the multiplicative combinations of the ρth order formed out of these elements. Let $A_1, A_2 \dots$ be those which do contain a, and let $B_1, B_2 \dots$ be those which do not contain a.

Then we may write

$$S_\rho = \alpha_1 A_1 + \alpha_2 A_2 + \dots + \beta_1 B_1 + \beta_2 B_2 + \dots.$$

But by hypothesis

$$(aS_\rho) = 0 = (aA_1) = (aA_2) = \text{etc.}$$

Hence multiplying the assumed equation by a we deduce

$$\beta_1 (aB_1) + \beta_2 (aB_2) + \dots = 0.$$

Now (aB_1), (aB_2), etc. are different multiplicative combinations of a, b, c, etc. of the $(\rho+1)$th order. Hence they are independent, and by hypothesis they do not vanish.

Accordingly the above equation requires $\beta_1 = 0 = \beta_2 = \text{etc.}$

Hence $$S_\rho = \alpha_1 A_1 + \alpha_2 A_2 + \text{etc.} = (aS_{\rho-1}).$$

Corollary. If $(e_1 e_2 \dots e_\sigma)$ $(\sigma < \rho)$ be a simple element of the σth order, and if σ equations hold of the type $e_\lambda S_\rho = 0$ $(\lambda = 1, 2 \dots \sigma)$, then $S_\rho = (e_1 e_2 \dots e_\sigma S_{\rho-\sigma})$; where $S_{\rho-\sigma}$ is an element of the $(\rho-\sigma)$th order.

Prop. II. If A denote a regional element of the σth order, and B denote a regional element of the ρth order $(\rho < \sigma)$ such that the subregion A contains the subregion B, then A can be written (BC); where C is a regional element of the $(\sigma-\rho)$th order. For let the subregion B be defined by the ρ independent elements $a_1, a_2 \dots a_\rho$. Then to these independent elements $\sigma-\rho$ other independent elements $a_{\rho+1}, a_{\rho+2} \dots a_\sigma$ can be added such that the σ elements $a_1, a_2 \dots a_\sigma$ define the region A. But

$$A = \Delta (a_1 a_2 \dots a_\rho a_{\rho+1} \dots a_\sigma) = \Delta (BC') = (BC);$$

where C' stands for the product $(a_{\rho+1} a_{\rho+2} \dots a_\sigma)$, and $C = \Delta C'$.

Corollary. It follows from the two foregoing theorems that the combinatorial product of two subregions is zero if they possess one or more elements in common.

If they possess no common subregion their product is the region which contains them both.

Prop. III. If A_ρ and A_σ be two regional elements of orders ρ and σ respectively, and if $\rho + \sigma = \nu + \gamma$, then we can write $A_\rho = (C_\gamma B_{\rho-\gamma})$ and $A_\sigma = (C_\gamma B_{\sigma-\gamma})$, where C_γ is a regional element of the γth order and $B_{\rho-\gamma}$ and $B_{\sigma-\gamma}$ are regional elements of the $(\rho-\gamma)$th and $(\sigma-\gamma)$th orders.

For the subregions A_ρ and A_σ must contain in common a subregion of at least $\gamma - 1$ dimensions. Hence we are at liberty to assume the regional element C_γ as a common factor both to A_ρ and A_σ.

Prop. IV. All the elements of the derived manifold of the $(\nu-1)$th order are simple. For let A and B be two simple elements of the $(\nu-1)$th order. Then, since $(\nu-1)+(\nu-1)=\nu+(\nu-2)$, we may assume by the previous proposition a regional element C the $(\nu-2)$th order which is a common factor of A and B.

Hence $A=(aC)$, and $B=(bC)$, where a and b are of the first order.

Thus $$A+B=(a+b)\,C.$$

But $a+b$ is some element of the first order, call it c.

Hence $$A+B=cC.$$

But cC is simple. Hence the sum of any number of simple elements of the $(\nu-1)$th order is a simple element.

NOTE. All the propositions of this chapter are substantially to be found in the *Ausdehnungslehre von* 1862. The application of Combinatorial Multiplication to the theory of Determinants is investigated by R. F. Scott, cf. *A Treatise on the Theory of Determinants*, Cambridge, 1880. Terms obeying the combinatorial law of multiplication are called by him 'alternate numbers.'

CHAPTER II.

Regressive Multiplication.

98. Progressive and Regressive Multiplication. (1) According to the laws of combinatorial multiplication just explained the product of two extensive magnitudes S_ρ and S_σ respectively of the ρth and σth order must necessarily be null, if $\rho+\sigma$ be greater than ν, where the original manifold is of $\nu-1$ dimensions. Such products can therefore never occur, since every term of any equation involving them would necessarily be null.

In the case $\rho+\sigma>\nu$ we are accordingly at liberty to assign a fresh law of multiplication to be denoted by the symbol $S_\rho S_\sigma$. Let multiplication according to this new law (to be defined in § 100) be termed 'regressive,' and in contradistinction let combinatorial multiplication be called progressive.

Thus if $\rho+\sigma<\nu$, the product $S_\rho S_\sigma$ is formed according to the progressive law already explained. Such products will be called progressive products. If $\rho+\sigma>\nu$, the product $S_\rho S_\sigma$ will be formed according to the regressive law. Such products will be called regressive products. If $\rho+\sigma=\nu$, the product $S_\rho S_\sigma$ may be conceived indifferently as formed according to the progressive or regressive law.

(2) In this last case it is to be noted that $S_\rho S_\sigma$ must necessarily be of the form $\alpha(e_1e_2\ldots e_\nu)$, where $e_1, e_2\ldots e_\nu$ are ν independent reference elements of the original manifold. Since therefore such products can only represent a numerical multiple of a given product, we are at liberty to assume them to be merely numerical.

Thus for example we may assume

$$(e_1e_2\ldots e_\nu)=1, \text{ and } (S_\rho S_\sigma)=\alpha;$$

where it is to be remembered that $\rho+\sigma=\nu$.

Let a product which is merely numerical be always enclosed in a bracket, as thus $(e_1e_2\ldots e_\nu)$; other products will be enclosed in brackets where convenient, but numerical products invariably so.

99. Supplements. (1) Corresponding to any multiplicative combination E_μ of the μth order ($\mu<\nu$) of the elements $e_1, e_2\ldots e_\nu$, there exists [cf. § 65 (4)] a multiplicative combination $E_{\nu-\mu}$ of the $(\nu-\mu)$th order which contains those elements as factors which are omitted from E_μ. Let it be assumed that

$$(e_1 e_2 \ldots e_\nu) = \pm 1.$$

Hence $$(E_\mu E_{\nu-\mu}) = \pm (e_1 e_2 \ldots e_\nu) = \pm 1.$$

We may notice that if $E'_{\nu-\mu}$ be any other multiplicative combination of the $(\nu-\mu)$th order, then $(E_\mu E'_{\nu-\mu}) = 0$.

(2) The 'supplement' of any multiplicative combination E_μ of the reference elements $e_1, e_2 \ldots e_\nu$ and of the μth order is that multiplicative combination $E_{\nu-\mu}$ of the $(\nu-\mu)$th order which contains those reference elements omitted from E_μ multiplied in such succession that

$$(E_\mu E_{\nu-\mu}) = 1.$$

Let the supplement of E_μ be denoted by $|E_\mu$.

(3) Then if $E_{\nu-\mu}$ contain the same elements as $|E_\mu$ but multiplied in any succession, $E_{\nu-\mu}$ will be called the multiplicative combination supplementary to E_μ.

Then since $(E_\mu E_{\nu-\mu}) = \pm 1$, we see that $|E_\mu = (E_\mu E_{\nu-\mu}) E_{\nu-\mu}$; where $(E_\mu E_{\nu-\mu})$ is treated as a numerical multiplier of $E_{\nu-\mu}$.

The fundamental equations satisfied by $|E_\mu$ are

$$(E_\mu | E_\mu) = 1, \text{ and } (E_\mu' | E_\mu) = 0;$$

where E_μ' is any multiplicative combination of the μth order other than E_μ.

(4) The analogy of the above definitions leads us in the extreme cases to define

$$|(e_1 e_2 e_3 \ldots e_\nu) = 1, \text{ and } |1 = (e_1 e_2 \ldots e_\nu).$$

Since $(e_1 e_2 \ldots e_\nu) = 1$, it follows from these definitions that, $|1 = 1$.

(5) Let the supplement of a sum of multiplicative combinations of a given order be defined to be the sum of the supplements. This definition is consistent with that of subsection (2), since [cf. § 94 (2)] the sum of different multiplicative combinations is not a multiplicative combination of the reference elements.

Thus $$|(E_\mu + E_\mu' + \ldots) = |E_\mu + |E_\mu' + \ldots$$

Let this definition be assumed to apply to the special case where E_μ is repeated α times.

Thus $|(E_\mu + E_\mu + \ldots \text{ to } \alpha \text{ terms}) = |E_\mu + |E_\mu + \ldots \text{ to } \alpha \text{ terms} = \alpha |E_\mu$.

Hence $$|(\alpha E_\mu) = \alpha |E_\mu.$$

Now let $\mu = \nu$, and $E_\mu = E_\nu = 1$. Then the above equation becomes $|\alpha = \alpha$.

Also finally $$|(\alpha E_\mu + \alpha' E_\mu' + \text{etc.}) = \alpha |E_\mu + \alpha' |E_\mu' + \text{etc.}$$

(6) The symbol $|$ may be considered as denoting an operation on the terms following it. It will be called the operation of taking the supplement.

This operation is distributive in reference to addition and also in reference to the product of a numerical factor and an extensive magnitude. For $|(A+B)=|A+|B$, and $|(\alpha . A)=|\alpha . |A$.

(7) Let the symbol $||A$ denote the supplement of the supplement of A. If A be an extensive magnitude of the μth order, then

$$||A=(-1)^{\mu(\nu-\mu)} A.$$

For with the notation used above,

$$|E_\mu=(E_\mu E_{\nu-\mu}) E_{\nu-\mu}, \text{ and } |E_{\nu-\mu}=(E_{\nu-\mu} E_\mu) E_\mu.$$

Hence from (5) $||E_\mu=(E_\mu E_{\nu-\mu})|E_{\nu-\mu}=(E_\mu E_{\nu-\mu})(E_{\nu-\mu} E_\mu) E_\mu$.

But $\quad (E_{\nu-\mu} E_\mu)=(-1)^{\mu(\nu-\mu)}(E_\mu E_{\nu-\mu})$; and $(E_\mu E_{\nu-\mu})=\pm 1$.

Hence $\quad ||E_\mu=(-1)^{\mu(\nu-\mu)} E_\mu$.

But $\quad A=\Sigma \alpha E_\mu$; and therefore $||A=(-1)^{\mu(\nu-\mu)} A$.

(8) It must finally be noted that the supplement of an extensive magnitude must be taken to refer to a definite set of reference elements of the original manifold, and that it has no signification except in relation to such a set.

(9) The following notation for the operation of the symbol $|$ on products will be observed. The symbol will be taken to operate on all succeeding letters of a product up to the next dot; thus $a|bcd$ means that $|(bcd)$ is to be multiplied into a; and $a|bc.d$ means that $|(bc)$ is to be multiplied into a and d into this product. Also a second $|$ will be taken to act as a dot in limiting the operation of a former $|$: thus $|A|B$ means that $|B$ is multiplied into $|A$, and it does not mean $|(A|B)$.

Again, $|$ placed before a bracket will be taken to act only on the magnitude inside the bracket: thus $|(AB)C$ means that C is multiplied into $|(AB)$. Johnson's notation with dots might be employed [cf. § 20]: thus $|ab.c..d$ would mean that c is multiplied into $|(ab)$ and d into this product.

100. Definition of Regressive Multiplication. (1) If A_ρ and A_σ be extensive magnitudes of the ρth and σth orders, where $\rho+\sigma>\nu$; then $|A_\rho$ and $|A_\sigma$ are extensive magnitudes of the $(\nu-\rho)$th and $(\nu-\sigma)$th orders, and $(\nu-\rho)+(\nu-\sigma)<\nu$. Hence $|A_\rho$ and $|A_\sigma$ can be multiplied together according to the progressive rule of multiplication, already explained.

Now the regressive product of A_ρ and A_σ will be so defined that the operation of taking the supplement may be distributive in reference to the factors of a product.

(2) Let the regressive product $A_\rho A_\sigma$ be defined to be an extensive magnitude, such that its supplement is $|A_\rho|A_\sigma$.

In symbols, $\quad |A_\rho A_\sigma=|A_\rho|A_\sigma$.

Since $|A_\rho|A_\sigma$ is of the $(2\nu-\rho-\sigma)$th order, the regressive product $A_\rho A_\sigma$ is an extensive quantity of the $(\rho+\sigma-\nu)$th order.

(3) If $\rho+\sigma=\nu$, then $A_\rho A_\sigma$ can be indifferently conceived either as progressive or as regressive. For if E_ρ, E_ρ', etc. are the multiplicative combinations of the reference elements of the ρth order, we can write

$$A_\rho = \Sigma\alpha_\rho E_\rho, \text{ and } A_\sigma = \Sigma\alpha_\sigma |E_\rho.$$

Hence $$(A_\rho A_\sigma) = \alpha_\rho\alpha_\sigma + \alpha_\rho'\alpha_\sigma' + \text{etc.},$$

since $$(E_\rho|E_\rho) = 1, \text{ and } (E_\rho|E_\rho') = 0.$$

Also $$|A_\rho = \Sigma\alpha_\rho |E_\rho \text{ and } |A_\sigma = \Sigma\alpha_\sigma ||E_\rho = (-1)^{\rho(\nu-\rho)}\alpha_\sigma E_\rho.$$

Hence $$(|A_\rho|A_\sigma) = (-1)^{\rho(\nu-\rho)}\{\alpha_\rho\alpha_\sigma(|E_\rho . E_\rho) + \alpha_\rho'\alpha_\sigma'(|E_\rho' . E_\rho') + \text{etc.}\}.$$

Now $$(|E_\rho . E_\rho) = (-1)^{\rho(\nu-\rho)}(E_\rho|E_\rho) = (-1)^{\rho(\nu-\rho)}.$$

Thus finally

$$(|A_\rho|A_\sigma) = \alpha_\rho\alpha_\sigma + \alpha_\rho'\alpha_\sigma' + \text{etc.} = |\{\alpha_\rho\alpha_\sigma + \alpha_\rho'\alpha_\sigma' + \text{etc.}\} = |(A_\rho A_\sigma).$$

(4) Again if $\rho+\sigma<\nu$, the product $A_\rho A_\sigma$ is progressive and the product $|A_\rho|A_\sigma$ is regressive.

But by definition of the regressive product $|A_\rho|A_\sigma$ we have

$$|\{|A_\rho|A_\sigma\} = ||A_\rho||A_\sigma.$$

Now $$||A_\rho = (-1)^{\rho(\nu-\rho)}A_\rho, \text{ and } ||A_\sigma = (-1)^{\sigma(\nu-\sigma)}A_\sigma.$$

Hence $$|\{|A_\rho|A_\sigma\} = (-1)^{\rho(\nu-\rho)+\sigma(\nu-\sigma)}A_\rho A_\sigma.$$

Therefore, taking the supplement of each side,

$$||\{|A_\rho|A_\sigma\} = (-1)^{\rho(\nu-\rho)+\sigma(\nu-\sigma)}|A_\rho A_\sigma.$$

Now $|A_\rho|A_\sigma$ is an extensive magnitude of the $(\nu-\rho-\sigma)$th degree. Hence

$$||\{|A_\rho|A_\sigma\} = (-1)^{(\nu-\rho-\sigma)(\rho+\sigma)}|A_\rho|A_\sigma.$$

Also $$(-1)^{(\nu-\rho-\sigma)(\rho+\sigma)} = (-1)^{\rho(\nu-\rho)+\sigma(\nu-\sigma)}.$$

Hence $$|A_\rho A_\sigma = |A_\rho|A_\sigma.$$

(5) Finally therefore in every case, whether the product $A_\rho A_\sigma$ be progressive or regressive, we deduce $|A_\rho A_\sigma = |A_\rho|A_\sigma$.

101. Pure and Mixed Products. (1) A product in which all the multiplications indicated are all progressive or all regressive (as the case may be) is called pure; if the multiplications are all progressive the product is called a pure progressive product; if all regressive, a pure regressive product.

Thus if A, B, C and D be extensive magnitudes, the product $AB . CD$ is a pure regressive product if the product of A and B is regressive, and that of C and D, and that of AB and CD.

(2) A product which is not pure is called 'mixed.' Thus if the product of A and B is regressive, and that of C and D is progressive, then the product $AB . CD$ is mixed.

(3) A pure product is associative.

This proposition is true by definition, if the pure product be progressive.

If the product (P) of magnitudes A, B, C, etc. be a pure regressive product, then the product of $|A$, $|B$, $|C$, etc. is a pure progressive product. But this product is associative.

Hence $$|P = |A \,.\, |B \,|C \ldots = |A \,|B \,|C \ldots .$$

Taking the supplements of both sides

$$||P = ||A \,.\, ||B \,||C \ldots = ||A \,||B \,||C \ldots ,$$

hence $$P = A \,.\, BC \ldots = ABC \ldots .$$

For instance, if the product $AB \,.\, CD$ be pure, we may write

$$AB \,.\, CD = ABCD.$$

A mixed product is not in general associative.

102. Rule of the Middle Factor. (1) We have now to give rules for the identification of that extensive magnitude of the $(\rho + \sigma - \nu)$th order which is denoted by the regressive product $A_\rho A_\sigma$. This will be accomplished by the following theorems.

(2) *Proposition* A. Let E_ρ, E_σ, E_τ be three multiplicative combinations of the reference units of the ρth, σth, and τth orders respectively, and let $\rho + \sigma + \tau = \nu$.

To prove that $$E_\rho E_\sigma \,.\, E_\rho E_\tau = (E_\rho E_\sigma E_\tau)\, E_\rho.$$

It will be noticed that the products $E_\rho E_\sigma$ and $E_\rho E_\tau$ are progressive, while the final product of $E_\rho E_\sigma$ and $E_\rho E_\tau$ is regressive; and also that $(E_\rho E_\sigma E_\tau)$ is either zero or ± 1.

Case I. Let $(E_\rho E_\sigma E_\tau) = 0$. Then since in this product there are only ν factors of the first order, one of the ν reference elements must be absent in order that another one may be repeated.

Let e_1 be the absent element. Then e_1 is contained neither in $E_\rho E_\sigma$ nor in $E_\rho E_\tau$. Hence it is contained both in $|(E_\rho E_\sigma)$ and in $|(E_\rho E_\tau)$.

Therefore $$|(E_\rho E_\sigma \,.\, E_\rho E_\tau) = |(E_\rho E_\sigma)\,|(E_\rho E_\tau) = 0.$$

Hence $$E_\rho E_\sigma \,.\, E_\rho E_\tau = 0 = (E_\rho E_\sigma E_\tau)\, E_\rho.$$

Case II. Let $(E_\rho E_\sigma E_\tau) = \pm 1$.

In this case no factor of the first order is repeated in the product $(E_\rho E_\sigma E_\tau)$.

Hence [cf. § 99 (3)] $|E_\sigma = (E_\sigma E_\rho E_\tau)\, E_\rho E_\tau$, $|E_\tau = (E_\tau E_\rho E_\sigma)\, E_\rho E_\sigma$.

Hence $$|(E_\tau E_\sigma) = |E_\tau \,|E_\sigma = (E_\tau E_\rho E_\sigma)(E_\sigma E_\rho E_\tau)\, E_\rho E_\sigma \,.\, E_\rho E_\tau \quad \ldots\ldots\ldots(1).$$

But from § 99 (3) $$|(E_\tau E_\sigma) = (E_\tau E_\sigma E_\rho)\, E_\rho \quad \ldots\ldots\ldots\ldots\ldots\ldots\ldots\ldots(2).$$

Also $(E_\tau E_\rho E_\sigma) = (-1)^{\tau(\rho+\sigma)} (E_\rho E_\sigma E_\tau)$, $(E_\sigma E_\rho E_\tau) = (-1)^{\rho\sigma} (E_\rho E_\sigma E_\tau)$,

and $$(E_\tau E_\sigma E_\rho) = (-1)^{\rho\sigma + \tau(\rho+\sigma)} (E_\rho E_\sigma E_\tau).$$

Therefore from equations (1) and (2), and remembering that $(E_\rho E_\sigma E_\tau) = \pm 1$, it follows that $$E_\rho E_\sigma \,.\, E_\rho E_\tau = (E_\rho E_\sigma E_\tau) E_\rho.$$

(3) *Proposition* B. If A_ρ, A_σ, A_τ be any simple extensive magnitudes of the ρth, σth and τth orders respectively, such that $\rho + \sigma + \tau = \nu$, then $$A_\rho A_\sigma \,.\, A_\rho A_\tau = (A_\rho A_\sigma A_\tau) A_\rho.$$

For let us assume that this formula holds for the case when the factors of the first order of A_ρ, A_σ and A_τ are composed out of a given set of ν independent elements $a_1, a_2 \dots a_\nu$. Then we shall show that the formula holds for products formed out of the set $a_1', a_2 \dots a_\nu$, where $a_1' = \Sigma \alpha_\mu a_\mu$ and replaces a_1.

Now a_1 may occur in A_ρ, A_σ, or A_τ.

Firstly assume that it occurs in A_σ. Let $A_\sigma = a_1 A_{\sigma-1}$; and let $A_\sigma' = a_1' A_{\sigma-1}$. Then A_σ' is what A_σ becomes when a_1' is everywhere substituted for a_1, and A_ρ and A_τ are unaltered by this substitution.

Thus $$A_\sigma' = a_1' A_{\sigma-1} = \Sigma \alpha_\mu (a_\mu A_{\sigma-1}).$$

Hence $$A_\rho A_\sigma' \,.\, A_\rho A_\tau = \Sigma \alpha_\mu (A_\rho a_\mu A_{\sigma-1} \,.\, A_\rho A_\tau).$$

But each product of the type $A_\rho a_\mu A_{\sigma-1} \,.\, A_\rho A_\tau$ is such that the factors of the first degree in A_ρ, $a_\mu A_{\sigma-1}$, and A_τ are composed of the set of elements $a_1, a_2 \dots a_\nu$. Accordingly by our assumption $$A_\rho a_\mu A_{\sigma-1} \,.\, A_\rho A_\tau = (A_\rho a_\mu A_{\sigma-1} A_\tau) A_\rho.$$

Hence $$A_\rho A_\sigma' \,.\, A_\rho A_\tau = \Sigma \alpha_\mu (A_\rho a_\mu A_{\sigma-1} A_\tau) A_\rho = (A_\rho \,.\, \Sigma \alpha_\mu a_\mu \,.\, A_{\sigma-1} A_\tau) A_\rho$$
$$= (A_\rho a_1' A_{\sigma-1} A_\tau) A_\rho = (A_\rho A_\sigma' A_\tau) A_\rho.$$

Secondly, it can be proved in exactly the same manner that if a_1 occurs in A_τ, so that when a_1' is substituted for a_1, A_τ becomes A_τ' and A_ρ and A_σ are unaffected, then $A_\rho A_\sigma \,.\, A_\rho A_\tau' = (A_\rho A_\sigma A_\tau') A_\rho$.

Thirdly, assume that a_1 occurs in A_ρ.

Let a_1 be changed into $a_1' = \alpha_1 a_1 + \alpha_2 a_2$, and let $A_\rho = (a_1 A_{\rho-1})$, and $A_\rho' = (a_1' A_{\rho-1})$.

If a_2 occur in $A_{\rho-1}$, then $$A_\rho' = \alpha_1 (a_1 A_{\rho-1}) + \alpha_2 (a_2 A_{\rho-1}) = \alpha_1 (a_1 A_{\rho-1}) = \alpha_1 A_\rho.$$

Accordingly A_ρ' is merely a multiple of A_ρ, and we deduce immediately that $$A_\rho' A_\sigma \,.\, A_\rho' A_\tau = \alpha^2 A_\rho A_\sigma \,.\, A_\rho A_\tau = \alpha^2 (A_\rho A_\sigma A_\tau) A_\rho$$
$$= (A_\rho' A_\sigma A_\tau) A_\rho'.$$

If a_2 does not occur in $A_{\rho-1}$, suppose that it occurs in A_σ. Let $A_\sigma = a_2 A_{\sigma-1}$.

Then $$A_\rho' A_\sigma = \alpha_1 (a_1 A_{\rho-1} a_2 A_{\sigma-1}) + \alpha_2 (a_2 A_{\rho-1} a_2 A_{\sigma-1})$$
$$= \alpha_1 (a_1 A_{\rho-1} a_2 A_{\sigma-1}) = \alpha_1 A_\rho A_\sigma.$$

And $$A_\rho' A_\tau = \alpha_1 A_\rho A_\tau + \alpha_2 (a_2 A_{\rho-1} A_\tau).$$

Hence $\quad A_\rho' A_\sigma \,.\, A_\rho' A_\tau = \alpha_1^2 A_\rho A_\sigma \,.\, A_\rho A_\tau + \alpha_1 \alpha_2 A_\rho A_\sigma \,.\, a_2 A_{\rho-1} A_\tau.$

Now $$A_\rho A_\sigma \,.\, A_\rho A_\tau = (A_\rho A_\sigma A_\tau)\, A_\rho.$$

And $\quad A_\rho A_\sigma \,.\, a_2 A_{\rho-1} A_\tau = a_1 A_{\rho-1} a_2 A_{\sigma-1} \,.\, a_2 A_{\rho-1} A_\tau = - a_2 A_{\rho-1} a_1 A_{\sigma-1} \,.\, a_2 A_{\rho-1} A_\tau$

$$= - (a_2 A_{\rho-1} a_1 A_{\sigma-1} A_\tau)\, a_2 A_{\rho-1} = (a_1 A_{\rho-1} a_2 A_{\sigma-1} A_\tau)\, a_2 A_{\rho-1} = (A_\rho A_\sigma A_\tau)\, a_2 A_{\rho-1}.$$

Therefore by substitution,

$$A_\rho' A_\sigma \,.\, A_\rho' A_\tau = \alpha_1 (A_\rho A_\sigma A_\tau)\, \{\alpha_1 A_\rho + \alpha_2 a_2 A_{\rho-1}\}.$$

But $\quad (A_\rho' A_\sigma A_\tau) = \alpha_1 (A_\rho A_\sigma A_\tau)$, and $\alpha_1 A_\rho + \alpha_2 a_2 A_{\rho-1} = A_\rho'$.

Hence finally $$A_\rho' A_\sigma \,.\, A_\rho' A_\tau = (A_\rho' A_\sigma A_\tau)\, A_\rho'.$$

But by repeated substitutions for a_1, a_1', etc. of the type $a_1' = \alpha_1 a_1 + \alpha_2 a_2$, $a_1'' = \beta_1 a_1' + \beta_3 a_3$, and so on, a_1 is finally replaced by any arbitrary element $\Sigma \alpha_\mu a_\mu$.

Thus if any element of the set $a_1, a_2 \dots a_\nu$ be replaced by an arbitrary element, the formula still holds. Hence by successive substitution the ν elements $a_1, a_2 \dots a_\nu$ can be replaced by ν other elements $b_1, b_2 \dots b_\nu$.

But if E_ρ, E_σ, E_τ be simple magnitudes formed by products of the reference elements $e_1, e_2 \dots e_\nu$, the formula, $E_\rho E_\sigma \,.\, E_\rho E_\tau = (E_\rho E_\sigma E_\tau)\, E_\rho$, has been proved to hold by proposition A. Therefore if A_ρ, A_σ, A_τ be simple magnitudes formed by products of any set of elements $a_1, a_2 \dots a_\nu$ the formula,

$$A_\rho A_\sigma \,.\, A_\rho A_\tau = (A_\rho A_\sigma A_\tau)\, A_\rho, \text{ holds.}$$

(4) *Corollary.* It is easy to see that the formula still holds if A_σ and A_τ be compound. But it does not hold if A_ρ is compound.

Proposition B is the foundation of all the formulae in this algebra. The following important formulae given by propositions C and D can be deduced from it.

(5) *Proposition* C. $A_\rho A_\sigma \,.\, A_\rho A_\tau = (A_\rho A_\sigma A_\tau)\, A_\rho$, when $\rho + \sigma + \tau = 2\nu$.

In this case the products $A_\rho A_\sigma$ and $A_\rho A_\tau$ are both regressive. Hence the products $|A_\rho|A_\sigma$ and $|A_\rho|A_\tau$ are both progressive, and

$$(\nu - \rho) + (\nu - \sigma) + (\nu - \tau) = \nu.$$

Hence by proposition B, $|A_\rho|A_\sigma \,.\, |A_\rho|A_\tau = (|A_\rho|A_\sigma|A_\tau)\,|A_\rho$.

Therefore by taking supplements of both sides

$$A_\rho A_\sigma \,.\, A_\rho A_\tau = (A_\rho A_\sigma A_\tau)\, A_\rho.$$

(6) *Proposition* D. $A_\rho A_\sigma \,.\, A_\sigma A_\tau = (A_\rho A_\sigma A_\tau)\, A_\sigma$,

and $$A_\rho A_\tau \,.\, A_\sigma A_\tau = (A_\rho A_\sigma A_\tau)\, A_\tau;$$

where $\rho + \sigma + \tau = \nu$ or 2ν.

These formulae follow immediately from propositions B and C.

For
$$A_\rho A_\sigma . A_\sigma A_\tau = (-1)^{\rho\sigma} A_\sigma A_\rho . A_\sigma A_\tau$$
$$= (-1)^{\rho\sigma} (A_\sigma A_\rho A_\tau) A_\sigma$$
$$= (A_\rho A_\sigma A_\tau) A_\sigma.$$
Similarly for the other formula.

(7) These formulae may all be included in one rule, which we will call the *rule of the middle factor*, given in the following proposition.

Proposition E. Let A_ρ and A_σ be two simple extensive magnitudes of the ρth and σth order such that $\rho + \sigma = \nu + \gamma$. Then the regions A_ρ and A_σ have a common region of at least $\gamma - 1$ dimensions. Let C_γ be this common region. Then we may write (cf. § 97 Prop. III.) $A_\rho = B_{\rho-\gamma} C_\gamma$, and $A_\sigma = C_\gamma B_{\sigma-\gamma}$. And it is easy from the foregoing propositions to prove that
$$A_\rho A_\sigma = B_{\rho-\gamma} C_\gamma . A_\sigma = (B_{\rho-\gamma} A_\sigma) C_\gamma$$
$$= A_\rho . C_\gamma B_{\sigma-\gamma} = (A_\rho B_{\sigma-\gamma}) C_\gamma.$$
These formulae embody the rule of the middle factor.

103. Extended Rule of the Middle Factor. (1) But this rule in its present form is not very easily applicable in most cases. Thus suppose that the complete manifold be of three dimensions, so that $\nu = 4$, and let $A_\rho = pqr$, and $A_\sigma = st$; where p, q, r, s, t are elements of the complete manifold. Then to find the product $pqr . st$, the rule directs us to find the element x which the line st must have in common with the plane pqr and to write either pqr in the form uvx or st in the form xz; and then
$$pqr . st = uvx . st = (uvst) x, \text{ and } pqr . st = pqr . xz = (pqrz) x.$$
But no rule has yet been given to express x in terms of p, q, r, s, t.

This defect is remedied by the following proposition embodied in equations (1) and (2) of the next subsection, which we will call the 'extended rule of the middle factor.'

(2) *Proposition* F. Let A_ρ and B_σ be simple extensive magnitudes of the ρth and σth orders respectively, and let $\rho + \sigma = \nu + \gamma$, where γ must be less than ν. Let $C_\gamma^{(1)}$, $C_\gamma^{(2)}$, etc. denote the multiplicative combinations of the γth order which can be formed out of the factors of the first order of A_ρ. Then we may write
$$A_\rho = A_{\rho-\gamma}^{(1)} C_\gamma^{(1)} = A_{\rho-\gamma}^{(2)} C_\gamma^{(2)} = \text{etc.},$$
where $A_{\rho-\gamma}^{(1)}$, $A_{\rho-\gamma}^{(2)}$, etc. are extensive magnitudes of the $(\rho - \gamma)$th order.

Then according to the extended rule of the middle factor
$$A_\rho B_\sigma = (A_{\rho-\gamma}^{(1)} B_\sigma) C_\gamma^{(1)} + (A_{\rho-\gamma}^{(2)} B_\sigma) C_\gamma^{(2)} + \text{etc.} \quad\ldots\ldots\ldots\ldots\ldots(1).$$
Similarly let $D_\gamma^{(1)}$, $D_\gamma^{(2)}$, etc. be the multiplicative combinations of the γth order formed out of the factors of the first order of B_σ. Then we may write
$$B_\sigma = D_\gamma^{(1)} B_{\sigma-\gamma}^{(1)} = D_\gamma^{(2)} B_{\sigma-\gamma}^{(2)} = \text{etc.},$$
where $B_{\sigma-\gamma}^{(1)}$, $B_{\sigma-\gamma}^{(2)}$, etc. are extensive magnitudes of the $(\sigma - \gamma)$th order.

Then according to the extended rule of the middle factor

$$A_\rho B_\sigma = (A_\rho B^{(1)}_{\sigma-\gamma}) D^{(1)}_\gamma + (A_\rho B^{(2)}_{\sigma-\gamma}) D^{(2)}_\gamma + \text{etc.} \quad\ldots\ldots\ldots\ldots(2).$$

Equations (1) and (2) form the extended rule of the middle factor which has now to be proved.

Let $a_1, a_2 \ldots a_\rho$ be the ρ factors of A_ρ, and let $b_1, b_2 \ldots b_\sigma$ be the σ factors of B_σ.

Let $\nu - \rho$ other elements $a_{\rho+1}, a_{\rho+2} \ldots a_\nu$ be added to $a_1 \ldots a_\rho$, so as to form a set of ν independent elements.

Then we may write B_σ in the form

$$\beta_1 B^{(1)}_\sigma + \beta_2 B^{(2)}_\sigma + \text{etc.};$$

where $B^{(1)}_\sigma$, $B^{(2)}_\sigma$, etc. are the multiplicative combinations of the σth order of the elements $a_1, a_2 \ldots a_\nu$; and any number of the coefficients β_1, β_2, etc. may be zero.

Also, remembering that $(\rho - \gamma) + \sigma = \nu$, let the index-notation be so arranged that $B^{(1)}_\sigma$ contains those a's which do not appear in $A^{(1)}_{\rho-\gamma}$, and $B^{(2)}_\sigma$ contains those which do not appear in $A^{(2)}_{\rho-\gamma}$, and so on.

Then it may be noted that to every magnitude $A^{(\mu)}_{\rho-\gamma}$ there corresponds a magnitude $B^{(\mu)}_\sigma$, but not necessarily conversely.

Furthermore it is obvious that when $\lambda \neq \mu$

$$\{A^{(\lambda)}_{\rho-\gamma} B^{(\mu)}_\sigma\} = 0.$$

Then $$A_\rho B_\sigma = \Sigma \beta_\mu (A_\rho B^{(\mu)}_\sigma) = \Sigma \beta_\mu (A^{(\mu)}_{\rho-\gamma} C^{(\mu)}_\gamma\} . B^{(\mu)}_\sigma).$$

Now $C^{(\mu)}_\gamma$ must represent a subregion contained in the subregion $B^{(\mu)}_\sigma$; since $C^{(\mu)}_\gamma$ is a product of γ of the a's which do not appear in $A^{(\mu)}_{\rho-\gamma}$, and $B^{(\mu)}_\sigma$ is a product of all those a's which do not appear in $A^{(\mu)}_{\rho-\gamma}$. Hence by the rule of the middle factor

$$A^{(\mu)}_{\rho-\gamma} C^{(\mu)}_\gamma . B^{(\mu)}_\sigma = (A^{(\mu)}_{\rho-\gamma} B^{(\mu)}_\sigma) C^{(\mu)}_\gamma.$$

Also since $A^{(\mu)}_{\rho-\gamma} B^{(\lambda)}_\sigma = 0$, we deduce

$$\beta_\mu (A^{(\mu)}_{\rho-\gamma} B^{(\mu)}_\sigma) = \sum_\lambda \beta_\lambda (A^{(\mu)}_{\rho-\gamma} B^{(\lambda)}_\sigma) = A^{(\mu)}_{\rho-\gamma} . \sum_\lambda \beta_\lambda B^{(\lambda)}_\sigma = (A^{(\mu)}_{\rho-\gamma} B_\sigma).$$

Hence finally $$A_\rho B_\sigma = \sum_\mu (A^{(\mu)}_{\rho-\gamma} B_\sigma) C^{(\mu)}_\gamma;$$

which is the equation (1) of the enunciation. An exactly similar proof yields equation (2).

(3) The following formulae are important special examples of this extended rule of the middle factor.

Let $\nu = 3$, the complete manifold being therefore of two dimensions.

Then $$pq . rs = (prs) q - (qrs) p = (pqs) r - (pqr) s \quad\ldots\ldots\ldots\ldots(3).$$

Let $\nu = 4$, the complete manifold being therefore of three dimensions.

Then $$pqr . st = st . pqr = (pqrt) s - (pqrs) t$$
$$= (pqst) r + (rpst) q + (qrst) p \quad\ldots\ldots\ldots\ldots(4).$$

And $$pqr \,.\, stu = -\, stu \,.\, pqr = (pqrs)\, tu + (pqru)\, st + (pqrt)\, us$$
$$= (pstu)\, qr + (rstu)\, pq + (qstu)\, rp \;\; \ldots\ldots\ldots\ldots\ldots\ldots (5).$$

(4) Take the supplements of these formulae.

When $\nu = 3$, the supplement of a magnitude of the first order is a magnitude of the second order. Let P, Q, R, S be magnitudes of the second order such that $P = |p$, $Q = |q$, etc.

Then by taking the supplement of (3) we deduce
$$PQ \,.\, RS = (PRS)\, Q - (QRS)\, P = (PQS)\, R - (PQR)\, S \ldots\ldots (3').$$

Again let $\nu = 4$; then the supplement of a magnitude of the first order is one of the third order.

Let P, Q, R, S, T, U be put for $|p$, $|q$, $|r$, $|s$, $|t$, $|u$; and let the supplements of equations (4) and (5) be taken.

Then $$PQR \,.\, ST = ST \,.\, PQR = (PQRT)\, S - (PQRS)\, T$$
$$= (PQST)\, R + (RPST)\, Q + (QRST)\, P \;\; \ldots\ldots\ldots (4').$$

And
$$PQR \,.\, STU = -\, STU \,.\, PQR = (PQRS)\, TU + (PQRU)\, ST + (PQRT)\, US$$
$$= (PSTU)\, QR + (RSTU)\, PQ + (QSTU)\, RP \ldots\ldots\ldots\ldots\ldots\ldots (5').$$

In fact by taking supplements any formula involving magnitudes of the first order is converted into one involving planar elements, i.e. magnitudes of the $(\nu - 1)$th order; where the complete manifold is of $\nu - 1$ dimensions.

104. Regressive Multiplication independent of Reference Elements. (1) The rule of the middle factor and the extended rule disclose the fact that the regressive product of two magnitudes A and B is independent of the special reference elements in the original manifold which were chosen for defining the operation of taking the supplement. Accordingly regressive multiplication is an operation independent of any special reference elements or of their intensities, though such elements are used in its definition for the sake of simplicity. Also it is independent of the fact that the product of the ν reference elements was taken to be unity for simplicity of explanation. Thus the product may be assumed to have any numerical value Δ which may be convenient [cf. § 98 (2)].

It would have been possible to define regressive multiplication by means of the rule of the middle factor. It would then have been necessary to prove that it is a true multiplication, namely that it is distributive in reference to addition.

(2) It is useful to bear in mind the following summary of results respecting the multiplication of two regions P_ρ and P_σ, of the ρth and σth orders respectively:

If $\rho + \sigma < \nu$, then $P_\rho P_\sigma$ is progressive and represents the containing region [cf. § 65 (6)] of the two regions P_ρ and P_σ; unless P_ρ and P_σ overlap, and in this case the progressive product $P_\rho P_\sigma$ is zero.

If $\rho + \sigma > \nu$, then $P_\rho P_\sigma$ is regressive and represents the complete region common both to P_ρ and P_σ; unless P_ρ and P_σ overlap in a region of order greater than $\rho + \sigma - \nu$, and in this case $P_\rho P_\sigma$ is zero.

If $\rho + \sigma = \nu$, then $(P_\rho P_\sigma)$ is a mere number and can be considered either as progressive or regressive.

The only formulae which in practice it is necessary to retain in the memory are the extended rule of the middle factor [cf. § 103] and proposition G of § 105.

105. PROPOSITION G. If $a_1, a_2 \ldots a_\rho$ be ρ points in a region of $\nu - 1$ dimensions ($\nu > \rho$), and if $B_1, B_2 \ldots B_\rho$ be ρ planar elements, then

$$(a_1 a_2 \ldots a_\rho . B_1 B_2 \ldots B_\rho) = \begin{vmatrix} (a_1 B_1), & (a_1 B_2) & \ldots & (a_1 B_\rho) \\ (a_2 B_1), & (a_2 B_2) & \ldots & (a_2 B_\rho) \\ \ldots & \ldots & \ldots & \ldots \\ (a_\rho B_1), & (a_\rho B_2) & \ldots & (a_\rho B_\rho) \end{vmatrix}.$$

For assume that the formula is true for the number $\rho - 1$ respectively of points and of planar elements, to prove that it is true for the number ρ; where $\rho < \nu$.

Let $\Delta_{\sigma\tau}$ denote the minor of the element $(a_\sigma B_\tau)$ of the above determinant.

Now the product of the $\rho + 1$ regional elements of the ρth order $(a_1 a_2 \ldots a_\rho)$, $B_1, B_2 \ldots B_\rho$ is a pure regressive product and is therefore associative.

Hence $(a_1 a_2 \ldots a_\rho . B_1 B_2 \ldots B_\rho) = \{(a_1 a_2 \ldots a_\rho . B_1) B_2 B_3 \ldots B_\rho\}$

$$= (a_1 B_1)(a_2 a_3 \ldots a_\rho . B_2 B_3 \ldots B_\rho) + (a_2 B_1)(a_1 a_3 \ldots a_\mu . B_2 B_3 \ldots B_\rho)$$
$$+ \ldots + (a_\rho B_1)(a_1 a_2 \ldots a_{\rho-1} . B_2 B_3 \ldots B_\rho).$$

But since the theorem holds for the number $\rho - 1$, $a_2 a_3 \ldots a_\rho . B_2 B_3 \ldots B_\rho = \Delta_{11}$, with ρ similar equations.

Hence

$$(a_1 a_2 \ldots a_\rho . B_1 B_2 \ldots B_\rho) = (a_1 B_1) \Delta_{11} + (a_2 B_1) \Delta_{21} + \ldots + (a_\rho B_1) \Delta_{\rho 1}$$
$$= \begin{vmatrix} (a_1 B_1), & (a_1 B_2) & \ldots & (a_1 B_\rho) \\ (a_2 B_1), & (a_2 B_2) & \ldots & (a_2 B_\rho) \\ \ldots & \ldots & \ldots & \ldots \\ (a_\rho B_1), & (a_\rho B_2) & \ldots & (a_\rho B_\rho) \end{vmatrix}.$$

But when $\rho = 2$,

$$(a_1 a_2 . B_1 B_2) = (a_1 a_2 . B_1) B_2$$
$$= \{(a_1 B_1) a_2 - (a_2 B_1) a_1\} B_2$$
$$= (a_1 B_1)(a_2 B_2) - (a_2 B_1)(a_1 B_2).$$

Therefore the theorem is true universally.

106. Müller's Theorems*. (1) Let A, B and C be three regional elements of the κth, ρth and σth orders respectively. Also let $\kappa+\rho+\sigma=\nu+\tau$, where τ is positive and not zero. Then the products $A.BC$, ABC and ACB are mixed products. Hence in general [cf. § 101] $A.BC$, ABC, and ACB are not congruent. The conditions will now be investigated which are necessary that an addition relation of the form,

$$A.BC = \lambda ABC + \mu ACB \quad \text{.......................(i)},$$

may exist.

(2) It will be only necessary to consider the proofs for the cases when BC is a progressive product, and consequently when the product of BC into A, namely $A.BC$, is regressive. For if it has been proved in this case that under certain conditions the above equation (i) holds, then by taking supplements

$$|A.|B|C = \lambda|A|B|C + \mu|A|C|B.$$

Let $A' = |A$, $B' = |B$, $C' = |C$. Then $B'C'$ is regressive, and the product of $B'C'$ into A', namely $A'.B'C'$, is progressive.

Also the conditions which hold between A, B, C can be interpreted as conditions between A', B', C'. Hence it follows that, when $B'C'$ is regressive, under certain conditions

$$A'.B'C' = \lambda A'B'C' + \mu A'C'B'.$$

(3) Let the ρ points which are the factors of B be $b_1, b_2, \dots b_\rho$, so that $B = b_1b_2 \dots b_\rho$. Also let the σ points which are the factors of C be $c_1, c_2, \dots c_\sigma$, so that $C = c_1c_2 \dots c_\sigma$. Let the multiplicative combinations of the μth order ($\mu < \rho$) formed out of $b_1, b_2, \dots b_\rho$ be written B_μ, B_μ', B_μ'', and so on. Also let $B_{\rho-\mu}$, $B'_{\rho-\mu}$, $B''_{\rho-\mu}$, and so on, be the complementary multiplicative combinations of the $(\rho-\mu)$th order formed out of $b_1, b_2, \dots b_\rho$, so that

$$B = (B_\mu B_{\rho-\mu}) = (B_\mu' B'_{\rho-\mu}) = (B_\mu'' B''_{\rho-\mu}) = \text{etc.}$$

Let this convention hold for any number μ; and also a similar convention for the multiplicative combinations formed out of $c_1, c_2, \dots c_\sigma$, which are the factors of C.

(4) Assume that the product BC is progressive. Then the multiplication symbolized by the dot in $A.BC$ is regressive. Also

$$BC = b_1b_2 \dots b_\rho c_1c_2 \dots c_\sigma = D, \text{ say.}$$

Let the convention of the previous subsection apply to the multiplicative combinations formed out of the $\rho+\sigma$ factors of the first order which compose D.

Then by the extended rule of the middle factor [cf. § 103]

$$A.BC = AD = \Sigma (AD_{\rho+\sigma-\tau}) D_\tau \quad \text{....................(ii)},$$

* Cf. Emil Müller, 'Ueber das gemischte Product,' *Mathematische Annalen*, vol. XLVIII. 1897.

where D_τ, D_τ', etc., are the multiplicative combinations of the τth order formed out of $b_1, b_2, \dots b_\rho, c_1, c_2, \dots c_\sigma$. If τ be less than both ρ and σ, some of the multiplicative combinations D_τ, D_τ', etc., contain only b's, some only c's, and some both b's and c's. If τ be less than ρ and greater than σ (assuming $\rho > \sigma$), then some of the D_τ's contain only b's, and some contain both b's and c's; but none contain only c's. If τ be greater than both ρ and σ, then all the D_τ's must contain both b's and c's.

(5) Let the products BC, AB, AC be progressive. Then

$$\rho + \sigma < \nu, \quad \kappa + \rho < \nu, \quad \kappa + \sigma < \nu.$$

By the extended rule of the middle factor

$$ABC = \Sigma (ABC_{\sigma-\tau})\, C_\tau, \quad ACB = \Sigma (ACB_{\rho-\tau})\, B_\tau.$$

Hence if a relation of the form of equation (i) holds, no D_τ's must exist which contain both b's and c's. But this condition can only hold when $\tau = 1$.

Hence the condition is that

$$\kappa + \rho + \sigma = \nu + 1.$$

Also, remembering that $c_\mu BC^{(\mu)}_{\sigma-1} = (-1)^\rho Bc_\mu C^{(\mu)}_{\sigma-1} = (-1)^\rho BC$, equation (ii) becomes

$$A \,.\, BC = (-1)^\rho \Sigma (ABC^{(\mu)}_{\sigma-1})\, c_\mu + \Sigma (AB^{(\mu)}_{\rho-1} C) b_\mu.$$

And

$$ABC = \Sigma (ABC^{(\mu)}_{\sigma-1})\, c_\mu, \quad ACB = \Sigma (ACB^{(\mu)}_{\rho-1})\, b_\mu = (-1)^{\sigma(\rho-1)} \Sigma (AB^{(\mu)}_{\rho-1} C)\, b_\mu.$$

Hence $$A \,.\, BC = (-1)^\rho ABC + (-1)^{\sigma(\rho-1)} ACB \quad \text{..............(iii)}.$$

This is the required equation of the form of equation (i).

(6) Let the products BC, AB, AC be regressive.

Then $$\rho + \sigma > \nu, \quad \kappa + \rho > \nu, \quad \kappa + \sigma > \nu.$$

Taking the supplements, $|A$, $|B$, $|C$ are of orders $(\nu - \kappa)$, $(\nu - \rho)$, $(\nu - \sigma)$ respectively, and $|B|C$, $|A|B$, $|A|C$ are progressive.

Hence in order that a relation of the required form may hold, by the previous subsection,

$$(\nu - \kappa) + (\nu - \rho) + (\nu - \sigma) = \nu + 1,$$

that is, $$\kappa + \rho + \sigma = 2\nu - 1.$$

Also from equation (iii)

$$|A \,.\, |B|C = (-1)^{\nu-\rho} |A|B|C + (-1)^{(\nu-\sigma)(\nu-\rho-1)} |A|C|B.$$

Hence $$A \,.\, BC = (-1)^{\nu-\rho} ABC + (-1)^{(\nu-\sigma)(\nu-\rho-1)} ACB \quad \text{...........(iv)}.$$

(7) Let the products BC and AB be progressive; and let the product AC be regressive.

Then $$\kappa + \rho + \sigma = \nu + \tau, \quad \kappa + \rho < \nu, \quad \rho + \sigma < \nu, \quad \kappa + \sigma > \nu.$$

Hence $$\sigma > \tau, \quad \kappa > \tau, \quad \rho < \tau.$$

By the extended rule of the middle factor

$$ABC = \Sigma\,(ABC_{\sigma-\tau})\,C_\tau,\ \ AC = \Sigma\,(AC_{\nu-\kappa})\,C_{\kappa+\sigma-\nu}.$$

Hence $$ACB = \Sigma\,(AC_{\nu-\kappa})\,C_{\kappa+\sigma-\nu}\,B.$$

Accordingly if a relation of the form of equation (i) holds, the D_τ's [cf. subsection (4)] must consist of two classes only, namely those composed only of c's, and those which contain all the b's.

But this is only possible if

$$\rho = 1.$$

In this case B is of the first order and will be written b.

Then remembering that

$$C_\tau b C_{\sigma-\tau} = (-1)^\tau\, bC,\ \ bC_{\tau-1}\,C_{\sigma-\tau+1} = bC,$$

and that $$\kappa + \sigma - \nu = \tau - 1,\ \ \sigma - \tau + 1 = \nu - \kappa,$$

equation (ii) takes the form

$$A\,.\,BC = (-1)^\tau\,\Sigma\,(AbC_{\sigma-\tau})\,C_\tau + \Sigma\,(AC_{\sigma-\tau+1})\,bC_{\tau-1},$$
$$= (-1)^\tau ABC + (-1)^{\tau-1} ACB \ldots\ldots\ldots\ldots\text{(v)}.$$

This is the required equation of the form of equation (i).

(8) Let the products BC and AB be regressive and the product AC be progressive. This case can be deduced from subsection (7) by the method of subsection (6).

The necessary condition for the existence of the required addition relation is

$$\rho = \nu - 1.$$

Then from the assumptions it follows that

$$\kappa > 1,\ \sigma > 1,\ \kappa + \sigma < \nu,\ \kappa + \nu - 1 + \sigma = \nu + \tau.$$

Also $$A\,.\,BC = (-1)^{\nu-\tau} ABC + (-1)^{\nu-\tau-1} ACB \ldots\ldots\ldots\ldots\text{(vi)}.$$

(9) Let the products BC and AC be progressive, and the product AB be regressive.

Now $$A\,.\,BC = (-1)^{\rho\sigma} A\,.\,CB.$$

Hence this case can be deduced from that of subsection (7).

The necessary condition is that

$$\sigma = 1.$$

Then $$A\,.\,CB = (-1)^\tau ACB + (-1)^{\tau-1} ABC.$$

Hence $$A\,.\,BC = (-1)^{\rho+\tau-1} ABC + (-1)^{\rho+\tau} ACB \ldots\ldots\ldots\ldots\text{(vii)}.$$

(10) Let the products BC and AC be regressive, and the product AB be progressive.

Then from the previous subsection

$$\sigma = \nu - 1,$$

And $$A\,.\,BC = (-1)^{2\nu-\rho-\tau-1} ABC + (-1)^{2\nu-\rho-\tau} ACB$$
$$= (-1)^{\rho+\tau-1} ABC + (-1)^{\rho+\tau} ACB \ldots\ldots\ldots\ldots\text{(viii)}.$$

(11) Let the product BC be progressive, and the products AB and AC be regressive.

Then $\quad AB = \Sigma (AB_{\nu-\kappa}) B_{\rho+\kappa-\nu}, \; AC = \Sigma (AC_{\nu-\kappa}) C_{\sigma+\kappa-\nu}.$

Hence $\; ABC = \Sigma (AB_{\nu-\kappa}) B_{\rho+\kappa-\nu} C, \; ACB = \Sigma (AC_{\nu-\kappa}) C_{\sigma+\kappa-\nu} B.$

Thus the D_τ's of subsection (4) equation (ii) must either contain all the b's or all the c's; and thus the $D_{\rho+\sigma-\tau}$'s of the same equation must contain only b's or only c's. Hence the $D_{\rho+\sigma-\tau}$'s are of the first order, that is to say, $\rho + \sigma - \tau = 1$. But $\kappa + \rho + \sigma = \nu + \tau$.

Hence the required condition is that

$$\kappa = \nu - 1.$$

Then $$AB = \Sigma (Ab_\mu) B^{(\mu)}_{\rho-1}, \; AC = (Ac_\mu) C^{(\mu)}_{\sigma-1};$$

where $$B^{(\mu)}_{\rho-1} b_\mu = B, \; C^{(\mu)}_{\sigma-1} c_\mu = C.$$

Thus $$ABC = \Sigma (Ab_\mu) B^{(\mu)}_{\rho-1} C, \; ACB = \Sigma (Ac_\mu) C^{(\mu)}_{\sigma-1} B.$$

Now $$B^{(\mu)}_{\rho-1} Cb_\mu = (-1)^\sigma B^{(\mu)}_{\rho-1} b_\mu C = (-1)^\sigma BC,$$

and $$C^{(\mu)}_{\sigma-1} Bc_\mu = (-1)^\rho C^{(\mu)}_{\sigma-1} c_\mu B = (-1)^\rho CB = (-1)^{\rho\sigma+\rho} BC.$$

Hence by comparing with equation (ii) of subsection (4)

$$A \,.\, BC = (-1)^\sigma ABC + (-1)^{\rho(\sigma+1)} ACB \quad \text{...............(ix).}$$

(12) Let the product BC be regressive, and the products AB and AC be progressive.

Then from the previous subsection the condition is

$$\kappa = 1.$$

Also $$A \,.\, BC = (-1)^{\nu-\sigma} ABC + (-1)^{(\nu-\rho)(\nu-\sigma+1)} ACB \text{.........(x).}$$

(13) It has nowhere been assumed in the foregoing reasoning that A is simple. Accordingly A may be compound.

107. Applications and Examples. (1) The condition that an element x may lie in a subregion P_ρ of $\rho - 1$ dimensions is, $xP_\rho = 0$. This equation may therefore be regarded as the equation of the subregion.

(2) The supplementary equation is, $|x\,|P_\rho = 0$. The product of $|x$ and $|P_\rho$ is regressive, and the equation indicates that $|x$ and $|P_\rho$ overlap in a regional element of an order greater than the excess of the orders of $|x$ and $|P_\rho$ above ν. Now the order of $|x$ is $\nu - 1$, and the order of $|P_\rho$ is $\nu - \rho$. Hence the order of the common region is greater than

$$(\nu - 1) + (\nu - \rho) - \nu,$$

that is, is greater than $\nu - \rho - 1$. But the subregional element $|P_\rho$ is only of order $\nu - \rho$. Hence $|P_\rho$ must be contained in the plane $|x$. This is the signification of the supplementary equation.

(3) The supplementary equation can be regarded as the original equation and written in the form

$$X_{\nu-1}P_\rho = 0;$$

where $X_{\nu-1}$ is a planar element, and P_ρ is a subregional element of the ρth order. The preceding proof shows that this equation is the condition that the plane $X_{\nu-1}$ contains the subregion P_ρ.

The supplementary equation is now $|X_{\nu-1}|P_\rho = 0$, and signifies that the point $|X_{\nu-1}$ lies in the region $|P_\rho$ of $\nu-\rho-1$ dimensions.

(4) The theory of duality also applies, and $xP_\rho = 0$ can be regarded as the condition that the subregion P_ρ contains the given point x; and the equation, $X_{\nu-1}P_\rho = 0$, as the condition that the subregion P_ρ is contained in the given plane $X_{\nu-1}$.

(5) In the previous subsection it has been assumed that P_ρ is a regional element, that is to say, is simple. Now let S_ρ be a compound extensive magnitude of the ρth order. Then in general it is impossible to satisfy the equation $xS_\rho = 0$, except by the assumption that $x = 0$.

For xS_ρ is an extensive magnitude of the $\rho+1$th order; but this manifold is defined by $\frac{\nu!}{(\nu-\rho-1)!\,(\rho+1)!}$ independent units (cf. § 94). Hence if $xS_\rho = 0$, the coefficient of each of these units, as it appears in the expression xS_ρ, must vanish. Thus there are $\frac{\nu!}{(\nu-\rho-1)!\,(\rho+1)!}$ equations to be satisfied. But in $x\,(=\Sigma\xi e)$ there are only $\nu-1$ unknowns, namely, the ratios of $\xi_1, \xi_2 \ldots \xi_\nu$. But if ρ be any one of the numbers $2, 3 \ldots \nu-2$,

$$\frac{\nu!}{(\nu-\rho-1)!\,(\rho+1)!} \geqq \nu.$$

In these cases the requisite equations cannot be satisfied. If $\rho = 1$, then S_ρ is a point and must be simple: the equation $xS_\rho = 0$ then means that $x \equiv S_\rho$.

If $\rho = \nu-1$, then S_ρ is a planar element and must be simple (cf. § 97, Prop. IV).

(6) Let $P_{\nu-1}$ be the planar element

$$\pi_1 e_2 e_3 \ldots e_\nu - \pi_2 e_1 e_3 \ldots e_\nu + \pi_3 e_1 e_2 e_4 \ldots e_\nu - \pi_4 e_1 e_2 e_3 e_5 \ldots e_\nu + \ldots + (-1)^{\nu-1}\pi_\nu e_1 e_2 \ldots e_{\nu-1}.$$

Then $\pi_1, \pi_2 \ldots \pi_\nu$ are the co-ordinates of the planar element $P_{\nu-1}$ with respect to the reference elements $e_1, e_2 \ldots e_\nu$.

Also if x be $\Sigma\xi e$, then

$$(xP_{\nu-1}) = (\pi_1\xi_1 + \pi_2\xi_2 + \ldots + \pi_\nu\xi_\nu)(e_1 e_2 \ldots e_\nu).$$

Hence the equation $(xP_{\nu-1}) = 0$, is equivalent to the usual equation of a plane, namely,

$$\pi_1\xi_1 + \pi_2\xi_2 + \ldots + \pi_\nu\xi_\nu = 0.$$

And conversely $P_{\nu-1}$, as defined in this subsection, is a planar element in the plane which is defined by the equation

$$\pi_1\xi_1 + \dots + \pi_\nu\xi_\nu = 0.$$

(7) Another simple method of obtaining a slightly different form of a planar element corresponding to the plane

$$\pi_1\xi_1 + \dots + \pi_\nu\xi_\nu = 0,$$

is found by means of § 73 (2). The point in which the plane cuts the straight line e_1e_ρ is by that article $\frac{e_1}{\pi_1} - \frac{e_\rho}{\pi_\rho}$. Hence by multiplying the $\nu - 1$ such points which lie on the $\nu - 1$ such straight lines meeting in e_1, a planar element in the plane is found to be

$$\prod_{\rho=2}^{\rho=\nu} \left(\frac{e_1}{\pi_1} - \frac{e_\rho}{\pi_\rho}\right).$$

Hence

$$\prod_{\rho=2}^{\rho=\nu} \left(\frac{e_1}{\pi_1} - \frac{e_\rho}{\pi_\rho}\right) \equiv \pi_1 e_2 e_3 \dots e_\nu - \pi_2 e_1 e_3 \dots e_\nu + \dots + (-1)^{\nu-1} \pi_\nu e_1 e_2 \dots e_{\nu-1}.$$

Therefore by multiplying out the left-hand side and comparing the coefficients of the term $e_2e_3 \dots e_\nu$ on the two sides,

$$(-1)^{\nu-1} \pi_1\pi_2 \dots \pi_\nu \prod_{\rho=2}^{\rho=\nu} \left(\frac{e_1}{\pi_1} - \frac{e_\rho}{\pi_\rho}\right)$$
$$= \pi_1 e_2 e_3 \dots e_\nu - \pi_2 e_1 e_3 \dots e_\nu + \dots + (-1)^{\nu-1} \pi_\nu e_1 e_2 \dots e_{\nu-1}.$$

This factorization of the right-hand side of the above equation into a product of $\nu - 1$ points forms another proof of § 97, Prop. IV.

(8) Among special applications of these theorems we may notice that the condition that x may lie on the straight line joining a and b is

$$xab = 0;$$

the condition that x may lie in the two dimensional region abc is

$$xabc = 0;$$

the condition that x may lie in the three dimensional region $abcd$ is

$$xabcd = 0.$$

(9) Let the complete manifold be of more than two dimensions so that the multiplication of linear elements is progressive. The multiplication of a planar and linear element together is necessarily regressive.

Then two lines ab and cd intersect if $abcd = 0$. For this is the condition that a, b, c and d lie in the same subregion of two dimensions.

The point where a line ab intersects a given plane $P_{\nu-1}$ is $P_{\nu-1} . ab$. But by § 103 (2) (the extended rule of the middle factor)

$$P_{\nu-1} . ab = (P_{\nu-1}b)\, a - (P_{\nu-1}a)\, b.$$

If the line lie entirely in the plane, $(P_{\nu-1}b)=0$, and $(P_{\nu-1}a)=0$; hence

$$P_{\nu-1}\,.\,ab=0.$$

If the planar element be written as the product $c_1c_2\ldots c_{\nu-1}$, then the point of intersection of the line ab with it can be written $c_1c_2\ldots c_{\nu-1}\,.\,ab$. And by § 103 (2)

$$c_1c_2\ldots c_{\nu-1}\,.\,ab=(c_1c_2\ldots c_{\nu-2}ab)\,c_{\nu-1}+(-1)^\nu(c_{\nu-1}c_1\ldots c_{\nu-3}ab)\,c_{\nu-2}+\ldots$$
$$+(-1)^\nu(c_2c_3\ldots c_{\nu-1}ab)\,c_1.$$

The last form exhibits the fact that the point of intersection lies in the plane $c_1c_2\ldots c_{\nu-1}$; while the form $(P_{\nu-1}b)\,a-(P_{\nu-1}a)\,b$ exhibits the fact that the point of intersection lies on the straight line ab.

(10) Two planar elements $P_{\nu-1}$ and $Q_{\nu-1}$ must intersect in a region of $\nu-3$ dimensions, or in other words the extensive magnitude $P_{\nu-1}\,.\,Q_{\nu-1}$ is a regional element of the $(\nu-2)$th order. Let such subregions be called subplanes. The magnitudes denoted by $\lambda P_{\nu-1}+\mu Q_{\nu-1}$ for varying values of the ratio λ/μ are planes containing the subplane $P_{\nu-1}\,.\,Q_{\nu-1}$, common to $P_{\nu-1}$ and $Q_{\nu-1}$.

In regions of three dimensions straight lines and subplanes are identical.

(11) If four given planes $P_{\nu-1}$, $Q_{\nu-1}$, $R_{\nu-1}$, $S_{\nu-1}$ contain a common subplane $L_{\nu-2}$, then the four points of intersection of any straight line with these planes form a range with a given anharmonic ratio.

For let $P_{\nu-1}=L_{\nu-2}\,a$, $Q_{\nu-1}=L_{\nu-2}\,b$, $R_{\nu-1}=L_{\nu-2}\,c$, $S_{\nu-1}=L_{\nu-2}\,d$.

Let pq be any line, and assume that p lies in $P_{\nu-1}$ and q in $Q_{\nu-1}$.

Then $L_{\nu-2}p=\varpi P_{\nu-1}$, and $L_{\nu-2}q=\rho Q_{\nu-1}$. Also let pq intersect $R_{\nu-1}$ and $S_{\nu-1}$ in r and s.

Then $$r=R_{\nu-1}\,.\,pq=(R_{\nu-1}\,q)\,p-(R_{\nu-1}\,p)\,q$$
$$=-(L_{\nu-2}qc)\,p+(L_{\nu-2}pc)\,q=-\rho\,(Q_{\nu-1}\,c)\,p+\varpi\,(P_{\nu-1}\,c)\,q.$$

Similarly $$s=-\rho\,(Q_{\nu-1}\,d)\,p+\varpi\,(P_{\nu-1}\,d)\,q.$$

Hence the anharmonic ratio $(pq,\,rs)=\dfrac{(Q_{\nu-1}\,c)\,(P_{\nu-1}\,d)}{(Q_{\nu-1}\,d)\,(P_{\nu-1}\,c)}$.

This ratio is the same for all lines pq; it can also be expressed as $(R_{\nu-1}\,b)\,(S_{\nu-1}\,a)/(S_{\nu-1}\,b)\,(R_{\nu-1}\,a)$.

(12) If $R_{\nu-1}=\lambda P_{\nu-1}+\mu Q_{\nu-1}$, and $S_{\nu-1}=\lambda' P_{\nu-1}+\mu' Q_{\nu-1}$, then $c=\lambda a+\mu b$, and $d=\lambda' a+\mu' b$.

Also since $(aP_{\nu-1})=0=(bQ_{\nu-1})$, we have

$$(pq,\,rs)=\frac{\lambda\,(Q_{\nu-1}\,a)\,.\,\mu'\,(P_{\nu-1}b)}{\lambda'\,(Q_{\nu-1}\,a)\,.\,\mu\,(P_{\nu-1}\,b)}=\frac{\lambda\mu'}{\lambda'\mu}.$$

We also notice that $|\,P_{\nu-1}$, $|\,Q_{\nu-1}$, $|\,R_{\nu-1}$, and $|\,S_{\nu-1}$ are four collinear points with the same anharmonic ratio, $\lambda\mu'/\lambda'\mu$, as the four planes.

NOTE. In developing the theory of Regressive Multiplication the *Ausdehnungslehre von* 1862 has been closely adhered to.

CHAPTER III.

SUPPLEMENTS.

108. SUPPLEMENTARY REGIONS. (1) The supplement of a regional element P_ρ of the ρth order is a regional element $|P_\rho$ of the $(\nu-\rho)$th order [cf. § 65 (4) and § 99]. The two subregions P_ρ and $|P_\rho$ are called supplementary. In particular $|x$ is the supplementary plane of the point x, and x the supplementary point of the plane $|x$.

(2) If P_ρ be expressed as the product of ρ points $p_1, p_2, \dots p_\rho$, then taking the supplement

$$|P_\rho = |p_1|p_2 \dots |p_\rho.$$

Hence if P_ρ be the containing region of the ρ independent points, then P_ρ is the common region of the ρ supplementary planes of those points.

(3) If P_ρ and P_ρ' be two regional elements both of the ρth order, then $(P_\rho | P_\rho')$ is merely numerical.

Hence $(P_\rho|P_\rho') = |(P_\rho|P_\rho') = (|P_\rho||P_\rho') = (-1)^{\rho(\nu-\rho)}(|P_\rho . P_\rho') = (P_\rho'|P_\rho)$.

(4) Thus if y lies in the supplementary plane of x, then $(y|x) = 0 = (x|y)$. Hence x lies in the supplementary plane of y.

(5) *Definition.* Points which lie each in the supplementary plane of the other will be called mutually normal points.

If the points $x \, (= \Sigma \xi e)$ and $y \, (= \Sigma \eta e)$ be mutually normal, then

$$(x|y) = (\xi_1\eta_1 + \xi_2\eta_2 + \dots \xi_\nu\eta_\nu) = 0.$$

(6) A point x does not in general lie in its own supplementary plane, unless it lies on the quadric

$$(x|x) = 0 = \xi_1^2 + \xi_2^2 + \dots + \xi_\nu^2.$$

Let points which lie in their own supplementary planes be called self-normal; and let the quadric which is the locus of such points be called the self-normal quadric.

109. NORMAL SYSTEMS OF POINTS. (1) All the points normal to a given point x_1 lie in the plane $|x_1$. Let x_2 be any such point, and let x_3 lie in the subregion $|x_1|x_2$, and x_4 in the subregion $|x_1|x_2|x_3$; and so on; and finally let x_ν be the point $|x_1|x_2 \dots |x_{\nu-1}$. Then assuming that none of these points are self-normal, we have deduced a system of ν independent mutually normal points, starting with any arbitrary point x_1.

(2) *Definition.* Let a system of ν independent mutually normal elements be called a normal system.

(3) The intensities of the normal system of points as denoted by $x_1, x_2 \ldots x_\nu$ are arbitrary.

Definition. Let any point p be said to be denoted at its normal intensity when $(p\,|\,p) = 1$. Note that the normal intensity of a point is not necessarily its unit intensity.

(4) Then if $x_1, x_2, \ldots x_\nu$ be a normal system of elements at their normal intensities, the following equations are satisfied

$$(x_1\,|\,x_1) = (x_2\,|\,x_2) = \text{etc.} = 1, \text{ and } (x_\rho\,|\,x_\sigma) = 0, \text{ where } \rho \neq \sigma.$$

If $x_\rho = \xi_{1\rho}e_1 + \xi_{2\rho}e_2 + \ldots \xi_{\nu\rho}e_\nu$, these equations can be written

$$\xi_{1\rho}^2 + \xi_{2\rho}^2 + \ldots + \xi_{\nu\rho}^2 = 1,$$

with $\nu - 1$ other similar equations,

and
$$\xi_{1\rho}\xi_{1\sigma} + \xi_{2\rho}\xi_{2\sigma} + \ldots + \xi_{\nu\rho}\xi_{\nu\sigma} = 0,$$

with $\frac{1}{2}\nu(\nu-1) - 1$ other similar equations.

(5) Also by § 97, Prop. I., the following equations hold

$$|x_1 = \lambda_1 x_2x_3 \ldots x_\nu, \quad |x_2 = \lambda_2 x_1x_3 \ldots x_\nu, \quad \ldots, \quad |x_\nu = \lambda_\nu x_1x_2 \ldots x_\nu.$$

Hence
$$(x_1\,|\,x_1) = \lambda_1\,(x_1x_2 \ldots x_\nu) = 1.$$

Therefore
$$\lambda_1 = \frac{1}{(x_1x_2 \ldots x_\nu)} = -\lambda_2 = \lambda_3 = \ldots = (-1)^{\nu-1}\lambda_\nu.$$

Also since $(x_1\,|\,x_1)$ is merely numerical, then by § 99 (5)

$$(x_1\,|\,x_1) = |(x_1\,|\,x_1) = (\,|x_1\,||\,x_1) = \frac{1}{(x_1x_2 \ldots x_\nu)^2}(x_2x_3 \ldots x_\nu\,|\,x_2x_3 \ldots x_\nu).$$

Hence by § 105 and by the previous subsection of the present article

$$(x_1\,|\,x_1) = \frac{1}{(x_1x_2 \ldots x_\nu)^2}\begin{vmatrix} (x_2\,|\,x_2), & (x_2\,|\,x_3), & \ldots & (x_2\,|\,x_\nu) \\ (x_3\,|\,x_2), & (x_3\,|\,x_3), & \ldots & (x_3\,|\,x_\nu) \\ \ldots & \ldots & \ldots & \ldots \\ (x_\nu\,|\,x_2), & (x_\nu\,|\,x_3), & \ldots & (x_\nu\,|\,x_\nu) \end{vmatrix} = \frac{1}{(x_1x_2 \ldots x_\nu)^2}.$$

Hence $(x_1x_2 \ldots x_\nu)^2 = 1$, and therefore $(x_1x_2 \ldots x_\nu) = \pm 1$.

Now if x_ρ be at its normal intensity, then (cf. § 89) $-x_\rho$ is also at its normal intensity. Hence by properly choosing the signs of $x_1, x_2, \ldots x_\nu$, we can make $(x_1x_2 \ldots x_\nu) = 1$. Thus finally with this convention

$$|x_1 = x_2x_3 \ldots x_\nu, \quad |x_2 = -x_1x_3 \ldots x_\nu, \quad \ldots \quad |x_\nu = (-1)^{\nu-1}x_1x_2 \ldots x_{\nu-1}.$$

(6) Hence the operator | bears the same relation to the normal system $x_1, x_2 \ldots x_\nu$ at normal intensities as it does to the original reference-elements. Accordingly in the operation of taking the supplement the original reference-elements may be replaced by any normal system at normal intensities.

110. Extension of the definition of Supplements. (1) This possibility of replacing the original reference-elements by other elements in the operation of taking the supplement suggests an extended* conception of the operation.

In the original definition the terms $e_1, e_2 \dots e_\nu$ represent the reference-elements at their normal intensities as well as at their unit intensities. But suppose now, as a new definition which is allowable by § 109 (3) and (6), that the normal intensities of these reference-elements are $\epsilon_1, \epsilon_2 \dots \epsilon_\nu$. Then by hypothesis [cf. § 109 (3)]

$$|\epsilon_1 e_1 = \epsilon_2 e_2 \,.\, \epsilon_3 e_3 \dots \epsilon_\nu e_\nu, \quad |\epsilon_2 e_2 = -\,\epsilon_1 e_1 \,.\, \epsilon_3 e_3 \dots \epsilon_\nu e_\nu,$$

and so on.

Also it must be assumed that $\epsilon_1 \epsilon_2 \dots \epsilon_\nu (e_1 e_2 \dots e_\nu) = 1$.

Let $$\epsilon_1 \epsilon_2 \dots \epsilon_\nu = \Delta = \frac{1}{(e_1 e_2 \dots e_\nu)}.$$

Then $$|e_1 = \frac{\Delta}{\epsilon_1^2} e_2 e_3 \dots e_\nu, \quad |e_2 = \frac{-\Delta}{\epsilon_2^2} e_1 e_3 \dots e_\nu, \text{ and so on.}$$

(2) This extended definition in no way alters the fundamental properties of the operation denoted by $|$. For this operation has been proved to be referred to an indefinite number of normal sets of points and cannot therefore be dependent on the symbolism by which we choose to denote one set of them.

Thus it follows that the symbol $|$ obeys the distributive law both for multiplication and addition. Also $||P_\rho = (-1)^{\rho(\nu-\rho)} P_\rho$, where P_ρ is of the ρth order.

But $$(e_1 | e_1) = \frac{1}{\epsilon_1^2}, \quad (e_2 | e_2) = \frac{1}{\epsilon_2^2}, \dots (e_\nu | e_\nu) = \frac{1}{\epsilon_\nu^2}.$$

Also it is not necessary that $\epsilon_1, \epsilon_2 \dots \epsilon_\nu$ should all be real; thus any number of their squares may be conceived as being negative.

(3) The self-normal quadric is defined by the equation $(x|x) = 0$; that is by $$\xi_1^2/\epsilon_1^2 + \xi_2^2/\epsilon_2^2 + \dots \xi_\nu^2/\epsilon_\nu^2 = 0.$$
If $\epsilon_1, \epsilon_2 \dots \epsilon_\nu$ be all real, this quadric is purely imaginary: but if some of them be pure imaginaries, this quadric is real. Since only the ratios of $\epsilon_1, \epsilon_2 \dots \epsilon_\nu$ are required for defining the self-normal quadric, it is allowable when convenient to define, $\epsilon_1 \epsilon_2 \dots \epsilon_\nu = 1$. Hence in this case $(e_1 e_2 \dots e_\nu) = 1$.

(4) The equation of the supplementary plane of any point x is $(y|x) = 0$; that is, if $x = \Sigma \xi e$ and $y = \Sigma \eta e$, the equation

$$\xi_1 \eta_1/\epsilon_1^2 + \xi_2 \eta_2/\epsilon_2^2 + \dots + \xi_\nu \eta_\nu/\epsilon_\nu^2 = 0.$$

But this is the equation of the polar plane of x [cf. § 78 (1)].

Hence the method of supplements is simply a symbolic application of the theory of reciprocal polars and its extension to linear elements and to other regional elements in manifolds of more than three dimensions.

* This extension is not given by Grassmann.

(5) Normal sets of elements are obviously sets of polar reciprocal elements forming a self-conjugate set with respect to the self-normal quadric.

In future it will be better to speak of taking the supplement with respect to an assumed self-normal quadric, rather than with respect to a particular set of normal elements.

111. Different Kinds of Supplements. (1) It may be desirable to take supplements with respect to various quadrics. The operation of taking the supplement with respect to one quadric is different from the operation of taking it with respect to another. If one operation be denoted by the symbol $|$, let another be denoted by the symbol I. Then $|P$ and $\mathrm{I}P$ denote different extensive magnitudes. But the operator I possesses all the properties which have been proved to belong to the operator $|$.

Also if the supplement is taken with respect to a third quadric, the operator might be denoted by I_1 and so on.

(2) Confining ourselves to two operations of taking the supplement, denoted by $|$ and I, we see that the two self-normal quadrics are denoted by

$$(x\,|x) = 0, \text{ and } (x\,\mathrm{I}x) = 0.$$

But [cf. § 83 (6)] in general two quadrics possess one and only one system of ν distinct self-conjugate points.

Let $e_1, e_2 \dots e_\nu$ be these points and let $\epsilon_1, \epsilon_2 \dots \epsilon_\nu$ be their normal intensities with respect to the operation $|$, and $\epsilon_1', \epsilon_2' \dots \epsilon_\nu'$ those with respect to the operation I.

Then
$$\epsilon_1\epsilon_2 \dots \epsilon_\nu = \frac{1}{(e_1e_2 \dots e_\nu)} = \epsilon_1'\epsilon_2' \dots \epsilon_\nu' = \Delta.$$

Hence
$$(x\,|x) = \Delta\left(\frac{\xi_1^2}{\epsilon_1^2} + \frac{\xi_2^2}{\epsilon_2^2} + \dots + \frac{\xi_\nu^2}{\epsilon_\nu^2}\right),$$

and
$$(x\,\mathrm{I}x) = \Delta\left(\frac{\xi_1^2}{\epsilon_1'^2} + \frac{\xi_2^2}{\epsilon_2'^2} + \dots + \frac{\xi_\nu^2}{\epsilon_\nu'^2}\right).$$

Also
$$|e_1 = \frac{\Delta}{\epsilon_1^2} e_2e_3 \dots e_\nu, \text{ and } \mathrm{I}e_1 = \frac{\Delta}{\epsilon_1'^2} e_2e_3 \dots e_\nu.$$

112. Normal Points and Straight Lines. (1) The following propositions can easily be seen to be true for mutually normal points with respect to any quadric.

On any straight line one point and only one point can be found normal to a given point, unless every point on the line is normal to the given point.

If a be the given point and bc the given line, this point is

$$bc\,|a = (b\,|a)\,c - (c\,|a)\,b,$$

unless
$$bc\,|a = 0 = (a\,|b) = (a\,|c).$$

(2) There are two exceptional self-normal points on every straight line (viz. the points in which the line cuts the self-normal quadric), but in general these self-normal points are normal to no other points on the line. If however these two self-normal points coincide, so that the line is tangential, then this double point is normal to every other point on the line.

(3) It follows from the harmonic properties of poles and polars that the pairs of normal points on a line form a system of points in involution, with the self-normal points as foci.

(4) This harmonic theorem can be proved thus: let a_1, a_2 be the two self-normal points of any line; then $(a_1 | a_1) = 0 = (a_2 | a_2)$.

Let $\lambda a_1 + \mu a_2$ and $\lambda' a_1 + \mu' a_2$ be any pair of normal points. Then

$$(\lambda a_1 + \mu a_2) | (\lambda' a_1 + \mu' a_2) = 0.$$

Hence $$(\lambda\mu' + \lambda'\mu)(a_1 | a_2) = 0.$$

Hence $$\lambda/\mu = -\lambda'/\mu'.$$

113. Mutually normal regions. (1) Two regions P_ρ and P_σ, where ρ and σ denote the orders of P_ρ and P_σ respectively, are called mutually normal, or normal to each other, if every pair of points p_ρ and p_σ respectively in P_ρ and P_σ are mutually normal.

(2) Let P_ρ be defined by the points $p_\rho^{(1)}, p_\rho^{(2)}, \dots p_\rho^{(\rho)}$, and P_σ by the points $p_\sigma^{(1)}, p_\sigma^{(2)}, \dots p_\sigma^{(\sigma)}$. Then any point on P_σ must lie on the intersection of the supplementary planes of $p_\rho^{(1)}, p_\rho^{(2)}, \dots p_\rho^{(\rho)}$. Similarly any point on P_ρ must lie on the intersection of the supplementary planes of $p_\sigma^{(1)}, p_\sigma^{(2)}, \dots p_\sigma^{(\sigma)}$. Thus the condition that P_σ and P_ρ should be mutually normal is that P_σ should be contained in $|P_\rho$, or that P_ρ should be contained in $|P_\sigma$. Either condition is sufficient to secure the satisfaction of the other.

(3) If P_ρ and P_σ be mutually normal, then

$$\sigma \leqq \nu - \rho, \text{ that is } \nu \geqq \rho + \sigma.$$

(4) If $\nu = \rho + \sigma$, then $P_\sigma \equiv | P_\rho$. Hence the supplementary regions are mutually normal. The supplementary region of P_ρ will be called the complete normal region of P_ρ, or (where there is no risk of mistake) the normal region of P_ρ. Thus the supplementary plane of a point is its normal region.

(5) In any subregion P_ρ (of the ρth order) ρ mutually normal points can be found, of which any assumed point in P_ρ (which is not self-normal) is one. For [cf. § 78 (9)] take x_1 to be any point in P_ρ, then $|x_1$ intersects P_ρ in a region of the $(\rho-1)$th order. Take x_2 to be any point (not self-normal) in this region. Then $|x_2$ intersects this region in a subregion of the $(\rho-2)$th order; take x_3 (not self-normal) in this subregion of the $(\rho-2)$th order, and so on.

If however P_ρ lie in the tangent plane to the self-normal quadric at one of the self-normal points lying in P_ρ, then this self-normal point must be one of any set of ρ mutually normal points in P_ρ. For the supplementary plane of the self-normal point by hypothesis contains P_ρ, hence the supplementary plane of any point in P_ρ contains the self-normal point. Thus proceeding as above in the choice of x_1, x_2, etc., the last point, x_ρ, chosen must be the self-normal point.

(6) To find the subregion (if any) of the highest order normal to P_ρ which is necessarily contained in P_σ; where ρ and σ are respectively the orders of P_ρ and P_σ.

Any region normal to P_ρ is contained in $|P_\rho$. Now $|P_\rho$ and P_σ do not necessarily intersect unless $(\nu - \rho) + \sigma > \nu$, that is, unless $\sigma > \rho$.

Assume $\sigma > \rho$. Then [cf. § 65 (5)] $|P_\rho$ necessarily overlaps P_σ in a subregion of the $\{(\nu - \rho) + \sigma - \nu\}$th order, that is, of the $(\sigma - \rho)$th order. Every point in this subregion is necessarily normal to P_ρ; and hence this subregion of the $(\sigma - \rho)$th order, contained in P_σ, is normal to P_ρ. If the intersection of P_σ and $|P_\rho$ is not of a higher order than $(\sigma - \rho)$, the regional element $P_\sigma|P_\rho$ defines it; thus if $P_\sigma|P_\rho$ be not zero, it is the subregion of P_σ normal to P_ρ.

(7) By subsection (5) ρ mutually normal points can be found in P_ρ and $(\sigma - \rho)$ mutually normal points can be found in the intersection of P_σ and $|P_\rho$ $(\sigma > \rho)$. Also by the previous subsection each point of the one set is normal to each point of the other set. Thus the σ points form a mutually normal set. Hence it is easy to see that, given two subregions P_σ and P_ρ $(\sigma > \rho)$, σ mutually normal points can be found in them, and of these ρ (or any less number) can be chosen in P_ρ and the remainder in P_σ; also that any one point in P_ρ can be chosen arbitrarily to be one of these points or (if ρ points are to be taken in P_ρ) any one point in the intersection of P_σ and $|P_\rho$.

114. Self-normal Elements. (1) Every element in a subregion defined by ρ independent self-normal elements mutually normal to each other is itself self-normal.

For if $a_1, a_2, \ldots a_\rho$ be such elements,

$$(a_1|a_1) = 0 = (a_2|a_2) = \text{etc.} = (a_1|a_2) = \text{etc.}$$

Hence $\{(\lambda_1 a_1 + \lambda_2 a_2 + \ldots + \lambda_\rho a_\rho)|(\lambda_1 a_1 + \lambda_2 a_2 + \ldots + \lambda_\rho a_\rho)\} = 0$.

Also any two elements of such a subregion are normal to each other.

For $\quad \{(\lambda_1 a_1 + \lambda_2 a_2 + \ldots + \lambda_\rho a_\rho)|(\mu_1 a_1 + \mu_2 a_2 + \ldots + \mu_\rho a_\rho)\} = 0.$

(2) Accordingly such a subregion is itself a complete generating region [cf. § 79] of the quadric, $(x|x) = 0$; or is contained in one.

But from § 79 the generating regions of this quadric are, in general, of $\nu/2-1$ or $(\nu-1)/2-1$ dimensions according as ν is even or odd.

Hence sets of $\nu/2$ or $(\nu-1)/2$ (as ν is even or odd) self-normal and mutually normal elements can be found.

(3) Also by § 80 a set of conjugate co-ordinates $j_1, j_2 \ldots, k_1, k_2 \ldots$ can be found all self-normal and all mutually normal except in pairs, *i.e.* $(j_1|k_1)$ is not zero, nor $(j_2|k_2)$ and so on. But $(j_1|j_1)=(j_1|j_2)=\text{etc.}=(j_1|k_2)=\text{etc.}=0$.

If ν be even, ν such co-ordinates can be found which define the complete manifold; but if ν be odd, $\nu-1$ can be found, and one co-ordinate element e remains over, which can be assumed to be normal to the $\nu-1$ other elements, but not self-normal.

(4) Firstly let ν be even. Let $e_1, e_2, \ldots e_\nu$ be a set of normal elements, $\epsilon_1, \epsilon_2 \ldots \epsilon_\nu$ being their normal intensities according to the notation of § 110. Then by § 80, we may assume

$$j_1=\lambda_1(\epsilon_1 e_1+i\epsilon_2 e_2), \qquad k_1=\lambda_1(\epsilon_1 e_1-i\epsilon_2 e_2),$$
$$j_2=\lambda_2(\epsilon_3 e_3+i\epsilon_4 e_4), \qquad k_2=\lambda_2(\epsilon_3 e_3-i\epsilon_4 e_4),$$
$$\ldots\ldots\ldots\ldots\ldots\ldots\ldots\ldots\ldots\ldots$$
$$j_{\frac{\nu}{2}}=\lambda_{\frac{\nu}{2}}(\epsilon_{\nu-1}e_{\nu-1}+i\epsilon_\nu e_\nu), \quad k_{\frac{\nu}{2}}=\lambda_{\frac{\nu}{2}}(\epsilon_{\nu-1}e_{\nu-1}-i\epsilon_\nu e_\nu).$$

Hence $j_1k_1=-2i\lambda_1^2\epsilon_1\epsilon_2 e_1 e_2$, with $\frac{\nu}{2}-1$ other similar equations.

Thus
$$(j_1k_1j_2k_2\ldots j_{\frac{\nu}{2}}k_{\frac{\nu}{2}})=(-2i)^{\frac{\nu}{2}}\lambda_1^2\lambda_2^2\ldots\lambda_{\frac{\nu}{2}}^2.$$

Again
$$|j_1=\lambda_1(\epsilon_1|e_1+i\epsilon_2|e_2)=\frac{\lambda_1\Delta}{\epsilon_1\epsilon_2}(\epsilon_2 e_2 e_3\ldots e_\nu-i\epsilon_1 e_1 e_3\ldots e_\nu).$$

But,
$$j_2k_2j_3k_3\ldots j_{\frac{\nu}{2}}k_{\frac{\nu}{2}}=(-2i)^{\frac{\nu}{2}-1}\epsilon_3\epsilon_4\ldots\epsilon_\nu\lambda_2^2\lambda_3^2\ldots\lambda_{\frac{\nu}{2}}^2 e_3e_4\ldots e_\nu;$$

hence
$$j_1j_2k_2j_3k_3\ldots j_{\frac{\nu}{2}}k_{\frac{\nu}{2}}=(-2i)^{\frac{\nu}{2}-1}\frac{\Delta}{\epsilon_1\epsilon_2}\lambda_1\lambda_2^2\lambda_3^2\ldots\lambda_{\frac{\nu}{2}}^2(\epsilon_1e_1e_3\ldots e_\nu+i\epsilon_2e_2e_3\ldots e_\nu)$$
$$=(-2)^{\frac{\nu}{2}-1}i^{\frac{\nu}{2}}\lambda_2^2\lambda_3^2\ldots\lambda_{\frac{\nu}{2}}^2|j=-\tfrac{1}{2}\lambda_1^{-2}(j_1k_1j_2k_2\ldots j_{\frac{\nu}{2}}k_{\frac{\nu}{2}})|j.$$

(5) Now let $\lambda_1, \lambda_2, \ldots \lambda_{\frac{\nu}{2}}$ be so chosen that

$$\lambda_1=\lambda_2=\ldots=\lambda_{\frac{\nu}{2}}=-\frac{i}{\sqrt{2}}.$$

Then
$$j_1=\frac{1}{\sqrt{2}}(\epsilon_2e_2-i\epsilon_1e_1), \qquad k_1=\frac{1}{\sqrt{2}}(\epsilon_2e_2+i\epsilon_1e_1),$$
$$j_2=\frac{1}{\sqrt{2}}(\epsilon_4e_4-i\epsilon_3e_3), \qquad k_2=\frac{1}{\sqrt{2}}(\epsilon_4e_4+i\epsilon_3e_3),$$
$$\ldots\ldots\ldots\ldots\ldots\ldots\ldots\ldots\ldots\ldots$$
$$j_{\frac{\nu}{2}}=\frac{1}{\sqrt{2}}(\epsilon_\nu e_\nu-i\epsilon_{\nu-1}e_{\nu-1}), \quad k_{\frac{\nu}{2}}=\frac{1}{\sqrt{2}}(\epsilon_\nu e_\nu+i\epsilon_{\nu-1}e_{\nu-1}).$$

And
$$(j_1k_1j_2k_2\ldots j_{\frac{\nu}{2}}k_{\frac{\nu}{2}}) = (-2i)^{\frac{\nu}{2}}\left(\frac{-1}{2}\right)^{\frac{\nu}{2}} = i^{\frac{\nu}{2}}.$$

Hence $j_1j_2k_2j_3k_3\ldots j_{\frac{\nu}{2}}k_{\frac{\nu}{2}} = i^{\frac{\nu}{2}}|j_1$. Similarly $j_2j_1k_1j_3k_3\ldots j_{\frac{\nu}{2}}k_{\frac{\nu}{2}} = i^{\frac{\nu}{2}}|j_2$, and so on.

When $\lambda_1, \lambda_2, \ldots \lambda_{\frac{\nu}{2}}$ have been chosen as above, the conjugate self-normal elements will be said to be in their standard normal form.

When the self-normal elements are in this form

$$j_1k_1 = i\epsilon_1\epsilon_2e_1e_2\,;\; (j_1\,|\,k_1) = (k_1\,|\,j_1) = i^{-\frac{\nu}{2}}(k_1j_1j_2k_2\ldots j_{\frac{\nu}{2}}k_{\frac{\nu}{2}}) = -1.$$

(6) Secondly, let ν be odd, and let $e, e_1, e_2, \ldots e_{\nu-1}$ be the set of normal elements with normal intensities $\epsilon, \epsilon_1, \ldots \epsilon_{\nu-1}$.

Let the standard normal forms of the conjugate self-normal elements be

$$j_1 = \frac{1}{\sqrt{2}}(\epsilon_2e_2 - i\epsilon_1e_1), \qquad k_1 = \frac{1}{\sqrt{2}}(\epsilon_2e_2 + i\epsilon_1e_1),$$

$$j_2 = \frac{1}{\sqrt{2}}(\epsilon_4e_4 - i\epsilon_3e_3), \qquad k_2 = \frac{1}{\sqrt{2}}(\epsilon_4e_4 + i\epsilon_3e_3),$$

....

$$j_{\frac{\nu-1}{2}} = \frac{1}{\sqrt{2}}(\epsilon_{\nu-1}e_{\nu-1} - i\epsilon_{\nu-2}e_{\nu-2}), \quad k_{\frac{\nu-1}{2}} = \frac{1}{\sqrt{2}}(\epsilon_{\nu-1}e_{\nu-1} + i\epsilon_{\nu-2}e_{\nu-2}).$$

Then $j_1k_1 = i\epsilon_1\epsilon_2e_1e_2$, with similar equations.

Hence
$$(ej_1k_1j_2k_2\ldots j_{\frac{\nu-1}{2}}k_{\frac{\nu-1}{2}}) = i^{\frac{\nu-1}{2}}\epsilon_1\epsilon_2\ldots\epsilon_{\nu-1}(ee_1\ldots e_{\nu-1})\,i^{\frac{\nu-1}{2}}\epsilon^{-1}.$$

Also $ej_1j_2k_2\ldots j_{\frac{\nu-1}{2}}k_{\frac{\nu-1}{2}} = -i^{\frac{\nu-1}{2}}\epsilon^{-1}|j_1$, with similar equations for the other elements.

And
$$(j_1\,|\,k_1) = (k_1\,|\,j_1) = -1.$$

115. Self-normal Planes. (1) Let a be a self-normal element; now the region $|a$ contains all the self-normal elements which are normal to a. Hence $|a$ contains all the generating regions of the quadric which contain a. Therefore $|a$ is the tangent plane to the quadric at a.

(2) The plane-equation of the quadric is, $(X\,|\,X) = 0$, where X is any planar element. For this equation is the condition that the region X contains its supplementary element $|X$.

A tangent plane X, for which $(X\,|\,X) = 0$, will be called a self-normal plane.

116. Complete Region of Three Dimensions. (1) The application of these formulæ to a manifold of three dimensions is important. Consider a skew quadrilateral $j_1j_2k_1k_2$ formed by generators of the self-normal quadric; so that j_1j_2 and k_1k_2 are two generators of one system, and j_1k_2 and j_2k_1 are two generators of the other system.

A self-conjugate tetrahedron $e_1e_2e_3e_4$ can be found such that if x be the point $\Sigma\xi e$, the self-normal quadric is

$$\xi_1^2/\epsilon_1^2 + \xi_2^2/\epsilon_2^2 + \xi_3^2/\epsilon_3^2 + \xi_4^2/\epsilon_4^2 = 0,$$

the normal intensities of e_1, e_2, e_3, e_4 are then ϵ_1, ϵ_2, ϵ_3, ϵ_4, and

$$(\epsilon_1\epsilon_2\epsilon_3\epsilon_4) = \Delta = \frac{1}{(e_1e_2e_3e_4)}.$$

(2) Assume $j_1 = \frac{1}{\sqrt{2}}(\epsilon_2e_2 - i\epsilon_1e_1),\ k_1 = \frac{1}{\sqrt{2}}(\epsilon_2e_2 + i\epsilon_1e_1),$

$$j_2 = \frac{1}{\sqrt{2}}(\epsilon_4e_4 - i\epsilon_3e_3),\ k_2 = \frac{1}{\sqrt{2}}(\epsilon_4e_4 + i\epsilon_3e_3).$$

Hence $(j_1k_1j_2k_2) = i^2 = -1.$

Also $j_1j_2k_2 = -|j_1,\ j_2j_1k_1 = -|j_2,\ k_1k_2j_2 = -|k_1,\ k_2k_1j_1 = -|k_2.$

Thus $|j_1j_2 = j_1j_2k_2 . j_2j_1k_1 = -(j_1k_1j_2k_2)j_1j_2 = j_1j_2;$

and similarly $|k_1k_2 = k_1k_2.$

Also $|j_1k_2 = -j_1k_2,\ |j_2k_1 = -j_2k_1.$

(3) Hence for a generator (G) of one system of the self-normal quadric $|G = G$, and for a generator G' of the other system $|G' = -G'$. Let the system of generators to which G belongs be called the positive system, and that to which G' belongs be called the negative system.

117. Inner Multiplication. (1) The product of one extensive magnitude (such as P_ρ) into the supplement of another extensive magnitude (such as $|P_\sigma$) is of frequent occurrence; and the rules for its transformation deserve study. These rules are of course merely a special application of the general rules of progressive and regressive multiplication, which have been explained above.

(2) This product $P_\rho|P_\sigma$ may also be regarded from another point of view. Since $P_\rho|(P_\sigma + P_\sigma') = P_\rho|P_\sigma + P_\rho|P_\sigma'$, we may conceive [cf. § 19] the symbol $|$ not as an operation on P_σ but as the mark of a special sort of multiplication between P_ρ and P_σ. Let this species of multiplication be called 'Inner Multiplication,' and let the product $P_\rho|P_\sigma$ be termed the inner product of P_ρ and P_σ. In distinction to Inner Multiplication Progressive and Regressive Multiplication are called Outer Multiplication.

(3) It is obvious that inner products and inner multiplication must be understood to refer to a definitely assumed self-normal quadric; and further that, corresponding to different self-normal quadrics, there can be different sorts of inner multiplication. But general formulæ for the transformation of such products can be laid down.

118. ELEMENTARY TRANSFORMATIONS. (1) Let P_ρ and P_σ be extensive magnitudes, simple or complex, of the ρth and σth orders respectively.

Then $\nu-\sigma$ is the order of $|P_\sigma$. The product $P_\rho|P_\sigma$ is progressive if $\rho+(\nu-\sigma)<\nu$, that is, if $\rho<\sigma$; and is regressive, if $\rho>\sigma$.

(2) If $\rho<\sigma$; $$P_\rho|P_\sigma=(-1)^{\rho(\nu-\sigma)}|P_\sigma.P_\rho;$$
and hence $$|(P_\rho|P_\sigma)=(-1)^{\rho(\nu-\sigma)}||P_\sigma.|P_\rho.$$
But by § 99 (7), $$||P_\sigma=(-1)^{\sigma(\nu-\sigma)}P_\sigma.$$
Therefore finally, $$|(P_\rho|P_\sigma)=(-1)^{(\rho+\sigma)(\nu-\sigma)}P_\sigma|P_\rho.$$

(3) If $\rho>\sigma$; then $$|(P_\rho|P_\sigma)=|P_\rho.||P_\sigma=(-1)^{\sigma(\nu-\sigma)}(|P_\rho.P_\sigma)$$
$$=(-1)^{\sigma(\nu-\sigma)}(-1)^{\sigma(\nu-\rho)}(P_\sigma|P_\rho)=(-1)^{\sigma(\rho+\sigma)}(P_\sigma|P_\rho).$$

(4) If $\rho=\sigma$; then $(P_\rho|P_\sigma)$ is merely numerical: write P_ρ' instead of P_σ.
Then $$(P_\rho|P_\rho')=|(P_\rho|P_\rho')=(P_\rho'|P_\rho).$$

119. RULE OF THE MIDDLE FACTOR. (1) The extended rule of the middle factor can be applied to transform $P_\rho|Q_\sigma$, where P_ρ and Q_σ are simple magnitudes of the ρth and σth orders respectively. In the first place assume that $\rho>\sigma$. Let the multiplicative combinations of the σth order formed out of the factors of the first order of P_ρ be $P_\sigma^{(1)}$, $P_\sigma^{(2)}$, etc., and let
$$P_\rho=P_\sigma^{(1)}P_{\rho-\sigma}^{(1)}=P_\sigma^{(2)}P_{\rho-\sigma}^{(2)}=\text{etc.}$$
Then from the extended rule of the middle factor, we deduce
$$P_\rho|Q_\sigma=(P_\sigma^{(1)}|Q_\sigma)P_{\rho-\sigma}^{(1)}+(P_\sigma^{(2)}|Q_\sigma)P_{\rho-\sigma}^{(2)}+\text{etc.}\ldots\ldots\ldots\ldots\ldots(1).$$

(2) Secondly, assume that $\rho<\sigma$.
Then $$(P_\rho|Q_\sigma)=(-1)^{(\rho+\sigma)(\nu-\rho)}|(Q_\sigma|P_\rho).$$
Now let $Q_\rho^{(1)}$, $Q_\rho^{(2)}$, etc. be the multiplicative combinations of the ρth order formed out of the factors of the first order of Q_σ, and let
$$Q_\sigma=Q_\rho^{(1)}Q_{\sigma-\rho}^{(1)}=Q_\rho^{(2)}Q_{\sigma-\rho}^{(2)}=\text{etc.}$$
Then by equation (1) of the first case
$$(Q_\sigma|P_\rho)=(Q_\rho^{(1)}|P_\rho)Q_{\sigma-\rho}^{(1)}+(Q_\rho^{(2)}|P_\rho)Q_{\sigma-\rho}^{(2)}+\text{etc.}$$
$$=(P_\rho|Q_\rho^{(1)})Q_{\sigma-\rho}^{(1)}+(P_\rho|Q_\rho^{(2)})Q_{\sigma-\rho}^{(2)}+\text{etc.}$$
Hence $$(-1)^{(\rho+\sigma)(\nu-\rho)}(P_\rho|Q_\sigma)=(P_\rho|Q_\rho^{(1)})|Q_{\sigma-\rho}^{(1)}+(P_\rho|Q_\rho^{(2)})|Q_{\sigma-\rho}^{(2)}+\text{etc.}\ldots(2).$$
The formulæ of equations (1) and (2) will be called the rule of the middle factor for inner multiplication.

120. IMPORTANT FORMULA. (1) The rule of the middle factor does not apply when both factors are of the same order. But the transformation in this case is given by § 105. For if each factor be of the ρth order, then
$$(a_1a_2\ldots a_\rho|b_1b_2\ldots b_\rho)=(a_1a_2\ldots a_\rho.|b_1|b_2\ldots|b_\rho)$$
$$=\begin{vmatrix}(a_1|b_1), & (a_1|b_2), & \ldots\ldots & (a_1|b_\rho)\\ (a_2|b_1), & (a_2|b_2), & \ldots\ldots & (a_2|b_\rho)\\ \ldots & \ldots & \ldots & \ldots\\ (a_\rho|b_1), & (a_\rho|b_2), & \ldots\ldots & (a_\rho|b_\rho)\end{vmatrix}\ldots\ldots\ldots(3).$$

Important special cases of this formula are

$$(a_1a_2 \dots a_\rho \,|\, a_1a_2 \dots a_\rho) = \begin{vmatrix} (a_1|a_1), & (a_1|a_2), & \dots\dots & (a_1|a_\rho) \\ (a_2|a_1), & (a_2|a_2), & \dots\dots & (a_2|a_\rho) \\ \dots & \dots & \dots & \dots \\ (a_\rho|a_1), & (a_\rho|a_2), & \dots\dots & (a_\rho|a_\rho) \end{vmatrix}$$

$$(a_1a_2 \,|\, b_1b_2) = (a_1 \,|\, b_1)(a_2 \,|\, b_2) - (a_1 \,|\, b_2)(a_2 \,|\, b_1);$$

$$(a_1a_2 \,|\, a_1a_2) = (a_1 \,|\, a_1)(a_2 \,|\, a_2) - (a_1 \,|\, a_2)^2.$$

(2) Also if the complete region be of $\nu - 1$ dimensions, the products $(a_1a_2 \dots a_\nu)$ and $(b_1b_2 \dots b_\nu)$, although merely numerical, may each be conceived as progressive products. The proof of § 105 still holds in this case, and therefore

$$(a_1a_2 \dots a_\nu)(b_1b_2 \dots b_\nu) = (a_1a_2 \dots a_\nu \,|\, b_1b_2 \dots b_\nu)$$
$$= \begin{vmatrix} (a_1|b_1), & (a_1|b_2), & \dots\dots & (a_1|b_\nu) \\ (a_2|b_1), & (a_2|b_2), & \dots\dots & (a_2|b_\nu) \\ \dots & \dots & \dots & \dots \\ (a_\nu|b_1), & (a_\nu|b_2), & \dots\dots & (a_\nu|b_\nu) \end{vmatrix}.$$

This is the ordinary rule for the multiplication of two determinants.

121. INNER MULTIPLICATION OF NORMAL REGIONS. If A, B, C be three mutually normal regions, (so that the multiplication ABC must be pure progressive), then

$$(ABC \,|\, ABC) = (A \,|\, A)(B \,|\, B)(C \,|\, C) = (AB \,|\, AB)(C \,|\, C).$$

This theorem can easily be proved independently; but we will deduce it at once from the formula for $(a_1a_2 \dots a_\rho \,|a_1a_2 \dots a_\rho)$ of § 120.

For let $A = a_1a_2 \dots a_\rho$, $B = b_1b_2 \dots b_\sigma$, $C = c_1c_2 \dots c_\tau$; then each of the groups

$$(a_1, a_2 \dots a_\rho), \ (b_1, b_2 \dots b_\sigma), \ (c_1, c_2 \dots c_\tau)$$

may be conceived [cf. § 113 (5)] to consist of mutually normal elements. But since A, B, C are mutually normal regions, it follows that the whole set of $\rho + \sigma + \tau$ elements are mutually normal.

Hence

$$(ABC \,|\, ABC) = (a_1a_2 \dots a_\rho b_1b_2 \dots b_\sigma c_1c_2 \dots c_\tau \,|a_1 \dots b_1 \dots c_1 \dots c_\tau)$$
$$= (a_1 \,|\, a_1)(a_2 \,|\, a_2) \dots (b_1 \,|\, b_1) \dots (b_\sigma \,|\, b_\sigma)(c_1 \,|\, c_1) \dots (c_\tau \,|\, c_\tau).$$

Also $(A \,|A) = (a_1 \,|a_1)(a_2 \,|a_2) \dots (a_\rho \,|a_\rho)$, and $(B \,|\, B)$, $(C \,|\, C)$ and $(AB \,|\, AB)$ are equal to similar expressions. Hence the theorem follows.

122. GENERAL FORMULA FOR INNER MULTIPLICATION. (1) Equations (1) and (2) of § 119 can be extended so as to prove two more general formulæ which include both them and equation (3) of § 120.

Consider the product $P_{\rho+\sigma} \,|\, Q_\rho Q_\tau$, where $P_{\rho+\sigma}$, Q_ρ, Q_τ represent simple magnitudes of the $(\rho + \sigma)$th, ρth, and τth orders respectively.

In the first place assume that $\sigma > \tau$.

Then $$P_{\rho+\sigma}|Q_\rho Q_\tau = P_{\rho+\sigma}(|Q_\rho|Q_\tau).$$

But since $\sigma > \tau$, the product is a pure regressive product and is therefore associative. Hence

$$P_{\rho+\sigma}|Q_\rho Q_\tau = P_{\rho+\sigma}|Q_\rho \,.\, |Q_\tau.$$

Now let $P_\rho^{(1)}$, $P_\rho^{(2)}$, etc. be the multiplicative combinations of the ρth order formed out of the factors of the first order of $P_{\rho+\sigma}$, and let $P_\sigma^{(1)}$, $P_\sigma^{(2)}$, etc. be the multiplicative combinations of the σth order formed out of the factors of $P_{\rho+\sigma}$, so that

$$P_{\rho+\sigma} = P_\rho^{(1)} P_\sigma^{(1)} = P_\rho^{(2)} P_\sigma^{(2)} = \text{etc.}$$

Hence by equation (1), $P_{\rho+\sigma}|Q_\rho = \sum_\lambda (P_\rho^{(\lambda)}|Q_\rho) P_\sigma^{(\lambda)}$.

Therefore finally $$P_{\rho+\sigma}|.\, Q_\rho Q_\tau = \sum_\lambda (P_\rho^{(\lambda)}|Q_\rho) P_\sigma^{(\lambda)}|Q_\tau \;\ldots\ldots\ldots\ldots\ldots\ldots(4).$$

(2) Secondly, let $\sigma < \tau$. Let $Q_\rho Q_\tau = Q_{\rho+\tau}$, and let $Q_\rho^{(1)}$, $Q_\rho^{(2)}$, etc. be the multiplicative combinations of the ρth order formed out of the factors of the first order of $Q_{\rho+\tau}$, and let $Q_\tau^{(1)}$, $Q_\tau^{(2)}$ be those of the τth order, so that

$$Q_{\rho+\tau} = Q_\rho^{(1)} Q_\tau^{(1)} = Q_\rho^{(2)} Q_\tau^{(2)} = \text{etc.}$$

Also let $P_{\rho+\sigma} = P_\rho P_\sigma$, where P_ρ is of the ρth order, and P_σ of the σth order.

Then $$P_{\rho+\sigma}|Q_{\rho+\tau} = (-1)^{(\sigma+\tau)(\nu-\rho-\sigma)}|(Q_{\rho+\tau}|P_{\rho+\sigma}).$$

But by equation (4) $Q_{\rho+\tau}|P_\rho P_\sigma = \sum_\lambda (Q_\rho^{(\lambda)}|P_\rho) Q_\tau^{(\lambda)}|P_\sigma$.

Hence

$$|(Q_{\rho+\tau}|P_\rho P_\sigma) = \sum_\lambda (Q_\rho^{(\lambda)}|P_\rho)|(Q_\tau^{(\lambda)}|P_\sigma) = (-1)^{(\sigma+\tau)(\nu-\sigma)} \sum_\lambda (P_\rho|Q_\rho^{(\lambda)}) P_\sigma|Q_\tau^{(\lambda)}.$$

Therefore finally, $$P_\rho P_\sigma|Q_{\rho+\tau} = (-1)^{\rho(\sigma+\tau)} \sum_\lambda (P_\rho|Q_\rho^{(\lambda)}) P_\sigma|Q_\tau^{(\lambda)} \;\ldots\ldots(5).$$

(3) Equations (4) and (5) are more general than equations (1), (2) and (3) but the readiness with which the equations first found can be recovered from the extended rule of the middle factor makes them to be of the greater utility.

The theory of Inner Multiplication and the above formulæ are given in Grassmann's *Ausdehnungslehre* von 1862.

123. Quadrics. (1) The theory of quadrics can be investigated by the aid of this notation. Let the quadric be chosen as the self-normal quadric for the operation $|$. Then the equation of the quadric is $(x|x) = 0$.

Let the reference points $e_1, e_2 \ldots e_\nu$ be any ν independent elements, not necessarily mutually normal. Then if $x = \Sigma\xi e$, the equation of the quadric according to the notation of Book III., Chapter III. is written

$$(\alpha \big)\!\big(x)^2 = \alpha_{11}\xi_1^2 + \ldots + 2\alpha_{12}\xi_1\xi_2 + \ldots = 0.$$

But $$(x|x) = (e_1|e_1)\,\xi_1^2 + \ldots + 2(e_1|e_2)\,\xi_1\xi_2 + \ldots.$$

Hence we may write, $(e_1|e_1) = \alpha_{11}$, $(e_2|e_2) = \alpha_{22}$, etc., $(e_1|e_2) = \alpha_{12}$, etc.

(2) Since by § 120 $(e_1e_2 \ldots e_\nu)^2 = (e_1e_2 \ldots e_\nu \,|e_1e_2 \ldots e_\nu)$

$$= \begin{vmatrix} (e_1 |e_1), & (e_1|e_2), \ldots & (e_1 |e_\nu) \\ (e_2 |e_1), & (e_2 |e_2), \ldots & (e_2 |e_\nu) \\ \ldots\ldots & \ldots\ldots & \ldots\ldots \\ (e_\nu |e_1), & (e_\nu |e_2), \ldots & (e_\nu |e_\nu) \end{vmatrix},$$

it follows that $(e_1e_2 \ldots e_\nu)^2$ is the discriminant of the quadratic expression $(x \,|x)$.

Since $(e_1e_2 \ldots e_\nu)$ cannot vanish ($e_1, e_2, \ldots e_\nu$ being independent), it follows that the quadric cannot be conical.

(3) The equation of the polar plane of any point x becomes $(x \,|y) = 0$.

The plane-equation of the quadric is $(X_{\nu-1} \,|X_{\nu-1}) = 0$; where $X_{\nu-1}$ is a planar element.

The equation of the polar point of any plane $X_{\nu-1}$, becomes

$$(X_{\nu-1} \,|Y_{\nu-1}) = 0.$$

(4) Let bx be any line drawn through a given point b; and let this line intersect the quadric in the point $\lambda b + \mu x$.

Then $$\lambda^2 (b \,|b) + 2\lambda\mu (b \,|x) + \mu^2 (x \,|x) = 0 \qquad\qquad (1).$$

This quadratic for λ/μ in general gives two points on the quadric.

(5) These points coincide if $(b \,|b)(x \,|x) - (b \,|x)^2 = 0$.

This is the equation of the tangent quadric cone with vertex b.

But $$(b \,|b)(x \,|x) - (b \,|x)^2 = (bx \,|bx).$$

Hence this cone can be written $(bx \,|bx) = 0$.

(6) The identity $(b \,|b)(x \,|x) - (b \,|x)^2 = (bx \,|bx)$, can be written

$$(\alpha \between b)^2 (\alpha \between x)^2 - \{(\alpha \between b \between x)\}^2 = \Sigma\, (\beta_\rho \xi_\sigma - \beta_\sigma \xi_\rho)(\beta_\lambda \xi_\mu - \beta_\mu \xi_\lambda)(e_\rho e_\sigma \,|e_\lambda e_\mu).$$

Also $(e_\rho e_\sigma \,|e_\lambda e_\mu) = (e_\rho \,|e_\lambda)(e_\sigma \,|e_\mu) - (e_\rho \,|e_\mu)(e_\sigma \,|e_\lambda) = \alpha_{\rho\lambda}\alpha_{\sigma\mu} - \alpha_{\rho\mu}\alpha_{\sigma\lambda}$.

(7) The roots of the quadratic equation (1) are

$$\frac{\lambda_1}{\mu_1} = \frac{-(b \,|x) + \sqrt{\{(b \,|x)^2 - (x \,|x)(b \,|b)\}}}{(b \,|b)},$$

and $$\frac{\lambda_2}{\mu_2} = \frac{-(b \,|x) - \sqrt{\{(b \,|x)^2 - (x \,|x)(b \,|b)\}}}{(b \,|b)}.$$

(8) If a_1 and a_2 are the points $\lambda_1 b + \mu_1 x$, and $\lambda_2 b + \mu_2 x$, then the anharmonic ratio $(a_1a_2,\, xb)$ is

$$\frac{-(b \,|x) - \sqrt{\{(b \,|x)^2 - (x \,|x)(b \,|b)\}}}{-(b \,|x) + \sqrt{\{(b \,|x)^2 - (x \,|x)(b \,|b)\}}} = \frac{(b \,|x) + \sqrt{\{-(bx \,|bx)\}}}{(b \,|x) - \sqrt{\{-(bx \,|bx)\}}}.$$

(9) Firstly let $(b \,|x)^2 < (x \,|x)(b \,|b)$.

Let θ be such that $\cos^2\theta = \dfrac{(b \,|x)^2}{(x \,|x)(b \,|b)}$; and let $\rho = (a_1a_2,\, xb)$.

Then $$\rho = \frac{\cos\theta + i\sin\theta}{\cos\theta - i\sin\theta} = e^{2i\theta}.$$

Also $$\sin^2\theta = \frac{(bx \mid bx)}{(x \mid x)(b \mid b)}.$$

Therefore we deduce the group of equations

$$\theta = \frac{1}{2i}\log\rho = \cos^{-1}\frac{(b \mid x)}{\sqrt{\{(x \mid x)(b \mid b)\}}} = \sin^{-1}\sqrt{\frac{(bx \mid bx)}{\{(x \mid x)(b \mid b)\}}} \quad \ldots(2).$$

(10) Secondly, let $$(b \mid x)^2 > (x \mid x)(b \mid b).$$

Put $$\cosh^2\theta = \frac{(b \mid x)^2}{(x \mid x)(b \mid b)}.$$

Then $$\rho = \frac{\cosh\theta + \sinh\theta}{\cosh\theta - \sinh\theta} = e^{2\theta}.$$

Also $$\sinh^2\theta = \frac{-(bx \mid bx)}{(x \mid x)(b \mid b)}.$$

Hence we deduce the group of equations

$$\theta = \frac{1}{2}\log\rho = \cosh^{-1}\frac{(b \mid x)}{\sqrt{\{(x \mid x)(b \mid b)\}}} = \sinh^{-1}\sqrt{\frac{-(bx \mid bx)}{\{(x \mid x)(b \mid b)\}}} \ldots(3).$$

(11) If $(bx \mid bx)$ be positive for every pair of elements b and x, it follows from § 82 (4) that the quadric, $(x \mid x) = 0$, is imaginary.

If the quadric be a closed real quadric and b and x both lie within it, or if both lie without it and the line bx cut the quadric in real points, then it follows from the same article that $(bx \mid bx)$ is necessarily negative.

124. PLANE-EQUATION OF A QUADRIC. (1) Taking the supplement of the equation, $(bx \mid bx) = 0$, and writing B instead of $\mid b$ and X instead of $\mid x$, we find the equation $(BX \mid BX) = 0$, which can also be written

$$(B \mid B)(X \mid X) - (B \mid X)^2 = 0.$$

This equation [cf. § 84 (4)] is the plane equation of the degenerate quadric enveloped by sub-planes lying in the plane B and touching the quadric.

(2) Again, by a proof similar to that in § 123, let B and X be any two planes, and let the two planes through their intersection BX which touch the quadric be A_1 and A_2. Also let ρ be the anharmonic ratio of the range $\{BX, A_1A_2\}$.

Then if $$(B \mid X)^2 < (B \mid B)(X \mid X),$$

$$\theta = \frac{1}{2i}\log\rho = \cos^{-1}\frac{(B \mid X)}{\sqrt{\{(X \mid X)(B \mid B)\}}} = \sin^{-1}\sqrt{\frac{(BX \mid BX)}{\{(X \mid X)(B \mid B)\}}}.$$

And if $$(B \mid X)^2 > (B \mid B)(X \mid X),$$

$$\theta = \frac{1}{2}\log\rho = \cosh^{-1}\frac{(B \mid X)}{\sqrt{\{(X \mid X)(B \mid B)\}}} = \sinh^{-1}\sqrt{\frac{-(BX \mid BX)}{\{(X \mid X)(B \mid B)\}}}.$$

(3) Again, let L_ρ be any subregion of $\rho - 1$ dimensions which touches the quadric. This condition requires that L_ρ should lie in the tangent plane

to the quadric at some point b, and should contain b. We can prove that the condition to be satisfied by L_ρ is, $(L_\rho \mid L_\rho) = 0$.

For let $l_1, l_2, \ldots l_\rho$, be assumed to be ρ mutually normal points on L_ρ, which is possible according to § 113 (5).

Then by § 120 (1) $(L_\rho \mid L_\rho) = (l_1 \mid l_1)(l_2 \mid l_2) \ldots (l_\rho \mid l_\rho)$.

Hence if $(L_\rho \mid L_\rho) = 0$, then one at least of the points $l_1, l_2, \ldots l_\rho$ must be self-normal. Assume that l_ρ is the self-normal point b. Then the remaining points $l_1, l_2, \ldots l_{\rho-1}$ all lie on the plane $\mid b$, which is the tangent plane of b. Thus Plucker's conception of the line equation of a quadric in three dimensions can be generalized for any subregion in any number of dimensions*.

(4) Consider the four subregions $B_\rho, X_\rho, A_\rho, A_\rho'$, of $\rho - 1$ dimensions which lie in the same subregion of ρ dimensions. Then considering this containing subregion as a complete region we see that $B_\rho, X_\rho, A_\rho, A_\rho'$ have the properties of planes in this region.

Let A_ρ and A_ρ' both contain the subregion of $\rho - 2$ dimensions in which B_ρ and X_ρ intersect. So that $A_\rho = \lambda B_\rho + \mu X_\rho$, and $A_\rho' = \lambda' B_\rho + \mu' X_\rho$. Then the four subregions $B_\rho, X_\rho, A_\rho, A_\rho'$ form a range with a definite anharmonic ratio $\lambda\mu'/\lambda'\mu$; let this ratio be called ρ. Let A_ρ and A_ρ' touch the self-normal quadric.

Then λ/μ and λ'/μ' are the roots of the quadratic

$$\lambda^2 (B_\rho \mid B_\rho) + 2\lambda\mu (B_\rho \mid X_\rho) + \mu^2 (X_\rho \mid X_\rho) = 0.$$

Hence, as before, if $\quad (B_\rho \mid X_\rho)^2 < (B_\rho \mid B_\rho)(X_\rho \mid X_\rho)$,

then
$$\theta = \frac{1}{2i} \log \rho = \cos^{-1} \frac{(B_\rho \mid X_\rho)}{\sqrt{\{(B_\rho \mid B_\rho)(X_\rho \mid X_\rho)\}}}.$$

And if $\quad (B_\rho \mid X_\rho)^2 > (B_\rho \mid B_\rho)(X_\rho \mid X_\rho)$,

then
$$\theta = \frac{1}{2} \log \rho = \cosh^{-1} \frac{(B_\rho \mid X_\rho)}{\sqrt{\{(B_\rho \mid B_\rho)(X_\rho \mid X_\rho)\}}}.$$

It is to be noticed that the formulæ for $\sin\theta$ and $\sinh\theta$ do not hold unless ρ be unity or $\nu - 1$.

* As far as I am aware this generalized form of Plucker's line-equation has not been given before.

CHAPTER IV.

DESCRIPTIVE GEOMETRY.

125. APPLICATION TO DESCRIPTIVE GEOMETRY. An extensive manifold of $\nu - 1$ dimensions is a positional manifold of $\nu - 1$ dimensions with other properties superadded. These further properties have in general no meaning for a positional manifold merely as such. But yet it is often possible conveniently to prove properties of all positional manifolds by reasoning which introduces the special extensive properties of extensive manifolds. This is due to the fact that the calculus of extension and some of the properties of extensive manifolds are capable of a partial interpretation which construes them merely as directions to form 'constructions' in a positional manifold. Ideally a construction is merely an act of fixing attention upon a certain aggregate of elements so as to mark them out in the mind apart from all others; physically, it represents some operation which makes the constructed objects evident to the senses. Now an extensive magnitude of any order, say the ρth, may be interpreted as simply representing the fact of the construction of the subregion of $\rho - 1$ dimensions which it represents. This interpretation leaves unnoticed that congruent products may differ by a numerical factor, and that, therefore, extensive magnitudes must be conceived as capable of various intensities. Accordingly, in all applications of the Calculus to Positional Manifolds by the use of this interpretation it will be found that the congruence of products is the sole material question, and that their intensities can be left unnoticed; except when the products are numerical and are the coefficients of elements of the first order which have intensities in positional manifolds. The sign of congruence, viz. $\equiv$ [cf. § 64 (2)], rather than that of equality is adapted to this type of reasoning. Also supplements never explicitly appear, since they answer to no mental process connected with this type of reasoning.

126. EXPLANATION OF PROCEDURE. (1) In the present chapter and in the succeeding one some applications of the calculus to positional manifolds are given. Except in § 130 on Projection, the manifolds are of two dimensions,

and the investigations form an example of the application of the Calculus to Descriptive Geometry of Two Dimensions. Other applications of this type have already been given in §§ 106, 107.

(2) In two dimensional complete regions the only products are of two points which produce a linear element, of two linear elements which produce a point, and of a point and a linear element which produce a numerical quantity. If a product yields an extensive magnitude, the act of using such a product is equivalent to the claim to be able to construct that subregion which the magnitude represents. Thus the product ab represents the indefinitely produced line joining ab, and the use of the product is the equivalent to drawing the line. Similarly if L and L' are two linear elements in a plane, the use of LL' is equivalent to the claim to be able to identify the point of intersection of the two lines L and L', which by hypothesis have been constructed. Thus the representation of a point by a product of certain assumed points is the construction of that point by drawing straight lines joining the assumed points and is the point of intersection of lines thus drawn.

127. Illustration of Method. The method of reasoning in the application of this algebra to Descriptive Geometry is exemplified by the proof of the following theorem*.

If abc and def be two coplanar triangles, and if s be a point such that sd, se, sf cut the sides bc, ca, ab respectively in three collinear points, then sa, sb, sc cut the sides ef, fd, de respectively in three collinear points.

For by hypothesis

$$(sd \,.\, bc)\,(se \,.\, ca)\,(sf \,.\, ab) = 0.$$

Hence by the extended rule of the middle factor

$$\{(sdc)\,b - (sdb)\,c\}\,\{(sea)\,c - (sec)\,a\}\,\{(sfb)\,a - (sfa)\,b\} = 0.$$

Multiplying out and dividing by the numerical factor (abc),

$$(sdc)\,(sea)\,(sfb) - (sdb)\,(sec)\,(sfa) = 0.$$

The symmetry of this condition as between the triangles abc and def proves the proposition.

128. von Staudt's Construction. (1) Let a, c, b represent any three points in a two-dimensional region, which are not collinear.

In ac assume any point d arbitrarily, and in cb assume any point e.

Since the intensities of a, c and b are quite arbitrary, we may assume that

$$d = a + \delta c, \quad e = b + \epsilon c,$$

where δ and ϵ are any assumed numerical magnitudes.

Then it is to be proved that any point x on ac (*i.e.* of the form $a + \xi c$) can be exhibited as a product of the assumed points, or in other words can be constructed. This construction to be given is due to von Staudt†.

* Due, I believe, to H. M. Taylor.

† *Geometrie der Lage*, 1847.

(2) Firstly, consider the following products, or in other words make the constructions symbolized by them:

$$q = ae \,.\, db, \quad q_1 = qc \,.\, de, \quad p_1 = q_1 b \,.\, ac.$$

Thus $$p_1 = ae \,.\, db \,.\, c \,.\, de \,.\, b \,.\, ac.$$

Then p_1 is the point $a + 2\delta c$.

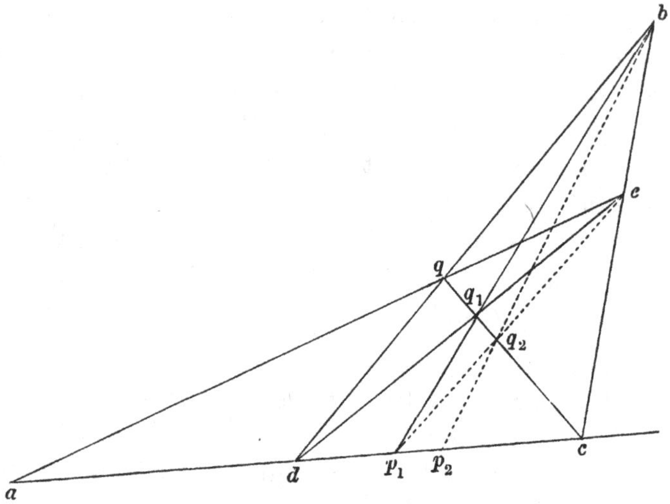

FIG. 1.

For $$q \equiv (ab + \epsilon ac)(ab + \delta cb) = \delta ab \,.\, cb + \epsilon ac \,.\, ab + \epsilon\delta ac \,.\, cb$$
$$= \delta\,(acb)\,b + \epsilon\,(acb)\,a + \epsilon\delta\,(acb)\,c = -(abc)\,\{\epsilon a + \delta b + \epsilon\delta c\}$$
$$\equiv \epsilon a + \delta b + \epsilon\delta c;$$

where the numerical factor $-(abc)$ has been dropped for brevity. This will be done in future without remark.

$$q_1 = (\epsilon ac + \delta bc) \,.\, de = (\epsilon ac + \delta bc)(ab + \epsilon ac + \delta cb)$$
$$= \epsilon ac \,.\, ab + \delta\epsilon ac \,.\, cb + \delta bc \,.\, ab + \delta\epsilon bc \,.\, ac$$
$$= \epsilon\,(acb)\,a + \delta\epsilon\,(acb)\,c + \delta\,(acb)\,b + \delta\epsilon\,(acb)\,c \equiv \epsilon a + \delta b + 2\delta\epsilon c.$$
$$p_1 = q_1 b \,.\, ac \equiv \epsilon\,\{a + 2\delta c\}\,b \,.\, ac \equiv \epsilon\,(abc)\,\{a + 2\delta c\} \equiv a + 2\delta c.$$

Hence p_1 is the point $a + 2\delta c$.

(3) Again, substitute d for a and p_1 for d in the above product. The new lines in the figure are represented by dotted lines. Then since $p_1 \equiv d + \delta c$, we obtain the point

$$p_2 = de \,.\, p_1 b \,.\, c \,.\, p_1 e \,.\, b \,.\, dc \equiv d + 2\delta c \equiv a + 3\delta c.$$

Similarly by substituting p_1 for d and p_2 for p_1 in this construction, we find $p_3 \equiv a + 4\delta c$, and so on successively. Thus finally if ν be any positive integer, we find $p_\nu \equiv a + (\nu + 1)\,\delta c$.

(4) Secondly, consider the point $e' = qp_1 . bc \equiv b - \epsilon c$ (cf. fig. 2). Make the following construction,

$$r_1 = ae' . de, \quad p_{\frac{1}{2}} = r_1 b . ac = ae' . de . b . ac.$$

Now

$$r_1 \equiv (ab - \epsilon ac)(ab + \epsilon ac + \delta cb) \equiv \epsilon ab . ac + \delta ab . cb - \epsilon ac . ab - \delta\epsilon ac . cb$$

$$\equiv -2\epsilon ac . ab + \delta ab . cb - \delta\epsilon ac . cb \equiv -2\epsilon(acb)a + \delta(acb)b - \delta\epsilon(acb)c;$$

and $p_{\frac{1}{2}} \equiv -(acb)\{2\epsilon ab + \delta\epsilon cb\} . ac \equiv a + \frac{1}{2}\delta c.$

(5) Similarly by substituting $p_{\frac{1}{2}}$ instead of d in this construction we obtain

$$p_{\frac{1}{2^2}} \equiv a + \frac{1}{2^2}\delta c;$$

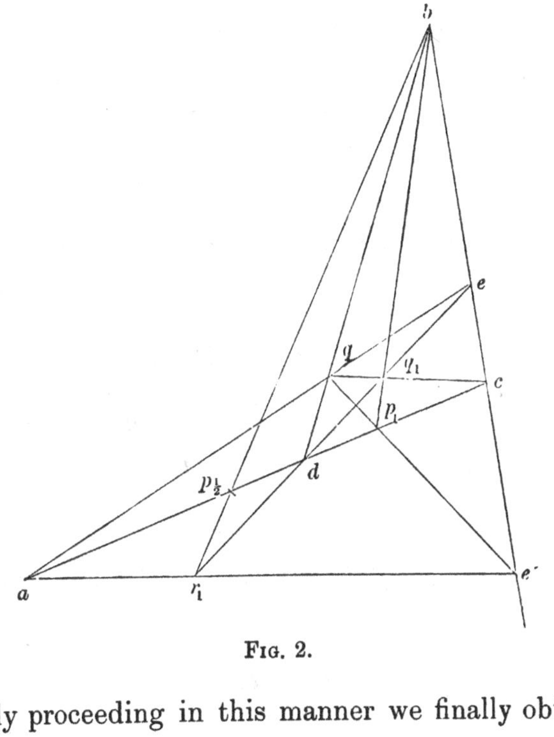

FIG. 2.

and by continually proceeding in this manner we finally obtain if ν be any positive integer

$$p_{\frac{1}{2^\nu}} \equiv a + \frac{1}{2^\nu}\delta c.$$

(6) Then if μ be any other positive integer, the construction of the first figure can be applied μ times starting with $p_{\frac{1}{2^\nu}}$ instead of with d. Thus the point

$$p_{\frac{\mu}{2^\nu}} \equiv a + \frac{\mu}{2^\nu}\delta c,$$

can be constructed.

(7) Thirdly, make the following construction (cf. fig. 3):

$$q' = ab \,.\, de',\ d' = q'e \,.\, ac = ab \,.\, de' \,.\, e \,.\, ac.$$

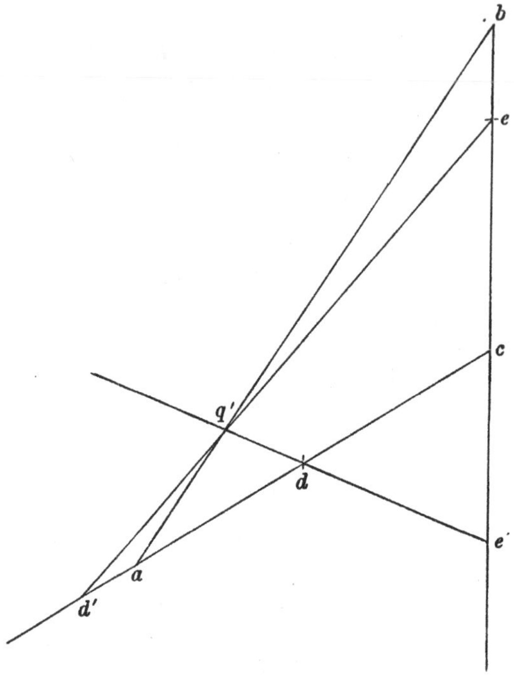

FIG. 3.

Now $q' \equiv ab\,\{ab - \epsilon ac + \delta cb\} \equiv -\,\epsilon ab \,.\, ac + \delta ab \,.\, cb \equiv -\,(abc)\,\{\epsilon a + \delta b\}$,

$$d' \equiv (\epsilon ae + \delta be)\, ac \equiv (\epsilon ab + \epsilon^2 ac + \delta\epsilon bc)\, ac$$

$$\equiv \epsilon ab \,.\, ac + \delta\epsilon bc \,.\, ac \equiv \epsilon\,(abc)\,\{a - \delta c\} \equiv a - \delta c.$$

(8) In this construction if we substitute for d any constructed point of the form

$$p_{\frac{\mu}{2^\nu}} \equiv a + \frac{\mu}{2^\nu}\,\delta c,$$

we obtain

$$p'_{\frac{\mu}{2^\nu}} \equiv a - \frac{\mu}{2^\nu}\,\delta c.$$

Thus all points of the form $\left(a \pm \frac{\mu}{2^\nu}\,\delta c\right)$ can now be constructed.

(9) Fourthly, let p, p' and p'' be three constructed points, and let

$$p \equiv a + \varpi\delta c,\ \ p' \equiv a + \varpi'\delta c,\ \ p'' \equiv a + \varpi''\delta c.$$

Then $$p' \equiv p + (\varpi' - \varpi)\,\delta c.$$

Now in the first construction substitute p for a and p' for d. Then we obtain

$$p_1' \equiv p + 2(\varpi' - \varpi)\,\delta c \equiv a + (2\varpi' - \varpi)\,\delta c.$$

Similarly by substituting p_1' for a, and p'' for d, we obtain

$$p_2' \equiv a + (2\varpi'' - 2\varpi' + \varpi)\,\delta c;$$

and so on by successive substitutions.

(10) But any positive number ξ, rational or irrational, can be expressed to any approximation desired in the scale of 2, as the radix of notation, in the form

$$\beta_0 + \frac{\beta_1}{2} + \frac{\beta_2}{2^2} + \frac{\beta_3}{2^3} + \text{etc.};$$

where β_0 is the integer next below ξ and β_1, β_2, etc. are either unity or zero.

If the series is finite any point of the form $a \pm \xi\delta c$ can be constructed in a finite number of constructions; if the series is infinite it can be constructed in an infinite number of constructions; and (since the series is convergent) this means that in a finite (but sufficiently large) number of constructions we can construct a point $a \pm \xi'\delta c$, where $\xi \sim \xi'$ is less than any assigned finite number however small.

Thus any point $a + \xi c$ on the line ac can be constructed, and similarly any point on the line bc can be constructed; and it is sufficiently easy to see that any point $a + \xi b + \eta c$ can be constructed.

This type of construction can easily be extended to a projective manifold of any dimensions.

129. Grassmann's Constructions. (1) Grassmann's constructions* in a complete region of two dimensions have for their ultimate object to construct the point $a + \psi(\xi_1, \xi_2)(a_1 + a_2)$, where $\psi(\xi_1, \xi_2)$ is any rational integral homogeneous function of ξ_1, ξ_2, and a, a_1, a_2 are any three given not collinear points, provided also that the point $a + \xi_1 a_1 + \xi_2 a_2$ and also certain points of the form $a + \alpha_1 a_1 + \alpha_2 a_2$ are given, where the α's are known coefficients. In order to accomplish this end the constructions are given for the following series of points,

$$a + \xi_1(a_1 + a_2),\ a + \xi_2(a_1 + a_2),\ a + \xi_1\xi_2(a_1 + a_2),$$

$$a + \xi_1^{\nu}(a_1 + a_2)\ \{\text{where } \nu \text{ is any positive integer}\},\ a + \xi_2^{\nu}(a_1 + a_2),$$

$$a + \frac{\gamma_1\xi_1 + \gamma_2\xi_2}{\gamma_1 + \gamma_2}(a_1 + a_2)\ \{\text{where } \gamma_1 + \gamma_2 \text{ is not zero}\}.$$

Then finally a construction is deduced for

$$a + \frac{\gamma\xi_1^{\nu}\xi_2^{\mu} + \gamma'\xi_1^{\nu'}\xi_2^{\mu'} + \dots}{\gamma + \gamma' + \dots}(a_1 + a_2),$$

when $\gamma + \gamma' + \dots$ is not zero, and $\mu, \nu, \mu', \nu', \dots$ are positive integers, and $\mu + \nu = \mu' + \nu' = \dots$.

* Cf. *Ausdehnungslehre von* 1862, §§ 325—329.

(2) Let a, a_1, a_2 denote any three elements forming a reference triangle in the two dimensional region; let $x = a + \xi_1 a_1 + \xi_2 a_2$; and let $d = a + a_1 + a_2$.

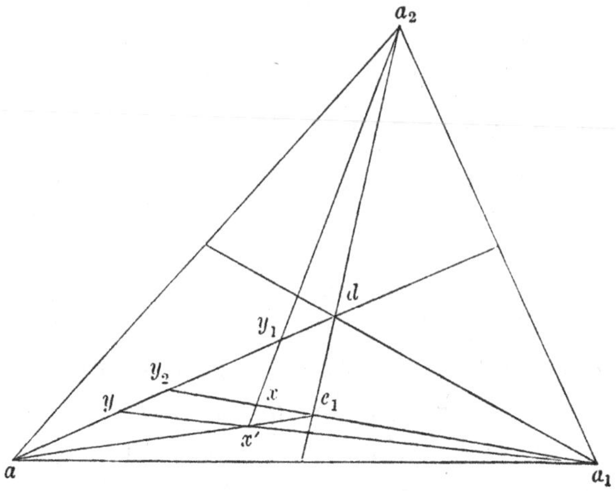

FIG. 4.

Firstly, make the following constructions (fig. 4),

$$y_1 = xa_2 \,.\, ad,\;\; y_2 = xa_1 \,.\, ad.$$

Then $$y_1 \equiv (aa_2 + \xi_1 a_1 a_2)(aa_1 + aa_2) \equiv -(aa_1a_2)\{a + \xi_1(a_1 + a_2)\}$$
$$\equiv a + \xi_1(a_1 + a_2).$$

Note that in future numerical factors which do not involve ξ_1 or ξ_2 will be dropped without remark.

Similarly $$y_2 = a + \xi_2(a_1 + a_2).$$

(3) Secondly, make the following constructions (fig. 4),

$$e_1 = y_2a_1 \,.\, a_2d,\;\; x' = e_1a \,.\, y_1a_2,\;\; y = x'a_1 \,.\, ad.$$

Then $$e_1 \equiv (aa_1 + \xi_2 a_2a_1)(a_2a + a_2a_1)$$
$$\equiv -(aa_1a_2)\,a - (aa_1a_2)\,a_1 - \xi_2(aa_1a_2)\,a_2$$
$$\equiv a + a_1 + \xi_2 a_2.$$

And $$x' \equiv (a_1a + \xi_2a_2a)(aa_2 + \xi_1a_1a_2)$$
$$\equiv -(aa_1a_2)\,a - \xi_1(aa_1a_2)\,a_1 - \xi_1\xi_2(aa_1a_2)\,a_2$$
$$\equiv a + \xi_1a_1 + \xi_1\xi_2a_2.$$

And $$y \equiv (aa_1 + \xi_1\xi_2a_2a_1)(aa_1 + aa_2)$$
$$\equiv (aa_1a_2)\{a + \xi_1\xi_2(a_1 + a_2)\} \equiv a + \xi_1\xi_2(a_1 + a_2)$$
$$\equiv y_2a_1 \,.\, a_2d \,.\, a \,.\, y_1a_2 \,.\, a_1 \,.\, ad.$$

(4) Now substitute y_1 for y_2 in the above construction (fig. 5). We obtain

$$y_1' = y_1a_1 \,.\, a_2d \,.\, a \,.\, y_1a_2 \,.\, a_1 \,.\, ad \equiv a + \xi_1^2(a_1 + a_2).$$

Similarly we construct

$$y_2' \equiv y_2 a_1 \,.\, a_2 d \,.\, a \,.\, y_2 a_2 \,.\, a_1 \,.\, ad \equiv a + \xi_2^2 (a_1 + a_2).$$

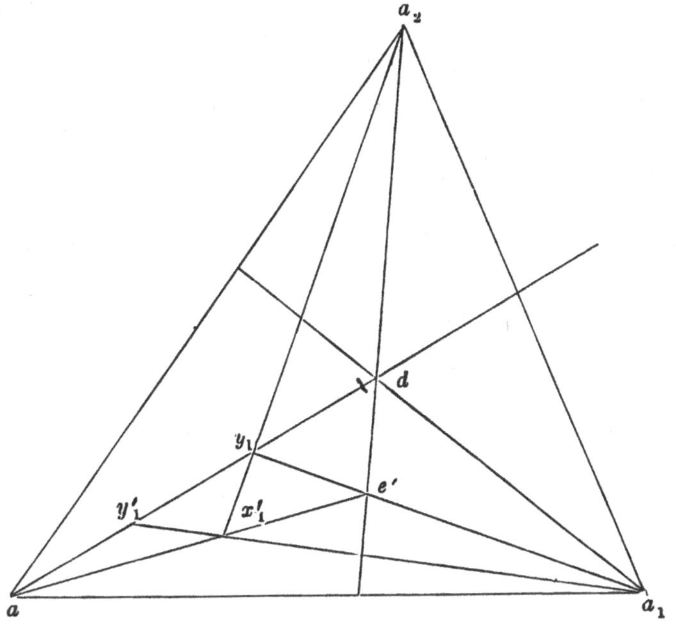

FIG. 5.

It is also obvious that in the constructions a_1 and a_2 can be interchanged. Thus

$$y \equiv y_2 a_2 \,.\, a_1 d \,.\, a \,.\, y_1 a_1 \,.\, a_2 \,.\, ad,$$

and
$$y_1' \equiv y_1 a_2 \,.\, a_1 d \,.\, a \,.\, y_1 a_1 \,.\, a_2 \,.\, ad,$$

and
$$y_2' \equiv y_2 a_2 \,.\, a_1 d \,.\, a \,.\, y_2 a_1 \,.\, a_2 \,.\, ad.$$

Also in the construction in subsection (3) for y from y_1 and y_2, y_1 and y_2 can be interchanged, thus giving two fresh forms of the construction, namely,

$$y \equiv y_1 a_1 \,.\, a_2 d \,.\, a \,.\, y_2 a_2 \,.\, a_1 \,.\, ad, \text{ and } y \equiv y_1 a_2 \,.\, a_1 d \,.\, a \,.\, y_2 a_1 \,.\, a_2 \,.\, ad.$$

(5) Let the symbol (ξ_1^ν) stand for the point $a + \xi_1^\nu (a_1 + a_2)$, and similarly let (ξ_2^ν) stand for the point $a + \xi_2^\nu (a_1 + a_2)$, and let $(\xi_1^\nu \xi_2^\mu)$ stand for the point $a + \xi_1^\nu \xi_2^\mu (a_1 + a_2)$.

Now substitute the point (ξ_1^ν) for y_2 in the first construction given for y, then we obtain

$$(\xi_1^{\nu+1}) \equiv (\xi_1^\nu) a_1 \,.\, a_2 d \,.\, a \,.\, y_1 a_2 \,.\, a_1 \,.\, ad.$$

(6) Again, let pR_1 denote that the point p has been multiplied successively by the factors a_1, $a_2 d$, a, $y_1 a_2$, a_1, ad, so that pR_1 stands for the point $pa_1 \,.\, a_2 d \,.\, a \,.\, y_1 a_2 \,.\, a_1 \,.\, ad$. In order to avoid misconception it may be mentioned that R_1 is *not* the product $a_1 \,.\, a_2 d \,.\, a \,.\, y_1 a_2 \,.\, a_1 \,.\, ad$; for in pR_1 the first factor a_1 is multiplied into the point p. Also pR_1 is itself a point: let $(pR_1) R_1$ be denoted by pR_1^2, and so on.

By applying this notation to the construction for $(\xi_1^{\nu+1})$ in terms of ξ_1^{ν}, we see that when ν is a positive integer, $(\xi_1^{\nu}) \equiv y_1 R_1^{\nu-1}$.

Since $y_1 a_2 \equiv x a_2$, R_1 may be conceived to stand for the set of factors a_1, $a_2 d$, a, $x a_2$, a_1, ad successively multiplied on to a point. Also $y_1 \equiv x a_2 \,.\, ad$.

Hence $$(\xi_1^{\nu}) \equiv x a_2 \,.\, ad \,.\, R_1^{\nu-1}.$$

Thus the point $a + \xi_1^{\nu}(a_1 + a_2)$ is exhibited as a product in which x occurs ν times.

(7) Similarly interchanging the suffixes 1 and 2, let R_2 stand for the set of factors a_2, $a_1 d$, a, $x a_1$, a_2, ad successively multiplied on to a point. Also $y_2 \equiv x a_1 \,.\, ad$.

Hence $$(\xi_2^{\mu}) \equiv x a_1 \,.\, ad \,.\, R_2^{\mu-1}.$$

Thus the point $a + \xi_2^{\mu}(a_1 + a_2)$ is exhibited as a product in which x occurs μ times.

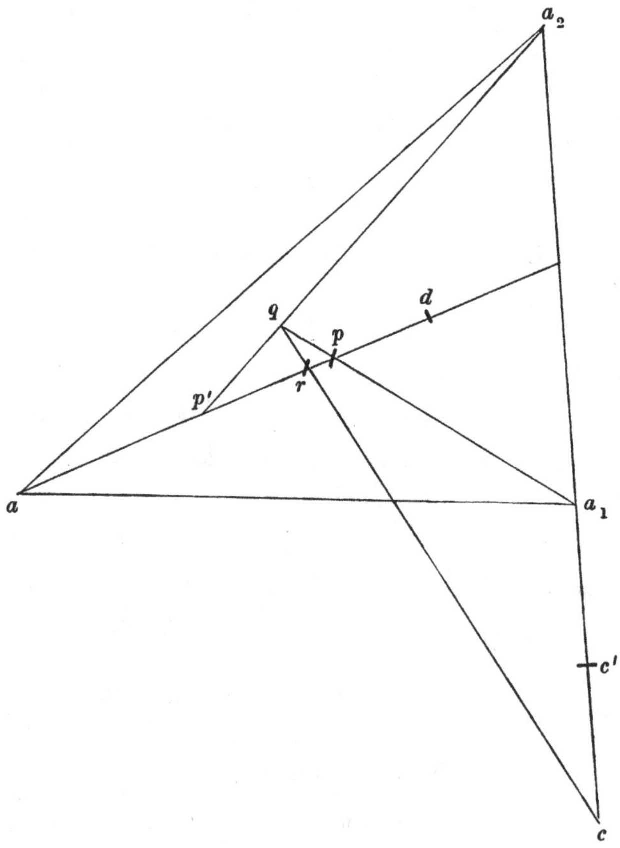

FIG. 6.

(8) Again, in any of the constructions for y (say the first) substitute (ξ_1^{ν}) and (ξ_2^{μ}) for y_1 and y_2, say, for example, (ξ_1^{ν}) for y_1 and (ξ_2^{μ}) for y_2.

Then $(\xi_1^\nu \xi_2^\mu) \equiv (\xi_2^\mu) a_1 . a_2 d . a . (\xi_1^\nu) a_2 . a_1 . ad.$

Hence the point $a + \xi_1^\nu \xi_2^\mu (a_1 + a_2)$ is represented as a product in which x occurs $(\mu + \nu)$ times.

(9) Finally, let p and p' be any two points $a + \varpi (a_1 + a_2)$ and $a + \varpi' (a_1 + a_2)$ on the line ad; and let c denote any point $\gamma_1 a_1 - \gamma_2 a_2$ on the line $a_1 a_2$. Make the constructions (fig. 6)

$$q \equiv pa_1 . p'a_2, \quad r \equiv qc . ad \equiv pa_1 . p'a_2 . c . ad.$$

Then
$$q \equiv pa_1 . p'a_2 \equiv (aa_1a_2) \{a + \varpi' a_1 + \varpi a_2\},$$
$$r \equiv \{\gamma_1 aa_1 - \gamma_2 aa_2 + (\gamma_1 \varpi + \gamma_2 \varpi') a_2 a_1\} \{aa_1 + aa_2\}$$
$$\equiv (aa_1a_2) \{(\gamma_1 + \gamma_2) a + (\gamma_1 \varpi + \gamma_2 \varpi')(a_1 + a_2)\}$$
$$\equiv a + \frac{\gamma_1 \varpi + \gamma_2 \varpi'}{\gamma_1 + \gamma_2} (a_1 + a_2).$$

Similarly, let p'' denote a third point $a + \varpi'' (a_1 + a_2)$, and let c' denote the point $(\gamma_1 + \gamma_2) a_1 - \gamma_3 a_2$. Make the construction

$$r' \equiv ra_1 . p''a_2 . c' . ad \equiv a + \frac{\gamma_1 \varpi + \gamma_2 \varpi' + \gamma_3 \varpi''}{\gamma_1 + \gamma_2 + \gamma_3} (a_1 + a_2).$$

And so on for any number of points p, p', p'', etc.

Thus if any number of points of the form $(\xi_1^\nu \xi_2^\mu)$ have been constructed, then the point

$$z \equiv a + \frac{\gamma \xi_1^\nu \xi_2^\mu + \gamma' \xi_1^{\nu'} \xi_2^{\mu'} + \dots}{\gamma + \gamma' + \dots} (a_1 + a_2),$$

can be exhibited in the form of a product.

(10) Hence the numerical product

$$(aa_1z) = \frac{\gamma \xi_1^\nu \xi_2^\mu + \gamma' \xi_1^{\nu'} \xi_2^{\mu'} + \dots}{\gamma + \gamma' + \dots} (aa_1a_2).$$

It will be observed that x occurs in this product $\{(\mu + \nu) + (\mu' + \nu') + \dots\}$ times, and that therefore if x be written in the form $\eta a + \eta_1 a_1 + \eta_2 a_2$, then the product is a *homogeneous* function of η, η_1, η_2 of degree $\{(\mu + \nu) + (\mu' + \nu') + \dots\}$.

But let ρ be the greatest of the numbers $(\mu + \nu), (\mu' + \nu')$, etc., then it is easily verified that the homogeneous function represented by the product is any required homogeneous function of degree ρ multiplied by η to the power: $[\{(\mu + \nu) + (\mu' + \nu') + \dots\} - \rho]$, and also by some constant numerical factor.

If however we keep x in the form $a + \xi_1 a_1 + \xi_2 a_2$, then the most general rational integral algebraic function (not necessarily homogeneous) of ξ_1 and ξ_2 can be exhibited in the form of a product; or if $\phi(\xi_1, \xi_2)$ be the function, it can be represented by a point

$$a + \frac{\phi(\xi_1, \xi_2)}{\Sigma \gamma} (a_1 + a_2),$$

which is constructed as a product of the point x and fixed points, partly arbitrarily chosen and partly chosen to suit the special function.

130. PROJECTION. (1) *Definition.* Let the complete region be of $\nu-1$ dimensions, and let x, y, etc., be elements on any given plane A of this region. Let e be any given point not on this plane and let B be any other given plane. Then the lines ex, ey, etc., intersect the plane B in elements x', y', etc.: the assemblage of elements x', y', etc., on the plane B is called the projection on B from the vertex e of the assemblage of elements x, y, etc., on the plane A.

Definition. Two assemblages of elements x, y, etc., on the plane A and x', y', etc. on the plane A', which is not necessarily distinct from A, are called mutually projective, if one assemblage can be derived from the other by a series of projections.

If one figure can be deduced from another by a single projection, the two figures are obviously in perspective.

(2) These constructions can be symbolized by products: thus the projection of x on to the plane B from the vertex e is $x' = xe\,.\,B$. Let the projected points always be assumed to be at intensities which are deduced from the intensities of the original points according to this formula.

Since $(\lambda x + \mu y)\,e\,.\,B = \lambda xe\,.\,B + \mu ye\,.\,B$, it follows that any range of elements on a line is projected into a homographic range.

(3) PROPOSITION I. Let any subregional element in the plane A be denoted by the product $x_1, x_2 \ldots x_\rho$, where ρ is less than $\nu-2$; also let $x_1', x_2' \ldots x_\rho'$ be the projections of the points $x_1, x_2, \ldots x_\rho$ on to the plane B from the vertex e, so that for instance $x_1' = x_1 e\,.\,B$; then it will be proved that*

$$x_1'x_2' \ldots x_\rho' = (eB)^{\rho-1}\, x_1x_2 \ldots x_\rho e\,.\,B \equiv x_1x_2 \ldots x_\rho e\,.\,B.$$

In other words, if X_ρ be any subregional element of the ρth order, and X_ρ' be the corresponding subregional element formed by the projected points, then

$$X_\rho' = (eB)^{\rho-1}\, X_\rho e\,.\,B \equiv X_\rho e\,.\,B.$$

Thus X_ρ' will be called the projection of X_ρ, and the above equation forms the universal formula for the projection of elements of any order.

(4) In order to prove this formula the following notation will be useful. Let $x_1x_2 \ldots (x_\sigma) \ldots x_\rho e$ denote that the elements $x_1, x_2 \ldots x_\rho, e$, *with the exception of* x_σ, are multiplied together in the order indicated. Then the extended rule of the middle factor gives the transformations

$$x_1x_2 \ldots x_\rho e\,.\,B = \sum_{\sigma=1}^{\sigma=\rho} (-1)^{\sigma-1}\,(x_\sigma B)\, x_1x_2 \ldots (x_\sigma) \ldots x_\rho e + (-1)^\rho\,(eB)\, x_1x_2 \ldots x_\rho.$$

Also $$x'_{\rho+1} = x_{\rho+1}\, e\,.\,B = (x_{\rho+1} B)\, e - (eB)\, x_{\rho+1}.$$

But $$x_1x_2 \ldots (x_\sigma) \ldots x_\rho e\,.\,e = 0,$$

and $$x_1x_2 \ldots (x_\sigma) \ldots x_\rho e\,.\,x_{\rho+1} = -\,x_1x_2 \ldots (x_\sigma) \ldots x_\rho x_{\rho+1} e.$$

* This formula has not, I think, been stated before.

Hence by multiplication and rearrangement of factors it follows that

$$\{x_1x_2 \dots x_\rho e \,.\, B\}\, x'_{\rho+1} = (-1)^\rho\, (eB)\, (x_{\rho+1} B)\, x_1x_2 \dots x_\rho e$$

$$+ \sum_{\sigma=1}^{\sigma=\rho} (-1)^{\sigma-1}\, (eB)\, (x_\sigma B)\, x_1x_2 \dots (x_\sigma) \dots x_{\rho+1}e - (-1)^\rho\, (eB)^2\, x_1x_2 \dots x_{\rho+1}$$

$$= (eB) \left[\sum_{\sigma=1}^{\sigma=\rho+1} (-1)^{\sigma-1}\, (x_\sigma B)\, x_1x_2 \dots (x_\sigma) \dots x_{\rho+1}\, e + (-1)^{\rho+1}\, (eB) x_1x_2 \dots x_{\rho+1} \right]$$

$$= (eB)\, x_1x_2 \dots x_{\rho+1}e \,.\, B.$$

Hence by successively applying this theorem we deduce

$$x_1'x_2' \dots x_\rho' = (eB)^{\rho-1}\, x_1x_2 \dots x_\rho e \,.\, B \equiv x_1x_2 \dots x_\rho e \,.\, B.$$

(5) It is obvious that the relation between a point x and its projection x' is reciprocal; that is, if x' be the projection of x on B from vertex e, then x is the projection of x' on A from vertex e.

For
$$x'e \,.\, A \equiv \{(xB)\, e - (eB)\, x\}\, e \,.\, A \equiv -(eB)\, xe \,.\, A$$
$$\equiv (eB)\, (eA)\, x \equiv x,$$
since $(xA) = 0$, by hypothesis.

Thus two figures are projective if they can both be projected into the same figure.

(6) Proposition II. Any three collinear points are projective with any other three collinear points.

This is the same as the proposition that any two homographic ranges are projective.

Firstly, let the two lines L and L', on which the points respectively lie, be intersecting, so that the complete region is of two dimensions. Let a, b, c and a', b', c' be the two sets of three points on L and L' respectively.

Take e and e' any two points on aa'. Construct the points $eb \,.\, e'b'$ and $ec \,.\, e'c'$: call them the points b'', c''. Construct the point $aa' \,.\, b''c'' = a''$.

Then we have evidently

$$a = a''e \,.\, L, \quad b = b''e \,.\, L, \quad c = c''e \,.\, L,$$

and
$$a' = a''e' \,.\, L', \; b' = b''e' \,.\, L', \; c' = c''e' \,.\, L'.$$

Thus the collinear points a'', b'', c'' can be projected both into a, b, c and a', b', c'. Hence a, b, c and a', b', c' are projective.

(7) It may be noticed that if the regressive multiplications are defined for a complete region of three dimensions and the ranges abc and $a'b'c'$ be coplanar, then the above results must be written

$$a = a''e \,.\, Ld, \quad b = b''e \,.\, Ld, \quad c = c''e \,.\, Ld,$$

and
$$a' = a''e' \,.\, L'd, \; b' = b''e' \,.\, L'd, \; c' = c''e' \,.\, L'd;$$

where d is any point not in the plane of the straight lines L and L', and e and e' being both on aa' are in the plane of L and L'.

And more generally, if the regressive multiplication be defined for a complete region of $\nu-1$ dimensions, let $D_{\nu-3}$ be any extensive magnitude of the $(\nu-3)$rd order which does not intersect the two dimensional regions containing L and L', then $LD_{\nu-3}$ and $L'D_{\nu-3}$ can be taken as the planes of the two projections, so that

$$a = a''e \,.\, LD_{\nu-3}, \text{ etc., and } a' = a''e' \,.\, L'D_{\nu-3}, \text{ etc.}$$

(8) Secondly, let the lines containing a, b, c and of a', b', c' be not intersecting. Take any point p on abc and p' on $a'b'c'$. Construct the line pp', and on it take any three points a'', b'', c''. Then a, b, c and a'', b'', c'' are projective, also a', b', c' and a'', b'', c'' are projective. Hence a, b, c and a', b', c' are projective.

(9) PROPOSITION III. If any ρ points in a subregion of $\rho-2$ dimensions (with only one addition relation) are projective with any other ρ points in another subregion of $\rho-2$ dimensions (with only one addition relation), then any $\rho+1$ points (with only one addition relation) in a subregion of $\rho-1$ dimensions are projective with any other $\rho+1$ points (with only one addition relation) in another subregion of $\rho-1$ dimensions.

For let $a_1, a_2, \ldots a_{\rho+1}$ and $b_1, b_2, \ldots b_{\rho+1}$ be any two sets of $\rho+1$ points in regions of $\rho-1$ dimensions.

Since the $\rho+1$ points $a_1 \ldots a_{\rho+1}$ are contained in a subregion of $\rho-1$ dimensions, $a_\rho a_{\rho+1}$ must intersect the subregion of the independent points $a_1, a_2, \ldots a_{\rho-1}$ in some point c; and similarly $b_\rho b_{\rho+1}$ must intersect the subregion of the independent points $b_1, b_2, \ldots b_{\rho-1}$ in some point d.

Now by hypothesis a series of projections can be made which transforms $b_1, b_2 \ldots b_{\rho-1}, d$ into $a_1, a_2 \ldots a_{\rho-1}, c$. Assume that such a series transforms b_ρ and $b_{\rho+1}$ into b_ρ' and $b'_{\rho+1}$.

Let $A_{\rho-1}$ stand for the subregional element $a_1a_2 \ldots a_{\rho-1}$, and let $D_{\nu-\rho}$ denote the product of any $\nu-\rho$ independent points which do not lie in $A_{\rho-1}$, where $\nu-1$ is the number of dimensions of the complete region. Then $A_{\rho-1}D_{\nu-\rho}$ is a planar element.

Again, c, b_ρ', $b'_{\rho+1}$ are collinear and so are c, a_ρ, $a_{\rho+1}$, hence b_ρ', $b'_{\rho+1}$, a_ρ, $a_{\rho+1}$ lie in the same two dimensional region. Therefore $a_\rho b_\rho'$ and $a_{\rho+1}b'_{\rho+1}$ intersect in some point e.

Let $D_{\nu-\rho}$ be so chosen that e does not lie in the plane $A_{\rho-1}D_{\nu-\rho}$: also let $D_{\nu-\rho}$ contain a_ρ and $a_{\rho+1}$. Then it cannot contain b_ρ' and $b'_{\rho+1}$, since it does not contain e.

Project on to the plane $A_{\rho-1}D_{\nu-\rho}$ from the vertex e. The points $a_1, a_2, \ldots a_{\rho-1}, c$ are unchanged, being already in that plane, also b_ρ' is projected into a_ρ, and $b'_{\rho+1}$ into $a_{\rho+1}$.

Hence the proposition is proved.

(10) It has already been proved that three collinear points can be projected into any other three collinear points; it follows that any ρ points in a

subregion of $\rho-2$ dimensions are projective with any other ρ points in another subregion of $\rho-2$ dimensions.

(11) PROPOSITION IV. The least number of separate projections required can also be easily determined. For we have proved in the course of subsection (9) that if $\phi(\rho)$ projections are required for two sets of ρ points in subregions of $\rho-2$ dimensions, then $\phi(\rho)+1$ projections are required for two sets of $\rho+1$ points in subregions of $\rho-1$ dimensions. We have therefore only to determine the least number requisite to project three collinear elements a, b, c into three other collinear elements a', b', c'.

The construction given above in the second and general case may be simplified thus. Join ab'. Project from any point e on bb'. Then a is unaltered, b becomes b' and c becomes some point c''. Now project a, b' c'' from the point of intersection of aa' and $c'c''$. Then a becomes a', b' is unaltered, c'' becomes c'. Hence two projections are in general requisite.

Thus three projections are requisite for four points in a two dimensional region, and $\rho-1$ projections for ρ points in a region of $\rho-2$ dimensions.

(12) These constructions still hold if the two sets of ρ points are both in the same subregion of $\rho-2$ dimensions. In such a case the same series of projections which transforms one set of ρ elements into another set of ρ elements may be conceived as applied to every point of the subregion. Thus every point of the subregion is transformed into some other point of the same subregion.

(13) PROPOSITION V. It will now be proved that the most general type of such a projective transformation is equivalent to the most general type of linear transformation which transforms every point of the given subregion into another point of that subregion.

If x' be the point into which any point x is finally projected, the relation between x' and x can be written in the form

$$x' = xe_1 . B_1 . e_2 . B_2 . e_3 . B_3 \dots e_{\rho-1} . B_{\rho-1}.$$

It is obvious therefore that x can be conceived as transformed into x' by some linear transformation. The only question is, whether it is of the most general type.

Now in the most general type of linear transformation, as applied to a region of $\rho-2$ dimensions, $\rho-1$ elements must remain unchanged. Let $a_1, a_2, \dots a_{\rho-1}$ be these elements, and let any other point x be represented by $\xi_1 a_1 + \xi_2 a_2 + \dots + \xi_{\rho-1} a_{\rho-1}$.

Then if x_1 be the transformed element corresponding to x, we have $x_1 = \alpha_1 \xi_1 a_1 + \alpha_2 \xi_2 a_2 + \dots + \alpha_{\rho-1} \xi_{\rho-1} a_{\rho-1}$, where $\alpha_1, \alpha_2, \dots \alpha_{\rho-1}$ are the constants which in conjunction with the fixed points define by their ratios the linear transformation.

Hence if a given point g is to be transformed into a given point d, where $g = \Sigma\gamma a$, and $d = \Sigma\delta a$, we must have, $\alpha_1 = \delta_1/\gamma_1$, $\alpha_2 = \delta_2/\gamma_2$, $\dots$ $\alpha_{\rho-1} = \delta_{\rho-1}/\gamma_{\rho-1}$.

Accordingly if the $\rho-1$ unchanged points are arbitrarily assumed, it is not possible by a linear transformation to transform more than one arbitrarily assumed point into another arbitrarily assumed point.

But it is possible by a series of projections to transform the ρ points $a_1, a_2, \dots a_{\rho-1}, c$ into the ρ points $a_1, a_2, \dots a_{\rho-1}, d$.

Hence the general type of projection is equivalent to the general type of linear transformation.

(14) It is to be noticed that a linear transformation can be conceived as transforming all the points of the complete region of (say) $\nu-1$ dimensions. But these points cannot be projectively transformed without considering the region of $\nu-1$ dimensions as a subregion in a containing region of ν dimensions. This fresh conception is of course always possible without in any way altering the intrinsic properties of the original region of $\nu-1$ dimensions.

The Theory of Linear Transformation in connection with this Calculus is resumed in Chapter VI. of this Book.

CHAPTER V.

DESCRIPTIVE GEOMETRY OF CONICS AND CUBICS.

131. GENERAL EQUATION OF A CONIC. (1) The following investigation concerning conics and cubics is in substance with some extensions a reproduction of Grassmann's applications of the Calculus of Extension to this subject*. In places the algebra is handled differently and alternative proofs are given for the sake of illustration.

A quadric surface in a complete region of two dimensions will be called a conic. It will also in this chapter be called a curve in order to agree with the usual nomenclature of Geometry.

(2) The complete region is of two dimensions: the product of three points or of three linear elements or of a point and a linear element is purely numerical. Also the product of three linear elements, being a pure progressive product, is associative; thus if L_1, L_2, L_3 be the linear elements, $(L_1L_2L_3) = (L_1 . L_2L_3)$. Also if p and q be points, then, since L_1L_2 is a point, (L_1L_2pq) is the product of the three points L_1L_2, p, q. Hence

$$(L_1L_2pq) = (L_1L_2 . pq).$$

(3) The equation, $$(xaBcDex) = 0,$$

where a, c, e are any points and B and D are any linear elements, is evidently, since x occurs twice, of the second degree in the three co-ordinates of x.

For let e_1, e_2, e_3 be the three reference points, and let $x = \xi_1e_1 + \xi_2e_2 + \xi_3e_3$. Also let the fixed points and lines be written in the form

$$a = \alpha_1e_1 + \alpha_2e_2 + \alpha_3e_3, \quad B = \beta_1e_2e_3 + \beta_2e_3e_1 + \beta_3e_1e_2,$$

and so on; where α_1, α_2, α_3, etc., are given numerical coefficients. Then the given equation, after multiplying the various expressions for the points and lines, takes the form

$$(e_1e_2e_3)^3 (\lambda_1 \xi_1^2 + \dots + 2\lambda_{23}\xi_2\xi_3 + \dots) = 0.$$

Hence, dividing out the numerical factor $(e_1e_2e_3)^3$, the given equation is equivalent to a single numerical equation of the second degree defining a quadric locus.

* Cf. *Ausdehnungslehre von* 1862, and *Crelle*, vols. XXXI, XXXVI, LII.

Write ϖ_x for the expression $(xaBcDex)$, then the following transformations by the aid of subsection (2) are obviously seen to be true:

$$\varpi_x = \{xaBcD \,.\, ex\} = \{xaBc\,(D \,.\, ex)\} = -\{xaBc \,.\, exD\}$$
$$= -\{c \,.\, axB \,.\, exD\} = \{c \,.\, exD \,.\, axB\} = -\{xeDcBax\};$$

where it is to be remembered in proving the transformations that xa is a linear element, xaB is a point, $xaBc$ is a linear element, $xaBcD$ is a point.

(4) From $\varpi_x = -\{c \,.\, axB \,.\, exD\} = 0$, it is obvious that a and e are points on the conic. In general c is not on the conic, for the points c, (acB), and (ecD) are not in general collinear.

(5) Points in which B and D meet the curve. Suppose that B meets the curve in the point p, and let $B = pq$. Also substitute p for x in the expression ϖ_x.

Now $$apB = ap \,.\, pq = (apq)\,p = (aB)\,p.$$

Therefore $$\varpi_p = -\{c \,.\, apB \,.\, epD\} = (aB)\,(cp \,.\, epD) = 0.$$

This involves either (i) that $(aB) = 0$, or (ii) that $(cp \,.\, epD) = 0$.

(i) Let $(aB) = 0$. Then $axB = (aB)\,x - (xB)\,a = -(xB)\,a$. Hence

$$\varpi_x = -(xB)\,(ca \,.\, exD) = 0.$$

Therefore the curve splits up into the two straight lines

$$(xB) = 0, \text{ and } (ca \,.\, exD) = 0.$$

Similarly if $(eD) = 0$, the curve becomes the two straight lines

$$(xD) = 0, \text{ and } (ce \,.\, axB) = 0.$$

These special cases in which the conic degenerates into two straight lines will not be further considered.

(ii) Let $(cp \,.\, epD) = 0$. Then $\{cp \,.\, epD\} = \{c\,(p \,.\, epD)\}$.

But $$p \,.\, epD = p\,\{(eD)\,p - (pD)\,e\} = -(pD)\,pe.$$

Hence $$(cp \,.\, epD) = (pD)\,(cep) = 0,$$

so that p lies in D or in the line ce.

Accordingly the two points in which B intersects the curve are $B \,.\, D$ and $B \,.\, ce$.

Similarly the points in which D intersects the curve are $B \,.\, D$ and $D \,.\, ca$.

(6) Let $g = B \,.\, D$, $b = B \,.\, ce$, $d = D \,.\, ca$.

Then $$b = (Be)\,c - (Bc)\,e, \quad d = (Da)\,c - (Dc)\,a.$$

Hence $$eb \,.\, ad = (Be)\,(Da)\,\{ec \,.\, ac\} = (Be)\,(Da)\,(eac)\,c \equiv c.$$

Also we may write $B = bg$, $D = dg$.

Hence the equation becomes

$$\{(xa \,.\, bg)\,(eb \,.\, ad)\,(dg \,.\, ex)\} = 0;$$

where a, b, d, e, g are five given points on the curve and x is a variable point.

(7) Conversely, if we take any five points a, b, d, e, g and write,

$$\{(xa \,.\, bg)\,(eb \,.\, ad)\,(dg \,.\, ex)\} = 0,$$

then the above reasoning shows that the five points are on the curve which is the locus of x. But only one conic can be drawn through five points; therefore by properly choosing the five points this equation can be made to represent any conic section, and is therefore the general equation of the second degree.

(8) If we perform the constructions indicated by the products on the left-hand side (cf. fig. 1), we see that the equation is a direct expression of Pascal's theorem, which is thereby proved.

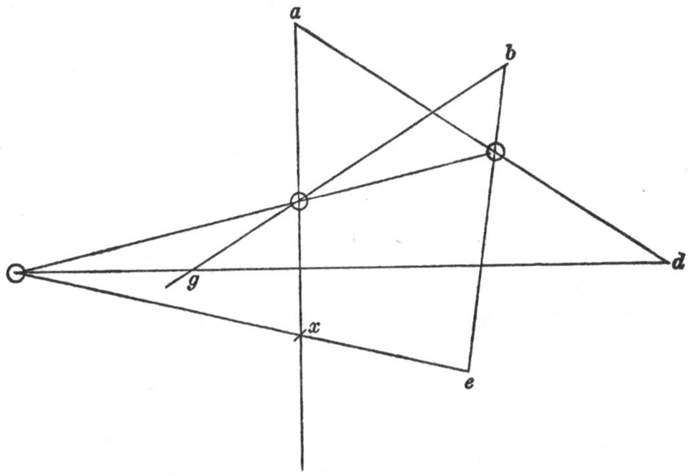

FIG. 1.

(9) Perform the operation of taking the supplement on the equation, and write X for the linear element $|x$, A for $|a$, and so on. Then

$$|\{(xa \,.\, bg)(eb \,.\, ad)(dg \,.\, ex)\} = \{(XA \,.\, BG)(EB \,.\, AD)(DG \,.\, EX)\}$$
$$= 0.$$

This is the general tangential equation of a conic [cf. § 107 (4)]: hence from subsection (7) it follows that A, B, D, E, G are tangents; and the equation is a direct expression of Brianchon's Theorem.

132. FURTHER TRANSFORMATIONS. (1) These results can be obtained by a different method which forms an instructive illustration of the algebra.

The following series of transformations follow immediately from the extended rule of the middle factor:

$$axB = (aB)\,x - (xB)\,a;$$

hence, $$axBcDx = \{(aB)\,xc \,.\, D - (xB)\,ac \,.\, D\}\,x$$
$$= (aB)(xD)\,cx - (xB)(aD)\,cx + (xB)(cD)\,ax.$$

Now $(aB)(xD) - (xB)(aD) = x\,[(aB)\,D - (aD)\,B] = x\,[a \,.\, DB] = (xa \,.\, DB).$

Hence, $$axBcDx = (xa \,.\, DB)\,cx + (xB)(cD)\,ax,$$

and, $$(axBcDxe) = (xa \,.\, DB)(cxe) + (xB)(cD)(axe).$$

Thus the equation of the curve, $\varpi_x = 0$, can be written

$$(xa \,.\, DB)(cxe) + (xB)(axe)(cD) = 0.$$

(2) To find where B meets the curve, put $(xB) = 0$. Then either

$$(xa \,.\, DB) = 0, \text{ or } (cxe) = 0.$$

Thus either x is the point BD or it is the point ceB; therefore these are the points where B meets the curve.

Similarly the points where D meets the curve are BD and caD.

(3) Obviously the points a and e lie on the curve.

(4) If $(cD) = 0$, the curve degenerates into the two straight lines

$$(xa \,.\, BD) = 0, \; (xce) = 0.$$

Similarly if $(cB) = 0$, the curve becomes the two straight lines,

$$(xe \,.\, BD) = 0, \; (xca) = 0.$$

(5) To find the second point in which any line through the point a cuts the curve.

Let L be the line, then $(aL) = 0$. Let x be the required point in L, then

$$xa \equiv L.$$

Hence $$(xaBcDex) \equiv (LBcDex) = 0.$$

Hence x is incident in the linear element $LBcDe$, also x is incident in L. Therefore $$x \equiv LBcDeL.$$

(6) It is to be noticed that a apart from L does not appear explicitly in this expression for x. Hence the theorem can be stated thus:

If a be any variable point on the line L, the conic through the five points a, BD, ceB, e, caD passes through the fixed point $LBcDeL$.

(7) The conditions that T should be the tangent at a are $(aT) = 0$, and $a \equiv TBcDeT$.

(8) The general expression ϖ_x is susceptible of a very large number of transformations of which the following is a type:

$$xa \,.\, bg = (xbg)\,a - (abg)\,x, \quad eb \,.\, ad = (ebd)\,a - (eba)\,d, \quad dg \,.\, ex = (dgx)\,e - (dge)\,x.$$

Hence $$\{(xa \,.\, bg)(eb \,.\, ad)(dg \,.\, ex)\} = (eba)(dge)(xbg)(adx) - (eba)(ade)(xbg)(dgx) + (abg)(ebd)(dgx)(aex) - (abg)(eba)(dgx)(dex).$$

(9) The equation,

$$(xa_1B_1a_2B_2 \ldots a_{n-1}B_{n-1}a_nx) = 0,$$

represents a conic. Hence the following theorem due to Grassmann:

'If all the sides of an n-sided polygon pass through n fixed points respectively, and $n-1$ of the corners move on $n-1$ fixed lines respectively, the nth corner moves on a conic section.'

133. Linear Construction of Cubics. The first linear constructions satisfied by any point of a cubic were given by Grassmann* in 1846; and the theory was extended and enlarged by him in 1848 and 1856†. An indefinite number of such linear constructions of increasing complexity can be successively written down by the aid of the calculus. The simplest types are given by

$$(xaAa_1 \, . \, xbBkCb_1 \, . \, xc) = 0 \quad \text{.........................(1)},$$

$$(xaAa_1 \, . \, xbBb_1 \, . \, xc) = 0 \quad \text{.........................(2)},$$

$$\left.\begin{array}{r}(x \, . \, xaBcD \, . \, xa_1B_1c_1D_1) = 0 \\ (xaBcDxD_1c_1B_1a_1x) = 0\end{array}\right\} \quad \text{.........................(3)},$$

$$(xaA \, . \, xbB \, . \, xcC) = 0 \quad \text{.........................(4)}.$$

The two equations, marked (3), are alternative forms of the same equation. It is to be noted that none of these constructions give a method of discovering points on a cubic; but that, given a point x on a cubic, the constructions can be made. Thus a point x on the cubic will be said to satisfy the corresponding construction, but not to be found by it.

134. First Type of Linear Construction of the Cubic. (1) To investigate the construction

$$(xaAa_1 \, . \, xbBkCb_1 \, . \, xc) = 0.$$

This equation asserts that if the three lines $xaAa_1$, $xbBkCb_1$, xc are concurrent, the locus of x is a cubic. Let y be the point of concurrence; then the construction is exemplified in figure 2.

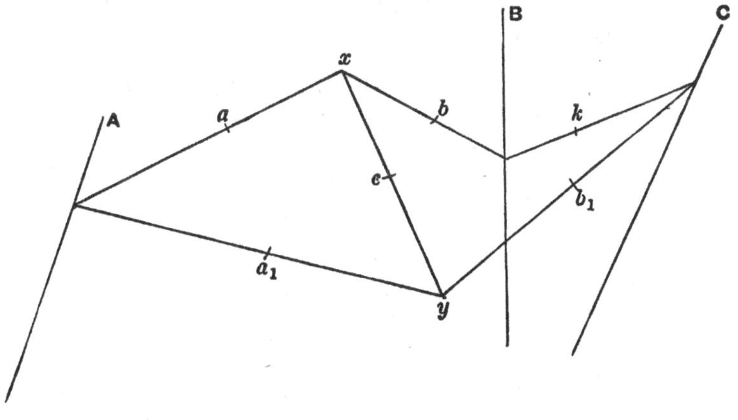

Fig. 2.

(2) It has now to be proved that any cubic can be represented by this construction. This will be proved by shewing that by a proper choice

* Cf. *Crelle's Journal*, vol. xxxi.

† Cf. *Crelle's Journal*, vols. xxxvi. and lii.

of the fixed lines and points of the construction the cubic may be made to pass through any nine arbitrarily assumed points. Thus we proceed to investigate the solution of the following problem: Given any nine arbitrarily assumed points in a plane to find a linear construction satisfied by any point of the cubic passing through them.

But previously to the direct solution of this problem in § 135 some properties of the expression $(xaAa_1 . xbBkCb_1 . xc)$ must be investigated.

(3) Let ϖ_x stand for the product $(xaAa_1 . xbBkCb_1 . xc)$.

Then $$\varpi_x = -(xaAa_1 . xc . xbBkCb_1).$$

Now put $$p = xaAa_1 . xc, \quad q = xbB.$$

Then $$\varpi_x = -(p . qkCb_1) = (pb_1Ckq).$$

It is easily proved that $(pb_1Ckq) = -(qkCb_1p)$.

(4) To find the particular positions of x for which $p = 0$, or $q = 0$.

Now $p = 0$, when $x = a$, and when $x = c$.

Also by § 105 $$p = (xaAa_1c)\,x - (xaAa_1x)\,c$$
$$= \{(xA)(aa_1c) - (aA)(xa_1c)\}\,x - (xA)(aa_1x)\,c.$$

Hence all the points x for which $p = 0$ (except $x = c$) must satisfy (in order to make the coefficient of c zero) either $(xA) = 0$, or $(xaa_1) = 0$.

If $(xA) = 0$, then, since the coefficient of x must also be zero, $(xa_1c) = 0$. Hence $x \equiv a_1cA$; and thus a_1cA is another of the required values of x for which p vanishes.

If $(xaa_1) = 0$, then $(xA)(aa_1c) - (aA)(xa_1c) = 0$. The only point on the line aa_1 which satisfies this equation is the point a. For if $\lambda a + \mu a_1$ be substituted for x, the equation reduces to $\mu(a_1A)(aa_1c) = 0$; and hence, $\mu = 0$.

Hence the three values of x for which $p = 0$ are a, c, a_1cA.

The only value of x for which $q = 0$ is $x = b$.

(5) To investigate the values of x for which $p \equiv x$. These are included among the points satisfying the equation $px = 0$. Though this equation for x is also satisfied by the points just found which make $p = 0$.

Now $$px = -(xA)(aa_1x)\,cx.$$

Hence if x lie in A, i.e. if $(xA) = 0$, or if x lie in aa_1, i.e. if $(xaa_1) = 0$, then $p \equiv x$. But the points a and a_1cA must be excluded, as involving $p = 0$.

(6) The points for which $q \equiv x$ are given by $qx = 0$, excluding the point b for which $q = 0$.

Now $$qx = xbBx = (xB)\,xb.$$

Hence if $(xB) = 0$, then $q \equiv x$.

Thus either of the points AB or aa_1B substituted for x in the expressions p and q make
$$p \equiv x \equiv q.$$

(7) Hence if x be either of these points

$$\varpi_x = (pb_1Ckq) \equiv (xb_1Ckx).$$

Now $$(xb_1Ckx) = (xb_1k)(xC).$$

Therefore $(xb_1Ckx) = 0$, implies either $(xb_1k) = 0$, or $(xC) = 0$.

Hence if the points AB and aa_1B lie on the cubic they must lie either on b_1k or C. Thus if A, B, C be concurrent, the point of concurrence lies on the cubic.

This analysis of the equation will enable us easily to follow Grassmann's solution of the problem.

135. Linear Construction of Cubic through nine arbitrary points. (1) Let the nine given points be $a, b, c, d, e, f, g, h, i$; and let the cubic be $(xaAa_1 . xbBkCb_1 . xc) = 0$. Then the curve obviously goes through the points a, b, c.

Let a_1cA, which lies on the cubic [cf. § 134 (4)], be the point d; and let A, B, C be concurrent in the point e, which is therefore on the cubic by § 134 (7). Hence we may write $A = de$.

Let the point aa_1B lie on b_1k and therefore be on the cubic by § 134 (7): let it be the point f, so that $(fb_1k) = 0$. Hence both e and f lie on B; therefore we may write $B = ef$. Also $a_1cA \equiv d$ now becomes $a_1c . de \equiv d$; hence d is the point of intersection of a_1c and de and therefore $(a_1cd) = 0$. And $(a_1aB) \equiv f$ becomes $a_1a . ef \equiv f$; hence $(a_1af) = 0$. Therefore, since a_1 lies both in cd and in af, we may assume $a_1 = af . cd$.

It has been assumed that no three of a, c, d, e, f, are collinear; for otherwise some of these equations become nugatory.

(2) We have k, C, b_1 still partially at disposal: the conditions to be satisfied by them being only,

$$(Ce) = 0, \quad (fb_1k) = 0.$$

Now let
$$g_1 = gaAa_1 . gc = ga(de)(af . cd) . gc,$$
$$h_1 = haAa_1 . hc = ha(de)(af . cd) . hc,$$
$$i_1 = iaAa_1 . ic = ia(de)(af . cd) . ic,$$
$$g_2 = gbB \qquad = gb . ef,$$
$$h_2 = hbB \qquad = hb . ef,$$
$$i_2 = ibB \qquad = ib . ef.$$

Thus the six points $g_1, h_1, i_1, g_2, h_2, i_2$ can be obtained by linear constructions from the nine given points $a, b, c, \dots i$. We proceed to choose k, C, b_1, so that the following equations hold [cf. § 134 (3)]

$$(g_2kCb_1g_1) = 0, \quad (h_2kCb_1h_1) = 0, \quad (i_2kCb_1i_1) = 0.$$

(3) Let C and k be chosen, if possible, to satisfy the equation

$$(i_1 b_1 C k i_2) = 0$$

without conditioning b_1. Then for this purpose we must write $i_1 b_1 C \equiv i_1$; that is to say, C must be assumed to pass through i_1. But e lies in C, hence we must assume $C = ei_1$.

Hence k is given by $(ki_1 i_2) = 0$. Further, except in the special case in which $(fei_1) = 0$, k and b_1 are also [cf. subsection (1)] related by $(fkb_1) = 0$. Thus $k \equiv i_1 i_2 . fb_1$, where b_1 is as yet any arbitrarily assumed point.

(4) The remaining equations can be written

$$(kg_2 Cg_1 b_1) = 0, \quad (kh_2 Ch_1 b_1) = 0.$$

Hence k must also be such that the three lines kf, kg_2Cg_1, kh_2Ch_1 intersect in the same point b_1; also k lies in $i_1 i_2$. Therefore k is one of the points in which $i_1 i_2$ intersects the cubic curve,

$$(xf . xg_2 Cg_1 . xh_2 Ch_1) = 0.$$

(5) But this curve is formed of three straight lines. For if x be any point in C, then

$$xg_2 C \equiv x, \quad xh_2 C \equiv x,$$

and hence $$(xf . xg_2 Cg_1 . xh_2 Ch_1) \equiv (xf . xg_1 . xh_1) = 0.$$

Thus C is part of the locus.

Now $g_2 (= gbB)$ and $h_2 (= hbB)$ both lie on B, also f lies on B. Thus if x be any point on B,

$$xf \equiv B, \quad xg_2 Cg_1 \equiv BCg_1, \quad xh_2 Ch_1 \equiv BCh_1.$$

Hence $$(xf . xg_2 Cg_1 . xh_2 Ch_1) \equiv (B . BCg_1 . BCh_1) = 0.$$

Thus B is part of the locus.

(6) Hence the remainder of the locus is another straight line.

To find this required line, let $y = xf . xg_2 Cg_1$.

Then $$(y . xh_2 Ch_1) = (xh_2 Ch_1 y) = -(yh_1 Ch_2 x).$$

Hence the equation of the three straight lines is

$$(xf . xg_2 Cg_1 . xh_2 Ch_1) = \{xf (xg_2 Cg_1) h_1 Ch_2 x\} = 0.$$

This equation is satisfied by any value of x for which

$$xf (xg_2 Cg_1) h_1 = 0;$$

that is, by any value of x making xf and $xg_2 Cg_1$ intersect in h_1; that is, if x satisfies $(xfh_1) = 0$, and $(xg_2 Cg_1 h_1) = 0$. But $(xg_2 Cg_1 h_1) = (h_1 g_1 Cg_2 x)$; hence x must lie in the intersection of fh_1 and $h_1 g_1 Cg_2$. Therefore

$$x = h_1 g_1 Cg_2 . fh_1.$$

Similarly another point on the third line is $g_1 h_1 Ch_2 . fg_1$. Hence the required line which completes the locus is

$$(h_1 g_1 Cg_2 . fh_1)(g_1 h_1 Ch_2 . fg_1).$$

(7) Put $$K = (h_1g_1Cg_2 . fh_1)(g_1h_1Ch_2 . fg_1).$$
Then k must lie in i_1i_2 [by subsection (3)] and in B or C or K.

Now the assumed equation of the cubic is
$$(xaAa_1 . xbBkCb_1 . xc) = 0.$$

Assume that k lies in B. Then $xbBk = (xbk) B$.

Hence the equation of the cubic becomes
$$(xbk)(xaAa_1 . BCb_1 . xc) = 0.$$

Accordingly the cubic degenerates into the straight line bk and a conic section; and cannot therefore be made to pass through any nine arbitrarily assumed points.

Assume that k lies in C. Then $xbBkC = (xbBC) k$.

Hence the equation of the cubic becomes
$$(xbBC)(xaAa_1 . kb_1 . xc).$$

Thus in this case also the cubic degenerates in a conic section and a straight line, namely, BCb.

Therefore the only possibility left is that k lie in K. It will be shown that this assumption allows the cubic to be of the general type by showing that the cubic passes through the nine arbitrarily assumed points.

Hence let it be assumed that $k = i_1i_2K$.

Accordingly with these assumptions the equations
$$(g_1b_1Ckg_2) = 0, \quad (h_1b_1Ckh_2) = 0, \quad (i_1b_1Cki_2) = 0,$$
are satisfied.

Again, b_1 has been determined, for by subsection (4) it is the point of intersection of
$$kf, \quad kg_2Cg_1, \quad kh_2Ch_1;$$
hence $$b_1 = kg_2Cg_1 . kf.$$

(8) Finally, therefore, it has been proved that the equation,
$$xaAa_1 . xbBkCb_1 . xc = 0,$$
denotes a curve passing through the nine arbitrarily assumed points $a, b, c, d, e, f, g, h, i$, provided that A, B, C, a_1, b_1, k are determined by the linear constructions,
$$A = de, \quad B = ef, \quad C = ei_1, \quad a_1 = af . cd, \quad k = i_1i_2K, \quad b_1 = kg_2Cg_1 . kf;$$
where
$$g_1 = ga(de)(af . cd) . gc, \quad h_1 = ha(de)(af . cd) . hc, \quad i_1 = ia(de)(af . cd) . ic,$$
$$g_2 = gb . ef, \quad h_2 = hb . ef, \quad i_2 = ib . ef,$$
and $$K = (h_1g_1Cg_2 . fh_1)(g_1h_1Ch_2 . fg_1).$$

(9) This linear construction satisfied by any point x on the cubic represents the general property of any ten-cornered figure $x, a, b, c, d, e, f, g, h, i$, inscribed in a cubic. It is the analogue for cubics of Pascal's Theorem for conics.

136. SECOND TYPE OF LINEAR CONSTRUCTION OF THE CUBIC. (1) Equation (2) § 133, namely

$$(xaAa_1 \,.\, xbBb_1 \,.\, xc) = 0$$

is a simplified form of (1), which has just been discussed. It can be derived from (1) by putting $k = b_1$. For in this case

$$xbBb_1Cb_1 = (xbBb_1b_1)\,C - (Cb_1)\,xbBb_1$$
$$= -(Cb_1)\,xbBb_1 \equiv xbBb_1.$$

(2) Hence as in § 134 (4) and (7) the points a, b, c, a_1cA, AB, aa_1B are seen to lie on the curve.

Similarly, from the symmetry of the equation, b_1cB, bb_1A are seen to be points on the curve.

Also it is easy to see that $aa_1 \,.\, bb_1$ is a point on the curve.

Let these nine points be denoted by $a, b, c, d, e, f, g, h, k$ respectively; so that

$$d \equiv a_1cA,\quad e \equiv AB,\quad f \equiv aa_1B,\quad g \equiv b_1cB,\quad h \equiv bb_1A,\quad k \equiv aa_1 \,.\, bb_1.$$

(3) To prove that the cubic denoted by this equation is of the general type.

Take any cubic, and inscribe in it any quadrilateral $khef$ as in the figure 3.

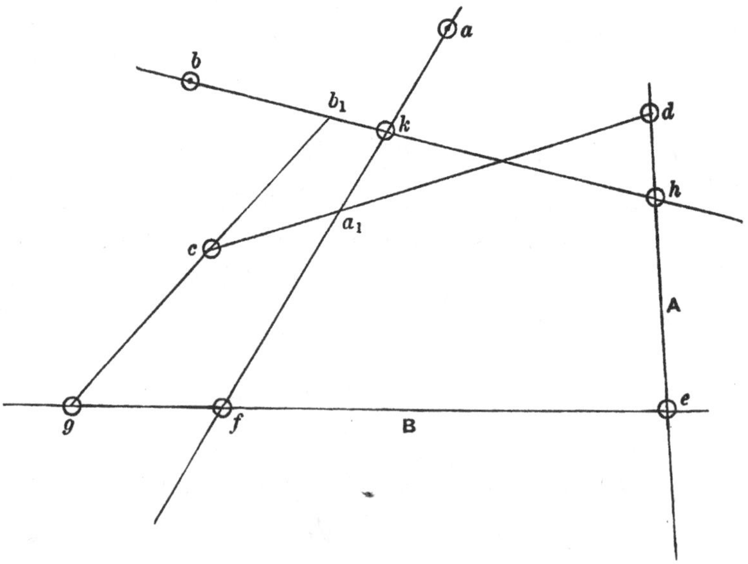

FIG. 3.

Let the side kh meet the curve again in b, the side he meet the curve again in d, the side ef in g, the side fk in a. Assume c to be any other point of the curve not collinear with any two of the other points. Then the assumed points on it determine the cubic. Join cd cutting kf in a_1, and cg cutting hk in b_1. Then if $fe \equiv B, he \equiv A$, the equation

$$xaAa_1 \,.\, xbBb_1 \,.\, xc = 0,$$

has been proved to represent a cubic through the nine points. Hence by a proper choice of constants the equation can represent any cubic.

(4) The construction represented by this equation is exemplified in figure 4.

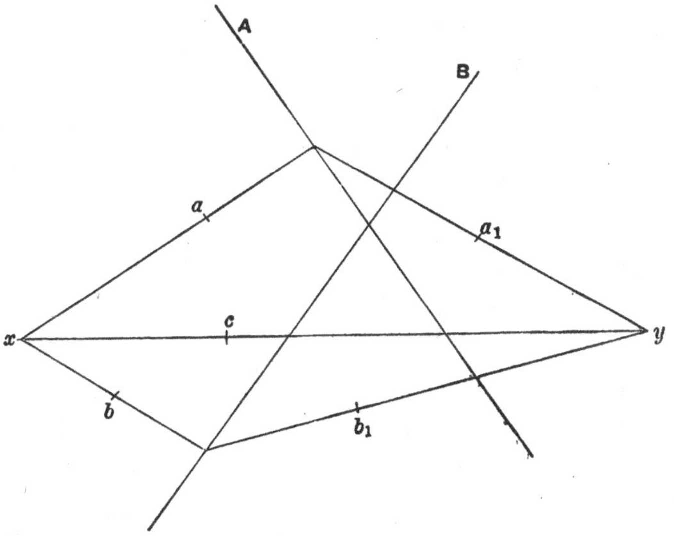

FIG. 4.

137. THIRD TYPE OF LINEAR CONSTRUCTION OF THE CUBIC. (1) The equation (3) of § 133 is

$$\varpi_x = (xaBcDxD_1c_1B_1a_1x) = 0.$$

The points a and a_1 obviously lie on the curve.

To discover other points on the curve, notice that by § 132 (1)

$$axBcDx = (xa \,.\, DB)\, cx + (xB)\,(cD)\, ax.$$

Hence $\varpi_x = (xa \,.\, DB)\,(cxD_1c_1B_1xa_1) + (xB)\,(cD)\,(axD_1c_1B_1xa_1)$.........(A).

But $(cxD_1c_1B_1xa_1) = 0$ is, by § 131 (4) and (5), a conic through the five points c, a_1, B_1D_1, $c_1a_1D_1$, c_1cB_1.

Also $(axD_1c_1B_1xa_1) = 0$ is, by § 131, a conic through the five points a, a_1, B_1D_1, $c_1a_1D_1$, c_1aB_1.

Hence the points B_1D_1 and $c_1a_1D_1$ lie on both conics and therefore also on the cubic.

But $$\varpi_x = xa_1B_1c_1D_1xDcBax.$$

Hence BD and caD are also points on the curve,

(2) As a verification notice that, if $x = BD$, then

$$xaB \equiv x, \quad xaBcD \equiv xcD \equiv x, \quad xaBcDx \equiv xx = 0:$$

also, if $x = ca \,.\, D$, then

$$xa \equiv ca, \quad xaBc \equiv caBc \equiv ca, \quad xaBcDx \equiv caDx \equiv xx = 0.$$

(3) To find where D cuts the curve a third time; note that $axBcD$ is a point in D; hence if x be also in D, $axBcDx \equiv D$, excluding the case when $axBcDx$ is zero.

Hence, by substituting D for $axBcDx$ in the equation of the curve, we see that x satisfies $(DD_1c_1B_1a_1x) = 0$, and $(Dx) = 0$: therefore $x = DD_1c_1B_1a_1D$.

Hence D cuts the curve in the three points BD, caD, $DD_1c_1B_1a_1D$, and similarly D_1 cuts the curve in the three points B_1D_1, $c_1a_1D_1$, D_1DcBaD_1.

(4) The two conics $(cxD_1c_1B_1xa_1) = 0$, and $(axD_1c_1Bxa_1) = 0$, have been proved to intersect in the three points a_1, B_1D_1, $c_1a_1D_1$. The fourth point of intersection is $caD_1c_1B_1a_1 \,.\, ca$; since by § 132 (5) this is the point in which the line ca meets either conic.

Hence the three points in which the line ca meets the cubic are a, caD, $caD_1c_1B_1a_1 \,.\, ca$. Similarly the three points in which the line c_1a_1 meets the cubic are a_1, $c_1a_1D_1$, $c_1a_1DcBa \,.\, c_1a_1$.

(5) It is easy to obtain expressions for the three points in which the line BDa cuts the cubic. Two of the points are already known, namely, a and BD. To find the third notice that from equation (A) of subsection (1), the required point is the point, other than a, in which the line cuts the conic $(xaD_1c_1B_1a_1x) = 0$. By § 132 (5) this is the point $(BDa)\, D_1c_1B_1a_1\, (BDa)$, which can also be written

$$BDaD_1c_1B_1a_1 \,.\, BDa.$$

Thus the three points in which BDa meets the cubic are a, BD, $BDaD_1c_1B_1a_1 \,.\, BDa$. Similarly the three points in which $B_1D_1a_1$ meets the cubic are a_1, B_1D_1, $B_1D_1a_1DcBa \,.\, B_1D_1a_1$.

(6) To find the three points in which B cuts the curve, notice that if $(xB) = 0$, then from equation (A) of subsection (1) the equation of the curve reduces to

$$(xa \,.\, DB)\,(cxD_1c_1B_1xa_1) = 0.$$

Hence either $xa \,.\, DB = 0$, and $x = BD$, which has been already discovered; or $(xcD_1c_1B_1a_1x) = 0$. Therefore the two remaining points in which B meets the cubic are the points in which B meets the conic $(xcD_1c_1B_1a_1x) = 0$.

These points can be immediately expressed, if $B = B_1$. In this case the cubic becomes

$$(xaBcDxD_1c_1Ba_1x) = 0;$$

and it will be proved [cf. subsection 13] that the equation still represents any cubic curve.

The points where B meets the conic, $(xcD_1c_1Ba_1x) = 0$, have been proved in § 131 (5) to be BD_1 and cc_1B. Hence B meets the simplified cubic in the three points BD, BD_1, cc_1B.

(7) The transformation

$$\varpi_x = -x \,.\, xaBcD \,.\, xa_1B_1c_1D_1$$

is established as follows.

Let $x_1 = xaBcDxD_1c_1B_1$, then $\varpi_x = x_1a_1x = -x_1 \,.\, xa_1$; since the product of three points is associative.

Let $X_2 = xaBcDxD_1c_1$, then $\varpi_x = -X_2B_1 \,.\, xa_1 = X_2 \,.\, xa_1B_1$; since the product of three linear elements is associative.

Let $x_3 = xaBcDxD_1$, then $\varpi_x = x_3c_1 \,.\, xa_1B_1 = -x_3 \,.\, xa_1B_1c_1$.

Let $X_4 = xaBcDx$, then $\varpi_x = -X_4D_1 \,.\, xa_1B_1c_1 = X_4 \,.\, xa_1B_1c_1D_1$.

Hence $\varpi_x = (xaBcD)\, x \,.\, xa_1B_1c_1D_1 = -x \,.\, xaBcD \,.\, xa_1B_1c_1D_1$.

The previous results can be easily obtained by means of this form of the equation.

(8) The geometrical meaning of the equation is that x, $xaBcD$, and $xa_1B_1c_1D_1$ are collinear. This property is shown in the annexed figure 5.

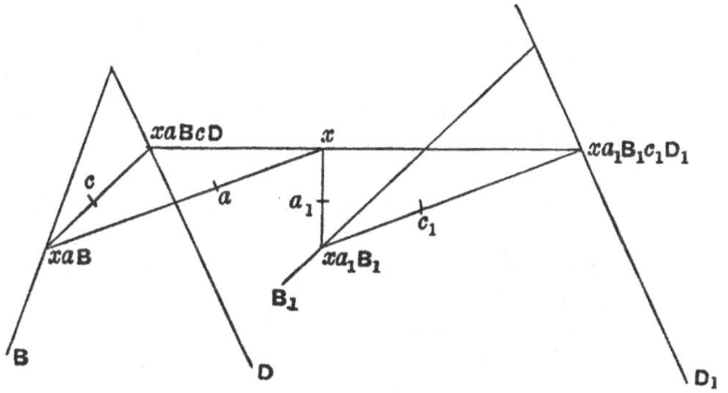

FIG. 5.

Hence if two variable triangles have a common variable vertex, and two sides, one of each triangle, which meet in the common vertex lie in the same straight line, and if also the four remaining sides pass respectively through four fixed points, and the four remaining vertices lie respectively on four fixed straight lines, then the locus of the common vertex is a cubic.

(9) The four lines $A\ (= ca)$, D, $A_1\ (= c_1a_1)$, D_1 have a special relation to the cubic

$$(xaBcDxD_1c_1B_1a_1x) = 0,$$

in addition to the fact that the points caD and $c_1a_1D_1$ both lie on the curve [cf. subsection (1)].

For suppose that the lines A, D, A_1, D_1 are arbitrarily assumed. Then the points $AD\ (= e)$ and $A_1D_1\ (= e_1)$ are determined.

Also suppose that the remaining points in which A and A_1 cut the curve are arbitrarily assumed on these lines, namely [cf. subsection (4)],

$$f(=AD_1c_1B_1a_1A),\ f_1(=A_1DcBaA_1),\ a \text{ and } a_1.$$

Thus a, a_1, f, f_1 are supposed to be known, and the equations $f=AD_1c_1B_1a_1A$ and $f_1=A_1DcBaA_1$ partially determine c_1 and B_1, and c and B, which are the remaining unknowns.

Again the arbitrarily assumed lines D and D_1 are supposed to meet the curve in two arbitrarily assumed points $e\,(=AD)$ and $e_1\,(=A_1D_1)$. Let two other points k and k_1 in which D and D_1 respectively meet the curve be arbitrarily assumed, so that [cf. subsection (3)] we may assume

$$k=DD_1c_1Ba_1D, \text{ and } k_1=D_1DcBaD_1.$$

Then the remaining points in which D and D_1 respectively meet the curve are [cf. subsection (3)] BD and B_1D_1. Call these points g and g_1. It will now be shown that g and g_1 are both determined by the previous assumptions of the eight points $a, a_1, e, e_1, f, f_1, k, k_1$; and that accordingly the group of four lines A, D, A_1, D_1 must bear some special relation to the cubic curve which passes through the eight assumed points.

(10) For if L_1 and L_2 are linear elements and p_1 and p_2 are points, the extended rule of the middle factor gives,

$$L_1L_2p_1=(L_1p_1)\,L_2-(L_2p_1)\,L_1, \text{ and } p_1p_2L_1=(p_1L_1)\,p_2-(p_2L_1)\,p_1.$$

Remembering these formulæ we see that

$$fa_1=AD_1c_1B_1a_1Aa_1=-(Aa_1)\,AD_1c_1B_1a_1\equiv AD_1c_1B_1a_1;$$

$$fa_1B_1\equiv AD_1c_1B_1a_1B_1=-(a_1B_1)\,AD_1c_1B_1\equiv AD_1c_1B_1;$$

$$fa_1B_1c_1\equiv AD_1c_1B_1c_1=-(B_1c_1)\,AD_1c_1\equiv AD_1c_1.$$

Hence $$AD_1c_1\,.\,fa_1\,.\,B_1=0.$$

Similarly $$DD_1c_1\,.\,ka_1\,.\,B_1=0.$$

Hence B_1 passes through the points $AD_1c_1\,.\,fa_1\,(=p)$ and $DD_1c_1\,.\,ka_1\,(=q)$.

Therefore we may write $B_1=(AD_1c_1\,.\,fa_1)\,(DD_1c_1\,.\,ka_1)=pq$.

Hence $g_1=B_1D_1=(AD_1c_1\,.\,fa_1)\,(DD_1c_1\,.\,ka_1)\,D_1=pqD_1=(pD_1)\,q-(qD_1)\,p$.

Now $$(pD_1)=(AD_1c_1\,.\,fa_1\,.\,D_1)=-(AD_1c_1\,.\,D_1\,.\,fa_1)$$
$$=(c_1D_1)\,(AD_1\,.\,fa_1)=(c_1D_1)\,(AD_1fa_1).$$

And $(qD_1)=(DD_1c_1\,.\,ka_1\,.\,D_1)=-(DD_1c_1\,.\,D_1\,.\,ka_1)=(c_1D_1)\,(DD_1ka_1)$.

Hence $$g_1\equiv(AD_1fa_1)\,q-(DD_1ka_1)\,p.$$

But $$p=AD_1c_1\,.\,fa_1=(AD_1fa_1)\,c_1-(c_1fa_1)\,AD_1;$$

and $$q=DD_1c_1\,.\,ka_1=(DD_1ka_1)\,c_1-(c_1ka_1)\,DD_1.$$

Also $(c_1fa_1)=-(A_1f)$, and $(c_1ka_1)=-(A_1k)$ by subsection (9).

Thus $$g_1\equiv(A_1k)\,(AD_1fa_1)\,DD_1-(A_1f)\,(DD_1ka_1)\,AD_1.$$

Hence the position of g_1 is completely determined by the arbitrarily assumed elements.

Similarly the position of g is completely determined.

(11) Hence ten points on the cubic are now known. The cubic is therefore independent of the positions of c and c_1 on A and A_1; except that c must not coincide with a or AD, nor c_1 with a_1 or A_1D_1, in which cases some of the previous equations become nugatory.

(12) We will now prove that the specialized form of equation introduced in subsection (6), namely

$$(xaBcDxD_1c_1Ba_1x) = 0,$$

where $(cD_1) = 0 = (c_1D)$ represents the most general form of cubic.

The three points in which D cuts the curve are [cf. subsection (3)], BD, caD, $DD_1c_1Ba_1D$.

But since $(c_1D) = 0$, $DD_1c_1Ba_1D \equiv DBa_1D \equiv DB$.

Hence D touches the curve at BD and cuts it again in caD. Similarly D_1 touches the curve at BD_1 and cuts it again in $c_1a_1D_1$.

Also [cf. subsection (6)] B cuts the curve in the points BD, BD_1, cc_1B.

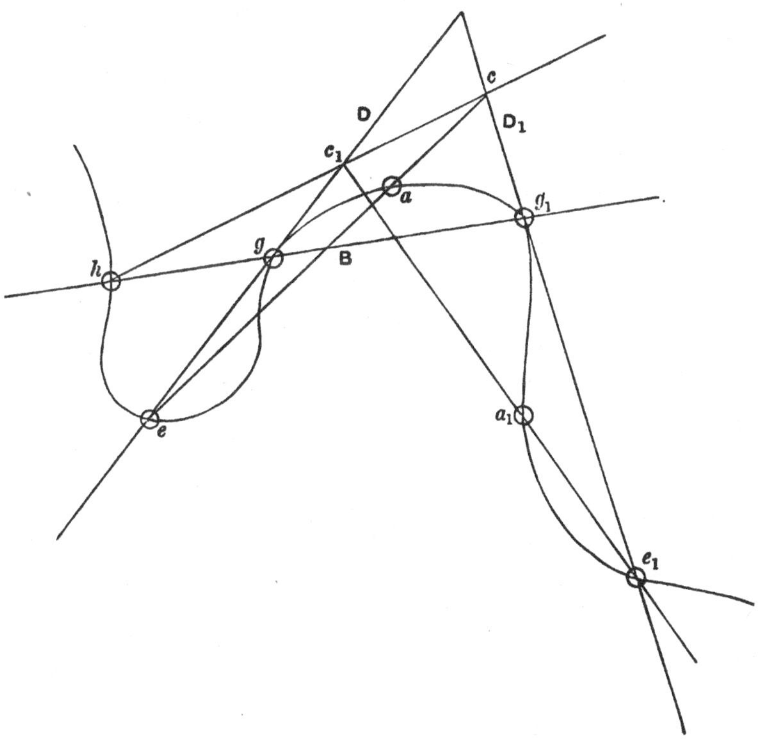

FIG. 6.

(13) Now (cf. fig. 6) take any cubic curve and draw the lines D and D_1 tangents to it at any two points g and g_1. Join gg_1 by the line B which cuts the curve in another point h. Through h draw any line cutting D in c_1 and D_1 in c. The tangents D and D_1 cut the curve again in two points e and e_1.

Now join ec; this line cuts the curve in two points. Call one of the two a. Similarly call one of the two points, in which e_1c_1 cuts the curve, a_1. Then by construction $h \equiv cc_1B$, $e \equiv caD$, $e_1 \equiv c_1a_1D_1$.

Now the tangents D, D_1 at g and g_1 and the points h, e, e_1, a, a_1 completely determine the cubic.

But $(xaBcDxD_1c_1Ba_1x) = 0$ is a cubic satisfying these conditions. Hence this equation represents the assumed cubic.

138. Fourth Type of Linear Construction of the Cubic. (1) The equation (4) of § 133 is

$$(xaA \,.\, xbB \,.\, xcC) = 0;$$

and it represents the fact that the points xaA, xbB, xcC are collinear. The construction is shown in figure 7.

It will be shown that any cubic can be thus described.

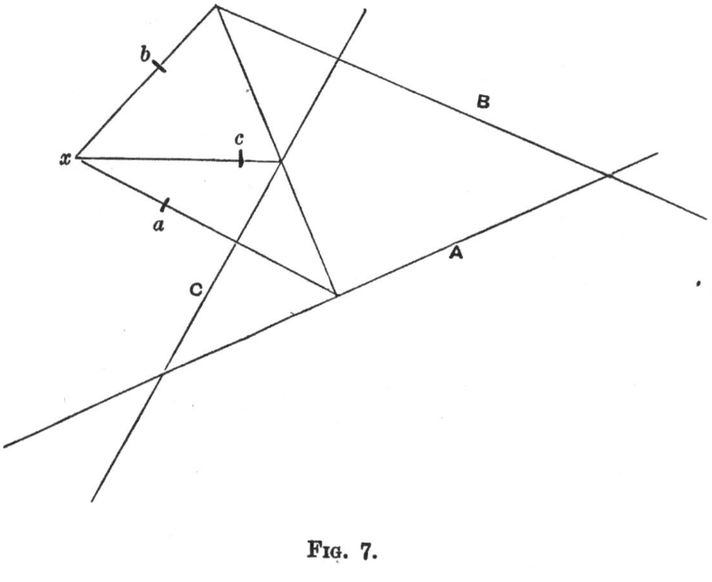

Fig. 7.

(2) To find where A cuts the cubic, note that if x lies in A, $xaA \equiv x$. Hence $(xaA \,.\, xbB \,.\, xcC) \equiv (x \,.\, xbB \,.\, xcC) \equiv (xB)\, xb \,.\, xcC$

$$\equiv (xB)(xC)(xbc);$$

where the sign of congruence means that only constant factors have been dropped.

Therefore the three points in which A cuts the cubic are AB, AC, bcA. Hence by symmetry, BC, caB, abC also lie on the cubic. Also obviously a, b, c lie on the cubic. Thus the two triangles respectively formed by a, b, c as vertices, and by A, B, C as sides are both inscribed in the cubic and their corresponding sides, namely A and bc, B and ca, C and ab, intersect also on the cubic.

(3) We have to prove that, given any triangle abc inscribed in a cubic, a triangle A, B, C always exists with these properties relatively to abc and the cubic.

Take a, b, c any three points on a given cubic, not collinear. Let bc cut the cubic again in f, ca in g, ab in h.

Let a, b, c be the reference triangle, and let ξ, η, ζ be the co-ordinates of any point x. Then we can write $x = \xi a + \eta b + \zeta c$.

Let A, B, C be any straight lines through f, g, h. Then, since any numerical multiples of A, B, and C can be substituted for them, we may write

$$A = \lambda bc + \gamma_1 ca + \beta_1 ab,$$
$$B = \gamma_2 bc + \mu ca + \alpha_2 ab,$$
$$C = \beta_3 bc + \alpha_3 ca + \nu ab;$$

where λ, μ, ν are at our disposal and $\beta_1, \gamma_1, \gamma_2, \alpha_2, \beta_3, \alpha_3$ are known from the equations, $f \equiv bcA$, $g \equiv caB$, $h \equiv abC$ and from the fact that one of the letters with each subscript can be assumed arbitrarily without affecting anything except the intensities of A, B, C, which are immaterial.

Now
$$xaA = (xA)\,a - (aA)\,x$$
$$= (abc)\,\{(\lambda\xi + \gamma_1\eta + \beta_1\zeta)\,a - \lambda\,(\xi a + \eta b + \zeta c)\}$$
$$= (abc)\,\{(\gamma_1\eta + \beta_1\zeta)\,a - \lambda\eta b - \lambda\zeta c\}.$$

Similarly
$$xbB = (abc)\,\{-\mu\xi a + (\alpha_2\zeta + \gamma_2\xi)\,b - \mu\zeta c\},$$
$$xcC = (abc)\,\{-\nu\xi a - \nu\eta b + (\beta_3\xi + \alpha_3\eta)\}.$$

Hence, $(xaA\,.\,xbB\,.\,xcC) = 0$ can be written as the ordinary algebraic equation,

$$\begin{vmatrix} \gamma_1\eta + \beta_1\zeta, & -\lambda\eta, & -\lambda\zeta, \\ -\mu\xi, & \alpha_2\zeta + \gamma_2\xi, & -\mu\zeta, \\ -\nu\xi, & -\nu\eta, & \beta_3\xi + \alpha_3\eta, \end{vmatrix} = 0.$$

This becomes on expanding the determinant

$$(\gamma_1\eta + \beta_1\zeta)(\alpha_2\zeta + \gamma_2\xi)(\beta_3\xi + \alpha_3\eta) - \mu\nu\eta\zeta\,(\gamma_1\eta + \beta_1\zeta)$$
$$- \nu\lambda\zeta\xi\,(\alpha_2\zeta + \gamma_2\xi) - \lambda\mu\xi\eta\,(\beta_3\xi + \alpha_3\eta) - 2\lambda\mu\nu\xi\eta\zeta = 0.$$

This is the equation to a cubic through the six points a, b, c, f, g, h: it is required to determine λ, μ, ν so that it may be the given cubic through these points.

The given cubic is determined by any other three points on it f_1, g_1, h_1 forming another triangle. Now λ, μ, ν can be so determined that the above equation is satisfied by the co-ordinates of these points. For by substituting successively the co-ordinates we find three linear equations to determine λ, μ, ν, each of the form

$$\frac{P_1}{\lambda} + \frac{P_2}{\mu} + \frac{P_3}{\nu} = \frac{P_4}{\lambda\mu\nu} + P_5,$$

where $P_1, P_2, \ldots, P_5$ do not contain λ, μ, ν.

Now put σ for $\lambda\mu\nu$, and solve these three linear equations for $\lambda^{-1}, \mu^{-1}, \nu^{-1}$. Then we may assume

$$\frac{1}{\lambda} = \frac{H_1}{\sigma} + K_1, \quad \frac{1}{\mu} = \frac{H_2}{\sigma} + K_2, \quad \frac{1}{\nu} = \frac{H_3}{\sigma} + K_3;$$

where $H_1, H_2, \ldots K_3$ do not contain λ, μ, ν.

Hence multiplying and replacing $\lambda\mu\nu$ by σ, an equation of the form,

$$p_0\sigma^3 + p_1\sigma^2 + p_2\sigma + p_3 = 0,$$

is found; where p_0, p_1, p_2, p_3 do not contain λ, μ, ν. Thus there are three values of σ, one of which must be real. Hence there are three systems of values of λ, μ, ν; and one system must consist of real values. Thus three systems of values can be found for A, B, C; and one of these systems must make A, B, C to be real lines.

Thus three triangles, of which one must be real, can be found related to a, b, c and to the given cubic in the required manner.

Let A, B, C be one of these triangles. Then

$$(xaA \,.\, xbB \,.\, xcC) = 0$$

is the given cubic.

The above proof of the required proposition is different from that which is given by Grassmann*.

139. CHASLES' CONSTRUCTION. (1) Another construction for a cubic given by Chasles†, without knowledge of Grassmann's results or methods is represented by

$$xeDpEdF \,.\, xfB \,.\, xdC = 0,$$

where $$(Ff) = 0 = (Bd).$$

(2) It is easy to prove‡ the following relations:

The points d, e, f, BC, CF lie on the curve.

The third point in which de cuts the curve is

$$deDpEdF\,(deC)\,Bf\,(de).$$

The third point in which ef cuts the curve is

$$efDpEdF\,(efB)\,Cd\,(ef).$$

The third point in which BCf cuts the curve is

$$FCdEpDe\,(BCf).$$

* *Crelle*, vol. LII.

† *Comptes Rendus*, vol. XXXVI., 1853.

‡ Cf. Grassmann, *loc. cit.*

The third point in which BCd cuts the curve is

$$BFdEpDe\,(BCd).$$

Also if we put $a \equiv deDpEdF\,(deC)\,Bf\,(de)$,

$$b \equiv efDpEdF\,(efB)\,Cd\,(ef),\quad c \equiv FC,\ A \equiv BCf,$$

$$a_1 \equiv cdEpDeAc\,(de),\ b_1 \equiv BFdEpDeBc\,(ef);$$

then the given cubic can be expressed by the construction

$$xaAa_1 \,.\, xbBb_1 \,.\, xc = 0.$$

CHAPTER VI.

MATRICES.

140. INTRODUCTORY. The leading properties of Matrices, that is of Linear Transformations, can be easily expressed by the aid of the Calculus of Extension. A complete investigation into the theory of Matrices will not be undertaken in this chapter: the subject will only be taken far enough to explain the method here employed and prove the results required in the subsequent investigations in the theory of Extensive Manifolds. Grassmann treated the subject in his *Ausdehnungslehre* of 1861 apparently in ignorance of Cayley's classical memoir on Matrices*. An exposition and amplification of Grassmann's methods was given by Buchheim†. The present chapter is in its greater part little more than a free translation of Grassmann's own writing, amplified by the aid of Buchheim's paper; except that Grassmann and Buchheim do not explicitly consider the case of a matrix operating on an extensive magnitude of an order higher than the first; and that the treatment here given of symmetrical matrices is new, and also that of skew matrices. I have also ventured in § 146 to distinguish between latent regions and semi-latent regions: in the ordinary nomenclature both would be called latent regions.

141. DEFINITION OF A MATRIX. (1) Let $e_1, e_2 \dots e_\nu$ be any ν reference elements in a complete region of $\nu - 1$ dimensions. Let the symbol

$$\frac{a_1, a_2 \dots a_\nu}{e_1, e_2 \dots e_\nu},$$

prefixed to any product of some or all of these elements, be defined to denote the operation of replacing the element e_1 by a_1, the element e_2 by a_2, and so on. Thus if $e_\kappa, e_\lambda \dots e_\rho$ be any of the original reference elements,

$$\frac{a_1, a_2 \dots a_\nu}{e_1, e_2 \dots e_\nu} e_\kappa e_\lambda \dots e_\rho = a_\kappa a_\lambda \dots a_\rho;$$

* *Phil. Trans.* vol. CXLVIII., 1858; and *Collected Mathematical Papers*, vol. II., no. 152.

† *Proc. London Math. Soc.* vol. XVI. 1885.

so that in this instance the symbol of operation has transformed the product $e_\kappa e_\lambda \dots e_\rho$ into the product $a_\kappa a_\lambda \dots a_\rho$. Let ϕ be put for the symbol $\dfrac{a_1, a_2 \dots a_\nu}{e_1, e_2 \dots e_\nu}$. The convention with respect to the operator ϕ will be the same as that with respect to the operator $|$ which is stated in § 99 (9). Then $\phi e_1 = a_1$, $\phi e_2 = a_2$, $\phi e_1 e_2 = a_1 a_2$, and so on.

(2) It follows from this definition that ϕ is distributive in reference to multiplication. For $\phi e_1 \phi e_2 = a_1 a_2 = \phi e_1 e_2$, and so on.

(3) Furthermore let ϕ be defined to be distributive in reference to addition, so that if $E_1, E_2 \dots E_\rho$ be regional elements of the σth order formed by the multiplicative combinations of the σth order formed out of the reference elements [cf. § 94 (1)], then

$$\phi(\alpha_1 E_1 + \alpha_2 E_2 + \dots + \alpha_\rho E_\rho) = \alpha_1 \phi E_1 + \alpha_2 \phi E_2 + \dots + \alpha_\rho \phi E_\rho.$$

For example, if $x = \xi_1 e_1 + \xi_2 e_2 + \dots + \xi_\nu e_\nu$, then

$$\begin{aligned} \phi x &= \xi_1 \phi e_1 + \xi_2 \phi e_2 + \dots + \xi_\nu \phi e_\nu \\ &= \xi_1 a_1 \ + \xi_2 a_2 \ + \dots + \xi_\nu a_\nu. \end{aligned}$$

(4) The operator ϕ—called by Grassmann a quotient—may be identified with Cayley's matrix. For assume

$$\begin{aligned} a_1 &= \alpha_{11} e_1 + \alpha_{21} e_2 + \dots + \alpha_{\nu 1} e_\nu, \\ a_2 &= \alpha_{12} e_1 + \alpha_{22} e_2 + \dots + \alpha_{\nu 2} e_\nu, \\ &\dots\dots\dots\dots\dots\dots\dots\dots \\ a_\nu &= \alpha_{1\nu} e_1 + \alpha_{2\nu} e_2 + \dots + \alpha_{\nu\nu} e_\nu. \end{aligned}$$

Then

$$\begin{aligned} \phi x = &(\alpha_{11}\xi_1 + \alpha_{12}\xi_2 + \dots + \alpha_{1\nu}\xi_\nu) e_1 \\ &+ (\alpha_{21}\xi_1 + \alpha_{22}\xi_2 + \dots + \alpha_{2\nu}\xi_\nu) e_2 \\ &+ \dots\dots\dots\dots\dots\dots\dots\dots \\ &+ (\alpha_{\nu 1}\xi_1 + \alpha_{\nu 2}\xi_2 + \dots + \alpha_{\nu\nu}\xi_\nu) e_\nu. \end{aligned}$$

Hence if we put $\phi x = \eta_1 e_1 + \eta_2 e_2 \dots + \eta_\nu e_\nu$, then with the usual notation for matrices,

$$(\eta_1, \eta_2 \dots \eta_\nu) = \left(\begin{array}{|c|} \alpha_{11}, \ \alpha_{12} \dots \alpha_{1\nu} \\ \alpha_{21}, \ \alpha_{22} \dots \alpha_{2\nu} \\ \dots\dots\dots\dots \\ \alpha_{\nu 1}, \ \alpha_{\nu 2} \dots \alpha_{\nu\nu} \end{array} \right)\!\!\left(\xi_1, \xi_2 \dots \xi_\nu \right).$$

Thus we may identify ϕ with the matrix $\|\alpha_{\rho\sigma}\|$.

(5) It will be convenient to call the elements $a_1, a_2 \dots a_\nu$ the elements of the numerator of the matrix, and $e_1, e_2 \dots e_\nu$ those of the denominator. The elements of the denominator must necessarily be independent, if the matrix is to have a meaning.

142. SUMS AND PRODUCTS OF MATRICES. (1) If E_σ denote any regional element of the σth order, say $e_1e_2 \dots e_\sigma$, then

$$\frac{\lambda a_1, \lambda a_2 \dots \lambda a_\nu}{e_1, e_2 \dots e_\nu} E_\sigma = \lambda^\sigma a_1 a_2 \dots a_\sigma = \lambda^\sigma \frac{a_1, a_2 \dots a_\nu}{e_1, e_2 \dots e_\nu} E_\sigma.$$

If ϕ denote the matrix $\frac{a_1, a_2 \dots a_\nu}{e_1, e_2 \dots e_\nu}$, then the matrix $\frac{\lambda a_1, \lambda a_2 \dots \lambda a_\nu}{e_1, e_2 \dots e_\nu}$ will be said to be the matrix ϕ multiplied by λ, and will be written symbolically $\lambda\phi$. This convention agrees with the ordinary notation, and will cause no confusion when the matrix is operating on elements of the first order, but must be abandoned when the matrix operates on regional elements of order greater than unity.

(2) If two matrices, operating on ν independent elements of the first order, give the same result in each case, then they give the same result whatever extensive magnitude they operate on.

For let $c_1, c_2 \dots c_\nu$ be any ν independent elements, ϕ and χ the two matrices.

Assume $$\phi c_1 = \chi c_1, \ \phi c_2 = \chi c_2, \dots \phi c_\nu = \chi c_\nu.$$

Then any extensive magnitude A_σ of the σth order can be written as the sum of terms of which $\lambda c_1 c_2 \dots c_\sigma$ is a type. Hence

$$\phi A_\sigma = \Sigma \lambda \phi c_1 c_2 \dots c_\sigma = \Sigma \lambda \,.\, \phi c_1 \,.\, \phi c_2 \,.\, \phi c_3 \dots \phi c_\sigma = \Sigma \lambda \,.\, \chi c_1 \,.\, \chi c_2 \,.\, \chi c_3 \dots \chi c_\sigma = \chi A_\sigma.$$

Thus the two matrices ϕ and χ must be considered as equivalent, and their equivalence may be expressed by $\phi = \chi$.

(3) If $c_1, c_2 \dots c_\nu$ be any ν independent elements, and if the matrix ϕ, originally given as $\frac{a_1, a_2 \dots a_\nu}{e_1, e_2 \dots e_\nu}$, give the results $\phi c_1 = d_1$, $\phi c_2 = d_2 \dots \phi c_\nu = d_\nu$, then ϕ can also be written in the form $\frac{d_1, d_2 \dots d_\nu}{c_1, c_2 \dots c_\nu}$. For if A be any extensive magnitude, it follows that $\phi A = \frac{d_1, d_2 \dots d_\nu}{c_1, c_2 \dots c_\nu} A$.

Hence any matrix can be written in a form in which any ν independent elements form its denominator.

(4) The sum of numerical multiples of matrices operating on any element of the first order can be replaced by a single matrix operating on the same element. For it can be seen that

$$\left\{\alpha \frac{a_1, a_2 \dots a_\nu}{e_1, e_2 \dots e_\nu} + \beta \frac{b_1, b_2 \dots b_\nu}{e_1, e_2 \dots e_\nu} + \dots\right\} x = \frac{\alpha a_1 + \beta b_1 + \dots, \ \alpha a_2 + \beta b_2 + \dots, \ \dots}{e_1 \qquad , \qquad e_2 \qquad , \ \dots} x.$$

But if the extensive magnitude operated on be of order greater than the first, then this theorem is not true. For example consider the product e_1e_2. Then

$$\left\{\frac{a_1, a_2 \dots}{e_1, e_2 \dots} + \frac{b_1, b_2 \dots}{e_1, e_2 \dots}\right\} e_1 e_2 = a_1 a_2 + b_1 b_2.$$

But in general $a_1a_2 + b_1b_2$ is not a single force, and cannot therefore be derived from e_1e_2 by the operation of a single matrix.

(5) A numerical multiplier can be conceived as a matrix. For if $x = \Sigma\xi e$, then $\lambda x = \Sigma\xi\lambda e$, where λ is some number. Hence λ may be conceived as the matrix $\dfrac{\lambda e_1, \lambda e_2 \dots \lambda e_\nu}{e_1, e_2 \dots e_\nu}$.

If A_σ be an extensive magnitude of the σth order, then

$$\frac{\lambda e_1, \lambda e_2 \dots \lambda e_\nu}{e_1, e_2 \dots e_\nu} A_\sigma = \lambda^\sigma A_\sigma.$$

Also from subsection (4) if ϕ be any matrix, λ any number, and x an element of the first order, then $(\phi + \lambda) x$ can be written χx, where χ is a single matrix.

(6) Let ϕ and χ be two matrices and A any extensive magnitude. Then the expression $\phi\chi A$ is defined to mean that the transformation $\chi A = B$ is first effected and then the transformation ϕB.

The combined operation $\phi\chi$ can itself be represented by a single matrix. For let $e_1, e_2 \dots e_\nu$ be the independent reference elements, and let

$$\chi e_1 = a_1,\ \chi e_2 = a_2 \dots \chi e_\nu = a_\nu;$$

and let
$$\phi a_1 = b_1,\ \phi a_2 = b_2 \dots \phi a_\nu = b_\nu.$$

Then the matrix which replaces e_1 by b_1, e_2 by $b_2 \dots e_\nu$ by b_ν, is equivalent to the complex operation $\phi\chi$.

(7) The operator $\phi\chi$, when operating on an element of the first order, may be conceived as a product [cf. § 19] of two matrices. For let ψ be a third matrix, and a any extensive magnitude of the first order. Then

$$\phi(\chi + \psi) a = \phi(\chi a + \psi a) = \phi\chi a + \phi\psi a = (\phi\chi + \phi\psi) a.$$

Hence the two operators $\phi(\chi + \psi)$ and $\phi\chi + \phi\psi$ are equivalent.

It is to be noticed that the sum of the matrices is another matrix and the product of matrices is another matrix. It will be convenient to speak of the product of two matrices when the matrices are operating on a magnitude of an order greater than the first. In this case the matrices have not a sum [cf. subsection (4)], and therefore strictly speaking have not a product [cf. § 19].

The product of three matrices is associative; that is $\phi\chi \cdot \psi A = \phi \cdot \chi\psi A$. For the meaning of $\phi\chi \cdot \psi A$ is that a single matrix ϕ_1 is substituted for the product $\phi\chi$, and the meaning of $\phi \cdot \chi\psi A$ is that a single matrix χ_1 is substituted for the product $\chi\psi$; and then the equation asserts that $\phi_1\psi A = \phi\chi_1 A$. Now let $e_1, e_2 \dots e_\nu$ be the ν reference elements; then, taking a typical element only, let $\psi e_\rho = a_\rho$, $\chi a_\rho = b_\rho$, $\phi b_\rho = c_\rho$. Hence

$$\phi_1 a_\rho = c_\rho, \text{ and } \chi_1 e_\rho = b_\rho.$$

Therefore $\phi_1\psi e_\rho = \phi_1 a_\rho = c_\rho$, and $\phi\chi_1 e_\rho = \phi b_\rho = c_\rho$.

Thus $\phi_1\psi e_\rho = \phi\chi_1 e_\rho$; and, since this is true for every reference element, $\phi_1\psi A = \phi\chi_1 A$, where A is any extensive magnitude.

143. ASSOCIATED DETERMINANT. If the matrix ϕ can be written in the alternative forms $\dfrac{a_1, a_2 \dots a_\nu}{e_1, e_2 \dots e_\nu}$ and $\dfrac{d_1, d_2 \dots d_\nu}{c_1, c_2 \dots c_\nu}$, then the ratios $\dfrac{(a_1 a_2 \dots a_\nu)}{(e_1 e_2 \dots e_\nu)}$ and $\dfrac{(d_1 d_2 \dots d_\nu)}{(c_1 c_2 \dots c_\nu)}$ are equal.

For let $c_1 = \gamma_{11}e_1 + \gamma_{12}e_1 + \dots + \gamma_{1\nu}e_\nu$, with $\nu - 1$ other similar equations.

Then $d_1 = \phi c_1 = \gamma_{11}\phi e_1 + \gamma_{12}\phi e_2 + \dots + \gamma_{1\nu}\phi e_\nu = \gamma_{11}a_1 + \gamma_{12}a_2 + \dots + \gamma_{1\nu}a_\nu$ with $\nu - 1$ similar equations.

Hence $(c_1 c_2 \dots c_\nu) = \Delta\,(e_1 e_2 \dots e_\nu)$, where Δ stands for the determinant

$$\begin{vmatrix} \gamma_{11}, & \gamma_{12} & \cdots & \gamma_{1\nu} \\ \gamma_{21}, & \gamma_{22} & \cdots & \gamma_{2\nu} \\ \cdots & \cdots & \cdots & \cdots \\ \gamma_{\nu 1}, & \gamma_{\nu 2} & \cdots & \gamma_{\nu\nu} \end{vmatrix}.$$

Similarly $$(d_1 d_2 \dots d_\nu) = \Delta\,(a_1 a_2 \dots a_\nu).$$

Finally therefore, $$\frac{(d_1 d_2 \dots d_\nu)}{(c_1 c_2 \dots c_\nu)} = \frac{(a_1 a_2 \dots a_\nu)}{(e_1 e_2 \dots e_\nu)}.$$

(2) If with the notation of § 141 (4) the matrix be

$$\begin{pmatrix} \alpha_{11}, & \alpha_{12} & \cdots & \alpha_{1\nu} \\ \alpha_{21}, & \alpha_{22} & \cdots & \alpha_{2\nu} \\ \cdots & \cdots & \cdots & \cdots \\ \alpha_{\nu 1}, & \alpha_{\nu 2} & \cdots & \alpha_{\nu\nu} \end{pmatrix},$$

then the ratio $(a_1 a_2 \dots a_\nu)/(e_1 e_2 \dots e_\nu)$ is the determinant

$$\begin{vmatrix} \alpha_{11}, & \alpha_{12} & \cdots & \alpha_{1\nu} \\ \alpha_{21}, & \alpha_{22} & \cdots & \alpha_{2\nu} \\ \cdots & \cdots & \cdots & \cdots \\ \alpha_{\nu 1}, & \alpha_{\nu 2} & \cdots & \alpha_{\nu\nu} \end{vmatrix}.$$

144. NULL SPACES OF MATRICES. (1) If the ν elements which form the numerator of a matrix are not independent, so that one or more relations exist between them, then the matrix can always be reduced to the form in which one or more of the elements of the numerator are null.

For let the matrix ϕ be $\dfrac{a_1, a_2 \dots a_\nu}{e_1, e_2 \dots e_\nu}$; and let $a_1, a_2 \dots a_\mu$ be independent, while the remaining $\nu - \mu$ elements of the numerator are expressible in terms of the preceding μ elements; so that we may assume $\nu - \mu$ equations of the form

$$a_{\mu+\rho} = \alpha_{\rho 1}a_1 + \alpha_{\rho 2}a_2 + \dots + \alpha_{\rho\mu}a_\mu.$$

Let $c_{\mu+1}, c_{\mu+2} \ldots c_\nu$ be defined by $(\nu-\mu)$ equations of the type,

$$c_{\mu+\rho} = a_{\rho 1}e_1 + a_{\rho 2}e_2 + \ldots + a_{\rho\mu}e_\mu - e_{\mu+\rho}$$
$$= d_{\mu+\rho} - e_{\mu+\rho};$$

where $$\phi d_{\mu+\rho} = a_{\mu+\rho}.$$

Then it is easily seen that $e_1, e_2 \ldots e_\mu, c_{\mu+1}, c_{\mu+2} \ldots c_\nu$ are ν independent elements. Hence these elements can be chosen to form the denominator of the matrix.

But $$\phi c_{\mu+\rho} = \phi d_{\mu+\rho} - \phi e_{\mu+\rho} = a_{\mu+\rho} - a_{\mu+\rho} = 0.$$

Hence the matrix takes the form $\dfrac{a_1, a_2 \ldots a_\mu, 0, 0 \ldots 0}{e_1, e_2 \ldots e_\mu, c_{\mu+1}, c_{\mu+2} \ldots c_\nu}$.

(2) In this case the associated determinant is zero. The region

$$(c_{\mu+1}, c_{\mu+2} \ldots c_\nu),$$

of $\nu-\mu-1$ dimensions, is called the null space of the matrix; and the matrix is said to be of nullity $\nu-\mu$. Thus if the associated determinant vanish, the matrix is of nullity other than zero.

Any point in the null space is said to be destroyed by the matrix, and will be called a null point of the matrix. Any point x is transformed by the matrix into a point in the region $(e_1, e_2 \ldots e_\mu)$. This region $(e_1, e_2 \ldots e_\mu)$ is said to be the space or region preserved by the matrix.

(3) Sylvester* has enunciated the theorem that the nullity of the product of two matrices is not less than the greater of their nullities, but not greater than the sum of the two nullities. The following proof is due to Buchheim†.

Let ϕ be a matrix of nullity α and let N_α be its null space; and let χ be a matrix of nullity β and let N_β be its null space. Also let $P_{\nu-\alpha}$ and $P_{\nu-\beta}$ be the spaces preserved by ϕ and χ respectively. Then if N_α and $P_{\nu-\beta}$ intersect in a region T_δ of $\delta-1$ dimensions, the nullity of the matrix $\phi\chi$ is $\beta+\delta$. For to find the nullity of $\phi\chi$ we have only to find the most general region which χ transforms into T_δ, since any point in this latter region is destroyed by ϕ. Now if T_δ and N_β be taken as co-ordinate regions [cf. § 65 (3)], any point in the region of $\beta+\delta-1$ dimensions, defined by the co-ordinate elements lying in T_δ and N_β, is transformed by χ into a point in T_δ. Thus the nullity of $\phi\chi$ is $\beta+\delta$, and the null space of $\phi\chi$ is the region defined by the co-ordinate points lying in T_δ and N_β. Hence the nullity of $\phi\chi$ is not less than β, being equal to β if N_α and $P_{\nu-\beta}$ do not intersect. Also it is immediately evident that the nullity of $\phi\chi$ is at least equal to the nullity of ϕ: for if χx lie in N_α, then $\phi\chi x = 0$. Hence the nullity of $\phi\chi$ is not less than α.

* Cf. 3 *Johns Hopkins Circulars* 33, "On the three Laws of Motion in the World of Universal Algebra."

† Cf. *Phil. Mag.* Series 5, vol. 18, November, 1884.

Again, to prove that the nullity of $\phi\chi$ is less than $\alpha+\beta$, note that if $\alpha+\beta>\nu$ the theorem is obvious. For a matrix of nullity ν would destroy all space. Assume therefore $\alpha+\beta<\nu$. Now δ is greatest when N_α is contained in $P_{\nu-\beta}$, since $\alpha<\nu-\beta$; hence the greatest possible value of δ is α. Thus the greatest possible value of the nullity of $\phi\chi$ is $\alpha+\beta$.

(4) Buchheim extends Sylvester's theorem. For if $\alpha+(\nu-\beta)<\nu$, that is, if $\alpha<\beta$, then in general N_α and $P_{\nu-\beta}$ do not intersect. In this case there is no region T_δ. Thus if $\alpha<\beta$, the nullity of $\phi\chi$ is in general β. Again, if $\alpha+(\nu-\beta)>\nu$, that is, if $\alpha>\beta$, then in general N_α and $P_{\nu-\beta}$ intersect in a region of $\alpha-\beta-1$ dimensions; thus $\delta=\alpha-\beta$, and in general the nullity of $\phi\chi$ is α.

Thus in general the nullity of $\phi\chi$ is equal to the greater of the two nullities of ϕ and χ; but if special conditions are fulfilled, it may have any greater value up to the sum of the two nullities.

145. LATENT POINTS. (1) If a point x is such that, ϕ being a given matrix,

$$\phi x = \rho x,$$

then x is called a latent point of the matrix, and the ordinary algebraic quantity ρ is called a latent root.

The transformation due to the matrix does not alter the position of a latent point x, it merely changes its intensity.

(2) Let the latent point x be expressed in the form $\Sigma\xi e$. Also let $(\phi-\rho)e_1=c_1$, $(\phi-\rho)e_2=c_2$, and so on.

Hence
$$(\phi-\rho)x = 0 = \Sigma\xi(\phi-\rho)e = \Sigma\xi c.$$

Therefore $c_1, c_2 \ldots c_\nu$ are not independent, and thus $(c_1c_2\ldots c_\nu)=0$.

This equation can also be written $\Pi[(\phi-\rho)e]=0$, that is

$$\{(\phi e_1-\rho e_1)(\phi e_2-\rho e_2)\ldots(\phi e_\nu-\rho e_\nu)\}=0.$$

This is an equation of the νth degree in ρ, of which the first term is $(-1)^\nu(e_1\ldots e_\nu)\rho^\nu$, and the last term is $(\phi e_1 . \phi e_2 \ldots \phi e_\nu)$. The roots of this equation in ρ are the latent roots of the matrix.

(3) From § 142 (4) and (5) with the notation of § 141 (4), $\phi-\rho$ is the matrix

$$\left(\begin{array}{llll} \alpha_{11}-\rho, & \alpha_{12} & \ldots & \alpha_{1\nu} \\ \alpha_{21} & , \alpha_{22}-\rho & \ldots & \alpha_{2\nu} \\ \ldots & \ldots & \ldots & \ldots \\ \alpha_{\nu1} & , \alpha_{\nu2} & \ldots & \alpha_{\nu\nu}-\rho \end{array}\right).$$

Hence the equation for the latent roots is

$$\left|\begin{array}{llll} \alpha_{11}-\rho, & \alpha_{12} & \ldots & \alpha_{1\nu} \\ \alpha_{21} & , c_{22}-\rho & \ldots & \alpha_{2\nu} \\ \ldots & \ldots & \ldots & \ldots \\ \alpha_{\nu1} & , \alpha_{\nu2} & \ldots & \alpha_{\nu\nu}-\rho \end{array}\right| = 0.$$

(4) If all the roots of this equation are unequal, then ν, and only ν, latent points exist, one corresponding to each root, and these points form an independent system. These propositions are proved in the following three subsections.

(5) There is at least one latent point corresponding to any root ρ_1 of the equation giving the latent roots. For let

$$(\phi - \rho_1) e_1 = c_1, \ (\phi - \rho_1) e_2 = c_2 \dots (\phi - \rho_1) e_\nu = c_\nu.$$

Then since $(c_1 c_2 \dots c_\nu) = 0$, a relation holds such as $\gamma_1 c_1 + \gamma_2 c_2 + \dots + \gamma_\nu c_\nu = 0$.

Hence $\gamma_1 (\phi - \rho_1) e_1 + \gamma_2 (\phi - \rho_1) e_2 + \dots + \gamma_\nu (\phi - \rho_1) e_\nu = 0$;

this becomes $\phi \{\gamma_1 e_1 + \gamma_2 e_2 + \dots + \gamma_\nu e_\nu\} = \rho_1 \{\gamma_1 e_1 + \gamma_2 e_2 + \dots + \gamma_\nu e_\nu\}$.

Hence the point $\gamma_1 e_1 + \gamma_2 e_2 + \dots + \gamma_\nu e_\nu$ is a latent point corresponding to the root ρ_1 of the equation.

(6) A system of ν such points, one corresponding to each root, form a system of independent elements. For let $a_1, a_2 \dots a_\nu$ be the ν latent points; then, if they are not all independent, at least two of them are independent, otherwise the ν points could not be distinct.

Assume that the μ points $a_1, a_2 \dots a_\mu$ are independent, and that another point a_σ can be expressed in terms of them, by the relation

$$a_\sigma = \alpha_1 a_1 + \alpha_2 a_2 + \dots + \alpha_\mu a_\mu.$$

Then $$\phi a_\sigma = \alpha_1 \phi a_1 + \alpha_2 \phi a_2 + \dots + \alpha_\mu \phi a_\mu,$$

that is, $$\rho_\sigma a_\sigma = \alpha_1 \rho_1 a_1 + \alpha_2 \rho_2 a_2 + \dots + \alpha_\mu \rho_\mu a_\mu.$$

Multiply the first equation by ρ_σ and subtract from this equation, then

$$0 = (\rho_1 - \rho_\sigma) \alpha_1 a_1 + (\rho_2 - \rho_\sigma) \alpha_2 a_2 + \dots + (\rho_\mu - \rho_\sigma) \alpha_\mu a_\mu.$$

Since none of the latent roots, $\rho_1, \rho_2 \dots \rho_\sigma$ are equal, this forms one relation between $a_1, a_2 \dots a_\mu$ contrary to the hypothesis. But at least two of the latent points must be independent, hence they are all independent.

(7) Two latent points cannot belong to the same latent root. For assume that a_1 and a_1' are two distinct points such that $\phi a_1 = \rho_1 a_1$, $\phi a_1' = \rho_1 a_1'$. Let $a_2, a_3 \dots a_\nu$ be latent points corresponding to the remaining $\nu - 1$ roots. Then $a_1, a_2 \dots a_\nu$ form an independent system. Hence a_1' can be written in the form $\alpha_1 a_1 + \alpha_2 a_2 + \dots + \alpha_\nu a_\nu$.

Hence $$\phi a_1' = \alpha_1 \phi a_1 + \alpha_2 \phi a_2 + \dots + \alpha_\nu \phi a_\nu$$
$$= \rho_1 \alpha_1 a_1 + \rho_2 \alpha_2 a_2 + \dots + \rho_\nu \alpha_\nu a_\nu.$$

But $$\phi a_1' = \rho_1 a_1' = \rho_1 \alpha_1 a_1 + \rho_1 \alpha_2 a_2 + \dots + \rho_1 \alpha_\nu a_\nu.$$

Therefore $$(\rho_2 - \rho_1) \alpha_2 a_2 + (\rho_3 - \rho_1) \alpha_3 a_3 + \dots + (\rho_\nu - \rho_1) \alpha_\nu a_\nu = 0.$$

Accordingly there is a relation between $a_2, a_3 \dots a_\nu$, which has been proved to be impossible.

Hence there is only one latent point corresponding to each latent root, when the latent roots are all unequal.

146. Semi-Latent Regions. (1) Let the region defined by μ latent points of a matrix with unequal latent roots be called a semi-latent region of the $(\mu-1)$th species. The number indicating the species of a semi-latent region is thus equal to the dimensions of the region when all the latent roots are unequal.

(2) Let $e_1, e_2 \dots e_\nu$ be the ν latent points of a matrix with unequal latent roots $\rho_1, \rho_2 \dots \rho_\nu$. Then the region defined by $e_1, e_2 \dots e_\mu$ $(\mu < \nu)$ is a semi-latent region. The characteristic property of a semi-latent region is that if x be any element in the region, then ϕx is an element of the same region; for if $x = \xi_1 e_1 + \dots + \xi_\mu e_\mu$, then

$$\phi x = \rho_1 \xi_1 e_1 + \rho_2 \xi_2 e_2 + \dots + \rho_\mu \xi_\mu e_\mu.$$

And if X be any regional element incident in the semi-latent region, then ϕX is a regional element incident in the same semi-latent region. In particular if $L = \lambda e_1 e_2 \dots e_\mu$, then

$$\phi L = \lambda \rho_1 \rho_2 \dots \rho_\mu e_1 e_2 \dots e_\mu = \rho_1 \rho_2 \dots \rho_\mu L.$$

(3) It is also important to notice that

$$\begin{aligned}\phi x &= \rho_1 x + (\rho_2 - \rho_1)\, \xi_2 e_2 + \dots + (\rho_\mu - \rho_1)\, \xi_\mu e_\mu \\ &= \rho_1 x + x',\end{aligned}$$

where x' is a point in the semi-latent region $e_2 \dots e_\mu$, excluding e_1. Thus x' lies in a semi-latent region of the $(\mu-2)$th species, whereas x lies in a semi-latent region of the $(\mu-1)$th species.

147. The Identical Equation. (1) If $\rho_1, \rho_2 \dots \rho_\nu$ be the latent roots of a matrix, no two being equal and none vanishing, and if $a_1, a_2 \dots a_\nu$ be the corresponding latent points, then it follows from above that the matrix can be written in the form

$$\frac{\rho_1 a_1,\ \rho_2 a_2 \dots \rho_\nu a_\nu}{a_1,\ a_2 \dots a_\nu}.$$

(2) If ϕ be the matrix, let ϕ^2 denote the matrix $\phi\phi$, ϕ^3 the matrix $\phi\phi\phi$, and so on. Also any point x can be written $\xi_1 a_1 + \xi_2 a_2 + \dots + \xi_\nu a_\nu$.

Hence $$\phi x = \rho_1 \xi_1 a_1 + \rho_2 \xi_2 a_2 + \dots + \rho_\nu \xi_\nu a_\nu.$$

Hence $$\phi x - \rho_1 x = (\rho_2 - \rho_1)\, \xi_2 a_2 + \dots + (\rho_\nu - \rho_1)\, \xi_\nu a_\nu.$$

Again $$\phi(\phi x - \rho_1 x) = \phi^2 x - \rho_1 \phi x = \rho_2 (\rho_2 - \rho_1)\, \xi_2 a_2 + \dots + \rho_\nu (\rho_\nu - \rho_1)\, \xi_\nu a_\nu.$$

Hence $$\phi^2 x - (\rho_1 + \rho_2)\, \phi x + \rho_1 \rho_2 x = (\rho_3 - \rho_2)(\rho_3 - \rho_1)\, \xi_3 a_3 + \dots + (\rho_\nu - \rho_2)(\rho_\nu - \rho_1)\, \xi_\nu a_\nu.$$

Proceeding in this way, we finally prove that

$$(\phi - \rho_1)(\phi - \rho_2) \dots (\phi - \rho_\nu)\, x = 0,$$

whatever element x may be.

(3) The equation may be written

$$(\phi - \rho_1)(\phi - \rho_2) \dots (\phi - \rho_\nu) = 0,$$

that is, $$\phi^\nu - (\rho_1 + \rho_2 + \dots \rho_\nu)\phi^{\nu-1} + \dots (-1)^\nu \rho_1\rho_2 \dots \rho_\nu = 0.$$

This is called the identical equation satisfied by the matrix ϕ. A similar equation is satisfied by any matrix, though the above proof has only been given for the case when all the roots are unequal and none vanish.

148. The Latent Region of a repeated latent root. (1) In the case when the equation giving the latent roots has equal roots, assume that α_1 of the roots are ρ_1, α_2 are ρ_2, ... α_μ are ρ_μ, where $\rho_1, \rho_2 \dots \rho_\mu$ are the μ distinct roots of the equation. Then

$$\alpha_1 + \alpha_2 + \dots + \alpha_\mu = \nu.$$

(2) Then subsections (5) and (6) of § 145 still hold, proving that at least one latent point corresponds to each distinct root, and that the μ latent points which therefore certainly exist are independent.

(3) Consider now the root ρ_1 which occurs α_1 times, where α_1 is greater than unity.

Let $e_1, e_2 \dots e_\nu$ be ν reference elements, and for brevity write

$$\rho_1 e_1 - \phi e_1 = e_1', \quad \rho_1 e_2 - \phi e_2 = e_2', \quad \dots \rho_1 e_\nu - \phi e_\nu = e_\nu'.$$

Then since ρ_1 is a latent root of the matrix ϕ, $(e_1' e_2' \dots e_\nu') = 0$. Hence $e_1', e_2' \dots e_\nu'$ are not independent [cf. § 145 (2)].

(4) Assume that $\nu - \beta_1$ of them and no more are independent, so that there are β_1 relations of the type

$$\lambda_{\sigma 1} e_1' + \lambda_{\sigma 2} e_2' + \dots + \lambda_{\sigma\nu} e_\nu' = 0 \quad \text{.....................(1)},$$

where σ is an integer less than or equal to β_1 and equation (1) denotes the σth relation of that type.

Let $$a_{1\sigma} = \lambda_{\sigma 1} e_1 + \lambda_{\sigma 2} e_2 + \dots + \lambda_{\sigma\nu} e_\nu.$$

Then $$(\rho_1 - \phi)\, a_{1\sigma} = \lambda_{\sigma 1} e_1' + \dots + \lambda_{\sigma\nu} e_\nu' = 0.$$

Hence $$\phi a_{1\sigma} = \rho_1 a_{1\sigma}.$$

Thus corresponding to each relation of the type (1) existing between $e_1', e_2' \dots e_\nu'$, there is a latent point, such as $a_{1\sigma}$, corresponding to the root ρ_1. Hence, since β_1 relations have been assumed to exist, there are β_1 latent points of the type $a_{1\sigma}$. Furthermore all these points are independent. For if not, the relations of the type (1) are not independent.

(5) The region of $\beta_1 - 1$ dimensions defined by $a_{11}, a_{12} \dots a_{1\beta_1}$ is such that if x be any point in it, $\phi(x) = \rho_1 x$.

This region is therefore such that every point in it is a latent point, corresponding to the root ρ_1. Let it be called the latent region of the matrix corresponding to ρ_1.

(6) The number β_1 cannot be greater than α_1. For let $a_{11}, a_{12} \dots a_{1\beta_1}$, defining the latent region corresponding to ρ_1, be chosen to be β_1 of the reference elements $e_1, e_2 \dots e_\nu$. Thus let $a_{11} = e_1$, $a_{12} = e_2$ and so on. Let $\bar{e}_\sigma$ stand for $(\rho - \phi) e_\sigma$, e_σ' for $(\rho_1 - \phi) e_\sigma$. Then the equation $(\bar{e}_1\bar{e}_2 \dots \bar{e}_\nu) = 0$, contains the factor $(\rho - \rho_1)^{\alpha_1}$.

But $(\rho - \phi) a_{1\sigma} = (\rho - \rho_1) a_{1\sigma} = (\rho - \rho_1) e_\sigma$, when $\sigma \leqq \beta_1$.

Hence the equation becomes $(\rho - \rho_1)^{\beta_1} (e_1 e_2 \dots e_{\beta_1} \bar{e}_{\beta_1+1} \dots \bar{e}_\nu) = 0$.

Therefore $$\beta_1 \overline{\overline{<}} \alpha_1.$$

149. THE FIRST SPECIES OF SEMI-LATENT REGIONS. (1) If $\beta_1 < \alpha_1$, then $(e_1 e_2 \dots e_{\beta_1} \bar{e}_{\beta_1+1} \dots \bar{e}_\nu) = 0$, is satisfied by the root ρ_1 which occurs $\alpha_1 - \beta_1$ times.

Hence $$(e_1 e_2 \dots e_{\beta_1} e'_{\beta+1} \dots e_\nu') = 0.$$

Thus the ν elements $e_1, e_2 \dots e_{\beta_1}, e'_{\beta+1} \dots e_\nu'$ are not independent. It is known that the β_1 elements $e_1, e_2 \dots e_{\beta_1}$ at least are independent. Assume that $\nu - \gamma_1$ only are independent $(\nu - \gamma_1 > \beta_1)$. Then γ_1 relations hold of the type

$$\kappa_{\sigma 1} e_1 + \kappa_{\sigma 2} e_2 + \dots + \kappa_{\sigma, \beta_1} e_{\beta_1} + (\mu_{\sigma, \beta_1+1} e'_{\beta_1+1} + \dots + \mu_{\sigma\nu} e_\nu') = 0 \dots\dots (2),$$

where σ is put successively equal to $1, 2 \dots \gamma_1$.

Since $e_1, e_2 \dots e_{\beta_1}$ are independent, in each relation of type (2), all the coefficients $\mu_{\sigma, \beta_1+1} \dots \mu_{\sigma, \nu}$ cannot vanish, nor can it be possible to eliminate all the elements $e'_{\beta_1+1} \dots e_\nu'$ between these relations and thus to find a relation between $e_1 \dots e_{\beta_1}$.

Thus if we assume γ_1 elements of the type

$$b_{1\sigma} = \mu_{\sigma, \beta_1+1} e_{\beta_1+1} + \dots + \mu_{\sigma\nu} e_\nu,$$

then these elements of the type $b_{1\sigma}$ are mutually independent, and are also independent of $e_1, e_2 \dots e_{\beta_1}$. Also

$$(\rho_1 - \phi) b_{1\sigma} = \mu_{\sigma, \beta_1+1} e'_{\beta_1+1} + \dots + \mu_{\sigma\nu} e_\nu' = -\kappa_{\sigma 1} e_1 - \dots - \kappa_{\sigma\beta_1} e_{\beta_1}.$$

The coefficients $\kappa_{\sigma 1} \dots \kappa_{\sigma\beta_1}$ cannot all vanish: for otherwise $b_{1\sigma}$ would belong to the latent region corresponding to ρ_1, which by supposition is only of $\beta_1 - 1$ dimensions.

Let $a'_{1\sigma}$ stand for $\kappa_{\sigma 1} e_1 + \dots \kappa_{\sigma\beta_1} e_{\beta_1}$, then $a'_{1\sigma}$ is a point in the latent region corresponding to ρ_1: and

$$\phi b_{1\sigma} = \rho_1 b_{1\sigma} + a'_{1\sigma}.$$

(2) Thus γ_1 independent elements, $b_{11}, \dots b_{1\gamma_1}$, satisfying an equation of this type [cf. § 146 (3)] have been proved to exist, defining a region of $\gamma_1 - 1$ dimensions. Also by the same reasoning as in § 148 (6) it is proved that $\beta_1 + \gamma_1 \overline{\overline{<}} \alpha_1$.

(3) The β_1 independent points of the latent region of the type $a_{1\sigma}$ corresponding to the root ρ_1 and the γ_1 points of the type $b_{1\sigma}$, just found, together define a region of $\beta_1 + \gamma_1 - 1$ dimensions, which will be called the

semi-latent region of the first species corresponding to the root ρ_1. This definition is in harmony with the definition of semi-latent regions given in § 146 for the case where all the latent roots are unequal. For let x be any point in this semi-latent region, then ϕx is another point in the same region; let X be any regional element incident in this region, then ϕX is a regional element incident in the same region. And if L be a regional element denoting the semi-latent region itself, then $\phi L = \rho_1^{\beta_1+\gamma_1} L$. Also we can write, $\phi x = \rho_1 x + y$, where y belongs to the latent region (that is, to the semi-latent region of the zero species).

It should be noticed that by definition the semi-latent region of the first species corresponding to any given repeated root contains the latent region corresponding to that root.

(4) The region defined by the points of typical form $b_{1\sigma}$ in subsection (1) is contained within the region defined by $e_{\beta_1+1} \dots e_\nu$; while the latent region is defined by $e_1, e_2 \dots e_{\beta_1}$. Hence the region defined by the points

$$b_{1\sigma} \, (\sigma = 1, 2 \dots \gamma_1)$$

does not intersect the latent region.

But from subsection (1), $\phi b_{1\sigma} = \rho_1 b_{1\sigma} + a'_{1\sigma}$, where $a'_{1\sigma}$ lies in the latent region. Now it can be proved that the γ_1 points $a'_{1\sigma} (\sigma = 1, 2 \dots \gamma_1)$ are independent. For if a relation of the type, $\Sigma \xi_\sigma a'_{1\sigma} = 0$, holds between them, then by writing $(\phi - \rho_1) b_{1\sigma}$ for $a'_{1\sigma}$, we have

$$\Sigma \xi_\sigma (\phi - \rho_1) b_{1\sigma} = 0, \text{ that is } (\phi - \rho_1) \Sigma \xi_\sigma b_{1\sigma} = 0.$$

Hence the point $\Sigma \xi_\sigma b_{1\sigma}$ lies in the latent region, and therefore the region defined by $b_{1\sigma} (\sigma = 1, 2 \dots \gamma_1)$ must intersect the latent region, contrary to what has been proved above.

Hence the γ_1 points $a'_{1\sigma} (\sigma = 1, 2 \dots \gamma_1)$ are independent. But they all lie in the latent region which is defined by β_1 points. Hence $\beta_1 \geqq \gamma_1$.

150. THE HIGHER SPECIES OF SEMI-LATENT REGIONS. (1) Semi-latent regions of the second and of higher species can be successively deduced by an application of the same reasoning as that of § 149 (1).

Thus to deduce, when $\beta_1 + \gamma_1 < \alpha_1$, the semi-latent region of the second species, corresponding to the repeated root ρ_1, take as before β_1 of the reference elements, namely $e_1, e_2 \dots e_{\beta_1}$, in the latent region, which is assumed to be of $\beta_1 - 1$ dimensions, and take γ_1 of the reference elements, namely

$$e_{\beta_1+1}, e_{\beta_1+2} \dots e_{\beta_1+\gamma_1},$$

in the semi-latent region of the first species (but not in the latent region), so that the $\beta_1 + \gamma_1$ reference elements thus assumed define the complete semi-latent region of the first species. Then, if $\sigma \leqq \beta_1$, $\phi e_\sigma = \rho_1 e_\sigma$, and hence $(\rho - \phi) e_\sigma = (\rho - \rho_1) e_\sigma$. Also, if $\sigma > \beta_1$ and $\leqq \beta_1 + \gamma_1$, $\phi e_\sigma = \rho_1 e_\sigma + a_\sigma$, where a_σ lies in the latent region and is therefore dependent on $e_1, e_2 \dots e_{\beta_1}$. Hence $(\rho - \phi) e_\sigma = (\rho - \rho_1) e_\sigma - a_\sigma$.

Thus $\{(\rho-\phi)e_1 \,.\, (\rho-\phi)e_2 \ldots (\rho-\phi)e_{\beta_1+\gamma_1}\} = (\rho-\rho_1)^{\beta_1+\gamma_1} e_1e_2 \ldots e_{\beta_1+\gamma_1}$.

Hence the equation for the latent roots, namely, $\prod_{\sigma=1}^{\nu}\{(\rho-\phi)e_\sigma\}=0$, can be written

$$(\rho-\rho_1)^{\beta_1+\gamma_1}\Big[e_1e_2 \ldots e_{\beta_1+\gamma_1} \prod_{\sigma=\beta_1+\gamma_1+1}^{\nu}\{(\rho-\phi)e_\sigma\}\Big]=0.$$

But the equation for the latent roots has by hypothesis the factor $(\rho-\rho_1)^{\alpha_1}$, where $\beta_1+\gamma_1<\alpha_1$, thus the expression

$$\Big[e_1e_2 \ldots e_{\beta_1+\gamma_1} \prod_{\sigma=\beta_1+\gamma_1+1}^{\nu}\{(\rho-\phi)e_\sigma\}\Big]$$

contains the factor $(\rho-\rho_1)^{\alpha_1-\beta_1-\gamma_1}$.

Thus writing ρ_1 for ρ we see that the ν points

$$e_1, e_2 \ldots e_{\beta_1+\gamma_1}, (\rho_1-\phi)e_{\beta_1+\gamma_1+1}, \ldots, (\rho_1-\phi)e_\nu$$

are not independent.

Assume that $\nu-\delta_1$ only are independent ($\nu-\delta_1>\beta_1+\gamma_1$), so that there are δ_1 independent relations of the type

$$\kappa_{\sigma1}e_1+\kappa_{\sigma2}e_2+\ldots+\kappa_{\sigma,\beta_1+\gamma_1}e_{\beta_1+\gamma_1} - \{\mu_{\sigma,\beta_1+\gamma_1+1}(\rho_1-\phi)e_{\beta_1+\gamma_1+1}+\ldots+\mu_{\sigma\nu}(\rho_1-\phi)e_\nu\}=0;$$

where σ is put successively $1, 2 \ldots \delta_1$. All the μ's cannot vanish simultaneously in any typical relation; and all the terms of the type

$$(\rho_1-\phi)e_\tau \; (\tau>\beta_1+\gamma_1)$$

cannot be simultaneously eliminated between the δ_1 relations, so as to leave a relation between the independent elements $e_1, e_2 \ldots e_{\beta_1+\gamma_1}$.

Now assume $\qquad c_{1\sigma}=\mu_{\sigma,\beta_1+\gamma_1+1}e_{\beta_1+\gamma_1+1}+\ldots+\mu_{\sigma\nu}e_\nu.$

Also note that the point $\kappa_{\sigma1}e_1+\kappa_{\sigma2}e_2+\ldots+\kappa_{\sigma,\beta_1+\gamma_1}e_{\beta_1+\gamma_1}$ ($=b_{1\sigma}$ say) lies in the semi-latent region of the first species. Hence the above typical relation takes the form

$$\phi c_{1\sigma}=\rho_1 c_{1\sigma}+b_{1\sigma}.$$

(2) Also by the same reasoning as in § 148 (6), it follows that

$$\beta_1+\gamma_1+\delta_1 \leqq \alpha_1.$$

(3) Also by the same reasoning as in § 149 (4) it follows that the region defined by the δ_1 points $c_{1\sigma}$ ($\sigma=1, 2 \ldots \delta_1$) does not intersect the semi-latent region of the first species. Also, as before, the δ_1 points of the type $b_{1\sigma}$ are independent and the subregion defined by them (lying in the semi-latent region of the first species) does not intersect the latent region; for otherwise some point of the type $\Sigma\xi_{1\sigma}c_{1\sigma}$ lies in the semi-latent region of the first order, contrary to the assumption that this semi-latent region is only of the $\beta_1+\gamma_1-1$ dimensions. Thus $\delta_1 \leqq \gamma_1$.

(4) If $\beta_1=\alpha_1$, then only a latent region exists corresponding to the repeated root α_1 and no semi-latent region. If $\beta_1<\alpha_1$ and $\beta_1+\gamma_1=\alpha_1$, then no semi-latent region of a species higher than the first exists. If $\beta_1+\gamma_1<\alpha_1$, and $\beta_1+\gamma_1+\delta_1=\alpha_1$, then no semi-latent region of a species higher than the

second exists. If $\beta_1 + \gamma_1 + \delta_1 < \alpha_1$, then by similar reasoning a semi-latent region of the third species exists, and so on till α_1 independent points have been introduced defining the complete series of semi-latent regions corresponding to the root ρ_1.

Also from subsection (3) and from § 149 (4) it follows that if $\alpha_1 > \mu\beta_1$, where μ is an integer, then in addition to the latent region at least μ semi-latent regions of the successive species must exist*.

(5) It follows from (3) and § 149 (4) that a matrix can always be written thus

$$\frac{\rho_1 a_{1\sigma},\ \rho_1 b_{1\sigma} + a_{1\sigma},\ \rho_1 c_{1\sigma} + b_{1\sigma},\ \ldots}{a_{1\sigma},\quad b_{1\sigma} \quad,\quad c_{1\sigma} \quad,\ \ldots},$$

where only those typical terms are exhibited which correspond to the latent root ρ_1.

151. THE IDENTICAL EQUATION. (1) Suppose that the number of different groups of points of the types $a_{1\sigma}$, $b_{1\sigma}$, $c_{1\sigma}$, and so on corresponding to a latent root ρ_1 is τ_1. Then

$$(\phi - \rho_1)\, a_{1\sigma} = 0,\ (\phi - \rho_1)^2\, b_{1\sigma} = 0,\ (\phi - \rho_1)^3\, c_{1\sigma} = 0,$$

and if $p_{1\sigma}$ be a point in the τ_1th group, that is in the semi-latent region of the $(\tau_1 - 1)$th species (but not in that of the $(\tau - 2)$th species), then $(\phi - \rho_1)^{\tau_1} p_{1\sigma} = 0$. Let the region defined by all these points be called the semi-latent region of the matrix corresponding to ρ_1.

(2) Now all the points of the different types thus found, corresponding to all the latent roots, are independent, and may be taken as a reference system.

Hence if $\tau_2, \tau_3 \ldots \tau_\mu$ be the corresponding numbers relating to the other latent roots, and x be any point, then

$$(\phi - \rho_1)^{\tau_1} (\phi - \rho_2)^{\tau_2} \ldots (\phi - \rho_\mu)^{\tau_\mu} x = 0.$$

Thus any matrix satisfies the identical equation

$$(\phi - \rho_1)^{\tau_1} (\phi - \rho_2)^{\tau_2} \ldots (\phi - \rho_\mu)^{\tau_\mu} = 0.$$

(3) Since $\tau_1 < \alpha_1$, $\tau_2 < \alpha_2 \ldots \tau_\mu < \alpha_\mu$, it follows that any matrix satisfies the equation

$$(\phi - \rho_1)^{\alpha_1} (\phi - \rho_2)^{\alpha_2} \ldots (\phi - \rho_\mu)^{\alpha_\mu} = 0.$$

Thus the equation of § 147 is proved for the case of equal roots. But in this case the matrix satisfies an equation of an order lower than the νth.

152. THE VACUITY OF A MATRIX. (1) A null space [cf. § 144] can only exist if a matrix has a zero latent root. The null space, or null region, is the latent region corresponding to the zero latent root.

* This theorem does not seem to have been noticed before: nor do I think that the relations $\gamma_1 \leqq \beta_1$, $\delta_1 \leqq \gamma_1$, etc. have been previously explicitly stated.

(2) If the zero latent root occur α times, then the matrix is said to be of vacuity α. Thus by definition the vacuity of a matrix is not less than its nullity. Let the semi-latent regions corresponding to the zero root be called also the vacuous regions of the matrix. Thus if b be a point in a vacuous region of the first species, $\phi b = a$, where a is a point in the null region; also if c be a point in the vacuous region of the second species, $\phi c = b$, where b is a point in the vacuous region of the first species; and so on.

(3) Assume that δ independent points, and no more, can be found in the vacuous regions of the first species defining a subregion which does not intersect the null region. Let $d_1, d_2 \dots d_\delta$, be these points, and let the β points $b_1, b_2 \dots b_\beta$ define the null region. Then any point x in the vacuous region of the first species can be written $\Sigma \xi d + \Sigma \eta b$.

Also by § 150 (3), $\delta < \beta$; and we may assume consistently with the previous assumptions, $\phi d_1 = \lambda_1 b_1$, $\phi d_2 = \lambda_2 b_2$, ... $\phi d_\delta = \lambda_\delta b_\delta$.

Hence $$\phi x = \phi \Sigma \xi d + \phi \Sigma \eta b = \Sigma \xi \phi d = \sum_{\rho=1}^{\delta} \xi_\rho \lambda_\rho b_\rho.$$

Thus any point in the vacuous region of the first species is transformed into a point in the subregion of the null region defined by $b_1, b_2 \dots b_\delta$. Call this subregion the subregion of the null region associated with the vacuous region of the first species.

153. Symmetrical Matrices*. (1) In general, if x and y be any two elements and ϕ any matrix, $(x \mid \phi y)$ is not equal to $(y \mid \phi x)$.

In order to obtain the conditions which must hold for these expressions to be equal, let the matrix be $\dfrac{a_1, a_2, \dots a_\nu}{e_1, e_2, \dots e_\nu}$, where, according to the notation of § 141 (4), $a_\rho = \alpha_{1\rho} e_1 + \alpha_{2\rho} e_2 + \dots + \alpha_{\nu\rho} e_\nu$.

In other words the matrix is
$$\left(\begin{array}{cccc} \alpha_{11}, & \alpha_{12}, & \dots & \alpha_{1\nu} \\ \alpha_{21}, & \alpha_{22}, & \dots & \alpha_{2\nu} \\ \dots & \dots & \dots & \dots \\ \alpha_{\nu 1}, & \alpha_{\nu 2}, & \dots & \alpha_{\nu\nu} \end{array}\right).$$

Then, supposing that $e_1, e_2, \dots e_\nu$ are a set of normal elements at unit normal intensities [cf. §§ 109 (3) and 110 (1)],
$$(e_\rho \mid \phi e_\sigma) = (e_\rho \mid a_\sigma) = \alpha_{\rho\sigma} (e_\rho \mid e_\rho) = \alpha_{\rho\sigma},$$
and
$$(e_\sigma \mid \phi e_\rho) = (e_\sigma \mid a_\rho) = \alpha_{\sigma\rho} (e_\sigma \mid e_\sigma) = \alpha_{\sigma\rho}.$$

Hence, if the required condition holds, $\alpha_{\rho\sigma} = \alpha_{\sigma\rho}$.

(2) Thus the matrix with the desired property is a matrix symmetrical about its leading diagonal when the elements of the denominator form a

* Symmetrical matrices are considered by Grassmann [cf. *Ausdehnungslehre von* 1862, § 391]; but his use of supplements implicitly implies a purely imaginary, self-normal quadric. Hence his conclusions are limited to those of subsection (7).

normal system (at unit normal intensities) with respect to the quadric chosen as the self-normal quadric.

Let such matrices be called symmetrical with respect to the normal systems, or, more shortly, symmetrical matrices.

(3) If μ out of the ν latent roots of a symmetrical matrix be distinct and not zero, so that at least μ points $c_1, c_2, \dots c_\mu$, can be found with the property $\phi c_\rho = \gamma_\rho c_\rho$, then the μ points $c_1, c_2, \dots c_\mu$ corresponding to different latent roots $\gamma_1, \gamma_2, \dots \gamma_\mu$ are mutually normal.

For let $x = \xi_1 c_1 + \xi_2 c_2 + \dots + \xi_\mu c_\mu$, and $y = \eta_1 c_1 + \eta_2 c_2 + \dots + \eta_\mu c_\mu$.

Then $$(y \mid \phi x) = (\Sigma \eta c \mid \Sigma \xi \gamma c) = \Sigma (\xi_\rho \eta_\sigma \gamma_\rho + \xi_\sigma \eta_\rho \gamma_\sigma)(c_\rho \mid c_\sigma),$$

and $$(x \mid \phi y) = (\Sigma \xi c \mid \Sigma \eta \gamma c) = \Sigma (\xi_\rho \eta_\sigma \gamma_\sigma + \xi_\sigma \eta_\rho \gamma_\rho)(c_\rho \mid c_\sigma).$$

Hence $(y \mid \phi x) = (x \mid \phi y)$ gives $\Sigma (\xi_\rho \eta_\sigma - \xi_\sigma \eta_\rho)(\gamma_\rho - \gamma_\sigma)(c_\rho \mid c_\sigma) = 0$.

Now let all the ξ's, except ξ_ρ, and all the η's, except η_σ, vanish; and it follows that $(c_\rho \mid c_\sigma) = 0$.

Hence $c_1, c_2 \dots c_\mu$ are mutually normal.

(4) Let c_1', c_1'', etc., be other points in the latent region of the root γ_1, so that $\phi c_1' = \gamma_1 c_1'$, etc.: then the same proof shows that c_1' is normal to all of $c_2, \dots c_\mu$, and so on. Hence the latent region corresponding to γ_1 is normal to the latent region corresponding to ρ_2, and so on.

(5) In the same way it can be proved that the whole semi-latent region corresponding to any latent root γ_1 is normal to the whole semi-latent region corresponding to any other latent root γ_2. For let d_1 be any point in the semi-latent region of γ_1 of the first species.

Then $$\phi d_1 = \gamma_1 d_1 + \lambda_1 c_1, \quad \phi c_2 = \gamma_2 c_2.$$

Hence $(c_2 \mid \phi d_1) = \gamma_1 (c_2 \mid d_1)$, by (3) and (4).

Also $(d_1 \mid \phi c_2) = \gamma_2 (d_1 \mid c_2) = \gamma_2 (c_2 \mid d_1)$.

But $(c_2 \mid \phi d_1) = (d_1 \mid \phi c_2)$, by hypothesis. Hence $(\gamma_1 - \gamma_2)(c_2 \mid d_1) = 0$, and $\gamma_1 \neq \gamma_2$ by hypothesis. Therefore $(c_2 \mid d_1) = 0$. Hence the semi-latent region of the first species corresponding to γ_1 is normal to the latent region corresponding to γ_2. Similarly the semi-latent region of the first species corresponding to γ_2 is normal to the latent region corresponding to γ_1.

Again d_1 and d_2 lying respectively in the semi-latent regions of the first species corresponding respectively to γ_1 and to γ_2 are normal to each other.

For $(d_1 \mid \phi d_2) = (d_1 \mid (\gamma_2 d_2 + \lambda_2 c_2)) = \gamma_2 (d_1 \mid d_2)$, and $(d_2 \mid \phi d_1) = \gamma_1 (d_2 \mid d_1)$.

Thus $(d_1 \mid \phi d_2) = (d_2 \mid \phi d_1)$ gives $(\gamma_1 - \gamma_2)(d_1 \mid d_2) = 0$; and hence $(d_1 \mid d_2) = 0$.

Similarly if f_1 be another point in the semi-latent region of the second species of the root γ_1, such that $\phi f_1 = \gamma_1 f_1 + \mu_1 d_1$, then the same proof shows that f_1 is normal to c_2, d_2 and f_2; and so on.

Hence the semi-latent regions of different roots are mutually normal.

(6) Again consider the equation

$$(c_1 \mid \phi d_1) = (d_1 \mid \phi c_1).$$

This becomes $\gamma_1 (c_1 \mid d_1) + \lambda_1 (c_1 \mid c_1) = \gamma_1 (c_1 \mid d_1).$

Hence $\lambda_1 (c_1 \mid c_1) = 0.$

Thus if c_1 does not lie on the self-normal quadric, $\lambda_1 = 0$.

Now suppose that the latent region defined by $c_1, c_1', c_1''\dots$ does not touch the self-normal quadric. Then it is always possible in an infinite number of ways to choose $c_1, c_1', c_1''\dots$ to be mutually normal and none of them self-normal. Also the most general form for d_1 is such that

$$\phi d_1 = \gamma_1 d_1 + \lambda_1 c_1 + \lambda_1' c_1' + \dots.$$

Then $(c_1 \mid \phi d_1) = \gamma_1 (c_1 \mid d_1) + \lambda_1 (c_1 \mid c_1) = (d_1 \mid \phi c_1) = \gamma_1 (c_1 \mid d_1).$

Hence $\lambda_1 = 0$, similarly $\lambda_1' = 0$, $\lambda_1'' = 0$, and so on. Hence d_1 lies in the latent region, and no semi-latent regions of the first or higher species exist corresponding to the root γ_1.

(7) It is a well-known proposition that the roots of the equation

$$\begin{vmatrix} \alpha_{11} - \rho, & \alpha_{12}, & \dots \alpha_{1\nu} \\ \alpha_{21}, & \alpha_{22} - \rho, & \dots \alpha_{2\nu} \\ \dots & \dots & \dots \\ \alpha_{\nu 1}, & \alpha_{\nu 2}, & \dots \alpha_{\nu\nu} - \rho \end{vmatrix} = 0,$$

are all real; provided that $\alpha_{\rho\sigma} = \alpha_{\sigma\rho}$, where all the quantities $\alpha_{\rho\sigma}$ are real.

Hence it follows that the latent regions of symmetrical matrices are all real. For if γ_1 be one of the real roots, the equation $\phi x = \gamma_1 x$, determines the ratios of the co-ordinates of x by real linear equations. If the self-normal quadric be imaginary owing to all the normal intensities being real [cf. § 110 (3)], a latent region, being real, cannot touch it. Hence in this case there can be no latent self-normal point, such that $\phi c_\rho = \gamma_\rho c_\rho$. Hence from above there are no semi-latent regions. Thus finally in this case a complete real normal system of the type $c_1, c_1', c_1''\dots c_2, c_2'\dots, c_3\dots, c_\mu, c_\mu', c_\mu''\dots$ can be found defining the latent regions of γ_1, γ_2, etc.; each element being at unit normal intensity.

(8) If the latent region defined by $c_1, c_1', c_1''\dots$ touches the self-normal quadric (assumed real), but is not part of a generating region, take c_1 to be the point of contact, and take $c_1', c_1''\dots$ to be mutually normal elements on the tangent plane at c_1, but not self-normal [cf. § 113 (5)].

Then the general form of d_1 is such that, $\phi d_1 = \gamma_1 d_1 + \lambda_1 c_1 + \lambda_1' c_1' + \dots.$

Hence $(c_1 \mid \phi d_1) = \gamma_1 (c_1 \mid d_1) = (d_1 \mid \phi c_1) = \gamma_1 (c_1 \mid d_1).$

Also $(c_1' \mid \phi d_1) = \gamma_1 (c_1' \mid d_1) + \lambda_1' (c_1' \mid c_1') = (d_1 \mid \phi c_1') = \gamma_1 (c_1' \mid d_1).$

Hence $\lambda_1' = 0$, similarly $\lambda_1'' = 0$, and so on.

Thus d_1 is such that $\phi d_1 = \gamma_1 d_1 + \lambda_1 c_1$.

There can only be one independent point d_1 satisfying this equation. For if d_1' be another point such that $\phi d_1' = \gamma_1 d_1' + \lambda_1' c_1$, then it was proved in § 149 (4) that if c_1, d_1, d_1' are independent, then $\lambda_1 c_1$ is independent of $\lambda_1' c_1$, whereas here they are the same point; which is impossible.

(9) If the latent region of the root γ_1 contain a real generating region of $\rho - 1$ dimensions of the self-normal quadric, let the points $c_{11}, c_{12}, \dots c_{1\rho}$ be chosen to be mutually normal points in this generating region [cf. §§ 79 and 80], and let the remaining points of the latent region be mutually normal and normal to $c_{11} \dots c_{1\rho}$, but not self-normal. Let these remaining points be $c_1, c_1', c_1'' \dots$.

Let d_1 be any point in the semi-latent region of the first species, but not in the latent region.

Then
$$\phi d_1 = \gamma_1 d_1 + \sum_{\kappa=1}^{\kappa=\rho} \lambda_\kappa c_{1\kappa} + \mu_1 c_1 + \mu_1' c_1' + \dots.$$

Hence $(c_1 | \phi d_1) = \gamma_1 (c_1 | d_1) + \mu_1 (c_1 | c_1) = (d_1 | \phi c_1) = \gamma_1 (c_1 | d_1).$

Hence $\mu_1 = 0$. Similarly $\mu_1' = 0$, $\mu_1'' = 0$, and so on.

Hence
$$\phi d_1 = \gamma_1 d_1 + \sum_{\kappa=1}^{\kappa=\rho} \lambda_\kappa c_{1\kappa}.$$

Thus in the semi-latent region of the first species the subregion of highest dimensions not necessarily intersecting the latent region cannot be of higher dimensions than the real generating region contained in the latent region [cf. § 149 (4)]. Similarly for the semi-latent regions of higher species.

154. Symmetrical Matrices and Supplements. (1) A one to one correspondence of points to planes is given by the operation which transforms the reference elements $e_1, e_2 \dots e_\nu$ into the planes $A_1, A_2 \dots A_\nu$, where [cf. § 97 Prop. IV.]

$$A_1 = \alpha_{11} E_1 + \alpha_{21} E_2 + \alpha_{31} E_3 + \dots + \alpha_{\nu 1} E_\nu,$$

and so on.

Then the element $x\,(= \Sigma \xi e)$ is transformed into the plane $X\,(= \Sigma \xi A)$.

Now let $e_1, e_2 \dots e_\nu$ be a normal system, so that $E_1 = |e_1$, and so on; then

$$A_1 = |(\alpha_{11} e_1 + \alpha_{21} e_2 + \dots) = |a_1, \text{ say.}$$

Similarly $A_2 = |(\alpha_{12} e_1 + \alpha_{22} e_2 + \dots) = |a_2$, and so on.

Also let ϕ denote the matrix $\dfrac{a_1, a_2 \dots a_\nu}{e_1, e_2 \dots e_\nu}$.

Then the type of one to one correspondence of points to planes, which we have been considering, can be denoted by $X = |\phi x$.

Similarly this type of correspondence could be denoted by $\phi | x$; but $|\phi$ and $\phi|$ are in general different operations.

(2) If to every point there corresponds a plane and to every plane there corresponds a point, then the matrix ϕ has no vacuity. In this case $\dfrac{e_1, e_2 \dots e_\nu}{a_1, a_2 \dots a_\nu}$

is a determinate matrix; denote it by ϕ^{-1}. Then if $X = |\phi x$, $|X = \phi x$, and $x = \phi^{-1}|X$.

In general the transformations $|\phi$ and $\phi^{-1}|$ are different: thus $|\phi x$ is different from $\phi^{-1}|x$.

(3) If the latent roots of ϕ are all unequal, then the operations $|\phi$ and $\phi^{-1}|$ can only be identical when the ν latent points of ϕ form a normal system, that is, when the matrix is symmetrical; and when, in addition, the product of the latent roots of the matrix is unity.

For let $c_1, c_2 \dots c_\nu$ be the latent points of ϕ, so that ϕ can be written

$$\frac{\gamma_1 c_1, \gamma_2 c_2 \dots \gamma_\nu c_\nu}{c_1, c_2 \dots c_\nu}.$$

Then $|\phi c_1 = \gamma_1 |c_1$. Hence $|\phi c_1 = \phi^{-1}|c_1$, becomes $\gamma_1 |c_1 = \phi^{-1}|c_1$, that is $\gamma_1\phi|c_1 = |c_1$.

Let $C_1 = c_2c_3 \dots c_\nu$, $C_2 = c_1c_3 \dots c_\nu$, ..., $C_\nu = c_1c_2 \dots c_{\nu-1}$.

Assume that $$|c_1 = \lambda_1 C_1 + \lambda_2 C_2 + \dots + \lambda_\nu C_\nu.$$

Then $$\gamma_1\phi\,|\,c_1 = \gamma_1\lambda_1\phi C_1 + \gamma_1\lambda_2\phi C_2 + \dots + \gamma_1\lambda_\nu\phi C_\nu.$$

But by § 141 (1) $\phi C_1 = \gamma_2\gamma_3 \dots \gamma_\nu C_1$, $\phi C_2 = \gamma_1\gamma_3 \dots \gamma_\nu C_2$, etc.

Hence $$\gamma_1\phi|c_1 = \gamma_1\gamma_2 \dots \gamma_\nu\lambda_1 C_1 + \gamma_1^2\gamma_3 \dots \gamma_\nu\lambda_2 C_2 + \dots + \gamma_1^2\gamma_2 \dots \gamma_{\nu-1}\lambda_\nu C_\nu$$
$$= |c_1 = \lambda_1 C_1 + \lambda_2 C_2 + \dots + \lambda_\nu C_\nu.$$

Hence since $\gamma_1, \gamma_2 \dots \gamma_\nu$ are all unequal, $\gamma_1\gamma_2 \dots \gamma_\nu = 1$, $\lambda_2 = 0$, $\lambda_3 = 0, \dots \lambda_\nu = 0$.

Thus $|c_1 = \lambda_1 C_1$; and similarly for $|c_2$, $|c_3$, etc. Accordingly the latent points of the matrix form a normal system, and the product of the latent roots is unity.

(4) Conversely if the matrix be a symmetrical matrix with unequal latent roots of which the product is unity, then $|\phi$ and $\phi^{-1}|$ are the same operations.

For let $c_1, c_2 \dots c_\nu$ be the latent points, $\gamma_1, \gamma_2 \dots \gamma_\nu$ the latent roots of ϕ. Then $c_1, c_2 \dots c_\nu$ are the latent points and $\gamma_1^{-1}, \gamma_2^{-1} \dots \gamma_\nu^{-1}$ are the latent roots of ϕ^{-1}.

Also $$|\phi c_1 = \gamma_1 |c_1, \text{ and } \phi\,|\phi c_1 = \gamma_1\phi\,|c_1 = \gamma_1\gamma_2 \dots \gamma_\nu |c_1 = |c_1.$$

Hence $|\phi c_1 = \phi^{-1}|c_1$. Similarly for the other latent points.

Thus finally $$|\phi x = \phi^{-1}|x.$$

(5) It is obvious that in this case the operation $|\phi$ is equivalent to the operation of taking the supplements with respect to some quadric with respect to which $c_1, c_2 \dots c_\nu$ form a self-normal system. Let I denote this operation; let $\epsilon_1, \epsilon_2 \dots \epsilon_\nu$ be the normal intensities of $c_1, c_2 \dots c_\nu$ with respect to this operation; and let $\delta_1, \delta_2 \dots \delta_\nu$ denote the normal intensities of $c_1, c_2 \dots c_\nu$ with respect to the operation $|$. Also put

$$\Delta' = \epsilon_1\epsilon_2 \dots \epsilon_\nu, \quad \Delta = \delta_1\delta_2 \dots \delta_\nu.$$

Then $$|c_1 = \frac{\Delta}{\delta_1^2} c_2 c_3 \dots c_\nu, \quad \mathrm{I}c_1 = \frac{\Delta'}{\epsilon_1^2} c_2 c_3 \dots c_\nu.$$

But $$\mathrm{I}c_1 = |\phi c_1 = \gamma_1 |c_1 = \frac{\gamma_1 \Delta}{\delta_1^2} c_2 c_3 \dots c_\nu.$$

Thus $\gamma_1 = \frac{\delta_1^2}{\epsilon_1^2} \frac{\Delta'}{\Delta}$. Similarly $\gamma_2 = \frac{\delta_2^2}{\epsilon_2^2} \frac{\Delta'}{\Delta}$, and so on.

Hence $\gamma_1 \gamma_2 \dots \gamma_\nu = 1 = \frac{\Delta'^{\nu-2}}{\Delta^{\nu-2}}$, therefore $\Delta = \Delta'$.

Hence $$\gamma_1 = \frac{\delta_1^2}{\epsilon_1^2}, \quad \gamma_2 = \frac{\delta_2^2}{\epsilon_2^2}, \quad \dots \quad \gamma_\nu = \frac{\delta_\nu^2}{\epsilon_\nu^2}.$$

Thus the symmetrical matrix ϕ, with unequal roots of product unity, has been expressed in the form $|\mathrm{I}$; so that $\phi x = |\mathrm{I}x$.

The latent points of the matrix are the one common system of self-normal points of the two self-normal quadrics corresponding to $|$ and I; and the relations between the latent roots and normal intensities are given above.

155. Skew Matrices. (1) The matrix $\frac{a_1, a_2 \dots a_\nu}{e_1, e_2 \dots e_\nu} (= \phi)$ has important properties in the special case when

$$a_1 = * + \alpha_{21} e_2 + \alpha_{31} e_3 + \dots + \alpha_{\nu 1} e_\nu,$$
$$a_2 = \alpha_{12} e_1 + * + \alpha_{32} e_3 + \dots + \alpha_{\nu 2} e_\nu,$$
etc.,

where $\alpha_{12} + \alpha_{21} = 0, \dots, \alpha_{\rho\sigma} + \alpha_{\sigma\rho} = 0 \dots$, and $e_1, e_2 \dots e_\nu$ form a normal system at unit normal intensities. Let such a matrix be called a skew matrix.

Then $$(e_\rho | \phi e_\rho) = 0, \; (e_\sigma | \phi e_\rho) = \alpha_{\sigma\rho} = -\alpha_{\rho\sigma} = -(e_\rho | \phi e_\sigma).$$

Thus $(x | \phi x) = 0$ (A), and $(x | \phi y) + (y | \phi x) = 0$ (B), whatever points x and y may be.

(2) Any latent point c_1 of this matrix, such that $\phi c_1 = \gamma_1 c_1$, where γ_1 is not zero, is self-normal. For from equation (A) $(c_1 | \phi c_1) = \gamma_1 (c_1 | c_1) = 0$.

Again, putting c_1 and c_2 for x and y in equation (B), where c_2 is another latent point,

$$(\gamma_1 + \gamma_2)(c_1 | c_2) = 0.$$

Hence either $\gamma_1 + \gamma_2 = 0$, or $(c_1 | c_2) = 0$.

(3) Assume that there are no repeated roots. The self-normal quadric contains generating regions of dimensions $\frac{\nu}{2} - 1$ or $\frac{\nu - 1}{2} - 1$, according as ν be even or odd (cf. § 79).

If ν be even, $\frac{\nu}{2}$ mutually normal elements $j_1, j_2, \dots j_{\frac{\nu}{2}}$ can be found on the quadric, defining one generating region, and $k_1, k_2 \dots k_{\frac{\nu}{2}}$ mutually normal elements defining another generating region. Also any element such as j_ρ can be made normal to all the k's, except k_ρ, and conversely k_ρ is normal to all the j's, except j_ρ (cf. § 80).

Then $j_1, j_2 \dots j_{\frac{\nu}{2}}$, $k_1, k_2 \dots k_{\frac{\nu}{2}}$ can be chosen as the latent points of the matrix. If $\gamma_1, \gamma_2 \dots \gamma_{\frac{\nu}{2}}$ be the latent roots corresponding to $j_1, j_2 \dots j_{\frac{\nu}{2}}$, then by subsection (2) $-\gamma_1, -\gamma_2 \dots -\gamma_{\frac{\nu}{2}}$ are the latent roots corresponding to $k_1, k_2 \dots k_{\frac{\nu}{2}}$.

Hence if $$x = \xi_1 j_1 + \xi_2 j_2 + \dots + \xi_{\frac{\nu}{2}} j_{\frac{\nu}{2}} + \eta_1 k_1 + \eta_2 k_2 + \dots + \eta_{\frac{\nu}{2}} k_{\frac{\nu}{2}},$$

then $$\phi x = \gamma_1 \xi_1 j_1 + \gamma_2 \xi_2 j_2 + \dots - \gamma_1 \eta_1 k_1 - \gamma_2 \eta_2 k_2 - \dots .$$

Thus $$(x \mid \phi x) = (\gamma_1 \xi_1 \eta_1 - \gamma_1 \xi_1 \eta_1)(j_1 \mid k_1) + \dots = 0.$$

(4) If ν be odd, $\frac{\nu-1}{2}$ mutually normal elements of the type j_ρ can be found, and $\frac{\nu-1}{2}$ of the type k_ρ, and an element e, not on the quadric, normal to all the j's and all the k's.

Let these be the latent points of the matrix, then the element e must be a null point of the matrix.

If $x = \zeta e + \Sigma \xi j + \Sigma \eta k$, then $\phi x = \Sigma \gamma \xi j - \Sigma \gamma \eta k$; and $(x \mid \phi x) = 0$.

(5) Assume that there are repeated roots. Let the roots γ_1 and γ_2 be both repeated, and neither zero. Let d_1 and d_2 be in the semi-latent regions of the first species (and not in the latent regions) corresponding to γ_1 and γ_2 respectively, f_1 and f_2 in the semi-latent regions of the second species (and not in the semi-latent regions of the first species), and so on. Let $c_1, c_1', \dots$ be in the latent region of γ_1, and $c_2, c_2', \dots$ in that of γ_2.

Then we may assume [cf. § 150 (5)],

$$\phi c_1 = \gamma_1 c_1, \quad \phi d_1 = \gamma_1 d_1 + \lambda_1 c_1, \quad \phi f_1 = \gamma_1 f_1 + \mu_1 d_1, \text{ and so on.}$$

Hence by equation (B), $(c_1 \mid \phi c_1') + (c_1' \mid \phi c_1) = 2\gamma_1 (c_1 \mid c_1') = 0$.

Hence c_1 and c_1' are mutually normal as well as being self-normal. Thus the latent region of a repeated root is a subregion of some generating region of the self-normal quadric.

(6) Again from equation (B), $(c_1' \mid \phi d_1) + (d_1 \mid \phi c_1') = 2\gamma_1 (c_1' \mid d_1) = 0$.

Hence the semi-latent region of the first species corresponding to γ_1 is normal to the latent region corresponding to γ_1.

Also by equation (A), $(d_1 \mid \phi d_1) = \gamma_1 (d_1 \mid d_1) = 0$.

Hence $(d_1 \mid d_1) = 0$. Therefore each point in the semi-latent region of the first species is self-normal. Further if d_1' be another point in this semi-latent region,

$$(d_1 \mid \phi d_1') + (d_1' \mid \phi d_1) = 2\gamma_1 (d_1 \mid d_1') = 0.$$

Thus $(d_1 | d_1') = 0$. Hence the semi-latent region of the first species is a subregion of a generating region of the self-normal quadric; and therefore the latent region and semi-latent region of the first species are together contained in the same generating region.

(7) The same proof applies to semi-latent regions of higher species. Hence the complete semi-latent region (which contains the latent region) corresponding to a repeated root is a subregion of a generating region of the self-normal quadric.

(8) The same proof shows that the complete semi-latent region of one repeated root γ_1 is normal to the complete semi-latent region of another repeated root γ_2 unless $\gamma_1 + \gamma_2 = 0$.

(9) Again assume that the matrix is of vacuity α and of nullity β. Let c be any null point of the matrix, and c_1 any latent point corresponding to the non-vanishing root γ_1.

Then $\phi c = 0$, hence $(c_1 | \phi c) = 0$. Thus by equation (B)

$$(c_1 | \phi c) + (c | \phi c_1) = \gamma_1 (c | c_1) = 0.$$

Hence the null region is normal to the latent regions of all the other latent roots.

Similarly the null region can be proved to be normal to all the semi-latent regions of the other latent roots.

(10) Let d be a point in the vacuous region of the first species: assume $\phi d = \lambda c$, where c is a null point.

Then $(c | \phi d) + (d | \phi c) = \lambda (c | c) = 0$, by equation (B).

Hence either $\lambda = 0$, and d is in the null space; or $(c | c) = 0$, that is to say, c is self-normal. Hence the subregion of the null region associated with the vacuous region of the first species is self-normal.

Also from equation (A), $(d | \phi d) = \lambda (c | d) = 0$.

Hence, assuming that λ is not zero, $(c | d) = 0$, that is to say, d is normal to c.

Again let c' be any other null point, then $(c' | \phi d) + (d | \phi c') = \lambda (c | c') = 0$.

Hence, assuming $\lambda \neq 0$, c is normal to every other null point.

(11) Similar theorems apply to vacuous regions of higher species.

BOOK V.

EXTENSIVE MANIFOLDS OF THREE DIMENSIONS.

CHAPTER I.

SYSTEMS OF FORCES.

156. NON-METRICAL THEORY OF FORCES. (1) The general theory of extensive manifolds, apart from the additional specification of the Theory of Metrics, has received very little attention. It is proposed here to investigate the properties of Extensive Manifolds of three dimensions, thereby on the one hand illustrating the development of one type of formulæ of the Calculus of Extension, and on the other hand discussing properties which are important from their connection with Geometry*.

(2) Since in this case four independent points define the complete region the simple extensive magnitudes are only of three orders, the point, the linear element, the planar element. Also the only complex extensive magnitudes are systems of linear elements. A linear element,—in that (α) it is an intensity associated with a straight line, (β) it is directed along the line, so as to be capable of two opposite senses, (γ) it is to be combined with other linear elements on the same line by a mere addition of the intensities [cf. § 95 (1)],—has so far identical properties with a force acting on a rigid body. Only in an extensive manifold no metrical ideas with respect to distance have been introduced. The other properties of a linear element, whereby it is defined by two points and is combined with other linear elements on other lines form a generalization of the properties of a force so as to avoid the introduction of any notion of distance. It will be noticed that the theorem respecting the combination of Forces known as Leibnitz's theorem expresses the aspect of the properties of forces which are here generalized. The parallelogram of forces is without meaning at this stage of our investigations: for the idea of a parallelogram depends on the Euclidean (or equivalent) axioms concerning parallel lines, and such axioms presuppose metrical conceptions with respect to distance which have not yet been enunciated.

* The formulæ and proofs of propositions in this book are, I believe, new. Many of the propositions are well-known; but I believe that they have hitherto been obtained in connection with *Metrical* Geometry, either Euclidean or non-Euclidean.

(3) We shall therefore use the term force as equivalent to linear element, meaning by it the generalized conception here developed apart from metrical considerations. It will be found that very few of the geometrical properties of ordinary mechanical forces are lost by this generalization.

Also, when no confusion will arise, plane will be used for planar element. The context will always shew the exact meaning of the term.

157. Recapitulation of Formulæ. (1) It will be useful to recapitulate the leading formulæ of the Calculus of Extension in the shape in which they appear, when the complete manifold is of three dimensions.

(2) The product of four points is merely numerical. The product of a linear element and planar element is the point of intersection of the line and plane. The product of two planar elements is a linear element in the line of intersection of the two planes. Thus a linear element can be conceived either as the product of two points or as the product of two planar elements. The product of three planar elements is a point. The product of three points a planar element. The product of a linear element and a point is a planar element. The product of two linear elements is merely numerical.

(3) The formulæ for regressive multiplication are [cf. § 103 (3) and (4)]

$$abc\,.\,de = de\,.\,abc = (abce)\,d - (abcd)\,e = (abde)\,c + (cade)\,b + (bcde)\,a \quad \ldots(1).$$

Thus five points a, b, c, d, e are connected by the equation

$$(bcde)\,a - (acde)\,b + (abde)\,c - (abce)\,d + (abcd)\,e = 0 \quad \ldots\ldots(2).$$

Again
$$abc\,.\,def = (abcf)\,de + (abcd)\,ef + (abce)\,fd$$
$$= (adef)\,bc + (bdef)\,ca + (cdef)\,ab = -def\,.\,abc \quad \ldots\ldots (3).$$

By taking supplements, we deduce that these formulæ still hold when planar elements A, B, C, D, E, F are substituted for the points a, b, c, d, e, f.

(4) Also from § 105 there come the group of formulæ, B_1, B_2, B_3, B_4 being planar elements,

$$(a_1a_2\,.\,B_1B_2) = (a_1B_1)(a_2B_2) - (a_1B_2)(a_2B_1) \quad \ldots\ldots(4);$$

$$(a_1a_2a_3\,.\,B_1B_2B_3) = \begin{vmatrix} (a_1B_1), & (a_1B_2), & (a_1B_3), \\ (a_2B_1), & (a_2B_2), & (a_2B_3), \\ (a_3B_1), & (a_3B_2), & (a_3B_3), \end{vmatrix} \quad \ldots\ldots(5);$$

$$(a_1a_2a_3a_4)(B_1B_2B_3B_4) = (a_1a_2a_3a_4\,.\,B_1B_2B_3B_4) =$$

$$\begin{vmatrix} (a_1B_1), & (a_1B_2), & (a_1B_3), & (a_1B_4), \\ (a_2B_1), & (a_2B_2), & (a_2B_3), & (a_2B_4), \\ (a_3B_1), & (a_3B_2), & (a_3B_3), & (a_3B_4), \\ (a_4B_1), & (a_4B_2), & (a_4B_3), & (a_4B_4), \end{vmatrix} \quad \ldots\ldots(6).$$

(5) Also from equation (4) a useful formula may be deduced by putting $B_1 = bcc_1$, $B_2 = bcc_2$. Then from equation (4)

$$a_1a_2 . (bcc_1)(bcc_2) = (a_1bcc_1)(a_2bcc_2) - (a_1bcc_2)(a_2bcc_1).$$

But from § 102, $(bcc_1)(bcc_2) = (bcc_1c_2)\, bc.$

Therefore $a_1a_2 . (bcc_1)(bcc_2) = (a_1a_2bc)(bcc_1c_2).$

Hence finally, $(a_1a_2bc)(bcc_1c_2) = (a_1bcc_1)(a_2bcc_2) - (a_1bcc_2)(a_2bcc_1)$ ......(7).

This equation can be written in another form by putting F for the force bc. Then

$$(a_1a_2F)(c_1c_2F) = (a_1c_1F)(a_2c_2F) - (a_1c_2F)(a_2c_1F) \quad \text{............(7')}.$$

158. Inner Multiplication. (1) If a be any point, then $|a$ is a planar element; and if A be any planar element, then $|A$ is a point. If F be a simple linear element, then $|F$ is a simple linear element; and if S be a system of linear elements, then $|S$ is a system of linear elements.

(2) Again [cf. § 99 (7)], $||a = -a$, $||A = -A$, $||F = F$ ....(8).

Also (cf. § 118), and hence
$$\left.\begin{aligned} |(abc\,|de) &= ||de\,.\,|abc = (de\,|abc), \\ |(de\,|abc) &= ||(abc\,|de) = -(abc\,|de) \end{aligned}\right\} \quad \text{...............(9)}.$$

Also and hence
$$\left.\begin{aligned} |(abc\,|d) &= -|abc\,.\,d = (d\,|abc), \\ |(d\,|abc) &= ||(abc\,|d) = (abc\,|d) \end{aligned}\right\} \quad \text{...................(10)}.$$

Also hence
$$\left.\begin{aligned} |(ab\,|c) &= -|ab\,.\,c = -(c\,|ab); \\ |(c\,|ab) &= -||(ab\,|c) = (ab\,|c) \end{aligned}\right\} \quad \text{...................(11)}.$$

Finally and
$$\left.\begin{aligned} (a\,|b) = (b\,|a), \text{ and } (ab\,|cd) &= (cd\,|ab), \\ (abc\,|def) &= (def\,|abc) \end{aligned}\right\} \quad \text{............(12)}.$$

(3) Again from the extended rule of the middle factor (cf. § 119),
$$\left.\begin{aligned} abc\,|de &= (ab\,|de)\,c + (bc\,|de)\,a + (ca\,|de)\,b. \\ \text{And } de\,|abc &= |(abc\,|de) = (de\,|bc)\,|a + (de\,|ca)\,|b + (ab\,|de)\,|c \end{aligned}\right\} \quad \text{......(13)}.$$

$$\left.\begin{aligned} \text{Again } abc\,|d &= (a\,|d)\,bc + (b\,|d)\,ca + (c\,|d)\,ab. \\ \text{And } d\,|abc &= |(abc\,|d) = (d\,|a)\,|bc + (d\,|b)\,|ca + (d\,|c)\,|ab \end{aligned}\right\} \quad \text{...(14)}.$$

$$\left.\begin{aligned} \text{Again } ab\,|c &= (a\,|c)\,b - (b\,|c)\,a. \\ \text{And } c\,|ab &= -|(ab\,|c) = (c\,|b)\,|a - (c\,|a)\,|b \end{aligned}\right\} \quad \text{...........(15)}.$$

(4) Again from § 120, $(ab\,|cd) = (a\,|c)(b\,|d) - (a\,|d)(b\,|c)$ ....(16),

$$(abc\,|def) = \begin{vmatrix} (a\,|d), & (a\,|e), & (a\,|f) \\ (b\,|d), & (b\,|e), & (b\,|f) \\ (c\,|d), & (c\,|e), & (c\,|f) \end{vmatrix} \quad \text{....................(17)},$$

$$(abcd\,|efgh) = \begin{vmatrix} (a\,|e), & (a\,|f), & (a\,|g), & (a\,|h), \\ (b\,|e), & (b\,|f), & (b\,|g), & (b\,|h), \\ (c\,|e), & (c\,|f), & (c\,|g), & (c\,|h), \\ (d\,|e), & (d\,|f), & (d\,|g), & (d\,|h), \end{vmatrix} \quad \text{...............(18)}.$$

(5) It is unnecessary to reproduce the special forms of the more general but less useful formulæ in § 122. These eighteen formulæ of the present and the preceding articles are the fundamental formulæ which will be appealed to as known. They are all immediate consequences either of the extended rule of the middle factor or of the formula of § 105.

159. ELEMENTARY PROPERTIES OF A SINGLE FORCE. (1) A force can be represented as a product of any two points in its line. This is a simple corollary of § 95.

(2) A system of forces lying in one plane is equivalent to a single force. This is a corollary of § 97, Prop. IV.

(3) A force can be resolved into the sum of two forces on lines concurrent with it and coplanar with it. For let a be the point of concurrence, then ab can be chosen to represent the given force. Two points c and d can be found on the other lines respectively, such that $b = \lambda c + \mu d$. Hence $ab = \lambda ac + \mu ad$. Thus ab is resolved as required.

(4) Any force can be resolved into the sum of two forces, of which one passes through a given point and one lies in a given plane, which does not contain the point.

For consider the plane P through the given force and the given point. It cuts the given plane in a line concurrent with the force, and through the point of concurrence a line can be drawn in P through the given point: then two forces can be found by (3) along these lines of which the sum is equivalent to the given force.

Thus if a be any given point, A any given plane, F any given force, then we can write,

$$F = ap + AP.$$

160. ELEMENTARY PROPERTIES OF SYSTEMS OF FORCES. (1) The letter S will only be used to denote a system of forces. Two congruent systems of forces (i.e. of the types S and λS) will be spoken of as the same system at different intensities. If F_1, F_2, etc. be any number of forces, then $S = \Sigma F$ represents the most general type of system.

(2) If a be any given point and A any given planar element not containing a, any system of forces (S) can be written

$$S = ap + AP.$$

For by § 159 (4), $F_1 = ap_1 + AP_1$, $F_2 = ap_2 + AP_2$, etc.

Hence $S = F_1 + F_2 + \ldots = a(p_1 + p_2 + \ldots) + A(P_1 + P_2 + \ldots) = ap + AP$.

Hence any system can always be represented by two forces of which one lies in a given plane, and one passes through a given point not lying in the plane.

(3) The mention of p and P can be avoided by means of the formula

$$(aA)\,S = a\,.\,AS + aS\,.\,A.$$

This can be proved as follows. From (2) of this article

$$S = ap + AP.$$

Multiplying by a, we have $aS = a\,.\,AP = (aP)\,A - (aA)\,P$.

Hence $$aS\,.\,A = -(aA)\,PA = (aA)\,AP.$$

Again multiplying by A, we have $AS = A\,.\,ap = (Ap)\,a - (Aa)\,p$.

Hence $$a\,.\,AS = (aA)\,ap.$$

The required formula follows at once.

(4) It follows from (2) that any system S can be expressed in the form

$$S = ab + cd.$$

For we may write cd instead of AP in the expression for S. It will be proved in § 162 (2) that one of the two lines, say ab, can be assumed arbitrarily.

(5) If e_1, e_2, e_3, e_4 be any four independent elements, then [cf. § 96 (1)] S can be written

$$S = \pi_{12}e_1e_2 + \pi_{23}e_2e_3 + \pi_{31}e_3e_1 + \pi_{14}e_1e_4 + \pi_{24}e_2e_4 + \pi_{34}e_3e_4.$$

Hence any system can be represented as six forces along the edges of any given tetrahedron.

When e_1, e_2, e_3, e_4 are unit reference elements, π_{12}, etc. will be called the co-ordinates of the system S.

161. CONDITION FOR A SINGLE FORCE. (1) If S be any system of forces, (SS) is not in general zero. For by § 160 (4), S may be written $ab + cd$; hence $(SS) = 2\,(abcd)$.

Thus (SS) only vanishes when $(abcd) = 0$, i.e. when ab and cd intersect. But in this case $ab + cd$ can be combined into a single force.

Thus $(SS) = 0$, is the required condition that S may reduce to a single force.

(2) If $S = ap + AP$, then

$$(SS) = 2\,(ap\,.\,AP) = 2\,(aA)\,(pP) - 2\,(aP)\,(pA).$$

If $$S = \pi_{12}e_1e_2 + \pi_{23}e_2e_3 + \pi_{31}e_3e_1 + \pi_{14}e_1e_4 + \pi_{24}e_2e_4 + \pi_{34}e_3e_4,$$

then $$\tfrac{1}{2}\,(SS) = \pi_{12}\pi_{34} + \pi_{23}\pi_{14} + \pi_{31}\pi_{24}.$$

(3) If S reduce to a single force, $|S$ reduces to a single force. For if $(SS) = 0$, then $|(SS) = 0$, that is $(|S\,|S) = 0$.

162. CONJUGATE LINES. (1) When a system S is reduced to the sum of two forces ab and cd, then the lines ab and cd are called conjugate lines, and the forces ab and cd are called conjugate forces with respect to the system. Also ab will be called conjugate to cd, and *vice versa*.

(2) To prove that in general any line ab has one and only one conjugate with respect to any system S, not a single force.

For if $S = \lambda ab + \mu cd$, then $S - \lambda ab$ is a single force.

Hence $\{(S - \lambda ab)(S - \lambda ab)\} = 0$; that is $(SS) - 2\lambda(abS) = 0$.

Therefore $\lambda = \frac{1}{2}\frac{(SS)}{(abS)}$; and hence $S - \frac{1}{2}\frac{(SS)}{(abS)} ab$ represents the force conjugate to λab. Since only one value of λ has been found, there is only one such force; and if (abS) be not zero, there is always one such force. Similarly if any line be symbolized by AB, its conjugate with respect to S is $S - \frac{1}{2}\frac{(SS)}{(ABS)} AB$.

(3) If two lines ab and cd intersect, their conjugates with respect to any system S intersect.

For by multiplication

$$\left\{S - \frac{1}{2}\frac{(SS)}{(abS)} ab\right\}\left\{S - \frac{1}{2}\frac{(SS)}{(cdS)} cd\right\} = \frac{1}{4}\frac{(SS)^2(abcd)}{(abS)(cdS)} = 0,$$

since by hypothesis $(abcd) = 0$.

163. NULL LINES, PLANES AND POINTS. (1) If L be any force, and $(LS) = 0$, then the line L is called a null line of the system S.

Note that L can be written in the two forms ab and AB; the product (abS) is a pure progressive product; the product (ABS) is a pure regressive product.

If F be any force, then (FS) is called the moment of S about the force F.

(2) The assemblage of null lines of any given system S will be called the linear complex* defined by the system S.

(3) If a be any point, then the planar element aS defines a plane containing a, which is called the null plane of the point a with respect to the system S.

If A be any plane, then the point AS lies in A and is called the null point of the plane A with respect to the system S.

* Linear Complexes were first invented and studied by Plücker, cf. *Phil. Trans.* vol. 155, 1865, and his book *Neue Geometrie des Raumes*, 1868. The theory of Linear Complexes is developed in Clebsch and Lindemann's *Vorlesungen über Geometrie*, vol. 2, 1891; also (among other places) in Kœnig's *La Géométrie Reglée*, Paris, 1895, and in Dr Rudolf Sturm's *Liniengeometrie*, 3 vols., Leipzig, 1892, 1893, 1896. The chief advances in Line Geometry, since Plücker, are due to Klein. Buchheim first pointed out the possibility of applying Grassmann's *Ausdehnungslehre* to the investigation of the Linear Complex, cf. *On the Theory of Screws in Elliptic Space*, *Proc. of London Math. Soc.* vols. XV, XVI, and XVII, 1884 and 1886.

164. PROPERTIES OF NULL LINES. (1) All the null lines of S which pass through any point a lie in the null plane of a; and conversely all the null lines which lie in any plane A pass through its null point. For if ab be any null line of S through a, then $(abS) = 0 = (aS \,.\, b)$. Hence b lies on the plane aS.

Similarly if AB be any null line of S in A, then $(ABS) = 0 = (AS \,.\, B)$. Hence B contains the point AS.

(2) If a lie on the null plane of b, then b lies on the null plane of a.

For
$$(bS \,.\, a) = 0 = -(aS \,.\, b).$$

It is obvious that in this case ab is a null line.

(3) If any null line L of a system of forces intersect any line ab, it intersects its conjugate.

For by hypothesis,
$$(SL) = 0 = (abL).$$

Hence
$$L\left(S - \frac{1}{2}\frac{(SS)}{(abS)}ab\right) = 0.$$

Also obviously any line intersecting each of two conjugates is a null line.

(4) The conjugates of all lines through a given point a lie in the null plane of a.

For let ab be any line through a. Then the plane through a and the conjugate of ab is defined by $a\left\{S - \frac{1}{2}\frac{(SS)}{(abS)}ab\right\}$, that is, by aS.

It follows as a corollary that $aS \,.\, bS$ represents the line conjugate to ab. For this conjugate lies in the line of intersection of the null planes of a and b. Thus
$$aS \,.\, bS \equiv S - \frac{1}{2}\frac{(SS)}{(abS)}ab.$$

(5) If the system do not reduce to a single force, no two points have the same null plane and no two planes have the same null point.

For if x and y be two points such that $xS = yS$, then putting $x = y + z$, $zS = 0$. Hence by § 97, Prop. I., $S = zp$; and therefore S reduces to a single force, contrary to the assumption. Thus no two points with the same null plane exist.

If X and Y be two planes with the same null point, then $XS = YS$. Hence by taking supplements $|X|S = |Y|S$. But $|S$ is a system of forces, and hence the points $|X$ and $|Y$ cannot have the same null planes with regard to it unless $|S$ reduce to a single force. Hence from § 161 (3) X and Y cannot have the same null points with regard to S, unless S reduce to a single force.

(6) The relations between planes and their null points and between points and their null planes can be expressed in terms of ordinary algebraic equations involving their coordinates*. For let

$$x = \xi_1 e_1 + \xi_2 e_2 + \xi_3 e_3 + \xi_4 e_4,$$

and $$X = \lambda_1 e_2 e_3 e_4 - \lambda_2 e_1 e_3 e_4 + \lambda_3 e_1 e_2 e_4 - \lambda_4 e_1 e_2 e_3.$$

Then the equation, either of a plane through x, or of a point on X, is

$$(xX) = (\lambda_1 \xi_1 + \lambda_2 \xi_2 + \lambda_3 \xi_3 + \lambda_4 \xi_4)(e_1 e_2 e_3 e_4) = 0.$$

Also [cf. § 160 (5)] let S be the system

$$\alpha_{12} e_1 e_2 + \alpha_{34} e_3 e_4 + \alpha_{13} e_1 e_3 + \alpha_{42} e_4 e_2 + \alpha_{14} e_1 e_4 + \alpha_{23} e_2 e_3.$$

Then by simple multiplication $xS = (\alpha_{23}\xi_1 - \alpha_{13}\xi_2 + \alpha_{12}\xi_3)\, e_1 e_2 e_3 + \text{etc.}$

Hence the co-ordinates $\lambda_1, \lambda_2, \lambda_3, \lambda_4$ of the null plane of x can be written,

$$\begin{aligned}
\sigma\lambda_1 &= \quad * \quad + \alpha_{34}\xi_2 + \alpha_{42}\xi_3 + \alpha_{23}\xi_4, \\
\sigma\lambda_2 &= \alpha_{43}\xi_1 + \quad * \quad + \alpha_{14}\xi_3 + \alpha_{31}\xi_4, \\
\sigma\lambda_3 &= \alpha_{24}\xi_1 + \alpha_{41}\xi_2 + \quad * \quad + \alpha_{12}\xi_4, \\
\sigma\lambda_4 &= \alpha_{32}\xi_1 + \alpha_{13}\xi_2 + \alpha_{21}\xi_3 + \quad * \quad ,
\end{aligned}$$

where we assume $\alpha_{12} + \alpha_{21} = 0 = \alpha_{13} + \alpha_{31} = \text{etc.}$

Again by simple multiplication, we find

$$XS = (* + \alpha_{21}\lambda_2 + \alpha_{31}\lambda_3 + \alpha_{41}\lambda_4)(e_1 e_2 e_3 e_4)\, e_1 + \text{etc.}$$

Hence the co-ordinates $\xi_1, \xi_2, \xi_3, \xi_4$ of the null point of X are given by

$$\begin{aligned}
\sigma'\xi_1 &= \quad * \quad + \alpha_{21}\lambda_2 + \alpha_{31}\lambda_3 + \alpha_{41}\lambda_4, \\
\sigma'\xi_2 &= \alpha_{12}\lambda_1 + \quad * \quad + \alpha_{32}\lambda_3 + \alpha_{42}\lambda_4, \\
\sigma'\xi_3 &= \alpha_{13}\lambda_1 + \alpha_{23}\lambda_2 + \quad * \quad + \alpha_{43}\lambda_4, \\
\sigma'\xi_4 &= \alpha_{14}\lambda_1 + \alpha_{24}\lambda_2 + \alpha_{34}\lambda_3 + \quad * \quad .
\end{aligned}$$

(7) Thus, if the reference elements be normal points at unit normal intensities, a skew matrix [cf. § 155] in a complete region of three dimensions operating on x can be symbolized by $|xS$.

165. Lines in Involution. (1) A system of forces can always be found so that five given lines are null lines with respect to it. But if six lines are null lines with respect to some system, their co-ordinates must satisfy a condition.

For let $L_1, L_2, L_3, L_4, L_5, L_6$ be any six independent lines. Then [cf. § 96 (2)] we may write any system S,

$$S = \xi_1 L_1 + \xi_2 L_2 + \xi_3 L_3 + \xi_4 L_4 + \xi_5 L_5 + \xi_6 L_6.$$

Assume that $(L_1 S) = 0 = (L_2 S) = (L_3 S) = (L_4 S) = (L_5 S).$

* Cf. Clebsch and Lindemann, *Vorlesungen über Geometrie*, vol. II. pp. 41 et seq.

Then the five ratios $\xi_1 : \xi_2 : \xi_3 : \xi_4 : \xi_5 : \xi_6$ are determined by the five equations

$$\begin{aligned}
* \quad + \xi_2 (L_1L_2) + \xi_3 (L_1L_3) + \xi_4 (L_1L_4) + \xi_5 (L_1L_5) + \xi_6 (L_1L_6) &= 0,\\
\xi_1 (L_2L_1) + \quad * \quad + \xi_3 (L_2L_3) + \xi_4 (L_2L_4) + \xi_5 (L_2L_5) + \xi_6 (L_2L_6) &= 0,\\
\cdots\cdots\cdots\cdots\cdots\cdots\cdots\cdots\cdots\cdots\cdots\cdots\cdots\cdots\cdots\cdots\cdots\cdots &\\
\xi_1 (L_5L_1) + \xi_2 (L_5L_2) + \xi_3 (L_5L_3) + \xi_4 (L_5L_4) + \quad * \quad + \xi_6 (L_5L_6) &= 0.
\end{aligned}$$

Hence S is completely determined. Therefore one and only one system of forces can in general be found such that the five lines $L_1, L_2 \ldots L_5$ are null lines with respect to it.

(2) If L_6 be also a null line with respect to the same system then eliminating ξ_1, ξ_2, etc. from the six equations of condition, we find

$$\begin{vmatrix} *, & (L_1L_2), & (L_1L_3), \ldots & (L_1L_6) \\ (L_2L_1), & *, & (L_2L_3), \ldots & (L_2L_6) \\ \cdots & \cdots & \cdots & \cdots \\ (L_6L_1), & (L_6L_2), & \cdots\cdots, & * \end{vmatrix} = 0;$$

where it is to be noticed that $(L_1L_2) = (L_2L_1)$.

(3) *Definition.* Six lines which are null lines with respect to the same system are said to be in involution; and each is said to be in involution with respect to the other five.

Thus the propositions of the preceding article can be stated thus:

The lines through a given point in involution with five given lines lie in a plane, cf. § 164 (1).

The lines in a given plane in involution with five given lines are concurrent, cf. § 164 (1).

Again, a linear complex may be conceived as defined by five independent lines belonging to it.

166. Reciprocal Systems. (1) Two systems of forces S and S' are said to be reciprocal* if $(SS') = 0$.

It is obvious that a force on a null line of any system is a force reciprocal to the system.

(2) If two systems be reciprocal, the null lines of one system taken in pairs are conjugates with respect to the other system.

For let S and S' be the two systems. Then $(SS') = 0$. Let ab be a null line of S, its conjugate with respect to S' is $S' - \frac{1}{2}\frac{(S'S')}{(abS')}ab$.

* Reciprocal systems of mechanical forces were first studied by Sir R. S. Ball, cf. *Transactions of the Royal Irish Academy*, 1871 and 1874, vol. 25, and *Phil. Trans.* (London), vol. 164, 1874, and his book *Theory of Screws* (1876), ch. III. The theory of systems of forces for non-Euclidean Geometry was first worked out by Lindemann in his classical memoir, *Mechanik bei Projectiven Maasbestimmung*, Math. Annal. vol. VII, 1873. The most complete presentment of Sir R. S. Ball's Theory of Screws is given by H. Gravelius, *Theoretische Mechanik*, Berlin, 1889.

But $$S\left\{S' - \tfrac{1}{2}\frac{(S'S')}{(abS')}ab\right\} = (SS') - \tfrac{1}{2}\frac{(S'S')}{(abS')}(abS) = 0.$$

Hence the conjugate of ab with respect to S' is a null line of S.

It is to be noted that there are conjugates of either system which are not null lines of the other.

167. Formulæ for Systems of Forces. (1) The following formulæ are obvious extensions of the standard formulæ of § 157, remembering the distributive law of multiplication.

From equation (1), § 157,

Also $$\left.\begin{aligned} abc\,.\,S = S\,.\,abc &= (abS)\,c + (caS)\,b + (bcS)\,a.\\ Sc\,.\,de &= (Sce)\,d - (Scd)\,e \end{aligned}\right\}\quad\ldots\ldots\ldots\ldots(19).$$

From equation (3), $abc\,.\,dS = (adS)\,bc + (bdS)\,ca + (cdS)\,ab$ ....(20).

By taking supplements, and replacing $|S$ by S, we see that the formulæ hold when planar elements replace the points.

(2) To prove that, if a be any point and S any system of forces

$$\left.\begin{aligned} S\,.\,aS = aS\,.\,S &= \tfrac{1}{2}(SS)\,a,\\ S\,.\,AS = AS\,.\,S &= \tfrac{1}{2}(SS)\,A \end{aligned}\right\}\quad\ldots\ldots\ldots\ldots(21).$$

For let $$S = bc + de.$$

Then $S\,.\,aS = de\,.\,abc + bc\,.\,ade = -(abcd)\,e + (abce)\,d - (abde)\,c + (acde)\,b$
$= (bcde)\,a$; from § 157, equation (2).

Also $(SS) = 2\,(bcde)$. Hence $S\,.\,aS = \tfrac{1}{2}(SS)\,a$.

The second formula follows by taking supplements.

(3) To prove that

$$\left.\begin{aligned} aS\,.\,bS &= (abS)\,S - \tfrac{1}{2}(SS)\,ab,\\ AS\,.\,BS &= (ABS)\,S - \tfrac{1}{2}(SS)\,AB \end{aligned}\right\}\quad\ldots\ldots\ldots\ldots(22).$$

For let $S = \lambda ab + cd$. Then $aS\,.\,bS = acd\,.\,bcd = (abcd)\,cd$.

But $$(abcd) = (abS),\ \text{and}\ cd = S - \tfrac{1}{2}\frac{(SS)}{(abS)}ab.$$

Hence $$aS\,.\,bS = (abS)\,S - \tfrac{1}{2}(SS)\,ab.$$

This forms another proof of the corollary to § 164 (4).

(4) From equations (21) and (22) it is easily proved that

$$\left.\begin{aligned} aS\,.\,bS\,.\,cS &= \tfrac{1}{2}(SS)\,\{(bcS)\,a + (caS)\,b + (abS)\,c\} = \tfrac{1}{2}(SS)\,S\,.\,abc\,;\\ AS\,.\,BS\,.\,CS &= \tfrac{1}{2}(SS)\,S\,.\,ABC \end{aligned}\right\}\ldots(23).$$

Also from equation (22), $$\left.\begin{aligned} aS\,.\,bS\,.\,S &= \tfrac{1}{2}(SS)\,(abS),\\ AS\,.\,BS\,.\,S &= \tfrac{1}{2}(SS)\,(ABS) \end{aligned}\right\}\quad\ldots\ldots\ldots\ldots(24).$$

(5) To prove that if a be any point and S and S' any two systems of forces, then

$$\left.\begin{aligned} S\,.\,aS' + S'\,.\,aS &= (SS')\,a, \\ S\,.\,AS' + S'\,.\,AS &= (SS')\,A \end{aligned}\right\} \quad\ldots\ldots\ldots\ldots\ldots\ldots\ldots\ldots(25).$$

For in equations (21) write $S + S'$ instead of S.

Then $$(S+S')\,.\,a\,(S+S') = \tfrac{1}{2}\,\{(S+S')\,(S+S')\}\,a.$$

Hence by multiplying out both sides,

$$S\,.\,aS + S'\,.\,aS' + S\,.\,aS' + S'\,.\,aS = \tfrac{1}{2}\,(SS)\,a + \tfrac{1}{2}\,(S'S')\,a + (SS')\,a.$$

But $S\,.\,aS = \frac{1}{2}\,(SS)\,a$, and $S'\,.\,aS' = \frac{1}{2}\,(S'S')\,a$. Hence the required result.

Similarly from equation (22) we can prove

$$\left.\begin{aligned} aS\,.\,bS' + aS'\,.\,bS &= (abS)\,S' + (abS')\,S - (SS')\,ab, \\ AS\,.\,BS' + AS'\,.\,BS &= (ABS)\,S' + (ABS')\,S - (SS')\,AB \end{aligned}\right\} \quad\ldots\ldots(26).$$

CHAPTER II.

GROUPS OF SYSTEMS OF FORCES.

168. SPECIFICATIONS OF A GROUP. (1) If $S_1, S_2, \ldots S_6$ be any six independent [cf. § 96 (2)] systems of forces, then any system can be written in the form $\lambda_1 S_1 + \lambda_2 S_2 + \ldots + \lambda_6 S_6$. Let $\lambda_1, \lambda_2, \ldots \lambda_6$ be called the co-ordinates of S as referred to the six systems.

Definitions. The assemblage of systems, found from the expression $\lambda_1 S_1 + \lambda_2 S_2$ by giving the ratio $\lambda_1 : \lambda_2$ all possible values, will be called a 'dual group' of systems. The assemblage of systems, found from the expression $\lambda_1 S_1 + \lambda_2 S_2 + \lambda_3 S_3$ by giving the ratios $\lambda_1 : \lambda_2 : \lambda_3$ all possible values, will be called a 'triple group' of systems.

The assemblage, found from $\lambda_1 S_1 + \lambda_2 S_2 + \lambda_3 S_3 + \lambda_4 S_4$ by giving the ratios $\lambda_1 : \lambda_2 : \lambda_3 : \lambda_4$ all possible values, will be called a 'quadruple group.' The assemblage, found from $\lambda_1 S_1 + \lambda_2 S_2 + \lambda_3 S_3 + \lambda_4 S_4 + \lambda_5 S_5$ by giving the ratios $\lambda_1 : \lambda_2 : \lambda_3 : \lambda_4 : \lambda_5$ all possible values, will be called a 'quintuple group.'

(2) A dual group will be said to be of one dimension, a triple group of two dimensions, and so on.

It is obvious that a group of $\rho - 1$ dimensions ($\rho = 2, 3, 4, 5$) can be defined by any ρ independent systems belonging to it; and also that not more than ρ independent systems can be found belonging to it.

(3) Again, if the co-ordinates $\lambda_1, \lambda_2, \ldots \lambda_6$ of any system S satisfy a linear equation of the form,

$$\alpha_1\lambda_1 + \alpha_2\lambda_2 + \alpha_3\lambda_3 + \alpha_4\lambda_4 + \alpha_5\lambda_5 + \alpha_6\lambda_6 = 0,$$

then S belongs to a given quintuple group.

For by eliminating λ_6, we can write

$$\alpha_6 S = \lambda_1(\alpha_6 S_1 - \alpha_1 S_6) + \lambda_2(\alpha_6 S_2 - \alpha_2 S_6) + \lambda_3(\alpha_6 S_3 - \alpha_3 S_6) \\ + \lambda_4(\alpha_6 S_4 - \alpha_4 S_6) + \lambda_5(\alpha_6 S_5 - \alpha_5 S_6).$$

Hence $\alpha_6 S_1 - \alpha_1 S_6$, $\alpha_6 S_2 - \alpha_2 S_6$, etc., define a quintuple group to which S belongs.

Similarly it can be proved that if the co-ordinates $\lambda_1 \ldots \lambda_6$ satisfy two linear equations $\Sigma\alpha\lambda = 0$, $\Sigma\beta\lambda = 0$, then the system must belong to a certain quadruple group: if the co-ordinates satisfy three linear equations, the system must belong to a certain triple group: and if four linear equations, to a certain dual group.

(4) Hence a dual group may be conceived as defined by two systems belonging to it, or by four linear equations connecting the co-ordinates of any system belonging to it.

And generally, a group of $\rho - 1$ dimensions ($\rho = 2, 3, 4, 5$) is defined by ρ independent systems belonging to it, or by $6 - \rho$ linear equations connecting the co-ordinates of any system belonging to it.

169. Systems Reciprocal to Groups. (1) *Definition.* A system of forces, which is reciprocal to every system of a group, is said to be reciprocal to the group.

If a system S' be reciprocal to ρ independent systems, $S_1, S_2, \ldots S_\rho$, of a group of $\rho - 1$ dimensions, it is reciprocal to the group.

For any system of the group is $S = \lambda_1 S_1 + \ldots + \lambda_\rho S_\rho$.

Hence $$(SS') = \lambda_1 (S_1 S') + \ldots + \lambda_\rho (S_\rho S').$$

But by hypothesis $(S_1 S') = 0 = (S_2 S') = \ldots = (S_\rho S')$. Hence $(SS') = 0$.

(2) All the systems reciprocal to a given group of $\rho - 1$ dimensions form a group of $5 - \rho$ dimensions.

For let $E_1, E_2, \ldots E_6$ be any six independent reference forces.

Then any system can be written

$$S = \lambda_1 E_1 + \lambda_2 E_2 + \ldots + \lambda_6 E_6.$$

If this system be reciprocal to the ρ independent systems $S_1, S_2, \ldots S_\rho$ which define the given group, then the following ρ equations hold:

$$\lambda_1 (E_1 S_1) + \lambda_2 (E_2 S_1) + \ldots + \lambda_6 (E_6 S_1) = 0,$$
$$\lambda_1 (E_1 S_2) + \lambda_2 (E_2 S_2) + \ldots + \lambda_6 (E_6 S_2) = 0,$$
$$\ldots\ldots\ldots\ldots\ldots\ldots\ldots\ldots\ldots\ldots\ldots$$
$$\lambda_1 (E_1 S_\rho) + \lambda_2 (E_2 S_\rho) + \ldots + \lambda_6 (E_6 S_\rho) = 0.$$

Hence by § 168 (4) the group of reciprocal systems is of $(5 - \rho)$ dimensions, and is therefore defined by any $(6 - \rho)$ independent systems belonging to it.

(3) *Definition.* Let this group of reciprocal systems be called the group reciprocal to the given group; and let the two groups be called reciprocal.

It is to be noted that there is only one system reciprocal to a quintuple group; or in other words, the reciprocal group is of no dimensions.

170. COMMON NULL LINES AND DIRECTOR FORCES. (1) *Definition.* A line which is a null line of every system of a group is called a 'common null line of the group.'

It is obvious that if a line be a null line of ρ independent systems of a group of $(\rho - 1)$ dimensions, it is a common null line of the group.

Definition. Those systems of forces of a group which are simple, that is, which reduce to single forces, are called 'director forces of the group'; and the lines, on which they lie, are called 'director lines of the group.'

(2) Since the null lines of a system are the lines of forces reciprocal to the system, it follows that the common null lines of a group must be the director lines of the reciprocal group; and conversely.

(3) Let $S_1, S_2, \ldots S_\rho$ define a group of $\rho - 1$ dimensions, and let $S'_{\rho+1}$, $S'_{\rho+2}, \ldots S_6'$ define the reciprocal group.

Call the first group G, the second group G'.

Then if $\lambda_1 S_1 + \lambda_2 S_2 + \ldots + \lambda_\rho S_\rho$ be a director force of G, we must have $(\lambda_1 S_1 + \ldots + \lambda_\rho S_\rho)(\lambda_1 S_1 + \ldots + \lambda_\rho S_\rho) = 0$.

Hence $$\lambda_1^2 (S_1 S_1) + 2\lambda_1\lambda_2 (S_1 S_2) + \ldots + \lambda_\rho^2 (S_\rho S_\rho) = 0.$$

Let this equation be called the director equation of the group G.

If $\alpha_1 : \alpha_2 : \ldots : \alpha_\rho$ be a system of values of the ratios $\lambda_1 : \lambda_2 : \ldots : \lambda_\rho$ which satisfy this equation, then $\alpha_1 S_1 + \alpha_2 S_2 + \ldots + \alpha_\rho S_\rho$ is a director line of G and a null line of G'.

Similarly if $\lambda_{\rho+1} S'_{\rho+1} + \ldots + \lambda_6 S_6'$ be a director line of G', the λ's must satisfy the equation

$$\lambda^2_{\rho+1} (S'_{\rho+1} S'_{\rho+1}) + 2\lambda_{\rho+1}\lambda_{\rho+2} (S'_{\rho+1} S'_{\rho+2}) + \text{etc.} = 0;$$

and the director line of G' is a null line of G.

(4) A common null line of the group G is a null line of any one of its director forces F. But the null lines of a single force are the lines intersecting it. Accordingly each common null line of a group intersects all the director lines and conversely.

171. QUINTUPLE GROUPS. (1) Let a quintuple group be defined by the five systems S_1, S_2, S_3, S_4, S_5 and let S_6' be the system which forms the reciprocal group.

The director equation is

$$\lambda_1^2 (S_1 S_1) + 2\lambda_1\lambda_2 (S_1 S_2) + \ldots + \lambda_5^2 (S_5 S_5) = 0.$$

If $\alpha_1 : \alpha_2 : \alpha_3 : \alpha_4 : \alpha_5$ satisfies this equation, then $\alpha_1 S_1 + \alpha_2 S_2 + \alpha_3 S_3 + \alpha_4 S_4 + \alpha_5 S_5$ is a director line of the quintuple group; and accordingly is a null line of S_6'.

Hence the director lines of a quintuple group form a linear complex defined by the system S_6' [cf. § 163 (2)].

Thus conversely a linear complex may be said to be defined, not only by any five independent lines belonging to it [cf. § 165 (3)], but also by any five independent systems of the group reciprocal to S_6'.

(2) Also if $e_1e_2e_3e_4$ be the four co-ordinate points and any system S be denoted by $\pi_{12}e_1e_2 + \pi_{34}e_3e_4 + \pi_{13}e_1e_3 + \pi_{42}e_4e_2 + \pi_{14}e_1e_4 + \pi_{23}e_2e_3$, then a linear complex is defined by the two equations

$$\Sigma\alpha\pi = 0 \qquad (1),$$

and

$$\pi_{12}\pi_{34} + \pi_{13}\pi_{42} + \pi_{14}\pi_{23} = 0 \qquad (2),$$

where the α's are given coefficients.

For the first equation secures that the variable system S belong to a given quintuple group, and the second that it be a director force of the group. Then by subsection (1) the lines, on which these director forces lie, form a linear complex.

(3) The system reciprocal to the quintuple group given by equation (1) can easily be expressed. For let this equation be written at length in the form,

$$\alpha_{34}\pi_{12} + \alpha_{12}\pi_{34} + \alpha_{42}\pi_{13} + \alpha_{13}\pi_{42} + \alpha_{23}\pi_{14} + \alpha_{14}\pi_{23} = 0.$$

Then the system, $S_6' = \alpha_{12}e_1e_2 + \alpha_{34}e_3e_4 + \alpha_{13}e_1e_3 + \alpha_{42}e_4e_2 + \alpha_{14}e_1e_4 + \alpha_{23}e_2e_3$, is reciprocal to any system S, whose co-ordinates satisfy equation (1). Therefore S_6' is the required system. All the lines of the linear complex are null lines of S_6'.

(4) In general a quintuple group has no common null line. But if the reciprocal system reduce to a single force, then this line is the common null line of the group. The linear complex is in this case called a special linear complex. It consists of the assemblage of lines which intersect the line of the reciprocal force.

172. Quadruple and Dual Groups. (1) Let S_1 and S_2 define a dual group and S_3', S_4', S_5', S_6' the reciprocal quadruple group. Let the dual group be called G and the quadruple group G'.

The director equation of G is

$$\lambda_1^2(S_1S_1) + 2\lambda_1\lambda_2(S_1S_2) + \lambda_2^2(S_2S_2) = 0.$$

This equation is a quadratic in λ_1/λ_2, and has in general two roots, real or imaginary. Let α_1/α_2 and β_1/β_2 be the roots, assumed unequal [cf. subsection (9) below].

Then $\alpha_1S_1 + \alpha_2S_2$ and $\beta_1S_1 + \beta_2S_2$ are the only two director forces of the dual group G.

Thus a dual group has in general two and only two director forces; and a quadruple group has two and only two common null lines.

Another statement of this proposition is that two systems of forces have one and only one common pair of conjugate lines.

(2) Also the common null lines of a dual group are the lines intersecting the two director lines of the group; and the director lines of a quadruple group are the lines intersecting the two common null lines of the group.

(3) *Definition.* The assemblage of common null lines of a dual group is called the 'congruence' defined by the group.

Thus the lines of a congruence are lines intersecting two given lines. The lines indicated by the director equation of the group G', namely

$$\lambda_3^2 (S_3'S_3') + 2\lambda_3\lambda_4 (S_3'S_4') + \ldots + \lambda_6^2 (S_6'S_6') = 0,$$

form the congruence defined by the group G.

(4) Through any point one and only one line of a congruence can in general be drawn.

To find the line through any point x of the congruence defined by the group G, notice that it must lie in the null planes of x with respect to any two systems S_1 and S_2 of the group. Hence $xS_1 . xS_2$ is the common null line through x.

Similarly in any plane X one and only one line of the congruence lies. This line is $XS_1 . XS_2$.

(5) The equation, $xS_1 . xS_2 = 0$, implies that x is on one of the two director lines of G.

For if a_1a_2 and b_1b_2 are the director lines, and $S_1 = \lambda_1 a_1a_2 + \mu_1 b_1b_2$, $S_2 = \lambda_2 a_1a_2 + \mu_2 b_1b_2$, then $xS_1 . xS_2 = (\lambda_1\mu_2 - \lambda_2\mu_1)\, xa_1a_2 . xb_1b_2$.

Hence, assuming that the director lines are not co-planar, either $xa_1a_2 = 0$, or $xb_1b_2 = 0$.

Similarly the equation, $XS_1 . XS_2 = 0$, implies that the plane X contains one of the director lines.

If $xS_1 . xS_2 = 0$, and $XS_1 . XS_2 = 0$, then the theorems of subsection (4) do not hold.

(6) If the congruence be defined as the assemblage of the director lines of the quadruple group G', the line belonging to it which lies in any plane or passes through any point can be determined thus:

Lemma. If L denote a single force the two equations, $(abL) = 0$, $(bcL) = 0$, imply the equation $(caL) = 0$ and that L lies in the plane abc. But if L denote a system which is not a single force then the three equations cannot coexist. For the equations $(abL) = 0$ and $(bcL) = 0$ imply that b is the null point of the plane abc with respect to L. Hence ca cannot be a null line (assuming that abc is not zero), unless L represent a single force lying in the plane abc.

Now let abc represent any given plane, and let $\lambda_3 S_3' + \lambda_4 S_4' + \lambda_5 S_5' + \lambda_6 S_6'$ represent any system of the group G'. Then it follows from the Lemma that the three equations,

$$\lambda_3 (bcS_3') + \lambda_4 (bcS_4') + \lambda_5 (bcS_5') + \lambda_6 (bcS_6') = 0,$$
$$\lambda_3 (caS_3') + \lambda_4 (caS_4') + \lambda_5 (caS_5') + \lambda_6 (caS_6') = 0,$$
$$\lambda_3 (abS_3') + \lambda_4 (abS_4') + \lambda_5 (abS_5') + \lambda_6 (abS_6') = 0,$$

are the three conditions that this system may represent the director line in the plane abc.

Hence the system of the group G' which can be written in the form

$$\left|\begin{matrix} S_3', & S_4', & S_5', & S_6' \\ (bcS_3'), & (bcS_4'), & (bcS_5'), & (bcS_6') \\ (caS_3'), & (caS_4'), & (caS_5'), & (caS_6') \\ (abS_3'), & (abS_4'), & (abS_5'), & (abS_6') \end{matrix}\right|$$

is the director force of the group which lies in the plane abc.

(7) Similarly the line of the congruence, which passes through any point ABC, where A, B, C are planes, is found by substituting A, B, C for a, b, c respectively in the above expression.

(8) Again, if the plane abc contain one of the two common null lines of G', then every line lying in it and passing through its point of intersection with the other common null line must be a director line.

Hence the above expression for the single director line lying in the plane abc must be nugatory.

Accordingly the conditions, that the plane abc may contain one of the two common null lines of G', are

$$\left\|\begin{matrix} (bcS_3'), & (bcS_4'), & (bcS_5'), & (bcS_6') \\ (caS_3'), & (caS_4'), & (caS_5'), & (caS_6') \\ (abS_3'), & (abS_4'), & (abS_5'), & (abS_6') \end{matrix}\right\| = 0.$$

Similarly the conditions, that the point ABC may lie on one of the common null lines, is found by replacing the points a, b, c by the planes A, B, C in the above conditions.

(9) An exceptional type of dual group arises, when the director equation has two equal roots. In this case, with the notation of subsection (1), if S_1 and S_2 be any two systems of the group,

$$(S_1S_1)(S_2S_2) = (S_1S_2)^2.$$

A group of this type will be called a parabolic group.

There is only one director force in the group. Let it be D, and substitute D for S_1 in the above equation. Then, since $(DD) = 0$, the equation reduces to $(DS_2) = 0$. Hence the director line is a common null line of all the other systems of the group; in other words, the director force is reciprocal to every other system of the group.

The null plane of a point on the director line is the same for each system of the group, and contains the director line. For, if S be any system of the group and D the director force, any other system of the group can be written $\lambda D + \mu S$. Hence, if x be any point on the line D,

$$x(\lambda D + \mu S) = \mu xS \equiv xS.$$

Since the director line is a common null line of the group, the plane xS contains the director line.

Similarly the null point of a plane containing the director line is the same for each system of the group, and lies on the director line.

The theorems of subsection (4) still hold. For, if x be any point not on the director line, the common null lines of the group through x must intersect the director force D; and therefore must pass through the common null point of the plane xD. Hence there is only one such line through x, and there is always one such line. Also, if S_1 and S_2 be any two systems of the group, the common null line through x is $xS_1 . xS_2$.

The theorem of subsection (5) still holds. For, if ab be the director force, any system of the group can be written in the form $ac + bd$.

Now $$xab . x(ac + bd) = (xabc)\, xa + (xabd)\, xb.$$

Hence, $xab . x(ac + bd) = 0$, implies $(xabc) = 0 = (xabd)$. Therefore x must lie on the line ab.

Now, if $S_1 = ac + bd$, any other system S_2 of the group can be written in the form $\lambda ab + \mu S_1$.

Hence $$xS_1 . xS_2 = xS_1 . x(\lambda ab + \mu S_1) = \lambda xS_1 . xab.$$

Now λ is not zero, if S_2 be different from S_1. Hence, $xS_1 . xS_2 = 0$, implies, $xab . xS_1 = 0$.

(10) If N be any line not intersecting the director force D of a parabolic group, then one and only one system of the group can be found for which N is a null line.

For let S be any system of the group. Then $\lambda D + \mu S$ is any other system. If N is a null line of this system

$$\lambda (ND) + \mu (NS) = 0.$$

Now by hypothesis (ND) is not zero. Hence the system $D(NS) - S(ND)$ has N for a null line. And no other system has N for a null line.

If $D = e_1e_2$, and $N = e_3e_4$, then the conjugate with respect to

$$D(NS) - S(ND)$$

of the line e_1e_3 must intersect both D and N. Hence $D(NS) - S(ND)$ can be written in the form

$$\lambda e_1e_3 + \mu ab,$$

where e_1e_3 is any given line intersecting D and N, and a lies on D and b on N.

173. Anharmonic Ratio of Systems. (1) The null points of any given plane with respect to the systems of a dual group are collinear. For let the two systems S_1 and S_2 define the group, and let S be any third system of the group. Also let A be any plane.

Then $S = \lambda_1 S_1 + \lambda_2 S_2$, also the null point of A with respect to S is $AS = \lambda_1 AS_1 + \lambda_2 AS_2$. Hence AS, AS_1, AS_2 are collinear.

(2) The anharmonic ratio of the four null points of any plane with respect to four systems of a dual group is the same for all planes and depends only on the four systems. For let S_1, S_2, $\lambda_1 S_1 + \lambda_2 S_2$, $\mu_1 S_1 + \mu_2 S_2$ be the four systems. The four null points of any plane A are AS_1, AS_2, $\lambda_1 AS_1 + \lambda_2 AS_2$, $\mu_1 AS_1 + \mu_2 AS_2$. The anharmonic ratio of these four points, taking the first two and the last two as conjugates, is $\lambda_1\mu_2/\lambda_2\mu_1$. This ratio is independent of A.

(3) Similarly the four null planes of any point a with respect to the four systems have the same line of intersection, and their anharmonic ratio is also $\lambda_1\mu_2/\lambda_2\mu_1$.

(4) *Definitions.* Let this ratio be called the anharmonic ratio of the four systems. If the anharmonic ratio be -1, the four systems are said to be harmonic; and one pair are harmonic conjugates to the other pair. Pairs of systems, harmonically conjugate to the two systems S_1 and S_2, are said to form an involution, of which S_1 and S_2 are the foci.

The anharmonic ratio of the four systems $\lambda_1 S_1 + \lambda_2 S_2$, $\lambda_1' S_1 + \lambda_2' S_2$, $\lambda_1'' S_1 + \lambda_2'' S_2$, $\lambda_1''' S_1 + \lambda_2''' S_2$ is

$$(\lambda_1\lambda_2'' - \lambda_2\lambda_1'')(\lambda_1'\lambda_2''' - \lambda_2'\lambda_1''')/(\lambda_1\lambda_2''' - \lambda_2\lambda_1''')(\lambda_1'\lambda_2'' - \lambda_2'\lambda_1'').$$

(5) There is one and only one system belonging to a dual group which is reciprocal to a given system of the group. For if $\alpha_1 S_1 + \alpha_2 S_2$ be any given system, and $\lambda_1 S_1 + \lambda_2 S_2$ a system of the dual group reciprocal to it, then

$$\lambda_1 \{\alpha_1 (S_1 S_1) + \alpha_2 (S_1 S_2)\} + \lambda_2 \{\alpha_1 (S_1 S_2) + \alpha_2 (S_2 S_2)\} = 0.$$

And this equation determines $\lambda_1 : \lambda_2$ uniquely. Thus a dual group can be divided into pairs of reciprocal systems. Each director force is its own reciprocal system.

But if the group be parabolic [cf. § 172 (9)], the director force is the only system of the group reciprocal to any of the other systems. For, if S be any system and D the director force, any other system can be written $\lambda D + \mu S$. If this system be reciprocal to S, $\lambda (DS) + \mu (SS) = 0$. But $(DS) = 0$, and (SS) is not zero. Hence $\mu = 0$.

(6) A pair of reciprocal systems of a dual group are harmonic conjugates to the two director forces of the group.

For let D_1 and D_2 be the two director forces, and $\lambda_1 D_1 + \lambda_2 D_2$ and $\mu_1 D_1 + \mu_2 D_2$ be the two reciprocal systems.

Then $$(\lambda_1\mu_2 + \lambda_2\mu_1)(D_1 D_2) = 0.$$

Hence (assuming that the director lines do not intersect),

$$\lambda_1/\lambda_2 = -\mu_1/\mu_2.$$

The two reciprocal systems can therefore be written $\lambda_1 D_1 + \lambda_2 D_2$, $\lambda_1 D_1 - \lambda_2 D_2$, and are harmonic conjugates to D_1 and D_2.

(7) Hence systems S_1, S_2, S_3, etc., belonging to one dual group form an assemblage of systems in involution with their reciprocal systems S_1', S_2', S_3', etc., belonging to the same dual group. The foci of the involution are the director forces.

The dual group will be called elliptic or hyperbolic according as these foci are imaginary or real.

(8) Since a single system uniquely defines a linear complex, we can also speak of the anharmonic ratio of four linear complexes which have the same congruence in common. An assemblage of complexes with the same congruence in common contains two and only two special complexes. These are the foci of an involution in which each complex corresponds to its reciprocal complex, that is, to the complex of the assemblage which is defined by a system reciprocal to its own.

These theorems respecting linear complexes are merely other statements of the theorems proved above.

174. Self-Supplementary Dual Groups. (1) Let the operation of taking the supplement be assumed to refer to any given quadric.

The system $|S$ will be called the supplementary system of S, where S is any system. Also S and $|S$ define a dual group. This dual group has the property that the supplement of any system belonging to it also belongs to the group.

For if $S' = \lambda S + \mu\,|S$, then $|S' = \lambda\,|S + \mu S$. Let the group be called 'self-supplementary.'

(2) A self-supplementary group is obviously in general determined by any one system belonging to it. For if S be known, S and $|S$ in general determine the group.

But if S' be of the form $\lambda S \pm \lambda\,|S$, then $|S' = \pm S'$, hence S' and $|S'$ do not determine the group. A system S', such that $|S' = \pm S'$, is called a self-supplementary system.

(3) If two generators of the same system of any quadric are conjugate lines with respect to any system of forces, then the generators of that system of generators taken in pairs are all conjugate lines with respect to that system of forces. Let S be the system of forces, D_1 and D_2 the two generators which are conjugate with respect to S; and let G be any third generator of the same system of generators. We require to prove that $\left(S - \frac{1}{2}\frac{(SS)}{(GS)}G\right)$ is also a generator of the quadric.

Now let the operation of taking supplements be performed in reference to this quadric. Then $|D_1 = \pm D_1$, and $|D_2 = \pm D_2$, where both the upper signs or both the under signs are to be taken [cf. § 116 (3)]. Hence since S can be written $\lambda_1 D_1 + \lambda_2 D_2$, we have $|S = \pm S$. Also $|G = \pm G$.

Therefore $$|\left(S-\frac{1}{2}\frac{(SS)}{(GS)}G\right)=|S-\frac{1}{2}\frac{(SS)}{(GS)}|G$$
$$=\pm\left(S-\frac{1}{2}\frac{(SS)}{(GS)}G\right).$$

Accordingly the conjugate of G is a generator [cf. § 116 (3)].

(4) Conversely it is obvious that if S be self-supplementary, that is, if $|S=\pm S$, then the conjugate of any generator G belonging to one of the two systems of the self-normal quadric is another generator of the same system of generators as G.

For $$|\left(S-\frac{1}{2}\frac{(SS)}{(GS)}G\right)=\pm\left(S-\frac{1}{2}\frac{(SS)}{(GS)}G\right).$$

It is obvious that if $|S=S$, the generator must be of the positive system; if $|S=-S$, the generator must be of the negative system.

(5) In general the director lines of a self-supplementary group are supplementary to each other.

For if the group be defined by S and $|S$, the director equation is
$$(\lambda^2+\mu^2)(SS)+2\lambda\mu(S|S)=0.$$

Let the roots of this equation be α/β and β/α, then the director forces D_1 and D_2 are $D_1=\alpha S+\beta|S$, $D_2=\beta S+\alpha|S$. Hence $|D_1=D_2$, and $|D_2=D_1$.

(6) But if we choose two director forces so that each lies on the self-normal quadric, that is, so that $|D_1=\pm D_1$, and $|D_2=\pm D_2$ (making the same choice of both ambiguities), then any system $S=\lambda D_1+\mu D_2$ belonging to the group is self-supplementary. Hence these exceptional groups cannot be defined by two systems of the form S and $|S$. Therefore the above reasoning fails.

Also if $(SS)=\pm(S|S)$, the roots of the director equation are equal; and the group is parabolic [cf. § 172 (9)]. If $(SS)=(S|S)$, the director force is $S-|S$, and is self-supplementary, and belongs to the negative system of generating lines: if $(SS)=-(S|S)$, the director force is $S+|S$, and belongs to the positive system. This is the most general type of self-supplementary parabolic group, in which each system is not self-supplementary.

(7) In general there is one and only one self-supplementary system of each type (positive and negative) in each self-supplementary dual group.

For if the group be defined by S and $|S$, where S is any system, or by S' and $|S'$, where $|S'$ is any other system of the group, then any two pairs of self-supplementary systems of the two types belonging to the group are $S\pm|S$, and $S'\pm|S'$.

But if $S'=\lambda S+\mu|S$, then $|S'=\lambda|S+\mu S$; and hence
$$S'\pm|S'=(\lambda\pm\mu)(S\pm|S)\equiv S\pm|S.$$

Thus all such pairs of systems are identical.

(8) Any system S which is not self-supplementary has in general two and only two conjugate lines which are supplementary. The system obviously has one pair of such conjugate lines, namely, the director lines of the group S and $|S$. It has no more, for if possible let D and $|D$ be two such lines which are not the director lines of the group S, $|S$.

Then $$S = \lambda D + \mu|D;$$

hence $$|S = \lambda|D + \mu D.$$

Accordingly D and $|D$ must be director lines of the group S, $|S$, which by hypothesis is not the case.

This proposition does not hold, if the group $(S, |S)$ be parabolic.

(9) This proposition may also be stated thus: Any system has in general one and only one pair of conjugate lines which are polar reciprocal to each other with reference to a given quadric.

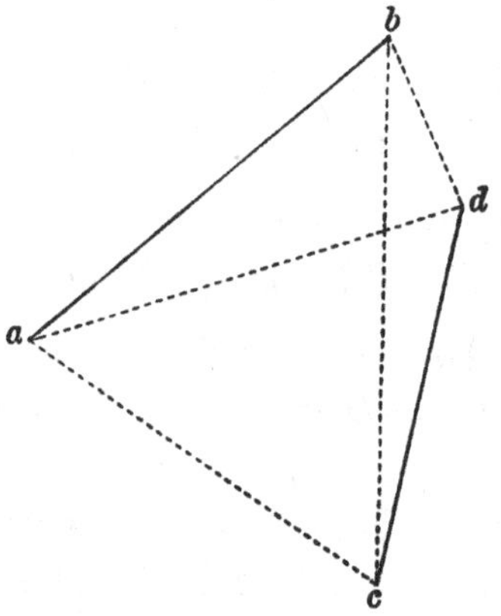

Let ab and cd be this pair of conjugates for any system S. Let ab and cd meet the quadric in a, b and c, d. Then ad, ac and bd, bc are generating lines of the quadric. But these lines are also null lines of the system S.

Hence in general [cf. § 175 (12) and (13)] any linear* complex has four lines which are generators of any given quadric, two belonging to one system of generators and two belonging to the other system.

(10) The proposition can easily be extended to self-supplementary systems with respect to the given quadric. For if S be any system, then $S \pm |S$ is the general type of a self-supplementary system. But the director lines of the group S and $|S$ are supplementary, and they are conjugate lines of $S \pm |S$ which belong to the dual group.

The discussion of self-supplementary systems, and of systems such that $(SS) \pm (S|S) = 0$, is resumed in § 175 (8) to (13).

* Cf. Clebsch and Lindemann, *Vorlesungen über Geometrie*, vol. II.

175. Triple Groups. (1) The reciprocal group of a triple group is another triple group. Let S_1, S_2, S_3 define any triple group G, and let S_4', S_5', S_6' define the reciprocal group G'. The director equation of G, namely, the condition that $\lambda_1 S_1 + \lambda_2 S_2 + \lambda_3 S_3$ reduce to a single force, is

$$\lambda_1^2 (S_1 S_1) + \lambda_2^2 (S_2 S_2) + \lambda_3^2 (S_3 S_3) + 2\lambda_1\lambda_2 (S_1 S_2) + 2\lambda_2\lambda_3 (S_2 S_3) + 2\lambda_3\lambda_1 (S_3 S_1) = 0.$$

This equation is also the condition that the line $\lambda_1 S_1 + \lambda_2 S_2 + \lambda_3 S_3$ be a common null line of the group G'.

(2) The condition that x may lie on a common null line of G is,

$$(xS_1 . xS_2 . S_3) = 0.$$

For $xS_1 . xS_2$ is a common null line of S_1 and S_2, and the given condition secures that it be also a null line of S_3.

(3) But the equation, $(xS_1 . xS_2 . S_3) = 0$, is the equation of a quadric surface.

Hence the common null lines of a triple group G are generators of a quadric surface. The director lines therefore, which are null lines of the triple group G', must also be generators of a quadric surface. Furthermore every null line intersects every director line, and conversely. Thus it follows that the quadric surfaces, on which the null lines of G and of G' lie, must be the same surface; and that the null lines of G are generators of one system on the surface, and the director lines of G (i.e. the null lines of G') are generators of the other system on the surface. Let the two systems of generators be called respectively the null system and the director system with respect to the given group.

(4) Hence a triple group G defines a quadric surface. The only other triple group which defines the same surface is the reciprocal group G'. The director system of generators with respect to G is the null system with respect to G', and *vice versa*.

(5) Conversely, any quadric surface defines a pair of reciprocal triple groups.

For take any three generators of the same system belonging to this quadric. Let G_1, G_2, G_3 be forces along them. Then G_1, G_2, G_3 define a triple group, and its associated quadric must contain the three lines G_1, G_2, G_3. But there is only one quadric which contains three given lines. Hence the associated quadric is the given quadric.

(6) The condition that the plane abc may contain a director line of the group G is

$$\begin{vmatrix} (bcS_1), & (bcS_2), & (bcS_3) \\ (caS_1), & (caS_2), & (caS_3) \\ (abS_1), & (abS_2), & (abS_3) \end{vmatrix} = 0.$$

For assume that $\lambda_1 S_1 + \lambda_2 S_2 + \lambda_3 S_3$ is a single force lying in the plane abc. Then, by the lemma of § 172 (6), the three following equations are the necessary and sufficient conditions,

$$\lambda_1 (bcS_1) + \lambda_2 (bcS_2) + \lambda_3 (bcS_3) = 0,$$
$$\lambda_1 (caS_1) + \lambda_2 (caS_2) + \lambda_3 (caS_3) = 0,$$
$$\lambda_1 (abS_1) + \lambda_2 (abS_2) + \lambda_3 (abS_3) = 0.$$

But these equations require the given condition.

Accordingly this is also the condition that abc may touch the associated quadric and contain a common null line of the group.

(7) Similarly the condition that the point ABC, where A, B, C are planar elements, may lie on the associated quadric is found by replacing a, b, c in the above condition by A, B and C.

(8) If the supplements of G and G' be taken with respect to the associated quadric, then from § 174 (3) and (4) every system belonging to G or G' is self-supplementary; and conversely all self-supplementary systems with respect to a given quadric must belong to one of the two associated groups of the quadric.

For any system S of group G we may assume $|S = S$; then for any system S' of G' we have $|S' = -S'$.

(9) Corresponding to each director line of a triple group, one parabolic dual subgroup can be found with that line as director line.

For let F_1, F_2, F_3 be any three director lines of the triple group, and let F_1 be the given director line. Then any system S of the triple group can be written

$$S = \lambda_1 F_1 + \lambda_2 F_2 + \lambda_3 F_3.$$

Now, if the subgroup (F_1, S) is parabolic, $(F_1 S) = 0$. Hence the required condition is

$$\lambda_2 (F_1 F_2) + \lambda_3 (F_1 F_3) = 0.$$

Thus the subgroup defined by F_1 and $(F_1 F_3) F_2 - (F_1 F_2) F_3$ is parabolic with F_1 as director line.

(10) Let the quadric defined by the triple group be self-supplementary. Hence by the previous subsection, if $S_1 = (F_1 F_3) F_2 - (F_1 F_2) F_3$, the dual group defined by F_1 and S_1 is parabolic and such that each system S is self-supplementary. If $|F_1 = F_1$, then $|S = S$; and if $|F_1 = -F_1$, $|S = -S$. Corresponding to each generator of either system there is one such parabolic self-supplementary dual group [cf. subsection (12), below].

(11) The most general type of self-supplementary parabolic group, in which each system is not self-supplementary, is the type defined by a generator, G, of the self-normal quadric and a self-supplementary system S; such that, either $|G = G$, and $|S = -S$, or, $|G = -G$, and $|S = S$ [cf. § 174 (7)].

For firstly let G, the director line of the parabolic subgroup, be such that $|G=G$; and let S_1 be any other system of the group. Then by § 174 (6)

$$|S_1 = \alpha G - S_1.$$

Also by hypothesis, $(GS_1)=0$.

Then any system S of the group can be written $\lambda G + \mu S_1$.

Hence $$|S = \lambda\, |G + \mu\, |S = (\lambda + \mu\alpha)\, G - \mu S_1.$$

Thus, if S be self-supplementary, that is, if $|S \equiv S$, then, $\lambda + \mu\alpha = -\lambda$; that is, $\lambda = -\frac{1}{2}\mu\alpha$.

Hence the system $S = S_1 - \frac{1}{2}\alpha G$, is such that $|S = -S$.

Accordingly the self-supplementary parabolic group can be defined by G and S; where $|G = G$, $|S = -S$.

Similarly if $|G = -G$, then the self-supplementary system S belonging to the group is such that $|S = S$.

Thus corresponding to any generator G of the self-normal quadric there are an infinite number of such parabolic self-supplementary groups, since any self-supplementary system S of the opposite denomination (positive or negative) to G will with G define such a group.

(12) It is evident [cf. § 174 (3) and (4)] that any self-supplementary system S has as null lines all the generators of the self-normal quadric of the opposite denomination. It also has as null lines two generators of the same denomination.

For we may write $S = \alpha_1 D_1 + \alpha_2 D_2$, where D_1 and D_2 are two generators of the same denomination as S. Let D_3 be a third such generator. Then any self-normal system of the same denomination as S can be written in the form

$$\lambda_1 D_1 + \lambda_2 D_2 + \lambda_3 D_3.$$

This system is a generator (D) if

$$\lambda_2\lambda_3 (D_2D_3) + \lambda_3\lambda_1 (D_3D_1) + \lambda_1\lambda_2 (D_1D_2) = 0.$$

Also D is a null line of S, if $(DS)=0$, that is, if

$$(\lambda_1\alpha_1 + \lambda_2\alpha_2)(D_1D_2) + \lambda_3 \{\alpha_1 (D_1D_3) + \alpha_2 (D_2D_3)\} = 0.$$

These two equations give two solutions for the set of ratios of λ_1 to λ_2 to λ_3. Hence S has two null lines among the generators of the same denomination [cf. subsection (10) above, and also § 174 (10)].

(13) Now let N be a generator of the opposite denomination to the self-supplementary system S; and let D and D' be the two generators of the same denomination as S, which are null lines of S according to the previous subsection.

Then D and D' necessarily intersect N. Also the parabolic group defined by N and S is of the type discussed in subsection (11). But D and D' and N must be common nulls of this group. Also no other generators of the quadric can be null lines of any system of the group, other than N and S. For consider the system $\lambda N + \mu S$. Then every generator of the D type intersects N, but only D and D' are null lines of S. Accordingly only D and D' of the generators of this type are null lines of $\lambda N + \mu S$. Again, all the generators of the N type are null lines of S; but no generator of this type, except N, intersects N. Hence N is the only null line belonging to the generators of this type.

Hence any system S', not self-supplementary, which is such that $(S'S') \pm (S' | S') = 0$, has two generators of one system and one generator of the other system as null lines. This proposition should be compared with that of § 174 (9).

(14) Thus, summing up and repeating, any quadric has in general two generators only of one system and two generators only of the other system, which are null lines of any system of forces S. But, as exceptional cases, either all the generators of one system and two only of the other system are null lines of S; or one generator only of one system and two only of the other system are null lines of S.

176. Conjugate sets of Systems in a Triple Group. (1) Any two systems S_1, S_2 of the triple group G define a subgroup. It is possible to find one and only one system S belonging to G which is reciprocal to the whole subgroup S_1, S_2.

For let S be such a system and let S_3 be any third independent system so that S_1, S_2, S_3 define G. Then we may write $S = \lambda_1 S_1 + \lambda_2 S_2 + \lambda_3 S_3$.

Hence by hypothesis

$$\lambda_1 (S_1 S_1) + \lambda_2 (S_1 S_2) + \lambda_3 (S_1 S_3) = 0,$$

$$\lambda_1 (S_2 S_1) + \lambda_2 (S_2 S_2) + \lambda_3 (S_2 S_3) = 0.$$

Thus the ratios $\lambda_1 : \lambda_2 : \lambda_3$ are completely determined, and therefore S is completely determined. The reciprocal system S can be written in the form

$$\begin{vmatrix} S_1, & S_2, & S_3 \\ (S_1 S_1), & (S_1 S_2), & (S_1 S_3) \\ (S_1 S_2), & (S_2 S_2), & (S_2 S_3) \end{vmatrix}.$$

This system does not belong to the dual subgroup (S_1, S_2), it the coefficient of S_3 does not vanish; that is, if

$$(S_1 S_1)(S_2 S_2) - (S_1 S_2)^2$$

be not zero; that is, if the subgroup (S_1, S_2) be not parabolic. In subsections (2), (3), (4), following, the subgroups will be assumed to be not parabolic.

(2) Also in the subgroup defined by S_1, S_2 we may choose S_1 and S_2 so as to be reciprocal [cf. § 173 (5)]. Thus three systems S_1, S_2, S_3 can be found, belonging to the triple group G, such that each system is reciprocal

to the subgroup formed by the other two. And one of these systems, say S_1, can be chosen arbitrarily out of the systems of the group G; and then S_2 and S_3 can be chosen in a singly-infinite number of ways out of the dual subgroup of G which is reciprocal to S_1.

Definition. Let such a set of three mutually reciprocal systems of a group G be called a 'conjugate' set of the group.

(3) If S_1, S_2, S_3 be a conjugate set of systems, then XS_1, XS_2, XS_3 are three conjugate points lying in the plane X with respect to the associated quadric of G.

For the director lines of the group (S_1, S_2) are generators of the quadric G; and it has been proved [cf. § 173 (6)] that the line joining the points XS_1 and XS_2 intersects these director lines in two points d_1 and d_2 such that the range formed by (d_1, d_2, XS_1, XS_2) is harmonic. But d_1 and d_2 are on the quadric G. Hence by the harmonic properties of poles and polars XS_1 is on the polar of XS_2, and XS_2 on the polar of XS_1.

Similarly for XS_2 and XS_3, and for XS_1 and XS_3. Hence the three points XS_1, XS_2, XS_3 are three mutually conjugate points on the plane X.

(4) An analogous proof shows that xS_1, xS_2, xS_3 are conjugate planes through the point x.

CHAPTER III.

INVARIANTS OF GROUPS.

177. DEFINITION OF AN INVARIANT. (1) Let $S_1, S_2, \dots S_\rho$ define a group G of $\rho - 1$ dimensions, and let $S_1', S_2', \dots S_\rho'$ be any ρ systems belonging to this group G. Then there must exist ρ equations of the typical form

$$S_\mu' = \lambda_{\mu 1} S_1 + \lambda_{\mu 2} S_2 + \dots + \lambda_{\mu\rho} S_\rho.$$

Also let Δ denote the determinant

$$\begin{vmatrix} \lambda_{11}, & \lambda_{12}, & \dots & \lambda_{1\rho} \\ \lambda_{21}, & \lambda_{22}, & \dots & \lambda_{2\rho} \\ \dots & \dots & \dots & \dots \\ \lambda_{\rho 1}, & \lambda_{\rho 2}, & \dots & \lambda_{\rho\rho} \end{vmatrix}.$$

Then, if Δ be not zero, the systems $S_1', S_2', \dots S_\rho'$ are independent [cf. § 96 and § 63 (4)] systems.

(2) Let $\phi(S_1, S_2, \dots S_\rho)$ be any function of the ρ systems $S_1, S_2, \dots S_\rho$ formed by multiplications and additions of $S_1, S_2, \dots S_\rho$ and of given points, forces, and planar elements. Let $\phi(S_1', S_2', \dots S_\rho')$ denote the same function only with $S_1', S_2', \dots S_\rho'$ substituted respectively for $S_1, S_2, \dots S_\rho$.

Then if $\phi(S_1', S_2', \dots S_\rho') = \Delta^\lambda \phi(S_1, S_2, \dots S_\rho)$, λ being an integer, $\phi(S_1, S_2, \dots S_\rho)$ is called an invariant of the group G.

The effect of substituting any other ρ independent systems of the group G for $S_1, S_2, \dots S_\rho$ in an invariant of the group is to reproduce the original function multiplied by a numerical factor which does not vanish.

178. THE NULL INVARIANTS OF A DUAL GROUP. (1) Let S_1 and S_2 define a dual group, and let $S = \lambda S_1 + \mu S_2$, $S' = \lambda' S_1 + \mu' S_2$, $\Delta = \lambda\mu' - \lambda'\mu$. Then the expressions $xS_1 . xS_2$ and $XS_1 . XS_2$, where x is any point and X is any planar element, are invariants of the group. Call them the Null Invariants.

For $\quad xS . xS' = \Delta xS_1 . xS_2$, and $XS . XS' = \Delta XS_1 . XS_2$.

It has already been proved [cf. § 172 (4)] that these expressions denote respectively the common null line of the group through the point x, and the common null line of the group in the plane X.

179. THE HARMONIC INVARIANTS OF A DUAL GROUP. (1) Another important invariant of the group is $xS_1 . S_2 - xS_2 . S_1$. Call this expression the Harmonic Point Invariant of the group; let it be denoted by $H(x)$.

This expression is easily proved to be an invariant by direct substitution. It represents a point. It must be noticed that the intensity to be ascribed to $H(x)$ depends on the special pair of systems (S_1, S_2) which is chosen to define the group.

It is obvious that $H(\lambda x + \mu x') = \lambda H(x) + \mu H(x')$.

(2) Similarly if X be any planar element, $XS_1 . S_2 - XS_2 . S_1$ is an invariant of the group. Call this expression the Harmonic Plane Invariant; and let it be denoted by $H(X)$. It represents a planar element. The intensity of $H(X)$ depends on the special pair of systems which define it.

Also $$H(\lambda X + \mu X') = \lambda H(X) + \mu H(X').$$

(3) If S_1 and S_2 be a pair of reciprocal systems, it is obvious from § 167 (5), equation 25, that

$$xS_1 . S_2 + xS_2 . S_1 = 0.$$

Hence in this case $H(x) = 2xS_1 . S_2 = -2xS_2 . S_1$.

Similarly $H(X) = 2XS_1 . S_2 = -2XS_2 . S_1$.

These expressions only hold when S_1 and S_2 are reciprocal.

(4) To find the relation between the points x and $H(x)$, and between the planes X and $H(X)$.

Let the common null line through x meet the director lines of the group in d_1 and d_2; and let the two director lines be written d_1e_1 and d_2e_2.

Then S_1 and S_2, which will be assumed to be reciprocal systems, can be written in the forms [cf. § 173 (6)]

$$S_1 = d_1e_1 + d_2e_2,\quad S_2 = \lambda(d_1e_1 - d_2e_2).$$

Also we may write $x = \xi_1 d_1 + \xi_2 d_2$.

Hence by multiplication $xS_1 = \xi_1 d_1 d_2 e_2 + \xi_2 d_2 d_1 e_1$,

$$xS_1 . S_2 = \lambda\xi_1 d_1 d_2 e_2 . d_1 e_1 - \lambda\xi_2 d_2 d_1 e_1 . d_2 e_2 = \lambda(d_1e_1d_2e_2)\{\xi_1 d_1 - \xi_2 d_2\}.$$

Also $$(S_1S_1) = 2(d_1e_1d_2e_2),\quad (S_2S_2) = -2\lambda^2(d_1e_1d_2e_2).$$

Therefore $H(x) = 2xS_1 . S_2 = \sqrt{\{-(S_1S_1)(S_2S_2)\}}(\xi_1 d_1 - \xi_2 d_2)$.

But $\xi_1 d_1 + \xi_2 d_2$, $\xi_1 d_1 - \xi_2 d_2$, d_1, d_2 form a harmonic range. Hence $H(x)$ lies on the common null line of the group through x, and is the harmonic conjugate of x with respect to the two points in which the null line meets the director lines.

(5) Similarly $H(X) = 2XS_1 . S_2 = \sqrt{\{-(S_1S_1)(S_2S_2)\}}(\xi_1 D_1 - \xi_2 D_2)$; where D_1 and D_2 are two planes both containing the common null line in the plane X, and respectively containing the two director lines; and $X = \xi_1 D_1 + \xi_2 D_2$.

Hence $H(X)$ contains the common null line of the group which lies in X, and is the harmonic conjugate of X with reference to the two planes containing the null line and the two director lines.

(6) Let $H\{H(x)\}$ be written $H^2(x)$, and let $H^3(x)$ denote $H\{H^2(x)\}$, and so on.

Then it has been proved in (4) and (5) that if d_1 and d_2 lie on the director lines of the group, and $x = \xi_1 d_1 + \xi_2 d_2$,

$$H(x) = \sqrt{\{-(S_1 S_1)(S_2 S_2)\}}\,(\xi_1 d_1 - \xi_2 d_2).$$

It follows that $H^2(x) = -(S_1 S_1)(S_2 S_2)(\xi_1 d_1 + \xi_2 d_2) = -(S_1 S_1)(S_2 S_2)\,x$.

Therefore $H^2(x) \equiv x$, and generally $H^\lambda(x) \equiv x$, or $\equiv H(x)$, according as λ is an even or an odd integer.

Similarly $H^2(X) = -(S_1 S_1)(S_2 S_2)\,X$; and hence $H^2(X) \equiv X$.

(7) If the group be parabolic [cf. § 172 (9)], then $H(x)$ is the null point (common to all the systems) of the plane through x and the single director line. For let D be the director force and S any other system of the group, then $(DS) = 0$.

Hence by subsection (3), $\quad H(x) = 2xD\,.\,S.$

Thus $H(x)$ is the null point of the plane xD with respect to S.

Accordingly all the points of the type $H(x)$ are concentrated on the director line; and, if $(xyD) = 0$, then $H(x) \equiv H(y)$.

Similarly $H(X)$ is the null plane of the point DX.

180. FURTHER PROPERTIES OF HARMONIC INVARIANTS. (1) If S_1 and S_2 are two reciprocal systems of the group, the null plane of x with respect to S_1 is the same as the null plane of $H(x)$ with respect to S_2. For by § 167 (2), since xS_1 and xS_2 are planar elements,

$$\left.\begin{aligned} H(x)\,S_2 &= \;\;\,2xS_1\,.\,S_2\,.\,S_2 = \;\;\,(S_2 S_2)\,xS_1 \\ \text{and}\quad H(x)\,S_1 &= -2xS_2\,.\,S_1\,.\,S_1 = -(S_1 S_1)\,xS_2\,. \end{aligned}\right\} \quad\ldots\ldots\ldots\ldots (1).$$

Similarly the null point of X with respect to S_1 is the same as the null point of $H(X)$ with respect to S_2.

$$\left.\begin{aligned} \text{For}\quad H(X)\,S_2 &= \;\;\,2XS_1\,.\,S_2\,.\,S_2 = \;\;\,(S_2 S_2)\,XS_1 \\ H(X)\,S_1 &= -2XS_2\,.\,S_1\,.\,S_1 = -(S_1 S_1)\,XS_2\,. \end{aligned}\right\} \quad\ldots\ldots\ldots\ldots (2).$$

(2) If S be any system of the dual group, to prove that

$$H(x)\,S = -H(xS), \quad H(X)\,S = -H(XS) \quad\ldots\ldots\ldots\ldots (3).$$

For let S' be the system reciprocal to S belonging to the group. Then we may write $H(x) = 2xS\,.\,S'$. Also from the second of equations (1) in subsection (1),

$$H(x)\,S = -(SS)\,xS'.$$

Again by § 167 (2) $H(xS) = 2xS\,.\,S\,.\,S' = (SS)\,xS' = -H(x)\,S.$

Similarly $\quad H(XS) = -H(X)\,S.$

(3) If the locus of x be the plane X, then the locus of $H(x)$ is the plane $H(X)$.

This proposition is obvious from the harmonic relation between x and $H(x)$ and between X and $H(X)$.

It can also be proved by means of the important transformation

$$XH(x) = xH(X) \quad\text{.............................(4)},$$

where x and X denote respectively any point and any plane.

For if S_1 and S_2 be any two reciprocal systems of the group, then remembering that the product of two planar elements and a force, or a system of forces, is a pure regressive product,

$$XH(x) = 2X.(xS_1.S_2) = 2X.xS_1.S_2 = -2xS_1.XS_2$$
$$= -2x(S_1.XS_2) = 2x(XS_1.S_2) = xH(X).$$

(4) If ab be a null line of any system S of the dual group, then $H(a)H(b)$ is also a null line of S.

For by hypothesis $(abS) = 0$. And by (2) of this article,

$$H(a)H(b)S = -H(a)H(bS).$$

But by (3) of this article and by § 179 (6),

$$H(a)H(bS) = bSH^2(a) \equiv bSa = 0.$$

Hence $$H(a)H(b)S = 0.$$

Since $H^2(x) \equiv x$, this proposition can also be stated thus, if $aH(b)$ be a null line of S, then $bH(a)$ is a null line of S.

(5) If S_1 and S_2 be reciprocal systems of the dual group and ab be a null line of S_2, then $H(a)H(b)$ is the conjugate of ab with respect to S_1.

This proposition will be proved [cf. § 164 (4)] by proving the important formula

$$H(a)H(b) = -2(S_2S_2)\,aS_1.bS_1 \quad\text{....................(5)};$$

where $$H(x) = 2xS_1.S_2.$$

For remembering that $(S_1S_2) = 0$, and $(abS_2) = 0$, and twice using equations (22) of § 167 (3),

$$H(a)H(b) = 4(aS_1)S_2.(bS_1)S_2 = 4(aS_1.bS_1.S_2)S_2 - 2(S_2S_2)\,aS_1.bS_1$$
$$= 4[\{(abS_1)S_1 - \tfrac{1}{2}(S_1S_1)ab\}S_2]S_2 - 2(S_2S_2)\,aS_1.bS_1$$
$$= -2(S_2S_2)\,aS_1.bS_1.$$

In connection with this proposition and that of subsection (4) the proposition of § 166 (2) should be referred to.

181. Formulæ connected with Reciprocal Systems. (1) A variety of formulæ connected with two reciprocal systems can be deduced from the preceding article.

Thus equation (22) of § 167 (3) can be written

$$(abS)S = \tfrac{1}{2}(SS)ab + aS.bS.$$

From this equation and from equation (5) of § 180 (5), it immediately follows that, if S_1 and S_2 be reciprocal and ab be a null line of S_2,

$$\left.\begin{aligned} 2(abS_1)S_1 &= (S_1S_1)\,ab - \frac{1}{(S_2S_2)}H(a)\,H(b). \\ \text{Similarly,}\quad 2(ABS_1)S_1 &= (S_1S_1)\,AB - \frac{1}{(S_2S_2)}H(A)\,H(B); \end{aligned}\right\}\ldots\ldots\ldots\ldots(1).$$

where AB is a null line of S_2.

(2) Also with the same assumptions as in (1), it follows from § 180 (5) that $aH(b)$ is a null line of S_1. Hence by the preceding subsection

$$2\{aH(b)S_2\}S_2 = (S_2S_2)\,aH(b) - \frac{1}{(S_1S_1)}H(a)\,H^2(a).$$

But by § 179 (6), $\quad H^2(b) = -(S_1S_1)(S_2S_2)\,b;$

also by an easy transformation

$$\{aH(b)S_2\} = (S_2S_2)(abS_1).$$

$$\left.\begin{aligned} \text{Hence}\quad 2(abS_1)S_2 &= aH(b) - bH(a). \\ \text{Similarly}\quad 2(ABS_1)S_2 &= AH(B) - BH(A). \end{aligned}\right\}\ldots\ldots\ldots\ldots(2).$$

(3) Also, since $(abS_2) = 0$, $(abS_2)S_2 = 0 = \frac{1}{2}(S_2S_2)\,ab + aS_2\,.\,bS_2$.

Hence $$ab = -\frac{2}{(S_2S_2)}aS_2\,.\,bS_2.$$

$$\left.\begin{aligned} &\text{Thus}\quad (abS_1)S_1 = \tfrac{1}{2}(S_1S_1)\,ab + aS_1\,.\,bS_1 = aS_1\,.\,bS_1 - \frac{(S_1S_1)}{(S_2S_2)}aS_2\,.\,bS_2. \\ &\text{Similarly, if } (ABS_2)=0,\quad (ABS_1)S_1 = AS_1\,.\,BS_1 - \frac{(S_1S_1)}{(S_2S_2)}AS_2\,.\,BS_2. \end{aligned}\right\}\ldots(3).$$

(4) Also, with the same assumptions, equations (26) of § 167 (5) become

$$\left.\begin{aligned} (abS_1)S_2 &= aS_1\,.\,bS_2 + aS_2\,.\,bS_1, \\ (ABS_1)S_2 &= AS_1\,.\,BS_2 + AS_2\,.\,BS_1. \end{aligned}\right\}\ldots\ldots\ldots\ldots(4).$$

182. Systems reciprocal to a Dual Group. (1) Let R be any system reciprocal to a whole dual group. Then R belongs to the reciprocal quadruple group. Also let S_1 and S_2 be two reciprocal systems of the dual group.

Then by equation (25) of § 167 (5) and remembering that $(RS_1) = 0 = (RS_2)$,

$$H(XR) = 2(XR)S_1\,.\,S_2 = -2(XS_1)R\,.\,S_2 = 2(XS_1)S_2\,.\,R = H(X)R.$$

Similarly, $$H(xR) = H(x)R.$$

(2) We may notice by comparison of this result with § 180 (2) that if S be any system of the dual group,

$$H(xS) = -H(x)S,\quad H(XS) = -H(X)S.$$

But if R be any system of the group reciprocal to the dual group,

$$H(xR) = H(x)R,\quad H(XR) = H(X)R.$$

183. THE POLE AND POLAR INVARIANTS OF A TRIPLE GROUP. (1) Let the triple group G be defined by three systems S_1, S_2, S_3. The same three systems taken in pairs define three dual subgroups. Let these dual subgroups be denoted by g_1, g_2, g_3; thus, let the group g_1 be defined by S_2, S_3, the group g_2 by S_3, S_1, and the group g_3 by S_1, S_2.

Let the harmonic invariants of the point x or of the plane X with respect to the groups g_1, g_2 and g_3 be denoted respectively by $H_1(x)$, $H_1(X)$, $H_2(x)$, $H_2(X)$, $H_3(x)$, $H_3(X)$.

(2) The expression

$$\frac{1}{\begin{vmatrix} (S_2S_2), & (S_2S_3) \\ (S_2S_3), & (S_3S_3) \end{vmatrix}} \cdot H_1(x) \cdot \begin{vmatrix} S_1, & S_2, & S_3 \\ (S_1S_2), & (S_2S_2), & (S_2S_3) \\ (S_1S_3), & (S_2S_3), & (S_3S_3) \end{vmatrix} \quad \text{.........(1)},$$

will be proved to be an invariant of the group, and will be called the Polar Invariant with respect to the group G. Similarly the expression

$$\frac{1}{\begin{vmatrix} (S_2S_2), & (S_2S_3) \\ (S_2S_3), & (S_3S_3) \end{vmatrix}} \cdot H_1(X) \cdot \begin{vmatrix} S_1, & S_2, & S_3 \\ (S_1S_2), & (S_2S_2), & (S_2S_3) \\ (S_1S_3), & (S_2S_3), & (S_3S_3) \end{vmatrix} \quad \text{.........(2)},$$

will be proved to be an invariant of the group, and will be called the Pole Invariant with respect to the group G.

(3) If R_1 be the system of the group G reciprocal to the subgroup g_1, then by properly choosing the intensity of R_1 we may write [cf. § 176 (1)]

$$\begin{vmatrix} (S_2S_2), & (S_2S_3) \\ (S_2S_3), & (S_3S_3) \end{vmatrix} R_1 = \begin{vmatrix} S_1, & S_2, & S_3 \\ (S_1S_2), & (S_2S_2), & (S_2S_3) \\ (S_1S_3), & (S_2S_3), & (S_3S_3) \end{vmatrix}.$$

Hence the polar invariant of x with respect to G is $H_1(x)\,R_1$, and the pole invariant of X with respect to G is $H_1(X)\,R_1$.

Let the polar invariant be denoted by $P(x)$ and the pole invariant by $P(X)$.

Then $$P(x) = H_1(x)\,R_1, \text{ and } P(X) = H_1(X)\,R_1.$$

(4) Another form for $P(x)$ and $P(X)$ can be found as follows.

We have $$H_1(x)\,S_2 = \{xS_2 \,.\, S_3 - xS_3 \,.\, S_2\}\,S_2 = \{(S_2S_3)\,x - 2xS_3 \,.\, S_2\}\,S_2$$
$$= (S_2S_3)\,xS_2 - (S_2S_2)\,xS_3.$$

Also $$H_1(x)\,S_3 = \{2xS_2 \,.\, S_3 - (S_2S_3)\,x\}\,S_3 = (S_3S_3)\,xS_2 - (S_2S_3)\,xS_3.$$

Hence from equation (2), $$\left.\begin{aligned} P(x) &= H_1(x)\,S_1 - (S_1S_3)\,xS_2 + (S_1S_2)\,xS_3. \\ P(X) &= H_1(X)\,S_1 - (S_1S_3)\,XS_2 + (S_1S_3)\,XS_3. \end{aligned}\right\} \text{...(3)}$$
Similarly

(5) The invariant property can easily be proved from this latter form. For write $x\{S_1, S_2, S_3\}$ for $P(x)$ as defined above, in order to bring out the

relations of $P(x)$ to the three systems S_1, S_2, S_3. Then it follows from the form for $P(x)$ given in equations (3) that

$$x\{S_1, S_2, S_3\} = -x\{S_1, S_3, S_2\} \qquad (a),$$

also

$$x\{S_1, S_2, S_2\} = 0 \qquad (b).$$

Furthermore $H_1(x)S_1 = \{2xS_2 . S_3 - (S_2S_3)x\} S_1$

$$= 2(S_1S_3)xS_2 - 2xS_2 . S_1 . S_3 - (S_2S_3)xS_1.$$

Hence $x\{S_1, S_2, S_3\} = (S_1S_3)xS_2 - 2xS_2 . S_1 . S_3 - (S_2S_3)xS_1 + (S_1S_2)xS_3$

$$= (S_1S_3)xS_2 - (S_2S_3)xS_1 - \{2xS_2 . S_1 - (S_1S_2)x\} S_3$$

$$= H_2(x)S_3 - (S_2S_3)xS_1 + (S_1S_3)xS_2$$

$$= x\{S_3, S_1, S_2\} \qquad (c).$$

Lastly $\quad x\{S_1 + S_1', S_2, S_3\} = x\{S_1, S_2, S_3\} + x\{S_1', S_2, S_3\} \qquad (d).$

Now let S, S', S'' be three systems of the group, such that

$$S = \lambda S_1 + \mu S_2 + \nu S_3,\quad S' = \lambda' S_1 + \mu' S_2 + \nu' S_3,\quad S'' = \lambda'' S_1 + \mu'' S_2 + \nu'' S_3;$$

and let Δ denote the determinant $\Sigma \pm \lambda\mu'\nu''$. Then from the equations (a), (b), (c), (d), which have just been proved, we deduce at once that

$$x\{S, S', S''\} = \Delta x\{S_1, S_2, S_3\}.$$

This proves that $P(x)$ is an invariant of the group.

An exactly similar proof shews that $P(X)$ is an invariant of the group.

Now that the invariant property is proved we may abandon the notation $x\{S_1, S_2, S_3\}$ for $P(x)$.

184. CONJUGATE SETS OF SYSTEMS AND THE POLE AND POLAR INVARIANTS. (1) Let R_1, R_2, R_3 be a set of conjugate systems of the group G. Then

$$(R_1R_2) = (R_2R_3) = (R_3R_1) = 0.$$

Also let g_1, g_2, g_3 denote the subgroups R_2R_3 and R_3R_1, and R_1R_2 respectively. Hence R_1 is reciprocal to the group g_1, and R_2 to the group g_2, and R_3 to the group g_3.

Then $P(x)$ and $P(X)$ take the simple forms $2xR_2 . R_3 . R_1$ and $2XR_2 . R_3 . R_1$. This follows at once from the forms for $P(x)$ and $P(X)$ given in § 183 (4), equation (3).

(2) It also follows that

$$P(x) = 2xR_2 . R_3 . R_1 = 2xR_1 . R_2 . R_3 = 2xR_3 . R_1 . R_2 = -2xR_3 . R_2 . R_1 = \text{etc.};$$

with similar transformations for $P(X)$.

(3) The equation, $aR_1 = P(b)$, can be solved for a. For let

$$P(b) = 2bR_2 . R_3 . R_1.$$

Then multiplying each side of the given equation by R_1,

$$aR_1 . R_1 = \tfrac{1}{2}(R_1R_1)a = P(b)R_1 = 2bR_2 . R_3 . R_1 . R_1 = (R_1R_1)bR_2 . R_3.$$

Hence

$$a = 2bR_2 . R_3 = -2bR_3 . R_2.$$

Also a condition holds. For

$$aR_2 = -2bR_3 \,.\, R_2 \,.\, R_2 = -(R_2R_2)\, bR_3.$$

Thus $$(abR_2) = -(baR_2) = (R_2R_2)(bbR_3) = 0.$$

Similarly $$(abR_3) = 0.$$

Accordingly ab is a common null line of the subgroup g_1.

185. Interpretation of $P(x)$ and $P(X)$. (1) $P(x)$ denotes a planar element, and $P(X)$ denotes a point.

To find the plane $P(x)$, write $P(x)$ in the form $2xR_2 \,.\, R_3 \,.\, R_1$, which is given in the last article.

Let the common null line through x of the subgroup g_1 intersect the director lines of g_1 in d_1 and d_1'. Then d_1 and d_1' are on the quadric G [cf. § 175 (4)]. Also the four points x, $2xR_2 \,.\, R_3$, d_1, d_1' form a harmonic range [cf. § 179 (4)]. Hence the point $2xR_2 \,.\, R_3$ lies on the polar plane of x with respect to this quadric.

But $P(x)$ is the null plane of this point with respect to R_1; and therefore the plane $P(x)$ passes through the point $2xR_2 \,.\, R_3$.

Now let R_1', R_2', R_3' be another set of conjugate systems of the group G. Then the same proof shews that the plane $P(x)$ passes through the point $2xR_2' \,.\, R_3'$; and that this point, $2xR_2' \,.\, R_3'$, lies in the polar plane of x with respect to the quadric G. Similarly for a third set of conjugate systems, such as R_1'', R_2'', R_3''.

Hence the plane $P(x)$ passes through the three (not collinear) points $2xR_2 \,.\, R_3$, $2xR_2' \,.\, R_3'$, $2xR_2'' \,.\, R_3''$.

Hence $P(x)$ denotes a planar element of the polar plane of x with respect to the quadric G. Similarly $P(X)$ denotes the pole of the plane X with respect to the quadric G.

(2) It follows as a corollary from subsection (1) and from § 176 (3) and (4) that we can express the angular points of tetrahedrons self-conjugate with respect to the quadric G, which have one face in a given plane.

For let X be the given plane, and R_1, R_2, R_3 a set of conjugate systems of the group G. Then by § 176 (3) XR_1, XR_2, XR_3 are three conjugate points in the plane X, and by the present article $P(X)$ is the pole of X. Hence these four points are the corners of a self-conjugate tetrahedron with one face in the plane X.

By taking different sets of conjugate systems an infinite number of such tetrahedrons may be found.

(3) Similarly we can express the four planes which are the faces of a self-conjugate tetrahedron with respect to G, of which one corner is at a given point x.

For, by the same reasoning as that just employed, the four planes are xR_1, xR_2, xR_3 and $P(x)$.

By taking different sets of conjugate systems an infinite number of such tetrahedrons may be found.

(4) The interpretations of $P(x)$ and of $P(X)$, which are given in (2) and (3), shew that $P^2(x)$ [i.e. $P\{P(x)\}$] must denote the point x, and that $P^2(X)$ must denote the plane X.

This result can also easily be proved by direct transformation.

(5) Again it follows from the interpretations of $P(x)$ and $P(X)$ that if y lie on $P(x)$, then x lies on $P(y)$; and that if Y contain $P(X)$, then X contains $P(Y)$.

This result can also be proved by direct transformation, namely the following equations hold

$$[P(x)\,y] = [P(y)\,x], \quad [P(X)\,Y] = [P(Y)\,X].$$

186. Relations between Conjugate Sets of Systems. (1) It follows from § 181 (3), equation (3), that if R_1, R_2, R_3 be a conjugate set of systems, and if $(abR_2) = 0$, then

$$(abR_1)\,R_1 = aR_1 \,.\, bR_1 - \frac{(R_1R_1)}{(R_2R_2)}\, aR_2 \,.\, bR_2.$$

Now if we take $aR_1 = P(b)$, then by § 184 (3) the condition $(abR_2) = 0$ is fulfilled; and $(aR_2) = -(R_2R_2)\,bR_3$.

Also $\qquad (abR_1) = -(baR_1) = -\{bP(b)\} = \{P(b)\,b\}$.

Hence finally if b be any point,

$$\{P(b)\,b\}\,R_1 = P(b)\,.\,bR_1 - (R_1R_1)\,bR_2\,.\,bR_3;$$

and, since b bears no special relation to R_1, by the cyclical interchange of suffixes,

$$\left.\begin{aligned}\{P(b)\,b\}\,R_2 &= P(b)\,.\,bR_2 - (R_2R_2)\,bR_3\,.\,bR_1,\\ \{P(b)\,b\}\,R_3 &= P(b)\,.\,bR_3 - (R_3R_3)\,bR_1\,.\,bR_2.\end{aligned}\right\} \quad \text{..........(1)}$$

(2) Similarly if B be any plane,

$$\left.\begin{aligned}\{P(B)\,B\}\,R_1 &= P(B)\,.\,BR_1 - (R_1R_1)\,BR_2\,.\,BR_3,\\ \{P(B)\,B\}\,R_2 &= P(B)\,.\,BR_2 - (R_2R_2)\,BR_3\,.\,BR_1,\\ \{P(B)\,B\}\,R_3 &= P(B)\,.\,BR_3 - (R_3R_3)\,BR_1\,.\,BR_2.\end{aligned}\right\} \quad \text{..........(2)}$$

(3) It is to be noticed that $P(B)$, BR_1, BR_2, BR_3 are the four angular points of a self-conjugate tetrahedron with respect to the quadric G. This tetrahedron has the plane of one face, namely B, arbitrarily chosen, but is otherwise definitely assigned by the conjugate set of systems R_1, R_2, R_3.

Similarly, $P(b)$, bR_1, bR_2, bR_3 are the four faces of a self-conjugate tetrahedron with respect to the quadric G. This tetrahedron has one angular point, namely b, arbitrarily chosen, but is otherwise definitely assigned by the set R_1, R_2, R_3.

(4) Let p_1, p_2, p_3, p denote the angular points of a self-conjugate tetrahedron with respect to the quadric G. Then one reciprocal set of systems with respect to the group G can be expressed by

$$\left.\begin{array}{l} pp_1 + \mu_1 p_2 p_3, \\ pp_2 + \mu_2 p_3 p_1, \\ pp_3 + \mu_3 p_1 p_2; \end{array}\right\} \qquad (3)$$

where μ_1, μ_2, μ_3 are given definite numbers.

Similarly if P_1, P_2, P_3, P denote planar elements in the faces of a self-conjugate tetrahedron, then one reciprocal set of systems can be expressed by

$$\left.\begin{array}{l} PP_1 + \lambda_1 P_2 P_3, \\ PP_2 + \lambda_2 P_3 P_1, \\ PP_3 + \lambda_3 P_1 P_2; \end{array}\right\} \qquad (4)$$

where $\lambda_1, \lambda_2, \lambda_3$ are given definite numbers.

(5) The proposition of the preceding subsection, symbolized in equations (3) and (4), may be enunciated as follows: Corresponding to any given set of conjugate systems of a group G, one and only one tetrahedron self-conjugate with respect to the quadric G can be found with three corners in a given plane, such that its opposite edges taken in pairs are respectively conjugate lines of the three systems of the conjugate set.

Also corresponding to any given set of conjugate systems of a group G, one and only one tetrahedron self-conjugate with respect to the quadric G can be found with one corner given, such that its opposite edges taken in pairs are respectively conjugate lines of the three systems of the conjugate set.

Such self-conjugate tetrahedrons will be said to be associated with the corresponding conjugate sets of systems, and *vice versa*.

(6) The group G' reciprocal to the group G is also a triple group, and defines the same quadric as G.

Now let p, p_1, p_2, p_3 be the four angular points of a tetrahedron which is self-conjugate with respect to this quadric. Also let the conjugate set of systems of the group G associated with this tetrahedron be

$$pp_1 + \mu_1 p_2 p_3, \quad pp_2 + \mu_2 p_3 p_1, \quad pp_3 + \mu_3 p_1 p_2.$$

Then it is obvious that the conjugate set of the reciprocal group G', associated with this tetrahedron, is

$$pp_1 - \mu_1 p_2 p_3, \quad pp_2 - \mu_2 p_3 p_1, \quad pp_3 - \mu_3 p_1 p_2.$$

For it follows from mere multiplication that any system of the last set is reciprocal to each system of the first set. Hence the three systems of the last set each belong to the group G'. Furthermore they obviously are reciprocal to each other, and therefore form a conjugate system of the group G'. And lastly, the form in which they are expressed shews them to be the conjugate set of systems associated with the tetrahedron p, p_1, p_2, p_3.

(7) Similarly an analogous proof shews that if P, P_1, P_2, P_3 be the four faces of a tetrahedron self-conjugate with respect to the quadric of G and G', and if the associated conjugate set of G be

$$PP_1 + \mu_1 P_2 P_3, \quad PP_2 + \mu_2 P_3 P_1, \quad PP_3 + \mu_3 P_1 P_2;$$

then the associated conjugate set of the group G' is

$$PP_1 - \mu_1 P_2 P_3, \quad PP_2 - \mu_2 P_3 P_1, \quad PP_3 - \mu_3 P_1 P_2.$$

187. THE CONJUGATE INVARIANT OF A TRIPLE GROUP. (1) If S_1, S_2, S_3 be any three systems of the group G, the equation of the quadric G is [cf. § 175 (3)]

$$(xS_1 \,.\, xS_2 \,.\, S_3) = 0.$$

(2) If x and y be any two points and $\lambda x + \mu y$ be a point on the quadric lying on the line joining them, then

$$\lambda^2 (xS_1 \,.\, xS_2 \,.\, S_3) + \lambda\mu \{(xS_1 \,.\, yS_2 \,.\, S_3) + (yS_1 \,.\, xS_2 \,.\, S_3)\} + \mu^2 (yS_1 \,.\, yS_2 \,.\, S_3) = 0.$$

Hence the condition that the points x and y should be conjugate is

$$(xS_1 \,.\, yS_2 \,.\, S_3) + (yS_1 \,.\, xS_2 \,.\, S_3) = 0.$$

If y be regarded as fixed, this is the equation of the polar plane of y.

(3) Let the expression $\frac{1}{2}\{xS_1 \,.\, yS_2 \,.\, S_3 + yS_1 \,.\, xS_2 \,.\, S_3\}$ be denoted by $G(xy)$. It will be proved to be an invariant of the group G, and will be called the Conjugate Invariant.

The equation $G(xx) = 0$, is the equation of the quadric G.

It follows from symmetry that, $G(xy) = G(yx)$.

(4) In order to prove the invariant property let us write $G(xy)$ in the form $xy\{S_1, S_2, S_3\}$.

Then obviously

$$xy\{S_1 + S_1', S_2, S_3\} = xy\{S_1, S_2, S_3\} + xy\{S_1', S_2, S_3\} \quad\ldots\ldots\ldots(1).$$

Also

$$xy\{S_1, S_2, S_3\} = -\,xy\{S_2, S_1, S_3\} \quad\ldots\ldots\ldots(2),$$

and

$$xy\{S_1, S_1, S_3\} = 0 \quad\ldots\ldots\ldots(3).$$

Furthermore $\quad yS_2 \,.\, S_3 = (S_2 S_3)\, y - yS_3 \,.\, S_2.$

Hence $xS_1 \,.\, yS_2 \,.\, S_3 = xS_1 \,.\, (yS_2 \,.\, S_3) = (S_2 S_3)\, xS_1 \,.\, y - xS_1 \,.\, yS_3 \,.\, S_2.$

Similarly $\quad yS_1 \,.\, xS_2 \,.\, S_3 = (S_2 S_3)\, yS_1 \,.\, x - yS_1 \,.\, xS_3 \,.\, S_2.$

Also $\quad xS_1 \,.\, y + yS_1 \,.\, x = xyS_1 + yxS_1 = 0.$

Therefore $\quad xy\{S_1, S_2, S_3\} = -\,xy\{S_1, S_3, S_2\} = xy\{S_3, S_1, S_2\} \quad\ldots\ldots\ldots (4).$

Now let S, S', S'' be any three systems of the group G, such that

$$S = \lambda S_1 + \mu S_2 + \nu S_3, \quad S' = \lambda' S_1 + \mu' S_2 + \nu' S_3, \quad S'' = \lambda'' S_1 + \mu'' S_2 + \nu'' S_3.$$

Also let Δ stand for the determinant $\Sigma \pm \lambda\mu'\nu''$.

Then from equations (1), (2), (3), (4) it follows that

$$xy\{S, S', S''\} = \Delta xy\{S_1, S_2, S_3\}.$$

This proves the invariant property of $xy\{S_1, S_2, S_3\}$. This expression for the conjugate invariant may now be abandoned in favour of $G(xy)$, in which the special systems used do not appear.

(5) Let R_1, R_2, R_3 be a conjugate set of systems of the group; so that

$$(R_1R_2) = (R_2R_3) = (R_3R_1) = 0.$$

Then $(yR_1 . xR_2 . R_3) = -(yR_1 . R_2 . xR_3) = (yR_2 . R_1 . xR_3)$

$$= -(yR_2 . R_3 . xR_1) = (xR_1 . yR_2 . R_3).$$

Hence $G(xy) = \frac{1}{2}\{(xR_1 . yR_2 . R_3) + (yR_1 . xR_2 . R_3)\}$

$$= (xR_1 . yR_2 . R_3) = (xR_1 . R_3 . yR_2) = \{(xR_1 . R_3) R_2 . y\}.$$

But $$xR_1 . R_3 . R_2 = \tfrac{1}{2} P(x).$$

Therefore $$G(xy) = \tfrac{1}{2}\{P(x) y\} = \tfrac{1}{2}\{P(y) x\}.$$

The equation $G(xx) = 0$, can be written in the form $\{P(x) x\} = 0$.

(6) Similarly the condition that the plane X touches the quadric G is

$$(XS_1 . XS_2 . S_3) = 0.$$

The condition that the planes X and Y are conjugate is

$$(XS_1 . YS_2 . S_3) + (YS_1 . XS_2 . S_3) = 0.$$

The expression $\frac{1}{2}\{(XS_1 . YS_2 . S_3) + (YS_1 . XS_2 . S_3)\}$ can be proved to be an invariant of the group and will also be called the conjugate invariant, and denoted by $G(XY)$. Also

$$G(XY) = G(YX).$$

Furthermore if R_1, R_2, R_3 be a set of conjugate systems of the group, then

$$G(XY) = (XR_1 . YR_2 . R_3).$$

Also $$G(XY) = \tfrac{1}{2}\{P(X) Y\} = \tfrac{1}{2}\{P(Y) X\}.$$

Therefore the plane-equation of the quadric G is

$$\{P(X) X\} = 0.$$

(7) Also from § 183 (3), $P(x) = H_1(x) R_1$, hence

$$G(xy) = \tfrac{1}{2}\{P(x) y\} = \tfrac{1}{2}\{H_1(x) R_1 y\} = \tfrac{1}{2}\{H_1(x) . yR_1\}$$

$$= \tfrac{1}{2}\{H_1(yR_1) x\}, \text{ from § 180 (3)}$$

$$= \tfrac{1}{2}\{H_1(y) R_1 x\}, \text{ from § 182 (1)}$$

$$= \tfrac{1}{2}\{H_1(y) xR_1\} = \tfrac{1}{2}\{H_1(xR_1) y\}.$$

Similarly $G(XY) = \frac{1}{2}\{H_1(X) YR_1\} = \frac{1}{2}\{H_1(YR_1) X\}$

$$= \tfrac{1}{2}\{H_1(Y) XR_1\} = \tfrac{1}{2}\{H_1(XR_1) Y\}.$$

(8) Corresponding to each director force D_1 of the triple group, there is one parabolic subgroup [cf. § 175 (9)]. Let S_2 be any system of this subgroup, and let $H_3(x)$ and $H_3(X)$ be the harmonic invariants with respect to this subgroup of a point x and of a plane X. Then we will prove that $H_3(x)$ [cf. § 179 (7)] is the point of contact of that tangent line from x to the

quadric, which intersects D_1; also that $H_3(X)$ is the tangent plane containing that tangent line to the quadric, which lies in the plane X and intersects the line D_1.

For let y be another point, and S_3 any system of the triple group which does not belong to the given parabolic subgroup. Then, according to subsection (2), if $\lambda x + \mu y$ be a point on the quadric,

$$\lambda^2 (xD_1 . xS_2 . S_3) + \lambda\mu \{(xD_1 . yS_2 . S_3) + (yD_1 . xS_2 . S_3)\} + \mu^2 (yD_1 . yS_2 . S_3) = 0.$$

Now let
$$y = H_3(x) = 2xD_1 . S_2.$$

Then by § 179 (7), $yD_1 = 0$. Hence $(yD_1 . yS_2 . S_3) = 0$; and
$$(yD_1 . xS_2 . S_3) = 0.$$

Also $yS_2 = 2xD_1 . S_2 . S_2 = xD_1 (S_2 S_2)$. Hence $(xD_1 . yS_2 . S_3) = 0$.

Thus the equation reduces to $\lambda^2 = 0$. Hence the two points, in which the line $xH_3(x)$ meets the quadric, coincide at the point $H_3(x)$. Similarly for the second part of the theorem which concerns $H_3(X)$.

Another mode of stating the propositions of this subsection is that, all quadrics with a given parabolic subgroup touch along the director line of that subgroup.

Hence the equation of any one of a group of quadrics which touch along a common generator can be immediately written down.

188. TRANSFORMATIONS OF $G(pp)$ AND $G(PP)$. (1) Let a point p on a plane X be conceived as the null point of X with respect to some system of the group; it is proved in subsection (3) below that, if $G(X, X)$ be not zero, p may be any point on the plane X. Hence p can be written $X(\lambda_1 R_1 + \lambda_2 R_2 + \lambda_3 R_3)$; where R_1, R_2, R_3 are three reciprocal systems of the group.

Then writing $\lambda_1 XR_1 + \lambda_2 XR_2 + \lambda_3 XR_3$ for the second p in $G(pp)$, and then using § 187 (7)

$$G(pp) = \lambda_1 G(p, XR_1) + \lambda_2 G(p, XR_2) + \lambda_3 G(p, XR_3)$$
$$= \tfrac{1}{2} \lambda_1 H_1(XR_1) . pR_1 + \tfrac{1}{2} \lambda_2 H_2(XR_2) . pR_2 + \tfrac{1}{2} \lambda_3 H_3(XR_3) . pR_3.$$

But R_1 is reciprocal to the dual group g_1, and therefore by § 182 (1) $H_1(XR_1) = H_1(X) R_1$; similarly $H_2(XR_2) = H_2(X) R_2$, $H_3(XR_3) = H_3(X) R_3$.

Hence

$$G(pp) = \tfrac{1}{2} \lambda_1 H_1(X) R_1 . pR_1 + \tfrac{1}{2} \lambda_2 H_2(X) R_2 . pR_2 + \tfrac{1}{2} \lambda_3 H_3(X) R_3 . pR_3.$$

But [cf. § 167, equation (21)]

$$H_1(X) R_1 . pR_1 = H_1(X) R_1 . R_1 . p = \tfrac{1}{2} H_1(X) p . (R_1 R_1).$$

Similarly $H_2(X) R_2 . pR_2 = \tfrac{1}{2} H_2(X) p . (R_2 R_2)$,

and
$$H_3(X) R_3 . pR_3 = \tfrac{1}{2} H_3(X) p . (R_3 R_3).$$

Hence

$$G(pp) = \tfrac{1}{2} \{\lambda_1 (R_1 R_1) H_1(X) p + \lambda_2 (R_2 R_2) H_2(X) p + \lambda_3 (R_3 R_3) H_3(X) p\}.$$

Again $\frac{1}{2}H_1(X)p = \frac{1}{2}H_1(X).X(\lambda_1 R_1 + \lambda_2 R_2 + \lambda_3 R_3)$.

And by § 187 (7) $\frac{1}{2}H_1(X).XR_1 = G(XX)$.

Also $\frac{1}{2}H_1(X).XR_2 = XR_2.R_3.XR_2 = 0$; similarly $\frac{1}{2}H_1(X).XR_3 = 0$.

Therefore $\frac{1}{2}H_1(X)p = \lambda_1 G(XX)$. Similarly

$$\frac{1}{2}H_2(X)p = \lambda_2 G(XX),\ \frac{1}{2}H_3(X)p = \lambda_3 G(XX).$$

Thus finally $G(pp) = G(XX)\{\lambda_1^2(R_1R_1) + \lambda_2^2(R_2R_2) + \lambda_3^2(R_3R_3)\}$.

(2) Now if $G(pp) = 0$, p is a point on the section of the quadric G made by the plane X. But $G(pp) = 0$ involves either $G(XX) = 0$, or

$$\lambda_1^2(R_1R_1) + \lambda_2^2(R_2R_2) + \lambda_3^2(R_3R_3) = 0.$$

If $G(XX) = 0$, the plane touches the quadric and therefore contains one director line of the group and one common null line. The null points of this plane in respect to the various systems of the group must lie on this common null line.

If $G(XX)$ be not zero, then

$$\lambda_1^2(R_1R_1) + \lambda_2^2(R_2R_2) + \lambda_3^2(R_3R_3) = 0.$$

But this is the director equation of the group. Hence the director forces of the group are in general the only systems in respect to which the null points of any plane—not a tangent plane—lie on the quadric.

(3) If $G(XX)$ be zero, then by (2) the three points XR_1, XR_2, XR_3 are collinear and lie on the common null line. The point of contact of the plane is $P(X)$, which is also collinear with the three points.

If $G(XX)$ be not zero, the three points XR_1, XR_2, XR_3 form a triangle on the plane. Hence any point on the plane can be represented by $X(\lambda_1 R_1 + \lambda_2 R_2 + \lambda_3 R_3)$, that is by XS where S is any system of the group. Also $P(X)$ does not lie on the plane X.

(4) Similarly if $P = x(\lambda_1 R_1 + \lambda_2 R_2 + \lambda_3 R_3)$,

then $$G(PP) = G(xx)\{\lambda_1^2(R_1R_1) + \lambda_2^2(R_2R_2) + \lambda_3^2(R_3R_3)\}.$$

The plane P is a tangent plane of the quadric G, if $G(PP) = 0$.

But this equation involves either $G(xx) = 0$,

or $$\lambda_1^2(R_1R_1) + \lambda_2^2(R_2R_2) + \lambda_3^2(R_3R_3) = 0.$$

If $G(xx) = 0$, the point x lies on the quadric and therefore is contained in one director line of the group and one common null line. The null planes of this point with respect to the various systems of the group all touch the quadric (since $G(PP) = 0$), hence they all contain the common null line through the point.

If $G(xx)$ be not zero, then $\lambda_1^2(R_1R_1) + \lambda_2^2(R_2R_2) + \lambda_3^2(R_3R_3) = 0$. But this is the director equation of the group. Hence the director forces of the group are in general the only systems of the group with respect to which the null planes of any point—not lying on the quadric—touch the quadric.

If $G(xx)$ be zero, the three planes xR_1, xR_2, xR_3 are collinear, and contain the common null line through x. The tangent plane at x is $P(x)$ and also contains this null line.

If $G(xx)$ be not zero, the three planes xR_1, xR_2, xR_3 are not collinear. Hence any plane through x can be represented in the form

$$x(\lambda_1 R_1 + \lambda_2 R_2 + \lambda_3 R_3).$$

Also $P(x)$ does not contain the point x.

(5) Let p, p_1, p_2, p_3 be the corners of a self-conjugate tetrahedron of a quadric; and let a conjugate set of systems of one of the two triple groups defined by the quadric be

$$R_1 = pp_1 + \mu_1 p_2 p_3,\quad R_2 = pp_2 + \mu_2 p_3 p_1,\quad R_3 = pp_3 + \mu_3 p_1 p_2.$$

Also let any point x be defined by $\xi p + \xi_1 p_1 + \xi_2 p_2 + \xi_3 p_3$.

Then $$xR_1 = \mu_1 \xi pp_2 p_3 + \mu_1 \xi_1 p_1 p_2 p_3 + \xi_2 pp_1 p_2 + \xi_3 pp_1 p_3,$$

and $$xR_2 = -\mu_2 \xi pp_1 p_3 - \xi_1 pp_1 p_2 + \mu_2 \xi_2 p_1 p_2 p_3 + \xi_3 pp_2 p_3.$$

Hence $xR_2 \,.\, R_3 = (pp_1 p_2 p_3)\{\mu_2 \mu_3 \xi p_1 - \xi_1 p + \mu_2 \xi_2 p_3 - \mu_3 \xi_3 p_2\}$.

Therefore finally $G(xx) = (pp_1 p_2 p_3)^2 \{\mu_1 \mu_2 \mu_3 \xi^2 + \mu_1 \xi_1^2 + \mu_2 \xi_2^2 + \mu_3 \xi_3^2\}$.

Accordingly the equation of the quadric, $G(xx) = 0$, can be written in the form

$$\xi^2 + \frac{\xi_1^2}{\mu_2 \mu_3} + \frac{\xi_3^2}{\mu_3 \mu_1} + \frac{\xi_1^2}{\mu_1 \mu_2} = 0.$$

(6) Conversely let the equation,

$$\alpha \xi^2 + \alpha_1 \xi_1^2 + \alpha_2 \xi_2^2 + \alpha_3 \xi_3^2 = 0,$$

be the equation of the quadric; where x is the point $\xi p + \xi_1 p_1 + \xi_2 p_2 + \xi_3 p_3$.

Also let one group of the two reciprocal groups, defined by this quadric, be defined by the conjugate set

$$pp_1 + \mu_1 p_2 p_3,\quad pp_2 + \mu_2 p_3 p_1,\quad pp_3 + \mu_3 p_1 p_2.$$

Then by comparison with (5) we find

$$\frac{\alpha}{\mu_1 \mu_2 \mu_3} = \frac{\alpha_1}{\mu_1} = \frac{\alpha_2}{\mu_2} = \frac{\alpha_3}{\mu_3}.$$

Hence $$\frac{\alpha^3}{\mu_1^3 \mu_2^3 \mu_3^3} = \frac{\alpha_1 \alpha_2 \alpha_3}{\mu_1 \mu_2 \mu_3}.$$

Therefore $$\mu_1 \mu_2 \mu_3 = \pm \sqrt{\frac{\alpha^3}{\alpha_1 \alpha_2 \alpha_3}}.$$

Finally $$\mu_1 = \pm \lambda \alpha \alpha_1,\quad \mu_2 = \pm \lambda \alpha \alpha_2, \text{ and } \mu_3 = \pm \lambda \alpha \alpha_3;$$

where all the upper signs are to be chosen or all the lower signs, and $1/\lambda$ is put for $+\sqrt{(\alpha \alpha_1 \alpha_2 \alpha_3)}$.

Thus one group associated with the quadric is defined by the three systems

$$pp_1 + \lambda\alpha\alpha_1 p_2 p_3, \quad pp_2 + \lambda\alpha\alpha_2 p_3 p_1, \quad pp_3 + \lambda\alpha\alpha_3 p_1 p_2.$$

The reciprocal group is defined by the three systems

$$pp_1 - \lambda\alpha\alpha_1 p_2 p_3, \quad pp_2 - \lambda\alpha\alpha_2 p_3 p_1, \quad pp_3 - \lambda\alpha\alpha_3 p_1 p_2.$$

If $\alpha\alpha_1\alpha_2\alpha_3$ be positive, λ is real; if negative, λ is imaginary.

CHAPTER IV.

MATRICES AND FORCES.

189. LINEAR TRANSFORMATIONS IN THREE DIMENSIONS. (1) Let a real linear transformation of elements in a complete region of three dimensions be denoted by the matrix ϕ, as in Book IV, Chapter VI. All the matrices considered will be assumed to be of zero nullity [cf. § 144 (2)].

(2) The following notation respecting latent and semi-latent regions, which agrees with and extends that in §§ 145 to 150, will be used.

The region, such that for each point x in it $\phi x = \gamma x$, is called the latent region corresponding to the latent root γ of the matrix. Any subregion of this latent region will be called a latent subregion corresponding to the latent root γ.

(3) The region, such that for each point x in it $\phi x = \gamma x + y$, where y is a point in an included latent region, is called a semi-latent region of the first species. Here two cases arise. Firstly, if γ be a repeated root, then [cf. § 149] there may be a semi-latent region of the first species corresponding to the root γ. Similarly [cf. § 150], there may be semi-latent subregions of species higher than the first.

Secondly, let γ_1 and γ_2 be two latent roots, of which either or both may be repeated; and let e_1, e_2 be a pair of latent points belonging to the two roots respectively.

Then [cf. § 146] any point $x = \xi_1 e_1 + \xi_2 e_2$ is transformed according to the rule,

$$\phi x = \gamma_1 \xi_1 e_1 + \gamma_2 \xi_2 e_2 = \gamma_1 x + \delta_2 e_2 = \gamma_2 x + \delta_1 e_1.$$

The region defined by the assemblage of independent latent points corresponding to both γ_1 and γ_2, that is by the points e_1, e_1', e_1'', etc., e_2, e_2', e_2'', etc., is called the semi-latent region of the first species corresponding to the two roots conjointly. If neither of the roots be repeated, such a region is necessarily a straight line.

(4) A subregion of a semi-latent region of the αth species, which is not contained in a semi-latent region of a lower species, and which is such that

any point x in it is transformed into a point in the same subregion, is called a semi-latent subregion of the αth species.

The semi-latent regions, or subregions, which are of most importance in the present investigation, are straight lines. A semi-latent straight line is necessarily of the first species; so that its species need not be mentioned. It necessarily contains at least one latent point: let e_1 be a latent point on such a line, and x any point on it.

Then the transformation of x takes place according to the law

$$\phi x = \gamma_2 x + \delta_1 e_1.$$

If γ_2 be not the latent root corresponding to e_1, then another latent point e_2 exists on the line, and the line is a semi-latent subregion with respect to the two latent roots γ_1 and γ_2 conjointly.

If γ_2 be equal to γ_1, the latent root corresponding to e_1, then, either $\delta_1 = 0$, and all the points on the line are latent, and the line is a latent region (or subregion); or δ_1 is not zero, and there is only one latent point on the line. Thus, if $\gamma_2 = \gamma_1$, the line is either a latent or semi-latent subregion corresponding to the root γ_1.

(5) A semi-latent plane is at most of the second species. The transformation takes place according to the law

$$\phi x = \gamma x + x';$$

where x is any point on the plane, and x' is some point (depending in general on x) on a semi-latent line lying in the plane.

If the plane be semi-latent of the first species, it necessarily contains lines which are semi-latent subregions of the first species. For let x', above, be a latent point, so that $\phi x' = \gamma x'$, then the point $\lambda x + \mu x'$ is transformed according to the law

$$\phi(\lambda x + \mu x') = \lambda(\gamma x + x') + \mu\gamma x' = \lambda\gamma x + (\lambda + \mu\gamma) x'.$$

Hence any point on the line xx' is transformed into another point on the line xx'; and therefore the line is a semi-latent subregion.

190. Enumeration of Types of Latent and Semi-Latent Regions*. (1) Let the four latent roots $\gamma_1, \gamma_2, \gamma_3, \gamma_4$, of the matrix ϕ be distinct. Then [cf. § 145 (5), (6), (7)] there are four and only four independent latent points, one corresponding to each root. Let these points be e_1, e_2, e_3, e_4. The only semi-latent regions of the first species are the six edges of the tetrahedron $e_1e_2e_3e_4$; and each edge is semi-latent with respect to two roots

* This enumeration has, I find, been made by H. Grassmann (the younger) in a note to §§ 377—390 of the *Ausdehnungslehre von* 1862 in the new edition, edited by F. Engel (cf. note at end of this chapter). He gives interesting applications to Euclidean Space. He also refers to von Staudt, *Beiträge zur Geometrie der Lage*, 3rd Ed., 1860, for a similar enumeration made by different methods.

conjointly. The only semi-latent regions of the second species are the four faces of the tetrahedron $e_1e_2e_3e_4$; and each face is semi-latent with respect to three roots conjointly. All the manifold necessarily belongs to a semi-latent region of the third species; therefore such species need not be further considered.

The enumeration will be made without regard to the difference of type which arises according as the roots are real or imaginary.

(2) Let there be three distinct latent roots of the matrix. Let these roots be γ_1, γ_3, γ_4; and let the root γ_1 be the repeated root.

There are necessarily [cf. § 148 (2)] three latent points e_1, e_3, e_4, corresponding respectively to the three roots. There is a line e_1e_2 which is, either (Case I.) a latent region corresponding to the root γ_1, or (Case II.) a semi-latent region corresponding to the root γ_1.

Case I. The line e_1e_2 is a latent region corresponding to the root γ_1, e_3 is the latent point corresponding to γ_3, and e_4 is the latent point corresponding to γ_4. The semi-latent regions of the first species are, the plane $e_1e_2e_3$ corresponding to the roots γ_1 and γ_3 conjointly, the plane $e_1e_2e_4$ corresponding to the roots γ_1 and γ_4 conjointly, the line e_3e_4 corresponding to the roots γ_3 and γ_4 conjointly. Any line through e_3, which intersects e_1e_2, is a semi-latent subregion corresponding to the roots γ_1 and γ_3 conjointly; any line through e_4, which intersects e_1e_2, is a semi-latent subregion corresponding to the two roots γ_1 and γ_4 conjointly.

The complete manifold forms a semi-latent region of the second species corresponding to all the roots conjointly. All planes through the line e_3e_4 are semi-latent subregions of the second species corresponding to all the roots conjointly.

Case II. The point e_1 is the sole latent point corresponding to the root γ_1, the points e_3 and e_4 are the latent points corresponding to the roots γ_3 and γ_4. The semi-latent regions of the first species are, the line e_1e_2 corresponding to the root γ_1, the line e_1e_3 corresponding to the roots γ_1 and γ_3 conjointly, the line e_1e_4 corresponding to the roots γ_1 and γ_4 conjointly, the line e_3e_4 corresponding to the roots γ_3 and γ_4 conjointly. The semi-latent regions of the second species are the planes $e_1e_2e_3$, $e_1e_2e_4$, $e_1e_3e_4$; the roots to which they correspond need not be mentioned.

(3) Let the matrix have a triple latent root γ_1; and let γ_4 be the other root.

There are necessarily two latent points e_1 and e_4 corresponding to these roots respectively. There is [cf. §§ 149, 150] a plane $e_1e_2e_3$ which is, either (Case I.) a latent region corresponding to the root γ_1, or (Case II.) a semi-latent region of the first species corresponding to the root γ_1, or (Case III.) a semi-latent region of the second species corresponding to the root γ_1.

According to § 150 (5), the points e_1, e_2, e_3 can always be so chosen that

$$\phi e_3 = \gamma_1 e_3 + \delta_2 e_2,\quad \phi e_2 = \gamma_1 e_2 + \delta_1 e_1,\quad \phi e_1 = \gamma_1 e_1;$$

where δ_2 cannot vanish unless δ_1 also vanishes.

Thus, in Case I. $\delta_1 = 0 = \delta_2$; in Case II. $\delta_1 = 0$, and δ_2 is not zero; in Case III. neither δ_1 nor δ_2 vanishes.

Any point $x = \Sigma \xi e$ is transformed according to the rule

$$\phi x = (\gamma_1 \xi_1 + \delta_1 \xi_2)\, e_1 + (\gamma_1 \xi_2 + \delta_2 \xi_3)\, e_2 + \gamma_1 \xi_3 e_3 + \gamma_4 \xi_4 e_4.$$

Thus in Case I., $\phi x = \gamma_1 x + (\gamma_4 - \gamma_1)\, \xi_4 e_4$ ....(A).

In Case II., $\phi x = \gamma_1 x + \delta_2 \xi_3 e_2 + (\gamma_4 - \gamma_1)\, \xi_4 e_4$ ....(B).

In Case III., $\phi x = \gamma_1 x + \delta_1 \xi_2 e_1 + \delta_2 \xi_3 e_2 + (\gamma_4 - \gamma_1)\, \xi_4 e_4$................(C).

Case I. The plane $e_1e_2e_3$ is the latent region corresponding to the triple root γ_1, and e_4 is the latent point corresponding to the root γ_4. Any line in the plane $e_1e_2e_3$ is a latent subregion corresponding to the root γ_1.

The complete manifold is the semi-latent region of the first species corresponding to the two roots conjointly. Any line through e_4 is a semi-latent subregion corresponding to the two roots conjointly.

Case II. The latent regions are, the line e_1e_2 corresponding to the root γ_1, and the point e_4 corresponding to the root γ_4. The semi-latent regions of the first species are, the plane $e_1e_2e_3$ corresponding to the root γ_1, and the plane $e_1e_2e_4$. There is some one point [cf. § 149 and § 150 (5)] in the line e_1e_2 (the point e_2, according to the present notation), such that any line through it in the plane $e_1e_2e_3$ is a semi-latent subregion corresponding to the root γ_1. Also any line through e_4 in the plane $e_1e_2e_4$ is a semi-latent subregion corresponding to the roots γ_1 and γ_4 conjointly. The semi-latent region of the second species is the complete manifold. Any plane through e_2e_4 is a semi-latent subregion of the second species; for from equation (B)

$$e_2e_4\phi x \equiv e_2e_4x.$$

Case III. The only latent points are, the point e_1 corresponding to the root γ_1, and the point e_4 corresponding to the root γ_4. The semi-latent regions of the first species are, the line e_1e_2 corresponding to the root γ_1, and the line e_1e_4 corresponding to the roots γ_1 and γ_4 conjointly. The semi-latent regions of the second species are the planes $e_1e_2e_3$, and the planes $e_1e_2e_4$.

(4) Let there be only one latent root to the matrix, which occurs quadruply. Let γ_1 be this root, and let e_1 be the latent point corresponding to it which necessarily exists [cf. § 148 (2)]. Then [cf. § 150 (5)] it is always possible to find three other points e_2, e_3, e_4, such that

$$\phi e_4 = \gamma_1 e_4 + \delta_3 e_3,\quad \phi e_3 = \gamma_1 e_3 + \delta_2 e_2,\quad \phi e_2 = \gamma_1 e_2 + \delta_1 e_1,\quad \phi e_1 = \gamma_1 e_1.$$

Then any point $x = \Sigma \xi e$ is transformed according to the rule

$$\begin{aligned}\phi x &= (\gamma_1 \xi_1 + \delta_1 \xi_2)\, e_1 + (\gamma_1 \xi_2 + \delta_2 \xi_3)\, e_2 + (\gamma_1 \xi_3 + \delta_3 \xi_4)\, e_3 + \gamma_1 \xi_4 e_4 \\ &= \gamma_1 x + \delta_1 \xi_2 e_1 + \delta_2 \xi_3 e_2 + \delta_3 \xi_4 e_3 \quad \text{......(D).}\end{aligned}$$

Five cases now arise according as, either (Case I.) δ_1, δ_2 and δ_3 all vanish; or (Case II.) δ_1 and δ_2 vanish, but δ_3 does not vanish; or (Case III.) δ_1 vanishes, but δ_2 and δ_3 do not vanish; or (Case IV.) δ_1, δ_2, δ_3 do not vanish; or (Case V.) δ_2 vanishes, but δ_1 and δ_3 do not vanish.

Thus in Case I., $\phi x = \gamma_1 x$ ....(E).

In Case II., $\phi x = \gamma_1 x + \delta_3 \xi_4 e_3$ ....(F).

In Case III., $\phi x = \gamma_1 x + \delta_2 \xi_3 e_2 + \delta_3 \xi_4 e_3$ ....(G).

In Case IV., $\phi x = \gamma_1 x + \delta_1 \xi_2 e_1 + \delta_2 \xi_3 e_2 + \delta_3 \xi_4 e_3$ ....(H).

In Case V., $\phi x = \gamma_1 x + \delta_1 \xi_2 e_1 + \delta_3 \xi_4 e_3$ ....(I).

Case I. Every point is latent; and the operation of the matrix is simply equivalent to a numerical multiplier. This case need not be further considered; and may generally be neglected in subsequent discussions.

Case II. The latent region is the plane $e_1e_2e_3$. The semi-latent region of the first species is the complete manifold. Every line through the latent point e_3 is a semi-latent subregion. All semi-latent planes must pass through e_3; and all planes through e_3 are semi-latent.

Case III. The latent region is the line e_1e_2. The semi-latent region of the first species is the plane $e_1e_2e_3$. Every line through e_2 and lying in the plane $e_1e_2e_3$ is a semi-latent subregion. The semi-latent region of the second species is the complete manifold. Every plane through the line e_2e_3 is a semi-latent subregion of the second species; for by equation (G)

$$e_2e_3\phi x \equiv e_2e_3x.$$

Case IV. The sole latent point is e_1. The semi-latent region of the first species is the line e_1e_2. The semi-latent region of the second species is the plane $e_1e_2e_3$. The semi-latent region of the third species is the complete manifold.

Case V. The latent region is the line e_1e_3. The semi-latent region of the first species is the complete manifold. Every point lies on some straight line which is a semi-latent subregion. For consider the condition that the point $x (= \Sigma\xi e)$ may lie on a semi-latent straight line through the latent point $a (= \alpha_1 e_1 + \alpha_3 e_3)$.

Now $\phi x = \gamma_1 x + \delta_1 \xi_2 e_1 + \delta_3 \xi_4 e_3 = \gamma_1 x + \lambda a$, by hypothesis.

Hence $$\delta_1 \xi_2 = \lambda \alpha_1, \quad \delta_3 \xi_4 = \lambda \alpha_3.$$

Thus x must lie on the plane of which the equation is

$$\alpha_3 \delta_1 \xi_2 = \alpha_1 \delta_3 \xi_4.$$

This plane passes through a; and all lines lying in it which pass through a are semi-latent. A planar element in this plane is $e_1e_3(\alpha_1\delta_3e_2 + \alpha_3\delta_1e_4)$.

(5) Let there be two distinct latent roots, and let each be once repeated. Let γ_1 and γ_3 be the two distinct latent roots; and let e_1 and e_3 be the two latent points, which certainly exist, corresponding to these roots respectively.

Then it is always possible to find two points e_2 and e_4, such that

$$\phi e_2 = \gamma_1 e_2 + \delta_1 e_1, \quad \phi e_4 = \gamma_3 e_4 + \delta_3 e_3.$$

Hence any point $x = \Sigma\xi e$ is transformed according to the rule

$$\phi x = (\gamma_1\xi_1 + \delta_1\xi_2)\, e_1 + \gamma_1\xi_2 e_2 + (\gamma_3\xi_3 + \delta_3\xi_4)\, e_3 + \gamma_3\xi_4 e_4.$$

Three cases now arise according as, either (Case I.) δ_1 and δ_3 both vanish; or (Case II.) δ_1 vanishes and δ_3 does not vanish; or (Case III.) neither δ_1 nor δ_3 vanishes.

The case, when δ_3 vanishes and δ_1 does not vanish, is not a type of case distinct from Case II.

Thus in Case I.,

$$\phi x = \gamma_1 x + (\gamma_3 - \gamma_1)(\xi_3 e_3 + \xi_4 e_4) = \gamma_3 x + (\gamma_1 - \gamma_3)(\xi_1 e_1 + \xi_2 e_2) \ldots\ldots \text{(J)}.$$

In Case II.,

$$\phi x = \gamma_1 x + \{(\gamma_3 - \gamma_1)\,\xi_3 + \delta_3\xi_4\}\, e_3 + (\gamma_3 - \gamma_1)\,\xi_4 e_4$$
$$= \gamma_3 x + (\gamma_1 - \gamma_3)(\xi_1 e_1 + \xi_2 e_2) + \delta_3\xi_4 e_3 \ldots\ldots\ldots \text{(K)}.$$

In Case III.,

$$\phi x = (\gamma_1\xi_1 + \delta_1\xi_2)\, e_1 + \gamma_1\xi_2 e_2 + (\gamma_3\xi_3 + \delta_3\xi_4)\, e_3 + \gamma_3\xi_4 e_4 \ldots\ldots\ldots\ldots \text{(L)}.$$

Case I. The latent regions are, the line e_1e_2 corresponding to the root γ_1, and the line e_3e_4 corresponding to the root γ_3. The semi-latent region of the first species is the complete manifold, and it corresponds to the two roots conjointly. Every line intersecting both e_1e_2 and e_3e_4 is a semi-latent subregion corresponding to the two roots conjointly.

Case II. The latent regions are, the line e_1e_2 corresponding to the root γ_1, and the point e_3 corresponding to the root γ_3. The semi-latent regions of the first species are, the line e_3e_4 corresponding to the root γ_3, and the plane $e_1e_2e_3$ corresponding to the roots γ_1 and γ_3 conjointly. Every line through e_3 in this plane is a semi-latent subregion corresponding to the two roots conjointly. The semi-latent region of the second species is the complete manifold. Every plane through e_3e_4 is a semi-latent subregion.

Case III. The only two latent points are, e_1 corresponding to the root γ_1, and e_3 corresponding to the root γ_3.

The semi-latent regions of the first species are, the line e_1e_2 corresponding to the root γ_1, the line e_3e_4 corresponding to the root γ_3, the line e_1e_3 corresponding to the roots γ_1 and γ_3 conjointly. The semi-latent regions of the second species are, the plane $e_1e_2e_3$ corresponding to the roots $\gamma_1, \gamma_1, \gamma_3$ conjointly, and the plane $e_1e_3e_4$ corresponding to the roots $\gamma_1, \gamma_3, \gamma_3$ conjointly. It is to be noticed that, in order to define the roots corresponding to the semi-latent regions of the second species, the repeated roots must be counted twice. The semi-latent region of the third species is the complete manifold.

191. MATRICES AND FORCES. (1) Let S denote any system of forces, and ϕ any matrix. Then ϕS is defined in § 141 (1) and (3), and denotes another system of forces.

If $\phi S \equiv S$, the system S is said to be a latent system of the matrix ϕ [cf. § 160 (1)].

If every system S, belonging to a group of systems G, is transformed into another system ϕS of the same group, the group G is said to be a semi-latent group of the matrix.

If every system of the group is latent, the group is said to be latent.

(2) The following properties of the transformation are immediately evident.

If $(SS') = 0$, then $(\phi S \phi S') = (\phi . SS') = 0$. Hence if S and S' are reciprocal, ϕS and $\phi S'$ are also reciprocal; and if S reduce to a single force, ϕS reduces to a single force.

If S belong to the group defined by $S_1, S_2, \dots S_\rho$, then ϕS belongs to the group $\phi S_1, \phi S_2, \dots \phi S_\rho$.

(3) Let $e_1e_2e_3e_4$ be any fundamental tetrahedron of reference; and let

$$S = \varpi_{12}e_1e_2 + \varpi_{23}e_2e_3 + \varpi_{31}e_3e_1 + \varpi_{14}e_1e_4 + \varpi_{24}e_2e_4 + \varpi_{34}e_3e_4,$$

and

$$\phi S = \varpi_{12}'e_1e_2 + \varpi_{23}'e_2e_3 + \varpi_{31}'e_3e_1 + \varpi_{14}'e_1e_4 + \varpi_{24}'e_2e_4 + \varpi_{34}'e_3e_4.$$

Then it is obvious [cf. § 141] that the coefficients $\varpi_{12}, \dots, \varpi_{34}$ are transformed into the coefficients $\varpi_{12}', \dots, \varpi_{34}'$ by a linear transformation represented by a matrix of the sixth order.

But the most general matrix of the sixth order contains thirty-six constants; whereas the matrix of the fourth order, from which the present transformation is derived, contains only sixteen constants. Accordingly relations must hold between the thirty-six constants reducing the number of independent constants to sixteen.

(4) An interpretation of these relations can be found as follows. In general the transformation of the sixth order yields only six latent systems: and in general a matrix of the fourth order has four latent points forming a tetrahedron. The six edges of this tetrahedron are latent forces. Hence in general the six latent systems are six single forces along the edges of a tetrahedron. The expression of these conditions yields twenty independent equations, which reduce the number of independent constants to sixteen.

(5) Let the corners of the tetrahedron $e_1e_2e_3e_4$ be the latent points of the matrix ϕ. This is always possible, when the roots $\gamma_1, \gamma_2, \gamma_3, \gamma_4$ are unequal.

Then the latent roots of the matrix of the sixth order, which transforms the co-ordinates of any system of forces, are given by the sextic

$$(\sigma - \gamma_1\gamma_2)(\sigma - \gamma_3\gamma_4)(\sigma - \gamma_2\gamma_3)(\sigma - \gamma_1\gamma_4)(\sigma - \gamma_3\gamma_1)(\sigma - \gamma_2\gamma_4) = 0.$$

(6) Apart from the special cases when some of the roots of the matrix ϕ are repeated, the only cases, for which latent systems exist other than forces along the six edges of the tetrahedron, arise when two roots of this sextic are equal [cf. § 145 (4) and § 148]. There are two types of such equality; namely, the type given by $\gamma_1\gamma_2 = \gamma_3\gamma_4$, and the type given by $\gamma_2\gamma_3 = \gamma_3\gamma_1$. The second type necessitates $\gamma_1 = \gamma_2$; and thus leads back to the special cases when the matrix ϕ possesses repeated latent roots.

(7) It is evident from (3) that, if $S_1, S_2, \dots S_6$ be any six independent systems, and any system S be defined by

$$S = \lambda_1 S_1 + \lambda_2 S_2 + \dots + \lambda_6 S_6,$$

and

$$\phi S = \lambda_1' S_1 + \lambda_2' S_2 + \dots + \lambda_6' S_6,$$

then $\lambda_1', \lambda_2', \dots \lambda_6'$, can be derived from $\lambda_1, \lambda_2, \dots \lambda_6$ by a linear transformation.

192. Latent Systems and Semi-Latent Groups. (1) A force on any semi-latent line is a latent force. For let $e_1 e_2$ be the semi-latent line, and e_1 the latent point on it. Then [cf. § 189 (4)] we may assume,

$$\phi e_2 = \gamma_2 e_2 + \delta_1 e_1, \quad \phi e_1 = \gamma_1 e_1.$$

Hence

$$\phi e_1 e_2 = \phi e_1 \phi e_2 = \gamma_1 \gamma_2 e_1 e_2.$$

(2) If one of two conjugate forces of a latent system be latent, the other force is also latent. For let $S = \lambda D_1 + \mu D_2$; and assume that $\phi S \equiv S$, $\phi D_1 \equiv D_1$. Then since

$$\phi S = \lambda \phi D_1 + \mu \phi D_2 \equiv \lambda D_1 + \mu D_2,$$

it follows that $\phi D_2 \equiv D_2$.

Also from subsection (1), $\phi D_1 = \gamma\gamma' D_1$, and $\phi D_2 = \gamma''\gamma''' D_2$; where γ, γ', γ'', γ''' are latent roots of the matrix, but not necessarily distinct roots. Thus $\gamma\gamma' = \gamma''\gamma'''$. This forms another proof of § 191 (6).

(3) Hence, if a latent system S (not a single force) exist and also a semi-latent line which is not a null line of S, then another semi-latent line not intersecting the first must exist, and such that the two lines are conjugates with respect to S.

(4) The null plane of a latent point with respect to a latent system is semi-latent.

For let e be the latent point, S the latent system, P a planar element in the null plane. Then

$$P = \lambda e S, \quad \phi P = \lambda \phi e S = \lambda \phi e \phi S \equiv e S \equiv P.$$

Conversely the null point of a semi-latent plane with respect to a latent system is latent.

(5) Hence from (2), (3), (4) it is easy to prove that, if a tetrahedron of latent points exist, every latent system must have two semi-latent lines as one pair of conjugates.

(6) If a group is semi-latent, its reciprocal group obviously is also semi-latent.

(7) In general a semi-latent group of $\rho - 1$ dimensions ($\rho \gtreqless 6$) contains ρ latent systems. This follows from § 191 (3) and (7).

(8) If no special relation holds between the latent roots of the matrix, then [cf. § 191 (4)] the only six latent systems are six single forces on the six edges of the tetrahedron formed by the four latent points. It follows that in general a semi-latent group of $(\rho - 1)$ dimensions must have ρ edges of the fundamental tetrahedron of the matrix as director lines.

Thus in general a dual group of the general type can be a semi-latent group, namely, any dual group with two non-intersecting edges of the fundamental tetrahedron as director lines. Also [cf. subsection (6)] in general a quadruple group of the general type can be a semi-latent group, namely, any quadruple group with two non-intersecting edges of the fundamental tetrahedron as common null lines.

But in general a triple group of the general type cannot be semi-latent. For the director lines of a triple group, which are generating lines of the same species of a quadric, do not intersect unless the quadric degenerate into a cone or into two planes. But there are not three non-intersecting edges of a tetrahedron.

(9) A matrix can always be constructed with four assigned latent roots, so that any given dual group (or any given quadruple group) is semi-latent. For it is only necessary to choose two of the latent points on one director line (or common null line), and two on the other director line (or common null line).

(10) Every semi-latent dual group, which is not parabolic and does not consist of single forces, contains at least two distinct latent systems, which are either the director forces or two reciprocal systems.

For, if D_1 and D_2 be the director forces of the group, then D_1 and D_2 remain the director forces of the group after transformation. Hence either $\phi D_1 \equiv D_1$, and $\phi D_2 \equiv D_2$; or $\phi D_1 \equiv D_2$, and $\phi D_2 \equiv D_1$.

In the first case the two director forces are the two latent systems.

In the second case, let $\phi D_1 = \alpha D_2$, $\phi D_2 = \beta D_1$. Let one of the latent systems be $S = \lambda D_1 + \mu D_2$.

Then
$$\phi S = \sigma S = \lambda \phi D_1 + \mu \phi D_2 = \lambda \alpha D_2 + \mu \beta D_1.$$
Hence
$$\sigma \lambda = \mu \beta, \quad \sigma \mu = \lambda \alpha.$$
Therefore
$$\sigma^2 = \alpha \beta.$$
Hence
$$\frac{\lambda}{\sqrt{\beta}} = \pm \frac{\mu}{\sqrt{\alpha}}.$$

Thus the two latent systems are

$$\sqrt{\beta}D_1 + \sqrt{\alpha}D_2 \text{ and } \sqrt{\beta}D_1 - \sqrt{\alpha}D_2;$$

and these systems are reciprocal [cf. § 173 (6)].

But reciprocal systems of a dual group of the general type are necessarily distinct.

(11) The director force of a semi-latent parabolic group is evidently latent.

Also [cf. § 172 (9)] the null plane of any point on the director line is the same for each system of the group. Hence the null plane of every latent point on the director line is semi-latent. Thus there must be as many semi-latent planes passing through the director line as there are null points on it, for a semi-latent parabolic group to be possible. And conversely there must be as many latent points on the director line as there are semi-latent planes through it.

(12) If a semi-latent line exists which does not intersect the director line of a semi-latent parabolic group, then the parabolic group contains at least one latent system in addition to its director force.

For let D be its director force and N the semi-latent line. Then [cf. § 172 (10)] one and only one system of the group exists, for which N is a null line. This system is $D(NS) - S(ND)$; where S is any other system of the group. But since the group is semi-latent, and the lines N and D are semi-latent, this system must be latent.

(13) Let $D = e_1e_2$, and $N = e_3e_4$; and let e_1 and e_3 be the latent points on D and N which certainly exist. Then e_1e_2 is a semi-latent line. Now by subsections (3) and (4), either e_1e_2 is not a null line of the latent system, and then its conjugate is also a latent line intersecting both D and N in two other null points; or e_1e_2 is a null line of the system.

In this last case e_2 is the null point with respect to the latent system of the plane e_1N, since the two null lines e_1e_2 and N intersect in it: also e_1 is the null point of the plane e_2D for all the systems of the group [cf. § 172 (9)], since the null lines e_1e_2 and D intersect in it.

(14) It follows from (12) that the only possibility, for the existence of a semi-latent dual group with only one latent system, is when all the semi-latent lines intersect one of their number. Then such a semi-latent group is parabolic, and the director force is the single latent system, and is on that one of the semi-latent lines intersected by all the rest.

Furthermore, if two non-intersecting semi-latent lines exist, and no latent system exists which is not a single force, then no semi-latent parabolic group with either of these lines as director line is possible. For by (12), such a group must contain a second latent system, and by hypothesis such a system does not exist.

(15) Every semi-latent triple group of the general type must contain at least three distinct independent latent systems, unless it contains a semi-latent dual group with only one latent system. This is obvious, remembering that the properties of semi-latent triple groups are particular cases of the properties of the linear transformation of points in a complete region of two dimensions [cf. § 145 (4) and § 148].

Also by the preceding subsections such a semi-latent dual group is parabolic. Hence, if for any matrix no semi-latent parabolic group with only one latent system exists, every semi-latent triple group must have three distinct latent systems.

(16) If a plane region of latent points exists, every director force D of a semi-latent triple group intersects this plane, and therefore has a latent point on it. Thus D and ϕD are either congruent or intersect. If they intersect, the triple group is not of the general type, since ϕD is also a director force. Hence the only possible type of semi-latent triple group is latent.

(17) If a line of latent points exists, this line is either a generator of any quadric or intersects it in two points. Now consider the quadric defined by any semi-latent triple group G. The reciprocal group G' is also semi-latent.

Firstly let the latent line intersect the quadric. Then by the same reasoning as in the previous subsection (16), two director forces of G and two of G' must be latent, assuming that G is of the general type.

Secondly let the latent line be a director line of G. Then as in (16) all the director forces of G' (null lines of G) are latent.

Thirdly let the latent line be a null line of G'. Then all the director forces of G are latent.

193. Enumeration of Types of Latent Systems and Semi-Latent Groups. (1) Let no two roots of the sextic of § 191 (5) be equal. The four latent roots of the matrix ϕ are unequal, and only four latent points e_1, e_2, e_3, e_4 exist. Then [cf. § 191 (6)] the only latent systems are the six forces on the edges of the tetrahedron $e_1e_2e_3e_4$.

By § 192 (10) the only semi-latent dual groups, not parabolic, have two edges of the tetrahedron as director lines.

No semi-latent parabolic group can exist, for by § 192 (12) and (14) a latent system of the group must exist which is not a director line; and there is no such system.

No semi-latent triple group of the general type can exist. For [cf. § 192 (15)] such a group must contain three latent systems, that is, three edges of the tetrahedron $e_1e_2e_3e_4$ as director lines. But three non-intersecting edges cannot be found.

The semi-latent quadruple and quintuple groups can be found by the use of § 192 (6). Thus it is not necessary to enumerate them.

(2) Let the four roots $\gamma_1, \gamma_2, \gamma_3, \gamma_4$ of the matrix be unequal, and [cf. § 191 (6)] let $\gamma_1\gamma_3 = \gamma_2\gamma_4$.

Case I. Let no other roots of the sextic of § 191 (5) be equal. Then, as in (1), the four latent points $e_1e_2e_3e_4$ form a tetrahedron. The latent systems are the six single forces on the edges of the tetrahedron, and any system of the type $\lambda e_1e_3 + \mu e_2e_4$. It can be seen by the use of § 192 (2) that no other system can be latent. The dual group defined by e_1e_3 and e_2e_4 is therefore latent. The dual groups defined by any system of the type $\lambda e_1e_3 + \mu e_2e_4$ together with e_1e_2, or e_3e_4, or e_1e_4, or e_3e_2 are semi-latent. They are parabolic groups. The only semi-latent dual groups, not parabolic, have two edges of the tetrahedron as director lines.

By § 192 (14) no semi-latent dual group exists with only one latent system.

The semi-latent triple groups of the general type are, all groups of the type defined by $\lambda e_1e_3 + \mu e_2e_4$, e_1e_2, e_3e_4; and all groups of the type defined by $\lambda e_1e_3 + \mu e_2e_4$, e_1e_4, e_2e_3.

The generality of these types is proved by § 175 (9) and (12). By § 192 (15) no other semi-latent triple group (of the general type) exists.

Case II. Let $\gamma_1\gamma_3 = \gamma_2\gamma_4$, and $\gamma_1\gamma_4 = \gamma_2\gamma_3$. Then $\gamma_1^2 = \gamma_2^2$, $\gamma_3^2 = \gamma_4^2$. Hence, excluding the case of equal roots which is discussed later [in subsection (5)], $\gamma_1 = -\gamma_2$, $\gamma_3 = -\gamma_4$.

As in Case I. there are only four latent points, which form a tetrahedron. The latent systems are, the six single forces on the edges of the tetrahedron, and [cf. § 191 (6)] any system of the type $\lambda e_1e_3 + \mu e_2e_4$, and any system of the type $\lambda e_1e_4 + \mu e_2e_3$.

The semi-latent dual groups are, the semi-latent group defined by e_1e_2 and e_3e_4, the latent group defined by e_1e_3 and e_2e_4, the latent group defined by e_1e_4 and e_2e_3, any group defined by systems of the types $\lambda e_1e_3 + \mu e_2e_4$ and $\lambda' e_1e_4 + \mu' e_2e_3$. This last type of semi-latent group is not parabolic, unless one of the four quantities λ, μ, λ', μ' vanishes.

The semi-latent triple groups of the general type are all groups defined by sets of three systems of the following types:

$$e_1e_2,\ e_3e_4,\ \lambda e_1e_3 + \mu e_2e_4,$$
$$e_1e_2,\ e_3e_4,\ \lambda e_1e_4 + \mu e_2e_3,$$
$$e_1e_4,\ e_2e_3,\ \lambda e_1e_3 + \mu e_2e_4,$$
$$e_1e_3,\ e_2e_4,\ \lambda e_1e_4 + \mu e_2e_3.$$

Thus there are four types of semi-latent triple groups for this case.

There appears to be a fifth type of group of which a typical specimen is the group defined by $\lambda e_1e_3 + \mu e_2e_4$, $\lambda' e_1e_4 + \mu' e_2e_3$, e_1e_3. But this group cannot be of the general type. For from § 175 (9) only one parabolic subgroup of a triple group (of the general type) exists with a given director line of the triple group as its director line. Whereas in the group above mentioned, e_1e_3 is the director line of two such subgroups.

By § 192 (15) no other semi-latent triple group of the general type can exist.

(3) Let $\gamma_1 = \gamma_2$. Then from § 190 (2) there are two cases to be considered; but two extra cases arise, when the relation $\gamma_1^2 = \gamma_3\gamma_4$ is satisfied. Thus there are four cases in all; in the first two cases it is assumed that $\gamma_1^2 \neq \gamma_3\gamma_4$. Let the notation of § 190 (2) be adopted.

Case I. [Cf. § 190 (2), Case I.] The line e_1e_2 is the latent region corresponding to the root γ_1. Then from § 190 (2) the latent systems are, e_1e_2, e_3e_4, any force of the type $e_3(\lambda e_1 + \mu e_2)$, any force of the type $e_4(\lambda e_1 + \mu e_2)$.

From § 192 (3) no other latent systems exist. Hence no latent systems exist, which are not single forces.

The semi-latent dual groups (not consisting entirely of single forces) are the group defined by e_1e_2, e_3e_4; and the groups of the type defined by $e_3(\lambda e_1 + \mu e_2)$, $e_4(\lambda' e_1 + \mu' e_2)$.

No semi-latent parabolic group exists [cf. § 192 (14)].

No semi-latent triple group of the general type exists; since three non-intersecting latent forces do not exist, and the only latent systems are single forces.

Case II. [Cf. § 190 (2), Case II.] The line e_1e_2 is the semi-latent region corresponding to the root γ_1. The latent systems are the forces e_1e_2, e_1e_3, e_1e_4, e_3e_4. It is impossible for any latent system (not a single force) to exist, since e_1e_2, e_1e_3, e_1e_4 are not coplanar, and cannot therefore all be null lines. Hence the theorem of § 192 (3) applies; and the truth of the statement can easily be seen, since $\lambda e_1e_2 + \mu e_3e_4$ is not latent.

The only semi-latent dual group, not consisting entirely of single forces, is that defined by e_1e_2, e_3e_4.

No semi-latent parabolic group can exist [cf. § 192 (11) and (14)].

No semi-latent triple group of the general type can exist [cf. § 192 (15)].

Case III. (Subcase of Case I.) Let $\gamma_1^2 = \gamma_3\gamma_4$; and let the arrangement of latent and semi-latent points and regions be that of Case I. Then, from Case I. and § 192 (3) the only latent systems are, e_1e_2, e_3e_4, any force of the type $e_3(\lambda e_1 + \mu e_2)$, any force of the type $e_4(\lambda e_1 + \mu e_2)$, and any system of the type $\lambda e_1e_2 + \mu e_3e_4$.

The semi-latent dual groups (not consisting entirely of single forces) are the (latent) group defined by e_1e_2, e_3e_4; any group of the type defined by $e_3(\lambda e_1 + \mu e_2)$, $e_4(\lambda' e_1 + \mu' e_2)$; the parabolic groups of the types defined by

$\lambda e_1e_2 + \mu e_3e_4$ together with either $e_3(\lambda' e_1 + \mu' e_2)$, or $e_4(\lambda' e_1 + \mu' e_2)$. No other semi-latent parabolic groups exist [cf. § 192 (14)]. The semi-latent triple groups are groups of the type defined by

$$\lambda e_1e_2 + \mu e_3e_4,\ e_3(\lambda' e_1 + \mu' e_2),\ e_4(\lambda'' e_1 + \mu'' e_2).$$

Case IV. (Subcase of Case II.) Let $\gamma_1^2 = \gamma_3\gamma_4$; and let the arrangement of latent and semi-latent points and regions be that of Case II. Then the latent systems are the forces e_1e_2, e_1e_3, e_1e_4, e_3e_4, and any system of the type $\lambda e_1e_2 + \mu e_3e_4$. By § 192 (3) no other latent systems exist.

The semi-latent dual groups, not consisting entirely of single forces, are, the (latent) group defined by e_1e_2, e_3e_4, all parabolic groups of the types defined by $\lambda e_1e_2 + \mu e_3e_4$ together with e_1e_3 or e_1e_4. No other semi-latent parabolic groups can exist [cf. § 192 (11) and (14)]. No semi-latent triple group of the general type can exist.

(4) Let $\gamma_1 = \gamma_2 = \gamma_3$. Then from § 190 (3) there are three cases to be considered.

Case I. [Cf. § 190 (3), Case I.] The plane $e_1e_2e_3$ is the latent region corresponding to the root γ_1. The latent systems are, any force in the plane $e_1e_2e_3$, any force through the point e_4.

No other latent system exists. For all the latent lines in the plane $e_1e_2e_3$ cannot be null lines, since all the null lines lying in a plane must pass through the null point. Thus some line bc lying in the plane $e_1e_2e_3$ may be assumed not to be a null line of any such latent system. Then by § 192 (3) the system must be of the form $\lambda e_4a + \mu bc$, where b and c lie in the plane e_1, e_2, e_3, and e_4a is any line through e_4. Let a be assumed to be the point in which this line meets the plane $e_1e_2e_3$.

Hence $$\phi(\lambda e_4a + \mu bc) = \gamma_1\gamma_4\lambda e_4a + \gamma_1^2\mu bc.$$

Therefore such systems are not latent.

The semi-latent dual groups are all groups of the type defined by e_4a and bc, where bc lies in the plane $e_1e_2e_3$.

No semi-latent parabolic group can exist [cf. § 192 (14)].

No semi-latent triple group of the general type can exist [cf. § 192 (16)].

Case II. [Cf. § 190 (3), Case II.] The line e_1e_2 is the latent region corresponding to the root γ_1; and the plane $e_1e_2e_3$ is the semi-latent region of the first species corresponding to the first root. Also e_2 and e_3 are such that $\phi e_3 = \gamma_1 e_3 + \delta_2 e_2$, where δ_2 is not zero. The latent forces are, e_1e_2, any force through e_2 in the plane $e_1e_2e_3$, any force through e_4 in the plane $e_1e_2e_4$. No other latent systems exist. For a similar proof to that in Case I. shows that a latent system, not a single force, cannot have two non-intersecting semi-latent lines as conjugate lines. Hence such a system must have all the semi-latent lines as null lines. Therefore e_2 must be the null point of the plane

$e_1e_2e_3$, and e_4 the null point of the plane $e_4e_1e_2$. But since the null line e_1e_2 does not go through e_4, which is the null point of the plane $e_1e_2e_4$, this is impossible.

The semi-latent dual groups (not entirely consisting of single forces) are all groups defined by a force through e_2 in the plane $e_1e_2e_3$ and a force through e_4 in the plane $e_1e_2e_4$.

No semi-latent parabolic group exists [cf. § 192 (14)].

No semi-latent triple group of the general type exists [cf. § 192 (15) and (17)].

Case III. [Cf. § 190 (3), Case III.] The only latent points are, the point e_1 corresponding to γ_1, and the point e_4 corresponding to γ_4.

The only latent systems are, the force e_1e_2, and the force e_1e_4. There can be no other latent system (not a single force). For by § 192 (4) the null point with respect to such a system of the semi-latent plane $e_1e_2e_3$ is e_1, and therefore the null point of the semi-latent plane $e_1e_2e_4$ is e_4. Hence the line e_1e_4 is not a null line of such a system; and therefore from § 192 (3) another semi-latent line, not intersecting e_1e_4, is required. But such a line does not exist.

There are no semi-latent dual groups, not consisting entirely of single forces. For the only possibility of such a group lies in the possibility of a semi-latent parabolic group with only its director force latent [cf. § 192 (14)]. But e_1e_4 cannot be the director force of such a group, since by § 192 (11) there must be as many semi-latent planes containing e_1e_4 as there are latent points on e_1e_4. But e_1 and e_4 are both latent points; while $e_1e_2e_4$ is the only semi-latent plane through $e_1e_2e_4$. Again e_1e_2 cannot be the director force, for by § 192 (11) there ought to be as many latent points on it as there are semi-latent planes through it. But there are two semi-latent planes through it, namely $e_1e_2e_3$ and $e_1e_2e_4$; and there is only one latent point on it.

There are no semi-latent triple groups of the general type [cf. § 192 (15)].

(5) Let there be only one latent root γ_1 of the matrix. Then [cf. § 190 (4)] there are five cases to be considered: but, of these, the first may be dismissed at once.

Case II. [Cf. § 190 (4), Case II.] The latent region is the plane $e_1e_2e_3$. The latent systems are of the type $\lambda e_3a + \mu bc$, where a is any point and the force bc is any force lying in the plane $e_1e_2e_3$.

If S be any one of these latent systems, $\phi S = \gamma_1^2 S$.

The semi-latent dual groups (not consisting entirely of single forces) are, all groups defined by the forces e_3a and bc, where a is any point and bc is any line lying in the plane $e_1e_2e_3$; and any parabolic group with a director force of the type e_3d, where d lies in the plane $e_1e_2e_3$. All the semi-latent dual groups are thus either latent, or else possess only one latent system.

In order to prove the above statements, first notice that, if S_1 and S_2 be the latent systems of a semi-latent group with two distinct latent systems, $\phi S_1 = \gamma_1^2 S_1$ and $\phi S_2 = \gamma_1^2 S_2$. Hence the group is latent, and hence the director forces are latent, if there are two of them.

Accordingly the director forces of a non-parabolic semi-latent group are of the type described.

Again [cf. § 192 (11)] the number of latent points on the director line of a semi-latent parabolic group is equal to the number of semi-latent planes passing through it. Hence a force of the type $e_3 d$ is the only possible director force of such a group [cf. § 190 (4), Case II.]. Now if d be any point on this director line, a system S of one of the parabolic groups with this line as director line can be written in the form $S = \lambda e_3 x + \mu d y$, where x and y are any points [cf. § 172 (9)]. Now the dual group defined by $e_3 d$ and S is semi-latent.

For [cf. § 190 (4), equation (F)],

$$\begin{aligned}\phi S &= \lambda \phi e_3 \phi x + \mu \phi d \phi y \\ &= \gamma_1 \lambda e_3 (\gamma_1 x + \delta e_3) + \gamma_1 \mu d (\gamma_1 y + \delta' e_3) \\ &= \gamma_1^2 S - \delta' \gamma_1 \mu e_3 d.\end{aligned}$$

Thus ϕS also belongs to the dual group.

Furthermore, if δ' be not zero (that is, if dy do not lie in the plane $e_1 e_2 e_3$), $e_3 d$ is the only latent system of the group. But if δ' be zero (that is, if dy do lie in the plane $e_1 e_2 e_3$), every system is latent and the group is therefore latent.

No semi-latent triple group of the general type exists [cf. § 192 (16)].

Case III. [Cf. § 190 (4), Case III.] The latent region is the line $e_1 e_2$. The latent systems are all forces in the plane $e_1 e_2 e_3$ through the point e_2. There are no other latent systems: for all the planes through $e_2 e_3$ are semi-latent; and their null points with respect to any latent system must be latent [cf. § 192 (4)]. But e_2 is the only latent point on all these planes.

There are no semi-latent dual non-parabolic groups (not consisting entirely of single forces). For there are evidently no semi-latent dual groups (not single forces) with two latent systems. There are semi-latent parabolic groups of the type defined by $e_2 a$ and $ab + \lambda e_2 k$; where a is any point on the plane $e_1 e_2 e_3$, but not on $e_1 e_2$ or $e_2 e_3$ [cf. § 192 (11)], b is any point on the plane $e_1 e_2 e_3$, k is any point, and λ is determined by a certain condition. In order to prove this, note that the form assumed is obvious from § 192 (11) and from the consideration, that the null point of the semi-latent plane $e_1 e_2 e_3$ is latent and lies on the director force $e_2 a$, and must therefore be e_2. Now let $a = \alpha_1 e_1 + \alpha_2 e_2 + \alpha_3 e_3$, where neither α_1 nor α_3 is zero; let $b = \beta_1 e_1 + \beta_2 e_2 + \beta_3 e_3$; let $k = \kappa_1 e_1 + \kappa_2 e_2 + \kappa_3 e_3 + \kappa_4 e_4$.

Then, using § 190 (4), equation (G),

$$\phi\,(ab + \lambda e_2 k) = \gamma_1{}^2\,(ab + \lambda e_2 k) + \delta_2\gamma_1\,(\alpha_3 e_2 b - \beta_3 e_2 a) + \lambda\gamma_1\delta_3\kappa_4 e_2 e_3.$$

Now $\phi\,(ab + \lambda e_2 k)$ must belong to the group defined by $ab + \lambda e_2 k$ and ae_2. Hence

$$\delta_2\gamma_1\,(\alpha_3 e_2 b - \beta_3 e_2 a) + \lambda\gamma_1\delta_3\kappa_4 e_2 e_3 \equiv e_2 a.$$

Hence $\delta_2\gamma_1\alpha_3 e_2 ab - \lambda\gamma_1\delta_3\kappa_4 e_2 e_3 a = 0.$

(It is easy to deduce from this equation another proof of the limitation of the position of a, namely that it is not to lie on $e_1 e_2$ or on $e_2 e_3$.)

Again $e_2 ab \equiv e_1 e_2 e_3 \equiv e_2 e_3 a = \alpha e_2 e_3 a,$ say.

Then $$\lambda = \frac{\delta_2\gamma_1\alpha_3\alpha}{\gamma_1\delta_3\kappa_4}.$$

These semi-latent parabolic groups have only one latent system.

There are no semi-latent triple groups of the general type [cf. § 192 (17)].

Case IV. [Cf. § 190 (4), Case IV.] The sole latent point is the point e_1. The only latent system is the force $e_1 e_2$.

The semi-latent dual groups are parabolic groups, with $e_1 e_2$ as axis, and with e_1 as the null point of the semi-latent plane $e_1 e_2 e_3$. Such semi-latent groups have only one latent system, namely the director force $e_1 e_2$. Also all such groups are not semi-latent; one condition has to be fulfilled, which will be investigated as follows.

Let $k = \kappa_1 e_1 + \kappa_2 e_2 + \kappa_3 e_3 + \kappa_4 e_4$, where κ_4 is not zero; let $a = \alpha_1 e_1 + \alpha_2 e_2$, where α_2 is not zero; and let $b = \beta_1 e_1 + \beta_2 e_2 + \beta_3 e_3$, where β_3 is not zero. Then a parabolic group of the specified type is defined by $e_1 a$ and $e_1 k + \lambda ab$. If k, a and b are given, then λ is determined by a condition, which is found as follows.

From § 190 (4), equation (H), remembering that $e_1 e_2 \equiv e_1 a \equiv e_2 a$,

$$\phi\,(e_1 k + \lambda ab) = \gamma_1{}^2\,(e_1 k + \lambda ab) + \xi e_1 a + \gamma_1\delta_3\kappa_4 e_1 e_3 + \lambda\gamma_1\delta_1\alpha_2 e_1 b\,;$$

where ξ need not be calculated.

But $\phi\,(e_1 k + \lambda ab)$ belongs to the parabolic group.

Thus $\gamma_1\delta_3\kappa_4 e_1 e_3 + \lambda\gamma_1\delta_1\alpha_2 e_1 b \equiv e_1 e_2.$

Hence $\gamma_1\delta_3\kappa_4 e_1 e_2 e_3 + \lambda\gamma_1\delta_1\alpha_2 e_1 e_2 b = 0.$

Also $e_1 e_2 b = \beta_3 e_1 e_2 e_3.$

Therefore $\gamma_1\delta_3\kappa_4 + \lambda\gamma_1\delta_1\alpha_2\beta_3 = 0.$

But $\gamma_1,\ \delta_1,\ \alpha_2,\ \beta_3,\ \kappa_4$ are not zero.

Hence $$\lambda = -\frac{\delta_3\kappa_4}{\delta_1\alpha_2\beta_3}.$$

No semi-latent triple group exists. For if such a group existed, $e_1 e_2$ must be a director line. But the reciprocal group must also be semi-latent, and therefore it must also contain $e_1 e_2$ as a director line.

Case V. [Cf. § 190 (4), Case V.] The latent region is the line e_1e_3. Those latent systems which are single forces are grouped in planes: thus corresponding to the latent point $a\ (=\alpha_1e_1+\alpha_3e_3)$, there are an infinite number of latent forces of the type ax; where x is any point on the plane of which the equation is

$$\alpha_3\delta_1\xi_2=\alpha_1\delta_3\xi_4.$$

Any point x lies on the line of a latent force.

For $$x\phi x=x\,(\delta_1\xi_2e_1+\delta_3\xi_4e_3).$$

Hence $$\phi\,(x\phi x)=\gamma_1^2x\,(\delta_1\xi_2e_1+\delta_3\xi_4e_3)=\gamma_1^2x\phi x.$$

Also the force e_1e_3 is latent.

The only latent systems, which are not single forces, are formed by any two latent forces as conjugate forces: thus, if x and y be any points, $x\phi x+y\phi y$ is a latent system. And if S be such system $\phi S=\gamma_1^2S$. Also it may be noted that e_1e_3 is a null line of all such systems.

There is no *à priori* impossibility in the existence of latent systems with all the semi-latent lines as null lines. The following investigation shows that such a latent system does not exist.

For, since e_1e_3, e_1e_4, e_2e_3 are to be null lines, systems, with all the semi-latent lines as null lines, must be of the form

$$\xi_{13}e_1e_3+\xi_{14}e_1e_4+\xi_{23}e_2e_3.$$

This form is found by assuming

$$S=\xi_{23}e_2e_3+\xi_{31}e_3e_1+\xi_{12}e_1e_2+\xi_{14}e_1e_4+\xi_{24}e_2e_4+\xi_{34}e_3e_4;$$

then the following equations must hold

$$0=(e_1e_3S)=(e_1e_4S)=(e_2e_3S).$$

Also any point $a\ (=\alpha_1e_1+\alpha_3e_3)$ on the line e_1e_3 is the null point of the semi-latent plane $e_1e_3\,(\alpha_1\delta_3e_2+\alpha_3\delta_1e_4)$, for all values of the ratio of α_1 to α_3.

Hence $$e_1e_3\,(\alpha_1\delta_3e_2+\alpha_3\delta_1e_4)\,.\,(\xi_{13}e_1e_3+\xi_{14}e_1e_4+\xi_{23}e_2e_3)\equiv\alpha_1e_1+\alpha_3e_3;$$

that is $$\alpha_1\delta_3\xi_{14}e_1+\alpha_3\delta_1\xi_{23}e_3\equiv\alpha_1e_1+\alpha_3e_3.$$

Hence systems with all the semi-latent lines as null lines must be of the type

$$\xi_{13}e_1e_3+\xi\,(\delta_1e_1e_4+\delta_3e_2e_3).$$

But by operating on such a system with the matrix, it is easy to see that such a system is not latent. For, if S be the system,

$$\phi S=\gamma_1^2S+2\xi\gamma_1\delta_1\delta_3e_1e_3.$$

Any dual group defined by two latent systems is latent, and has therefore (unless it be parabolic) two latent forces as director forces.

Any parabolic group with e_1e_3 as director line is semi-latent. Such a group is either latent, or has only one latent system (the director force). For, let a and b be any two (latent) points on e_1e_3, and let x and y be any other two points. Then $\phi a = \gamma_1 a$, $\phi b = \gamma_1 b$, $\phi x = \gamma_1 x + x'$, $\phi y = \gamma_1 y + y'$; where x' and y' are points on e_1e_3. Hence, if S be the system $ax + by$,

$$\phi S = \gamma_1^2 S + \lambda e_1 e_3;$$

since

$$ax' \equiv by' \equiv e_1 e_3.$$

Thus S and e_1e_3 define a semi-latent parabolic group. If S be latent (that is, if λ be zero), the group is latent.

The semi-latent triple groups (not entirely single forces) are of two kinds, which will be called, type I. and type II.

A semi-latent group of type I. is defined by any three non-intersecting latent forces, namely by $x\phi x$, $y\phi y$, $z\phi z$. Any such group is latent.

The groups of type II. are the groups reciprocal to those of type I. Thus e_1e_3, which is a common null line of every group of type I., is a common director line of every group of type II. Also, since every other semi-latent line intersects e_1e_3, this force is the only latent director force of any group of type II.

Again it has been proved that every semi-latent line is a null line of the system $\delta_1 e_1 e_4 + \delta_3 e_2 e_3$. Hence this system is reciprocal to every system of every semi-latent group of type I. Accordingly the semi-latent parabolic subgroup defined by e_1e_3, $\delta_1 e_1 e_4 + \delta_3 e_2 e_3$ is a subgroup of every reciprocal semi-latent group of type II. The force e_1e_3 is the only latent system of this common subgroup, since the system $\delta_1 e_1 e_4 + \delta_3 e_2 e_3$ is not latent. Thus a semi-latent group of type II. has only one latent system, namely e_1e_3. For, by § 175 (9), a director force (e_1e_3) of a triple group can only be a null line of one dual subgroup of systems; and e_1e_3 is a null line of every latent system.

Semi-latent groups of type II. are defined by e_1e_3, $\delta_1 e_1 e_4 + \delta_3 e_2 e_3$, and any system S. For the system S can be written

$$S = \varpi_{12} e_1 e_2 + \varpi_{13} e_1 e_3 + \varpi_{34} e_3 e_4 + \varpi_{23} e_2 e_3 + \varpi_{14} e_1 e_4 + \varpi_{24} e_2 e_4.$$

Hence $\phi S = \gamma_1^2 S + (\varpi_{23}\delta_1\gamma_1 + \varpi_{14}\delta_3\gamma_1 + \varpi_{24}\delta_1\delta_3)\, e_1 e_3 + \varpi_{24}\gamma_1 (\delta_1 e_1 e_4 + \delta_3 e_2 e_3)$.

Thus ϕS belongs to the triple group defined by S, e_1e_3 and $\delta_1 e_1 e_4 + \delta_3 e_2 e_3$.

The other semi-latent parabolic groups cannot belong to any semi-latent triple group. For the reciprocal to such a group must be another semi-latent triple group. But groups of type I. and II. are respectively reciprocal in pairs: so this reciprocal semi-latent group must contain another semi-latent parabolic subgroup. Thus two reciprocal triple groups would each contain the director line e_1e_3 in common. This is impossible for triple groups, not of a special kind.

(6) Let the roots of the matrix ϕ be equal in pairs; so that $\gamma_1 = \gamma_2$, and $\gamma_3 = \gamma_4$. Then [cf. § 190 (5)] there are three cases to be considered, as far as the distribution of latent and semi-latent points and regions is concerned.

But each of these three principal cases gives rise to another case, in which the additional relation $\gamma_1 = -\gamma_3$ is fulfilled. Thus there are six cases in all.

Case I. [Cf. § 190 (5), Case I.] The latent regions are the lines e_1e_2 and e_3e_4. The latent systems are the forces e_1e_2, e_3e_4, and any system of the quadruple group which has e_1e_2 and e_3e_4 as common null lines.

The semi-latent dual groups are defined by any two of these latent systems; since by § 192 (14) no semi-latent parabolic group exists with only one latent system.

The semi-latent triple groups are defined by any three of these latent systems.

There are only two types of semi-latent triple groups, of a general kind: let them be called type I. and type II.

Type I. consists of groups defined by three systems of the form, ab, $a'b'$, $a''b''$, where a, a', a'' lie on e_1e_2, and b, b', b'' lie on e_3e_4.

Type II. consists of groups defined by three systems of the form e_1e_2, e_3e_4, $\lambda ab + \mu a'b'$. Each group of type II. is reciprocal to a corresponding group in type I.

Groups defined by two forces of the form ab and $a'b'$, together with either e_1e_2 or e_3e_4, are not of the general kind; since in them the director force e_1e_2 (or e_3e_4) intersects the director forces ab and $a'b'$.

Case II. [Cf. § 190 (5), Case II.] The latent regions are the line e_1e_2 and the point e_3. The latent systems are, the forces e_1e_2, e_3e_4, and any force of the type $e_3(\lambda e_1 + \mu e_2)$. There can be no other latent systems; for the coplanar lines of the type $e_3(\lambda e_1 + \mu e_2)$ and e_1e_2 are not all concurrent and therefore cannot all be null lines. Hence by using § 192 (3) the proposition is easily proved.

There is a semi-latent dual group defined by e_1e_2 and e_3e_4.

There are semi-latent parabolic groups with any force of the type $e_3(\lambda e_1 + \mu e_2)$ as director line [cf. § 192 (14)].

Now by § 192 (11) the point $\lambda e_1 + \mu e_2$ is the null point with respect to the group of the plane $e_1e_2e_3$; and the point e_3 is the null point of the plane $(\lambda e_1 + \mu e_2)e_3e_4$; since both e_1e_2 and e_3e_4 must be common null lines of the group. Hence any other system of such a group must be of the form

$$S = e_3(\lambda e_1 + \mu e_2) + \xi e_2e_3 + \eta(\lambda e_1 + \mu e_2)e_4.$$

And by § 190 (5), equation (K), it follows that

$$\phi S = \gamma_1\gamma_3 S + \eta\delta_3\gamma_1(\lambda e_1 + \mu e_2)e_3.$$

Hence any parabolic dual group of the type defined by $e_3(\lambda e_1 + \mu e_2)$ and $\xi e_2e_3 + \eta(\lambda e_1 + \mu e_2)e_4$ is semi-latent; and all such semi-latent groups contain only one latent system.

There can be no semi-latent triple groups of the general type [cf. § 192 (17)].

Case III. [Cf. § 190 (5), Case III.] The only two latent points are e_1 and e_3. The latent systems are e_1e_2, e_3e_4, e_1e_3, and any system of the type $\lambda e_3e_1 + \mu(\gamma_1\delta_3e_2e_3 - \gamma_3\delta_1e_1e_4)$. For [cf. § 192 (3) and (4)] the point e_1 is the null point of the plane $e_1e_2e_3$, and the point e_3 is the null point of the plane $e_3e_4e_1$. Hence any possible latent system S (not a single force) must be of the form

$$\lambda e_3e_1 + \mu_{23}e_2e_3 + \mu_{14}e_1e_4.$$

Now [cf. § 190 (5), equation (L)]

$$\phi S = \gamma_1\gamma_3 S + (\gamma_3\delta_1\mu_{23} + \gamma_1\delta_3\mu_{14})\, e_1e_3 \quad\text{..................(1).}$$

Thus, if S be latent,

$$\mu_{23} = \gamma_1\delta_3\mu, \quad \mu_{14} = -\gamma_3\delta_1\mu.$$

The semi-latent dual groups (not consisting entirely of single forces) are, the group defined by e_1e_2 and e_3e_4; any parabolic group [cf. equation (1)] of the type defined by the director force e_1e_3 and a system of the form $\mu_{23}e_2e_3 + \mu_{14}e_1e_4$; any parabolic group of the type defined by e_1e_2 and $\lambda e_3e_1 + \mu(\gamma_1\delta_3e_2e_3 - \gamma_3\delta_1e_1e_4)$; any parabolic group of the type defined by e_3e_4 and $\lambda e_3e_1 + \mu(\gamma_1\delta_3e_2e_3 - \gamma_3\delta_1e_1e_4)$.

It can easily be proved that no other dual groups are semi-latent, as follows. The only possibility lies in the existence of a parabolic group with a single latent system. But, from § 192 (11) and (14), e_1e_3 is the only possible director force of such a group; and the group must have the form stated above. Further, by equation (1) above, the group stated is actually semi-latent.

In searching for triple groups of the general type, it is useful to notice that such a group cannot be defined by a director force D and two systems S and S', such that $(DS) = 0 = (DS')$. For by § 175 (9) in a triple group of the general type only one parabolic group with D as director force can exist.

The semi-latent triple groups with three latent systems are all groups of the type defined by e_1e_2, e_3e_4, $\lambda e_3e_1 + \mu(\gamma_1\delta_3e_2e_3 - \gamma_3\delta_1e_1e_4)$. Call such groups the semi-latent groups of type I.

No two groups of type I. can be reciprocal to each other, since all such groups have one pair of director forces in common, namely e_1e_2 and e_3e_4. Thus there must be another type of semi-latent groups. Call them the groups of type II. The only semi-latent subgroups, which a group of type II. can contain, are parabolic subgroups of the type defined by e_1e_3, $\mu_{23}e_2e_3 + \mu_{14}e_1e_4$. Now the condition that the system $\mu_{23}e_2e_3 + \mu_{14}e_1e_4$ may be reciprocal to the system $\lambda e_3e_1 + \mu(\gamma_1\delta_3e_2e_3 - \gamma_3\delta_1e_1e_4)$ is

$$\gamma_1\delta_3\mu_{14} - \gamma_3\delta_1\mu_{23} = 0.$$

Hence all groups of type II. contain the parabolic subgroup defined by e_1e_3, $\gamma_1\delta_3e_2e_3 + \gamma_3\delta_1e_1e_4$. There is no other latent system which, in conjunction with this subgroup, will define a triple group of the general type. Hence all groups of type II. contain only one latent system, namely e_1e_3.

Any system S, which has e_1e_2 and e_3e_4 as null lines, defines in conjunction with the systems e_1e_3, $\gamma_1\delta_3e_2e_3 + \gamma_3\delta_1e_1e_4$, a semi-latent triple group of type II.

For we may write

$$S = \varpi_{23}e_2e_3 + \varpi_{31}e_3e_1 + \varpi_{14}e_1e_4 + \varpi_{24}e_2e_4.$$

Then $\phi S = \gamma_1\gamma_3 S + (\varpi_{23}\delta_1\gamma_3 + \varpi_{14}\gamma_1\delta_3 + \varpi_{24}\delta_1\delta_3)\, e_1e_3 + \varpi_{24}(\gamma_1\delta_3e_2e_3 + \gamma_3\delta_1e_1e_4)$.

Hence ϕS belongs to the required group. There can be no other semi-latent triple groups. For any other semi-latent triple groups must contain a parabolic subgroup of the type e_1e_3, $\mu_{23}e_2e_3 + \mu_{14}e_1e_4$. But two such triple groups (of the general type) cannot be reciprocal to each other, since they both contain a common director force e_1e_3.

Case IV. (Subcase of Case I.) The latent and semi-latent points and regions are the same as in Case I.: but the additional relation $\gamma_1 = -\gamma_3$ is satisfied. Thus $\gamma_1^2 = \gamma_3^2$.

The latent systems are the forces e_1e_2, e_3e_4, any system of the type $\lambda e_1e_2 + \mu e_3e_4$, and any system of the quadruple group which has e_1e_2 and e_3e_4 as common null lines. Any latent system S is either such that $\phi S = \gamma_1^2 S = \gamma_3^2 S$; or such that $\phi S = \gamma_1\gamma_3 S = -\gamma_1^2 S = -\gamma_3^2 S$.

No semi-latent parabolic group exists with only one latent system [cf. § 192 (14)] Hence all semi-latent groups have their full number of latent systems.

The semi-latent dual groups are defined by any two of these latent systems. The semi-latent triple groups are defined by any three of these latent systems.

There are three types of semi-latent triple groups. Type I. and type II. are the same as in Case I.; and their groups are reciprocal in pairs. Type III. consists of groups defined by three latent systems of the form $\lambda e_1e_2 + \mu e_3e_4$, ab, $a'b'$. The groups of this type are reciprocal in pairs; since the group defined by $\lambda e_1e_2 + \mu e_3e_4$, ab, $a'b'$, is reciprocal to the group defined by $\lambda e_1e_2 - \mu e_3e_4$, ab', $a'b$.

Case V. (Subcase of Case II.) The latent and semi-latent points and regions are the same as in Case II.: but the additional relation $\gamma_1 = -\gamma_3$ is satisfied. The latent systems are, the forces e_1e_2, e_3e_4, any system of the type $\lambda e_1e_2 + \mu e_3e_4$, any force of the type $e_3(\lambda e_1 + \mu e_2)$.

There is a latent dual group defined by e_1e_2 and e_3e_4; semi-latent parabolic groups defined by latent systems of the type $\lambda e_1e_2 + \mu e_3e_4$ and $e_3(\lambda' e_1 + \mu' e_2)$; and semi-latent parabolic groups, with only one latent system, defined by systems of the type $e_3(\lambda e_1 + \mu e_2)$ and $\xi e_2e_3 + \eta(\lambda e_1 + \mu e_2)e_4$ [cf. Case II.].

There can be no semi-latent triple groups of the general type [cf. § 192 (17)].

Case VI. (Subcase of Case III.) The latent and semi-latent points and regions are the same as in Case III.: but the additional relation $\gamma_1 = -\gamma_3$ is satisfied.

The latent systems are the same as in Case III. with the additional set of latent systems of the form $\lambda e_1 e_2 + \mu e_3 e_4$.

The semi-latent dual groups (not consisting entirely of single forces) are, the (latent) group defined by $e_1 e_2$ and $e_3 e_4$; any parabolic group of the type defined by $e_1 e_3$ and $\mu_{23} e_2 e_3 + \mu_{14} e_1 e_4$; any parabolic group of the type defined by $e_1 e_2$ and $\lambda e_3 e_1 + \mu (\gamma_1 \delta_3 e_2 e_3 - \gamma_3 \delta_1 e_1 e_4)$; any parabolic group of the type defined by $e_3 e_4$ and $\lambda e_3 e_1 + \mu (\gamma_1 \delta_3 e_2 e_3 - \gamma_3 \delta_1 e_1 e_4)$; and any group of the type defined by $\lambda e_1 e_2 + \mu e_3 e_4$ and $\lambda' e_3 e_1 + \mu' (\gamma_1 \delta_3 e_2 e_3 - \gamma_3 \delta_1 e_1 e_4)$.

The semi-latent triple groups are simply those of type I. and type II. in Case III. For the only possibility of additional semi-latent triple groups (of a general kind) beyond those of Case III. lies in the semi-latent groups of the type defined by $\lambda e_1 e_2 + \mu e_3 e_4$, $\lambda' e_3 e_1 + \mu' (\gamma_1 \delta_3 e_2 e_3 - \gamma_3 \delta_1 e_1 e_4)$, $e_1 e_3$. But these triple groups are not of the general kind, since the director force $e_1 e_3$ is the director force of two parabolic subgroups belonging to such groups.

194. TRANSFORMATION OF A QUADRIC INTO ITSELF. (1) When a triple group is semi-latent, the matrix must transform the director generators of the associated quadric [cf. § 175 (4) and (5)] into director generators of the same quadric. Thus each point on the associated quadric is transformed into a point on the same quadric; and the quadric may be said to be transformed into itself by a direct transformation, the associated triple groups being semi-latent.

(2) There is another way in which a matrix may transform a quadric into itself, so that the associated triple groups are not semi-latent. For the generators of one system may be transformed into generators of the other system, and vice versa. Let this be called the skew transformation of a quadric into itself, and let the first method be called the direct transformation.

(3) If a matrix transforms a quadric into itself, by either direct or skew transformation, then every semi-latent line either has two distinct latent points on it, or touches the quadric.

For let any semi-latent line cut the quadric in the two distinct points p and q. Then ϕp and ϕq are also on the quadric and on the semi-latent line. Hence either $\phi p \equiv p$, $\phi q \equiv q$, in which case p and q are two distinct latent points on the line, or $\phi p \equiv q$, $\phi q \equiv p$. In the second case $\phi^2 p \equiv \phi q \equiv p$, and $\phi^2 q \equiv q$.

Now if e be the sole latent point on the line, and γ be the repeated latent root corresponding to e, then $\phi p = \gamma p + \lambda e$, and $\phi^2 p = \gamma^2 p + 2\lambda\gamma e \equiv p$. Hence $\lambda = 0$. Thus p is a latent point. Similarly q is a latent point. Thus the line is a latent region corresponding to the repeated root, and e is not the sole latent point on it.

But if the semi-latent line touches the quadric, this reasoning does not apply. The point of contact must be a latent point; and, as far as has been shown, it may be the only latent point on the line.

(4) If the semi-latent line does not touch the quadric, and if $\phi p \equiv q$, $\phi q \equiv p$, assume that e_1 and e_2 are the two distinct latent points on the line, and that γ_1 and γ_2 are the corresponding latent roots.

Also let $p = \lambda_1 e_1 + \lambda_2 e_2$. Then $\phi^2 p = \gamma_1^2 \lambda_1 e_1 + \gamma_2^2 \lambda_2 e_2 \equiv \lambda_1 e_1 + \lambda_2 e_2$. Hence $\gamma_1^2 = \gamma_2^2$. But if $\gamma_1 = \gamma_2$, then $\phi p \equiv p$, $\phi q \equiv q$, which is contrary to the assumption. Hence in the present case $\gamma_1 = -\gamma_2$; and $\phi p = \gamma_1 (\lambda_1 e_1 - \lambda_2 e_2) \equiv q$. Thus $(pq, e_1 e_2)$ form a harmonic range. Thus e_1 and e_2 are conjugate points with respect to the quadric.

(5) The polar reciprocal of a latent or semi-latent line is itself latent or semi-latent. For since the quadric is unaltered by transformation and the original line retains the same position, its polar reciprocal must also retain the same position.

Also if a point be latent, its polar plane must be semi-latent. Hence if the latent point be not on the quadric, at least one other latent point must exist on the semi-latent polar plane. Then the line joining these latent points is semi-latent; and its reciprocally polar line is also semi-latent.

195. Direct Transformation of Quadrics. (1) It follows from the enumeration of § 193, that the only cases in which semi-latent triple groups of the general type exist are those cases stated in § 193 (2), Cases I. and II.: in § 193 (3) Case III.: in § 193 (5), Case V.: in § 193 (6), Cases I. and III. and IV. and VI. In all these cases the relation, $\gamma\gamma' = \gamma''\gamma'''$, holds between the four latent roots of the matrix. In the two cases of § 193 (2) the four latent roots are distinct; and there are only four latent points, which form a tetrahedron. In Case III. of § 193 (3) two latent roots γ_1 and γ_2 are equal, and $\gamma_1^2 = \gamma_3\gamma_4$: also a latent line exists corresponding to the double root: this is really a subcase of Case I. in § 193 (2). In Case V. of § 193 (5) all the latent roots are equal, there is a latent line, and an infinite number of semi-latent lines intersecting the latent line. This is really a subcase of Case III. of § 193 (6), and will be discussed after that case.

In Cases I. and III. of § 193 (6), the latent roots are equal in pairs, namely $\gamma_1 = \gamma_2$, $\gamma_3 = \gamma_4$; and either (Case I.) two latent lines exist corresponding respectively to the two distinct roots; or (Case III.) only two latent points exist, one corresponding to each root. Case I. is a subcase of § 193 (2).

Case IV. of § 193 (6) is Case I. with the additional relation $\gamma_1 = -\gamma_3$. It is partly merely a subcase of (6) Case I.: but it also transforms other quadrics according to the type of (2) Case I.: thus it is partly a subcase of (2) Case II.

Case VI. of § 193 (6) is a subcase of Case III., and in no way differs from it in its properties with regard to the direct transformation of quadrics.

(2) The general type of direct transformation of quadrics is given by § 193 (2), Case I. Then the associated quadric of any group of the type defined by e_1e_4, e_2e_3, $\alpha e_1e_3 + \beta e_2e_4$ is transformed into itself by direct transformation. The reciprocal group, associated with the same quadric, is e_1e_2, e_3e_4, $\alpha e_1e_3 - \beta e_2e_4$, and this group is also semi-latent.

The semi-latent lines e_1e_4, e_2e_3 are generators of one system, and the semi-latent lines e_1e_2, e_3e_4 are generators of the other system. Hence the quadric has four of its generators semi-latent, two of one system and two of the other.

It follows that the semi-latent lines e_1e_3 and e_2e_4 are reciprocally polar to each other, so that the polar plane of any point on one contains the other.

All quadrics containing these four generators are transformed into themselves. For they are defined by either group of a pair of reciprocal triple groups of the types mentioned above.

(3) In Case II. of § 193 (2), the only difference from the general case is that an additional set of quadrics are transformed into themselves; namely the associated quadrics of groups of the type defined by e_1e_2, e_3e_4, $\alpha e_1e_4 + \beta e_2e_3$. The reciprocal group associated with the same quadric is e_1e_3, e_2e_4, $\alpha e_1e_4 - \beta e_2e_3$; and this group is also semi-latent.

(4) In Case III. of § 193 (3) the only difference from the general case is that any two points a_1 and a_2 on the line e_1e_2 can be substituted for e_1 and e_2. Then a_1e_3, a_2e_4, a_1e_4, a_2e_3 are generators of one of the quadrics; and e_1e_2 and e_3e_4 are reciprocal lines.

(5) Case I. of § 193 (6) is really only a subcase of that described in the subsection (3). Let a_1 and a_2 be any two latent points on the line e_1e_2; and let a_3 and a_4 be any two latent points on the line e_3e_4. Then the explanation of the previous subsection applies, substituting any tetrahedron such as $a_1a_2a_3a_4$ for the tetrahedron $e_1e_2e_3e_4$ in the previous subsection. Thus all quadrics which have the two latent lines e_1e_2 and e_3e_4 as generating lines are transformed into themselves by direct transformation. Also all such quadrics have two generators of each system latent, or semi-latent.

(6) Case IV. of § 193 (6) transforms into themselves the quadrics mentioned above both in (3) and in (5): that is, quadrics with the latent lines e_1e_2 and e_3e_4 as generating lines, and quadrics with e_1e_2 and e_3e_4 as reciprocal lines. The transformation may be represented as follows. Let x be any point; draw through x the line xpq intersecting e_1e_2 and e_3e_4 in p and q. Then x, ϕx, p, q form a harmonic range.

Let $S = e_1e_2 + e_3e_4$, $S' = e_1e_2 - e_3e_4$. Then [cf. § 179 (4)] we may put

$$\phi x = 2Sx \,.\, S' = H(x).$$

(7) In Cases III. and VI. of § 193 (6) the semi-latent triple groups belong to two types; and are reciprocal in pairs, one from each type. A typical specimen of type I. is defined by e_1e_2, e_3e_4, $\lambda e_3e_1 + \mu(\gamma_1\delta_3e_2e_3 - \gamma_3\delta_1e_1e_4)$: a typical specimen of type II. is defined by

$$e_1e_3,\ \gamma_1\delta_3e_2e_3 + \gamma_3\delta_1e_1e_4,\ \varpi_{23}e_2e_3 + \varpi_{31}e_3e_1 + \varpi_{14}e_1e_4 + \varpi_{24}e_2e_4.$$

Hence all quadrics, which are transformed into themselves, have two semi-latent generators of one system, namely e_1e_2 and e_3e_4; and *one* semi-latent generator of the other system, namely e_1e_3. All the quadrics, so transformed, touch each other along the generator e_1e_3; since [cf. § 187 (8)] the parabolic subgroup, defined by e_1e_3 and $\gamma_1\delta_3e_2e_3 + \gamma_3\delta_1e_1e_4$, is common to them all.

The systems of the type $\lambda e_3e_1 + \mu(\gamma_1\delta_3e_2e_3 - \gamma_3\delta_1e_1e_4)$ have only three null lines which are generators of the quadric associated with the triple group, defined by any one of them together with e_1e_2 and e_3e_4 [cf. § 175 (13)]. Thus consider the quadric defined by e_1e_2, e_3e_4, and $e_3e_1 + \nu(\gamma_1\delta_3e_2e_3 - \gamma_3\delta_1e_1e_4)$. Let it be called the quadric A; also for brevity write $S = \nu(\gamma_1\delta_3e_2e_3 - \gamma_3\delta_1e_1e_4)$. Then we have to prove that the system $S' = \lambda e_3e_1 + \mu(\gamma_1\delta_3e_2e_3 - \gamma_3\delta_1e_1e_4)$ has only three null lines which are generating lines of the quadric A; unless $\lambda\nu = \mu$. For take the quadric A as the self-supplementary quadric, then [cf. § 175 (8)] we may assume that

$$|e_1e_2 = e_1e_2,\ |e_3e_4 = e_3e_4,\ |\{e_3e_1 + S\} = \{e_3e_1 + S\},\ |e_1e_3 = -e_1e_3.$$

Hence $|S = 2e_3e_1 + S$. Also it is easy to see that $(e_1e_3S) = 0$.

Now $\nu S' = \lambda\nu e_3e_1 + \mu S$. Hence $\nu^2(S'S') = \mu^2(SS)$.

Again $\nu|S' = \lambda\nu|e_3e_1 + \mu|S = (2\mu - \lambda\nu)e_3e_1 + \mu S$.

Hence $\nu^2(S'|S') = \mu^2(SS) = \nu^2(S'S')$.

Thus $(S'|S') = (S'S')$. But [cf. § 175 (13)], this is the condition that S' may have only three null lines which are generators of the quadric A, assuming that $|S' \neq S'$.

(8) Case V. of § 193 (5) is a subcase of the case discussed in the previous subsection. The semi-latent triple groups belong to two types, such that the groups are reciprocal in pairs, one from each type. A typical group of type I. is defined by three latent forces $x\phi x$, $y\phi y$, $z\phi z$: a typical group of type II. is defined by e_1e_3, $\delta_1e_1e_4 + \delta_3e_2e_3$, S; where S is any system. Thus any quadric transformed into itself has the latent line e_1e_3 as a generator, and has no other latent or semi-latent generator of the same system; also all the generators of the opposite system are semi-latent [cf. § 192 (17)]. All quadrics which are transformed into themselves touch along the generator e_1e_3, since [cf. § 187 (8)] they have a common parabolic subgroup with e_1e_3 as director force.

(9) If the four latent roots $\gamma_1, \gamma_2, \gamma_3, \gamma_4$, satisfying the relation $\gamma_1\gamma_3 = \gamma_2\gamma_4$, be assigned, then a matrix can in general be constructed, which will transform the quadric into itself by direct transformation, and at the same time make an assigned system latent.

For let S be the assigned system. Then [cf. § 174 (9)] S has in general one pair of conjugate lines which are polar reciprocal with respect to the given quadric. Let e_1e_3 and e_2e_4 be these lines, cutting the quadric in the points e_1, e_2, e_3, e_4. Consider the matrix for which e_1, e_2, e_3, e_4 are the latent points corresponding to the latent roots $\gamma_1, \gamma_2, \gamma_3, \gamma_4$. The given system is obviously latent, since it is of the form $\alpha e_1e_3 + \beta e_2e_4$. Also all quadrics containing the four generators $e_1e_4, e_2e_3, e_1e_2, e_3e_4$ are transformed into themselves, and among them the given quadric.

(10) But if the system S has only three null lines, which are generators of the quadric to be transformed into itself; then the matrix must be of the type of § 193 (6) Case III., or must belong to one of the subcases. Then [cf. subsection (7)] with the notation of § 193 (6) Case III., let e_1, e_2, e_3, e_4 be so chosen that the three generators, which are null lines are, e_1e_2, e_1e_3, e_3e_4. The system S can be written $e_3e_1 + \alpha e_2e_3 + \beta e_1e_4$, where α and β are known, since S is known. Then $\gamma_1, \gamma_3, \delta_1, \delta_3$ must be so chosen that

$$\frac{\alpha}{\beta} = -\frac{\gamma_1\delta_3}{\gamma_3\delta_1}.$$

Also by an easy extension of subsection (7) [cf. § 175 (13) and § 187 (8)] the quadric is defined by three systems of the form

$$e_1e_2,\ e_3e_4,\ e_3e_1 + \lambda(\alpha e_2e_3 + \beta e_1e_4).$$

Hence the quadric is transformed into itself, at the same time as S is latent, by the operation of the matrix.

(11) Thus from (9) and (10), it is always possible to find a matrix which transforms directly a given quadric into itself, and keeps a given system of forces latent. And the matrix is not completely determined by these conditions.

196. Skew Transformation of Quadrics. (1) When a quadric is transformed into itself by a skew transformation, no generator can be semi-latent.

(2) If ϕ be a matrix which transforms a certain quadric into itself by a skew transformation, then the matrix ϕ^2 transforms the same quadric into itself by a direct transformation. It is useful to notice that the latent points of ϕ are also latent points of ϕ^2, though the converse is not necessarily true.

(3) Let the matrix ϕ have four distinct latent roots. Let $\gamma_1, \gamma_2, \gamma_3, \gamma_4$ be the distinct latent roots, and e_1, e_2, e_3, e_4 the corresponding latent points. Then the latent roots of ϕ^2 are $\gamma_1^2, \gamma_2^2, \gamma_3^2, \gamma_4^2$, and e_1, e_2, e_3, e_4 are latent points. Now either $\gamma_1^2, \gamma_2^2, \gamma_3^2, \gamma_4^2$ are distinct; or, two are equal, $\gamma_1^2 = \gamma_2^2$, so that $\gamma_1 = -\gamma_2$; or, they are equal in pairs, $\gamma_1^2 = \gamma_2^2$, $\gamma_3^2 = \gamma_4^2$.

Hence ϕ^2 is either of the type of § 193 (2), or of § 193 (3), Cases I. or III., or of § 193 (6), Case I. Case II. of § 193 (3) cannot occur because [cf. subsection (2)] the line e_1e_2 has two null points, e_1 and e_2, on it. Similarly the other cases of § 193 (6) cannot occur: § 193 (6) Case IV. is inconsistent with the roots being distinct.

But ϕ^2 transforms the quadric into itself by direct transformation. Hence § 193 (3), Case I. is impossible; and § 193 (2) and § 193 (6), Case I. both make semi-latent lines of ϕ to be generating lines of the quadric, which is impossible by subsection (1) above.

If the additional relation, $\gamma_1^4 = \gamma_3^2\gamma_4^2$, hold, then ϕ^2 is the type of matrix described in § 193 (3) Case III. The latent roots of ϕ are connected either by the relations $\gamma_1 = -\gamma_2 = \sqrt{\gamma_3\gamma_4}$, or by $\gamma_1 = -\gamma_2 = \sqrt{-\gamma_3\gamma_4}$. With the notation of § 195 (4), the points a_1 and a_2 are on a quadric transformed directly into itself by ϕ^2; and a_1e_3, a_1e_4, a_2e_3, a_2e_4 are generators of this quadric.

Hence, if ϕ transforms this quadric into itself by a skew transformation, a_1 and a_2 cannot be latent [cf. subsection (1)].

Hence, since they lie on a semi-latent line, $\phi a_1 \equiv a_2$, $\phi a_2 \equiv a_1$. Hence, by § 194 (4), (a_1a_2, e_1e_2) forms a harmonic range. Also $\phi a_1 = \gamma_1 a_2$, $\phi a_2 = \gamma_1 a_1$. Now for the quadric defined by the group $a_1a_2 + \lambda e_3e_4$, a_1e_3, a_2e_4, to be transformed by ϕ by a skew transformation, this group must be transformed by ϕ into the reciprocal group $a_1a_2 - \lambda e_3e_4$, a_2e_3, a_1e_4.

Now $\phi(a_1a_2 + \lambda e_3e_4) = -\gamma_1^2 a_1a_2 + \gamma_3\gamma_4\lambda e_3e_4$, $\phi a_1e_3 = \gamma_1\gamma_3 a_2e_3$, $\phi a_2e_4 = \gamma_1\gamma_4 a_1e_4$. Thus it is necessary that $\gamma_1^2 = \gamma_3\gamma_4$.

Hence a matrix with four distinct latent roots, related so that

$$\gamma_1 = -\gamma_2 = \sqrt{\gamma_3\gamma_4},$$

transforms into themselves by a skew transformation quadrics, passing through e_3 and e_4, with e_1e_2 and e_3e_4 as polar reciprocal lines, and with e_1 and e_2 as polar reciprocal points.

(4) Let the matrix ϕ have three distinct roots. Assume e_1, e_2, e_3, e_4 to be such that

$$\phi e_1 = \gamma_1 e_1,\ \phi e_2 = \gamma_1 e_2 + \delta_1 e_1,\ \phi e_3 = \gamma_3 e_3,\ \phi e_4 = \gamma_4 e_4;$$

where δ_1 may, or may not, be zero.

Hence $\phi^2 e_1 = \gamma_1^2 e_1$, $\phi^2 e_2 = \gamma_1^2 e_2 + 2\delta_1\gamma_1 e_1$, $\phi^2 e_3 = \gamma_3^2 e_3$, $\phi^2 e_4 = \gamma_4^2 e_4$.

Now four cases arise.

Case A. Let γ_1^2, γ_3^2, γ_4^2 be distinct. Then the matrix ϕ^2 is of the type described in § 193 (3). Hence it cannot transform a quadric by direct transformation into itself; except in Case III. But in § 193 (3) Case III. the lines of the type a_1e_3, a_2e_4, a_1e_4, a_2e_3 [cf. § 195 (4)] are generators of the transformed quadrics. But these lines are semi-latent lines of ϕ as well as of ϕ^2: and hence [cf. subsection (1)] this case must be rejected.

Case B. Let $\gamma_1^2 = \gamma_3^2$; so that $\gamma_1 = -\gamma_3$. Then the matrix ϕ^2 is of one of the types (Cases I. and II.) described in § 193 (4); either it is Case I. if δ_1 vanish; or it is Case II. if δ_1 do not vanish. In either case ϕ^2 cannot transform a quadric into itself by direct transformation.

Case C. Let $\gamma_3^2 = \gamma_4^2$; and δ_1 be not zero. Then ϕ^2 is of the type described in § 193 (6), Cases II. and V. The other cases of § 193 (6) cannot occur, since the three latent roots γ_1, γ_3, γ_4 are by hypothesis distinct; and the points e_3 and e_4 are both latent points of ϕ^2. Hence ϕ^2 cannot transform a quadric into itself by direct transformation.

Case D. Let $\gamma_3^2 = \gamma_4^2$; and $\delta_1 = 0$. Then ϕ^2 is of the type described in § 193 (6), Cases I. and IV. If ϕ^2 belongs to the type of § 193 (6) Case I., then by § 195 (5), ϕ^2 transforms into themselves all quadrics with e_1e_2 and e_3e_4 as generating lines. But these are semi-latent lines of ϕ. Hence by subsection (1), this case is impossible.

But if ϕ^2 belong to the type of § 193 (6) Case IV., so that

$$\gamma_1 = \gamma_2 = i\gamma_3 = -i\gamma_4,$$

then ϕ^2 transforms quadrics directly for which e_1e_2 and e_3e_4 are polar reciprocal lines. Thus since $\gamma_3^2 = \gamma_4^2 = -\gamma_1\gamma_2$, we have a subcase of the transformation considered in subsection (3). But it is the alternative case for which ϕ does not effect a skew transformation.

(5) Assume that ϕ has two distinct roots, one root γ_1 occurring triply.

Let e_1, e_2, e_3, e_4 be assumed so that

$$\phi e_1 = \gamma_1 e_1,\ \ \phi e_2 = \gamma_1 e_2 + \delta_1 e_1,\ \ \phi e_3 = \gamma_1 e_3 + \delta_2 e_2,\ \ \phi e_4 = \gamma_4 e_4;$$

where δ_1 and δ_2 may, or may not, vanish.

Then

$$\phi^2 e_1 = \gamma_1^2 e_1,\ \ \phi^2 e_2 = \gamma_1^2 e_2 + 2\delta_1\gamma_1 e_1,\ \ \phi^2 e_3 = \gamma_1^2 e_3 + 2\delta_2\gamma_1 e_2 + \delta_1\delta_2 e_1,\ \ \phi^2 e_4 = \gamma_4^2 e_4.$$

Let the point $e_3' = 2\gamma_1 e_3 - \delta_2 e_2$. Then e_1, e_2, e_3' are such that

$$\phi^2 e_1 = \gamma_1^2 e_1,\ \ \phi^2 e_2 = \gamma_1^2 e_2 + 2\delta_1\gamma_1 e_1,\ \ \phi^2 e_3' = \gamma_1^2 e_3' + 4\delta_2\gamma_1^2 e_2.$$

Case A. Let γ_4^2 be not equal to γ_1^2. Then ϕ^2 must be one of the three types described in § 193 (4). But in no one of the three cases of that article does the matrix transform a quadric into itself by direct transformation.

Case B. Let $\gamma_4 = -\gamma_1$. Then ϕ^2 must be one of the types described in § 193 (5). The matrix ϕ^2 is of the type of Case III. of § 193 (5), if δ_1 and δ_2 do not vanish: it is of the type of Case II., if δ_1 vanishes: it is of the type of Case I., if δ_1 and δ_2 both vanish. But in Cases II. and III. no quadric is transformed into itself by direct transformation. In Case I. the matrix ϕ^2 is merely the numerical multiplier γ_1^2. Hence every quadric is transformed into itself, since no point changes its position.

Then the matrix ϕ has two latent roots γ_1 and $-\gamma_1$. There is a latent plane $e_1e_2e_3$ corresponding to γ_1; and a latent point e_4, not on $e_1e_2e_3$, corresponding to $-\gamma_1$. These are the only latent regions. Hence, by § 194 (5), for all quadrics which are transformed into themselves by ϕ, e_4 and $e_1e_2e_3$ must be pole and polar. Also e_4 cannot lie on such a quadric, since it does not lie on its polar plane.

Now, if p be any latent point on the plane $e_1e_2e_3$, and $x = \lambda e_4 + \mu p$, then $\phi x = \gamma_1(\mu p - \lambda e_4)$.

Hence $\{e_4 p, x\phi x\}$ forms a harmonic range. Thus if x be a point on a quadric for which e_4 and $e_1e_2e_3$ are pole and polar, ϕx is also on the same quadric. Also the transformation is skew, since by § 192 (16) it cannot be direct. This is a subcase of the skew transformation of subsection (3), since

$$\gamma_1 = -\gamma_4 = \sqrt{\gamma_2\gamma_3}.$$

(6) Assume that ϕ has only one root. Let e_1, e_2, e_3, e_4 be assumed so that

$$\phi e_1 = \gamma_1 e_1,\ \phi e_2 = \gamma_1 e_2 + \delta_1 e_1,\ \phi e_3 = \gamma_1 e_3 + \delta_2 e_2,\ \phi e_4 = \gamma_1 e_4 + \delta_3 e_3.$$

Then
$$\phi^2 e_1 = \gamma_1^2 e_1,\ \phi^2 e_2 = \gamma_1^2 e_2 + 2\gamma_1\delta_1 e_1,$$
$$\phi^2 e_3 = \gamma_1^2 e_3 + 2\gamma_1\delta_2 e_2 + \delta_1\delta_2 e_1,\ \phi^2 e_4 = \gamma_1^2 e_4 + 2\gamma_1\delta_3 e_3 + \delta_2\delta_3 e_2.$$

Let
$$e_3' = 2\gamma_1 e_3 - \delta_2 e_2,\ e_4' = 2\gamma_1^2 e_4 - 2\gamma_1\delta_3 e_3 + \delta_2\delta_3 e_2.$$

Then
$$\phi^2 e_3' = \gamma_1^2 e_3' + 4\gamma_1^2\delta_2 e_2,\ \phi e_4' = \gamma_1^2 e_4' + 2\gamma_1^2\delta_3 e_3'.$$

Hence ϕ^2 is a matrix of one of the types described in § 193 (5). The case, when $\delta_1 = \delta_2 = \delta_3 = 0$, need not be considered: for then ϕ is a mere numerical multiplier. Thus [cf. § 195 (1)] the only case of this type in which ϕ^2 transforms a quadric into itself by direct transformation is that of § 193, Case V. Then δ_1 and δ_3 do not vanish, and $\delta_2 = 0$. In this case $e_3' \equiv e_3$; also [cf. § 195 (8)] the latent line e_1e_3 is a generator of all quadrics transformed by ϕ^2. But e_1e_3 is also a latent line of ϕ. Hence ϕ cannot transform these quadrics into themselves by a skew transformation.

(7) Assume that the latent roots of ϕ are equal in pairs, so that $\gamma_1 = \gamma_2$, and $\gamma_3 = \gamma_4$. Let e_1, e_2, e_3, e_4 be such that

$$\phi e_1 = \gamma_1 e_1,\ \phi e_2 = \gamma_1 e_2 + \delta_1 e_1,\ \phi e_3 = \gamma_3 e_3,\ \phi e_4 = \gamma_3 e_4 + \delta_3 e_3.$$

Then $\phi^2 e_1 = \gamma_1^2 e_1,\ \phi^2 e_2 = \gamma_1^2 e_2 + 2\gamma_1\delta_1 e_1,\ \phi^2 e_3 = \gamma_3^2 e_3,\ \phi^2 e_4 = \gamma_3^2 e_4 + 2\gamma_3\delta_3 e_3.$

Hence ϕ^2 belongs to the type described in § 193 (6). Of the six cases of this type only Cases I. and III. and IV. and VI. yield quadrics which are transformed into themselves by ϕ^2 with a direct transformation. In Cases I. and IV., $\delta_1 = 0 = \delta_3$, and [cf. § 195 (5) and (6)] either e_1e_2 and e_3e_4 are generating lines of such quadrics, or they are reciprocally polar lines to them. If they are reciprocally polar lines, the four semi-latent lines, joining the two pairs of points in which e_1e_2 and e_3e_4 meet any such quadric, are generating lines of the quadric. But the latent lines e_1e_2 and e_3e_4, and the semi-latent lines joining any point on e_1e_2 to any point on e_3e_4, are latent and semi-latent lines of ϕ as well as of ϕ^2. Hence ϕ does not transform any of these quadrics into themselves by a skew transformation.

In Cases III. and VI. of § 193 (6) neither δ_1 nor δ_3 vanishes. All quadrics transformed into themselves by ϕ^2 have [cf. § 195 (7)] the three lines e_1e_2, e_1e_3, e_3e_4 as generators, which are semi-latent with respect to ϕ as well as with respect to ϕ^2. Thus ϕ does not transform these quadrics into themselves by a skew transformation.

(8) Thus there is only one case of skew transformation, namely the case (including its subcase) when

$$\gamma_1 = -\gamma_2 = \sqrt{\gamma_3\gamma_4};$$

and the subcase arises when

$$\gamma_1 = \gamma_2 = \gamma_3 = -\gamma_4.$$

In the general case the lines e_1e_2 and e_3e_4 are polar reciprocal with respect to any quadric so transformed, the points e_1 and e_2 are polar reciprocal, and the points e_3 and e_4 are on the quadric (except in the subcase, when $\gamma_3 = \gamma_4$ and the line e_3e_4 is a latent region).

In the subcase the point e_4 and the latent plane $e_1e_2e_3$ are pole and polar with respect to all quadrics so transformed.

NOTE. Homersham Cox, *On the Application of Quaternions and Grassmann's Algebra to different kinds of Uniform Space, Trans. of Camb. Phil. Soc.*, 1882, points out the connection between a positional manifold and Descriptive Geometry of any dimensions [cf. Book III.], and applies it to Hamilton's theory of nets. Also he points out the special applicability of Outer Multiplication to Descriptive Geometry [cf. Chapter IV., Book IV.]; this had already been practically demonstrated by Grassmann in his papers in *Crelle's Journal* on Cubics. Further [in correction of note, p. 278] he applies the calculus in the manner of this book to deduce some elementary propositions concerning Linear Complexes; he finds the condition for reciprocal systems [cf. § 116 (1)], for null lines [cf. § 163 (1)], the director equations of dual and triple groups [cf. § 172 (1) and § 175 (1)], and the condition for a parabolic group [cf. 172 (9)]. He also finds a defining equation of Intensity [cf. note, p. 168], which depends on the distance between points. The bulk of this very suggestive paper is concerned with the Theory of Metrics.

BOOK VI.

THEORY OF METRICS.

CHAPTER I.

THEORY OF DISTANCE.

197. AXIOMS OF DISTANCE. (1) In a positional manifold, to which no additional properties have been assigned by definition, no relation between any two points can be stated without reference to other points on the manifold. Thus consider a straight line which is a one-dimensional positional manifold. If e_1 and e_2 represent the reference elements at unit intensity, any point p can be written $\xi_1 e_1 + \xi_2 e_2$. But ξ_1 is not the expression of a quantitive relation between p and e_1. For ξ_1 depends on ξ_2, and ξ_1/ξ_2 represents a relation of p to the terms e_1 and e_2. But even this does not properly represent a relation of the element represented by p to those represented by e_1 and e_2. For no determinate principles have been assigned by which the terms e_1 and e_2 should be considered to represent their corresponding elements at unit intensities. Thus the arbitrary assumption as to the intensities is included, when ξ_1/ξ_2 is considered as representing a quantitive relation of p to e_1 and e_2.

The only relations between points, which are independent of the intensities, are the anharmonic ratio between four points [cf. § 69 (1)], and functions of this anharmonic ratio.

(2) A spatial manifold will be defined to be a positional manifold, in which a quantitive relation between any two points is defined to exist. This quantitive relation will be called the distance, and the following axioms will be assumed to hold of it.

(3) Axiom I. Any two points in a spatial manifold define a single determinate quantity called their distance, which, when real, may be conceived as measuring the separation or distinction between the points. When the distance vanishes, the points are identical.

Axiom II. If p, q, r be three points on a straight line, and q lie between p and r [cf. § 90 (3)], then the sum of the distances between p and q and between q and r is equal to the distance between p and r.

Axiom III. If a, b, c be any three points in a spatial manifold, and the distances ab and bc be finite, then the distance ac is finite. Also if the distance ab be finite and the distance bc be infinite, then the distance ac is infinite. Also if the distances ab and bc be real, then the distance ac is also real.

(4) Let pq be any straight line through p; and assume some rule to exist, by which one of the two intercepts between p and any point q on the line can be considered as the intercept [cf. § 90] such that points on it lie *between* p and q; then points on this line on the same side of p as q are points which either lie between p and q or are such that q lies between them and p. It follows from axiom II. that all points between p and q are at a less distance from p than the distance pq, and that all points on the same side as q, but beyond q, are at a greater distance from p than is q. Also it is evident that there cannot be another point on pq on the same side of p as q and at the same distance as q. For if q' be such a point, then by axiom II. the distance qq' must be zero; and hence by axiom I. the points q and q' coincide.

(5) Hence the relation of a point q to a point p in a spatial manifold is completely determined by (α) the straight line through p on which q lies, (β) the determination of the side of p on which q lies, (γ) the distance of q from p. Thus any quantitive relation between points on a straight line must be expressible in terms of their distance.

198. Congruent Ranges of Points. (1) Two ranges $p, q, r, s, \ldots$ and $p', q', r', s', \ldots$ of the same number of points in a spatial manifold are called congruent when the following conditions hold. Let the points $p, q, r, s, \ldots$ and the points $p', q', r', s', \ldots$ be mentioned in order; also let the distance between p and q be equal to that between p' and q', and the distance between q and r equal to that between q' and r', and so on.

(2) It follows from this definition of congruent ranges and from axiom II. of distance, that the distance between any two points on one range is equal to that between the corresponding points on the other range.

(3) Also [cf. § 197 (5)] any quantitive relation between points in the first range, which can be expressed without reference to other points of the spatial manifold, is equal to the corresponding relation between the corresponding points of the second range. Such a relation in a positional manifold is the anharmonic ratio of a range of four points. Hence congruent ranges must be homographic [cf. § 70].

(4) Also conversely, if on two homographic ranges the distances between three points of one range are respectively equal to the corresponding distances between the three corresponding points on the other range, then the ranges are congruent. For let pqr and $p'q'r'$ be the two groups of three

points on the two ranges, and let s and s' be any other two corresponding points on the ranges respectively. Then the anharmonic ratio of $(pqrs)$ equals that of $(p'q'r's')$. But if the range $(p'q'r's'')$ be constructed congruent to the range $(pqrs)$, then by the previous part of the proposition the anharmonic ratio of $(pqrs)$ is equal to that of $(p'q'r's'')$. Hence the anharmonic ratio of $(p'q'r's')$ is equal to that of $(p'q'r's'')$. Hence s' and s'' coincide.

The proposition may be stated thus, if three points of one range are congruent to the three corresponding points of a homographic range, then the ranges are congruent.

199. Cayley's Theory of Distance. (1) Cayley has invented in his 'Sixth Memoir on Quantics'* a generalized expression for the distance between two points of a positional manifold. This work was extended and simplified by Klein†, who pointed out its connection with Non-Euclidean geometry.

(2) Consider in the first place a one-dimensional region. Let a_1 and a_2 be two arbitrarily assumed points on it. Then any three points x_1, x_2, and x_3 of the region can be written $\lambda_1 a_1 + \mu_1 a_2$, $\lambda_2 a_1 + \mu_2 a_2$, $\lambda_3 a_1 + \mu_3 a_2$.

The anharmonic ratios ρ_{23}, ρ_{13}, ρ_{12} respectively of the ranges $(x_2 x_3, a_1 a_2)$, $(x_1 x_3, a_1 a_2)$, $(x_1 x_2, a_1 a_2)$ are given by

$$\rho_{23} = \frac{\lambda_2 \mu_3}{\lambda_3 \mu_2}, \quad \rho_{13} = \frac{\lambda_1 \mu_3}{\lambda_3 \mu_1}, \quad \rho_{12} = \frac{\lambda_1 \mu_2}{\lambda_2 \mu_1}.$$

Hence
$$\log \rho_{12} + \log \rho_{23} = \log \rho_{13}.$$

Then if γ be some numerical constant, real or imaginary, we may define [cf. Klein, loc. cit.] $\frac{\gamma}{2} \log \rho_{12}$ as the distance between any two points x_1 and x_2; where the distance is conceived as a signless quantity, but the ordinary conventions may hold as to the sign of lengths according to the direction of measurement. But the definition and the resulting conventions require further examination according to the different cases, which may arise [cf. subsection (4) below].

(3) Let the point-pair a_1, a_2 be called the absolute point-pair. Let these points be either both real, or let their corresponding co-ordinates, referred to any real set of reference elements, be conjugate complex numbers.

Then for real points λ_1 and μ_1 are both real when a_1 and a_2 are real, and are conjugate imaginaries when a_1 and a_2 are conjugate imaginary points. Similarly for λ_2 and μ_2, and for λ_3 and μ_3.

* Cf. *Phil. Trans.* 1859, and *Collected Papers*, Vol. II. No. 158.

† 'Ueber die sogenannte Nicht-Euklidische Geometrie,' *Math. Annalen*, Bd. IV. 1871.

Hence when a_1 and a_2 are real, in order that the distance between real points which both lie on the same intercept between a_1 and a_2 may be real, γ must be real. When a_1 and a_2 are conjugate imaginary points, in order that the distance between real points, such as x_1 and x_2, may be real, γ must be a pure imaginary; for $\log \rho_{12}$ is a pure imaginary. Let $\frac{\gamma}{i}$ be written for γ in this case.

(4) Thus if the absolute point-pair, a_1 and a_2, be real, the distance between any two real points, x_1 and x_2, lying between them, is defined to be the real positive quantity $\frac{\gamma}{2}\log \rho_{12}$, where γ is some real number, and ρ_{12} is so chosen as to be greater than unity. Since $\rho_{12} > 1$, it follows that, with the notation of subsection (2),

$$\rho_{12} = \frac{\lambda_1\mu_2}{\lambda_2\mu_1} \text{ when } \frac{\lambda_1}{\mu_1} > \frac{\lambda_2}{\mu_2}, \text{ and } \rho_{12} = \frac{\lambda_2\mu_1}{\lambda_1\mu_2} \text{ when } \frac{\lambda_2}{\mu_2} > \frac{\lambda_1}{\mu_1}.$$

Assuming $\frac{\lambda_1}{\mu_1} > \frac{\lambda_2}{\mu_2}$, then ρ_{12} can be described either as the anharmonic ratio of the range (x_1x_2, a_1a_2), or as the anharmonic ratio of the range (x_2x_1, a_2a_1). Thus a_1 bears the same relation to x_1, as a_2 bears to x_2, in this definition of distance. The points a_1 and x_1 will be considered as lying on one side of x_2, and the point a_2 on the other side [cf. § 90]. Let this be called the Hyperbolic definition of distance. It is to be noticed that, with this definition a pair of points, not in the same intercept between a_1 and a_2, have not a real distance.

If the absolute point-pair be two conjugate imaginary points, the distance between two real points x_1 and x_2 is defined to be one of the two values of $\frac{\gamma}{2i}\log \rho_{12}$ which lies between o and $\pi\gamma$. The ambiguity as to which value is to be chosen is discussed later in § 204, and its determination is possible in two ways. Let this be called the Elliptic definition of distance.

(5) Let the limiting case be considered in which the absolute points are coincident at some point u. Let e and u be the two reference points in the one-dimensional manifold. Let $a_1 = \alpha_1 e + \beta_1 u$, $a_2 = \alpha_2 e + \beta_2 u$. Then, when a_1 and a_2 ultimately coincide with u, α_1/β_1 and α_2/β_2 ultimately vanish. Let any other points x and y be written, $x = e + \xi u$, $y = e + \eta u$.

Now putting Δ for $\alpha_1\beta_2 - \alpha_2\beta_1$,

$$\Delta x = (\beta_2 - \alpha_2\xi)\, a_1 - (\beta_1 - \alpha_1\xi)\, a_2,$$

$$\Delta y = (\beta_2 - \alpha_2\eta)\, a_1 - (\beta_1 - \alpha_1\eta)\, a_2.$$

Hence according to Cayley's definition, the distance between x and y is

$$\tfrac{1}{2}\gamma \log \frac{(\beta_2 - \alpha_2\xi)(\beta_1 - \alpha_1\eta)}{(\beta_1 - \alpha_1\xi)(\beta_2 - \alpha_2\eta)}.$$

Therefore expanding in powers of α_1/β_1 and α_2/β_2 and retaining only the lowest powers, the distance becomes

$$\tfrac{1}{2}\gamma\left(\frac{\alpha_2}{\beta_2}-\frac{\alpha_1}{\beta_1}\right)(\eta-\xi).$$

Now let γ increase as α_1/β_1 and α_2/β_2 decrease, so that $\gamma\left(\frac{\alpha_2}{\beta_2}-\frac{\alpha_1}{\beta_1}\right)$ remains finite and equal to δ, say. Then in the limit when a_1 and a_2 coincide with u, the distance between the points $e+\xi u$ and $e+\eta u$ is $\delta(\eta-\xi)$. Therefore this definition of distance is a special limiting case of the more general definition first explained. Let it be called the Parabolic definition.

200. KLEIN'S THEOREM. (1) It can be shown (cf. Klein, loc. cit.) that this definition of the distance between two points is the only possible definition, which is consistent with the propositions on congruent ranges in § 198 (3) and (4).

Let $p, p_1, p_2, \ldots$ and $p_1, p_2, p_3, \ldots$ be two congruent ranges. Then by definition the distance pp_1 = the distance p_1p_2 = etc. Also by § 198 (3) and (4) the ranges are homographic; therefore [cf. § 71 (1)] the first range can be transformed into the second by a linear transformation. Let a_1 and a_2 be the two points on the line which are unaltered by this transformation [cf. § 71 (2)], and firstly assume them to be distinct [cf. § 71 (4) and (5)]. Then [cf. § 71 (8)]

$$p=\xi a_1+\xi' a_2,\ p_1=\nu\xi a_1+\xi' a_2,\ \ldots,\ p_\rho=\nu^\rho\xi a_1+\xi' a_2,\ \ldots\ p_\sigma=\nu^\sigma\xi a_1+\xi' a_2,\ \ldots.$$

Thus the anharmonic ratio $(p_\rho p_\sigma,\ a_1a_2)$ is $\nu^{\sigma-\rho}$. But the distance between p_ρ and p_σ is $(\sigma-\rho)$ times the distance p_1p_2, which is any arbitrarily assumed distance λ. Accordingly if a_1 and a_2 be Cayley's absolute point-pair and $p_\rho=\theta a_1+\phi a_2$, $p_\sigma=\psi a_1+\chi a_2$, we obtain

$$\text{dist. } p_\rho p_\sigma=\lambda(\sigma-\rho)=\frac{\lambda}{\log\nu}\log\frac{\psi\phi}{\chi\theta}.$$

But this is the definition of distance already given in § 199 (2), as far as concerns integral multiples of an arbitrarily assumed length λ. But since λ is any length, it may be assumed to be small compared to all lengths which are the subjects of discourse. Thus the definition must hold for all lengths.

(2) Secondly, let the two points, a_1 and a_2, unaltered by the linear transformation, be coincident, and write u for either of them [cf. § 71 (7)]. Let e be any other reference point on the line; then if any point p be written in the form $e+\xi u$, it is transformed [cf. § 71 (7) equation D] into the point $e+(\xi+\delta)u$, where δ does not depend on ξ. Thus if the range $p, p_1, p_2 \ldots$ be transformed into the congruent range $p_1, p_2, p_3 \ldots$, the distance between the points p_ρ and p_σ is $(\sigma-\rho)$ times the distance between p and p_1. But [cf. § 71 (9)] $p_\rho=e+(\xi+\rho\delta)u\ (=e+\eta u$, say), and

$$p_\sigma=e+(\xi+\sigma\delta)u\ (=e+\eta' u,\ \text{say}).$$

Therefore the distance $p_\sigma p_\rho = \lambda(\sigma - \rho) = \frac{\lambda}{\delta}(\eta' - \eta)$, where λ is real. But this is the Parabolic definition of distance of § 199 (5).

201. Comparison with the Axioms of Distance. The only difficulty in reconciling the Cayley-Klein theory of distance with the axioms of § 197 (3) arises from axiom I. For in axiom I. the distance is said to relate two points of a spatial manifold, whereas the definition of distance of § 199 relates four points of the manifold, namely the two points of which the distance is defined and the two points forming the absolute. But the two points which form the absolute, if real, are at an infinite distance from every point of the spatial manifold. They may be considered as extreme, or limiting points, of the manifold. Thus the distance only relates two points arbitrarily chosen. Again if the absolute point-pair be imaginary, and the distance only relates real arbitrary points, the other points which enter into the definition are special points and are imaginary.

202. Spatial Manifolds of many dimensions. (1) Consider a spatial manifold of $\nu - 1$ dimensions, where $\nu > 2$. Assume that Cayley's definition of distance applies to every straight line in it.

Let the whole, or part, of the spatial manifold be such that Cayley's definition of distance, in the same one of its three forms, applies to any two real points in it; so that a real distance exists between them. Then such a manifold, or such a part of a manifold, will be called a Space of $\nu - 1$ dimensions. If the Space of $\nu - 1$ dimensions be not the complete spatial manifold, then there must not be a real distance between any point in Space and any point in the remaining part of the spatial manifold. Let the remaining real part of the spatial manifold be called Anti-space. Thus a spatial manifold is either such that its complete real portion forms Space; or it is such that its complete real portion is partly Space and partly Anti-space.

(2) Consider any triangle abc in the complete spatial manifold, the whole or some part of which forms the space considered. Let b and c be real, and a either real or imaginary. Let the distance between b and c be real and finite.

It follows from § 197, Axiom III., that if a be one of the points of the absolute point-pair of the line ab, it must also be one of the points of the absolute point-pair of the line ac. Hence all the points which form the absolute point-pairs of all straight lines must form either an entirely imaginary surface, or a closed surface; and the part of the spatial manifold, which forms Space, must lie within the surface. For [cf. § 82 (1)], when the absolute is real, every straight line through any point in space must cut the absolute in a pair of real points. Then the part of the manifold outside the closed surface is Anti-space.

(3) Also every straight line, containing points in the spatial manifold, must cut this surface, whether it be real or imaginary, in one point-pair. The only *algebraic* surface for which this is possible is a quadric.

Let it be assumed in future that the absolute point-pairs form a quadric, which is either entirely imaginary or real and closed. Let this quadric be called the Absolute.

(4) When the absolute is imaginary, the spatial manifold is called elliptic*. There are two forms of elliptic geometry; the polar form in which the symbols $+x$ and $-x$ represent the same point at opposite intensities [cf. § 89 (1)]; the antipodal form in which $+x$ and $-x$ represent different points [cf. § 89 (2)]. The discrimination between the two forms was first made by Klein.

When the absolute is real and closed, the spatial manifold is called hyperbolic. In hyperbolic geometry the symbols $+x$ and $-x$ represent the same point at opposite intensities.

Parabolic Space is a special limiting form which both Elliptic and Hyperbolic Space can assume, when the absolute degenerates into two coincident planes [cf. § 212 below].

(5) Let the distance between the two points a and b be written $D(ab)$ as an abbreviation for 'distance ab.'

203. Division of Space. In the polar form of elliptic geometry a plane does not divide space. For if x and y be any two points and L any plane, the straight line xy cuts the plane L in one point p only. But [cf. 90 (5)] there are two intercepts between x and y. Thus the plane L cuts one of the intercepts and does not cut the other. Hence it is always possible to join any two points by an intercept of a straight line which does not cut a given plane.

(2) But two planes do divide space. For it is possible to find two points, such that each of the two intercepts joining them cuts one of the two planes. For any straight line must cut the two planes in two points, say in p and q: on the straight line pq take two points x and y, one on each of the two intercepts joining p and q. Then the planes divide x from y in the way stated.

(3) In the antipodal form of elliptic geometry a plane does divide space. For any straight line cuts a plane L in two antipodal points. Then if one intercept between two points x and y contains only one point p on the plane, the other intercept must contain the antipodal point. Thus x and y are divided from each other by the plane. Points x and y, which are not divided

* Klein confines the term Elliptic to the Polar form of Elliptic Geometry. The Antipodal form is called by him Spherical Geometry.

from each other by the plane, must be such that one intercept between x and y does not cut the plane and the other intercept contains the two antipodal points of section, namely $\pm p$.

(4) In the hyperbolic geometry a plane does divide space. It might have been wrongly anticipated, since $\pm x$ represent the same point, that results analogous to those in the polar form hold. But in elliptic geometry space is the whole of the real part of the positional manifold; whereas in hyperbolic geometry space is only the part of the positional manifold within the closed absolute. Now no straight line lies completely within the absolute. Accordingly if one intercept, joining two points in space, itself lie completely in space, the other intercept passes out of space. Hence, ignoring points outside space, points joined by an intercept, lying entirely in space and cut by a plane, are divided from each other by that plane.

204. Elliptic Space. (1) Let the absolute be imaginary, so that the space is elliptic: let it be chosen to be the self-normal quadric [cf. Bk. IV., Ch. III.]. Then its equation can be written $(x|x)=0$. Also let it always be assumed that, when x represents any real point, $(x|x)$ is positive. Then from § 199 (4) and equation (2) of § 123 (9), the distance between any two points x_1 and x_2 is

$$D(x_1x_2)=\frac{\gamma}{2i}\log \rho_{12}=\gamma\cos^{-1}\frac{\pm(x_1|x_2)}{\sqrt{\{(x_1|x_1)(x_2|x_2)\}}}=\gamma\sin^{-1}\sqrt{\frac{(x_1x_2|x_1x_2)}{(x_1|x_1)(x_2|x_2)}};$$

where the inverse trigonometrical functions are to denote angles between 0 and π.

(2) If x and $-x$ represent the same point [cf. § 89 (1)], the ambiguity of sign must be determined so that $\pm(x_1|x_2)$ is positive; for this choice makes the distance of a point from itself to be zero. Hence, in the polar form of elliptic space, $D(x_1x_2)$ is not greater than $\frac{1}{2}\pi\gamma$.

(3) If x and $-x$ represent different points [cf. § 89 (2)], then the upper sign is to be chosen in determining the ambiguity. Thus if d be the distance between x and y, and d' the distance between $-x$ and y,

$$\cos\frac{d}{\gamma}=\frac{(x|y)}{\sqrt{\{(x|x)(y|y)\}}},\text{ and }\cos\frac{d'}{\gamma}=\frac{(-x|y)}{\sqrt{\{(x|x)(y|y)\}}}=\frac{-(x|y)}{\sqrt{\{(x|x)(y|y)\}}}.$$

Hence, in the antipodal form of elliptic space, $D(xy)$ is not greater than $\pi\gamma$.

205. Polar Form. (1) It is necessary, for the elucidation of the distance formula of the polar form of elliptic space, to investigate the circumstances under which $(x|z)$ and $(y|z)$ are of the same and of opposite signs.

Let the polar plane of x with respect to the absolute cut xy in x', and let that of y cut xy in y'. Let the closed [cf. § 65 (9)] oval line $xyx'y'$ of the figure represent the complete straight line xy. Any point z on this line can be written in the form $\alpha x + \xi x'$. For the sake of simplicity assume that α is positive and does not change, and that ξ alone varies as z shifts its position on

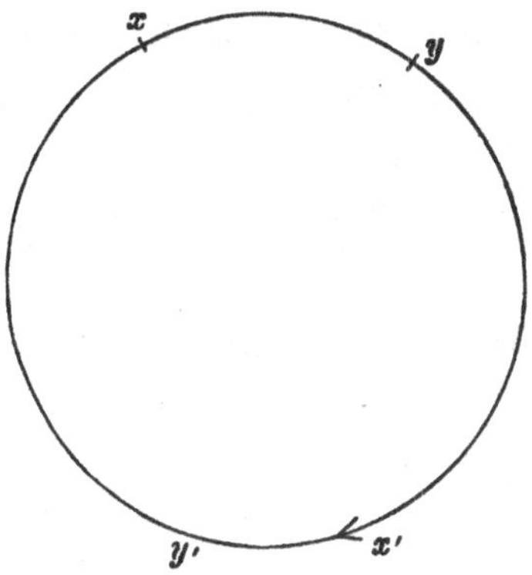

FIG. 1.

the line. Then for one of the two intercepts between x and x', ξ is positive; for the other, ξ is negative; let xyx' be the ξ positive intercept. Assume $y = \alpha x + \beta x'$, thus β is positive. Also we can write $y' = \beta(x' | x') x - \alpha(x | x) x'$ thus y' is on the ξ negative intercept. Also let it be noted that with the assumed form of y, $(x | y)$ [$= \alpha(x | x)$] is positive.

Now $$z = \alpha x + \xi x' = \lambda^2 \{\alpha^2 (x | x) + \xi\beta (x' | x')\} y + \lambda^2 (\alpha\beta - \alpha\xi) y',$$
where $$\lambda^{-2} = \alpha^2 (x | x) + \beta^2 (x' | x').$$

Accordingly $(x | z)$ is positive at all points of xy. And
$$(y | z) = \lambda^2 \{\alpha^2 (x | x) + \xi\beta (x' | x')\} (y | y).$$

Hence, remembering that as z moves from x' to y' in the direction of the arrow ξ changes gradually from $-\infty$ to $-\dfrac{\alpha^2 (x | x)}{\beta (x' | x')}$, we deduce that $(y | z)$ is positive when z is on the intercept $x'yxy'$ between x' and y', and is negative when z is on the other intercept.

(2) Secondly let z be any point not necessarily on the line xy. Now [cf. § 72 (5)] z can always be written in the form $z_1 + p$, where z_1 is on the line xy and p is on the subplane which is the intersection of the polars of x and y; and this representation is possible in one way only. Then $(x | z) = (x | z_1)$, and $(y | z) = (y | z_1)$.

(3) Hence, summing up the results of (1) and (2), we see that if z be separated from x and y by the polar planes of x and y, then $(x | z)$ and $(y | z)$ are necessarily of different signs, provided that $(x | y)$ is positive. But if z be not separated from x and y by the polar planes, then $(x | z)$ and $(y | z)$ are necessarily of the same sign, when $(x | y)$ is positive. Thus if $(y | z)$, $(z | x)$, $(x | y)$ are all of the same sign, they are all positive.

(4) Let the intercept between x and y on which x' and y' do not lie be called *the* intercept, while that intercept on which x' and y' do lie is called the polar intercept.

206. Length of Intercepts in Polar Form. (1) If x, y, z be three collinear points, it is as yet ambiguous as to which lies between the other two, since the straight line is a closed curve. The definition of distance has however really decided the question, as is shown by the following investigation.

(2) Let $(x|y)$ be positive, and firstly let z lie on the intercept [cf. § 205 (4)] between x and y.

Put $z = \lambda x + \mu y$; then λ, μ, $(x|z)$ and $(y|z)$ may be assumed to be positive.

Hence

$$\cos\frac{D(xz)}{\gamma} = \frac{(x|z)}{\sqrt{\{(x|x)(z|z)\}}}, \quad \sin\frac{D(xz)}{\gamma} = \sqrt{\frac{(xz|xz)}{(x|x)(z|z)}} = \mu\sqrt{\frac{(xy|xy)}{(x|x)(z|z)}};$$

and $$\cos\frac{D(zy)}{\gamma} = \frac{(y|z)}{\sqrt{\{(y|y)(z|z)\}}}, \quad \sin\frac{D(zy)}{\gamma} = \lambda\sqrt{\frac{(xy|xy)}{(x|x)(z|z)}}.$$

Thus $$\sin\frac{D(xz)+D(zy)}{\gamma} = \frac{\mu(y|z)+\lambda(x|z)}{(z|z)}\sqrt{\frac{(xy|xy)}{(x|x)(y|y)}} = \sin\frac{D(xy)}{\gamma}.$$

Also $$\cos\frac{D(xz)+D(zy)}{\gamma} = \frac{(x|z)(y|z)-\lambda\mu(xy|xy)}{(z|z)\sqrt{\{(x|x)(y|y)\}}} = \frac{(x|z)(y|z)+(zy|zx)}{(z|z)\sqrt{\{(x|x)(y|y)\}}}$$
$$= \frac{(x|y)}{\sqrt{\{(x|x)(y|y)\}}} = \cos\frac{D(xy)}{\gamma}.$$

Hence $$D(xz) + D(zy) = D(xy).$$

Thus when z lies on the intercept between x and y, as defined in § 205 (4), z lies between x and y according to the meaning of § 197, axiom II.

(3) If z lie on the intercept between y and x', then y lies on the intercept between x and z. Thus from subsection (2),

$$D(xy) + D(yz) = D(xz).$$

Similarly if z lie on the intercept between x and y', then x lies on the intercept between y and z, and

$$D(yx) + D(xz) = D(yz).$$

(4) If z lie between x' and y', then z' lies on the intercept between x and y, and each of the points x, y, z is separated from remaining two by the pair of polar planes of those two; so that each point lies on the polar intercept of the other two. Assume $(x|z)$ positive and $(y|z)$ negative: also let $z = \lambda x - \mu y$, where λ and μ are positive [cf. § 205 (1)].

Then $\cos\frac{D(xz)}{\gamma} = \frac{(x|z)}{\sqrt{\{(x|x)(z|z)\}}}$, $\sin\frac{D(xz)}{\gamma} = \mu\sqrt{\frac{(xy|xy)}{(x|x)(z|z)}}$,

$$\cos\frac{D(yz)}{\gamma} = \frac{-(y|z)}{\sqrt{\{(x|x)(z|z)\}}},\quad \sin\frac{D(yz)}{\gamma} = \lambda\sqrt{\frac{(xy|xy)}{(y|y)(z|z)}}.$$

Hence

$$\sin\frac{D(xy)+D(yz)}{\gamma} = -\frac{(y|z)}{(y|y)}\sqrt{\frac{(xy|xy)}{(x|x)(z|z)}} + \frac{\lambda(x|y)}{(y|y)}\sqrt{\frac{(xy|xy)}{(x|x)(z|z)}}$$
$$= \mu\sqrt{\frac{(xy|xy)}{(x|x)(z|z)}} = \sin\frac{D(xz)}{\gamma}.$$

Also

$$\cos\frac{D(xy)+D(yz)}{\gamma} = \frac{-(x|y)(y|z)-\lambda(xy|xy)}{(y|y)\sqrt{\{x|x)(z|z)\}}} = \frac{-(x|y)(y|z)-(zy|xy)}{(y|y)\sqrt{\{(x|x)(z|z)\}}}$$
$$= \frac{-(y|y)(z|x)}{(y|y)\sqrt{\{(x|x)(z|z)\}}} = -\cos\frac{D(xz)}{\gamma}.$$

Hence $$D(xy)+D(yz) = \pi\gamma - D(xz),$$

or $$D(yz)+D(zx)+D(xy) = \pi\gamma \quad\text{.....................(A).}$$

Hence no one of the points x, y or z lies between the other two according to the meaning of § 197, Axiom II.

(5) This difficulty in the reconciliation of the Polar form of Elliptic Geometry to the Axioms of Distance may be obviated as follows. The distance between two points must be specially associated with *the* intercept between them; since for the intercept only is the axiom II. of § 197 true. Let the distance between two points be also called the length of the intercept. Thus the intercept itself is considered as possessing a quantity of length.

(6) Again the polar intercept may also be considered as possessing a quantity of length. For, since $(x|x') = 0$,

$$D(xx') = \gamma\cos^{-1}0 = \tfrac{1}{2}\pi\gamma.$$

Also and similarly

$$D(yx') = D(xy') = \tfrac{1}{2}\pi\gamma - D(xy).$$

Hence $$D(x'y') = D(xx') - D(xy') = D(xy).$$

Thus $$D(yx') + D(x'y') + D(y'x) = \pi\gamma - D(xy).$$

Hence $\pi\gamma - D(xy)$ may be considered as the length of the polar intercept between x and y, since it is the sum of the lengths of its three parts.

Accordingly the whole length of the straight line may be considered to be $\pi\gamma$. This also agrees with equation (A) of subsection (4).

(7) The paradox of subsection (4) can now be explained. For each of the three points lies on the polar intercept between the other two: and the sum of the distances of any two from the third is in each case equal to the

length of the polar intercept. Thus axiom II. of § 197 ought to be amended into, the sum of the lengths of the parts which make up either the intercept, or the polar intercept, is equal to the length of the intercept, or of the polar intercept, as the case may be.

(8) Also if $\gamma \cos^{-1} \frac{(x|y)}{\sqrt{\{(x|x)(y|y)\}}}$ gives the length of the intercept between x and y, then $\gamma \cos^{-1} \frac{-(x|y)}{\sqrt{\{x|x)(y|y)\}}}$ gives the length of the polar intercept; and vice versa.

Let that intercept between x and y of which the length is

$$\cos^{-1} \frac{(x|y)}{\sqrt{\{(x|x)(y|y)\}}},$$

be called the intercept $(x|y)$, or $\overline{xy}$; and let the length of the intercept $(x|y)$ be called $\overline{xy}$. This name is useful in the ordinary case in which it is unknown and immaterial whether $(x|y)$ is positive or negative. If $(x|y)$ be positive, the intercept $(x|y)$ is *the* intercept between x and y according to § 205 (4).

(9) It is necessary in this connection to distinguish carefully between the points x and y, and the terms x and y by which they are symbolized [cf. § 14, *Definition*]. All congruent terms [cf. § 64 (2)] denote the same point (or regional element). Two points x and y divide the complete straight line into two intercepts. The sum of the lengths of the two intercepts is $\pi\gamma$. The length of the shortest intercept, which is the distance between the points x and y, is $D(xy)$. The length of the other (polar) intercept is $\pi\gamma - D(xy)$. The terms x and y, written in the form $(x|y)$ or $\overline{xy}$, define one of these intercepts. If $(x|y)$ be positive, this intercept is *the* intercept, and is of length $\overline{xy} = D(xy)$. If $(x|y)$ be negative, this intercept is the polar intercept, and is of length $\overline{xy} = \pi\gamma - D(xy)$. Let $x'' = -x$, $y'' = -y$. Then the terms x'' and y'' denote the same points as x and y. Also $(x''|y'') = (x|y)$; hence the intercept $(x''|y'')$ is the same as the intercept $(x|y)$. But the intercepts $(x''|y)$ and $(x|y'')$, which are the same intercept, are always the other intercept to the intercept $(x|y)$ or $(x''|y'')$.

Thus, summarizing and repeating the distinctions between $D(xy)$ and $\overline{xy}$;

$$D(xy) = D(x''y'') = D(x''y) = D(xy'');$$

$$\overline{xy} = \overline{x''y''}; \quad \overline{xy''} = \overline{x''y}; \quad \overline{xy} + \overline{x''y} = \pi\gamma;$$

$$\overline{xy} = D(xy), \text{ if } (x|y) \text{ be positive};$$

$$\overline{x''y} = D(xy), \text{ if } (x|y) \text{ be negative};$$

$$D(xy) \gtrless \tfrac{1}{2}\pi\gamma; \quad \overline{xy} \gtrless \pi\gamma.$$

Also the length of the intercept $(x|y)$ is written $\overline{xy}$. Then $\overline{xy}$ (as well as $(x|y)$) may also be taken as this name of the intercept. It is not often of much importance to know whether $\overline{xy} = D(xy)$ or $\pi\gamma - D(xy)$.

(10) If z be the point $x + \xi y$, then when ξ is positive z lies on the intercept $(x|y)$.

For
$$\cos\frac{\overline{xz}}{\gamma} = \frac{(x|x) + \xi(x|y)}{\sqrt{[(x|x)\{(x|x) + 2\xi(x|y) + \xi^2(y|y)\}]}}.$$

Hence as ξ changes gradually from 0 to $+\infty$, $\cos\frac{\overline{xz}}{\gamma}$ diminishes gradually from 1 to $\frac{(x|y)}{\sqrt{\{(x|x)(y|y)\}}}$, and this whether $(x|y)$ be positive or negative.

Thus $\overline{xz}$ gradually increases from 0 to $\overline{xy}$. Similarly at the same time $\overline{zy}$ gradually decreases from $\overline{xy}$ to 0. Hence z must lie in the intercept $(x|y)$.

207. Antipodal form. (1) In the antipodal form of elliptic geometry *the* intercept between x and y is that intercept which does not contain the

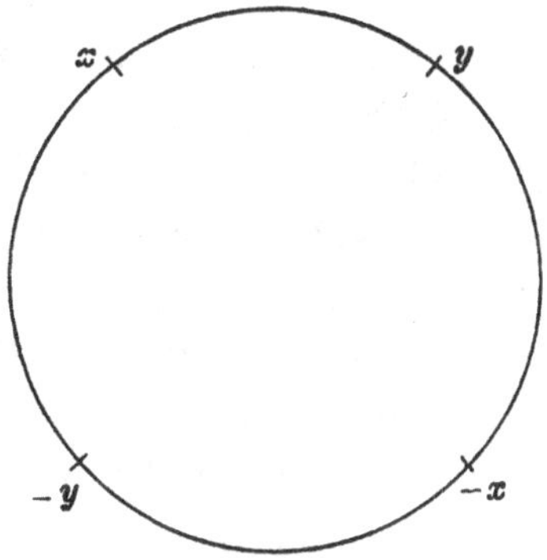

Fig. 2.

antipodal points $-x$ and $-y$; the intercept containing the antipodal points is called [cf. § 90 (6)] the antipodal intercept.

(2) Now by a proof similar to that in the previous article, if z lie in the intercept between x and y, $D(xz) + D(zx) = D(xy)$. If z lie in the intercept between y and $-x$, $D(xy) + D(yz) = D(xz)$. If z lie in the intercept between x and $-y$, $D(yx) + D(xz) = D(yz)$.

(3) If z lie in the intercept between $-x$ and $-y$, let z' $(= -z)$ be the antipodal point to z. Then z' lies in the intercept between x and y. Hence by subsection (1)
$$D(xz') + D(z'y) = D(xy).$$

But by § 204 (3),
$$D(xz) + D(xz') = \pi\gamma = D(yz) + D(yz').$$

Hence
$$D(xz) + D(zy) = 2\pi\gamma - D(xy).$$

Also
$$D(xy) + D(yz) = D(xy) + \pi\gamma - D(yz') = \pi\gamma + D(xz') = 2\pi\gamma - D(xz),$$

and
$$D(yx) + D(xz) = 2\pi\gamma - D(yz).$$

Thus no one of the three points x, y, z lies between the other two in the sense of axiom II. § 197. Accordingly this axiom is not literally satisfied; however the following explanations and additions shew that it is substantially satisfied.

(4) Analogously to the similar case of the polar form, let the distance between x and y be called the length of the intercept between x and y. Then the length of the intercept between y and $-x$ is $\pi\gamma - D(xy)$, and this is also the length of the intercept between $-y$ and x. The length of the intercept between $-x$ and $-y$ is $D(xy)$. Hence adding the three parts, the length of the antipodal intercept between x and y is $2\pi\gamma - D(xy)$.

Thus the length of the whole straight line is $2\pi\gamma$.

(5) The paradox of subsection (3) can now be explained. For each of the three points lies on the antipodal intercept between the other two: and the sum of the distances of any two from the third is in each case equal to the length of the antipodal intercept. Thus axiom II. § 197 ought to be amended into, The sum of the lengths of the parts which make up either the intercept or the antipodal intercept is equal to the length of the intercept or of the antipodal intercept, as the case may be.

208. Hyperbolic Space. (1) Secondly let the absolute quadric be real and closed. Then from § 199 (4) and equation (3) of § 123 (10), the distance $D(xy)$ between any two points x, and y within the quadric is

$$D(xy) = \tfrac{1}{2}\gamma \log \rho_{12} = \gamma \cosh^{-1} \frac{\pm (x \mid y)}{\sqrt{\{(x \mid x)(y \mid y)\}}} = \gamma \sinh^{-1} \sqrt{\frac{-(xy \mid xy)}{(x \mid x)(y \mid y)}}.$$

The ambiguity of sign must be determined so that $\pm(x \mid y)$ is positive. It has been proved in § 82 (9) that $(x \mid x)$ and $(y \mid y)$ are of the same sign: hence $\{(x \mid x)(y \mid y)\}$ is necessarily positive.

(2) The test as to which sign of the ambiguity is to be chosen is derived from the following lemma; which, it is useful to notice, applies to any closed quadric $(\alpha \between x^2) = 0$, and not solely to the absolute in its character of self-supplementary quadric.

Let e, x, y, z be four points within the quadric. Then, if $(e \mid x)$, $(e \mid y)$, $(e \mid z)$ are of one sign, also $(y \mid z)$, $(z \mid x)$, $(x \mid y)$ are of one sign.

For let the line xy cut the polar of e in e', and let xz cut it in e''. Then we may write $y = \lambda x + \eta e'$, $z = \mu x + \zeta e''$. Hence, since $(e \mid e') = 0 = (e \mid e'')$, $(e \mid y) = \lambda (e \mid x)$ and $(e \mid z) = \mu (e \mid x)$. Therefore from the hypothesis λ and μ are positive.

Again as η varies between $-\infty$ and $+\infty$, y takes all the positions on the line xe'. Also $(x \mid y) = \lambda (x \mid x) + \eta (e' \mid x)$. Hence $(x \mid y)$ is a linear function of the variable η; and thus as η varies, $(x \; y)$ can only change sign when it vanishes or is infinite. But when $(x|y)$ vanishes, y must lie on the polar plane of x, and this plane is entirely outside the quadric [cf. § 82 (6)]; similarly when $(x|y)$ is infinite, η is infinite and y coincides with e' which is outside the quadric since it lies on the polar plane of e.

Thus for all points y on that part of the line xe' which lies within the quadric, $(x|y)$ has the same sign. Now put $\eta = 0$. Hence $(x|y)$ has the same sign as $\lambda(x|x)$. But λ is positive. Thus $(x|y)$ has the same sign as $(x|x)$. Also $(x|x)$ has the same sign for all points within the quadric, say the positive sign. Hence $(x|y)$ is also positive. Thus the proposition is proved.

(3) Let $(x|x)$ be always assumed to be positive for points within the quadric: also let a point x within the quadric be said to be of standard sign when $(e|x)$ is positive, where e is any given point within the quadric chosen as a standard of reference. Then it follows from the above that for all points of standard sign within the quadric, $(x|y)$ is positive.

Thus the distance between two points x and y, within the quadric and of standard sign, is

$$D(xy) = \tfrac{1}{2}\gamma \log \rho_{12} = \gamma \cosh^{-1} \frac{(x|y)}{\sqrt{\{(x|x)(y|y)\}}} = \gamma \sinh^{-1} \sqrt{\frac{-(xy|xy)}{(x|x)(y|y)}}.$$

In future all symbols arbitrarily assumed to represent points within a real closed absolute will be assumed to represent them at standard sign.

(4) In hyperbolic space there is only one intercept between two points x and y which lies entirely within the space. Also if z lie within this intercept

$$D(xz) + D(zy) = D(xy).$$

Hence there is no ambiguity as to the application of axiom II. of § 197.

The distance between x and y will be called the length of the intercept between x and y.

The distance of any point from any point on the absolute is infinite. Thus the length of the part of any straight line within the spatial manifold is infinite.

209. THE SPACE CONSTANT. It is formally possible to assume that γ, instead of being an absolute constant, is constant only for each straight line; and accordingly is a function of any quantities which define the special straight line on which x_1 and x_2 lie. Such quantities can necessarily be expressed in terms of the co-ordinates of x_1 and x_2, since these points define the line x_1x_2. Hence the assumption of γ as a function of the co-ordinates of the straight line joining the points does not appear necessarily to offend against the axioms of § 197. Let the assumption be made that γ is constant and the same for all lines. Let γ be called the space-constant.

210. LAW OF INTENSITY IN ELLIPTIC AND HYPERBOLIC GEOMETRY. (1) The law of intensity (cf. Bk. III. ch. IV.) is also settled, if the assumption* be made that, when x_1 and x_2 are of the same intensity, $x_1 + x_2$ bisects the distance between x_1 and x_2; where for the polar form of elliptic geometry $(x_1 | x_2)$ is assumed to be positive, and for hyperbolic geometry x_1 and x_2 are both of standard sign. No special explanation is required for the antipodal form of elliptic geometry, since $x_1 + x_2$ is to bisect the distance between x_1 and x_2, and $x_1 - x_2$ is to bisect the distance between x_1 and $-x_2$.

Then by §§ 204 and 208,

$$\frac{\{x_1 | (x_1 + x_2)\}}{\sqrt{[(x_1 | x_1) \{(x_1 + x_2) | (x_1 + x_2)\}]}} = \frac{\{x_2 | (x_1 + x_2)\}}{\sqrt{[(x_2 | x_2) \{(x_1 + x_2) | (x_1 + x_2)\}]}}.$$

Hence $\{\sqrt{(x_1 | x_1)} - \sqrt{(x_2 | x_2)}\} \{\sqrt{(x_1 | x_1)(x_2 | x_2)} - (x_1 | x_2)\} = 0.$

Therefore either $(x_1 | x_1) = (x_2 | x_2)$, or $(x_1 | x_1)(x_2 | x_2) - (x_1 | x_2)^2 = 0$.

The second alternative is equivalent to $(x_1 x_2 | x_1 x_2) = 0$. This implies that the line $x_1 x_2$ touches the absolute [cf. § 123 (5)]; and this presupposes special positions for x_1 and x_2. In fact for such a case in elliptic geometry the line $x_1 x_2$ would then be imaginary; and in hyperbolic geometry x_1 and x_2 would lie outside the absolute.

Hence the alternative, $(x_1 | x_1) = (x_2 | x_2)$, must be adopted. Accordingly if the point x has a given intensity, $(x | x)$ is independent of the position of x. Thus with a proper choice of constants the intensity of x is $\sqrt{(x | x)}$; so that $(x | x) = 1$, when x is at unit intensity.

(2) Then, if x_1 and x_2 be at unit intensity and $(x_1 | x_2)$ be positive (except for antipodal elliptic space), the formulæ for the distance between them become,

$$d_{12} = \frac{\gamma}{2i} \log \rho_{12} = \gamma \cos^{-1} (x_1 | x_2) = \gamma \sin^{-1} \sqrt{(x_1 x_2 | x_1 x_2)};$$

and

$$d_{12} = \frac{\gamma}{2} \log \rho_{12} = \gamma \cosh^{-1} (x_1 | x_2) = \gamma \sinh^{-1} \sqrt{(- x_1 x_2 | x_1 x_2)};$$

according as the space is elliptic space or is hyperbolic space (of any number of dimensions), where in both cases x_1 and x_2 fulfil the condition

$$(x_1 | x_1) = 1 = (x_2 | x_2).$$

(3) As an illustration of these formulæ consider antipodal elliptic space of two dimensions.

Let the absolute be $(x | x) = \xi_1^2 + \xi_2^2 + \xi_3^2 = 0.$

Then the conditions, $(x | x) = 1 = (y | y)$, become

$$\xi_1^2 + \xi_2^2 + \xi_3^2 = 1 = \eta_1^2 + \eta_2^2 + \eta_3^2.$$

* This assumption is made by Homersham Cox, *loc. cit.*

And $$\cos\frac{d}{\gamma} = (x \mid y) = \xi_1\eta_1 + \xi_2\eta_2 + \xi_3\eta_3,$$

$$\sin\frac{d}{\gamma} = \sqrt{(xy \mid xy)} = \sqrt{\{(\xi_2\eta_3 - \xi_3\eta_2)^2 + (\xi_3\eta_1 - \xi_1\eta_3)^2 + (\xi_1\eta_2 - \xi_2\eta_1)^2\}}.$$

These are the formulæ of the ordinary Euclidean geometry of a sphere; where ξ_1, ξ_2, ξ_3 and η_1, η_2, η_3 are direction cosines.

211. Distances of Planes and of Subregions. (1) As yet only the distance between points has been defined. The same principles can easily be applied to planes.

For any planes X and Y can be expressed in terms of their polar points with respect to the absolute. Thus $X = \mid x$, and $Y = \mid y$. Hence, if the absolute be imaginary, $(X \mid X)$ and $(Y \mid Y)$ are necessarily of the same sign. If the absolute be real and closed, $(X \mid X)$ and $(Y \mid Y)$ are of the same sign, when x and y are either both within or both without the absolute. If x lie within the real closed absolute, the plane X contains [cf. § 82 (6)] no points lying in space, but only points in anti-space; but if x lie without the absolute, then [cf. § 82 (7)] the plane X contains points in space as well as points in anti-space.

Let X and Y be any two planes, and suppose that the plane $\lambda X + \mu Y$ touches the absolute quadric.

Then λ/μ must be one of the two roots λ_1/μ_1 and λ_2/μ_2 of the equation
$$\lambda^2 (X \mid X) + 2\lambda\mu (X \mid Y) + \mu^2 (Y \mid Y) = 0.$$

Let A_1 and A_2 be these two tangent planes; then the anharmonic ratio of the range $\{XY, A_1A_2\}$ is $\lambda_1\mu_2/\lambda_2\mu_1$, and this ratio is either real or of the form $e^{2i\phi}$, where ϕ is real. Let it be called ρ.

Then if ρ be real, the measure of the separation between X and Y can be defined to be $\frac{\kappa}{2}\log\rho$; and if ρ be imaginary, it can be defined to be $\frac{\kappa'}{2i}\log\rho$; where κ and κ' are constants.

There is no reason why either κ or κ' should necessarily be equal to the 'space-constant' γ. But there is no real loss of generality, and there is a gain in the interest of the analogy to ordinary geometry, if $\kappa = \gamma$, and $\kappa' = 1$. For it will be found that the hyperbolic measure of separation between planes can then be identified with the distance between two points; and the elliptic measure of separation can be considered as the angle between them, which is of no dimensions in length.

(2) Thus, [cf. § 124], it follows that the separation between two planes X and Y is that angle between 0 and π given by
$$\theta = \frac{1}{2i}\log\rho = \cos^{-1}\frac{\pm (X \mid Y)}{\sqrt{\{(X \mid X)(Y \mid Y)\}}} = \sin^{-1}\sqrt{\frac{(XY \mid XY)}{\{(X \mid X)(Y \mid Y)\}}},$$
when $$(X \mid Y)^2 < (X \mid X)(Y \mid Y).$$

And the separation is

$$d = \frac{\gamma}{2}\log\rho = \gamma\cosh^{-1}\frac{\pm(X\,|\,Y)}{\sqrt{\{(X\,|\,X)(Y\,|\,Y)\}}} = \gamma\sinh^{-1}\sqrt{\frac{-(XY\,|\,XY)}{\{(X\,|\,X)(Y\,|\,Y)\}}},$$

when $$(X\,|\,Y)^2 > (X\,|\,X)(Y\,|\,Y).$$

It must be noticed that the distinction between these two cases must not be identified simply with that between Elliptic and Hyperbolic Geometry as defined above. The trigonometrical functions must however always be adopted in Elliptic Geometry. This question will be considered in the succeeding chapters as far as it concerns Hyperbolic Geometry.

(3) Furthermore the ambiguity of sign is capable of being determined by exactly the same methods as obtained for points. But with respect to *planes*, in order to obtain an interesting extension of the ideas of ordinary geometry, the 'polar' form is invariably adopted, namely, $+X$ and $-X$ are considered as representing the same plane at opposite intensities.

(4) If the elliptic measure of distance between planes has to be adopted, the measure of separation of planes is called the angle between them.

The ambiguity of sign in the formula for the cosine of the angle leads to the definition that planes make two supplemental angles with each other, θ and $\pi - \theta$; and that of the two the acute angle is the measure of the separation of the planes.

(5) The law of intensity of planar elements is determined by the same principles as that of points. Let it be assumed that if X and Y be planar elements of the same sign and at the same intensity, then $X + Y$ bisects the distance between X and Y. Hence the defining equation of unit intensity can be written, $(X\,|\,X) = \delta$, where δ is a constant which will be determined later separately for Elliptic and Hyperbolic Geometry according to convenience. In Elliptic Geometry δ is always of the same sign: let it therefore be chosen to be unity. In Hyperbolic Geometry it is convenient to choose δ to be positive or negative according as the (real) plane does or does not cut the absolute: let it therefore be chosen to be ± 1.

(6) It is in general impossible to define one single measure of separation between any two subregions X_σ and Y_σ, of $\sigma - 1$ dimensions. But if they are both contained in the same subregion of σ dimensions, then considering the latter subregion as the complete region, X_σ and Y_σ have the properties of planes in regard to it. Also the absolute in this complete region may be taken to be the section of the absolute by the region.

Hence in this case, cf. § 124 (4), the measure of the separation of X_σ and Y_σ (with the conventions, already explained, determining ambiguities) is either

$$\cos^{-1}\frac{\pm(X_\sigma\,|\,Y_\sigma)}{\sqrt{\{(X_\sigma\,|\,X_\sigma)(Y_\sigma\,|\,Y_\sigma)\}}},\ \text{or}\ \gamma\cosh^{-1}\frac{\pm(X_\sigma\,|\,Y_\sigma)}{\sqrt{\{(X_\sigma\,|\,X_\sigma)(Y_\sigma\,|\,Y_\sigma)\}}}.$$

Definition. $|X_\sigma$ and $|Y_\sigma$ are called the absolute polar regions of X_σ and Y_σ.

It is obvious that the separation between two regions is equal to that between their absolute polar regions.

212. Parabolic Geometry. (1) If the parabolic definition of distance hold for every straight line, then every straight line must meet the absolute in two coincident points. Hence the absolute must be two coincident planes. It can be seen as follows that the elliptic and hyperbolic definitions for $\nu - 1$ dimensions both degenerate into the parabolic definition, when the absolute is conceived as transforming itself gradually into two coincident planes.

(2) Let the co-ordinate points $e_1, e_2, \ldots e_\nu$ be ν self-normal points, then the equation of the absolute takes the form,

$$\alpha_1\xi_1^2 + \alpha_2\xi_2^2 + \ldots + \alpha_\nu\xi_\nu^2 = 0.$$

Now conceive the form of the quadric to be gradually modified by $\alpha_2, \ldots \alpha_\nu$ diminishing, till they ultimately vanish, while α_1 remains finite. Then ultimately the equation of the quadric becomes $\alpha_1\xi_1^2 = 0$; that is to say, the quadric becomes two coincident planes, the equation of each plane being $\xi_1 = 0$. Also the $\nu - 1$ co-ordinate points $e_2, e_3, \ldots e_\nu$ lie in this plane, and the point e_1 without it.

Also, cf. § 123 (6), $\quad (xy\,|xy) = \Sigma\alpha_\rho\alpha_\sigma(\xi_\rho\eta_\sigma - \xi_\sigma\eta_\rho)^2.$

Assume that, as the quadric approaches its degenerate form,

$$\alpha_2 = \frac{\kappa_2}{\gamma},\quad \alpha_3 = \frac{\kappa_3}{\gamma},\ \ldots,\ \alpha_\nu = \frac{\kappa_\nu}{\gamma},$$

where the κ's are finite and γ is ultimately infinite.

Then ultimately,

$$(xy\,|xy) = \alpha_1\Sigma\frac{\kappa_\rho}{\gamma}(\xi_1\eta_\rho - \xi_\rho\eta_1)^2 + \Sigma\frac{\kappa_\rho\kappa_\sigma}{\gamma^2}(\xi_\rho\eta_\sigma - \xi_\sigma\eta_\rho)^2 = \alpha_1\Sigma\frac{\kappa_\rho}{\gamma}(\xi_1\eta_\rho - \xi_\rho\eta_1)^2.$$

Similarly $\quad (x\,|x) = \alpha_1\xi_1^2,\ (y\,|y) = \alpha_1\eta_1^2.$

Then if the geometry be elliptic and γ be the space-constant,

$$d = \gamma\sin^{-1}\sqrt{\frac{(xy\,|xy)}{(x\,|x)(y\,|y)}} = \gamma\sin^{-1}\Sigma\frac{\kappa_\rho(\xi_1\eta_\rho - \xi_\rho\eta_1)^2}{\gamma\alpha_1\xi_1\eta_1} = \Sigma\frac{\kappa_\rho}{\alpha_1}\frac{(\xi_1\eta_\rho - \xi_\rho\eta_1)^2}{\xi_1\eta_1}.$$

Now, since the geometry is elliptic, α_1 and $\kappa_2, \kappa_3, \ldots \kappa_\nu$ are all of the same sign. Put $\frac{\kappa_\rho}{\alpha_1} = \beta_\rho^2$.

Hence $$d = \Sigma\beta_\rho^2\frac{(\xi_1\eta_\rho - \xi_\rho\eta_1)^2}{\xi_1\eta_1}.$$

If the geometry be hyperbolic,

$$d = \gamma\sinh^{-1}\sqrt{\frac{-(xy\,|xy)}{(x\,|x)(y\,|y)}} = \Sigma - \frac{\kappa_\rho}{\alpha_1}\frac{(\xi_1\eta_\rho - \xi_\rho\eta_1)^2}{\xi_1\eta_1}.$$

Now, since the geometry is hyperbolic, the absolute is a real closed quadric; and hence [cf. § 82 (5)] α_1 must have one sign and $\kappa_2, \kappa_3, \dots \kappa_\nu$ another sign. Put $\frac{\kappa_\rho}{\alpha_1} = -\beta_\rho^2$.

Hence $$d = \Sigma\beta_\rho^2 \frac{(\xi_1\eta_\rho - \xi_\rho\eta_1)^2}{\xi_1\eta_1}.$$

(3) Thus as a limiting case both of Elliptic and Hyperbolic Geometry, we find a space with the distance between any two elements given by

$$d = \Sigma\beta_\rho^2 \frac{(\xi_1\eta_\rho - \xi_\rho\eta_1)^2}{\xi_1\eta_1};$$

where the $\nu - 1$ co-ordinate elements $e_2, e_3, \dots e_\nu$ lie on the absolute plane at an infinite distance.

213. Law of Intensity in Parabolic Geometry. (1) Let e_1 be the reference element not in the absolute plane, and let $u_2, u_3, \dots u_\nu$ be the reference elements in the absolute plane. Let it be assumed, as in § 210 (1), that, when x and y are of the same intensity, $x + y$ bisects the distance between x and y.

Now let $x = \xi_1 e_1 + \Sigma\xi u$, $y = \eta_1 e_1 + \Sigma\eta u$; then

$$x + y = (\xi_1 + \eta_1) e_1 + \Sigma (\xi + \eta) u.$$

Also the distance between x and $x + y$ is by § 212 (3)

$$\frac{\sqrt{\Sigma\beta_\rho^2 \{\xi_\rho(\xi_1 + \eta_1) - \xi_1(\xi_\rho + \eta_\rho)\}^2}}{\xi_1(\xi_1 + \eta_1)} = \frac{\sqrt{\Sigma\beta_\rho^2(\xi_\rho\eta_1 - \xi_1\eta_\rho)^2}}{\xi_1(\xi_1 + \eta_1)}.$$

Similarly the distance between $x + y$ and y is

$$\frac{\sqrt{\Sigma\beta_\rho^2(\xi_\rho\eta_1 - \xi_1\eta_\rho)^2}}{\eta_1(\xi_1 + \eta_1)}.$$

Hence since these distances are equal, $\xi_1(\xi_1 + \eta_1) = \eta_1(\xi_1 + \eta_1)$, and thence, $\xi_1 = \eta_1$.

(2) Hence the intensity of the point x is a function of ξ_1 only; but by § 85 (2) it must be a homogeneous function of the first degree. Thus the intensity of x is $\lambda\xi_1$, where λ is some constant; and, if e_1 be chosen to be at unit intensity, then $\lambda = 1$. Hence the absolute plane is the locus of zero intensity and the law of intensity explained in § 87 (4) must hold. And the expression for a point x at unit intensity is $e_1 + \Sigma\xi u$, where e_1 is at unit intensity.

Also the distance between the two points $e_1 + \Sigma\xi u$ and $e_1 + \Sigma\eta u$, both at unit intensity, is $\Sigma\beta_\rho^2(\xi - \eta)^2$.

Furthermore by properly choosing the intensities of $u_2, u_3, \dots u_\nu$, this expression for the distance can be reduced to $\Sigma(\xi - \eta)^2$. Thus* parabolic

* Cf. Riemann, *Ueber die Hypothesen, welche der Geometrie zu Grunde liegen*, Collected Mathematical Works.

space of $\nu-1$ dimensions can be interpreted to be simply an ordinary Euclidean space of that number of dimensions; where $e_1u_2, e_1u_3, \ldots e_1u_\nu$ are $\nu-1$ axes at right-angles, and $\xi_2, \xi_3 \ldots \xi_\nu$ are rectangular Cartesian co-ordinates. The interpretation of (the vectors) $u_2, u_3, \ldots u_\nu$ will be considered in Book VII.

HISTORICAL NOTE. An interesting critical 'Short History of Metageometry' is to be found in Chapter I. of *The Foundations of Geometry*, by Bertrand A. W. Russell, Cambridge, 1897. Klein also gives an invaluable short history of the subject in his lithographed *Vorlesungen über Nicht-Euklidische Geometrie*, Göttingen, 1893; he makes the important division of the subject into three periods. The following are the creative works of the ideas of the three periods.

First Period.

Lobatschewsky, *Geometrische Untersuchungen zur Theorie der Parallel-linien*, Berlin, 1840; translated by Prof. G. B. Halsted, Austin, Texas, 1891. Lobatschewsky's first publication of his discovery was in a discourse at Kasan, 1826 (cf. Halsted's preface); and subsequently in papers (Russian) published at Kasan between 1829 and 1830 (cited by Stäckel and Engel, cf. below).

John Bolyai, *The Science Absolute of Space*, 1832; translated by Prof. Halsted, 1891; also cf. German edition by Frischauf, cited below. The original is written in Latin, and is an appendix to a work on Geometry by his father, Wolfgang Bolyai.

Second Period.

Rieman, *Ueber die Hypothesen, welche der Geometrie zu Grunde liegen*, written 1854, Gesammelte Werke; translated by Clifford, cf. his *Collected Mathematical Papers.*

Helmholtz, *Ueber die thatsächlichen Grundlagen der Geometrie*, 1866, and *Ueber die Thatsachen, die der Geometrie zum Grunde liegen*, 1868; both in the *Wissenschaftliche Abhandlungen*, Vol. II.

Beltrami, *Saggio di Interpretazione della Geometria non-Euclidea*, Giornale di Matematiche, Vol. VI. 1868; translated into French by J. Hoüel in the *Annales Scientifiques de l'École Normale Supérieure*, Vol. VI. 1869.

Third Period.

Cayley, *Sixth Memoir upon Quantics*, Phil. Trans., 1859; and, *Collected Papers*, Vol. II., No. 158.

Klein, *Ueber die sogenannte Nicht-Euklidische Geometrie*, two papers, 1871, 1872, *Math. Annalen*, Vols. IV., VI.

Lindemann, *Mechanik bei Projectiven Maasbestimmung*, 1873, *Math. Annalen*, Vol. VII.

Lie, *Ueber die Grundlagen der Geometrie*, Leipziger Berichte, 1890.

A bibliography up to 1878 is given by G. B. Halsted, *American Journal of Mathematics*, Vols. I., II.

The following very incomplete list of a few out of the large number of writers on the subject may be useful:

Flye, Ste Marie, *Études analytiques sur la théorie des parallèles*, Paris, 1871.

M. L. Gérard, Thèse, *Sur la Géometrie Non-Euclidienne*, Paris, 1892.

Poincaré, *Théorie des Groupes Fuchsiennes*, Acta Mathematica, Vol. I., 1882.

Clebsch and Lindemann, *Vorlesungen über Geometrie*, Vol. II. Dritte Abtheilung, Leipzig, 1891.

Frischauf, *Elemente der Absoluten Geometrie nach Johann Bolyai*, Leipzig, 1876.

Killing, *Die Nicht-Euklidischen Raumformen in Analytischer Behandlung*, Leipzig, 1885.

Stäckel and Engel, *Die Theorie der Parallel-linien von Euklid bis auf Gauss*, Leipzig, 1895. This book contains a very useful bibliography of books on the Theory of Parallels from the year 1482 to the year 1837.

Veronese, cf. *loc. cit.* p. 161.

Burnside, *On the Kinematics of Non-Euclidean Space*, Proc. of Lond. Math. Soc., 1894.

Clifford, *Preliminary Sketch of Biquaternions*, Proc. of Lond. Math. Soc., 1873, and *Collected Mathematical Papers*.

Newcomb, *Elementary Theorems relating to the Geometry of a space of three dimensions and of uniform positive curvature in the fourth dimension*, Crelle, Vol. 33, 1877.

The philosophical questions suggested by the subject are considered by Russell, *Foundations of Geometry* (mentioned above); in this work references will be found to the previous philosophical writers on the subject.

The first application of an 'extraordinary' algebra to non-Euclidean Geometry was made for Elliptic Space by Clifford, *Sketch of Biquaternions, Proc. of London Math. Society*, Vol. IV. 1873, also reprinted in his *Collected Papers*; this algebra will be considered in Vol. II. of this work. The first applications of Grassmann's *Calculus of Extension to Non-Euclidean Geometry* were made independently, by Homersham Cox (cf. *loc. cit.* p. 346), to Hyperbolic and Elliptic Space, and by Buchheim to Elliptic Space; *On the Theory of Screws in Elliptic Space, Proc. London Math. Soc.*, 1884 and 1886, Vols. XV. XVI. XVII.

The idea of starting a 'pure' Metrical Geometry with a series of definitions referring to a Positional Manifold is obscurely present in Cayley's *Sixth Memoir on Quantics*; it is explicitly worked out by Homersham Cox (*loc. cit.*) and by Sir R. S. Ball, *On the Theory of Content, Trans. of Roy. Irish Academy*, Vol. XXIX. 1889. Sir R. S. Ball confines himself to three dimensions, and uses Grassmann's idea of the addition of points, but uses none of Grassmann's formulæ for multiplication. But the general idea of a pure science of extension, founded upon conventional definitions, which shall include as a special case the geometry of ordinary experience, is clearly stated in Grassmann's *Ausdehnungslehre von* 1844; and from a point of view other than that of a Positional Manifold it has been carefully elaborated by Veronese (*loc. cit.*).

Homersham Cox constructs a linear algebra [cf. § 22] analogous to Clifford's *Biquaternions*, which applies to Hyperbolic Geometry of two and three and higher dimensions. He also points out the applicability of Grassmann's Inner Multiplication for the expression of the distance formulæ both in Elliptic and Hyperbolic Space; and applies it to the metrical theory of systems of forces. His whole paper is most suggestive [cf. notes, p. 346 and at the end of this volume].

Buchheim states the distance formulæ for both Elliptic and Hyperbolic Space in the same form as they are given in this chapter, with unimportant variations in notation. He then deduces Clifford's theory of parallel lines; and proceeds to investigate the theory of screws in Elliptic and Hyperbolic Space of three dimensions. In his last paper he obtains an important theorem respecting the motion of a rigid body in Elliptic Space of $2\mu-1$ dimensions. Many of his results are deduced by the aid of Biquaternions, and of Cayley's Algebra of Matrices. A further account of his important papers is given in the note at the end of the volume.

CHAPTER II.

Elliptic Geometry.

214. Introductory. In the following application of the formulæ of the Calculus of Extension to the investigation of Elliptic Geometry the polar form will be exclusively considered. Most of the theorems and investigations apply, *mutatis mutandis*, to both forms. But each form requires its own special explanations, which though important geometrically are only remotely possessed of any algebraic interest. So to avoid prolixity one form is adhered to.

The space spoken of throughout this chapter will be of $\nu - 1$ dimensions where ν is any number. It is the merit of this Calculus that the general formulæ for $\nu - 1$ dimensions are as simple and short as those for two or for three dimensions.

215. Triangles. (1) Let the terms a, b, c denote three points; there are eight modes of associating the pairs of intercepts [cf. § 206 (8)] joining each pair of points; namely, using lengths as named, that by associating $\overline{bc}, \overline{ca}, \overline{ab}$; or $\pi\gamma - \overline{bc}, \pi\gamma - \overline{ca}, \pi\gamma - \overline{ab}$; or $\pi\gamma - \overline{bc}, \overline{ca}, \overline{ab}$; or $\overline{bc}, \pi\gamma - \overline{ca}, \pi\gamma - \overline{ab}$; and so on.

(2) Let the angle α between the two intercepts $\overline{ab}$ and $\overline{ac}$ be defined to be that angle (out of the two supplementary alternatives) given by [cf. § 211 (6)]

$$\cos \alpha = \frac{(ab \mid ac)}{\sqrt{\{(ab \mid ab)(ac \mid ac)\}}}.$$

Similarly for the angles β and γ.

Thus the angle between $\overline{ab}$ and $\pi\gamma - \overline{ca}$ is found by putting $-c$ for c in the above and is

$$\cos^{-1} \frac{-(ab \mid ac)}{\sqrt{\{(ab \mid ab)(ac \mid ac)\}}},$$

that is $\pi - \alpha$.

Let the angles α, β, γ be associated with the intercepts $\overline{bc}, \overline{ca}, \overline{ab}$; and let this system of intercepts and angles be called the triangle abc.

(3) Now $(ab\,|\,ac) = (a\,|\,a)(b\,|\,c) - (a\,|\,b)(a\,|\,c)$.

Also [cf. § 206 (8)]

$$\sin\frac{\overline{ab}}{\gamma} = \sqrt{\frac{(ab\,|\,ab)}{(a\,|\,a)(b\,|\,b)}},\quad \sin\frac{\overline{ac}}{\gamma} = \sqrt{\frac{(ac\,|\,ac)}{(a\,|\,a)(c\,|\,c)}};$$

and

$$\cos\frac{\overline{ab}}{\gamma} = \frac{(a\,|\,b)}{\sqrt{\{(a\,|\,a)(b\,|\,b)\}}},\quad \cos\frac{\overline{ac}}{\gamma} = \frac{(a\,|\,c)}{\sqrt{\{(a\,|\,a)(c\,|\,c)\}}}.$$

Hence

$$\cos\frac{\overline{bc}}{\gamma} = \cos\frac{\overline{ab}}{\gamma}\cos\frac{\overline{ac}}{\gamma} + \sin\frac{\overline{ab}}{\gamma}\sin\frac{\overline{ac}}{\gamma}\cos\boldsymbol{\alpha};$$

with similar formulæ for $\boldsymbol{\beta}$ and $\boldsymbol{\gamma}$.

(4) When $\boldsymbol{\alpha} = 0$, then c is collinear with a and b. Also $(ab\,|\,ac)$ is positive: hence we can write either $c = \xi a + b$, or $c = -\xi a + b$, where ξ is positive. In the first case by § 206 (9) c lies in the intercept $\overline{ab}$; in the second case, since $b = c + \xi a$, b lies in the intercept $\overline{ac}$.

Also

$$\cos\frac{\overline{bc}}{\gamma} = \cos\frac{\overline{ab} \sim \overline{ac}}{\gamma}.$$

Thus $\overline{bc} = \overline{ab} - \overline{ac}$ in the first case, and $\overline{bc} = \overline{ac} - \overline{ab}$ in the second case.

(5) Let a', b', c' stand for $-a$, $-b$, $-c$ respectively. Then

$$\cos\frac{\overline{b'c'}}{\gamma} = \frac{(b'\,|\,c')}{\sqrt{\{(b'\,|\,b')(c'\,|\,c')\}}} = \frac{(b\,|\,c)}{\sqrt{\{(b\,|\,b)(c\,|\,c)\}}} = \cos\frac{\overline{bc}}{\gamma}.$$

Thus $\overline{b'c'} = \overline{bc}$. Similarly $\overline{c'a'} = \overline{ca}$, $\overline{a'b'} = \overline{ab}$.

Again it is easy to see from (3) that the angle between $\overline{a'b'}$ and $\overline{a'c'}$ is $\boldsymbol{\alpha}$; and so on. Hence the triangle $a'b'c'$ is the same as the triangle abc, both in its sides and angles and angular points. The two are therefore identical.

(6) Consider the triangle $a'bc$, which by subsection (5) is the same as $ab'c'$. Its sides are easily seen to be related to those of abc as follows:

$$\overline{b'c'} = \overline{bc},\quad \overline{c'a} = \pi\gamma - \overline{ca},\quad \overline{ab'} = \pi\gamma - \overline{ab}.$$

Hence by subsection (3) its angles are $\boldsymbol{\alpha}$, $\pi - \boldsymbol{\beta}$, $\pi - \boldsymbol{\gamma}$.

Similarly the triangle $ab'c$, or $a'bc'$, has sides $\pi\gamma - \overline{bc}$, $\overline{ca}$, $\pi\gamma - \overline{ab}$, and angles $\pi - \boldsymbol{\alpha}$, $\boldsymbol{\beta}$, $\pi - \boldsymbol{\gamma}$.

And the triangle abc', or $a'b'c$, has sides $\pi\gamma - \overline{bc}$, $\pi\gamma - \overline{ca}$, $\overline{ab}$, and angles $\pi - \boldsymbol{\alpha}$, $\pi - \boldsymbol{\beta}$, $\boldsymbol{\gamma}$.

(7) Hence of the eight possible cases mentioned in subsection (1) only four can have angles associated with them in accordance with the convention of subsection (2). Accordingly three points will be said to define four triangles, where a triangle is taken to mean three determinate intercepts and three angles between each pair of intercepts. *The* triangle defined by the *terms* a, b, c will be taken to mean the triangle with the intercepts $(b\,|\,c)$, $(c\,|\,a)$, $(a\,|\,b)$ as sides, and will be called the triangle abc. The other triangles defined by the *points* a, b, c are the triangles $a'bc$ (or $ab'c'$), $ab'c$ (or $a'bc'$), abc' (or $a'b'c$).

There are two main cases to be considered: firstly when one of the four triangles defined by the points a, b, c has all its sides less than $\frac{1}{2}\pi\gamma$, that is to say, has the three lengths $D(bc)$, $D(ca)$, $D(ab)$ for its sides [cf. § 204 (2)]; secondly, when one at least of the sides of each of the four triangles is greater than $\frac{1}{2}\pi\gamma$.

(8) Case I. Let no one of a, b, c be divided from the other two by their polar planes, then [cf. § 205 (3)] $(b|c)$, $(c|a)$, $(a|b)$ may be assumed to be of the same sign; and this sign must be positive. Hence $\overline{bc} = D(bc)$, $\overline{ca} = D(ca)$, $\overline{ab} = D(ab)$. Thus one triangle (the triangle abc) is formed by the intercepts of the lengths $D(bc)$, $D(ca)$, $D(ab)$; each being less than $\frac{1}{2}\pi\gamma$.

Then by subsection (6) the other three triangles formed by the three points are (i) that formed by the intercepts $D(bc)$, $\pi\gamma - D(ca)$, $\pi\gamma - D(ab)$, with angles $\boldsymbol{\alpha}$, $\pi - \boldsymbol{\beta}$, $\pi - \boldsymbol{\gamma}$; (ii) that formed by the intercepts $\pi\gamma - D(bc)$, $D(ca)$, $\pi\gamma - D(ab)$, with angles $\pi - \boldsymbol{\alpha}$, $\boldsymbol{\beta}$, $\pi - \boldsymbol{\gamma}$; (iii) that formed by the intercepts $\pi\gamma - D(bc)$, $\pi\gamma - D(ca)$, $D(ab)$, with angles $\pi - \boldsymbol{\alpha}$, $\pi - \boldsymbol{\beta}$, $\boldsymbol{\gamma}$.

(9) Each of these last three triangles has two sides greater than $\frac{1}{2}\pi\gamma$. Let the triangle with each side less than $\frac{1}{2}\pi\gamma$ be called the principal triangle abc, let the other three be called the secondary triangles.

(10) Case II. Assume that a is divided from b and c by the polar planes of b and c. Then [cf. § 205 (3)] we may assume $(b|c)$ and $(a|b)$ to be positive, and $(a|c)$ negative. Hence $\overline{bc} = D(bc)$, $\overline{ca} = \pi\gamma - D(ca)$, $\overline{ab} = D(ab)$.

Also
$$(ab|ac)\ \{= (a|a)(b|c) - (a|b)(a|c)\} \text{ is positive;}$$
$$(bc|ba)\ \{= (b|b)(c|a) - (a|b)(b|c)\} \text{ is negative;}$$
$$(ca|cb)\ \{= (c|c)(a|b) - (b|c)(c|a)\} \text{ is positive.}$$

Thus, considering the triangle abc, the angles $\boldsymbol{\alpha}$ and $\boldsymbol{\gamma}$ are acute, and $\boldsymbol{\beta}$ is obtuse; and the obtuse angle is opposite to the side greater than $\frac{1}{2}\pi\gamma$. The other three triangles, defined by the *points* a, b, c, are (i) that formed by $D(bc)$, $D(ca)$, $\pi\gamma - D(ab)$, with angles $\boldsymbol{\alpha}$, $\pi - \boldsymbol{\beta}$, $\pi - \boldsymbol{\gamma}$. This triangle has one side, namely $\pi\gamma - D(ab)$, greater than $\frac{1}{2}\pi\gamma$, and one obtuse angle, $\pi - \boldsymbol{\gamma}$, opposite to it. (ii) The triangle formed by $\pi\gamma - D(bc)$, $D(ca)$, $D(ab)$, with angles $\pi - \boldsymbol{\alpha}$, $\pi - \boldsymbol{\beta}$, $\boldsymbol{\gamma}$. This triangle has one side, namely $\pi\gamma - D(bc)$, greater than $\frac{1}{2}\pi\gamma$, and one obtuse angle, namely $\pi - \boldsymbol{\alpha}$, opposite to it. (iii) The triangle formed by $\pi\gamma - D(bc)$, $\pi\gamma - D(ca)$, $\pi\gamma - D(ab)$, with the angles $\pi - \boldsymbol{\alpha}$, $\boldsymbol{\beta}$, $\pi - \boldsymbol{\gamma}$. This triangle has all its sides greater than $\frac{1}{2}\pi\gamma$, and all its angles obtuse.

(11) Thus in this case the points a, b, c define three triangles each with one side greater than $\frac{1}{2}\pi\gamma$, and one triangle with all its sides greater than $\frac{1}{2}\pi\gamma$. Call this case, the case with no principal triangle. This possibility respecting triangles in elliptic space of the polar form has apparently been overlooked. Let the set, of three triangles, each with one side greater than $\frac{1}{2}\pi\gamma$, be called the principal set.

216. FURTHER FORMULÆ FOR TRIANGLES. (1) The two typical transformations, from which the further formulæ connecting the sides and angles are deduced, are

$$(a|a)(abc|abc) = (ab|ab)(ac|ac) - (ab|ac)^2 \qquad \text{(i)};$$

and

$$(b|c)(abc|abc) = (bc|ba)(ca|cb) + (ab|ac)(bc|bc) \qquad \text{(ii)}.$$

Both of these formulæ can be proved by mere multiplication. Thus for instance [cf. § 120]

$$\begin{aligned}(bc|ba)(ca|cb) &+ (ab|ac)(bc|bc)\\ = \{(b|b)(c|a) - (a|b)(b|c)\}\{(c|c)(a|b) - (b|c)(c|a)\}&\\ + \{(a|a)(b|c) - (a|b)(c|a)\}\{(b|b)(c|c) - (b|c)^2\}&\\ = (b|c)\{2(b|c)(c|a)(a|b) + (a|a)(b|b)(c|c) - (a|a)(b|c)^2 - (b|b)(c|a)^2&\\ - (c|c)(a|b)^2\}&\\ = (b|c)(abc|abc).&\end{aligned}$$

(2) Since $\sin \boldsymbol{\alpha} = \sqrt{\{1 - \cos^2 \boldsymbol{\alpha}\}} = \dfrac{\sqrt{\{(ab|ab)(ac|ac) - (ab|ac)^2\}}}{\sqrt{\{(ab|ab)(ac|ac)\}}}$,

it follows from equation (i) of subsection (1) that

$$\sin \boldsymbol{\alpha} = \sqrt{\frac{(a|a)(abc|abc)}{(ab|ab)(ac|ac)}}.$$

But

$$\sin \frac{\overline{bc}}{\gamma} = \sqrt{\frac{(bc|bc)}{(b|b)(c|c)}}.$$

Hence

$$\frac{\sin \boldsymbol{\alpha}}{\sin \frac{\overline{bc}}{\gamma}} = \frac{\sin \boldsymbol{\beta}}{\sin \frac{\overline{ca}}{\gamma}} = \frac{\sin \boldsymbol{\gamma}}{\sin \frac{\overline{ab}}{\gamma}}$$

$$= \sqrt{\frac{(a|a)(b|b)(c|c)(abc|abc)}{(bc|bc)(ca|ca)(ab|ab)}}.$$

(3) From equation (ii) of subsection (1)

$$\sin \boldsymbol{\beta} \sin \boldsymbol{\gamma} \cos \frac{\overline{bc}}{\gamma} = \cos \boldsymbol{\beta} \cos \boldsymbol{\gamma} + \cos \boldsymbol{\alpha};$$

that is,

$$\cos \boldsymbol{\alpha} = -\cos \boldsymbol{\beta} \cos \boldsymbol{\gamma} + \sin \boldsymbol{\beta} \sin \boldsymbol{\gamma} \cos \frac{\overline{bc}}{\gamma},$$

with two similar equations.

(4) If a, b, c be at unit intensity then [cf. § 120 (1) and § 210 (2)]

$$(abc|abc) = \begin{vmatrix} 1, & \cos \frac{\overline{ab}}{\gamma}, & \cos \frac{\overline{ac}}{\gamma} \\ \cos \frac{\overline{ab}}{\gamma}, & 1, & \cos \frac{\overline{bc}}{\gamma} \\ \cos \frac{\overline{ac}}{\gamma}, & \cos \frac{\overline{bc}}{\gamma}, & 1 \end{vmatrix}.$$

This determinant is the square of the well known function, which in Spherical Trigonometry is sometimes called the Staudtian of the triangle.

(5) It is evident that the usual formulæ of Spherical Trigonometry, for example Napier's Analogies, hold for triangles in Elliptic Geometry. For these formulæ are mere algebraic deductions from the fundamental formulæ of § 215 (3) and of subsections (2) and (3) of this article.

(6) Let a circle be defined to be a curve line [cf. § 67 (4)] in a two-dimensional subregion, such that each point of it is at the same distance (its radius) from a point (its centre) in the subregion. Then it follows from subsection (2) that the perimeter of a circle of radius ρ is $2\pi\gamma \sin \frac{\rho}{\gamma}$.

For consider the chord pq, subtending an angle α at the centre. Draw cl perpendicular to pq.

Then, since by symmetry l is the middle point of pq,

$$\sin \frac{\overline{pq}}{2\gamma} = \sin \frac{\overline{pl}}{\gamma} = \sin \frac{\alpha}{2} \sin \frac{\overline{cp}}{\gamma} = \sin \frac{\alpha}{2} \sin \frac{\rho}{\gamma}.$$

Therefore when α is made small enough,

$$\overline{pq} = \alpha\gamma \sin \frac{\rho}{\gamma}.$$

Accordingly, assuming that the length of the arc of a curve is to be reckoned as ultimately equal to the chord joining its extremities, the circumference of the circle $= \Sigma \overline{pq}$, ultimately, $= \gamma \sin \frac{\rho}{\gamma} \Sigma\alpha = 2\pi\gamma \sin \frac{\rho}{\gamma}$.

217. Points inside a Triangle. (1) Consider the triangle abc, that is, the triangle with its sides formed by the intercepts $(b \mid c)$, $(c \mid a)$, $(a \mid b)$. Any point of the form $\lambda a + \mu b + \nu c$, where λ, μ, ν are of the same sign, will be said to be inside the triangle. Other points of this form will be said to be outside the triangle.

(2) To prove that any straight line, in the two dimensional subregion defined by a, b and c, cuts the sides of the triangle, either two internally and one externally, or all three externally.

Write $p = \lambda a + \mu b + \nu c$; and let px be any line through p and another point x in the two dimensional region. Without loss of generality we may consider that the complete manifold [cf. § 103 (3)] is the two-dimensional region defined by a, b, c.

Then
$$px \,.\, bc = \lambda ax \,.\, bc + \mu bx \,.\, bc + \nu cx \,.\, bc$$
$$= \{\lambda (xca) - \mu (xbc)\} b + \{\lambda (xab) - \nu (xbc)\} c;$$
$$px \,.\, ca = \{\mu (xab) - \nu (xca)\} c + \{\mu (xbc) - \lambda (xca)\} a;$$
$$px \,.\, ab = \{\nu (xbc) - \lambda (xab)\} a + \{\nu (xca) - \mu (xab)\} b.$$

Let
$$\theta_1 = \mu (xab) - \nu (xca),\quad \theta_2 = \nu (xbc) - \lambda (xab),$$
$$\theta_3 = \lambda (xca) - \mu (xbc).$$

Hence $px \,.\, bc = \theta_3 b - \theta_2 c$, $px \,.\, ca = \theta_1 c - \theta_3 a$, $px \,.\, ab = \theta_2 a - \theta_1 b$.

Now $px . bc$ is the point of intersection of px and bc; and if θ_2 and θ_3 are of the same sign, this point is external to the intercept $(b|c)$; and if θ_2 and θ_3 are of opposite sign, the point is within the intercept $(b|c)$. But θ_1, θ_2, θ_3 are either all three of the same sign, or two are of one sign and the third of the opposite sign. Hence the proposition is evident.

(3) Any line in the two dimensional region, which contains a point inside the triangle, cuts two of the sides internally and one externally; also conversely. With the notation of the previous subsection, assume that p lies within the triangle. Then λ, μ, ν may be assumed to be all positive. Also without any loss of generality, x may be assumed to be on the line bc, so that $(xbc) = 0$.

Then $\theta_1 = \mu(xab) - \nu(xca)$, $\theta_2 = -\lambda(xab)$, $\theta_3 = \lambda(xca)$. Hence, if (xab) and (xca) are of the same sign, θ_2 and θ_3 are of opposite signs; also, if (xab) and (xca) are of opposite signs, θ_1 is of opposite sign to both θ_2 and θ_3. Hence in either case the first part of the proposition is true.

To prove the converse, assume that the sides $(c|a)$ and $(a|b)$ are cut internally at the points $\alpha a + \gamma c$, $\alpha_1 a + \beta_1 b$; where α, γ, α_1, β_1 can be assumed to be all positive. Then any point on the straight line can be written in the form $\xi(\alpha a + \gamma c) + \eta(\alpha_1 a + \beta_1 b)$. Hence all points, for which ξ and η are of the same sign, lie within the triangle.

218. Oval Quadrics. (1) If three points a, b, c, lie within [cf. § 82 (1)] a closed quadric, $(\alpha \between x)^2 = 0$, then the quadric cuts all of the sides of one of the triangles defined by the *points* a, b, c externally.

For [cf. § 208 (2)] we may assume $(\alpha \between b \between c)$, $(\alpha \between c \between a)$, $(\alpha \between a \between b)$ to be all positive, when $(\alpha \between x)^2$ is positive, x being a point within the quadric. Now with this assumption as to the terms a, b, c, consider the triangle abc. Let any side bc cut the quadric in a point $\mu b + \nu c$. Then

$$\mu^2 (\alpha \between b)^2 + 2\mu\nu (\alpha \between b \between c) + \nu^2 (\alpha \between c)^2 = 0.$$

Thus the two roots for $\mu : \nu$ given by this equation are both negative. Hence any side $(b|c)$ of the triangle abc is cut by the quadric in two external points. It follows that the sides of any of the remaining three triangles defined by the points a, b, c are cut two internally and one externally.

(2) An oval* quadric is a quadric which cuts externally the sides of any principal [cf. § 215 (9)] triangle abc, of which the three angular points lie within it.

(3) Let a sphere be defined to be a surface locus contained in the complete manifold [§ 67 (1)], such that every point of it lies at a given distance (the radius) from a given point (the centre).

* Oval quadrics have not, as far as I am aware, been previously defined. In the special case of Euclidean space of three dimensions, ellipsoids and hyperboloids of two sheets are both closed quadrics; but only ellipsoids are oval quadrics.

A sphere is a closed quadric. For if e be the centre and ρ the radius, the equation of the sphere is

$$\frac{(e|x)^2}{(x|x)(e|e)} = \cos^2\frac{\rho}{\gamma}, \text{ that is, } (e|x)^2 - (e|e)(x|x)\cos^2\frac{\rho}{\gamma} = 0.$$

Now if y be a point at a distance from e less than ρ, then

$$\frac{(e|y)^2}{(y|y)(e|e)} > \cos^2\frac{\rho}{\gamma}; \text{ hence } (e|y)^2 - (y|y)(e|e)\cos^2\frac{\rho}{\gamma} \text{ is positive.}$$

Also there must be two real points on any line through y which lie on the surface. For, let any line through y cut the plane $|e$ in e', so that $(e|e') = 0$. Then any point z on this line can be written $y + \xi e'$. Hence

$$(e|z)^2 - (z|z)(e|e)\cos^2\frac{\rho}{\gamma} = \left\{(e|y)^2 - (y|y)(e|e)\cos^2\frac{\rho}{\gamma}\right\}$$
$$- 2\xi(y|e')(e|e)\cos^2\frac{\rho}{\gamma} - \xi^2(e'|e')(e|e)\cos^2\frac{\rho}{\gamma}.$$

Hence, since $\left\{(e|y)^2 - (y|y)(e|e)\cos^2\frac{\rho}{\gamma}\right\}$ is positive, it is always possible to find two real values of ξ for which $(e|z)^2 - (z|z)(e|e)\cos^2\frac{\rho}{\gamma} = 0$. Accordingly [cf. § 82 (1)] any point at a distance from the centre less than the radius is within the sphere, and for such points $(e|y)^2 - (y|y)(e|e)\cos^2\frac{\rho}{\gamma}$ is positive.

(4) A sphere of radius less than $\frac{1}{4}\pi\gamma$ is an oval quadric. For let e be the centre of the sphere, and x and y two points within it. Then by (3) the two intercepts $D(ex)$, and $D(ey)$ both lie within the sphere and are cut externally by it. Now let the intercepts $(e|x)$ and $(e|y)$ be these intercepts, so that $(e|x)$ and $(e|y)$ are both positive. Then the triangle exy has two sides cut externally by the sphere, and hence by (1) the third side $(x|y)$ is cut externally.

But $$\cos\frac{\overline{xy}}{\gamma} = \cos\frac{\overline{ex}}{\gamma}\cos\frac{\overline{ey}}{\gamma} + \sin\frac{\overline{ex}}{\gamma}\sin\frac{\overline{ey}}{\gamma}\cos\theta,$$

where θ is the angle at e of the triangle exy.

Hence $$\cos\frac{\overline{xy}}{\gamma} > \cos\frac{\overline{ex}}{\gamma}\cos\frac{\overline{ey}}{\gamma} - \sin\frac{\overline{ex}}{\gamma}\sin\frac{\overline{ey}}{\gamma}$$
$$> \cos\frac{\overline{ex} + \overline{ey}}{\gamma}.$$

Hence $\overline{xy} < \overline{ex} + \overline{ey} < \frac{1}{2}\pi\gamma$; since $\overline{ex}$ and $\overline{ey}$ are by hypothesis each less than $\frac{1}{4}\pi\gamma$.

Thus $\overline{xy} = D(xy)$. Hence that intercept joining any two points within the sphere, which is cut externally by the sphere, is the shortest intercept. Hence the sphere is an oval quadric.

(5) It is also evident by the proof of the preceding subsection that any sphere of radius greater than $\frac{1}{4}\pi\gamma$ is not an oval quadric. Hence also it is easy to prove that any oval quadric can be completely contained within some sphere of radius $\frac{1}{4}\pi\gamma$.

(6) Furthermore it follows from (1) and (4) that any three points lying within a sphere of radius $\frac{1}{4}\pi\gamma$ define a principal triangle.

219. Further Properties of Triangles. (1) Two angles of a principal triangle [cf. § 215 (9)] cannot be obtuse. For if possible let $\boldsymbol{\alpha}$ and $\boldsymbol{\beta}$ be both obtuse. Then from § 215 (3)

$$\cos\frac{\overline{bc}}{\gamma} = \cos\frac{\overline{ca}}{\gamma}\cos\frac{\overline{ab}}{\gamma} + \sin\frac{\overline{ab}}{\gamma}\sin\frac{\overline{ca}}{\gamma}\cos\boldsymbol{\alpha},$$

$$\cos\frac{\overline{ca}}{\gamma} = \cos\frac{\overline{bc}}{\gamma}\cos\frac{\overline{ab}}{\gamma} + \sin\frac{\overline{bc}}{\gamma}\sin\frac{\overline{ab}}{\gamma}\cos\boldsymbol{\beta}.$$

Hence both $\cos\frac{\overline{bc}}{\gamma} - \cos\frac{\overline{ca}}{\gamma}\cos\frac{\overline{ab}}{\gamma}$ and $\cos\frac{\overline{ca}}{\gamma} - \cos\frac{\overline{bc}}{\gamma}\cos\frac{\overline{ab}}{\gamma}$ are negative, since $\cos\boldsymbol{\alpha}$ and $\cos\boldsymbol{\beta}$ are negative. But $\cos\frac{\overline{bc}}{\gamma}$ and $\cos\frac{\overline{ca}}{\gamma}$ are both positive by hypothesis and one of them must be the greater, say $\cos\frac{\overline{bc}}{\gamma}$. Then $\cos\frac{\overline{bc}}{\gamma} - \cos\frac{\overline{ca}}{\gamma}\cos\frac{\overline{ab}}{\gamma}$ has the sign of $\cos\frac{\overline{bc}}{\gamma}$, and is therefore positive. Hence there cannot be two obtuse angles in a principal triangle. It has been proved [cf. § 215 (11)] that, if no principal triangle exist, the triangles of the principal set defined by a, b, c have each only one obtuse angle, while the remaining triangle has three obtuse angles.

(2) In any triangle abc [cf. § 215 (7)] if $\boldsymbol{\beta}$ and $\boldsymbol{\gamma}$ be both acute or both obtuse, the foot of the perpendicular from a on to bc falls within the intercept $(b\,|\,c)$; otherwise it falls without the intercept $(b\,|\,c)$.

For let $p = \lambda b + \mu c$ be the foot of this perpendicular. Then

$$(ap\,|\,bc) = \lambda\,(ab\,|\,bc) + \mu\,(ac\,|\,bc) = 0.$$

Hence we may write $p = (ca\,|\,cb)\,b + (ba\,|\,bc)\,c$.

Now $\boldsymbol{\beta}$ and $\boldsymbol{\gamma}$ are respectively acute or obtuse according as $(ba\,|\,bc)$ and $(ca\,|\,cb)$ are positive or negative [cf. § 215 (2)]. Hence the proposition.

(3) If the angles $\boldsymbol{\beta}$ and $\boldsymbol{\gamma}$ be both acute, the triangles abp and acp have $\boldsymbol{\beta}$ and $\boldsymbol{\gamma}$ respectively as angles (and not $\pi - \boldsymbol{\beta}$ and $\pi - \boldsymbol{\gamma}$), also the sum of their angles at a is equal to $\boldsymbol{\alpha}$. This proposition is easily seen to be true.

(4) The sum of the three angles of any principal triangle, or of a triangle from a principal set is greater than two right-angles.

Firstly, let the angle $\boldsymbol{\gamma}$ be a right-angle, and let $\boldsymbol{\alpha}$ and $\boldsymbol{\beta}$ be acute.

Then by one of Napier's Analogies,

$$\tan \tfrac{1}{2}(\alpha+\beta) = \frac{\cos \frac{1}{2}\dfrac{\overline{bc} \sim \overline{ca}}{\gamma}}{\cos \frac{1}{2}\dfrac{\overline{bc} + \overline{ca}}{\gamma}} \cot \tfrac{1}{2}\gamma = \frac{\cos \frac{1}{2}\dfrac{\overline{bc} \sim \overline{ca}}{\gamma}}{\cos \frac{1}{2}\dfrac{\overline{bc} + \overline{ca}}{\gamma}}.$$

Now since $\overline{bc}$ and $\overline{ca}$ are each less than $\frac{1}{2}\pi\gamma$, it follows that

$$\cos \tfrac{1}{2}\frac{\overline{bc} \sim \overline{ca}}{\gamma} > \cos \tfrac{1}{2}\frac{\overline{bc} + \overline{ca}}{\gamma}.$$

Hence $\alpha+\beta > \dfrac{\pi}{2}$. Hence $\alpha+\beta+\gamma > \pi$.

Secondly, let *abc* be any principal triangle or a triangle from a principal set. Then at least two of its angles are acute, say α and β. Draw a perpendicular *cd* from *c* on to the opposite side. Then *d* lies between *a* and *b* on the intercept $(a|b)$, and *abc* is divided into two right-angled triangles. Hence obviously from subsection (3) the theorem holds for the triangle *abc*.

220. Planes one-sided. (1) It has been proved in § 203 (1) that a plane does not divide space. An investigation of the meaning to be attached to the idea of the sides of a plane is therefore required.

Let two points *a* and *b* be said to be on the same side of a plane *P*, when the intercept $D(ab)$ does not contain the point of intersection of *ab* and *P*, that is to say, the point $ab \,.\, P$.

Conversely when the intercept $D(ab)$ does contain the point $ab \,.\, P$, let *a* and *b* be said to be on opposite sides of the plane.

(2) Suppose that *a* and *b* are on opposite sides of the plane, but that they each approach the plane along the line *ab* so that $D(ab)$ diminishes and ultimately vanishes. Then in the limit *a* and *b*, though coincident in position, both lie on the plane on opposite sides of it.

Thus a plane can be considered to have two sides in the sense, that at each point of the plane there may be considered to be two coincident points on opposite sides of the plane. This idea can obviously be extended to any surface.

(3) If *a* be any point on a plane *P*, then *a* and $-a$ may be considered as symbolizing the two coincident points on opposite sides of the plane.

For let *b* be any other point not on the plane; and assume, for example, that $(a|b)$ is positive. Write a' for $-a$. Then if *a* be considered to be on the same side of the plane as *b*, the intercept $(a|b)$ does not contain a' (by the definition of subsection (1)), the intercept $(a'|b)$ does not contain *a* (since $\overline{a'b}$ is the length of the long (polar) intercept between a' and *b*, namely, $\pi\gamma - D(ab)$), and the straight line is completed by the indefinitely small intercept $D(aa')$ which passes through the plane.

Thus if b be a given representation of the point b, which is taken as the standard representation, a is on the same side as the point b of any plane on which a lies when $(a|b)$ is positive, and is on the opposite side when $(a|b)$ is negative.

It must be carefully noticed that the choice of sides for a and $-a$ depends not only on the position of the point b, but also on the special term b which represents the point. For $-b$ represents the same point, and if $-b$ be taken as the standard representation, a and $-a$ would according to the above definition change sides of any plane on which they lie.

(4) Suppose that a sphere of radius $\frac{1}{4}\pi\gamma$ be described cutting the plane, and that attention be confined to points within this sphere. Then [cf. § 218 (6)] any three such points, a, b, c, define a principal triangle: let it be the triangle abc.

Now if a and b be on the same side of the plane, then c is on the same side of the plane as a or on the opposite side of it, according as c is on the same side as b or on the opposite side.

For the plane cuts the two dimensional region abc in a straight line, and by hypothesis this straight line cuts the intercept $(a|b)$ externally, hence by § 217 (2) and (3) it cuts the other two intercepts, $(a|c)$, $(b|c)$, both externally or both internally.

Thus when attention is confined to the space within this sphere, the ordinary ideas concerning the two sides of a plane hold good.

(5) But if the points a, b, c do not define a principal triangle, let the triangle abc be one of the principal set. Assume that $(a|b)$, $(a|c)$ are positive and that $(b|c)$ is negative. Now the straight line, in which any plane cuts the region abc, must cut the sides of the triangle abc either all externally or two internally and one externally.

If the line cut all the sides externally, it cuts $D(ab)$, $D(ac)$ externally and $D(bc)$ internally. Hence a and b are on the same side of the plane, also a and c; but b and c are on opposite sides.

If the line cut $(a|b)$, $(a|c)$ internally and $(b|c)$ externally, it cuts $D(ab)$, $D(ac)$, $D(bc)$ all internally. Hence any two of the three are on opposite sides of the plane to each other.

If the line cut $(a|b)$, $(b|c)$ internally and $(a|c)$ externally, it cuts $D(ab)$ internally, and $D(bc)$, $D(ac)$ externally. Hence c and b are on the same side of the plane, also c and a; but a and b are on opposite sides. Similarly if the line cut $(a|c)$, $(b|c)$ internally and $(a|b)$ externally, then c and b are on the same side of the plane, also b and a; but a and c are on opposite sides.

Hence, when three points do not form a principal triangle, the ordinary ideas concerning a plane dividing space cannot apply.

(6) It has been defined in (3) that if the point a lie on the plane P and b be another point not on the plane, then the term a symbolizes a point on the same side as the point b, when $(a|b)$ is positive.

Let c be another point on the plane so that the triangle abc is a principal triangle. Then by hypothesis $(b|c)$ is positive, and the term c symbolizes a point on P on the same side as the point b. Hence, assuming that the theorems of subsections (4) and (5) are to hold when two angular points are on the plane, a and c are on the same side of the plane when $(a|c)$ is positive.

(7) If a and c be two points on the plane P and on the same side of it, then the point $\lambda a + \mu c$ is defined to describe a straight line without cutting the plane, when any two neighbouring points of the line successively produced by the gradual variation of λ and μ are on the same side of the plane.

Suppose that λ varies from 1 to 0 as μ varies from 0 to 1, then the intercept $(a|c)$ is described without cutting the plane. Also every point on this intercept is on the same side as both a and c. But now starting from c let the moving point describe the other intercept without cutting the plane. Then λ must vary from 0 to -1 while μ varies from 1 to 0. But the final point reached is $-a$. Thus a moving point, starting from a and traversing a complete straight line drawn on the plane without cutting the plane, ends at $-a$, that is on the opposite side of the plane.

Again, if Q be another plane cutting P, and the subplane of intersection does not cut either P or Q, then when the moving point starting from a has made a complete circuit of a straight line lying in the subplane PQ it is on the opposite side both of P and Q to a.

In order to understand the full meaning of this property, consider for example space of three dimensions. Let the two sides of P at a be called the upper and under side, and the two sides of Q at a be called the right and left side. Let a dial with a pointer lie in the plane P at a with face upwards and pointer pointing to the right. Let the dial be carried round the straight line of intersection of the planes so that in neighbouring positions both face and pointer respectively look to the same sides of the two planes. Then, when the complete circuit has been made, the dial at a is face downwards and the pointer points to the left.

The property of planes proved in this subsection is expressed by saying that planes are one-sided. The discovery of this property of planes in the polar form of elliptic geometry is due to Klein*.

(8) The definition in subsection (7) of a straight line drawn on a plane without cutting it can obviously be applied to any curve-line drawn on the plane. Also by the method of (7) it is easy to prove that a point, starting from

* Cf. *Math. Annal.*, Vol. VI.

a and describing a closed curve on a plane P, returns to a or to $-a$ according as the closed curve cuts the subplane of intersection of P and the polar plane of a (that is, the subplane $P|a$) an even or an odd number of times.

221. Angles between Planes. (1) Since in Elliptic Geometry the absolute is imaginary, the separation [cf. § 211 (2)] between planes must necessarily be measured by the trigonometrical formula and not by the hyperbolic formula. The same applies to the separation between any two subregions, when the idea of a measure of separation between them can be applied [cf. § 211 (6)]. Let the measure of the separation between planes or between subregions (excluding points) be called the angle between them. Thus the angle between two planes X and Y is one of the two supplementary angles (less than π).

$$\cos^{-1} \frac{\pm (X|Y)}{\sqrt{\{(X|X)(Y|Y)\}}}.$$

(2) Let $< XY$ stand for that one of the two supplementary angles between X and Y which is defined by

$$\cos < XY = \frac{(X|Y)}{\sqrt{\{(X|X)(Y|Y)\}}}.$$

(3) The points $|X (= x)$ and $|Y (= y)$ are the absolute poles of the planes X and Y. The length $\overline{xy}$ of the intercept $(x|y)$ is given by

$$\cos \frac{\overline{xy}}{\gamma} = \frac{(x|y)}{\sqrt{\{(x|x)(y|y)\}}} = \cos < XY.$$

Hence
$$\frac{\overline{xy}}{\gamma} = < XY.$$

(4) If Z be a third plane, the angles between the subplanes XY and XZ are the two supplementary angles (less than π) defined by

$$\cos^{-1} \frac{\pm (XY|XZ)}{\sqrt{\{(XY|XY)(XZ|XZ)\}}}.$$

These angles are the same as those between the lines xy and xz, where $z = |Z$.

222. Stereometrical Triangles. (1) The angles which the planes A, B, C make with each other, and also the angles which the subplanes BC, CA, AB make with each other can be associated together by definition, so as to form what will be called a stereometrical triangle. Let the stereometrical triangle ABC be the association of the three angles $< BC$, $< CA$, $< AB$, with the three angles α, β, γ, defined by

$$\cos \alpha = \frac{(AB|AC)}{\sqrt{\{(AB|AB)(AC|AC)\}}},$$

with two similar equations for β and γ.

(2) Then if $a = |A$, $b = |B$, $c = |C$, the triangle abc is the 'polar' triangle of the stereometrical triangle ABC. Also the angles of the triangle abc are respectively equal to α, β, γ; while the sides of the triangle abc are respectively equal to $\gamma(< BC)$, $\gamma(< CA)$, $\gamma(< AB)$.

(3) Accordingly, corresponding to every formula for a triangle defined by three points there exists a formula for a stereometrical triangle defined by three planes. Thus the ordinary formulæ of Spherical Trigonometry, in ordinary three dimensional Euclidean space, are shown to hold for the relations between three planes of any number of dimensions in Elliptic Geometry.

(4) From § 215 (3) it follows that

$$\cos < BC = \cos < CA \cos < AB + \sin < CA \sin < AB \cos \alpha;$$

with two similar formulæ.

Now if the complete space be three dimensional, the subplanes BC, CA, AB are three straight lines meeting at a point; and thus α, β, γ correspond to the 'sides' of an ordinary three dimensional spherical triangle, while $< BC$, $< CA$, $< AB$, correspond to the angles.

Thus according to analogy the above formula ought to be

$$\cos < BC = - \cos < CA \cos < AB + \sin < CA \sin < AB \cos \alpha.$$

This difference of sign is explained by noting that the angles to be associated with the stereometrical triangle ABC were defined by convention in subsection (1); and that if the angles of the triangle ABC had been defined to be $\pi - < BC$, etc., and $\pi - \alpha$, etc., the signs of the formulæ obtained would have agreed, when the complete region is of three dimensions, with those of ordinary Spherical Trigonometry.

223. Perpendiculars. (1) Any two mutually normal [cf. § 108 (5)] points x, y are at the same distance from each other. For since $(x|y) = 0$, $\cos \frac{\overline{xy}}{\gamma} = 0$, and therefore $\overline{xy} = \frac{1}{2}\pi\gamma$. Such points may also be called quadrantal. The condition that two lines ab and ac should be at right-angles (or perpendicular) is $(ab|ac) = 0$.

Lines, or other subregions, which are perpendicular must be carefully distinguished from lines, or other subregions, which are mutually normal [cf. § 113 (1)].

(2) Let any region L_ρ of $\rho - 1$ dimensions be cut by a straight line ab in the point a; then, if $(\rho - 1)$ independent lines drawn through a in the region L_ρ be perpendicular to ab, all lines drawn through a in the region L_ρ are perpendicular to ab. For let $ap_1, ap_2, \ldots ap_{\rho-1}$ be the $(\rho - 1)$ independent lines.

Then by hypothesis $(ab|ap_1) = 0 = (ab \; ap_2) = \text{etc.}$

But $\lambda a + \Sigma\mu p$ represents any point in L_ρ. Hence any line through a is $(\mu_1 ap_1 + \mu_2 ap_2 + \ldots)$.

And $\{ab \mid (\mu_1 ap_1 + \mu_2 ap_2 + \ldots)\} = \mu_1 (ab \mid ap_1) + \mu_2 (ab \mid ap_2) + \ldots = 0$; which is the required condition of perpendicularity.

Then ab will be said to be perpendicular to the region L_ρ, or at right-angles to it.

(3) Any line perpendicular to the region L_ρ intersects the supplementary (or complete normal) region $|L_\rho$; and conversely, any line intersecting both L_ρ and $|L_\rho$ is perpendicular to both.

For, with the notation of the previous subsection, let ab be the line; and let b be the point on the line ab normal to a [cf. § 113 (5)], then b is normal to every point on L_ρ. For, if p be such a point, $(ab \mid ap) = 0$.

Hence $(a \mid a)(b \mid p) - (a \mid p)(a \mid b) = 0$. Hence $(a \mid a)(b \mid p) = 0$.

But $(a \mid a)$ is not in general zero. Hence we must have $(b \mid p) = 0$.

Hence b lies in $|L_\rho$; and therefore ab intersects $|L_\rho$.

(4) If P_ρ and P_σ be two regions normal to each other [cf. § 113], and if a be any point in P_ρ, then any line drawn through a in the region P_ρ is perpendicular to the region aP_σ.

For let a' be any other point in P_ρ, and b be any point in P_σ, then by hypothesis, $(a \mid b) = (a' \mid b) = 0$.

Hence $$(aa' \mid ab) = (a \mid a)(a' \mid b) - (a \mid b)(a' \mid a) = 0.$$

(5) Let two planes L and M intersect in the subplane LM, and a_1 be any point in LM. From a_1 draw $a_1 l$ in the plane L perpendicular to the subplane LM, and draw $a_1 m$ in the plane M perpendicular to the subplane LM, then the angle between L and M is equal to that between $a_1 l$ and $a_1 m$.

For in the subplane LM, which is of $\nu - 3$ dimensions (the space throughout this chapter being of $\nu - 1$ dimensions), we can find [cf. § 113 (5)] $\nu - 3$ other points $a_2, a_3, \ldots a_{\nu-2}$, so that $a_1, a_2, \ldots a_{\nu-2}$ are mutually normal. Also take l in the line $a_1 l$ to be the point normal to a_1. Then by subsection (3) l is normal to $a_1, a_2, \ldots a_{\nu-2}$; and therefore to every point in LM.

Similarly in the line $a_1 m$ let m be normal to every point in LM.

Then $$(a_1 \mid a_2) = (a_1 \mid a_3) = (a_\rho \mid a_\sigma) = \ldots = (a_\rho \mid a_{\nu-2}) = 0,$$

and $$(a_1 \mid l) = (a_2 \mid l) = \ldots = (a_{\nu-2} \mid l) = 0,$$

and $$(a_1 \mid m) = (a_2 \mid m) = \ldots = (a_{\nu-2} \mid m) = 0.$$

Also we may write $L = (a_1 a_2 \ldots a_{\nu-2} l)$, and $M = (a_1 a_2 \ldots a_{\nu-2} m)$.

Then from § 120 (1)

$$(L \mid L) = (a_1 \mid a_1)(a_2 \mid a_2) \ldots (l \mid l),\ (M \mid M) = (a_1 \mid a_1)(a_2 \mid a_2) \ldots (m \mid m),$$
$$(L \mid M) = (a_1 \mid a_1)(a_2 \mid a_2) \ldots (l \mid m).$$

Hence if θ be the angle between L and M, and ϕ between $a_1 l$ and $a_1 m$,

$$\cos\theta = \frac{(L\,|\,M)}{\sqrt{\{(L\,|\,L)(M\,|\,M)\}}} = \frac{(l\,|\,m)}{\sqrt{\{(l\,|\,l)(m\,|\,m)\}}} = \frac{(a_1 l\,|\,a_1 m)}{\sqrt{\{(a_1 l\,|\,a_1 l)(a_1 m\,|\,a_1 m)\}}} = \cos\phi.$$

Thus $$\theta = \phi.$$

Corollary. It is also obvious that $\theta = \phi = \dfrac{\overline{lm}}{\gamma}$.

(6) Any line perpendicular to any plane L also passes through its absolute pole [cf. subsection (3)].

Thus if any plane M include one perpendicular to L, then from any point of the subplane LM a perpendicular to L can be drawn lying in M. For, if M includes one perpendicular to L, it includes the pole $|L$. Then any line joining any point in LM to $|L$ must be perpendicular to L and must lie in M.

Also since $|L$ lies in M, then $|M$ lies in L. Hence this property is reciprocal. Such planes will be said to be at right angles.

It is obvious that, if two planes are at right angles, their poles are quadrantal.

(7) If two planes L and L' be each cut perpendicularly by a third plane M, it follows at once from the formulæ for stereometrical triangles investigated in § 222, that the angle between the subplanes LM and $L'M$ is equal to that between the planes L and L'.

224. Shortest Distances from Points to Planes. (1) The shortest distance from a point to a plane is the shortest intercept of the straight line through the point perpendicular to the plane.

For let x be the point, p the foot of the perpendicular, and q any other point on the plane. Let the terms x and p be so chosen [cf. § 206 (9)] that $\overline{xp} = D(xp)$; so that $\overline{xp}$ is the shorter of the two intercepts between x and p.

Then by § 215 (3), $\cos\dfrac{\overline{xq}}{\gamma} = \cos\dfrac{\overline{xp}}{\gamma}\cos\dfrac{\overline{pq}}{\gamma}$. Hence, if $\overline{pq}$ be greater than $\frac{1}{2}\pi\gamma$, $\overline{xq}$ is also greater than $\frac{1}{2}\pi\gamma$. Thus the points x, p, q must define a principal triangle. Let the terms x, p, q be so chosen that xpq is this principal triangle. Then from the above formula, $D(xq) > D(xp)$.

This length of the perpendicular will be called simply the distance of the point from the plane.

(2) It is obvious that the other intercept of the straight line xp is the longest intercept of a straight line drawn from x to the plane.

(3) The pole of the plane is easily seen to be the point which is further from the plane than any other point, namely at a distance $\frac{1}{2}\pi\gamma$.

(4) Let ρ be the distance of the point x from the plane L. Then $\frac{1}{2}\pi\gamma - \rho$ is the distance between x and the point $|L$.

Hence $\sin\frac{\rho}{\gamma} = \cos\left(\frac{\frac{1}{2}\pi\gamma - \rho}{\gamma}\right) = \pm\frac{(x \,||\, L)}{\{(x\,|x)\,(|L\,||\,L)\}^{\frac{1}{2}}} = \pm\frac{(xL)}{\{(x\,|x)\,(L\,|L)\}^{\frac{1}{2}}}$, where, as in the other cases, the ambiguity in sign is to be so determined as to make $\sin\frac{\rho}{\gamma}$ positive. With this understanding we may write

$$\sin\frac{\rho}{\gamma} = \frac{(xL)}{\{(x\,|x)\,(L\,|L)\}^{\frac{1}{2}}}.$$

225. Common Perpendicular of Planes. (1) The line joining the poles $|L$ and $|L'$ of any two planes L and L' is obviously [cf. § 223 (3)] perpendicular to both planes L and L'. Further, any point on the line $|L\,|L'$ is normal to any point on the subplane LL'. Let the line $|L\,|L'$ intersect the planes in l and l'. Let a be any point on the subplane LL'. Then it is easy to prove that al and al' are each perpendicular to the subplane LL'. Hence the angle between al and al' is equal to the angle λ between the planes. Accordingly in the triangle lal', the two angles at l and l' are right-angles, $\overline{al}$ and $\overline{al'}$ are each $\frac{1}{2}\pi\gamma$, and a is λ. Hence $\overline{ll'} = \lambda\gamma$.

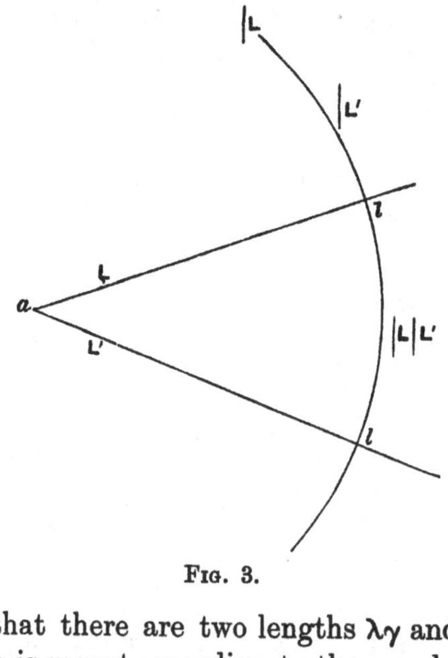

Fig. 3.

It is to be noted that there are two lengths $\lambda\gamma$ and $(\pi - \lambda)\,\gamma$ for $D\,(ll')$; the shortest of the two is meant according to the usual convention.

(2) It is easy to see that $D\,(ll')$ is greater than the distance of any point x in either plane from the other plane. For let x lie in L, and draw xp perpendicular to L'. Then xp passes through $|L'$. Also the distance from $|L'$ to p equals that from $|L'$ to l', both being $\frac{1}{2}\pi\gamma$; but that from $|L'$ to l is less than that from $|L'$ to x, since the line from $|L'$ to l is perpendicular to L. Hence $D\,(ll')$ is greater than $D\,(xp)$.

226. Distances from Points to Subregions*. (1) The least distance of a point a from a line bc can be found. For let p be the foot of the perpendicular from a to bc, and let b be any other point on bc. Then, by the same proof as in § 224 (1), the three points a, b, p define a principal triangle. Let this triangle be abp. Then, as in § 224 (1), $D(ap) < D(ab)$; and hence $D(ap)$ is the least distance which it is required to express.

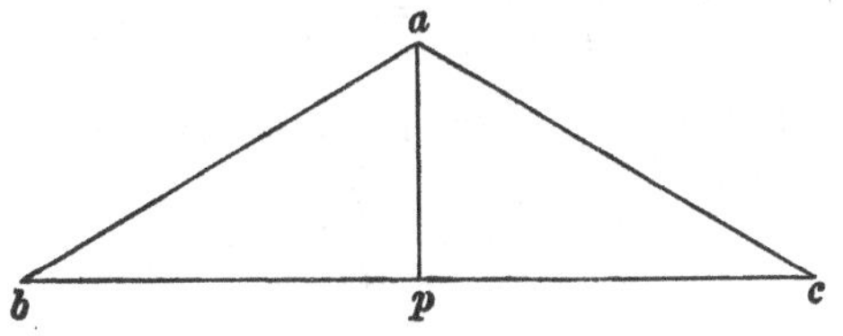

Fig. 4.

But $$\sin\frac{\overline{pa}}{\gamma} = \sin\frac{\overline{ab}}{\gamma}\sin\beta,$$

where β is the angle at b in the triangle abp, that is, the angle at b in the triangle abc, if the term c be properly chosen.

But $$\sin\frac{\overline{ab}}{\gamma} = \sqrt{\frac{(ab\,|ab)}{(a\,|a)(b\,|b)}},$$

and $$\sin\beta = \sqrt{\frac{\{(b\,|b)(abc\,|abc)\}}{\{(ab\,|ab)(bc\,|bc)\}}}.$$

Hence $$\sin\frac{\overline{pa}}{\gamma} = \sqrt{\frac{(abc\,|abc)}{(a\,|a)(bc\,|bc)}}.$$

Therefore, if F be the force bc, the distance (δ) of a from the line of F is given by

$$\sin\frac{\delta}{\gamma} = \sqrt{\frac{(aF\,|aF)}{(a\,|a)(F\,|F)}}.$$

(2) This formula can be extended to give the least distance of any point a from any subregion P_ρ of $\rho - 1$ dimensions.

For the argument of § 224 (1) still holds, and the least distance is evidently the length of the perpendicular ap from a to the subregion. One, and only one, such perpendicular line always exists, since [cf. §§ 72 (5) and 223 (3)] it intersects both P_ρ and $|P_\rho$.

Let F be a force on any line through p in the region P_ρ, and let $P_{\rho-2}$ be the subregion in P_ρ normal to F. Let $P_\rho = FP_{\rho-1}$.

Then $$\sin\frac{\overline{ap}}{\gamma} = \sqrt{\frac{(aF\,|aF)}{(a\,|a)(F\,|F)}}.$$

Now since ap is perpendicular to P_ρ it passes through the normal point to P_ρ in the region aP_ρ. Let p_ρ be this point. Then p_ρ is normal to every point in P_ρ.

* These formulæ, and the deductions from them in subsequent articles, have not been stated before, as far as I am aware.

Let $a = p_\rho + \lambda p$; where the two equations $(p_\rho|P_\rho) = 0$, and $(pF) = 0$, hold. Therefore $(aF) = (p_\rho F)$.

Also $(aP_\rho|aP_\rho) = (p_\rho FP_{\rho-2}|p_\rho FP_{\rho-2}) = (p_\rho F|p_\rho F)(P_{\rho-2}|P_{\rho-2})$
$= (aF|aF)(P_{\rho-2}|P_{\rho-2})$, (cf. § 121).

And $$(P_\rho|P_\rho) = (F|F)(P_{\rho-2}|P_{\rho-2}).$$

Therefore $$\frac{(aP_\rho|aP_\rho)}{(a|a)(P_\rho|P_\rho)} = \frac{(aF|aF)}{(a|a)(F|F)}.$$

Hence if δ be the distance of a from the subregion P_ρ, then

$$\sin\frac{\delta}{\gamma} = \sqrt{\frac{(aP_\rho|aP_\rho)}{(a|a)(P_\rho|P_\rho)}}.$$

(3) This formula includes all the other formulæ for distance from a point. For if P_ρ denote a point b, then it becomes

$$\sin\frac{\delta}{\gamma} = \sqrt{\frac{(ab|ab)}{(a|a)(b|b)}},$$

which is in accordance with § 204 (1).

And if P_ρ denote the plane L, then since (aL) is numerical,

$$(aL|aL) = (aL)^2;$$

hence the formula becomes

$$\sin\frac{\delta}{\gamma} = \frac{(aL)}{\sqrt{\{(a|a)(L|L)\}}};$$

in accordance with § 224 (4).

(4) Since ap is perpendicular to P_ρ, it also intersects $|P_\rho$ and is perpendicular to it. Let q be this point of intersection. Then $D(pq) = \frac{1}{2}\pi\gamma$, and $D(aq) = \frac{1}{2}\pi\gamma - \overline{ap}$.

Thus if δ' be the distance from a to $|P_\rho$,

$$\cos\frac{\delta'}{\gamma} = \sin\frac{\delta}{\gamma} = \sqrt{\frac{(aP_\rho|aP_\rho)}{(a|a)(P_\rho|P_\rho)}}.$$

But also by the same formula, replacing P_ρ by $|P_\rho$,

$$\sin\frac{\delta'}{\gamma} = \sqrt{\frac{(a|P_\rho|.\,a|P_\rho)}{(a|a)(P_\rho|P_\rho)}}.$$

Hence is obtained the formula

$$(aP_\rho|aP_\rho) + (a|P_\rho|.\,a|P_\rho) = (a|a)(P_\rho|P_\rho) \quad \text{..............(i).}$$

This formula is easily obtained by direct transformation by taking [cf. § 113 (7)] ρ mutually normal points in P_ρ and $\nu - \rho$ mutually normal points in $|P_\rho$ as reference points, as in § 229 following.

227. Shortest Distances between Subregions. (1) Let P_ρ and Q_σ be two non-intersecting subregions of the ρth and σth orders respectively, so that $\rho + \sigma < \nu$. A line, such that one of its intercepts is a maximum or a minimum distance between them, is perpendicular to both.

For let a, b, c be three points, and let $D(bc)$ be small. Then a, b and c define a principal triangle: let this triangle be the triangle abc. Then it is easy to prove from the formulæ [cf. §§ 215 and 216] that $D(ab) \sim D(ac)$ is a small quantity of the second order compared to $D(bc)$ when, and only when, the angle at b is a right angle.

The main proposition follows immediately from this lemma.

(2) Let $\rho > \sigma$; then the polar region, $|Q_\sigma$, of Q_σ, intersects P_ρ in a subregion of the $(\rho - \sigma)$th order at least.

Also any line pq, from a point p in this subregion to any point q in Q_σ, is perpendicular to Q_σ, and is of length $\frac{1}{2}\pi\gamma$. Accordingly such perpendiculars from P_ρ to Q_σ are of the greatest length possible for the shortest intercept of perpendiculars from P_ρ to Q_σ; but they are not necessarily perpendicular to P_ρ.

The polar region, $|P_\rho$, of P_ρ does not in general intersect Q_σ. Hence in general there are no such perpendiculars from Q_σ to P_ρ of length $\frac{1}{2}\pi\gamma$.

(3) Let $q_1, q_2, \ldots q_\sigma$ be σ independent points in Q_σ. Then any point x in Q_σ can be written $\Sigma\xi q$.

Hence [cf. § 226 (2)] the perpendicular δ from x to P_ρ is given by

$$\sin\frac{\delta}{\gamma} = \sqrt{\frac{\{(\Sigma\xi q)P_\rho|(\Sigma\xi q)P_\rho\}}{(\Sigma\xi q|\Sigma\xi q)(P_\rho|P_\rho)}}.$$

Write λ for $\sin\frac{\delta}{\gamma}$, and square both sides, and perform the multiplication; then

$$\lambda^2(P_\rho|P_\rho)\{\xi_1^2(q_1|q_1) + \xi_2^2(q_2|q_2) + \ldots + 2\xi_1\xi_2(q_1|q_2) + \ldots\}$$
$$= \xi_1^2(q_1P_\rho|q_1P_\rho) + \xi_2^2(q_2P_\rho|q_2P_\rho) + \ldots + 2\xi_1\xi_2(q_1P_\rho|q_2P_\rho) + \ldots\}.$$

If δ be a maximum or a minimum for variations of x in Q_σ, then λ is a maximum or minimum for variations of $\xi_1, \xi_2, \ldots \xi_\sigma$.

Hence
$$\frac{\partial\lambda}{\partial\xi_1} = 0 = \frac{\partial\lambda}{\partial\xi_2} = \ldots = \frac{\partial\lambda}{\partial\xi_\sigma}.$$

Thus by differentiation

$$\{(q_1P_\rho|q_1P_\rho) - \lambda^2(P_\rho|P_\rho)(q_1|q_1)\}\xi_1 + \{(q_1P_\rho|q_2P_\rho) - \lambda^2(P_\rho|P_\rho)(q_1|q_2)\}\xi_2 + \ldots = 0;$$

with $\sigma - 1$ other similar equations.

Thus, by eliminating $\xi_1, \xi_2, \ldots \xi_\sigma$, an equation is found for λ^2 of the form

$$\begin{vmatrix} \alpha_{11} - \lambda^2(q_1|q_1), & \alpha_{12} - \lambda^2(q_1|q_2), & \ldots, & \alpha_{1\sigma} - \lambda^2(q_1|q_\sigma), \\ \alpha_{21} - \lambda^2(q_2|q_1), & \alpha_{22} - \lambda^2(q_2|q_2), & \ldots, & \alpha_{2\sigma} - \lambda^2(q_2|q_\sigma), \\ \ldots & \ldots & \ldots & \ldots \\ \ldots & \ldots & \ldots & \ldots \end{vmatrix} = 0;$$

where $$\alpha_{11} = \frac{(q_1 P_\rho | q_1 P_\rho)}{(P_\rho | P_\rho)}, \quad \alpha_{12} = \alpha_{21} = \frac{(q_1 P_\rho | q_2 P_\rho)}{(P_\rho | P_\rho)},$$

with similar equations defining the other α's.

(4) Hence, if P_ρ and Q_σ be two subregions of the ρth and σth orders respectively ($\rho > \sigma$), there are in general σ common perpendiculars to the two subregions, which are the lines of maximum or minimum lengths joining them.

If P_ρ and Q_σ had been interchanged in the above reasoning, an equation of the ρth degree ($\rho > \sigma$) would have been found. But this equation would not merely determine the common perpendiculars to P_ρ and Q_σ. For, if δ be the length of the perpendicular from any point in P_ρ to Q_σ, then, with the notation of the previous subsection,

$$\lambda = \sin\frac{\delta}{\gamma}, \quad \frac{\partial\lambda}{\partial\xi_1} = \cos\frac{\delta}{\gamma} \cdot \frac{\partial\delta}{\gamma\partial\xi_1}.$$

Hence $\frac{\partial\lambda}{\partial\xi_1} = 0$, when $\delta = \frac{1}{2}\pi\gamma$, as well as when δ is a maximum or a minimum. Thus the infinite number of lines discussed in subsection (2) fulfil the conditions from which this equation of the ρth degree is derived.

(5) A formula can be found which determines the σ feet in Q_σ of these perpendiculars. For, if $q_1, q_2, \dots q_\sigma$, be these feet, then in the equation for λ^2 of subsection (3) the σ roots must be [cf. § 226 (2)]

$$\frac{(q_1 P_\rho | q_1 P_\rho)}{(q_1 | q_1)(P_\rho | P_\rho)}, \quad \frac{(q_2 P_\rho | q_2 P_\rho)}{(q_2 | q_2)(P_\rho | P_2)}, \text{ and so on.}$$

Hence equations must hold of the type $(q_1 P_\rho | q_2 P_\rho) = 0$, and of the type $(q_1 | q_2) = 0$.

Thus $q_1, q_2, \dots q_\sigma$ are the one common set of σ polar reciprocal points [cf. § 66 (6) and § 83 (6)] with respect to the sections by Q_σ of the two quadrics $(x|x) = 0$, and $(xP_\rho | xP_\rho) = 0$.

Thus the σ feet in Q_σ are mutually normal.

(6) These common perpendiculars all intersect $|P_\rho$ [cf. § 223 (3)]. These σ points of intersection with $|P_\rho$ are also mutually normal.

For any line joining P_ρ and Q_σ must lie in the region $P_\rho Q_\sigma$ of the $(\rho+\sigma)$th order defined by any $\rho+\sigma$ reference points, ρ from P_ρ and σ from Q_σ. Also the common perpendiculars, being perpendicular to P_ρ, all intersect the region $|P_\rho$ (of the $(\nu-\rho)$th order), and are perpendicular to it. Hence they all intersect the subregion, $P_\rho Q_\sigma P_\rho$, formed by the intersection of $|P_\rho$ with $P_\rho Q_\sigma$. But this subregion is of the σth order. Then the common perpendiculars of Q_σ and P_ρ are also common perpendiculars of Q_σ and $P_\rho Q_\sigma P_\rho$, since $P_\rho Q_\sigma | P_\rho$ is part of P_ρ. But by the previous subsection, if $p_1', p_2', \dots p_\sigma'$, be the σ feet in $P_\rho Q_\sigma | P_\rho$ of these perpendiculars, then

$p_1', p_2', \ldots p_\sigma'$ form a mutually normal set of points. Also they are the one common set of σ polar reciprocal points with respect to the sections by $P_\rho Q_\sigma | P_\rho$ of the two quadrics $(x|x)=0$, and $(xQ_\sigma|xQ_\sigma)=0$.

(7) Now, since $2\sigma < \nu$, the 2σ points $q_1, q_2, \ldots q_\sigma, p_1', p_2', \ldots p_\sigma'$ are in general independent. Hence the σ lines $p_1'q_1, p_2'q_2, \ldots$, cut P_ρ in σ independent elements $p_1, p_2, \ldots p_\sigma$, which define a subregion P_σ. Then by subsection (5) $p_1, p_2, \ldots p_\sigma$ are mutually normal. But they are also normal to $p_1', p_2', \ldots p_\sigma'$. Thus [cf. § 113] the σ lines of the σ common perpendiculars are mutually normal, so that any point on one is normal to any point on the other.

(8) The theorems of subsections (5) to (7) can be proved otherwise thus, assuming subsection (1) and that one common perpendicular exists between P_ρ and Q_σ. For, since this perpendicular (call it F_1) intersects P_ρ at right-angles, then [cf. § 223 (3)] any line drawn in P_ρ through the point of intersection intersects the region $|F_1$. But $\rho - 1$ independent such lines can be drawn. Thus $|F_1$ intersects P_ρ in a region of the $(\rho-1)$th order: similarly it intersects Q_σ in a region of the $(\sigma-1)$th order. Let these regions (both contained in $|F_1$) be called $P_{\rho-1}$ and $Q_{\sigma-1}$. Then by the original assumption $P_{\rho-1}$ and $Q_{\sigma-1}$ have a common perpendicular. Call it F_2. Then F_2 lies in $|F_1$ and is therefore normal to it.

Also $|F_2$ intersects $P_{\rho-1}$ and $Q_{\sigma-1}$ in two regions of the $(\rho-2)$th and $(\sigma-2)$th order; and so on. Hence $(\sigma < \rho)$ by continuing this process, σ common perpendiculars can be found, all mutually normal.

228. Spheres. (1) Let b be the centre and ρ the radius of a sphere; its equation is

$$(x|x)(b|b)\cos^2\frac{\rho}{\gamma} = (b|x)^2.$$

But [cf. § 110 (4)], $(b|x)=0$, is the equation of the polar plane of b with respect to the absolute.

Hence [cf. § 78 (2)] a sphere is a surface of the second degree, touching the absolute along the locus of contact of the tangent cone to the absolute with b as vertex.

(2) It is obvious that any point on a sphere of radius ρ and centre b is at a distance $\frac{1}{2}\pi\gamma - \rho$ from the polar plane of b, viz. from $|b$. But $|b$ may be any plane since b may be any point. Hence a sphere of radius ρ is the locus of points at constant distances, $\frac{1}{2}\pi\gamma - \rho$, from a plane.

A plane can be conceived to be the limiting case of a sphere of radius $\frac{1}{2}\pi\gamma$. For if b be the absolute pole of any plane, the equation of the plane is

$$(x|b)=0;$$

and this is the degenerate form of the equation of the sphere, when ρ is put equal to $\frac{1}{2}\pi\gamma$.

(3) Every line, perpendicular to a plane and passing through the pole of the plane with respect to a sphere, passes through the centre of the sphere.

For let b be the centre of the sphere, ρ its radius and let p be the pole of the plane with respect to the sphere.

The equation of the plane can be written

$$(x|p)(b|b)\cos^2\frac{\rho}{\gamma} - (x|b)(p|b) = 0.$$

Hence [cf. § 110 (4)] the absolute pole of the plane is

$$p(b|b)\cos^2\frac{\rho}{\gamma} - b(b|p).$$

But [cf. § 223 (3)] the perpendicular lines to the plane pass through the absolute pole. Hence the perpendicular through p to the plane is the line

$$p\left\{p(b|b)\cos^2\frac{\rho}{\gamma} - b(b|p)\right\};$$

that is, dropping numerical factors, the line pb. Accordingly this line passes through b.

Corollary. The perpendicular to a tangent plane of a sphere through its point of contact passes through the centre of the sphere.

(4) Let the length of a tangent line from any point x to the sphere, centre b, radius ρ, be τ. Let the line meet the sphere in p. Then considering the triangle xpb, by the last proposition, the angle at p is a right-angle. Hence the triangle xpb may be assumed to be a principal triangle.

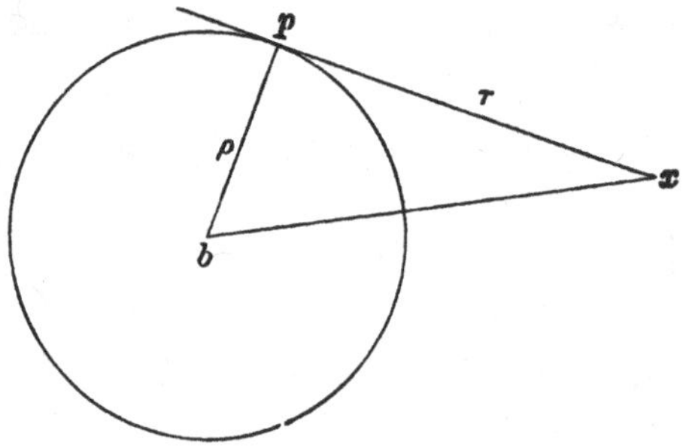

FIG. 5.

Hence $$\cos\frac{\overline{xb}}{\gamma} = \cos\frac{\overline{xp}}{\gamma}\cos\frac{\rho}{\gamma} = \cos\frac{\tau}{\gamma}\cos\frac{\rho}{\gamma}.$$

Therefore the lengths of all tangent lines from x to the sphere are equal.

Substituting for $\cos\frac{\overline{xb}}{\gamma}$, we find

$$(x|b)^2 = \cos^2\frac{\tau}{\gamma}\cos^2\frac{\rho}{\gamma}(x|x)(b|b).$$

Hence when τ is constant the locus of x is a sphere concentric with the original sphere.

In order that the tangent τ may be real, we must have

$$\frac{(x|b)^2}{\cos^2\frac{\rho}{\gamma}(x|x)(b|b)} < 1.$$

Hence $$\frac{(x|b)^2}{(x|x)(b|b)} < \cos^2\frac{\rho}{\gamma}.$$

Therefore the point x must be at a distance from b greater than ρ, that is to say, must be outside the sphere [cf. § 218 (3)].

(5) The intersection of any plane with a sphere is another sphere of $\nu - 3$ dimensions contained in the plane [cf. § 67 (1)].

For let L be the plane, and

$$(x|b)^2 - (x|x)(b|b)\cos^2\frac{\rho}{\gamma} = 0 \quad\text{.....................(i)}$$

be the equation of the sphere.

Then any point p on the perpendicular from b to L [cf. subsection (3)] can be written $b + \lambda|L$.

Hence the distance δ from p to any point x is given by

$$\cos^2\frac{\delta}{\gamma} = \frac{(x|p)^2}{(x|x)(p|p)} = \frac{\{(x|b) + \lambda(Lx)\}^2}{(x|x)\{(b|b) + 2\lambda(Lb) + \lambda^2(L|L)\}}.$$

Now let x lie on the locus of intersection of L with the sphere, then $(xL) = 0$, and x satisfies equation (i).

Hence $$\cos^2\frac{\delta}{\gamma} = \frac{(b|b)\cos^2\frac{\rho}{\gamma}}{(b|b) + 2\lambda(Lb) + \lambda^2(L|L)} \quad\text{.................(ii).}$$

Thus the distance of any given point on the line $b|L$ from any point on the locus of intersection of L and a sphere, centre b, is constant.

But this must hold for the point $L.b|L$, where the line $b|L$ intersects the plane L. Hence the locus of intersection is a sphere of $\nu - 3$ dimensions, with the point $L.b|L$ as centre.

The radius δ_1 of this sphere (of $\nu - 3$ dimensions) is easily proved by equation (ii) to be given by

$$\cos^2\frac{\delta_1}{\gamma} = \frac{(b|b)(L|L)\cos^2\frac{\rho}{\gamma}}{(b|b)(L|L) - (Lb)^2}.$$

(6) The locus of the intersection of any two spheres is contained in two planes, the radical planes.

For let $$(x|x)(b|b)\cos^2\frac{\rho}{\gamma} - (b|x)^2 = 0,$$

and $$(x|x)(c|c)\cos^2\frac{\sigma}{\gamma} - (c|x)^2 = 0,$$

be the equations of the two spheres.

Then two planes containing the locus of intersection are given by

$$(b\,|x)^2\,(c\,|c)\cos^2\frac{\sigma}{\gamma}-(c\,|x)^2\,(b\,|b)\cos^2\frac{\rho}{\gamma}=0\,;$$

that is, by the two equations,

$$(b\,|x)\,\sqrt{(c\,|c)}\cos\frac{\sigma}{\gamma}\pm(c\,|x)\,\sqrt{(b\,|b)}\cos\frac{\rho}{\gamma}=0.$$

Let these planes be called the radical planes.

(7) These planes are the loci from which equal tangents can be drawn to the spheres. Also from subsection (5), it follows that the locus of intersection of two spheres consists of two spheres of $\nu-3$ dimensions.

(8) Spheres cut each other at two angles of intersection, one corresponding to each radical plane.

For consider the radical plane

$$(b\,|x)\,\sqrt{(c\,|c)}\cos\frac{\sigma}{\gamma}-(c\,|x)\,\sqrt{(b\,|b)}\cos\frac{\rho}{\gamma}=0 \quad\text{............(iii).}$$

Then for points on this plane the equations of the two spheres can be written

$$\left.\begin{aligned}(b\,|x)&=\pm\sqrt{(x\,|x)\,(b\,|b)}\cos\frac{\rho}{\gamma},\\(c\,|x)&=\pm\sqrt{(x\,|x)\,(c\,|c)}\cos\frac{\sigma}{\gamma},\end{aligned}\right\}\text{....................(iv),}$$

where the same choice (upper or lower sign) determining the ambiguity is to be made for both equations.

Also the angle ω, at which the spheres cut each other along this plane, is the angle between the lines bx and cx through a point x on that part of the locus of intersection contained by the plane.

Hence
$$\cos\omega=\frac{(xb\,|xc)}{\sqrt{\{(xb\,|xb)\,(xc\,|xc)\}}}.$$

Hence, eliminating x by the use of equations (iii) and (iv),

$$\cos\omega=\frac{(b\,|c)-\sqrt{(b\,|b)\,(c\,|c)}\cos\frac{\rho}{\gamma}\cos\frac{\sigma}{\gamma}}{\sqrt{\{(b\,|b)\,(c\,|c)\}}\sin\frac{\rho}{\gamma}\sin\frac{\sigma}{\gamma}}$$

$$=\frac{\cos\frac{\delta}{\gamma}-\cos\frac{\rho}{\gamma}\cos\frac{\sigma}{\gamma}}{\sin\frac{\rho}{\gamma}\sin\frac{\sigma}{\gamma}}\quad\text{..................................(v),}$$

where δ is the length $\overline{bc}$ of that intercept between the points b and c defined by $(b\,|c)$.

Similarly for the other radical plane

$$(b\,|x)\,\sqrt{(c\,|\,c)}\cos\frac{\sigma}{\gamma}+(c\,|x)\,\sqrt{(b\,|\,b)}\cos\frac{\rho}{\gamma}=0,$$

the angle ω' of intersection between the spheres is given by

$$\cos\omega'=\frac{(b\,|\,c)+\sqrt{(b\,|\,b)\,(c\,|\,c)}\cos\frac{\rho}{\gamma}\cos\frac{\sigma}{\gamma}}{\sqrt{\{(b\,|\,b)\,(c\,|\,c)\}}\sin\frac{\rho}{\gamma}\sin\frac{\sigma}{\gamma}}$$

$$=\frac{\cos\frac{\delta}{\gamma}+\cos\frac{\rho}{\gamma}\cos\frac{\sigma}{\gamma}}{\sin\frac{\rho}{\gamma}\sin\frac{\sigma}{\gamma}}\quad\text{.............................(vi).}$$

Hence if δ' be the length of the other intercept between b and c, so that $\delta+\delta'=\pi\gamma$, then

$$\cos(\pi-\omega')=\frac{\cos\frac{\delta'}{\gamma}-\cos\frac{\rho}{\gamma}\cos\frac{\sigma}{\gamma}}{\sin\frac{\rho}{\gamma}\sin\frac{\sigma}{\gamma}}.$$

This equation exhibits the identity of type between the formulæ of the equations (v) and (vi).

Corollary. It may be noted as exemplifying equations (v) and (vi) that, if $\omega=\frac{\pi}{2}$, then $\cos\frac{\delta}{\gamma}=\cos\frac{\rho}{\gamma}\cos\frac{\sigma}{\gamma}$.

Hence $$\cos\omega'=2\cot\frac{\rho}{\gamma}\cot\frac{\sigma}{\gamma}.$$

This illustrates the fact that both parts of the intersection are not necessarily simultaneously real.

(9) Let it be assumed that $(b\,|\,c)$ is positive, so that $\delta=D(bc)<\frac{1}{2}\pi\gamma$. Also by definition, ρ and σ are both less than $\frac{1}{2}\pi\gamma$.

Then the two spheres have real or imaginary intersections on the corresponding radical planes, according as $\cos\omega$ and $\cos\omega'$ are numerically less or greater than unity.

Now $\cos\omega'$ is positive; and $\cos\omega'<1$,

if $$\cos\frac{D(bc)}{\gamma}+\cos\frac{\rho}{\gamma}\cos\frac{\sigma}{\gamma}<\sin\frac{\rho}{\gamma}\sin\frac{\sigma}{\gamma};$$

that is, if $$\cos\frac{D(bc)}{\gamma}<\cos\frac{\pi\gamma-\rho-\sigma}{\gamma};$$

that is, if $$D(bc)+\rho+\sigma>\pi\gamma\quad\text{..........................(vii).}$$

Also if $\cos\omega'$ be less than unity, $\cos\omega$ is necessarily numerically less than unity. Hence if ω' be real, ω is also real; and both parts of the intersection are real.

Thus the condition that the intersections of the spheres with both radical planes may be real is

$$D(bc) + \rho + \sigma > \pi\gamma.$$

(10) If $D(bc) + \rho + \sigma < \pi\gamma$, then one of the intersections is imaginary. The condition that the other may be real is

$$-1 < \frac{\cos\frac{D(bc)}{\gamma} - \cos\frac{\rho}{\gamma}\cos\frac{\sigma}{\gamma}}{\sin\frac{\rho}{\gamma}\sin\frac{\sigma}{\gamma}} < 1.$$

Hence
$$\cos\frac{D(bc)}{\gamma} > \cos\frac{\rho+\sigma}{\gamma},$$
$$< \cos\frac{\rho\sim\sigma}{\gamma}.$$

Therefore the condition that one intersection (at least) may be real is

$$\rho\sim\sigma < D(bc) < \rho + \sigma \ldots\ldots\ldots\ldots\text{(viii)}.$$

(11) It follows from the inequality (vii) of subsection (9) that two oval spheres cannot have two real intersections. For $D(bc) < \frac{1}{2}\pi\gamma$, and by § 218 (4) for oval spheres ρ and σ are both less than $\frac{1}{4}\pi\gamma$.

Hence
$$D(bc) + \rho + \sigma < \pi\gamma.$$

(12) Let the sphere, centre c, reduce to the plane L, so that we may write $|c = L$, and $\sigma = \frac{1}{2}\pi\gamma$.

Hence the angle ω at which the plane cuts the sphere, centre b, is given by

$$\sin\frac{\rho_1}{\gamma}\cos\omega = \frac{(bL)}{\{(b|b)(L|L)\}^{\frac{1}{2}}}.$$

If the plane cut the sphere at right-angles, $(bL) = 0$. Hence the plane contains the centre of the sphere.

(13) If the plane touches the sphere, $\omega = 0$. Hence the plane-equation [cf. § 78 (8)] of the sphere, of centre b and of radius ρ, is

$$(b|b)(L|L)\sin^2\frac{\rho}{\gamma} = (bL)^2.$$

If the sphere be defined as the locus of equal distance σ from the plane B, then remembering that $B = |b$, and that $\rho + \sigma = \frac{1}{2}\pi\gamma$, this equation becomes

$$(B|B)(L|L)\cos^2\frac{\sigma}{\gamma} = (B|L)^2.$$

Similarly the point equation of the sphere takes the two forms

$$(x|x)(b|b)\cos^2\frac{\rho}{\gamma} = (b|x)^2,$$

and
$$(x|x)(B|B)\sin^2\frac{\sigma}{\gamma} = (xB)^2.$$

229. PARALLEL SUBREGIONS. (1) Let P_ρ be a subregional element of the ρth order, then the locus of points x, which are at a given distance δ from the subregion P_ρ, by § 226 (2) is determined by the equation,

$$(x|x)(P_\rho|P_\rho)\sin^2\frac{\delta}{\gamma}=(xP_\rho|xP_\rho).$$

This is the equation of a quadric surface. In the special cases in which P_ρ is either a point or a plane, the surface reduces to a sphere.

(2) If P_ρ be neither a point nor a plane, *real* generating regions exist on this surface. For let $e_1, e_2, \dots e_\rho$ be ρ mutually normal points in P_ρ, each at unit normal intensity; and let $e_{\rho+1}, \dots e_\nu$ be $\nu-\rho$ mutually normal points in $|P_\rho$, each at unit normal intensity. Then $e_1, e_2, \dots e_\rho, e_{\rho+1}, \dots e_\nu$ form a set of ν mutually normal points at unit normal intensity. Let $e_1e_2 \dots e_\rho$ be written for P_ρ, then $(P_\rho|P_\rho)=1$. Also let x be written $\sum_1^\rho \xi e+\sum_{\rho+1}^\nu \eta e$. Then $(x|x)=\sum_1^\rho\xi^2+\sum_{\rho+1}^\nu\eta^2$.

Also

$$(xP_\rho|xP_\rho)=\sum_{\lambda=\rho+1}^{\lambda=\nu}\eta_\lambda^2(e_\lambda e_1\dots e_\rho|e_\lambda e_1\dots e_\rho)+\sum_{\lambda=\rho+1}^{\lambda=\nu}\sum_{\mu=\rho+1}^{\mu=\nu}\eta_\lambda\eta_\mu(e_\lambda e_1\dots e_\rho|e_\mu e_1\dots e_\rho),$$

where λ and μ are assumed to be unequal in the double summation. But $(e_\sigma|e_\sigma)=1$, for all values of σ; and $(e_\sigma|e_\tau)=0$, for all unequal values of σ and τ. Hence $(e_\lambda e_1\dots e_\rho|e_\lambda e_1\dots e_\rho)=1$, and $(e_\lambda e_1\dots e_\rho|e_\mu e_1\dots e_\rho)=0$.

Thus
$$(xP_\rho|xP_\rho)=\sum_{\rho+1}^{\nu}\eta_\lambda^2.$$

Hence the equation of the surface takes the form

$$\sin^2\frac{\delta}{\gamma}\sum_1^\rho\xi_\lambda^2-\cos^2\frac{\delta}{\gamma}\sum_{\rho+1}^\nu\eta_\lambda^2=0.$$

If $\rho<\frac{1}{2}\nu$, then [cf. § (80) (1) and (5)] noting the formation of conjugate sets of points from reciprocally polar sets, and remembering Sylvester's theorem [cf. § 82 (6)], it is evident that *real* generating regions defined by ρ points exist on the surface, that is, regions of $\rho-1$ dimensions.

If $\rho>\frac{1}{2}\nu$, then *real* generating regions defined by $\nu-\rho$ points exist on the surface, that is, regions of $\nu-\rho-1$ dimensions.

If $\rho=\frac{1}{2}\nu$ (ν even), then *real* generating regions defined by ρ points exist on the surface, that is, regions of $\rho-1$ dimensions.

(3) Let these real generating regions be said to be parallel to P_ρ. Thus a region parallel to P_ρ is by definition such that the distances from all points in it to P_ρ are equal, and has been proved to be of the type Q_ρ or $Q_{\nu-\rho}$, according as ρ or $\nu-\rho$ is least.

Also from § 226 (4) a surface of equal distance from P_ρ is also a surface of equal distance from $|P_\rho$. Thus all regions parallel to P_ρ are also parallel to $|P_\rho$, and conversely.

(4) Let the region Q_σ be parallel to P_ρ; where σ is equal to the least of the two ρ and $\nu-\rho$. Let q be any point in Q_σ, and let qp be the perpendicular from Q_σ to P_ρ. Then qp is also the perpendicular from p to Q_σ. For if not, let pq' be this perpendicular. Then $D(pq') < D(pq)$. Also from q', let $q'p'$ be drawn perpendicular to P_ρ. Then either p' coincides with p, or $D(p'q') < D(pq')$. Hence in any case $D(p'q') < D(pq)$. But, since the region Q_σ is parallel to P_ρ, $D(p'q') = D(pq)$. Thus pq must be a common perpendicular of P_ρ and Q_σ.

Thus, if for example ρ be less than $\nu-\rho$, so that $\sigma=\rho$, then P_ρ is parallel to Q_ρ. Thus P_ρ and Q_ρ are mutually parallel to each other. But the same proof does not shew that $|P_\rho$ is parallel to Q_ρ. For, by the preceding subsection, no region parallel to Q_ρ can be of an order greater than the ρth; and by hypothesis $\nu-\rho$ is greater than ρ. Also, if δ be the distance of P_ρ from Q_ρ, the entire region parallel to Q_ρ at a distance $\frac{1}{2}\pi\gamma-\delta$ from Q_ρ must be contained in $|P_\rho$. Hence a subregion P'_ρ of $|P_\rho$ of the ρth order is parallel to Q_ρ.

Accordingly a distinction must be drawn between the fact that one region is parallel to another region, and the fact that two regions are mutually parallel. Thus with the above notation, Q_ρ is parallel both to P_ρ and to $|P_\rho$; also $(\rho<\nu-\rho)$ P_ρ and Q_ρ are mutually parallel; but $|P_\rho$ and Q_ρ are not mutually parallel, though P_ρ' (a subregion of $|P_\rho$) and Q_ρ are mutually parallel. The feet of the perpendiculars from all points in Q_ρ to $|P_\rho$ must lie in this subregion P_ρ'. This agrees with § 227 (7): the perpendiculars found by the method of that article must be all equal.

(5) This theory of parallel regions is an extension* of Clifford's† theory of parallel lines in Elliptic Space of three dimensions. Consider a straight line L in a space of $\nu-1$ dimensions, $(\nu \geqq 4)$. Then the regions parallel to L are also straight lines, whatever be the dimensions of the space, provided that they are equal to or greater than 3. If the space be of three dimensions, then only two parallels to L can be drawn through any given point x, being the two generating lines of the quadric surface through x of equal distance from L. But if the space be of more than three dimensions, an indefinite number of parallels to L can be drawn through any given point. Also the tangent plane at x to the surface of equal distance from L which passes through x cuts this surface (a quadric) in another quadric of one lower dimension. Hence [cf. § 80 (8) and (12)] this quadric is a conical quadric formed by the parallels through x. Thus in a region of $\nu-1$ dimensions the parallels through x to a line L form a conical quadric of $\nu-3$ dimensions with x as vertex.

* Hitherto unnoticed as far as I am aware.

† Cf. *Preliminary Sketch of Biquaternions*, *Proc. of Lond. Math. Soc.*, vol. 4, 1873, reprinted in his *Collected Papers*.

CHAPTER III.

EXTENSIVE MANIFOLDS AND ELLIPTIC GEOMETRY.

230. INTENSITIES OF FORCES. (1) In considering the special metrical properties of extensive manifolds we shall confine ourself to three dimensions. The only regional elements in this case are planar elements and forces. The intensity of a planar element X is now taken to be [cf. § 211 (5)] $(X \mid X)$. The intensity of a force F has now to be determined.

(2) Let* it be defined that the intensity of the force xy is some function of the distance $\overline{xy}$ multiplied by the product of the intensities of x and y.

Thus assume that the intensity of xy is $\sqrt{\{(x \mid x)(y \mid y)\}}\, \phi(\overline{xy})$; where the function $\phi(\overline{xy})$ has now to be determined. Now let x and y be at unit intensity, and let α be any number, then $xy = x(y + \alpha x)$. Hence

$$(x \mid x) = 1 = (y \mid y);\ \{(y + \alpha x) \mid (y + \alpha x)\} = (y \mid y) + 2\alpha (x \mid y) + \alpha^2 (x \mid x)$$
$$= 1 + 2\alpha \cos\frac{\overline{xy}}{\gamma} + \alpha^2.$$

Accordingly the intensity of xy = the intensity of $x(y + \alpha x)$

$$= \left(1 + \alpha^2 + 2\alpha \cos\frac{\overline{xy}}{\gamma}\right)^{\frac{1}{2}} \phi\{\overline{x(y + \alpha x)}\}.$$

Therefore, $$\phi(\overline{xy}) = \left(1 + \alpha^2 + 2\alpha \cos\frac{\overline{xy}}{\gamma}\right)^{\frac{1}{2}} \phi\{\overline{x(y + \alpha x)}\}.$$

But $$\sin\frac{\overline{x(y + \alpha x)}}{\gamma} = \frac{\sin\dfrac{\overline{xy}}{\gamma}}{\left(1 + \alpha^2 + 2\alpha\cos\dfrac{\overline{xy}}{\gamma}\right)^{\frac{1}{2}}}.$$

Hence $$\frac{\phi(\overline{xy})}{\sin\dfrac{\overline{xy}}{\gamma}} = \frac{\phi\{\overline{x(y + \alpha x)}\}}{\sin\dfrac{\overline{x(y + \alpha x)}}{\gamma}}.$$

* This reasoning is very analogous to some reasoning in Homersham Cox's paper, cf. *loc. cit.* p. 370.

But $\overline{x(y+\alpha x)}$ can be made to be any length ($<\pi\gamma$) by choosing a suitable value for α.

Hence $\dfrac{\phi(\overline{xy})}{\sin\dfrac{\overline{xy}}{\gamma}} = \text{a constant} = 1$, say.

Therefore whatever points x and y are, the intensity of xy is

$$\{(x|x)(y|y)\}^{\frac{1}{2}} \sin\frac{\overline{xy}}{\gamma}.$$

(3) Hence the intensity of the force F is $(F|F)^{\frac{1}{2}}$. Thus if the force F be written PQ, where P and Q are planar elements, the intensity of F is $\{(P|P)(Q|Q)\}^{\frac{1}{2}} \sin \angle PQ$.

231. Relations between two forces. Let F and F' be any two forces. In general there are only two lines intersecting the four lines F, F', $|F, |F'$. These two lines [cf. § 223 (3)] are two common perpendiculars to F and F' [cf. § 227 (7) and (8)].

Let one perpendicular intersect F and F' in a and b respectively, and let the other intersect F and F' in c and d. Let $\overline{ab} = \delta$, and $\overline{cd} = \delta'$. Then one

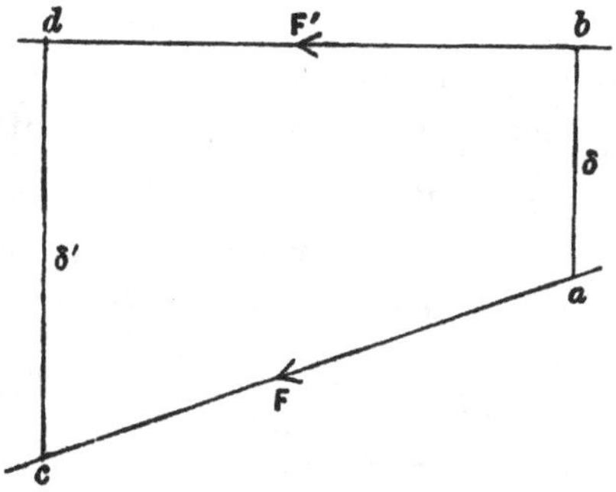

Fig. 6.

of the two is the shortest distance from one line to the other, and the other is the longest perpendicular distance. Also since ab intersects $F, F', |F, |F'$, $|ab$ intersects the same four lines. Hence cd is the line $|ab$, and ab is the line $|cd$. Thus $\overline{db} = \frac{1}{2}\pi\gamma = \overline{ac}$. Hence $(a|c) = 0 = (b|d) = (a|d) = (b|c)$.

(2) To prove that*

$$\frac{(FF')}{\{(F|F)(F'|F')\}^{\frac{1}{2}}} = \sin\frac{\delta}{\gamma}\sin\frac{\delta'}{\gamma}, \quad \frac{(F|F')}{\{(F|F)(F'|F')\}^{\frac{1}{2}}} = \cos\frac{\delta}{\gamma}\cos\frac{\delta'}{\gamma}.$$

Let
$$F = \lambda ac, \quad F' = \lambda' bd.$$

Then
$$\cos\frac{\delta}{\gamma} = \frac{(a|b)}{\{(a|a)(b|b)\}^{\frac{1}{2}}}, \quad \cos\delta' = \frac{(c|d)}{\{(c|c)(d|d)\}^{\frac{1}{2}}}.$$

Hence
$$\cos\frac{\delta}{\gamma}\cos\frac{\delta'}{\gamma} = \frac{(a|b)(c|d)}{\{(a|a)(b|b)(c|c)(d|d)\}^{\frac{1}{2}}}.$$

But $(F|F) = \lambda^2(ac|ac) = \lambda^2\{(a|a)(c|c) - (a|c)^2\} = \lambda^2(a|a)(c|c)$.

Similarly,
$$(F'|F') = \lambda'^2(b|b)(d|d).$$

* Cf. Homersham Cox, *loc. cit.*

Also $$(F|F') = \lambda\lambda'(ac|bd) = \lambda\lambda'(a|b)(c|d).$$

Therefore $$\cos\frac{\delta}{\gamma}\cos\frac{\delta'}{\gamma} = \frac{(F|F')}{\{(F|F)(F'|F')\}^{\frac{1}{2}}}.$$

(3) Again, from (1) we may write $cd = \mu|ab$.

Hence $$(cd|cd) = \mu^2(ab|ab).$$

Therefore [cf. § 204 (1)]

$$\mu = \frac{\sin\frac{\delta'}{\gamma}}{\sin\frac{\delta}{\gamma}}\left\{\frac{(c|c)(d|d)}{(a|a)(b|b)}\right\}^{\frac{1}{2}}.$$

Also $$(abcd) = \mu(ab|ab) = \mu(a|a)(b|b)\sin^2\frac{\delta}{\gamma}$$
$$= \sin\frac{\delta}{\gamma}\sin\frac{\delta'}{\gamma}\{(a|a)(b|b)(c|c)(d|d)\}^{\frac{1}{2}}.$$

But $$(FF') = \lambda\lambda'(acbd).$$

Hence assuming that the ambiguity of sign is so determined as to make both sides positive,

$$\frac{(FF')}{\{(F|F)(F'|F')\}^{\frac{1}{2}}} = \sin\frac{\delta}{\gamma}\sin\frac{\delta'}{\gamma}.$$

(4) If the lines F and F' intersect, either δ or δ' vanishes, say $\delta' = 0$.

Then $$(FF') = 0, \text{ and } \frac{(F|F')}{\{(F|F)(F'|F')\}^{\frac{1}{2}}} = \cos\frac{\delta}{\gamma}.$$

This agrees with § 211 (6).

232. Axes of a System of Forces. (1) A system of forces (S) can in general [cf. §§ 174 (9) and 175 (14)] be reduced in one way and in one way only to the form*,

$$S = a_1a_2 + \epsilon|a_1a_2.$$

Then the lines of the forces a_1a_2 and $\epsilon|a_1a_2$ are called the axes of the system (sometimes, the central axes), and the ratio of their intensities, namely ϵ $\left(\text{or } \frac{1}{\epsilon}\right)$, is called the parameter of the system.

Then $$(SS) = 2\epsilon(a_1a_2|a_1a_2),$$
and $$(S|S) = (1+\epsilon^2)(a_1a_2|a_1a_2).$$

(2) Let S denote the system $F + \epsilon|F$, and S' the system $F' + \eta|F'$. Also with the notation of § 231 let δ and δ' be the perpendicular distances between the lines F and F', reckoned algebraically as to sign.

* Cf. Homersham Cox, *loc. cit.*, p. 370.

Then $(SS') = (1+\epsilon\eta)(FF') + (\epsilon+\eta)(F|F')$

$$= \{(F|F)(F'|F')\}^{\frac{1}{2}} \left\{(1+\epsilon\eta)\sin\frac{\delta}{\gamma}\sin\frac{\delta'}{\gamma} + (\epsilon+\eta)\cos\frac{\delta}{\gamma}\cos\frac{\delta'}{\gamma}\right\};$$

and $(S|S') = (\epsilon+\eta)(FF') + (1+\epsilon\eta)F|F')$

$$= \{(F|F)(F'|F')\}^{\frac{1}{2}} \left\{(\epsilon+\eta)\sin\frac{\delta}{\gamma}\sin\frac{\delta'}{\gamma} + (1+\epsilon\eta)\cos\frac{\delta}{\gamma}\cos\frac{\delta'}{\gamma}\right\}.$$

(3) The simultaneous equations $(SS') = 0$, $(S|S') = 0$, in general secure that the axes of S and S' intersect at right angles. For from (2) unless either ϵ or η be ± 1,

$$\cos\frac{\delta}{\gamma}\cos\frac{\delta'}{\gamma} = 0 = \sin\frac{\delta}{\gamma}\sin\frac{\delta'}{\gamma}.$$

Thus $\delta = 0$, $\delta' = \frac{1}{2}\pi\gamma$; or vice versa. Hence F and F' intersect at right angles.

Therefore [cf. § 223 (3)] F intersects $|F'$ as well as F'; and F' intersects $|F$ as well as F. Also, since $(FF') = 0$, $(|F|F') = |(FF') = 0$. Thus $|F$ and $|F'$ intersect. Also these various pairs of lines [cf. § 223 (3)] intersect at right angles.

(4) Every dual group contains one pair of systems, and in general only one pair, such that their axes intersect at right angles.

Let S_1 and S_2 define the dual group, and let

$$S = \lambda_1 S_1 + \lambda_2 S_2,\quad S' = \mu_1 S_1 + \mu_2 S_2.$$

Then, $(SS') = 0$, becomes

$$\lambda_1\mu_1(S_1S_1) + (\lambda_1\mu_2 + \lambda_2\mu_1)(S_1S_2) + \lambda_2\mu_2(S_2S_2) = 0.$$

Similarly $(S|S') = 0$, becomes

$$\lambda_1\mu_1(S_1|S_1) + (\lambda_1\mu_2 + \lambda_2\mu_1)(S_1|S_2) + \lambda_2\mu_2(S_2|S_2) = 0.$$

Hence eliminating $\mu_1 : \mu_2$, the pair of systems are given by the quadratic for $\lambda_1 : \lambda_2$,

$$\lambda_1^2\{(S_1S_1)(S_1|S_2) - (S_1S_2)(S_1|S_1)\} - \lambda_2^2\{(S_2S_2)(S_1|S_2) - (S_1S_2)(S_2|S_2)\}$$
$$+ \lambda_1\lambda_2\{(S_1S_1)(S_2|S_2) - (S_2S_2)(S_1|S_1)\} = 0.$$

Let this pair of systems be called the central pair of the dual group, and let the points at which their axes intersect be called the centres of the group. There are [cf. subsection (3)] four centres to a dual group, forming a complete normal system of points.

If the group be not parabolic [cf. § 172 (9)], the two director forces D_1 and D_2 may be written for S_1 and S_2 in the above equation. The equation then becomes

$$(D_1D_2)\{\lambda_1^2(D_1|D_1) - \lambda_2^2(D_2|D_2)\} = 0.$$

Hence (considering only real groups) there are always two distinct roots to this equation. But, if D_1 and D_2 be both self-normal, this equation is an identity. For this exceptional case, cf. § 235 following.

If the group be parabolic. Let S_1 be any system, and replace S_2 by the single director force D. Then $(DS_1) = 0$, and the equation for $\lambda_1 : \lambda_2$ becomes

$$\lambda_1^2 (S_1 | D) + \lambda_1\lambda_2 (D | D) = 0.$$

Thus the central pair of the group are D and $(D|D) S_1 - (S_1|D) D$.

(5) To find* the locus of the axes of the systems of a dual group. Let the four centres be e, e_1, e_2, e_3, forming a complete normal system of unit points; also let ee_1 and $|ee_1$ be the axes of one central system, and ee_2 and $|ee_2$ of the other.

Let $S_1 = ee_1 + \epsilon_1 |ee_1$, $S_2 = ee_2 + \epsilon_2 |ee_2$, denote this central pair of systems of the group; and let ϵ_1 and ϵ_2 be called the principal parameters of the dual group. Any other system S' of the group can be written

$$S' = \lambda_1 S_1 + \lambda_2 S_2 = e(\lambda_1 e_1 + \lambda_2 e_2) + |e(\lambda_1\epsilon_1 e_1 + \lambda_2\epsilon_2 e_2).$$

Consider the system

$$S'' = (e + \zeta e_3)(\mu_1 e_1 + \mu_2 e_2) + \epsilon |(e + \zeta e_3)(\mu_1 e_1 + \mu_2 e_2).$$

It is a system of which a central axis intersects the line ee_3 at right angles [cf. § 223 (3)] in the point $e + \zeta e_3$; also its parameter is ϵ.

But we may assume $e_3 e_1 = |ee_2$, $e_2 e_3 = |ee_1$, $|e_3 e_1 = ee_2$, $|e_2 e_3 = ee_1$.

Hence
$$S'' = e\{(\mu_1 - \mu_2\zeta\epsilon) e_1 + (\mu_2 + \mu_1\zeta\epsilon) e_2\}$$
$$+ |e\{(\mu_1\epsilon - \mu_2\zeta) e_1 + (\mu_2\epsilon + \mu_1\zeta) e_2\}.$$

But in general we can make S'' and S' identical by putting

$$\mu_1 - \epsilon\mu_2\zeta = \lambda_1 \quad\text{.......(1)},$$
$$\mu_2 + \epsilon\mu_1\zeta = \lambda_2 \quad\text{.......(2)},$$
$$\epsilon\mu_1 - \mu_2\zeta = \epsilon_1\lambda_1 \quad\text{.......(3)},$$
$$\epsilon\mu_2 + \mu_1\zeta = \epsilon_2\lambda_2 \quad\text{.......(4)}.$$

Hence, by elimination, we find

$$(\epsilon^2 + 1)(\epsilon_1\lambda_1^2 + \epsilon_2\lambda_2^2) - \epsilon\{(\epsilon_1^2 + 1)\lambda_1^2 + (\epsilon_2^2 + 1)\lambda_2^2\} = 0 \quad\text{.......(5)}.$$

This is a quadratic to find ϵ; the two roots are reciprocals, namely ϵ and $\frac{1}{\epsilon}$, corresponding to the two axes of any system.

Again, let any point x on either axis of the system S' be

$$\xi e + \xi_1 e_1 + \xi_2 e_2 + \xi_3 e_3.$$

* Cf. Homersham Cox, *loc. cit.*

Then, assuming for example that x is on the axis $(e + \zeta e_3)(\mu_1 e_1 + \mu_2 e_2)$, by comparing with the original form of S'',

$$\frac{\xi_3}{\xi} = \zeta,\ \frac{\xi_2}{\xi_1} = \frac{\mu_2}{\mu_1} \quad \text{..............................(6).}$$

Also by elimination between equations (1), (2), (3), (4), (6)

$$(\epsilon_2 - \epsilon_1)\,\xi_1\xi_2\,(\xi^2 + \xi_3^2) = (1 - \epsilon_1\epsilon_2)\,(\xi_1^2 + \xi_2^2)\,\xi\xi_3 \quad \text{..............(7).}$$

This surface is the analogue in Elliptic space of the cylindroid. It is the locus of all the axes of systems of the dual group. All the central axes intersect at right angles the lines ee_3 and $|ee_3$ which are called the axes of the dual group.

(6) Equation (5) of the previous subsection can also be found thus. Assume that ab is the axis of the system S'.

Thus $\quad S' = e(\lambda_1 e_1 + \lambda_2 e_2) + |e(\lambda_1\epsilon_1 e_1 + \lambda_2\epsilon_2 e_2) = ab + \epsilon\,|ab.$

Then $\quad (S'S') = 2(\epsilon_1\lambda_1^2 + \epsilon_2\lambda_2^2)(ee_1\,|ee_1) = 2\epsilon\,(ab\,|ab)$;

and $\quad (S'\,|S') = \{(1 + \epsilon_1^2)\,\lambda_1^2 + (1 + \epsilon_2^2)\,\lambda_2^2\}\,(ee_1\,|ee_1) = (1 + \epsilon^2)\,(ab\,|ab).$

Thus finally

$$\frac{1+\epsilon^2}{2\epsilon} = \frac{(1+\epsilon_1^2)\,\lambda_1^2 + (1+\epsilon_2^2)\,\lambda_2^2}{2\,(\epsilon_1\lambda_1^2 + \epsilon_2\lambda_2^2)}.$$

This is equation (5) of the previous subsection.

233. Non-Axal Systems of Forces. (1) If a system of forces, S, be such that $(SS) = \pm(S\,|S)$, then [cf. § 174 (8) and § 175 (13)] S has not a pair of axes [cf. § 232 (1)]; provided that S be not self-supplementary [cf. § 235, following], in which case it has an infinite number of pairs of axes.

Such systems may be called non-axal. It will now be proved that all non-axal systems are imaginary.

(2) No real hyperbolic dual group can contain a real non-axal system. For let F and F' be the real director forces of this group, and let the non-axal system be $\lambda F + \lambda' F'$. Then by subsection (1)

$$\lambda^2 (F\,|F) + 2\lambda\lambda' \{(F\,|F') \mp (FF')\} + \lambda'^2 (F'\,|F') = 0.$$

Hence from § 231 (2), if δ and δ' are the lengths of the two common perpendiculars to F and F', this equation becomes

$$\lambda^2 (F\,|F) + 2\lambda\lambda' \{(F\,|F)(F'\,|F')\}^{\frac{1}{2}} \cos\frac{\delta \pm \delta'}{\gamma} + \lambda'^2 (F'\,|F') = 0.$$

But the roots of this equation are necessarily imaginary. Hence the four non-axal systems, which belong to any real hyperbolic group, are necessarily imaginary.

(3) But any real system must belong to some real hyperbolic groups. For [cf. § 162 (2)] the conjugate with respect to the system of any real line, not a null line, is a real line. Now the dual group with these two lines as director lines is a real hyperbolic group, and contains the real system.

(4) It therefore follows from (2) and (3) that all non-axal systems are imaginary.

Hence any real system S, for which $(SS) = \pm (S|S)$, is self-supplementary.

234. Parallel Lines. (1) An interesting case arises [cf. § 231] with regard to lines with a special relation discovered by Clifford*, and called by him the parallelism of lines [cf. § 229]. It is to be noted that the parallel lines of Hyperbolic Space [cf. Ch. IV. of this Book] do not exist (as real lines) in Elliptic Space, and conversely these parallel lines of Elliptic Space do not exist in Hyperbolic Space.

In general only two lines intersect the four lines F, F', $|F$, $|F'$. But if these four lines are generators of a quadric, then an infinite number of lines—namely, the generators of the opposite system—intersect them.

The two lines F and F' have then the peculiarity that an infinite number of common perpendiculars can be drawn. F and F' will then be proved to be mutually parallel according to the definitions of § 229 (3) and (4).

(2) Since the four lines are generators of the same quadric a relation [cf. § 175 (4) and (5)] of the form†,

$$\lambda F + \mu |F + \lambda' F' + \mu' |F' = 0$$

must exist.

Taking its supplement, it must be identical with

$$\lambda |F + \mu F + \lambda' |F' + \mu' F' = 0.$$

Hence $$\frac{\lambda}{\mu} = \frac{\mu}{\lambda} = \frac{\lambda'}{\mu'} = \frac{\mu'}{\lambda'}.$$

Accordingly $$\lambda = \pm \mu, \; \lambda' = \pm \mu'.$$

Firstly, let $\lambda = \mu$, $\lambda' = \mu'$. The condition becomes

$$\lambda (F + |F) + \lambda' (F' + |F') = 0.$$

Let the relation of F and F' be called 'right parallelism.'

Secondly, let $\lambda = -\mu$, $\lambda' = -\mu'$. The condition becomes

$$\lambda (F - |F) + \lambda' (F' - |F') = 0.$$

Let the relation of F and F' be called 'left parallelism.'

* Cf. *loc. cit.*, p. 370.

† This form of the relation between parallel lines was first given by Buchheim, *Proc. London Math. Society*, *loc. cit.*

(3) Consider the equation

$$\lambda(F+|F)+\lambda'(F'+|F')=0.$$

Multiply it successively by F and F', then

$$\lambda(F|F)+\lambda'\{(FF')+(F|F')\}=0,$$

and

$$\lambda\{(FF')+(F|F')\}+\lambda'(F'|F')=0.$$

Hence by eliminating $\lambda:\lambda'$,

$$(F|F)(F'|F')=\{(FF')+(F|F')\}^2.$$

Therefore
$$\frac{(FF')}{\sqrt{\{(F|F)(F'|F')\}}}+\frac{(F|F')}{\sqrt{\{(F|F)(F'|F')\}}}=\pm 1.$$

Similarly from the equation

$$\lambda(F-|F)+\lambda'(F'-|F')=0,$$

we deduce

$$\frac{(FF')}{\sqrt{\{(F|F)(F'|F')\}}}-\frac{(F|F')}{\sqrt{\{(F|F)(F'|F')\}}}=\pm 1.$$

But it has been proved in § 231, using its notation, that

$$\frac{(F|F')}{\sqrt{\{(F|F)(F'|F')\}}}=\cos\frac{\delta_1}{\gamma}\cos\frac{\delta_2}{\gamma},\quad \frac{(FF')}{\sqrt{\{(F|F)(F'|F')\}}}=\sin\frac{\delta_1}{\gamma}\sin\frac{\delta_2}{\gamma}.$$

Hence for right-parallelism, assuming that $\frac{\delta_1}{\gamma}$ and $\frac{\delta_2}{\gamma}$ are both acute angles but not necessarily both positive (with the usual conventions as to signs of lengths), $\cos\delta_1\cos\delta_2+\sin\delta_1\sin\delta_2=1$; therefore $\delta_1=\delta_2$.

For left-parallelism, $\cos\delta_1\cos\delta_2-\sin\delta_1\sin\delta_2=1$; therefore $\delta_1=-\delta_2$.

But δ_1 and δ_2 taken positively are the greatest and least perpendicular distances from one line to the other. Hence the lines are parallel according to § 229.

(4) Thus through any point b, a right-parallel line and a left-parallel line to any line F may be obtained by the following construction.

Draw ba perpendicular to F, and let the least of the two distances of b from F be δ, which is $\overline{ba}$. Find the polar line of ab, which must intersect F at right-angles in some point c. On this line take d and d' on opposite sides of c at distance δ from it. Then bd and bd' are respectively the right and left-parallel to F through b. It is to be noted that $|F$ is both a right and a left-parallel to F; and that a line parallel to F is parallel to $|F$.

(5) Since two parallel lines are generators of the same quadric [cf. subsection (2)], they are not coplanar.

235. Vector Systems*. (1) Any system (S) of the type $F \pm |F$ is called a vector system. Such a system has an infinite number of pairs of axes, consisting of all lines parallel to F taken in pairs. For let F' be any right or left-parallel to F. Then a relation exists of the form, $F \pm |F = \lambda(F' \pm |F')$. Accordingly $S = \lambda(F' \pm |F')$.

Let a system of the form $F + |F$ be called a right-vector system. If R be a right-vector system, $R = |R$, and $(R|R) = (RR)$: either of these equations is a sufficient test, if the system is known to be real [cf. § 233 (4)]. Let a system of the form $F - |F$ be called a left-vector system. If L be a left-vector system, $L = -|L$, and $(L|L) = -(LL)$: either of these equations is a sufficient test, if the system is known to be real. Vector systems are the self-supplementary systems of § 174 (2).

(2) The sum of two right-vector systems is a right-vector system, and the sum of two left-vector systems is a left-vector system; but the sum of a right-vector system and of a left-vector system is not a vector system†.

For let $R = a_1a_2 + |a_1a_2$, and $R' = b_1b_2 + |b_1b_2$, be two right-vector systems.

Then $$R + R' = (a_1a_2 + b_1b_2) + |(a_1a_2 + b_1b_2).$$

Now let $$a_1a_2 + b_1b_2 = c_1c_2 + \epsilon\,|c_1c_2.$$

Then $$R + R' = (1 + \epsilon)(c_1c_2 + |c_1c_2).$$

Accordingly $R + R'$ is a right-vector system.

A similar proof shows that the sum of two left-vector systems is a left-vector system. It is also obvious that another statement of the same proposition is that the dual group defined by two vector systems of the same name (right or left) contains only vector systems of that name.

(3) But if R is a right-vector system and L is a left-vector system, then $R + L$ is not a vector system.

For if it were, $$R + L \equiv |(R + L) \equiv R - L.$$

Hence $R \equiv L$. But a system cannot be both a right and a left vector system; since for such a system, $|S = S = -S$, which is impossible.

Any system‡ S can be written in the form $R + L$. For
$$S = \tfrac{1}{2}(S + |S) + \tfrac{1}{2}(S - |S);$$
and $\frac{1}{2}(S + |S)$ is a right-vector system, and $\frac{1}{2}(S - |S)$ is a left-vector system. This reduction is unique. For if $S = R + L = R' + L'$, then $R - R' = L' - L$. Hence a right-vector system would be equal to a left-vector system, which is impossible.

* This use of the word 'vector' seems to me to be very unfortunate. But an analogous use is too well established in connection with the kinematics of Elliptic space to be altered now. The theory of systems of forces is very analogous, as will be proved later, to the theory of motors and vectors investigated by Clifford; cf. *loc. cit.*, p. 370.

† Cf. Sir R. S. Ball, "On the Theory of Content," *Royal Irish Academy, Transactions*, 1889.

‡ Cf. Clifford, *loc. cit.*, p. 370.

(4) Any right-vector system and any left-vector system are reciprocal*. For let $R = a_1a_2 + |a_1a_2$, and $L = b_1b_2 - |b_1b_2$, then $(RL) = 0$.

236. VECTOR SYSTEMS AND PARALLEL LINES. (1) Let e_1 and e_2 be two unit quadrantal points: then the vector systems $e_1e_2 \pm |e_1e_2$ are called unit vector systems. If $R = e_1e_2 + |e_1e_2$, then

$$(RR) = 2(e_1e_2|e_1e_2) = 2(e_1|e_1)(e_2|e_2) = 2 = (R|R).$$

Also if $L = e_1e_2 - |e_1e_2$, then $(LL) = -2 = -(L|L)$.

(2) A vector system† possesses an infinite number of axes, it being possible to draw an axis of the system through any point; and this set of axes forms a set of parallels, right or left according as the vector system is right or left.

For let x be any other unit point and let p be another unit point on the right parallel to e_1e_2 and quadrantal to x.

Then by § 234 (2), $R = e_1e_2 + |e_1e_2 = \lambda(xp + |xp)$. Hence xp and $|xp$ are also axes of R.

Also $(R|R) = 2 = 2\lambda^2$. Hence $\lambda = \pm 1$. Thus a unit vector system, when expressed in terms of one pair of axes, is a unit vector system when expressed in terms of any other pair of axes.

(3) A simple expression‡ for a line drawn through a given point right or left-parallel to a given line can be found. For let a_1a_2 be the given line and x the given point. Consider the right-vector system $R = a_1a_2 + |a_1a_2$. Let xp be the required right-parallel to a_1a_2 drawn through x. Then

$$R = \lambda(xp + |xp).$$

Hence $$xR = \lambda x|xp = \lambda(x|p)|x - \lambda(x|x)|p;$$

therefore $$x|xR = -\lambda(x|x)\,xp \equiv xp.$$

Hence $x|xR$ is a force on the right-parallel to a_1a_2 drawn through x, where $R = a_1a_2 + |a_1a_2$.

Similarly if $L = a_1a_2 - |a_1a_2$, then $x|xL$ is the left-parallel to a_1a_2 drawn through x.

(4) It follows that, if any two lines are each parallels of the same name (right or left) to a third line, they are parallels of that name to each other. Let all the lines parallel (of the same name) to a given line be called a parallel set of lines.

(5) Any pair of conjugates of a vector system is a pair of parallel lines of an opposite denomination (right or left) to that of the system‡.

* Cf. Sir R. S. Ball, *Transactions R. I. A.*

† Cf. Clifford, *loc. cit.*, p. 370.

‡ Not previously published, as far as I am aware.

For let $$R = \lambda F + \mu F'.$$

Then $$R = \lambda F + \mu F' = |R = \lambda |F + \mu |F'.$$

Hence $$\lambda (F - |F) + \mu (F' - |F') = 0.$$

Thus by § 234 (2) F and F' are left-parallels, while R is a right-vector system.

It is to be noted that by § 234 (4) any pair of axes of a vector system, since they are reciprocally polar lines, are both right and left-parallels.

(6) Any* right-parallel set of lines and any left-parallel set of lines have one and only one pair of reciprocally polar lines in common [cf. § 234 (4)]. For let R and L be the associated vector systems of the two sets of parallels. Then they are necessarily reciprocal; also they have one and only one pair of common conjugates. These common conjugates are the lines F and F', where

$$F = R\sqrt{(LL)} + L\sqrt{\{-(RR)\}}; \quad F' = R\sqrt{(LL)} - L\sqrt{\{-(RR)\}}.$$

Hence $F' = |F$.

Also $$R \equiv (F + |F), \quad L \equiv (F - |F).$$

Thus F and $|F$ belong to the right-parallel set of R and to the left-parallel set of L.

(7) The common conjugates of two vector systems of the same denomination are a pair of imaginary generating lines of the absolute. This follows from § 235 (2).

237. Further Properties of Parallel Lines†. (1) If any straight line meet two parallel straight lines, it makes each exterior angle equal to the interior and opposite angle, or in other words the two interior angles equal to two right angles.

For let xp and yq be two parallel lines (say right-parallel); and let xy be any line meeting them. Then $\widehat{pxy}$, and $\widehat{qyx}$ are the two interior angles.

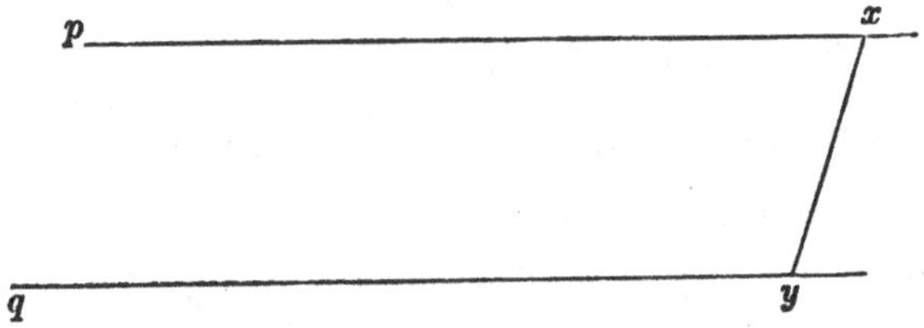

Fig. 7.

Let x, y, p, q be all unit points, and let $(x|p) = 0 = (y|q)$.

Then, by § 236 (2), it may be assumed that

$$xp + |xp = yq + |yq \quad \text{.............................(1).}$$

* Not previously published, as far as I am aware.

† Cf. Clifford, *loc. cit.*, p. 370.

Also $$\cos \widehat{yxp} = \frac{(xy \mid xp)}{\sqrt{\{(xy \mid xy)(xp \mid xp)\}}} = \frac{(xy \mid xp)}{\sqrt{(xy \mid xy)}}.$$

Similarly $$\cos \widehat{xyq} = \frac{(yx \mid yq)}{\sqrt{\{(xy \mid xy)(yq \mid yq)\}}} = -\frac{(xy \mid yq)}{\sqrt{\{xy \mid xy\}}}.$$

But from equation (1), multiplying by xy, we find $(xy \mid xp) = (xy \mid yq)$.

Hence $$\cos \widehat{yxp} + \cos \widehat{xyq} = 0, \text{ and } \widehat{yxp} + \widehat{xyq} = \pi.$$

A similar proof applies to left-parallel lines.

(2) Conversely, if two straight lines be such that every line intersecting both makes the two interior angles equal to two right angles, then the lines are parallel.

For let any line xy cut the lines xp, yq; so that $\widehat{yxp} = \widehat{xyq}'$.

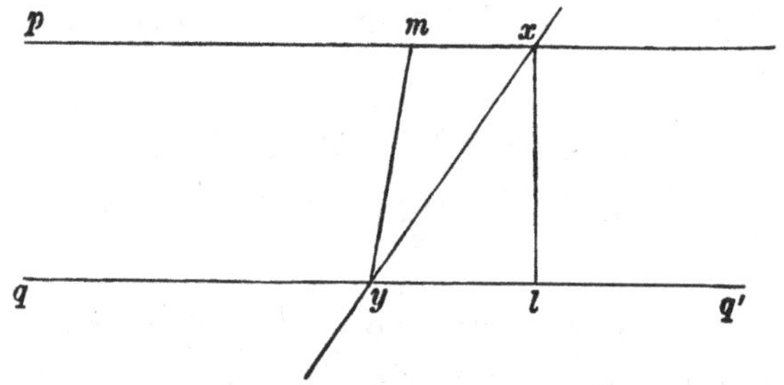

FIG. 8.

Draw xl and ym perpendicular to yq and xp respectively.

Then from § 216, considering the triangles xyl and xym,

$$\frac{\sin \frac{\overline{xl}}{\gamma}}{\sin \widehat{xyq}} = \frac{\sin \frac{\overline{xy}}{\gamma}}{\sin \frac{\pi}{2}} = \sin \frac{\overline{xy}}{\gamma} = \frac{\sin \frac{\overline{ym}}{\gamma}}{\sin \widehat{yxp}}.$$

Hence $$\overline{xl} = \overline{ym}.$$

Therefore the lines xp and yq are parallel.

(3) Parallelograms can be proved to exist in Elliptic Space: but they are not plane figures [cf. § 234 (5)].

For let ab and ac be any two lines intersecting at a. Then the right-

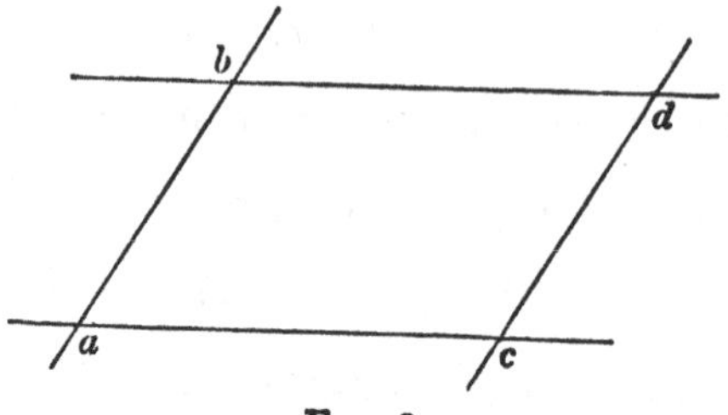

FIG. 9.

parallel through b to ac is, by § 236 (3), $F = b|b(ac + |ac)$. Similarly the left-parallel through c to ab is $F' = c|c(ab - |ab)$.

To prove that these lines intersect, we have to prove that $(FF') = 0$.

But it is easy to prove by multiplication and reduction that,

$$(FF') = b|\{b(ac + |ac)\} . c|\{c(ab - |ab)\} = 0.$$

Therefore the two parallels through b and c intersect in some point d. Therefore the opposite sides of the figure $abdc$ are parallel, one pair being right-parallels and the other pair being left-parallels.

Also if the angle $\widehat{cab}$ be θ, then $\widehat{abd} = \pi - \theta$, $\widehat{bdc} = \theta$, $\widehat{dca} = \pi - \theta$.

Further it is easy to prove that $\overline{ab} = \overline{cd}$, and $\overline{bd} = \overline{ac}$. Thus the opposite sides are equal. Hence if ac and bd be any two parallels and $\overline{ac} = \overline{bd}$, then ab and cd are parallels of opposite name (right or left) to ac and bd; and also $\overline{ab} = \overline{cd}$.

(4) Let ab, $a'b'$ be one pair of parallels, and let ac, $a'c'$ be another pair of the same name as the first pair: also let $\overline{ab} = \overline{a'b'}$, $\overline{ac} = \overline{a'c'}$, then bc and $b'c'$ are parallels of the same name, and $\overline{bc} = \overline{b'c'}$. For join aa', bb', cc'.

Then by (3) aa' and bb' are equal and parallels, of the opposite name to ab and $a'b'$; also aa' and cc' are equal and parallels of the opposite name to ac and $a'c'$. Hence [cf. § 236 (4)] bb' and cc' are equal and parallels of the opposite name to ab and $a'b'$. Hence bc and $b'c'$ are equal and parallels of the same name to ab and $a'b'$.

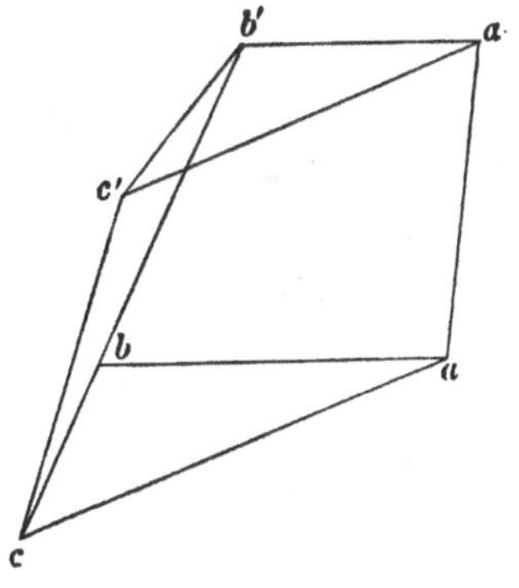

FIG. 10.

It is further obvious that the angle $\widehat{cab}$ is equal to the angle $\widehat{c'a'b'}$. Hence if from any point a' two parallels of the same name are drawn to any two lines ab, ac, the two pairs of intersecting lines contain the same angle.

238. Planes and Parallel Lines*. (1) One line, and only one line, belonging to a given parallel set of lines, lies in a given plane.

For let P be the given plane and e_1e_2 a line of the given parallel set.

Now if F be one of the set lying in P, $|F$ is also one of the set and passes through the point $|P$; and conversely. But one and only one of the set can be drawn through $|P$, hence one and only one of the set lies in P.

If p stand for $|P$, then by § 236 (3) the right-parallel $|F$, which passes through $|P$, is $p|pR$, where R stands for $e_1e_2 + |e_1e_2$. Hence

$$F = |p.|.|p|R = P|PR.$$

Thus the single right-parallel in the plane P is the line $P|PR$. Similarly for left-parallels.

(2) To any point p in a given plane P there corresponds one and only one point q in any other given plane Q, such that if any line through p be drawn in the plane P, the right-parallel line through q lies in the plane Q (or in other words the right-parallel in the plane Q passes through q).

For draw any two lines pp', pp'' in the plane P. Let their right parallels in the plane Q be qq', qq'' intersecting in q. Then q is the required point.

For take $$\overline{pp'} = \overline{qq'},\ \overline{pp''} = \overline{qq''}.$$

Then $$pp' + |pp' = qq' + |qq',\ pp'' + |pp'' = qq'' + |qq''.$$

Any other line through p and in the plane P can be written $pp' + \lambda pp''$. But from the above equations,

$$pp' + \lambda pp'' + |(pp' + \lambda pp'') = qq' + \lambda qq'' + |(qq' + \lambda qq'').$$

Hence the line $qq' + \lambda qq''$, which passes through q and lies in the plane Q, is the right-parallel to the line $pp' + \lambda pp''$.

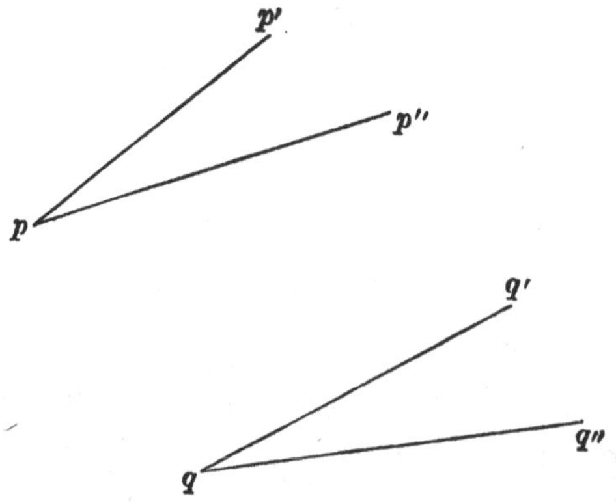

Fig. 11.

* These properties have not been stated before, as far as I am aware.

Similarly a unique point q_1 in the plane Q corresponds to the point p in the plane P with similar properties for left-parallels.

(3) With the construction of the preceding proposition (where $\overline{pp'} = \overline{qq'}$, $\overline{pp''} = \overline{qq''}$), it follows from § 237 (4) that $p'p''$ is the right-parallel to $q'q''$. Hence the points p' and q' in the planes P and Q correspond. Thus, given two corresponding points p and q, it is easy to find the point on one plane corresponding to any point on the other. For consider the point p' on P. Join pp' and draw qq' parallel to pp' and of the same length. Then q' corresponds to p'.

(4) The common perpendicular of two planes P and Q, namely $|PQ$, cuts the planes in two points p and q which are corresponding points both for

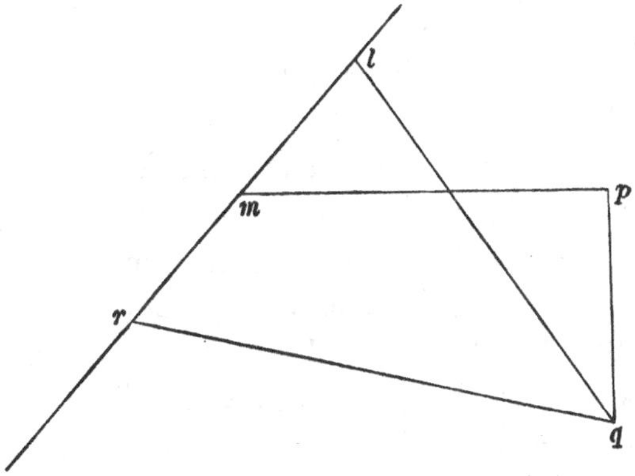

FIG. 12.

right and left-parallels. For in the plane P draw any line pm cutting the line PQ in m. Take two points l and r on PQ such that $\overline{lm} = \overline{mr} = \overline{pq}$. Join qr and ql. Then [cf. § 234 (4)] one of them (say qr) is a right-parallel to pm and the other ql is a left-parallel.

Accordingly, knowing that p and q are corresponding points, it is possible by (3) to construct the points q_1 and q_2 on Q corresponding to any point p' on P for right and left-parallels respectively.

CHAPTER IV.

Hyperbolic Geometry.

239. Space and Anti-space. (1) In hyperbolic geometry [cf. § 208] the absolute

$$(x\,|\,x) = 0,$$

is a real closed quadric.

If e be any point within such a quadric, then [cf. § 82 (6)] its polar plane does not cut the quadric in real points and the polar plane lies entirely without the quadric. Hence if e, e_1, e_2, ... $e_{\nu-1}$ form a normal system, and if e lie within the quadric, then the remaining points e_1, e_2, ... $e_{\nu-1}$ lie without it. Similarly [cf. § 82 (7)] if E, E_1, E_2, ... $E_{\nu-1}$ form a normal system of planes, and if E does not cut the quadric in real points, then E_1, E_2, ... $E_{\nu-1}$ must all cut the quadric in real points and include points within the quadric.

(2) Let that part of the complete spatial manifold of $\nu-1$ dimensions which is enclosed within the absolute be called Space [cf. § 202]. Let the part without the absolute be called Anti-space, or Ideal Space. Let a point within space be called spatial, a point in anti-space anti-spatial.

(3) A subregion may lie completely in anti-space, as far as its real elements are concerned, but cannot lie completely in space. Let a subregion which comprises spatial elements be called spatial, and a subregion which does not comprise spatial elements anti-spatial.

(4) Then a normal system of real elements e, e_1, ... $e_{\nu-1}$ consists of one spatial element e, and of $\nu-1$ anti-spatial elements. Let e be called the origin of this system.

A normal system of planes E, E_1, ... $E_{\nu-1}$ consists of one anti-spatial plane E, and of $\nu-1$ spatial planes.

If a plane P be spatial, its absolute pole $|P$ is anti-spatial; if the plane be anti-spatial, its absolute pole is spatial.

If an element p be spatial, its absolute polar $|p$ is anti-spatial; if p be anti-spatial, $|p$ is spatial.

If any subregion P_ρ, of $\rho-1$ dimensions, be spatial, the subregion $|P_\rho$ is anti-spatial; if P_ρ be anti-spatial, $|P_\rho$ is spatial.

240. INTENSITIES OF POINTS AND PLANES. (1) Let the absolute be referred to the ν normal elements $e, e_1, \dots e_{\nu-1}$, of which e is spatial.

Let α, $i\alpha_1$, $i\alpha_2, \dots i\alpha_{\nu-1}$ be the normal intensities [cf. § 110] of these elements; and let $i^{\nu-1}\Delta$ stand for $i^{\nu-1}\alpha\alpha_1\alpha_2\dots\alpha_{\nu-1}$. Then Δ is also real, where $\alpha, \alpha_1, \dots \alpha_{\nu-1}$ are real.

Also let $i^{\nu-1}\Delta(ee_1\dots e_{\nu-1}) = 1$.

Then
$$|e = \frac{i^{\nu-1}\Delta}{\alpha^2} e_1e_2\dots e_{\nu-1},\quad |e_1 = \frac{-i^{\nu-1}\Delta}{-\alpha_1^2} ee_2\dots e_{\nu-1},$$
$$|e_2 = \frac{i^{\nu-1}\Delta}{-\alpha_2^2} ee_1e_3\dots e_{\nu-1};\ \text{and so on.}$$

Hence if
$$x = \xi e + \xi_1 e_1 + \dots + \xi_{\nu-1}e_{\nu-1},$$
then
$$(x|x) = \left(\frac{\xi^2}{\alpha^2} - \frac{\xi_1^2}{\alpha_1^2} - \frac{\xi_2^2}{\alpha_2^2} - \dots - \frac{\xi^2_{\nu-1}}{\alpha^2_{\nu-1}}\right).$$

Thus [cf. § 82 (9)] if x be spatial and its co-ordinates real, $(x|x)$ is positive. This supposition will be adhered to.

(2) Any real plane is given by
$$L = \lambda e_1e_2\dots e_{\nu-1} - \lambda_1 ee_2\dots e_{\nu-1} + \lambda_2 ee_1e_3\dots e_{\nu-1} + \text{etc.};$$
where the ratios $\lambda : \lambda_1 : \lambda_2 :$ etc. are real.

But if $x\ (= \Sigma\xi e)$ be a real point with its co-ordinates real, we may suppose x to be the pole of L, and write $L = |x$
$$= i^{\nu-1}\Delta\left(\frac{\xi}{\alpha^2} e_1e_2\dots e_{\nu-1} - \frac{\xi_1}{\alpha_1^2} ee_2\dots e_{\nu-1} + \text{etc.}\right).$$

Hence
$$\lambda = i^{\nu-1}\frac{\Delta\xi}{\alpha^2},\quad \lambda_1 = i^{\nu-1}\frac{\Delta\xi_1}{\alpha_1^2},\ \text{etc.}$$

Therefore if ν be even, λ, λ_1, etc., are pure imaginaries, so that their ratios are real.

A plane will be considered to be in its standard form, when expressed in the form
$$L = i^{\nu-1}\lambda e_1e_2\dots e_{\nu-1} - i^{\nu-1}\lambda_1 ee_2\dots e_{\nu-1} + \text{etc.},$$
where λ, λ_1, etc. are real.

Then we can write $L = |x$, where the coefficients of x are real. Thus a real plane is, if ν be even, intensively imaginary [cf. § 88 (3)]; while a real point, spatial or anti-spatial, is always intensively real.

(3) If x be spatial, its intensity is unity when $(x|x) = 1$, and is real when $(x|x)$ is positive.

If x be anti-spatial, the intensity of x will be defined to be real when $(x|x)$ is negative, and to be unity when $(x|x) = -1$.

Thus in both cases the intensity is real when the coefficients are real.

The intensity [cf. § 211 (5)] of a plane L will be defined to be unity, when it is in the standard form $|x$, where x is at unit intensity.

Thus for anti-spatial planes at unit intensity, $|x$ is spatial, and

$$(L\,|\,L) = (|x\,.\,||x) = (x\,|\,x) = 1.$$

For spatial planes at unit intensity

$$(L\,|\,L) = -1.$$

(4) Thus for a spatial point $\Sigma\xi e$ at unit intensity

$$\frac{\xi^2}{\alpha^2} - \frac{\xi_1^2}{\alpha_1^2} - \dots - \frac{\xi^2_{\nu-1}}{\alpha^2_{\nu-1}} = 1.$$

For a spatial plane $\Sigma i^{\nu-1}\lambda E$ at unit intensity,

$$\alpha_1^2\lambda_1^2 + \alpha_2^2\lambda_2^2 + \dots + \alpha_{\nu-1}^2\lambda_{\nu-1}^2 - \alpha^2\lambda^2 = \Delta^2.$$

(5) But if the reference elements e, e_1, ... $e_{\nu-1}$ be at unit intensities, spatial and anti-spatial, then $\alpha = \alpha_1 = \dots = \alpha_{\nu-1} = 1$.

Hence [cf. subsection (1)] the point-equation of the absolute is

$$\xi^2 - \xi_1^2 - \dots - \xi^2_{\nu-1} = 0,$$

the plane-equation of the absolute is

$$\lambda_1^2 + \lambda_2^2 + \dots + \lambda^2_{\nu-1} - \lambda^2 = 0.$$

The intensity of a spatial point $\Sigma\xi e$ is $(\xi^2 - \xi_1^2 - \xi_2^2 - \dots - \xi^2_{\nu-1})^{\frac{1}{2}}$; the intensity of a spatial plane $\Sigma i^{\nu-1}\lambda E$ is $(\lambda_1^2 + \lambda_2^2 + \dots + \lambda^2_{\nu-1} - \lambda^2)^{\frac{1}{2}}$.

Also $$i^{\nu-1}(ee_1 \dots e_{\nu-1}) = 1;$$

and $|e = i^{\nu-1}e_1e_2 \dots e_{\nu-1}$, $|e_1 = i^{\nu-1}ee_2 \dots e_{\nu-1}$, $|e_2 = -i^{\nu-1}ee_1e_3 \dots e_{\nu-1}$, $|e_3 = i^{\nu-1}ee_1e_4 \dots e_{\nu-1}$, and so on.

This supposition will be adhered to, unless it is otherwise stated.

241. Distances of Points. (1) It will be seen that in the case, in which the line joining two points in anti-space does not cut the absolute in real points, the usual hyperbolic formula does not give a real distance between them. In this case it is convenient to use the Elliptic measure of distance. Thus any two points in anti-space (as well as any two points in space) are separated by a real distance, Elliptic or Hyperbolic. But a point in space cannot have a real distance of either type from a point in anti-space.

(2) Firstly, let two points x and y both be spatial, and of standard sign [cf. § 208 (3)]; then $\overline{xy}$ is given by

$$\cosh\frac{\overline{xy}}{\gamma} = \frac{(x\,|\,y)}{\sqrt{\{(x\,|\,x)\,(y\,|\,y)\}}};$$

and $\overline{xy}$, thus determined, is real. Also, since there can be no distinction between $D(xy)$ and $\overline{xy}$, the latter symbol will always be used for the distance.

(3) Secondly, let x and y both be anti-spatial, but let the line xy be spatial. Then if a_1 and a_2 be the points where the line meets the quadric, x and y lie together on the same intercept between a_1 and a_2. Also $(x|x)$ and $(y|y)$ are of the same negative sign. Hence the hyperbolic functions give a real distance. Thus $\overline{xy}$ is determined as a real quantity by the formula

$$\cosh \frac{\overline{xy}}{\gamma} = \frac{(x|y)}{\sqrt{\{(x|x)(y|y)\}}};$$

where the ambiguity of sign must be so determined that the right-hand side is positive.

(4) Thirdly, let x be spatial and y be anti-spatial. Then both the formulæ

$$\frac{\cosh}{\cos} \frac{\overline{xy}}{\gamma} = \frac{(x|y)}{\sqrt{\{(x|x)(y|y)\}}}$$

must make $\overline{xy}$ imaginary, since $(x|x)$ and $(y|y)$ are of opposite signs.

(5) Fourthly, let x and y both be anti-spatial, and let the line xy be anti-spatial. Then the two elements a_1 and a_2 in which xy meets the absolute are imaginary. Hence the elliptic law applies. Let the distance between x and y, determined by this law, be called the angular distance between the points, and denoted by $\angle xy$.

Then
$$\cos \angle xy = \frac{(x|y)}{\sqrt{\{(x|x)(y|y)\}}};$$

where the conventions of § 206 apply: so that, if x' stand for $-x$,

$$\cos \angle x'y = \frac{(x'|y)}{\sqrt{\{(x'|x')(y|y)\}}} = -\cos \angle xy.$$

Hence
$$\angle x'y + \angle xy = \pi.$$

(6) Thus, to conclude, if the line xy be spatial, and x and y be both spatial or both anti-spatial, then $\overline{xy}$ is real. If the line xy be anti-spatial, then $\angle xy$ is real. If x be spatial and y anti-spatial, both $\overline{xy}$ and $\angle xy$ are imaginary.

242. Distances of Planes. (1) Consider the formulæ for the separation between two planes P and Q.

Firstly, let both planes be spatial, and let the subplane PQ be spatial. Then $|P$ and $|Q$ are both anti-spatial, and $|P|Q$ is anti-spatial.

Hence $\angle |P|Q$ is real, and is given by

$$\cos \angle |P|Q = \frac{(|P||Q)}{\sqrt{\{(|P||P)(|Q||Q)\}}} = \frac{(P|Q)}{\sqrt{\{(P|P)(Q|Q)\}}}.$$

Hence the separation between P and Q is real, when determined by the elliptic formula. Let it be called the angle between P and Q, and denoted by $\angle PQ$.

Then
$$\cos \angle PQ = \frac{(P|Q)}{\sqrt{\{(P|P)(Q|Q)\}}}.$$

It is to be noticed that there are two angles $\angle PQ$ and $\pi - \angle PQ$, corresponding to the ambiguity of sign on the right-hand side.

(2) Secondly, let both planes be spatial, and let the subplane PQ be anti-spatial. Then $|P$ and $|Q$ are both anti-spatial, and $|P|Q$ is spatial.

Hence $\overline{|P|Q}$ is real, and is determined by
$$\cosh \frac{\overline{|P|Q}}{\gamma} = \frac{(|P\|Q)}{\sqrt{\{(|P\|P)(|Q\|Q)\}}},$$
$$= \frac{(P|Q)}{\sqrt{\{(P|P)(Q|Q)\}}}.$$

Hence the separation between P and Q is to be measured by the hyperbolic formula, and will be called the distance between the planes, and denoted by $\overline{PQ}$.

Then
$$\cosh \frac{\overline{PQ}}{\gamma} = \frac{(P|Q)}{\sqrt{\{(P|P)(Q|Q)\}}};$$
where as usual the terms P and Q are so chosen that $(P|Q)$ is positive.

(3) Thirdly, let P be spatial and Q be anti-spatial. Then $(P|P)$ and $(Q|Q)$ are of opposite signs. Hence both $\angle PQ$ and $\overline{PQ}$ are imaginary.

(4) Fourthly, let the planes P and Q both be anti-spatial. Then $|P$ and $|Q$ and $|P|Q$ are spatial, and $\overline{|P|Q}$ is real.

Hence
$$\cosh \frac{\overline{|P|Q}}{\gamma} = \frac{(|P\|Q)}{\sqrt{\{(|P\|P)(|Q\|Q)\}}}$$
$$= \frac{(P|Q)}{\sqrt{\{(P|P)(Q|Q)\}}}.$$

Hence $\overline{PQ}$ is real and $\angle PQ$ is imaginary.

Also
$$\cosh \frac{\overline{PQ}}{\gamma} = \frac{(P|Q)}{\sqrt{\{(P|P)(Q|Q)\}}};$$
where the terms P and Q are so chosen that $(P|Q)$ is positive.

243. SPATIAL AND ANTI-SPATIAL LINES. (1) If the elliptic measure for separation holds, then [cf. §§ 204 and 211]
$$\sin \angle xy = \sqrt{\frac{(xy|xy)}{\{(x|x)(y|y)\}}},$$
and
$$\sin \angle PQ = \sqrt{\frac{(PQ|PQ)}{\{(P|P)(Q|Q)\}}};$$
and if the hyperbolic measure holds, then [cf. §§ 208 and 211]
$$\sinh \frac{\overline{xy}}{\gamma} = \sqrt{\frac{-(xy|xy)}{\{(x|x)(y|y)\}}},$$
and
$$\sinh \frac{\overline{PQ}}{\gamma} = \sqrt{\frac{-(PQ|PQ)}{\{(P|P)(Q|Q)\}}}.$$

(2) Thus if xy be spatial, $(xy|xy)$ is negative. For if x and y be either both spatial or both anti-spatial, the proposition follows from the expression for $\sinh \frac{\overline{xy}}{\gamma}$. But if x be spatial and y anti-spatial, then

$$(xy|xy) = (x|x)(y|y) - (x|y)^2.$$

Now $(x|x)$ is positive and $(y|y)$ negative; hence again the proposition follows. But if xy be anti-spatial, then, from the expression for $\sin \angle xy$, $(xy|xy)$ is positive.

(3) Furthermore if x be anti-spatial and y be any point on the cone, $(xy|xy) = 0$, which envelopes the quadric, then $\overline{xy} = 0$ and $\angle xy = 0$. Hence any two points on a tangent line to the quadric are at zero distance from each other.

(4) Again, by similar reasoning, if the intersection of two spatial planes P and Q be spatial, $(PQ|PQ)$ is positive. If the intersection of two spatial planes be anti-spatial, $(PQ|PQ)$ is negative. If P be spatial and Q anti-spatial, $(PQ|PQ)$ is negative. Hence if PQ be spatial, $(PQ|PQ)$ is positive; if PQ be anti-spatial, $(PQ|PQ)$ is negative.

244. Distances of Subregions. (1) If two subregions P_ρ and Q_ρ, each of $\rho - 1$ dimensions, are contained in the same subregion (L) of ρ dimensions, then [cf. § 211 (6)] a single measure of the separation of P_ρ and Q_ρ can be assigned.

(2) Let the section of the absolute by L be real; and firstly let the intersection of P_ρ and Q_ρ be spatial. Then $\angle P_\rho Q_\rho$ is real, and

$$\cos \angle P_\rho Q_\rho = \frac{(P_\rho|Q_\rho)}{\sqrt{\{(P_\rho|P_\rho)(Q_\rho|Q_\rho)\}}}.$$

Secondly, let the intersection of P_ρ and Q_ρ be anti-spatial, but P_ρ and Q_ρ be both spatial. Then $\overline{P_\rho Q_\rho}$ is real, and

$$\cosh \frac{\overline{P_\rho Q_\rho}}{\gamma} = \frac{(P_\rho|Q_\rho)}{\sqrt{\{(P_\rho|P_\rho)(Q_\rho|Q_\rho)\}}}.$$

Thirdly, let P_ρ be spatial and Q_ρ be anti-spatial. Then $\overline{P_\rho Q_\rho}$ and $\angle P_\rho Q_\rho$ are both imaginary.

Fourthly, let P_ρ and Q_ρ be both anti-spatial. Then $\overline{P_\rho Q_\rho}$ is real, and

$$\cosh \frac{\overline{P_\rho Q_\rho}}{\gamma} = \frac{(P_\rho|Q_\rho)}{\sqrt{\{(P_\rho|P_\rho)(Q_\rho|Q_\rho)\}}}.$$

(3) Let the section of the absolute by L be imaginary. Then P_ρ and Q_ρ are anti-spatial, and we have a fifth case when $\angle P_\rho Q_\rho$ is real, and given by the formula of the first case.

245. Geometrical Signification. Geometrical meanings can be assigned to the co-ordinates of any spatial point x, at unit intensity and of standard sign, referred to a normal system $e, e_1, \dots e_{\nu-1}$ at unit intensities, of which e is the spatial origin.

Let $x = \xi e + \xi_1 e_1 + \dots + \xi_{\nu-1} e_{\nu-1}$, where $(x|x) = 1 = \xi^2 - \xi_1^2 - \dots - \xi_{\nu-1}^2$.

Then $$\cosh \frac{\overline{ex}}{\gamma} = (e|x) = \xi.$$

Let the angles, $\widehat{xee_1} = \lambda_1$, $\widehat{xee_2} = \lambda_2$, etc.

Then $$\cos \lambda_1 = \frac{(ee_1|ex)}{\sqrt{\{(ex|ex)(ee_1|ee_1)\}}}$$
$$= \frac{\xi_1}{\sinh \frac{\overline{ex}}{\gamma}}.$$

Hence $$\xi_1 = \sinh \frac{\overline{ex}}{\gamma} \cos \lambda_1, \quad \xi_2 = \sinh \frac{\overline{ex}}{\gamma} \cos \lambda_2, \text{ etc.}$$

Also the angles $\lambda_1, \lambda_2, \dots \lambda_{\nu-1}$ are connected by,
$$\Sigma \cos^2 \lambda = 1.$$

Similar geometrical interpretations hold for Elliptic Geometry.

246. Poles and Polars. (1) It will be noticed that the only case, when there is no real measure of separation between two points x and y, is when x is spatial and y is anti-spatial. In this case the point of intersection of xy and the polar of y is spatial. For this point is
$$xy|y = (x|y)\, y - (y|y)\, x = y', \text{ say.}$$
Then by simple multiplication we find
$$(y'|y') = (y|y)^2 (x|x) - (x|y)^2 (y|y).$$

But $(x|x)$ is positive and $(y|y)$ is negative. Hence $(y'|y')$ is positive, and therefore y' is spatial.

Also the term y' is of standard sign. For [cf. 208 (3)] x is by hypothesis of standard sign, and
$$(x|y') = (x|y)^2 - (x|x)(y|y) = -(xy|xy).$$

But xy is spatial; hence, by § 243 (2), $(xy|xy)$ is negative, and $(x\ y')$ is positive.

Similarly the point of intersection of xy and the polar of x is $x' = yx|x$; and x' is anti-spatial, since $|x$ is anti-spatial.

(2) Now $(y'|y') = (y|y)\{(x|x)(y|y) - (x|y)^2\} = (y|y)(xy|xy)$,

and $$(y'|x) = (x|y)^2 - (x|x)(y|y) = -(xy|xy).$$

Hence $$\cosh \frac{\overline{y'x}}{\gamma} = \frac{(y'|x)}{\sqrt{\{(x|x)(y'|y')\}}} = \sqrt{\frac{(xy|xy)}{(x|x)(y|y)}}.$$

Also since $(y|y)$, as well as $(xy|xy)$, is negative, $\overline{y'x}$ is real as given by this formula.

Similarly
$$\sinh\frac{\overline{y'x}}{\gamma} = \sqrt{\left\{\cosh^2\frac{\overline{y'x}}{\gamma} - 1\right\}}$$
$$= \frac{(x|y)}{\sqrt{\{-(x|x)(y|y)\}}}.$$

Also since x' and y are both anti-spatial, and $x'y$ is spatial, then $\overline{x'y}$ is real. And by a similar proof
$$\cosh\frac{\overline{x'y}}{\gamma} = \sqrt{\frac{(xy|xy)}{(x|x)(y|y)}} = \cosh\frac{\overline{y'x}}{\gamma}.$$

Hence
$$\overline{x'y} = \overline{y'x}.$$

(3) Let x and y be both anti-spatial; and let xy be spatial. Then $\overline{xy}$ is real.

Also $x' = yx|x$, and $y' = xy|y$ are both spatial points. For
$$(y'|y') = (y|y)^2(x|x) - (x|y)^2(y|y) = (y|y)(xy|xy).$$

Now $(y|y)$ and $(xy|xy)$ are, by hypothesis, both negative.

Hence $(y'|y')$ is positive, and y' is a spatial point. Similarly x' is a spatial point.

Also
$$(x'|y') = (x|y)^3 - (x|x)(y|y)(x|y) = -(x|y)(xy|xy).$$

Hence if the terms x and y be so chosen as to sign that $(x|y)$ is positive, $(x'|y')$ is also positive.

Now, since x' and y' are both spatial, $\overline{x'y'}$ is real; and
$$\cosh\frac{\overline{x'y'}}{\gamma} = \frac{(x'|y')}{\sqrt{\{(x'|x')(y'|y')\}}} = \frac{-(x|y)(xy|xy)}{+\sqrt{\{(x|x)(y|y)(xy|xy)^2\}}}.$$

Now, since $(xy|xy)$ is negative,
$$+\sqrt{\{(x|x)(y|y)(xy|xy)^2\}} = -(xy|xy)\sqrt{\{(x|x)(y|y)\}}.$$

Hence
$$\cosh\frac{\overline{x'y'}}{\gamma} = \frac{(x|y)}{\sqrt{\{(x|x)(y|y)\}}} = \cosh\frac{\overline{xy}}{\gamma}.$$

Therefore
$$\overline{x'y'} = \overline{xy}.$$

(4) Exactly in the same way let the plane P be spatial and Q anti-spatial, then $|P$ is anti-spatial and $|Q$ is spatial. Let the plane through PQ and $|P$ be called P' and that through PQ and $|Q$ be called Q'. Then Q' is obviously spatial, and P' can be proved to be anti-spatial.

Then P and Q' are two spatial planes with an anti-spatial intersection.

Hence
$$\cosh\frac{\overline{PQ'}}{\gamma} = \pm\frac{(P|Q')}{\sqrt{\{(P|P)(Q'|Q')\}}}.$$

But
$$Q' = PQ|Q = (P|Q)\,Q - (Q|Q)\,P.$$

Therefore $(P|Q') = (P|Q)^2 - (Q|Q)(P|P) = -(PQ|PQ)$,

and $$(Q'|Q') = (Q|Q)(PQ|PQ).$$

Hence $$\cosh\frac{\overline{PQ'}}{\gamma} = \sqrt{\frac{(PQ|PQ)}{(P|P)(Q|Q)}} = \cosh\frac{\overline{P'Q}}{\gamma}.$$

Similarly $$\sinh\frac{\overline{PQ'}}{\gamma} = \frac{(P|Q)}{\sqrt{\{-(P|P)(Q|Q)\}}} = \sinh\frac{\overline{P'Q}}{\gamma}.$$

Hence $$\overline{PQ'} = \overline{P'Q}.$$

247. Points on the Absolute. (1) A point u on the absolute is at an infinite distance from any other point. For $(u|u) = 0$, and hence

$$\cosh\frac{\overline{ux}}{\gamma} = \frac{(x|u)}{\sqrt{\{(x|x)(u|u)\}}} = \infty.$$

(2) To find a point u in which any spatial line xy cuts the absolute, put $u = x + \lambda y$.

Hence $$\lambda^2(y|y) + 2\lambda(x|y) + (x|x) = 0.$$

Now let ρ stand for the distance $\overline{xy}$, and let x and y be spatial points at unit intensity and of standard sign.

Hence $$\lambda^2 + 2\lambda\cosh\frac{\rho}{\gamma} + 1 = (\lambda + e^{\frac{\rho}{\gamma}})(\lambda + e^{-\frac{\rho}{\gamma}}) = 0.$$

Accordingly the two points, in which the line xy cuts the absolute, are $x - e^{-\frac{\rho}{\gamma}}y$ and $x - e^{\frac{\rho}{\gamma}}y$.

(3) In the same way if the line xy be spatial, but x and y be both anti-spatial at unit anti-spatial intensity, and $(x|y)$ be positive, then the points, in which xy cuts the absolute, are $x + e^{-\frac{\rho}{\gamma}}y$, $x + e^{\frac{\rho}{\gamma}}y$.

(4) Similarly let P, Q be two spatial planes, and PQ be anti-spatial: also let P and Q be at unit spatial intensity and let ρ be the distance between them (hyperbolic measure). Then the planes through PQ touching the absolute are $P + e^{-\frac{\rho}{\gamma}}Q$, and $P + e^{\frac{\rho}{\gamma}}Q$.

Also if P and Q be both anti-spatial at unit anti-spatial intensity, the tangent planes are $P - e^{-\frac{\rho}{\gamma}}Q$, and $P - e^{\frac{\rho}{\gamma}}Q$.

248. Triangles. (1) Consider a triangle abc, in which the measures for the separation of the angular points are all real. Then the cases which arise are (1) a, b, c all spatial; (2) a, b, c all anti-spatial, and bc, ca, ab all spatial; (3) a, b, c all anti-spatial, and bc, ca, ab all anti-spatial; (4) a, b, c all anti-spatial, and bc, ca, ab two spatial and one anti-spatial; (5) a, b, c all anti-spatial, and bc, ca, ab being one spatial and two anti-spatial.

(2) *Case I.* a, b, c all spatial. Let the triangle abc in this case be called a spatial triangle.

Let the angle between ab and ac be $\boldsymbol{\alpha}$, that between ba and bc be $\boldsymbol{\beta}$, and that between ca and cb be $\boldsymbol{\gamma}$.

To discriminate between $\boldsymbol{\alpha}$ and $\pi - \boldsymbol{\alpha}$, let $\boldsymbol{\alpha}$ be that angle which vanishes when b coincides with c; and similarly for $\boldsymbol{\beta}$ and $\boldsymbol{\gamma}$.

Thus, b and c being of standard sign according to the usual convention,

$$\cos \boldsymbol{\alpha} = \frac{(ab \,|\, ac)}{\sqrt{\{(ab \,|\, ab)\,(ac \,|\, ac)\}}}.$$

And
$$(ab \,|\, ac) = (a \,|\, a)\,(b \,|\, c) - (a \,|\, b)\,(a \,|\, c);$$

also
$$\sinh \frac{\overline{ab}}{\gamma} = \sqrt{\frac{-(ab \,|\, ab)}{(a \,|\, a)\,(b \,|\, b)}}, \quad \sinh \frac{\overline{ac}}{\gamma} = \sqrt{\frac{-(ac \,|\, ac)}{(a \,|\, a)\,(c \,|\, c)}}.$$

Hence
$$\cos \boldsymbol{\alpha} = -\frac{\cosh \frac{\overline{bc}}{\gamma} - \cosh \frac{\overline{ab}}{\gamma} \cosh \frac{\overline{ac}}{\gamma}}{\sinh \frac{\overline{ab}}{\gamma} \sinh \frac{\overline{ac}}{\gamma}}.$$

Finally,
$$\cosh \frac{\overline{bc}}{\gamma} = \cosh \frac{\overline{ab}}{\gamma} \cosh \frac{\overline{ac}}{\gamma} - \sinh \frac{\overline{ab}}{\gamma} \sinh \frac{\overline{ac}}{\gamma} \cos \boldsymbol{\alpha}.$$

(3) Also [cf. § 216 (1)]

$$\sin \boldsymbol{\alpha} = \frac{\sqrt{\{(ab \,|\, ab)\,(ac \,|\, ac) - (ab \,|\, ac)^2\}}}{\sqrt{\{(ab \,|\, ab)\,(ac \,|\, ac)\}}}$$
$$= \frac{\sqrt{\{(a \,|\, a)\,(abc \,|\, abc)\}}}{\sqrt{\{(ab \,|\, ab)\,(ac \,|\, ac)\}}}.$$

And
$$\sinh \frac{\overline{bc}}{\gamma} = \sqrt{\frac{-(bc \,|\, bc)}{(b \,|\, b)\,(c \,|\, c)}}.$$

Therefore
$$\frac{\sin \boldsymbol{\alpha}}{\sinh \frac{\overline{bc}}{\gamma}} = \sqrt{\frac{(a \,|\, a)\,(b \,|\, b)\,(c \,|\, c)\,(abc \,|\, abc)}{-(bc \,|\, bc)\,(ca \,|\, ca)\,(ab \,|\, ab)}}$$
$$= \frac{\sin \boldsymbol{\beta}}{\sinh \frac{\overline{ca}}{\gamma}} = \frac{\sin \boldsymbol{\gamma}}{\sin \frac{\overline{ab}}{\gamma}}.$$

(4) It is easily proved, exactly as in Elliptic Geometry [cf. § 216 (6)], that the perimeter of a spatial circle, with a spatial centre and of radius ρ, is $2\pi\gamma \sinh \frac{\rho}{\gamma}$. And that the length of an arc subtending an angle α at the centre is $\alpha\gamma \sinh \frac{\rho}{\gamma}$.

(5) *Case II.* The angular points a, b, c are anti-spatial, and the sides bc, ca, ab are spatial. Let the triangle abc in this case be called a semi-spatial triangle.

The distances between the sides bc, ca, ab must now be measured by the hyperbolic measure. Thus let α, β, γ be assumed to be lengths and not angles. Also adopt the conventions of Case I. Then by a similar proof

$$\cosh\frac{\overline{bc}}{\gamma} = \cosh\frac{\overline{ab}}{\gamma}\cosh\frac{\overline{ac}}{\gamma} - \sinh\frac{\overline{ab}}{\gamma}\sinh\frac{\overline{ac}}{\gamma}\cosh\frac{\alpha}{\gamma}.$$

Also
$$\frac{\sinh\frac{\alpha}{\gamma}}{\sinh\frac{\overline{bc}}{\gamma}} = \frac{\sinh\frac{\beta}{\gamma}}{\sinh\frac{\overline{ca}}{\gamma}} = \frac{\sinh\frac{\gamma}{\gamma}}{\sinh\frac{\overline{ab}}{\gamma}}$$
$$= \sqrt{\frac{(a|a)(b|b)(c|c)(abc|abc)}{(bc|bc)(ca|ca)(ab|ab)}}.$$

(6) *Case III.* The angular points a, b, c all anti-spatial, and bc, ca, ab also anti-spatial.

This case gives simply the ordinary formulæ of Elliptic Geometry [cf. § 215].

(7) *Case IV.* The angular points a, b, c are anti-spatial, the two sides ab, ac are spatial, and the third side bc is anti-spatial.

The sides ab and ac have a real measure of separation α, reckoned according to the hyperbolic formula, but the sides ba and bc, and the sides ca and cb have no real measure of separation.

Hence
$$\cos\angle bc = \cosh\frac{\overline{ab}}{\gamma}\cosh\frac{\overline{ac}}{\gamma} - \sinh\frac{\overline{ab}}{\gamma}\sinh\frac{\overline{ac}}{\gamma}\cosh\frac{\alpha}{\gamma}.$$

(8) *Case V.* The angular points a, b, c are anti-spatial, and ab, ac are anti-spatial and bc is spatial. Then α is real, and β and γ are imaginary; and α is measured by the elliptic formula. And

$$\cosh\frac{\overline{bc}}{\gamma} = \cos\angle ab\cos\angle ac + \sin\angle ab\sin\angle ac\cos\alpha.$$

There are no corresponding formulæ to be obtained by cyclic interchange; since β and γ are imaginary.

(9) The theory, given in § 217, of points inside a triangle holds without change for Hyperbolic Geometry.

249. PROPERTIES OF ANGLES OF A SPATIAL TRIANGLE. (1) Two angles of a spatial triangle cannot be obtuse. For if α and β be both obtuse, $\cos\alpha$ and $\cos\beta$ are both negative. Hence from § 248 (2)

$$\cosh\frac{\overline{ab}}{\gamma}\cosh\frac{\overline{ac}}{\gamma} < \cosh\frac{\overline{bc}}{\gamma}, \text{ and } \cosh\frac{\overline{bc}}{\gamma}\cosh\frac{\overline{ab}}{\gamma} < \cosh\frac{\overline{ac}}{\gamma}.$$

But $\cosh\frac{\overline{ab}}{\gamma}$ is necessarily greater than unity; hence these two inequalities are inconsistent.

(2) It follows from § 247 (2) that, when b and c are spatial points of standard sign, all points of the form $\lambda b + \mu c$, where λ/μ is positive, lie on the intercept between b and c; since the two points, in which xy cuts the absolute, both lie on that intercept for which λ/μ is negative. Hence it may be proved, exactly in the same way as in § 219 (2) dealing with Elliptic Geometry, that if in any triangle abc $\boldsymbol{\beta}$ and $\boldsymbol{\gamma}$ be both acute, the foot of the perpendicular from a on to bc falls within the intercept bc.

(3) The sum of the angles of any spatial triangle is less than two right-angles.

Firstly, let the angle $\boldsymbol{\gamma}$ be a right-angle. Then as in Elliptic Geometry, [cf. § 219 (4)].

$$\tan \tfrac{1}{2}(\boldsymbol{\alpha} + \boldsymbol{\beta}) = \frac{\cosh \frac{1}{2}\frac{\overline{bc} \sim \overline{ca}}{\gamma}}{\cosh \frac{1}{2}\frac{\overline{bc} + \overline{ca}}{\gamma}} \cot \tfrac{1}{2}\boldsymbol{\gamma}$$

$$= \frac{\cosh \frac{1}{2}\frac{\overline{bc} \sim \overline{ca}}{\gamma}}{\cosh \frac{1}{2}\frac{\overline{bc} + \overline{ca}}{\gamma}}.$$

Now
$$\cosh \frac{1}{2}\frac{\overline{bc} \sim \overline{ca}}{\gamma} < \cosh \frac{1}{2}\frac{\overline{bc} + \overline{ca}}{\gamma}.$$

Hence
$$\boldsymbol{\alpha} + \boldsymbol{\beta} < \frac{\pi}{2}.$$

Hence
$$\boldsymbol{\alpha} + \boldsymbol{\beta} + \boldsymbol{\gamma} < \pi.$$

Secondly, the theorem can be extended to any triangle by the reasoning of § 219 (4).

250. Stereometrical Triangles. (1) It is obvious by the theory of duality that a complete set of formulæ for stereometrical triangles [cf. § 222 (1)] can be set down, and that these can be ranged under eight cases just as in the case of ordinary triangles. It will be sufficient to obtain the results for the two most important cases.

(2) Firstly, let the planes A, B, C be spatial, and let the subplanes BC, CA, AB be also spatial.

For shortness put $BC = A_1$, $CA = B_1$, $AC = C_1$.

Let $\angle BC = \boldsymbol{\alpha}$, $\angle CA = \boldsymbol{\beta}$, $\angle AB = \boldsymbol{\gamma}$. Also $\angle B_1C_1$, $\angle C_1A_1$, $\angle A_1B_1$ are real.

Then if $|A = a$, $|B = b$, $|C = c$, the triangle abc is anti-spatial, and bc, ca, ab are anti-spatial. Hence from § 248 (6)

$$\cos \angle bc = \cos \angle ab \cos \angle ac + \sin \angle ab \sin \angle ac \cos \angle (ab)(ac).$$

But $\angle bc = \angle BC = \alpha$, $\angle ca = \angle CA = \beta$, $\angle ab = \angle AB = \gamma$. Also

$$\angle(ab)(ac) = \angle B_1C_1.$$

Hence, $$\cos\alpha = \cos\beta\cos\gamma + \sin\beta\sin\gamma\cos\angle B_1C_1.$$

(3) When the complete region is of two dimensions, this does not agree with the ordinary formula in Euclidean space for spherical trigonometry; and, as in the analogous case of Elliptic Geometry, the discrepancy is removed by replacing the angles by their supplements.

When the complete region is of three or more dimensions, we deduce, as in the case of Elliptic Geometry, that the 'Spherical Trigonometry' of Hyperbolic Space is the same as that of ordinary Euclidean Space. This theorem is due to J. Bolyai, as far as space of three dimensions is concerned: it is here extended to planes of any number of dimensions.

(4) Secondly, let the planes A, B, C be spatial, but let the subplanes A_1, B_1, C_1 be anti-spatial. Then the triangle abc has its three angular points anti-spatial, but its three sides bc, ca, ab spatial. Hence from § 248 (5)

$$\cosh\frac{\overline{bc}}{\gamma} = \cosh\frac{\overline{ab}}{\gamma}\cosh\frac{\overline{ac}}{\gamma} - \sinh\frac{\overline{ab}}{\gamma}\sin\frac{\overline{ac}}{\gamma}\cosh\frac{\overline{(ab)(ac)}}{\gamma}$$

Hence $$\cosh\frac{\alpha}{\gamma} = \cosh\frac{\beta}{\gamma}\cosh\frac{\gamma}{\gamma} - \sinh\frac{\beta}{\gamma}\sinh\frac{\gamma}{\gamma}\cosh\frac{\overline{B_1C_1}}{\gamma};$$

with two similar formulæ.

251. PERPENDICULARS. (1) The theory of normal points and of perpendiculars in Hyperbolic Geometry is much the same as in Elliptic Geometry (cf. § 223). The proofs of corresponding propositions will be omitted.

Any two mutually normal points satisfy, $(x|y) = 0$. If xy be spatial, then one point must be spatial and the other anti-spatial [cf. § 239 (4)]. In this case no real measure of distance exists between x and y. If xy be anti-spatial, then the elliptic measure holds, and $\angle xy = \frac{1}{2}\pi$.

The condition that two lines ab, ac should be at right-angles (or perpendicular) is $(ab|ac) = 0$. If a be spatial, the measure of distance between the lines is elliptic, and the angle between them is a right-angle. If a be anti-spatial, and both lines be anti-spatial, the measure of distance is elliptic and the angle is a right-angle. But if a be anti-spatial, ab be spatial, and ac be anti-spatial, there is no real measure of distance between the lines. It is impossible for two lines to be at right-angles when a is anti-spatial, and ab, ac both spatial.

(2) If a line ab cut any region L_ρ, of $\rho - 1$ dimensions in the point a, and if $\rho - 1$ independent lines drawn from a in L_ρ are perpendicular to ab, then all lines drawn from a in L_ρ are perpendicular to ab.

The line ab is then said to be perpendicular to the region L_ρ.

(3) Any line perpendicular to the region L_ρ intersects the supplementary (or complete normal) region $|L_\rho$; and conversely, any line intersecting both L_ρ and $|L_\rho$ is perpendicular to both.

(4) If P_ρ and P_σ be two regions normal to each other, and if a be any point in P_ρ, then any line drawn through a in the region P_ρ is perpendicular to the region aP_σ.

(5) Let two planes L and M intersect in the subplane LM, and let a_1 be any point in LM. From a_1 draw a_1l in the plane L perpendicular to the subplane LM, and draw a_1m in the plane M perpendicular to LM, then the separation between L and M is equal to that between a_1l and a_1m.

For as in the Elliptic Geometry,

$$\frac{(L|M)}{\sqrt{\{(L|L)(M|M)\}}} = \frac{(a_1l|a_1m)}{\sqrt{\{(a_1l|a_1l)(a_1m|a_1m)\}}}.$$

Hence if $\angle LM$ be real, then $\angle LM = \angle(a_1l)(a_1m)$; and if $\overline{LM}$ be real, then $\overline{LM} = \overline{(a_1l)(a_1m)}$.

(6) Any line, perpendicular to any plane L, also passes through its absolute pole.

If any plane M include one perpendicular to L, then from any point of the subplane LM a perpendicular to L can be drawn lying in M.

Also if $|L$ lies in M, then $|M$ lies in L; hence this property is reciprocal.

If two planes are at right-angles, their poles are mutually normal.

(7) Also if two planes L and L' be each cut perpendicularly by a third plane M, then the measure of separation between L and L' is the same as that between LM and $L'M$.

252. THE FEET OF PERPENDICULARS. (1) Let p be the foot of the perpendicular xp, drawn from any spatial point x to a spatial plane L. Then p is spatial.

For $p = x|L.L$; also put $l = |L$. Then it can easily be proved that

$$(p|p) = (L|L)(xl|xl).$$

Now since xl is spatial, $(xl|xl)$ is negative [cf. § 243 (2)]; and $(L|L)$ is negative, since the plane L is spatial. Hence $(p|p)$ is positive, and p is spatial.

This can be extended to any spatial subregion P_ρ by noticing that P_ρ has the property of a plane with respect to the region xP_ρ.

(2) If the plane L and the point x be both anti-spatial, then the perpendicular from x to L is spatial, since it passes through the spatial point $|L$.

(3) The line joining the poles $|L$ and $|L'$ of two planes L and L' is evidently the only common perpendicular to the two planes L and L'. It is anti-spatial, if LL' be spatial: it is spatial, if LL' be anti-spatial.

For let $l = |L$ and $l' = |L'$.

Then if LL' be spatial, by § 243 (4) $(LL'|LL')$ is positive. But $(ll'|ll') = (LL'|LL')$. Hence ll' is anti-spatial by § 243 (2).

If LL' be anti-spatial, by § 243 (4) $(LL'|LL')$ is negative, and hence $(ll'|ll')$ is negative. Therefore ll' is spatial.

(4) If $|LL' = (|L|L')$ be spatial and L be spatial, then $|LL'$ intersects L in a spatial point. For let d be this point.

Then
$$d = L|LL' = (L|L')|L - (L|L)|L'.$$
Hence
$$(d|d) = (L|L)(LL'|LL').$$
But $(L|L)$ is negative, and $(LL'|LL')$ is positive. Therefore d is spatial.

The theorem also follows immediately from subsections (1) and (3).

253. Distance between Planes. (1) To prove that the distance (hyperbolic or elliptic) between two planes is equal to the distance between the feet of their common perpendicular line.

For let L and L' be the two planes; and let $d = L|LL'$, $d' = L'|L'L$.

Then
$$d = (L|L')|L - (L|L)|L', \quad d' = (L'|L)|L' - (L'|L')|L.$$
Hence
$$(d|d') = (L|L')^3 - (L|L)(L'|L')(L|L')$$
$$= -(L|L')(LL'|LL').$$

Hence if LL' be anti-spatial and $(L|L')$ be positive, [cf. § 252 (3) and (4) and § 242 (2)]

$$\cosh\frac{\overline{dd'}}{\gamma} = \frac{-(L|L')(LL'|LL')}{\sqrt{\{(L|L)(L'|L')(LL'|LL')^2\}}}$$
$$= \frac{(L|L')}{\sqrt{\{(LL)(L'L')\}}} = \cosh\frac{\overline{LL'}}{\gamma}.$$

Hence $\overline{dd'}$ is the distance which has been defined as the measure of separation between the planes.

(2) Secondly if LL' be spatial, d and d' are anti-spatial and on an anti-spatial line [cf. § 252 (3) and (4) and § 242 (1)].

Then
$$\cos\angle dd' = \frac{(L|L')}{\sqrt{\{(L|L)(L'|L')\}}} = \cos\angle LL'.$$

Then $\angle dd'$ is the angle which has been defined as the measure of separation between the planes.

(3) Also [cf. § 211 (6)], when the distance formula can be applied to two subregions P_ρ and Q_ρ, each of $\rho - 1$ dimensions, these subregions are both contained in the same region of ρ dimensions; and therefore they have the properties of planes. Hence they possess a single common perpendicular; and, when P_ρ and Q_ρ are spatial and their common subregion anti-spatial, the length of this (spatial) perpendicular is the measure of separation between

the subregions; also when the common subregion is spatial, the angular length of this (anti-spatial) perpendicular is the angle between the subregions.

254. Shortest Distances. (1) The least distance from a spatial point x to a spatial plane L is the perpendicular distance $\overline{xp}$, where p is the foot of the perpendicular.

For let q be any other spatial point on the plane L. Then since [cf. § 251 (2)] the angle between px and pq is a right-angle,

$$\cosh \frac{\overline{xq}}{\gamma} = \cosh \frac{\overline{xp}}{\gamma} \cosh \frac{\overline{pq}}{\gamma}.$$

Hence $$\overline{xq} > \overline{xp}.$$

This length of the perpendicular will be called the distance of x from the plane L.

(2) To find this distance $\overline{xp}$, write l for $|L$, then

$$p = xl\,|\,l = (x\,|\,l)\,l - (l\,|\,l)\,x.$$

Hence $$(p\,|\,p) = (L\,|\,L)\,(xl\,|\,xl),$$

and $$(xp\,|\,xp) = (x\,|\,l)^2\,(xl\,|\,xl) = (xL)^2\,(xl\,|\,xl).$$

Thus $$\sinh \frac{\overline{xp}}{\gamma} = \sqrt{\frac{-(xp\,|\,xp)}{(x\,|\,x)(p\,|\,p)}} = \sqrt{\frac{(xL)^2}{-(x\,|\,x)(L\,|\,L)}} = \frac{\pm(xL)}{\sqrt{\{-(x\,|\,x)(L\,|\,L)\}}}.$$

This formula gives the distance from a spatial point x to a spatial plane L.

(3) The greatest hyperbolic distance from an anti-spatial point x to an anti-spatial plane L is the perpendicular distance $\overline{xp}$, where p is the foot of the perpendicular from x to L.

Let q be any other point on L such that xq is spatial: also xp is spatial from § 252 (2). But pq is anti-spatial.

Hence $$\cosh \frac{\overline{xq}}{\gamma} = \cos \widehat{pq} \cosh \frac{\overline{xp}}{\gamma}.$$

Hence $$\overline{xq} < \overline{xp}.$$

(4) It follows from (1) of this article and from § 253 that the length of the common perpendicular is the least distance between the spatial points of spatial planes with an anti-spatial intersection and that this least distance is what has been defined as the distance between the planes. The same holds for any two subregions of the same dimensions with a single measure of distance between them.

(5) A formula, analogous to the formula for Elliptic Geometry, in § 226 (1) and (2), can be found for the perpendicular distance of any point a from any subregion P_ρ, of $\rho - 1$ dimensions.

Case I. If a and P_ρ be both spatial, then

$$\sinh\frac{\delta}{\gamma}=\sqrt{\frac{-(aP_\rho|aP_\rho)}{(a|a)(P_\rho|P_\rho)}}.$$

Case II. If a and P_ρ be both anti-spatial, and aP_ρ be spatial

$$\sinh\frac{\delta}{\gamma}=\sqrt{\frac{-(aP_\rho|aP_\rho)}{(a|a)(P_\rho|P_\rho)}}.$$

Case III. If a and P_ρ be both anti-spatial, and aP_ρ be also anti-spatial

$$\sin\delta=\sqrt{\frac{(aP_\rho|aP_\rho)}{(a|a)(P_\rho|P_\rho)}}.$$

(6) To prove these formulæ first consider the distance of a from the straight line F. Let b and c be two spatial points on F and let ap be the perpendicular from a on F [fig. 1].

Then $$F=bc\equiv pb.$$

Also by hypothesis $(pa|bc)=0=(pa|pb)$, since $pb\equiv bc$.

Hence by formula (i) of § 216 (1)

$$(p|p)(pab|pab)=(pb|pb)(pa|pa)-(pa|pb)^2=(pb|pb)(pa|pa).$$

Hence $$\frac{(pa|pa)}{(p|p)(a|a)}=\frac{(pab|pab)}{(pb|pb)(a|a)}=\frac{(abc|abc)}{(bc|bc)(a|a)}.$$

Now, if the hyperbolic formula hold,

$$\sinh\frac{\delta}{\gamma}=\sqrt{\frac{-(pa|pa)}{(p|p)(a|a)}}=\sqrt{\frac{-(aF|aF)}{(F|F)(a|a)}};$$

and if the elliptic formula hold

$$\sin\delta=\sqrt{\frac{(pa|pa)}{(p|p)(a|a)}}=\sqrt{\frac{(aF|aF)}{(F|F)(a|a)}}.$$

In Case II the hyperbolic formula holds; since F is anti-spatial, and therefore the point, in which $|F$ meets the two dimensional region aF, must be spatial; remembering that the section of the absolute by aF is real. But ap passes through this point.

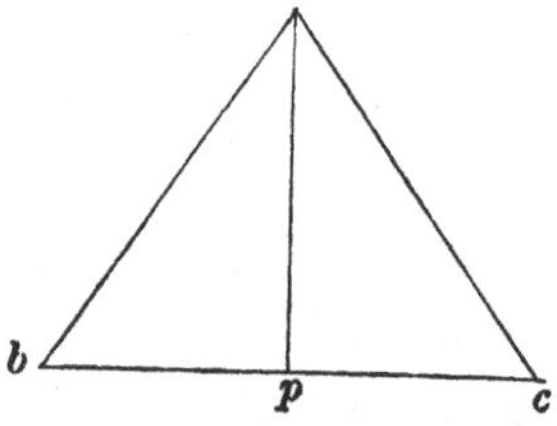

FIG. 1.

These formulæ may be extended to the general case of subregions of $\rho-1$ dimensions by exactly the same reasoning as that used for the analogous theorem of Elliptic Geometry in § 226 (2).

255. SHORTEST DISTANCES BETWEEN SUBREGIONS. (1) Let P_ρ and Q_σ be two non-intersecting sub-regions of the ρth and σth orders respectively, so that $\rho+\sigma<\nu$. A series of propositions concerning lines of maximum and minimum distance between P_ρ and Q_σ can be proved analogous to those for Elliptic Geometry in § 227. Let it be assumed throughout this article that

$\rho > \sigma$. Four different cases arise according as P_ρ and Q_σ are respectively spatial or anti-spatial. We will only consider here the single case in which P_ρ and Q_σ are both spatial.

(2) It can be proved, as in § 227 (1), that a line (spatial or anti-spatial) of maximum or minimum distance (hyperbolic or angular) between them is perpendicular to both.

(3) The polar regions $|P_\rho$ and $|Q_\sigma$ are both anti-spatial, and of the $(\nu-\rho)$th and $(\nu-\sigma)$th orders respectively. In general the region $|Q_\sigma$ intersects P_ρ in an anti-spatial subregion of the $(\rho-\sigma)$th order at least. The regions $|P_\rho$ and Q_σ do not in general intersect.

(4) Let $q_1, q_2, \dots q_\sigma$ be σ independent points in Q_σ. Then any point x in Q_σ can be written $\Sigma\xi q$.

Also write

$$\lambda^2 = \frac{\{(\Sigma\xi q)\, P_\rho \,|(\Sigma\xi q)\, P_\rho\}}{\{(\Sigma\xi q)\,|(\Sigma\xi q)\}\,(P_\rho\,|P_\rho)}.$$

If x be spatial, and xp be the perpendicular to P from x, then [cf. § 252 (1)] p is spatial. Hence by § 254 (5), Case I, $\sinh^2 \frac{\overline{xp}}{\gamma} = -\lambda^2$. If xp be anti-spatial, so that the elliptic measure of distance holds, $\sin^2 \angle xp = \lambda^2$.

Hence in all cases of lines of maximum or minimum length between P_ρ and Q_σ, σ conditions of the type, $\frac{\partial\lambda}{\partial\xi} = 0$, hold; where $\xi_1, \xi_2, \dots \xi_\sigma$ are successively put for ξ.

Thus by the same reasoning as in § 227 (3) a determinantal equation of the 6th degree is found for λ^2 of the form

$$\begin{vmatrix} \alpha_{11} - \lambda^2 (q_1\,|q_1), & \alpha_{12} - \lambda^2 (q_1\,|q_2), & \dots, & \alpha_{1\sigma} - \lambda^2 (q_1\,|q_\sigma), \\ \alpha_{21} - \lambda^2 (q_2\,|q_1), & \alpha_{22} - \lambda^2 (q_2\,|q_2), & \dots, & \alpha_{2\sigma} - \lambda^2 (q_2\,|q_2), \\ \dots & \dots & \dots & \dots \\ \dots & \dots & \dots & \dots \end{vmatrix} = 0;$$

where $$\alpha_{11} = \frac{(q_1 P_\rho\,|q_1 P_\rho)}{(P_\rho\,|P_\rho}, \quad \alpha_{12} = \alpha_{21} = \frac{(q_1 P_\rho\,|q_2 P_\rho)}{(P_\rho\,|P_\rho)},$$

with similar equations defining the other α's.

(6) Hence there are in general σ common perpendiculars to the two subregions P_ρ and Q_σ, $(\rho > \sigma)$.

If P_ρ and Q_σ had been interchanged in the above reasoning, so that x is a point in P_ρ, and

$$\lambda^2 = \frac{(xQ_\sigma\,|xQ_\sigma)}{(x\,|x)\,(Q_\sigma\,|Q_\sigma)},$$

then an equation of the ρth degree for λ^2 would have been found. But by the formula (i) of § 226 (4)

$$(xQ_\sigma|xQ_\sigma) + (x|Q_\sigma|.x|Q_\sigma) = (x|x)(Q_\sigma|Q_\sigma) \quad\text{..............(i).}$$

Now by subsection (3) above x may be supposed to lie in the region $P_\rho|Q_\sigma$, which is a subregion of P_ρ. Thus for all points x in this subregion, $x|Q_\sigma = 0$; and equation (i) becomes

$$(xQ_\sigma|xQ_\sigma) = (x|x)(Q_\sigma|Q_\sigma) \quad\text{.........................(ii).}$$

Now differentiate $\xi_1, \xi_2, \ldots \xi_\rho$ with respect to any variable θ, and put x' for $\Sigma \frac{d\xi}{d\theta} p$. Then equation (i) becomes, after differentiation,

$$(x'Q_\sigma|xQ_\sigma) + (x'|Q_\sigma|.x|Q_\sigma) = (x'|x)(Q_\sigma|Q_\sigma).$$

But
$$(x'|Q_\sigma|.x|Q_\sigma) = 0.$$

Hence
$$(x'Q_\sigma|xQ_\sigma) = (x'|x)(Q_\sigma|Q_\sigma) \quad\text{....................(iii);}$$

which holds for any point x in the subregion $P_\rho|Q_\sigma$, which has made any infinitesimal variation to the position $x + x'\delta\theta$ in the region P_ρ.

Thus differentiating λ^2, and using equations (ii) and (iii),

$$\frac{d\lambda^2}{d\theta} = 2\frac{(x'Q_\sigma|xQ_\sigma)(x|x) - (x'|x)(xQ_\sigma|xQ_\sigma)}{(x|x)^2(Q_\sigma|Q_\sigma)} = 0.$$

Thus the infinite number of lines drawn from any point in $P_\rho|Q_\sigma$ to Q_σ, which are not necessarily perpendicular to P_ρ, fulfil the conditions from which the equation of the ρth degree is derived. The analysis of this subsection could have been used for the corresponding subsection in Elliptic Geometry [cf. 227 (4)].

(7) It follows by the method of § 227 (5) that the σ feet in Q_σ of these σ perpendiculars are the one common set of σ polar reciprocal points with respect to the sections by Q_σ of the two quadrics $(x|x) = 0$, and $(xP_\rho|xP_\rho) = 0$.

(8) It follows by the method of § 227 (5) that the σ common perpendiculars all intersect $|P_\rho$; and that the σ points of intersection with $|P_\rho$ are mutually normal.

(9) It follows by the method of § 227 (7) that the σ lines of the perpendiculars are mutually normal; and that therefore they intersect P_ρ in σ mutually normal points, which define a subregion P_σ of the σth order.

(10) Also these theorems can be proved by the method of § 227 (8).

(11) One, and only one, of the σ perpendiculars is spatial. For consider a spatial point p in P_ρ and a spatial point q in Q_σ. Then the distance $\overline{pq}$ is real and finite; it varies continuously as p and q vary their positions continuously on P_ρ and Q_σ; and it approaches infinity as a limit, when p or q or both approach the absolute.

Hence there must be at least one position of pq, for which $\overline{pq}$ has a minimum value. Thus there is at least one spatial common perpendicular to P_ρ and Q_σ.

Let F be a force on the line of this perpendicular: then by (9) the remaining $\sigma - 1$ perpendiculars must lie in $|F$. Now $|F$ is anti-spatial [cf. § 239 (4)].

Hence the other $(\sigma - 1)$ perpendiculars are anti-spatial lines, and their lengths must be measured in angular measure.

256. RECTANGULAR RECTILINEAR FIGURES*. (1) Let attention be confined to rectilinear figures lying in a two-dimensional subregion. Then the straight lines of the figures have the properties of planes in this containing region. Let the two-dimensional region cut the absolute in a real section. Let all the rectilinear figures have all their corners spatial, unless otherwise stated.

(2) A rectangular quadrilateral (a rectangle) cannot exist. For in such a figure two opposite sides would have two common perpendiculars, contrary to § 253 (3).

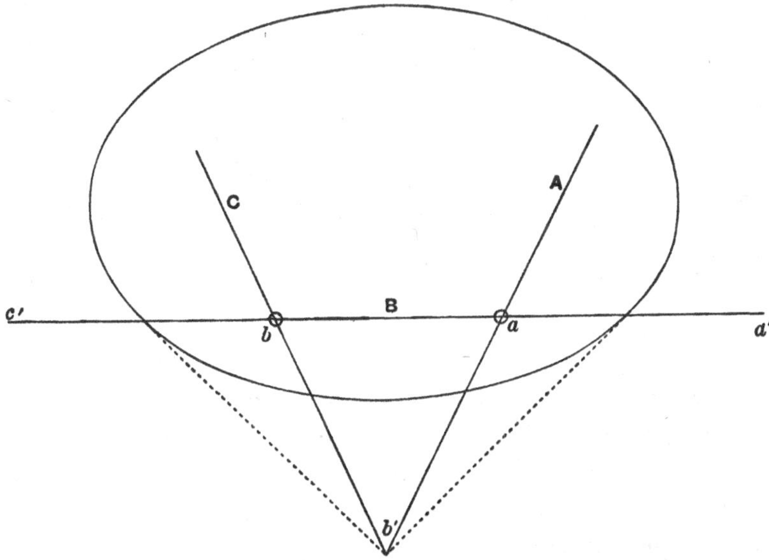

FIG. 2.

(3) Two alternate sides of any rectangular figure intersect on the (anti-spatial) pole of the included side.

Thus [see fig. 2] let A, B, and C be three consecutive sides of a rectangular figure, so that the angles at the intersections of A and B, and of B and C, are right-angles. Let the closed conic in the figure be the section of the absolute.

* These results have not been given before, as far as I am aware.

Then [cf. § 251 (3)] A and C must intersect in b', the pole of B with respect to the absolute.

Hence, corresponding to the rectangular spatial figure formed by the lines A, B, C, ..., there is the figure of which the anti-spatial corners a', b', c', ..., are the poles of the lines of the original figure. Let this be called the reciprocal figure.

Then in the reciprocal figure each corner, such as b', is normal to the two adjacent corners, such as a' and c': so that $(b' \mid a') = 0 = (b' \mid c')$.

(4) Let the point of intersection of A and B be a, and the point of intersection of B and C be b. Then $\overline{ab}$ is the side of the given figure corresponding to b'. Also by § 246 (3), $\overline{a'c'} = \overline{ab}$.

(5) A rectangular pentagon can be described as follows [cf. fig. 3]: take any two mutually normal anti-spatial points, c' and d'; and let a' be any third

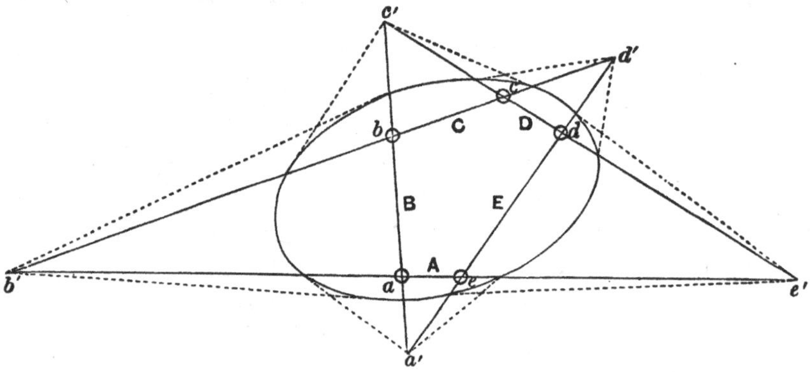

FIG. 3.

anti-spatial point, such that $a'c'$ and $a'd'$ are spatial. Let the line A be the polar of a', C of c', D of d'; let B be the line $a'c'$ and E the line $a'd'$. Then the lines A, B, C, D, E, taken in this order, form a rectangular pentagon with its corners spatial. Let A and C intersect in b', and A and D in e': then $a'b'c'd'e'$ is the reciprocal pentagon.

(6) The following formula holds for the rectangular pentagon, giving the length of any side $\overline{ae}$ in terms of the two adjacent sides $\overline{ab}$ and $\overline{de}$:

$$\cosh \frac{\overline{ae}}{\gamma} = \coth \frac{\overline{ab}}{\gamma} \coth \frac{\overline{de}}{\gamma} \quad \text{.........................(i).}$$

In order to prove this formula, assume that the two-dimensional region is the complete region with respect to which supplements are taken; so that we may write

$$b' = |a'c', \quad e' = |a'd'.$$

Then, by subsection (4),

$$\cosh^2 \frac{\overline{ae}}{\gamma} = \cosh^2 \frac{\overline{b'e'}}{\gamma} = \frac{(b' \mid e')^2}{(b' \mid b')(e' \mid e')}$$

$$= \frac{(a'c' \mid a'd')^2}{(a'c' \mid a'c')\,(a'd' \mid a'd')} = \frac{(a' \mid c')^2\,(a' \mid d')^2}{(a'c' \mid a'c')\,(a'd' \mid a'd')};$$

since $(c' \mid d') = 0$.

But $$\coth^2 \frac{\overline{ab}}{\gamma} = \coth^2 \frac{\overline{a'c'}}{\gamma} = \frac{(a' \mid c')^2}{-(a'c' \mid a'c')},$$

and $$\coth^2 \frac{\overline{de}}{\gamma} = \coth^2 \frac{\overline{a'd'}}{\gamma} = \frac{(a' \mid d')^2}{-(a'd' \mid a'd')}.$$

Hence $$\cosh^2 \frac{\overline{ae}}{\gamma} = \coth^2 \frac{\overline{ab}}{\gamma} \coth^2 \frac{\overline{de}}{\gamma}.$$

(7) The reciprocal figure of a rectangular hexagon can be decomposed into a pair of triangles conjugate with respect to the absolute. For, in figure 4, let the conic be the absolute: take any three anti-spatial points a', c', e', so that the three sides of the triangle $a'c'e'$ are spatial. Let A, C, E be the polars of these points; and let B be the line $a'c'$, D the line $c'e'$, F the line $e'a'$.

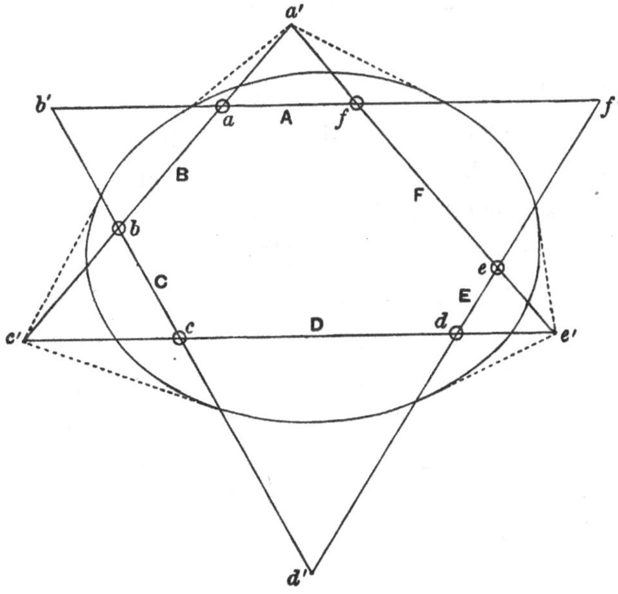

Fig. 4.

Then A, B, C, D, E, F, taken in this order, form a rectangular hexagon. Let A and B intersect in a', B and C in b, and so on. Thus $abcdef$ is the rectangular hexagon, and $a'b'c'd'e'f'$ is its reciprocal figure.

(8) The formulæ connecting the sides of a rectangular hexagon are simply the formulæ of § 248 (5) for a semi-spatial triangle.

For consider the semi-spatial triangle $a'c'e'$: let $\boldsymbol{\alpha}, \boldsymbol{\gamma}, \boldsymbol{\epsilon}$ be the measures of the separation between its sides.

Then [cf. § 248 (5)],

$$\cosh\frac{\overline{c'e'}}{\gamma} = \cosh\frac{\overline{a'c'}}{\gamma}\cosh\frac{\overline{a'e'}}{\gamma} - \sinh\frac{\overline{a'c'}}{\gamma}\sinh\frac{\overline{a'e'}}{\gamma}\cosh\frac{\boldsymbol{\alpha}}{\gamma};$$

and

$$\frac{\sinh\dfrac{\boldsymbol{\alpha}}{\gamma}}{\sinh\dfrac{\overline{c'e'}}{\gamma}} = \frac{\sinh\dfrac{\boldsymbol{\gamma}}{\gamma}}{\sinh\dfrac{\overline{a'e'}}{\gamma}} = \frac{\sinh\dfrac{\boldsymbol{\epsilon}}{\gamma}}{\sinh\dfrac{\overline{a'c'}}{\gamma}}.$$

But, by subsection (4), $\overline{cd} = \overline{c'e'}$, $\overline{ef} = \overline{e'a'}$, $\overline{ab} = \overline{a'c'}$; and by § 253 (3),

$$\boldsymbol{\alpha} = \overline{af},\ \boldsymbol{\gamma} = \overline{bc},\ \boldsymbol{\epsilon} = \overline{de}.$$

Hence the formulæ connecting the sides of the hexagon are

$$\cosh\frac{\overline{cd}}{\gamma} = \cosh\frac{\overline{ab}}{\gamma}\cosh\frac{\overline{ef}}{\gamma} - \sinh\frac{\overline{ab}}{\gamma}\sinh\frac{\overline{ef}}{\gamma}\cosh\frac{\overline{af}}{\gamma} \text{ ....(ii);}$$

and

$$\frac{\sinh\dfrac{\overline{ab}}{\gamma}}{\sinh\dfrac{\overline{de}}{\gamma}} = \frac{\sinh\dfrac{\overline{cd}}{\gamma}}{\sinh\dfrac{\overline{fa}}{\gamma}} = \frac{\sinh\dfrac{\overline{ef}}{\gamma}}{\sinh\dfrac{\overline{bc}}{\gamma}} \text{.......................(iii).}$$

257. Parallel Lines. (1) Two spatial straight lines in a subregion of two dimensions may intersect spatially, or non-spatially, or on the absolute.

In the first case let them be called secant*, in the second case non-secant, in the third case parallel. These parallel lines are not the analogues of parallel lines in Elliptic Geometry, cf. § 234.

(2) If the straight lines be secant, then by starting from a spatial point on either line the point of intersection can be reached after traversing a finite distance. This case is illustrated in figure 5.

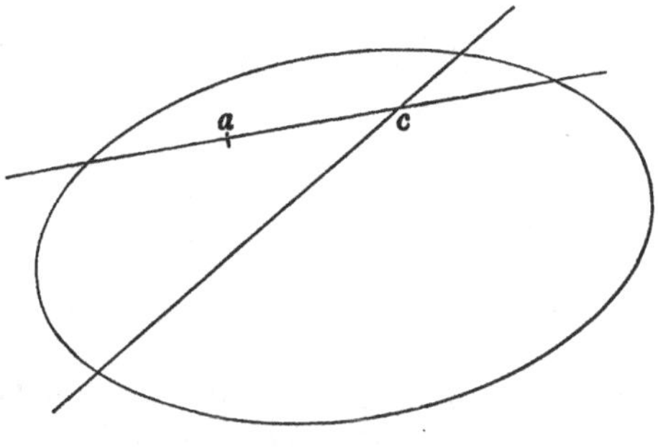

Fig. 5.

* Cf. Lobatschewsky and J. Bolyai (*loc. cit.*).

If the straight lines be non-secant, then the point of intersection has neither a real linear nor a real angular distance from a spatial point on either of the lines. This case is illustrated in figure 6.

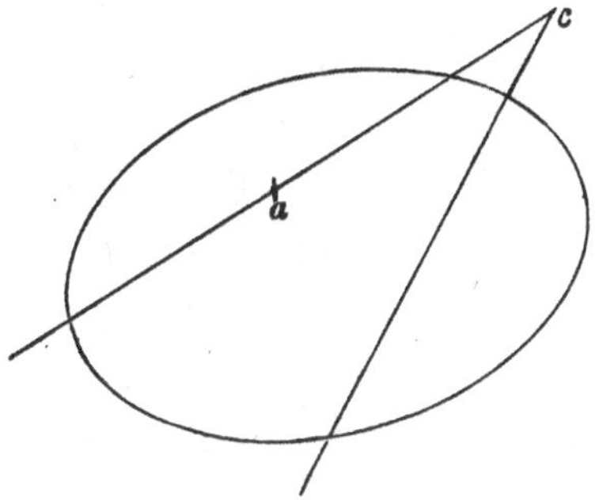

FIG. 6.

If the straight lines be parallel then the point of intersection is at an

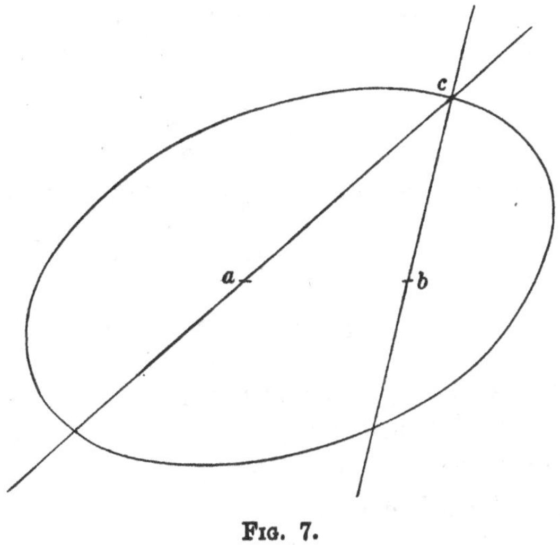

FIG. 7.

infinite distance from any spatial point on either of the lines. This case is illustrated in figure 7.

(3) Let the two straight lines ac, bc intersect at a point c on the absolute.

Then $$(ac \mid bc) = (a \mid b)(c \mid c) - (a \mid c)(b \mid c) = -(a \mid c)(b \mid c).$$

Hence if θ be the acute angle between ac and bc,

$$\cos\theta = \frac{-(ac \mid bc)}{\sqrt{\{(ac \mid ac)(bc \mid bc)\}}} = \frac{(a \mid c)(b \mid c)}{\sqrt{\{(a \mid c)^2 (b \mid c)^2\}}} = 1.$$

Hence $$\theta = 0.$$

Therefore the angle, which two parallel lines make with each other, is zero.

(4) Any spatial straight line meets the absolute in two points a_1 and a_2. Hence through any spatial point p two straight lines can be drawn in the plane parallel to the given straight line, namely the line pa_1 and the line pa_2.

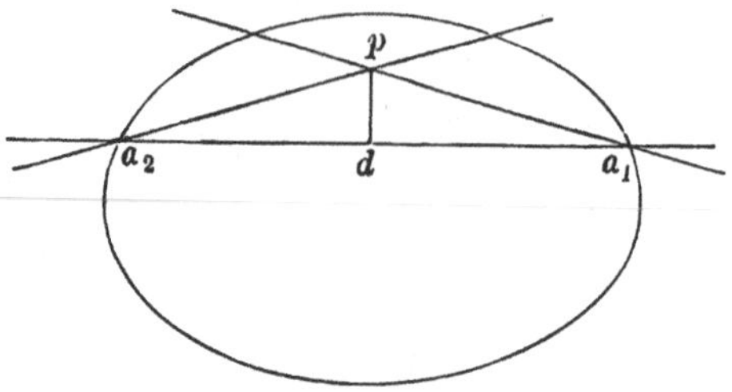

FIG. 8.

From p draw the perpendicular pd on to the line a_1a_2. The length between p and d is $\overline{pd}$. Let the angle between dp and pa_1 or pa_2 be called the 'angle of parallelism;' there is only one angle of parallelism, since it follows from the subsequent analysis that these angles are equal. Then the angle of parallelism is a function of $\overline{pd}$ only. For in the right-angled triangle pda_1, we have

$$\cos\angle a_1 = \sin\angle p \cosh\frac{\overline{pd}}{\gamma}.$$

But
$$\cos\angle a_1 = 1.$$

Hence
$$\sin\angle p = \operatorname{sech}\frac{\overline{pd}}{\gamma}.$$

This relation can also be written in either of the forms,

$$\cot\angle p = \sinh\frac{\overline{pd}}{\gamma},$$

and
$$\cot\frac{\angle p}{2} = e^{\overline{pd}/\gamma},$$

where e is here the base of Napierian logarithms.

Let the angle of parallelism corresponding to a perpendicular distance, ζ, from the given straight line be denoted by $\Pi(\zeta)$*.

Then the formula above becomes

$$\cot\frac{\Pi(\zeta)}{2} = e^{\frac{\zeta}{\gamma}}.$$

It is to be noticed that when $\zeta = 0$, $\Pi(\zeta) = \frac{1}{2}\pi$; as ζ increases, $\Pi(\zeta)$ diminishes; and when ζ is infinite, $\Pi(\zeta)$ is zero.

(5) It is possible to draw a straight line parallel to two secant straight lines. For let the two straight lines intersect at an angle α; draw cp bisecting the angle α; and produce it to p so that

$$\overline{cp} = \gamma \log\cot\frac{\alpha}{2}.$$

The perpendicular to cp through p is parallel to both the secant lines.

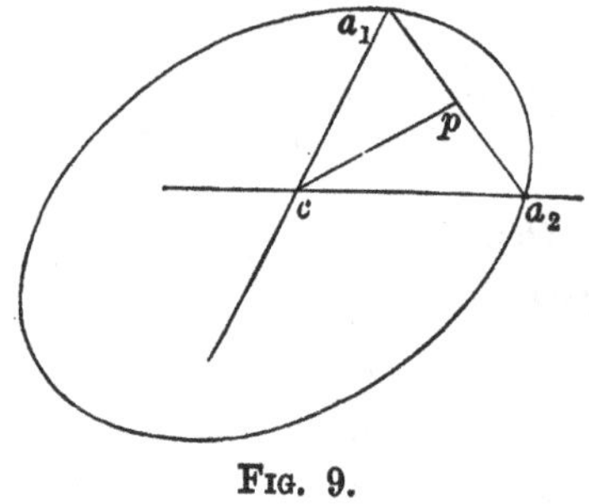

FIG. 9.

* Cf. Lobatschewsky, *loc. cit.*

258. Parallel Planes. (1) Let the complete region be of three dimensions, then the planes are ordinary two-dimensional planes, and the subplanes are lines.

Let two planes L and L' intersect in a spatial line LL', and let this line cut the absolute in the two points a_1 and a_2. Then through any two points p and p' in the planes L and L' respectively two pairs of parallel lines can be drawn, namely pa_1, $p'a_1$, and pa_2, $p'a_2$. Thus all the lines through a_1 in the two planes L and L' form one series of parallel lines distributed between the two planes; and all the lines through a_2 form another series. And both series are parallel to the line of intersection of the planes.

(2) If the line LL' touches the quadric, the points a_1 and a_2 coincide, and the two series of parallel lines coincide. The planes may then be called parallel.

The condition that LL' may touch the absolute quadric is

$$(LL' \mid LL') = 0.$$

Hence $\sin \angle LL' = 0$, and $\angle LL' = 0$. Therefore parallel planes are inclined to each other at a zero angle.

(3) The planes through any point p which are parallel to a given plane L envelope a cone, which has p for vertex and the section of the absolute by L for its section in the plane L.

Let pd be drawn perpendicular to L, and let ab be any tangent line to the absolute lying in the plane L and touching it at a.

Then the plane pab is one of the parallel planes through p. Let L' be

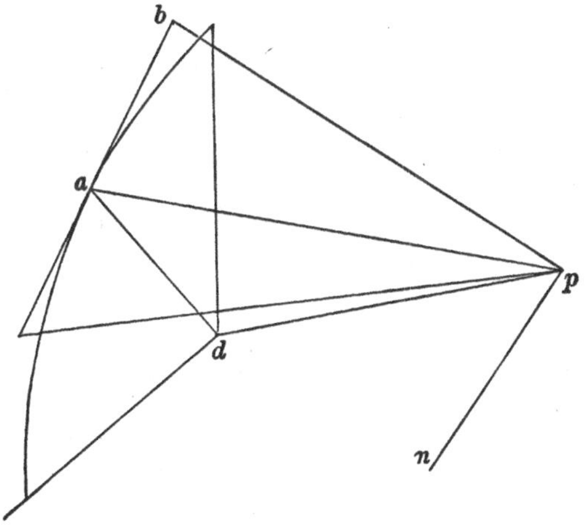

Fig. 10.

written for the plane pab. Now draw pn perpendicular to L'. Then pa, pd and pn are co-planar. For pd and pn pass through $|L$ and $|L'$. But the line $|L|L'$ is the normal subregion to the line ab. Hence $|LL'$ passes through a, since *ab* touches the absolute quadric. Hence the three lines *pn, pd, pa* are co-planar.

Now the two lines *da* and *pa* are parallel. Hence the angle $\angle apd$ is $\Pi(\overline{pd})$. But the angle $\angle apn = \frac{1}{2}\pi$. Hence $\angle dpn = \frac{1}{2}\pi - \Pi(\overline{pd})$. Thus if through any point *p*, distance ζ from any plane *L*, all the parallel planes to *L* are drawn, the normals to these planes form a cone of which all the generators make an angle $\frac{1}{2}\pi - \Pi(\zeta)$ with the perpendicular from *p* to *L*.

CHAPTER V.

Hyperbolic Geometry (*continued*).

259. The Sphere. (1) The equation of a sphere of radius ρ and of spatial centre b is

$$(x|x)(b|b)\cosh^2\frac{\rho}{\gamma} = (b|x)^2.$$

Every point of this sphere is spatial. For $(b|x)^2$ is positive and $(b|b)$ is positive; hence $(x|x)$ is positive [cf. § 240 (1)].

(2) The equation

$$\epsilon^2(x|x)(b|b) = (b|x)^2$$

represents an anti-spatial locus, when b is anti-spatial. For then $(b|b)$ is negative, and hence $(x|x)$ is negative [cf. § 240 (3)].

(3) The equation

$$-\epsilon^2(x|x)(b|b) = (b|x)^2$$

represents an anti-spatial locus, when b is spatial. For then $(b|b)$ is positive, and hence $(x|x)$ is negative.

But the equation represents a purely spatial locus, when b is anti-spatial. For then $(b|b)$ is negative, and hence $(x|x)$ is positive. Let ϵ be written $\sinh\frac{\sigma}{\gamma}$; then [cf. 254 (2)] σ is the distance of the spatial point x from the spatial plane $|b$.

(4) Thus, if the equation

$$\eta(x|x)(b|b) = (b|x)^2$$

be considered as the general form of equation of a sphere, there are two types of real spatial spheres; namely the type with spatial centre b, of which the equation is

$$(x|x)(b|b)\cosh^2\frac{\rho}{\gamma} = (b|x)^2;$$

and the type with anti-spatial centre b, of which the equation is

$$-(x|x)(b|b)\sinh^2\frac{\sigma}{\gamma} = (b|x)^2.$$

(5) This latter surface is the locus of points at a given distance σ from the spatial plane $B (= |b)$; and the equation can be written [cf. § 254 (2)].

$$-(x|x)(B|B)\sinh^2\frac{\sigma}{\gamma} = (xB)^2.$$

Let the spheres of this second type be also called Surfaces of Equal Distance; and let the plane B be called the Central Plane.

(6) The spatial sphere is a closed surface. For firstly, let the centre be spatial, and let it be taken as the origin, e, of a normal system of reference points with the notation of § 240.

Then $$(x|x) = \frac{\xi^2}{\alpha^2} - \frac{\xi_1^2}{\alpha_1^2} - \dots - \frac{\xi^2_{\nu-1}}{\alpha^2_{\nu-1}}, \quad (e|x) = \xi(e|e) = \frac{\xi}{\alpha^2}.$$

Hence the equation of the sphere is,

$$\frac{\xi^2}{\alpha^2} - \frac{\xi_1^2}{\alpha_1^2} - \dots - \frac{\xi^2_{\nu-1}}{\alpha^2_{\nu-1}} = \operatorname{sech}^2\frac{\rho}{\gamma}\frac{\xi^2}{\alpha^2};$$

that is, $$\tanh^2\frac{\rho}{\gamma}\frac{\xi^2}{\alpha^2} - \frac{\xi_1^2}{\alpha_1^2} - \dots - \frac{\xi^2_{\nu-1}}{\alpha^2_{\nu-1}} = 0.$$

Therefore the sphere is a closed surface [cf. § 82 (5)].

(7) Secondly, let the centre be the anti-spatial point, e_1, of this normal system of reference points.

Then $$(e_1|x) = \xi_1(e_1|e_1) = -\frac{\xi_1}{\alpha_1^2};$$

and $$-(e_1|e_1)\sinh^2\frac{\sigma}{\gamma} = \frac{\sinh^2\frac{\sigma}{\gamma}}{\alpha_1^2}.$$

Hence $$\frac{\xi^2}{\alpha^2} - \frac{\xi_1^2}{\alpha_1^2} - \dots - \frac{\xi^2_{\nu-1}}{\alpha^2_{\nu-1}} = \operatorname{cosech}^2\frac{\sigma}{\gamma}\frac{\xi_1^2}{\alpha_1^2}.$$

Therefore the equation of the sphere becomes

$$\frac{\xi^2}{\alpha^2} - \frac{\xi_1^2\coth^2\frac{\sigma}{\gamma}}{\alpha_1^2} - \frac{\xi_2^2}{\alpha_2^2} - \dots - \frac{\xi^2_{\nu-1}}{\alpha^2_{\nu-1}} = 0.$$

This is the equation of a closed surface [cf. § 82 (5)].

(8) It is to be noticed that this last closed surface touches the absolute along the real locus of $\nu - 3$ dimensions given by the equations

$$\xi_1 = 0, \quad \frac{\xi^2}{\alpha^2} - \frac{\xi_2^2}{\alpha_2^2} - \dots - \frac{\xi^2_{\nu-1}}{\alpha^2_{\nu-1}} = 0.$$

(9) If the centre of a sphere be spatial, it lies within the surface.

For let b be the centre of

$$(x|x)(b|b)\cosh^2\frac{\rho}{\gamma} - (b|x)^2 = 0.$$

Then if x be any point, the point $b + \lambda x$ lies on the surface, when

$$(b|b)^2 \sinh^2\frac{\rho}{\gamma} + 2\lambda (b|x)(b|b)\sinh^2\frac{\rho}{\gamma} + \lambda^2 \{(x|x)(b|b)\cosh^2\frac{\rho}{\gamma} - (b|x)^2\} = 0.$$

The roots of this quadratic for λ are real, if

$$(b|x)^2 (b|b)^2 \sinh^2\frac{\rho}{\gamma} - (b|b)^2 \{(x|x)(b|b)\cosh^2\frac{\rho}{\gamma} - (b|x)^2\} \text{ is positive;}$$

that is, if $$(b|x)^2 \cosh^2\frac{\rho}{\gamma} - (x|x)(b|b)\cosh^2\frac{\rho}{\gamma} \text{ is positive;}$$

that is, if $$-(bx|bx) \text{ is positive.}$$

But [cf. § 243 (2)] since the line bx is spatial, this condition is fulfilled. Hence [cf. § 82 (1)] b lies within the sphere.

Also if b be substituted in the expression

$$(x|x)(b|b)\cosh^2\frac{\rho}{\gamma} - (b|x)^2,$$

there results $(b|b)^2 \sinh^2\frac{\rho}{\gamma}$, which is positive.

Hence [cf. § 82 (9)] any point x, which makes this expression positive, lies within the surface.

(10) But if the centre b be anti-spatial, then the equation of the sphere is

$$-(x|x)(b|b)\sinh^2\frac{\sigma}{\gamma} - (b|x)^2 = 0.$$

Then if x be any point, the point $b + \lambda x$ lies on the surface, when

$$-(b|b)^2 \cosh^2\frac{\sigma}{\gamma} - 2(x|b)(b|b)\cosh^2\frac{\sigma}{\gamma} + \lambda^2 \{-(x|x)(b|b)\sinh^2\frac{\sigma}{\gamma} - (b|x)^2\} = 0.$$

The roots of this equation for λ are real, if

$$(x|b)^2 (b|b)^2 \cosh^2\frac{\sigma}{\gamma} - (b|b)^2 \{(x|x)(b|b)\sinh^2\frac{\sigma}{\gamma} + (b|x)^2\} \text{ is positive;}$$

that is, if $$(x|b)^2 - (x|x)(b|b) \text{ is positive;}$$

that is, if $$-(xb|xb) \text{ is positive.}$$

But xb may or may not be spatial; and therefore $(xb|xb)$ may be positive or negative.

Hence the anti-spatial centre of a sphere lies without the surface. But it is interesting to note that any spatial line, drawn through the (anti-spatial) centre, cuts the sphere in real points. Also substituting b in the expression

$$-(x|x)(b|b)\sinh^2\frac{\sigma}{\gamma} - (b|x)^2,$$

we obtain $-(b|b)^2 \cosh^2\frac{\sigma}{\gamma}$, which is negative. Hence any point x, which makes this expression positive, lies within the surface [cf. § 82 (9)]. Thus any spatial point on the central plane $|b$ lies within the surface. For, if x be such a point, $(x|b)=0$, and $-(x|x)(b|b)\sinh^2\frac{\sigma}{\gamma}$ is positive.

(11) It can be proved, exactly as in the case of Elliptic Geometry [cf. § 228 (3)], that the line, perpendicular to any plane and passing through its pole with respect to a sphere, passes through the centre of the sphere.

Hence it follows as a corollary, that the perpendicular to a tangent plane of a sphere through its point of contact passes through the centre of the sphere.

260. Intersection of Spheres. (1) The locus of the intersection of two spheres

$$\beta_1^2(x|x)=(b|x)^2,$$
$$\beta_2^2(x|x)=(c|x)^2,$$

lies on the two planes

$$\frac{(b|x)}{\beta_1}=\pm\frac{(c|x)}{\beta_2}.$$

These planes are respectively the absolute polar planes of the points

$$\frac{b}{\beta_1}\mp\frac{c}{\beta_2}.$$

But if b and c are both spatial and of standard sign, the point $\frac{b}{\beta_1}+\frac{c}{\beta_2}$ is spatial, and its polar plane is anti-spatial. This plane can only meet the spheres, which are entirely spatial, in imaginary points. If b and c are anti-spatial, one of the two points $\frac{b}{\beta_1}\mp\frac{c}{\beta_2}$ must be anti-spatial, and hence its polar plane spatial. The other point may or may not be anti-spatial.

These radical planes are perpendicular to the line bc, since bc passes through their poles.

(2) It can be proved, as in the Elliptic Geometry [cf. § 228 (4)] that the lengths of all tangent lines from any given point to a sphere with a spatial centre are equal. Also if ρ be the radius, and b the centre of the sphere, and τ the length of the tangent line from x, then

$$\cosh\frac{\tau}{\gamma}=\frac{(x|b)}{\cosh\frac{\rho}{\gamma}\sqrt{\{(x|x)(b|b)\}}}.$$

Also, by an easy modification of the proof and by reference to § 246 (2), we find when the centre b is anti-spatial, and the distance from the plane $|b$ is σ,

$$\cosh\frac{\tau}{\gamma}=\frac{(x|b)}{\sinh\frac{\sigma}{\gamma}\sqrt{\{-(x|x)(b|b)\}}}.$$

(3) The locus of points, from which equal tangents are drawn to two spheres with spatial centres b and c, and with radii ρ_1 and ρ_2, is given by

$$\frac{(x|b)}{\cosh\frac{\rho_1}{\gamma}\sqrt{(b|b)}} = \frac{(x|c)}{\cosh\frac{\rho_2}{\gamma}\sqrt{(c|c)}}.$$

This locus is the spatial radical plane.

The cases when b or c, or both, are anti-spatial can easily be discussed.

(4) The theorems of § 228 (5) also hold, with necessary alterations.

(5) The angles of intersection of two spheres can be investigated by the method of § 228 (8). Let the two spheres be

$$\epsilon\,(b|b)\,(x|x) = (b|x)^2.$$

and

$$\eta\,(c|c)\,(x|x) = (c|x)^2;$$

where ϵ stands for $\cosh^2\frac{\rho_1}{\gamma}$, if b be spatial, and for $-\sinh^2\frac{\sigma_1}{\gamma}$, if b be anti-spatial [cf. § 259 (4)]; and η stands for $\cosh^2\frac{\rho_2}{\gamma}$, or for $-\sinh^2\frac{\sigma_2}{\gamma}$, according as c is spatial, or anti-spatial.

Then it can easily be proved, as in the analogous theorem of Elliptic Geometry, that the angles of intersection, ω and ω', of the two spheres, which correspond to the two radical planes, are given by

$$\cos\omega = \frac{+\sqrt{\{\epsilon\eta\,(b|b)\,(c|c)\}} - (b|c)}{\sqrt{\{(\epsilon-1)(\eta-1)(b|b)(c|c)\}}},$$

and

$$\cos\omega' = \frac{-\sqrt{\{\epsilon\eta\,(b|b)\,(c|c)\}} - (b|c)}{\sqrt{\{(\epsilon-1)(\eta-1)(b|b)(c|c)\}}}.$$

Also let it be assumed that in all cases $(b|c)$ is positive. Four separate cases now arise.

(6) Firstly, let b and c be both spatial.

Then

$$\cos\omega = \frac{\cosh\frac{\rho_1}{\gamma}\cosh\frac{\rho_2}{\gamma} - \cosh\frac{\overline{bc}}{\gamma}}{\sinh\frac{\rho_1}{\gamma}\sinh\frac{\rho_2}{\gamma}},$$

and

$$\cos\omega' = \frac{-\cosh\frac{\rho_1}{\gamma}\cosh\frac{\rho_2}{\gamma} - \cosh\frac{\overline{bc}}{\gamma}}{\sinh\frac{\rho_1}{\gamma}\sinh\frac{\rho_2}{\gamma}}.$$

Since $\coth\frac{\rho_1}{\gamma}\coth\frac{\rho_2}{\gamma}$ is necessarily greater than unity, ω' is always imaginary.

The spheres have one real intersection, if

$$-1 < \cos\omega < 1;$$

that is, if $\rho_1 \sim \rho_2 < \overline{bc} < \rho_1 + \rho_2.$

(7) Secondly, let b and c be both anti-spatial; and let bc be anti-spatial. Let $b = |B$, and $c = |C$; so that B and C are the central planes of the spheres. Then, since bc is anti-spatial, BC is spatial.

Now
$$\cos\omega = \frac{\sinh\frac{\sigma_1}{\gamma}\sinh\frac{\sigma_2}{\gamma} - \cos\angle BC}{\cosh\frac{\sigma_1}{\gamma}\cosh\frac{\sigma_2}{\gamma}},$$

and
$$\cos\omega' = \frac{-\sinh\frac{\sigma_1}{\gamma}\sinh\frac{\sigma_2}{\gamma} - \cos\angle BC}{\cosh\frac{\sigma_1}{\gamma}\cosh\frac{\sigma_2}{\gamma}}.$$

Then ω and ω' are both necessarily real.

(8) Thirdly, let b and c be both anti-spatial; and let bc be spatial. Then BC is anti-spatial.

Then
$$\cos\omega = \frac{\sinh\frac{\sigma_1}{\gamma}\sinh\frac{\sigma_2}{\gamma} - \cosh\frac{\overline{BC}}{\gamma}}{\cosh\frac{\sigma_1}{\gamma}\cosh\frac{\sigma_2}{\gamma}},$$

$$\cos\omega' = \frac{-\sinh\frac{\sigma_1}{\gamma}\sinh\frac{\sigma_2}{\gamma} - \cosh\frac{\overline{BC}}{\gamma}}{\cosh\frac{\sigma_1}{\gamma}\cosh\frac{\sigma_2}{\gamma}}.$$

The angle ω is real, if

$$\overline{BC} < \sigma_1 + \sigma_2.$$

The angle ω' is real, if

$$\overline{BC} < \sigma_1 \sim \sigma_2.$$

The first condition secures a real intersection on one radical plane; the second condition secures a real intersection on both radical planes.

(9) Fourthly, let b be spatial, and c be anti-spatial. Let δ be the distance from b to the central plane C.

Then [cf. § 254 (2)],

$$\sinh\frac{\delta}{\gamma} = \pm\frac{(bC)}{\sqrt{\{-(b|b)(C|C)\}}}.$$

Hence,
$$\cos\omega = \frac{\cosh\frac{\rho_1}{\gamma}\sinh\frac{\sigma_2}{\gamma} - \sinh\frac{\delta}{\gamma}}{\sinh\frac{\rho_1}{\gamma}\cosh\frac{\sigma_2}{\gamma}};$$

and
$$\cos \omega' = \frac{-\cosh \frac{\rho_1}{\gamma} \sinh \frac{\sigma_2}{\gamma} - \sinh \frac{\delta}{\gamma}}{\sinh \frac{\rho_1}{\gamma} \cosh \frac{\sigma_2}{\gamma}}.$$

Then ω is real, if $\sigma_2 - \rho_1 < \delta < \sigma_2 + \rho_1$; and ω' is real, if $\delta < \rho_1 - \sigma_2$.

This condition for ω' includes the conditions for ω, since δ has been assumed to be positive.

(10) Now a spatial plane is a particular case of a sphere with an anti-spatial centre c, when $\sigma = 0$; the plane is then $|c$.

Hence from subsection (9) the plane L cuts the sphere, with spatial centre b, at an angle ω given by

$$\sinh \frac{\rho_1}{\gamma} \cos \omega = \frac{-(bL)}{\sqrt{\{-(b|b)(L|L)\}}}.$$

And from subsection (8), the plane L cuts the sphere, with anti-spatial centre b, at an angle ω given by

$$\cosh \frac{\sigma_1}{\gamma} \cos \omega = \frac{-(bL)}{\sqrt{\{(b|b)(L|L)\}}}.$$

Hence putting $\omega = 0$, the plane-equation of a sphere is

$$-(b|b)(L|L)\sinh^2 \frac{\rho_1}{\gamma} = (bL)^2,$$

when the centre is spatial; and is

$$(b|b)(L|L)\cosh^2 \frac{\sigma_1}{\gamma} = (bL)^2,$$

when the centre is anti-spatial.

261. LIMIT-SURFACES. (1) If b be on the absolute, the surface denoted by

$$\epsilon^2 (x|x) = (b|x)^2$$

is called a limit-surface. It must be conceived as a sphere of infinite radius. Since the centre is on the absolute, by § 259 (11) all the perpendiculars from points on to their polars with respect to the surface are parallel lines.

(2) Let a distance δ be measured from every point x on the above limit-surface along the normal xb, either towards or away from b. Let y be the point reached. Then x and λ can be eliminated from

$$x + \lambda b = y, \quad \cosh^2 \frac{\delta}{\gamma} = \frac{(x|y)^2}{(x|x)(y|y)}, \quad \epsilon^2 (x|x) = (b|x)^2.$$

The result, remembering that $(b|b) = 0$, is easily seen to be

$$\left\{\epsilon^2 \exp\left(\frac{2\delta}{\gamma}\right)(y|y) - (b|y)^2\right\} \left\{\epsilon^2 \exp\left(-\frac{2\delta}{\gamma}\right)(y|y) - (b|y)^2\right\} = 0.$$

The surface obtained by measuring towards b is therefore

$$\epsilon^2 \exp\left(-\frac{2\delta}{\gamma}\right)(y\,|\,y) - (b\,|\,y)^2 = 0.$$

The surface obtained by measuring from b is

$$\epsilon^2 \exp\left(\frac{2\delta}{\gamma}\right)(y\,|\,y) - (b\,|\,y)^2 = 0.$$

Both these surfaces are again limit-surfaces with b as centre.

(3) Now assume that the spatial origin e, of a normal system of unit reference points $e, e_1, \ldots e_{\nu-1}$, is on the surface.

Let ee_1 pass through the centre b. Then b is of the form $e \pm e_1$, say $e + e_1$. The equation of the surface becomes

$$\epsilon^2 (x\,|\,x) = \{x\,|\,(e + e_1)\}^2.$$

But since e is on the surface, we can put $x = e$ in this equation. Hence

$$\epsilon^2 = 1.$$

The equation now is

$$(x\,|\,x) = \{x\,|\,(e + e_1)\}^2.$$

This form, by its freedom from arbitrary constants, shows that all limit-surfaces are merely repetitions of the same surface differently placed.

262. Great Circles on Spheres. (1) Let any two-dimensional region, through the centre of a sphere and cutting the sphere in real points, be said to cut the sphere in a great circle. Accordingly a great circle is in general defined by two points on a sphere, since these two points and the centre of the sphere (if not collinear) are sufficient to define the two-dimensional region. The radius of the circle is the radius of the sphere, and the centre of the circle is the centre of the sphere. If the centre be anti-spatial, the circle is the surface of equal distance in the two-dimensional region from the line of intersection of the two-dimensional region with the polar plane of the centre. The two-dimensional region, since it contains the centre, is perpendicular to the polar plane of the centre, that is, to the central plane of the sphere.

(2) If the centre b be spatial, and two points pq on the surface define a great circle, then the length of the arc $\overline{pq}$ of the great circle [cf. § 248 (4)] is $\alpha\gamma \sinh \frac{\rho}{\gamma}$, where ρ is the radius of the circle, and α is the acute angle between pb and qb.

(3) Let the centre be anti-spatial. Consider any two points p and p' on a surface of equal distance σ from any given plane. Let the two perpendiculars from p and p' meet the given plane in q and q'. Then the length of

the arc pp', traced on the great circle joining p and p', can be found in terms of q and q'. For putting $\overline{qq'} = \delta$, it is easy to prove that

$$\cosh \frac{\overline{pp'}}{\gamma} = -\sinh^2 \frac{\sigma}{\gamma} + \cosh^2 \frac{\sigma}{\gamma} \cosh \frac{\delta}{\gamma}.$$

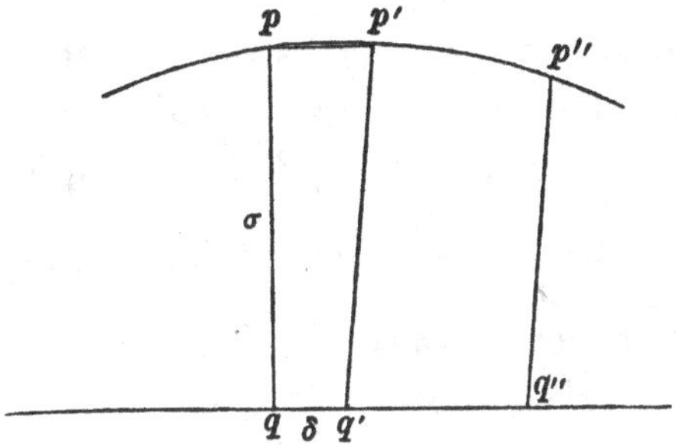

FIG. 1.

Hence when $\overline{pp'}$ and δ are small,

$$1 + \frac{1}{2}\frac{\overline{pp'}^2}{\gamma^2} = -\sinh^2 \frac{\sigma}{\gamma} + \cosh^2 \frac{\sigma}{\gamma}\left(1 + \frac{1}{2}\frac{\delta^2}{\gamma^2}\right).$$

Therefore $$\overline{pp'} = \delta \cosh \frac{\sigma}{\gamma}.$$

But ultimately, $\overline{pp'} = \text{arc } pp'$.

Therefore $$\frac{\text{arc } pp'}{\overline{qq'}} = \cosh \frac{\sigma}{\gamma}.$$

But if p'' be any point on the arc pp' prolonged to a finite distance, and if $p''q''$ be drawn perpendicular to the plane, it is obvious that

$$\frac{\text{arc } pp''}{\overline{qq''}} = \frac{\text{arc } pp'}{\overline{qq'}} = \cosh \frac{\sigma}{\gamma}.$$

(4) Let d be the centre of a spatial sphere of radius ρ, and let a, b, c be three points on the sphere. Let the acute angle between db and dc be α', that between dc and da be β', that between da and db be γ'; let the angle between the two-dimensional regions dab and dac be $\boldsymbol{\alpha}$, that between dab and dbc be $\boldsymbol{\beta}$, that between dbc and dca be $\boldsymbol{\gamma}$. Then the three two-dimensional regions can be conceived as planes in a three-dimensional region.

Hence by § 250 (2),

$$\cos \alpha' = \cos \beta' \cos \gamma' + \sin \beta' \sin \gamma' \cos \boldsymbol{\alpha}.$$

Now $\boldsymbol{\alpha}, \boldsymbol{\beta}, \boldsymbol{\gamma}$ are the angles of the curvilinear triangle formed by the great circles joining a, b, c. Also if $\widehat{bc}, \widehat{ca}, \widehat{ab}$ stand for the lengths of the arcs of great circles, by (2) of the present article,

$$\alpha' = \frac{\widehat{bc}}{\gamma} \operatorname{cosech} \frac{\rho}{\gamma}, \quad \beta' = \frac{\widehat{ca}}{\gamma} \operatorname{cosech} \frac{\rho}{\gamma}, \quad \gamma' = \frac{\widehat{ab}}{\gamma} \operatorname{cosech} \frac{\rho}{\gamma}.$$

Hence

$$\cos\frac{\widehat{bc}}{\gamma\sinh\frac{\rho}{\gamma}} = \cos\frac{\widehat{ca}}{\gamma\sinh\frac{\rho}{\gamma}}\cos\frac{\widehat{ab}}{\gamma\sinh\frac{\rho}{\gamma}} + \sin\frac{\widehat{ca}}{\gamma\sinh\frac{\rho}{\gamma}}\sin\frac{\widehat{ab}}{\gamma\sinh\frac{\rho}{\gamma}}\cos\alpha;$$

with similar equations.

Thus the relations between the lengths of the arcs, forming a triangle of great circles on a sphere of spatial centre, and the angles between them are the same as the relations between the sides and angles of a triangle in an Elliptic Space, of which the space constant is $\gamma\sinh\frac{\rho}{\gamma}$. Thus an Elliptic Space of $\nu-2$ dimensions can always be conceived as a sphere of radius ρ with spatial centre in Hyperbolic Space of $\nu-1$ dimensions, the great circles being the straight lines of the Elliptic Space, γ being the space constant of the Hyperbolic Space, and $\gamma\sinh\frac{\rho}{\gamma}$ that of the Elliptic Space.

(5) Let a sphere with anti-spatial centre be a surface of equal distance σ from a spatial plane; and let a, b, c be three points on the sphere, and a', b', c' be the feet of the perpendiculars from a, b, c on to the plane of equi-distance. Let a, b, c be joined by great circles of lengths $\widehat{bc}$, $\widehat{ca}$, $\widehat{ab}$.

Let α, β, γ be the angles of the curvilinear triangle abc; they are also the angles of the triangle $a'b'c'$, since the two-dimensional regions containing the great circles are perpendicular to the plane of equal distance.

Then [cf. § 248 (2)]

$$\cosh\frac{\overline{b'c'}}{\gamma} = \cosh\frac{\overline{c'a'}}{\gamma}\cosh\frac{\overline{a'b'}}{\gamma} - \sinh\frac{\overline{c'a'}}{\gamma}\sinh\frac{\overline{a'b'}}{\gamma}\cos\alpha.$$

But by subsection (3) of the present article, $\overline{b'c'} = \dfrac{\widehat{bc}}{\cosh\frac{\sigma}{\gamma}}$, with similar equations. Hence

$$\cosh\frac{\widehat{bc}}{\gamma\cosh\frac{\sigma}{\gamma}} = \cosh\frac{\widehat{ca}}{\gamma\cosh\frac{\sigma}{\gamma}}\cosh\frac{\widehat{ab}}{\gamma\cosh\frac{\sigma}{\gamma}} - \sinh\frac{\widehat{ca}}{\gamma\cosh\frac{\sigma}{\gamma}}\sinh\frac{\widehat{ab}}{\gamma\cosh\frac{\sigma}{\gamma}}\cos\alpha.$$

Thus the relations between the lengths of the arcs, forming a triangle of great circles on a sphere of equal distance σ from a spatial plane, and the angles between them are the same as the relations between the sides and angles of a triangle in a Hyperbolic Space, of which the space constant is $\gamma\cosh\frac{\sigma}{\gamma}$. Thus, since $\gamma\cosh\frac{\sigma}{\gamma}$ is always greater than γ, a Hyperbolic Space of $\nu-2$ dimensions can always be conceived as a spherical locus with anti-spatial centre in a Hyperbolic Space of $\nu-1$ dimensions and of smaller space constant.

(6) The relations between the sides and angles of a curvilinear triangle formed by great circles on a Limit-surface can be found either from (4) or (5) by making ρ or σ ultimately infinite. Then with the notation of (4) or (5)

$$\widehat{bc}^2 = \widehat{ca}^2 + \widehat{ab}^2 - 2\widehat{ca} . \widehat{ab} \cos \alpha.$$

Hence triangles formed by great circles on Limit-surfaces have the same geometry as triangles in ordinary Euclidean Space; for instance, the sum of the angles of any such triangle must equal two right-angles. Thus a Euclidean Space of $\nu - 2$ dimensions can be conceived as a Limit-surface in a Hyperbolic Space of $\nu - 1$ dimensions*.

263. Surfaces of Equal Distance from Subregions. (1) Let P_ρ be a spatial subregion of $\rho - 1$ dimensions, and let P_ρ be the regional element of the ρth order which represents it. Then locus of points x at the given distance δ from this subregion is by § 254 (5),

$$(x|x)(P_\rho|P_\rho) \sinh^2 \frac{\delta}{\gamma} + (xP_\rho|xP_\rho) = 0.$$

(2) Now take as reference elements ν normal points, of which the spatial origin e and $\rho - 1$ other points $e_1, e_2, \ldots e_{\rho-1}$ lie in P_ρ, and the remaining $\nu - \rho$ elements lie in $|P_\rho$. Also let e be at unit spatial intensity, and $e_1, e_2, \ldots e_{\nu-1}$ at unit anti-spatial intensity. Let $P_\rho = ee_1 \ldots e_{\rho-1}$. Then $(P_\rho|P_\rho) = (-1)^{\rho-1}$. Let

$$x = \xi e + \xi_1 e_1 + \ldots + \xi_{\rho-1} e_{\rho-1} + \eta_\rho e_\rho + \ldots + \eta_{\nu-1} e_{\nu-1}.$$

Then $$(xP_\rho|xP_\rho) = (-1)^\rho \Sigma\eta^2, \ (x|x) = \xi^2 - \xi_1^2 - \ldots - \eta^2_{\nu-1}.$$

Hence the equation of the surface of equal distance from P_ρ becomes

$$(\xi^2 - \xi_1^2 - \ldots - \xi^2_{\rho-1} - \eta_\rho^2 - \ldots - \eta^2_{\nu-1}) \sinh^2 \frac{\delta}{\gamma} = \Sigma\eta^2;$$

that is,

$$(\xi^2 - \xi_1^2 - \ldots - \xi^2_{\rho-1}) \tanh^2 \frac{\delta}{\gamma} - (\eta_\rho^2 + \ldots + \eta^2_{\nu-1}) = 0.$$

This is a closed surface with no real generating regions. Hence the parallel regions of Elliptic Space [cf. § 229] have no existence in Hyperbolic Space.

* The idea of a space of one type as a locus in a space of another type, and of dimensions higher by one, is due partly to J. Bolyai, and partly to Beltrami. Bolyai points out that the relations between lines formed by great circles on a two-dimensional limit-surface are the same as those of straight lines in a Euclidean plane of two dimensions. Beltrami proves, by the use of the pseudosphere, that a Hyperbolic space of any number of dimensions can be considered as a locus in Euclidean space of higher dimensions. There is an error, popular even among mathematicians misled by a useful technical phraseology, that Euclidean space is in a special sense flat, and that this flatness is exemplified by the possibility of an Euclidean space containing surfaces with the properties of Hyperbolic and Elliptic spaces. But the text shows that this relation of Hyperbolic to Euclidean space can be inverted. Thus no theory of the flatness of Euclidean space can be founded on it.

264. Intensities of Forces. (1) Consider an extensive manifold of three dimensions. The only regional elements are planar elements and forces. A spatial planar element X is at unit intensity when $(X|X) = -1$, and an anti-spatial planar element X is at unit intensity when $(X|X) = 1$ [cf. § 240 (3)].

(2) In order to determine the intensity of a force xy, let it be defined that the intensity of xy is some function of the distance $\overline{xy}$, or $\angle xy$ if the measure of distance be elliptic, multiplied by the product of the intensities of x and y. Then by the same reasoning as in § 230 (2) for Elliptic Space it can be proved that: (α) if x and y be both spatial, the intensity of xy is $\sqrt{\{(x|x)(y|y)\}} \sinh \frac{\overline{xy}}{\gamma}$, that is $\{-xy|xy\}^{\frac{1}{2}}$; where it is to be noticed that, by § 243 (2), $(xy|xy)$ is negative, when xy is spatial: (β) if x and y be both anti-spatial and xy be spatial, the same law of intensity holds as in (α): (γ) if xy be anti-spatial, the intensity of xy is $\sqrt{\{(x|x)(y|y)\}} \sin \widehat{xy}$, that is $(xy|xy)^{\frac{1}{2}}$. Hence the intensity of a spatial force F is $\{-F|F\}^{\frac{1}{2}}$, that of an anti-spatial force F is $\{F|F\}^{\frac{1}{2}}$.

(3) If P and Q be two planes the standard form of a real force of the type xy is iPQ. If the force be anti-spatial, its intensity is

$$\sqrt{\{(P|P)(Q|Q)\}} \sinh \frac{\overline{PQ}}{\gamma};$$

if the force be spatial, its intensity is $\sqrt{\{(P|P)(Q|Q)\}} \sin \widehat{PQ}$.

265. Relations between two Spatial Forces. (1) In general [cf. §§ 231 (1) and 255 (6)] there are only* two lines intersecting the four lines

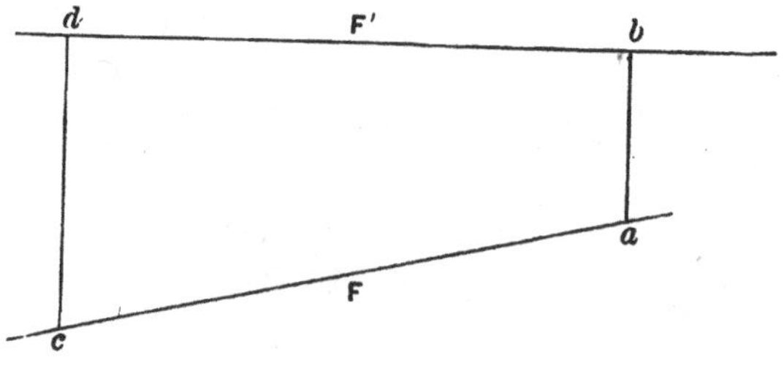

Fig. 2.

F, F', $|F$, $|F'$. Let these be the lines ab and cd. Then ab and cd are perpendicular to both F and F'; also each is the polar line of the other. One of the two must be spatial and the other anti-spatial. Assume ab spatial. Let $\overline{ab} = \delta$, and $\angle cd = \theta$. Then δ is the shortest distance between the lines, and θ will be called the angle between the lines.

* For the discussion of an exceptional case for imaginary lines see the corresponding discussion for Elliptic Space, cf. § 234, in which case the lines are real.

Let $$F = ac,\ F' = bd.$$

Then $$\cosh\frac{\delta}{\gamma} = \frac{(a\,|b)}{\sqrt{\{(a\,|a)(b\,|b)\}}},\ \cos\theta = \frac{(d\,|c)}{\sqrt{\{(d\,|d)(c\,|c)\}}},$$

also $(F\,|F') = (ac\,|bd) = (a\,|b)(c\,|d)$, since $(a\,|d) = 0 = (b\,|c)$.

Also $(F\,|F) = (ac\,|ac) = (a\,|a)(c\,|c)$, since $(a\,|c) = 0$.

And $$(F'\,|F') = (b\,|b)(d\,|d),\ \text{since } (b\,|d) = 0.$$

Hence $$\frac{(F\,|F')}{\sqrt{\{(F\,|F)(F'\,|F')\}}} = \frac{(a\,|b)(c\,|d)}{\sqrt{\{(a\,|a)(b\,|b)(c\,|c)(d\,|d)\}}} = \cos\theta\cosh\frac{\delta}{\gamma}.$$

(2) Again, let a' be the point normal to a on the line ab, and c' be the point normal to c on the line cd. Also let $(a'\,|a') = -(a\,|a)$, where a is assumed to be spatial, and $(c'\,|c') = (c\,|c)$.

Then $$\frac{b}{\sqrt{(b\,|b)}} = \frac{a\cosh\frac{\delta}{\gamma} + a'\sinh\frac{\delta}{\gamma}}{\sqrt{(a\,|a)}},\ \text{and}\ \frac{d}{\sqrt{-(d\,|d)}} = \frac{c\cos\theta + c'\sin\theta}{\sqrt{-(c\,|c)}}.$$

Hence

$$\frac{i\,(FF')}{\sqrt{\{(F\,|F)(F'\,|F')\}}} = \frac{i\,(acbd)}{\sqrt{\{(a\,|a)(b\,|b)(c\,|c)(d\,|d)\}}} = \frac{i\,(aa'cc')\sin\theta\sinh\frac{\delta}{\gamma}}{-(a\,|a)(b\,|b)}$$
$$= \pm\sin\theta\sinh\frac{\delta}{\gamma};$$

since [cf. § 240 (5)] $$i\,(aa'cc') = \pm(a\,|a)(b\,|b).$$

(3) If the forces intersect, the point of intersection is either spatial or anti-spatial or on the absolute. If the point of intersection be spatial, then $\delta = 0$. Hence $(FF') = 0$; and

$$\frac{(F\,|F')}{\sqrt{\{(F\,|F)(F'\,|F')\}}} = \cos\theta.$$

If the point of intersection be anti-spatial, then $\theta = 0$. Hence $(FF') = 0$; and

$$\frac{(F\,|F')}{\sqrt{\{(F\,|F)(F'\,|F')\}}} = \cosh\frac{\delta}{\gamma}.$$

If the point of intersection be on the absolute, so that the lines are parallel, then $\delta = 0 = \theta$. Hence $(FF') = 0$, and

$$(F\,|F')^2 = (F\,|F)(F'\,|F').$$

(4) Let two forces F and F' have a spatial intersection, and let their intensities be ρ and ρ'. Let the single force $F + F'$ be of intensity σ. Let θ be the angle between F and F'. Then

$$\sigma^2 = -(F + F')\,|(F + F') = \rho^2 + \rho'^2 \pm 2\rho\rho'\cos\theta.$$

The upper sign of the ambiguity must be chosen so that, when $\theta = 0$, $\sigma = \rho + \rho'$.

If F and F' be spatial but have an anti-spatial point of intersection, let δ be the shortest distance between F and F'. Then as before, if $F+F'$ be spatial,

$$\sigma^2 = \rho^2 + \rho'^2 + 2\rho\rho' \cosh \frac{\delta}{\gamma}.$$

But it is possible that, though F and F' are spatial, $F+F'$ may be anti-spatial, the intersection of F and F' being anti-spatial. In such a case

$$\sigma^2 = 2\rho\rho' \cosh \frac{\delta}{\gamma} - \rho^2 - \rho'^2.$$

266. Central axis of a System of Forces. (1) It has been proved in § 175 (14), also cf. § 232, that any system of forces has in general one and only one pair of conjugate lines, which are reciprocally polar with respect to a given quadric. Now let a system S have the two conjugate lines a_1a_2, a_3a_4, which are reciprocally polar with respect to the absolute. One of the two must be spatial, the other anti-spatial. Let a_1a_2 be spatial. Then $S = \lambda a_1a_2 + \mu a_3a_4$; and this form of reduction is unique. The line a_1a_2 will be called the central axis of the system. Let the points a_1, a_2, a_3, a_4 be so chosen at unit intensity, that

$$S = a_1a_2 + a_3a_4;$$

then $\overline{a_1a_2}$ $(=\delta)$ and $\angle\, a_3a_4 (=\alpha)$ will be called the parameters of the system.

A system, S, referred to its central axis may also be written in the form

$$a_1a_2 + i\epsilon \,|\, a_1a_2,$$

where ϵ is real.

Then $\quad (SS) = 2i\epsilon\,(a_1a_2 \,|\, a_1a_2)$, and $(S\,|\,S) = (1-\epsilon^2)\,(a_1a_2\,|\,a_1a_2)$.

(2) Let S denote the system $F + i\epsilon\,|\,F$, and S' the system $F' + i\eta\,|\,F'$. Also with the notation of § 265, let δ be the shortest distance between the lines and θ the angle between the two.

Then

$$\begin{aligned}(SS') &= (1-\epsilon\eta)\,(FF') + i\,(\epsilon+\eta)\,(F\,|\,F') \\ &= \{(F\,|\,F)\,(F'\,|\,F')\}^{\frac{1}{2}} \left\{\pm\, i\,(1-\epsilon\eta)\sin\theta\sinh\frac{\delta}{\gamma} + i\,(\epsilon+\eta)\cos\theta\cosh\frac{\delta}{\gamma}\right\}.\end{aligned}$$

And

$$\begin{aligned}(S\,|\,S') &= i\,(\epsilon+\eta)\,(FF') + (1-\epsilon\eta)\,(F\,|\,F') \\ &= \{(F\,|\,F)\,(F'\,|\,F')\}^{\frac{1}{2}} \left\{\pm\,(\epsilon+\eta)\sin\theta\sinh\frac{\delta}{\gamma} + (1-\epsilon\eta)\cos\theta\cosh\frac{\delta}{\gamma}\right\}.\end{aligned}$$

(3) The simultaneous equations $(SS') = 0$, $(S\,|\,S') = 0$, secure that the axes of S and S' intersect at right angles.

For from (2), unless ϵ or η be i, which is the case of an imaginary system analogous to the real vector system of Elliptic Geometry [cf. § 235], $(SS') = 0$ and $(S\,|\,S') = 0$ entail $\cos\theta\cosh\dfrac{\delta}{\gamma} = 0$, $\sin\theta\sinh\dfrac{\delta}{\gamma} = 0$; that is $\delta = 0$, $\theta = \dfrac{\pi}{2}$.

(4) Every dual group contains one pair of systems, and only one pair, such that their axes intersect at right angles. The proof is exactly the same as for the analogous theorem of Elliptic Geometry, cf. § 232 (4). Let this pair of systems be called the central systems of the group, and let the point in which their central axes intersect be the centre of the group.

(5) Dual groups with real director lines can be discriminated into three types according as, either (α) both director lines are spatial, or (β) both director lines are anti-spatial, or (γ) one director line is spatial and one is anti-spatial.

(6) To find the locus of the central axes of a dual group, let e be the centre, ee_1 and ee_2 the axes of the central system, and ee_3 a line perpendicular to the lines ee_1, ee_2. Also let $ee_1e_2e_3$ be a normal system at unit intensities. Let $S_1 = ee_1 + i\epsilon_1 | ee_1$, $S_2 = ee_2 + i\epsilon_2 | ee_2$, be the central systems of the group.

Now [cf. § 240 (5)] we may assune

$$e_3e_1 = -i\,|ee_2,\quad e_2e_3 = -i\,|ee_1,\quad |e_3e_1 = -iee_2,\quad |e_2e_3 = -iee_1.$$

Any other system S' of the group can be written

$$S' = \lambda_1 S_1 + \lambda_2 S_2 = e(\lambda_1 e_1 + \lambda_2 e_2) + i\,|\,.\,e(\lambda_1\epsilon_1 e_1 + \lambda_2\epsilon_2 e_2).$$

Then, as in § 232 (5), this system can be identified with the system

$$(e + \zeta e_3)(\mu_1 e_1 + \mu_2 e_2) + i\epsilon\,|\,.\,(e + \zeta e_3)(\mu_1 e_1 + \mu_2 e_2).$$

Hence all the central axes of systems of the group intersect ee_3 at right-angles. Let ee_3 be called the axis of the group.

The equation to find ϵ is

$$(\epsilon^2 - 1)\{\epsilon_1^2\lambda_1^2 + \epsilon_2^2\lambda_2^2\} - \epsilon\{(\epsilon_1^2 - 1)\lambda_1^2 + (\epsilon_2^2 - 1)\lambda_2^2\} = 0.$$

The locus of any point $\Sigma\xi e$ on an axis of any system of the group is

$$(\epsilon_1 - \epsilon_2)\,\xi_1\xi_2(\xi^2 - \xi_3^2) = (1 + \epsilon_1\epsilon_2)\,\xi\xi_3(\xi_1^2 + \xi_2^2).$$

267. Non-Axal Systems of Forces. (1) A system of forces, not self-supplementary, and such that $(SS) = \pm(S|S)$, is called a non-axal system [cf. § 233].

(2) All such systems are imaginary. For if S be real, then [cf. § 240 (5)] it is easily proved that (SS) is a pure imaginary, and that $(S|S)$ is real.

(3) Hence, from this article and from § 266 (3), any system S, such that $(SS) = \pm(S|S)$, whether it be self-supplementary or not, is imaginary.

(4) Accordingly the theorems of § 266 hold for all real systems of forces.

CHAPTER VI.

KINEMATICS IN THREE DIMENSIONS.

268. CONGRUENT TRANSFORMATIONS*. (1) A congruent transformation is a linear transformation, such that (α) the internal measure relations of any figure are unaltered by the transformation; e.g. if abc is transformed into $a'b'c'$, then $\overline{ab} = \overline{a'b'}$, and the angle between ab and ac is equal to that between $a'b'$ and $a'c'$, and similarly for the other sides and angles: (β) the transformation can be conceived as the result of another congruent transformation ρ times repeated, where ρ is any integer: (γ) real points are transformed into real points: and (δ) the intensities of points are unchanged by transformation.

It follows from (α) that points on the absolute must be transformed into points on the absolute.

Hence congruent transformations must transform the absolute into itself.

It follows from (α), (β) and (γ) that spatial points must be transformed into spatial points. For from (α) [cf. § 241 (1)], either all spatial points are transformed into spatial points, or all spatial points into anti-spatial points. Also from (β), since the integer ρ may be taken indefinitely large, a finite transformation can be considered as the result of ρ repetitions of an infinitesimal transformation. But an infinitesimal transformation must transform spatial points into spatial points. Hence the same holds for a finite transformation.

(2) Let the discussion be now confined to regions of three dimensions. To prove that a congruent transformation must transform the absolute by a direct transformation (cf. § 194).

For by (β) of (1) any congruent transformation can be conceived as the result of ρ repetitions of another congruent transformation. But an even number of applications of either a direct or a skew transformation of a quadric produces a direct transformation of the quadric. Hence every congruent transformation is a direct transformation of the absolute.

* The theory of congruent transformations is due to Klein, cf. *loc. cit.* p. 369; Buchheim has applied Grassman's algebra to this subject, cf. *loc. cit.* p. 370.

(3) By § 195 (2) to (6) among the latent points of a direct transformation of the general type there are the points of intersection of two conjugate polar lines with the quadric. Now in Elliptic Space the absolute is imaginary; and therefore the co-ordinates of the latent points on either one of the polar lines, referred to real reference points, form pairs of conjugate imaginaries. In Hyperbolic Space one polar line must be anti-spatial and one is spatial: the co-ordinates of the latent points on the anti-spatial polar line, referred to real reference points, are pairs of conjugate imaginaries: the latent points on the spatial polar line are two real points, and their co-ordinates are real.

Now either in Elliptic or in Hyperbolic Space let a_1, a_2, a_3, a_4 be the four above-mentioned latent points of a congruent transformation, and let a_1a_2 and a_3a_4 be conjugate polar lines. Also let a_3 and a_4 be imaginary points, then their co-ordinates are pairs of conjugate imaginaries. Hence if a_3 and a_4 be taken as reference points, the co-ordinates, η_3 and η_4, of a real point $y\,(=\eta_3a_3+\eta_4a_4)$ are conjugate imaginaries.

(4) Let the latent roots of the congruent matrix be $\alpha_1, \alpha_2, \alpha_3, \alpha_4$. Then y is changed into $\eta_3\alpha_3a_3+\eta_4\alpha_4a_4$, and by ($\gamma$) of (1), $\eta_3\alpha_3a_3+\eta_4\alpha_4a_4$ is a real point. Hence α_3 and α_4 must be conjugate imaginaries.

Similarly if a_1 and a_2 are imaginary points as in Elliptic Space, α_1 and α_2 are conjugate imaginaries; but if a_1 and a_2 are real points as in Hyperbolic Space, α_1 and α_2 must be real.

Hence in Elliptic Space the latent roots are two pairs of conjugate imaginaries, in Hyperbolic Space one pair are real and one pair are conjugate imaginaries.

(5) In Hyperbolic Space both the real latent roots α_1 and α_2 must be positive. For the given congruent matrix may be conceived, according to (β) of subsection (1), as the result of another congruent matrix twice applied. Let $\beta_1, \beta_2, \beta_3, \beta_4$ be the latent roots of this matrix. Then $\beta_1^2=\alpha_1$, and $\beta_2^2=\alpha_2$. But β_1 and β_2 are real by the same proof as that for α_1 and α_2; hence α_1 and α_2 are positive. Therefore the real roots of a congruent matrix in Hyperbolic Space are positive.

(6) By § 195, $\alpha_1\alpha_2=\alpha_3\alpha_4$. Hence in Elliptic Space we may put,

$$\alpha_1=\lambda e^{\frac{i\delta}{\gamma}},\quad \alpha_2=\lambda e^{-\frac{i\delta}{\gamma}},\quad \alpha_3=\lambda e^{\frac{i\delta'}{\gamma}},\quad \alpha_4=\lambda e^{-\frac{i\delta'}{\gamma}}.$$

And in Hyperbolic Space we may put,

$$\alpha_1=\lambda e^{\frac{\delta}{\gamma}},\quad \alpha_2=\lambda e^{-\frac{\delta}{\gamma}},\quad \alpha_3=\lambda e^{i\alpha},\quad \alpha_4=\lambda e^{-i\alpha}.$$

Also by (δ) of (1) the intensity of any point is unaltered by transformation. Now the intensity of $\eta_1a_1+\eta_2a_2$ is $\{\eta_1\eta_2\,(a_1|a_2)\}^{\frac{1}{2}}$, and the intensity of the transformed point is

$$\{\lambda^2\eta_1\eta_2\,(a_1\,|\,a_2)\}^{\frac{1}{2}}.$$

Hence $\lambda = 1$. Thus the latent roots in Elliptic Space take the form

$$a_1 = e^{\frac{i\delta}{\gamma}}, \quad a_2 = e^{-\frac{i\delta}{\gamma}}, \quad a_3 = e^{\frac{i\delta'}{\gamma}}, \quad a_4 = e^{-\frac{i\delta'}{\gamma}};$$

and in Hyperbolic Space

$$a_1 = e^{\frac{\delta}{\gamma}}, \quad a_2 = e^{-\frac{\delta}{\gamma}}, \quad a_3 = e^{i\alpha}, \quad a_4 = e^{-i\alpha}.$$

(7) The special type of direct transformation, with only three semi-latent lines [cf. § 195 (7)], cannot apply to Elliptic or Hyperbolic Space, so as to give a real congruent transformation. For, with the notation of § 195 (7), the points e_1, e_2, e_3, e_4 and the planes $e_1e_2e_3$ and $e_1e_3e_4$ are imaginary, both in Elliptic and Hyperbolic Space. But an imaginary plane always contains one straight line of real points. Hence the semi-latent plane $e_1e_2e_3$ contains one real line. But, since real points are transformed into real points, this line must be semi-latent. Also the semi-latent lines e_1e_2 and e_1e_3, which lie in this plane must be imaginary, since they are generators. Hence a third semi-latent line must lie in this plane; and this is impossible in this type of transformation.

(8) The theory of congruent transformations in Hyperbolic Space will first be discussed, cf. §§ 269 to 280, and then that of congruent transformations in Elliptic Space, cf. §§ 281 to 286.

269. ELEMENTARY FORMULAE. (1) Let a_1, a_2, a_3, a_4 be the latent points of a congruent transformation in Hyperbolic Space.

Let a_1a_2 and a_3a_4 be conjugate polar lines; and let them be called the axes of the transformation. Let a_1a_2 be spatial, and be called the spatial axis, or more shortly the axis; then a_3a_4 is anti-spatial, and may be called the anti-spatial axis. Thus a_1 and a_2 are real; and a_3 and a_4 are imaginary.

Then it at once follows that,

$$(a_1|a_1) = 0 = (a_2|a_2) = (a_3|a_3) = (a_4|a_4) = (a_1|a_3) = (a_1|a_4) = (a_2|a_3) = (a_2|a_4).$$

(2) Let e, e_1, e_2, e_3 be a normal system of elements at unit intensity, of which e is the spatial origin.

Let ee_1 be the line a_1a_2, and e_2e_3 the line a_3a_4.

Then by § 247, we may write

$$a_1 = e_1 + e, \quad a_2 = e_1 - e, \quad a_3 = e^{-\frac{i\pi}{4}} e_2 + e^{\frac{i\pi}{4}} e_3, \quad a_4 = e^{\frac{i\pi}{4}} e_2 + e^{-\frac{i\pi}{4}} e_3.$$

Hence $$(a_1|a_2) = (e_1|e_1) - (e|e) = -2;$$

and $$(a_3|a_4) = (e_2|e_2) + (e_3|e_3) = -2.$$

Also from § 240 (5) $$(ee_1e_2e_3) = i.$$

Hence $$(a_1a_2a_3a_4) = -4i(ee_1e_2e_3) = 4 = (a_1|a_2)(a_3|a_4).$$

(3) Again, by substituting for a_1, a_2, a_3, a_4 in the expression for any real point $x\,(=\xi_1 a_1 + \xi_2 a_2 + \xi_3 a_3 + \xi_4 a_4)$, we find

$$x = (\xi_1 - \xi_2)\, e + (\xi_1 + \xi_2)\, e_1 + (\xi_3 e^{-\frac{i\pi}{4}} + \xi_4 e^{\frac{i\pi}{4}})\, e_2 + (\xi_3 e^{\frac{i\pi}{4}} + \xi_4 e^{-\frac{i\pi}{4}})\, e_3.$$

But since x is real, the coefficients of e, e_1, e_2, e_3 are real. Hence ξ_3 and ξ_4 are conjugate imaginaries. Let $\xi_3 = \rho e^{i\theta}$, $\xi_4 = \rho e^{-i\theta}$.

Then $x = (\xi_1 - \xi_2)\, e + (\xi_1 + \xi_2)\, e_1 + 2\rho \cos\left(\theta - \frac{\pi}{4}\right) e_2 + 2\rho \cos\left(\theta + \frac{\pi}{4}\right) e_3$

$= \eta e + \eta_1 e_1 + \eta_2 e_2 + \eta_3 e_3$ (say).

Any spatial point x must satisfy the condition that $(x\,|\,x)$ be positive.

But
$$(x\,|\,x) = 2\xi_1\xi_2\,(a_1\,|\,a_2) + 2\xi_3\xi_4\,(a_3\,|\,a_4)$$
$$= -4\,(\xi_1\xi_2 + \xi_3\xi_4) = -4\,(\xi_1\xi_2 + \rho^2).$$

Hence for a spatial point $\xi_1\xi_2$ is negative, and $-\xi_1\xi_2 > \rho^2$.

(4) The congruent matrix transforms x into x', where

$$x' = \xi_1 e^{\frac{\delta}{\gamma}} a_1 + \xi_2 e^{-\frac{\delta}{\gamma}} a_2 + \rho e^{i(\theta+\alpha)} a_3 + \rho e^{-i(\theta+\alpha)} a_4$$
$$= (\xi_1 e^{\frac{\delta}{\gamma}} - \xi_2 e^{-\frac{\delta}{\gamma}})\, e + (\xi_1 e^{\frac{\delta}{\gamma}} + \xi_2 e^{-\frac{\delta}{\gamma}})\, e_1$$
$$+ 2\rho \cos\left(\theta + \alpha - \frac{\pi}{4}\right) e_2 + 2\rho \cos\left(\theta + \alpha + \frac{\pi}{4}\right) e_3$$
$$= \left(\eta \cosh\frac{\delta}{\gamma} + \eta_1 \sinh\frac{\delta}{\gamma}\right) e + \left(\eta \sinh\frac{\delta}{\gamma} + \eta_1 \cosh\frac{\delta}{\gamma}\right) e_1$$
$$+ (\eta_2 \cos\alpha + \eta_3 \sin\alpha)\, e_2 + (\eta_3 \cos\alpha - \eta_2 \sin\alpha)\, e_3.$$

270. Simple Geometrical Properties. (1) Consider any point x $(=\xi_1 a_1 + \xi_2 a_2)$ on the axis $a_1 a_2$. It is transformed into $x' = \xi_1 e^{\frac{\delta}{\gamma}} a_1 + \xi_2 e^{-\frac{\delta}{\gamma}} a_2$, which is again on the axis.

Furthermore,

$$\overline{xx'} = \frac{\gamma}{2} \log\{xx', a_1 a_2\} = \frac{\gamma}{2} \log \frac{\xi_1 e^{\frac{\delta}{\gamma}} \,.\, \xi_2}{\xi_2 e^{-\frac{\delta}{\gamma}} \,.\, \xi_1} = \frac{\gamma}{2} \log e^{\frac{2\delta}{\gamma}} = \delta.$$

Thus all points on the axis are transferred through the same distance δ.

(2) Again, consider any plane P through the axis $a_1 a_2$ and any point $x\,(=\xi_3 a_3 + \xi_4 a_4)$ on the anti-spatial axis $a_3 a_4$. Let the plane $a_1 a_2 a_3$ be called A_3, the plane $a_1 a_2 a_4$ be called A_4. Then A_3 and A_4 are the two planes through $a_1 a_2$ which touch the absolute in imaginary points a_3 and a_4. Also $P = \xi_3 A_3 + \xi_4 A_4$; and P is transformed into the plane $P' = \xi_3 e^{i\alpha} A_3 + \xi_4 e^{-i\alpha} A_4$.

Furthermore, $\angle PP' = \dfrac{1}{2i} \log (PP',\, A_3 A_4) = \dfrac{1}{2i} \log \dfrac{\xi_3 e^{i\alpha} \,.\, \xi_4}{\xi_4 e^{-i\alpha} \,.\, \xi_3} = \alpha.$

Hence every plane through the axis is rotated through the same angle α.

(3) The distance, ϖ, of any spatial point x from the axis is given by

$$\sinh\frac{\varpi}{\gamma} = \left\{\frac{(xa_1a_2 \,|\, xa_1a_2)}{-(x\,|\,x)(a_1a_2\,|\,a_1a_2)}\right\}^{\frac{1}{2}} = \frac{2\rho}{(x\,|\,x)^{\frac{1}{2}}};$$

where x is written in the form $\xi_1 a_1 + \xi_2 a_2 + \rho e^{i\theta} a_3 + \rho e^{-i\theta} a_4$. But the transformed point x' is $\xi_1 e^{\frac{\delta}{\gamma}} a_1 + \xi_2 e^{-\frac{\delta}{\gamma}} a_2 + \rho e^{i(\theta+\alpha)} a_3 + \rho e^{-i(\theta+\alpha)} a_4$; and it is obvious that its distance from the axis is the same as that of x. Hence the distance of a point from the axis is unaltered by the congruent transformation. Also it is easy to prove that

$$\cosh\frac{\varpi}{\gamma} = \left\{\frac{-4\xi_1\xi_2}{(x\,|\,x)}\right\}^{\frac{1}{2}} = \left\{\frac{-(xa_3a_4\,|\,xa_3a_4)}{4(x\,|\,x)}\right\}^{\frac{1}{2}}.$$

(4) To find the distance of the transformed point x' from the plane through x perpendicular to the axis a_1a_2 of the transformation.

This plane is represented by xa_3a_4. Now $xa_3a_4 = 2ixe_2e_3$; but $2ixe_2e_3$ is in the standard form of § 240 (2); hence xa_3a_4 is in the standard form. Now if ζ be the distance of x' from this plane, it is easily seen after some reduction that

$$\sinh\frac{\zeta}{\gamma} = \pm\frac{(x'xa_3a_4)}{\sqrt{\{-(x'\,|\,x')(xa_3a_4\,|\,xa_3a_4)\}}} = \cosh\frac{\varpi}{\gamma}\sinh\frac{\delta}{\gamma}.$$

Hence ζ is independent of α, and depends only on δ and on the distance of x from a_1a_2. Also when $\delta = 0$, $\zeta = 0$; and when $\varpi = 0$, ζ becomes δ in accordance with subsection (1).

(5) It is easily proved that

$$\cosh\frac{\overline{xx'}}{\gamma} = \frac{(x\,|\,x')}{\sqrt{\{(x\,|\,x)(x'\,|\,x')\}}} = \frac{\xi_1\xi_2\cosh\frac{\delta}{\gamma} + \rho^2\cos\alpha}{\xi_1\xi_2 + \rho^2} = \cosh^2\frac{\varpi}{\gamma}\cosh\frac{\delta}{\gamma} + \sinh^2\frac{\varpi}{\gamma}\cos\alpha.$$

271. TRANSLATIONS AND ROTATIONS. (1) Let the quantities δ and α be called the parameters of the transformation.

If $\alpha = 0$, the congruent transformation is called a translation through the distance δ with a_1a_2 as axis. The effect of the translation on a point x is that the transformed point x_1 is at the same distance as x from a_1a_2, is in the plane xa_1a_2, and is at a distance $\gamma\sinh^{-1}\left\{\cosh\frac{\varpi}{\gamma}\sinh\frac{\delta}{\gamma}\right\}$ from the plane xa_3a_4, where ϖ is the distance of x from a_1a_2, or in other words the plane $x_1a_3a_4$ is at a distance δ from the plane xa_3a_4, a_1a_2 being their common perpendicular.

(2) If $\delta = 0$, the congruent transformation is called a rotation through the angle α with a_1a_2 as axis. The effect of the rotation on the point x is that the transformed point x_2 is at the same distance ϖ from a_1a_2, and is in the plane xa_3a_4, and that the plane $x_2a_1a_2$ makes an angle α with the plane xa_1a_2.

(3) It is obvious that the general congruent transformation, axis a_1a_2, parameters (δ, α), in its effect on any point x is identical with the effect first of the translation, axis a_1a_2, parameters $(\delta, 0)$, bringing x to x_1, and then of the rotation, axis a_1a_2, parameters $(0, \alpha)$, bringing x_1 to x'; or first of the rotation bringing x to x_2, and then of the translation bringing x_2 to x'.

It is to be noticed that congruent transformations with the same axis are convertible as to order, but that when the axes of the transformations are different the order of operation affects the result.

It will be convenient for the future to use the letter K for the matrix representing a congruent transformation, so that Kx is the transformed position of x.

(4) A further peculiarity of translations and rotations is established by seeking the condition that a real plane, other than one of the faces of the tetrahedron $a_1a_2a_3a_4$ may remain unchanged in position, i.e. be semi-latent. Let x be any point in such a plane. Then x, Kx, K^2x, K^3x must all lie in the plane. Therefore $(xKxK^2xK^3x) = 0$.

But
$$x = \xi_1 a_1 + \xi_2 a_2 + \rho e^{i\theta} a_3 + \rho e^{-i\theta} a_4,$$
$$Kx = \xi_1 e^{\frac{\delta}{\gamma}} a_1 + \xi_2 e^{-\frac{\delta}{\gamma}} a_2 + \rho e^{i(\theta+\alpha)} a_3 + \rho e^{-i(\theta+\alpha)} a_4,$$
$$K^2x = \xi_1 e^{\frac{2\delta}{\gamma}} a_1 + \xi_2 e^{-\frac{2\delta}{\gamma}} a_2 + \rho e^{i(\theta+2\alpha)} a_3 + \rho e^{-i(\theta+2\alpha)} a_4,$$
$$K^3x = \xi_1 e^{\frac{3\delta}{\gamma}} a_1 + \xi_2 e^{-\frac{3\delta}{\gamma}} a_2 + \rho e^{i(\theta+3\alpha)} a_3 + \rho e^{-i(\theta+3\alpha)} a_4.$$

Therefore
$$\frac{(xKxK^2xK^3x)}{(a_1a_2a_3a_4)} = \xi_1\xi_2\rho^2 \begin{vmatrix} 1, & 1, & 1, & 1 \\ e^{\frac{\delta}{\gamma}}, & e^{-\frac{\delta}{\gamma}}, & e^{+i\alpha}, & e^{-i\alpha} \\ e^{\frac{2\delta}{\gamma}}, & e^{-\frac{2\delta}{\gamma}}, & e^{+2i\alpha}, & e^{-2i\alpha} \\ e^{\frac{3\delta}{\gamma}}, & e^{-\frac{3\delta}{\gamma}}, & e^{+3i\alpha}, & e^{-3i\alpha} \end{vmatrix}.$$

Now $\xi_1 = 0$, or $\xi_2 = 0$, or $\rho = 0$, each makes the plane to be one of the three real faces of the tetrahedron $(a_1a_2a_3a_4)$. Hence the determinant must vanish, if another real plane is semi-latent. But this condition can be written,

$$(e^{\frac{\delta}{\gamma}} - e^{-\frac{\delta}{\gamma}})(e^{\frac{\delta}{\gamma}} - e^{i\alpha})(e^{\frac{\delta}{\gamma}} - e^{-i\alpha})(e^{-\frac{\delta}{\gamma}} - e^{i\alpha})(e^{-\frac{\delta}{\gamma}} - e^{-i\alpha})(e^{i\alpha} - e^{-i\alpha}) = 0.$$

The only solutions, making δ and α real, of this equation are $\delta = 0$ or $\alpha = 0$.

Thus rotations and translations are the only congruent transformations by which planes do not change their positions. For rotations, planes perpendicular to the axis are rotated so as to remain coincident with themselves; for translations, planes containing the axis are translated so as to remain coincident with themselves. In other words, the only possible motions of a plane, which remains coincident with itself, are either rotations about a perpendicular axis, or translations along an axis lying in it.

Similarly it is obvious that rotations and translations are the only congruent transformations by which points—other than the four corners of the self-corresponding tetrahedron—do not alter their positions. For rotations these points are points on the axis, for translations they are points on the anti-spatial axis.

(5) The property, which discriminates a translation from a rotation, is that, as a translation is continually repeated, the distance between the original and final positions of any spatial point grows continually greater. For let the translation, axis a_1a_2, parameters $(\delta, 0)$, be repeated ν times, and let

$$x = \xi_1 a_1 + \xi_2 a_2 + \rho e^{i\theta} a_3 + \rho e^{-i\theta} a_4.$$

Then
$$K^\nu x = \xi_1 e^{\frac{\nu\delta}{\gamma}} a_1 + \xi_2 e^{\frac{\nu\delta}{\gamma}} a_2 + \rho e^{i\theta} a_3 + \rho e^{-i\theta} a_4.$$

Hence
$$(x \,|\, K^\nu x) = -4\xi_1\xi_2 \cosh\frac{\nu\delta}{\gamma} - 4\rho^2.$$

Therefore
$$\overline{xK^\nu x} = \gamma \cosh^{-1} \frac{-\left(4\xi_1\xi_2 \cosh\frac{\nu\delta}{\gamma} + 4\rho^2\right)}{(x\,|\,x)}.$$

Hence $\overline{xK^\nu x}$ grows continually greater as ν is increased. But in the case of a rotation, if the parameter α bear a commensurable ratio to four right angles, then after a certain number of repetitions every point coincides with its original position.

272. Locus of Points of Equal Displacement. (1) The locus of points, for which the distance of displacement [cf. § 270 (5)] in the general transformation, parameters (δ, α), is equal to a given length σ, is the quadric surface

$$2\xi_1\xi_2 \cosh\frac{\delta}{\gamma} + 2\rho^2 \cos\alpha = \cosh\frac{\sigma}{\gamma}\{2\xi_1\xi_2 + 2\rho^2\};$$

that is
$$(x\,|\,x)\left(\cosh\frac{\sigma}{\gamma} - \cos\alpha\right) = -\left(\cosh\frac{\delta}{\gamma} - \cos\alpha\right)(x\,|\,a_1)(x\,|\,a_2).$$

Also by referring to § 195 (2) we find that all quadrics of the form,

$$(x\,|\,x) + \mu^2 (x\,|\,a_1)(x\,|\,a_2) = 0,$$

remain coincident with themselves after the congruent transformation, axis a_1a_2. They are quadrics which touch the absolute at the ends of the axis a_1a_2. But by comparison with the quadric, which is the locus of points transferred through a given distance σ, we see that the system of quadrics found by varying σ is the same* as that found by varying μ.

* Cf. Sir R. S. Ball, "On the Theory of Content," *Transactions R. I. A.*, *loc. cit.* p. 370.

In fact we have

$$\mu^2 = \frac{\left(\cosh\frac{\delta}{\gamma} - \cos\alpha\right)}{\left(\cosh\frac{\sigma}{\gamma} - \cos\alpha\right)}.$$

(2) The sections of these quadrics by planes perpendicular to a_1a_2 are circles. For let pa_3a_4 be any such plane, and let $x = \zeta p + \zeta_3 a_3 + \zeta_4 a_4$. Then $(x|a_1) = \zeta(p|a_1)$, $(x|a_2) = \zeta(p|a_2)$; and $(xa_3a_4b) = \zeta(pa_3a_4b)$, where b is any fixed point.

Hence $$(x|a_1)(x|a_2) = \frac{(p|a_1)(p|a_2)}{(pa_3a_4b)^2}(xa_3a_4b)^2.$$

Therefore the section of the quadric,

$$(x|x) + \mu^2(x|a_1)(x|a_2) = 0,$$

by the plane pa_3a_4 is the intersection of the plane with the sphere

$$(x|x) + \frac{\mu^2(p|a_1)(p|a_2)}{(pa_3a_4b)}(xa_3a_4b)^2 = 0.$$

(3) The centre of this sphere is the point $|a_3a_4b$, that is the point where a_1a_2 meets $|b$. Hence the centre of the circle, which is the intersection of pa_3a_4 with the quadric [cf. § 260 (4)], is the point where a_1a_2 meets pa_3a_4. It is otherwise evident from § 270 (5), that all points on such a circle must receive equal displacements, and that the distances of their displaced positions from the plane of this circle must be the same for each point. Similarly the curves of intersection of the system of quadrics with planes through a_1a_2, that is of the form pa_1a_2, are lines of equal distance from a_1a_2, that is are circles with their anti-spatial centres on a_3a_4.

273. Equivalent Sets of Congruent Transformations. (1) Any congruent transformation (K) may be replaced by a combination of two transformations consisting first of a translation with its axis through any arbitrarily chosen spatial point b, and then of a rotation with its axis through Kb, where the given transformation changes b into Kb. First apply a translation, axis bKb and parameter $\overline{bKb}$; which is always a possible transformation. This translation converts b into Kb. The second transformation, which brings all the points into their final positions, must leave Kb unchanged. Hence the transformation must be a rotation with its axis through Kb.

(2) Applying the principles of § 2 75 below with regard to small displacements, it will be easy to see that any *small* transformation is equivalent to the combination of a rotation and a translation with their axes through any arbitrarily assigned point b.

274. Commutative Law. The operations K and $|$ performed successively on any point x are commutative, that is $|Kx = K|x$.

For let e, e_1, e_2, e_3 be a unit normal system, e being the spatial origin. Let the transformation K have ee_1 as axis, and δ, α as parameters.

Then if $$x = \xi e + \xi_1 e_1 + \xi_2 e_2 + \xi_3 e_3,$$

$$Kx = \left(\xi \cosh\frac{\delta}{\gamma} + \xi_1 \sinh\frac{\delta}{\gamma}\right) e + \left(\xi \sinh\frac{\delta}{\gamma} + \xi_1 \cosh\frac{\delta}{\gamma}\right) e_1$$
$$+ (\xi_2 \cos\alpha + \xi_3 \sin\alpha)\, e_2 + (\xi_3 \cos\alpha - \xi_2 \sin\alpha)\, e_3.$$

Also [cf. § 240 (5)] $|e = -ie_1e_2e_3$, $|e_1 = -iee_2e_3$, $|e_2 = +iee_1e_3$, $|e_3 = -iee_1e_2$;

and $$Ke = \cosh\frac{\delta}{\gamma}\, e + \sinh\frac{\delta}{\gamma}\, e_1,\quad Ke_1 = \sinh\frac{\delta}{\gamma}\, e + \cosh\frac{\delta}{\gamma}\, e_1,$$
$$Ke_2 = \cos\alpha\,.\,e_2 - \sin\alpha\,.\,e_3,\quad Ke_3 = \sin\alpha\,.\,e_2 + \cos\alpha\,.\,e_3.$$

Hence $$Kee_1 = KeKe_1 = ee_1, \text{ and } Ke_2e_3 = e_2e_3.$$

Thus $$|x = -i(\xi e_1 + \xi_1 e)\, e_2e_3 - i(\xi_3 e_2 - \xi_2 e_3)\, ee_1;$$

and $$K|x = -i(\xi Ke_1 + \xi_1 Ke)\, e_2e_3 - i(\xi_3 Ke_2 - \xi_2 Ke_3)\, ee_1.$$

Now $$\xi Ke_1 + \xi_1 Ke = \left(\xi \cosh\frac{\delta}{\gamma} + \xi_1 \sinh\frac{\delta}{\gamma}\right) e_1 + \left(\xi \sinh\frac{\delta}{\gamma} + \xi_1 \cosh\frac{\delta}{\gamma}\right) e,$$

and $$\xi_3 Ke_2 - \xi_2 Ke_3 = -(\xi_3 \sin\alpha + \xi_2 \cos\alpha)\, e_3 + (\xi_3 \cos\alpha - \xi_2 \sin\alpha)\, e_2.$$

Hence by substitution and comparison
$$|Kx = K|x.$$

275. Small Displacements. (1) Two finite congruent transformations, when successively applied, produce in general different results according to the order of operation. If however each transformation be small, and squares of small quantities be neglected, the order of operation is indifferent. This is a general theorem, which holds for any linear transformation whatever. Thus let $e_1e_2e_3e_4$ be any four reference points, and let x be transformed into Kx by the transformation

$$\xi_1' = (1 + \alpha_{11})\, \xi_1 + \alpha_{12}\xi_2 + \alpha_{13}\xi_3 + \alpha_{14}\xi_4,$$
$$\xi_2' = \alpha_{21}\xi_1 + (1 + \alpha_{22})\, \xi_2 + \alpha_{23}\xi_3 + \alpha_{24}\xi_4,$$

with two similar equations; where all letters of the form $\alpha_{\rho\rho}$ or $\alpha_{\rho\sigma}$ are small, and their squares and products are neglected. Again, let Kx be transformed into $K'Kx$ by the transformation

$$\xi_1'' = (1 + \beta_{11})\, \xi_1' + \beta_{12}\xi_2' + \beta_{13}\xi_3' + \beta_{14}\xi_4',$$

with three similar equations for ξ_2'', ξ_3'', ξ_4''; where all the letters of the form $\beta_{\rho\rho}$ or $\beta_{\rho\sigma}$ are small and their squares and products are neglected.

Substituting for ξ_1', ξ_2', etc. in the equations for ξ_1'', ξ_2'', etc. we find

$$\xi_1'' = (1 + \alpha_{11} + \beta_{11})\, \xi_1 + (\alpha_{12} + \beta_{12})\, \xi_2 + (\alpha_{13} + \beta_{13})\, \xi_3 + (\alpha_{14} + \beta_{14})\, \xi_4,$$

with three similar equations.

It is obvious from the form of these equations for the co-ordinates of $K'Kx$, that $K'Kx = KK'x$.

(2) Let a point x with reference to the normal system e, e_1, e_2, e_3 (spatial origin e) be written in the form $\eta e + \eta_1 e_1 + \eta_2 e_2 + \eta_3 e_3$.

Apply a congruent transformation, axis ee_1, parameters δ and α. Let x become Kx. Then, as in § 269 (4),

$$Kx = \left(\eta \cosh\frac{\delta}{\gamma} + \eta_1 \sinh\frac{\delta}{\gamma}\right) e + \left(\eta_1 \cosh\frac{\delta}{\gamma} + \eta \sinh\frac{\delta}{\gamma}\right) e_1$$
$$+ (\eta_2 \cos\alpha + \eta_3 \sin\alpha)\, e_2 + (\eta_3 \cos\alpha - \eta_2 \sin\alpha)\, e_3.$$

Assuming that δ and α are small, this equation becomes

$$Kx = \left(\eta + \frac{\delta}{\gamma}\eta_1\right) e + \left(\eta_1 + \frac{\delta}{\gamma}\eta\right) e_1 + (\eta_2 + \alpha\eta_3)\, e_2 + (\eta_3 - \alpha\eta_2)\, e_3.$$

(3) Accordingly if in any order small translations $\delta_1, \delta_2, \delta_3$ and small rotations $\alpha_1, \alpha_2, \alpha_3$ be applied with axes ee_1, ee_2, ee_3 respectively, and as the total result x becomes Kx, then

$$Kx = \left(\eta + \frac{\delta_1}{\gamma}\eta_1 + \frac{\delta_2}{\gamma}\eta_2 + \frac{\delta_3}{\gamma}\eta_3\right) e + \left(\eta_1 + \frac{\delta_1}{\gamma}\eta + \alpha_3\eta_2 - \alpha_2\eta_3\right) e_1$$
$$+ \left(\eta_2 + \frac{\delta_2}{\gamma}\eta + \alpha_1\eta_3 - \alpha_3\eta_1\right) e_2 + \left(\eta_3 + \frac{\delta_3}{\gamma}\eta + \alpha_2\eta_1 - \alpha_1\eta_2\right) e_3.$$

276. Small Translations and Rotations. (1) The result (K) of the three small translations of the preceding article by themselves is a translation, having as axis the line joining e and $\frac{\delta_1}{\gamma}e_1 + \frac{\delta_2}{\gamma}e_2 + \frac{\delta_3}{\gamma}e_3$, and as parameters $\sqrt{\{\delta_1^2 + \delta_2^2 + \delta_3^2\}}$ and 0.

For let
$$d = \frac{\delta_1}{\gamma}e_1 + \frac{\delta_2}{\gamma}e_2 + \frac{\delta_3}{\gamma}e_3.$$

Then
$$Ke = e + d, \text{ and } Kd = \frac{\delta_1^2 + \delta_2^2 + \delta_3^2}{\gamma^2}e + d.$$

Hence every point on ed is transferred to another point on ed; and any plane of the form edx is semi-latent. Therefore the resulting displacement is a translation with ed as axis. Let δ be the parameter. Then

$$\frac{\delta}{\gamma} = \sinh\frac{\delta}{\gamma} = \sqrt{\{-(eKe\,|\,eKe)\}} = \sqrt{-(ed\,|\,ed)}$$
$$= \sqrt{-(d\,|\,d)} = \sqrt{\frac{\delta_1^2 + \delta_2^2 + \delta_3^2}{\gamma^2}}.$$

(2) The result (K) of the three rotations by themselves is a rotation, having as axis the line joining e and $\alpha_1 e_1 + \alpha_2 e_2 + \alpha_3 e_3$, and as parameters 0 and $\sqrt{\{\alpha_1^2 + \alpha_2^2 + \alpha_3^2\}}$.

For let
$$a = \alpha_1 e_1 + \alpha_2 e_2 + \alpha_3 e_3.$$

Then $$Ke = e, \text{ and } Ka = a.$$

Hence every point on ea is unaltered by the resulting transformation. Therefore the transformation is a rotation with ea as axis.

To find its parameter, calculate the angle α between the planes iee_1a and $iK(ee_1a)$; this is the required parameter.

Let $A = iee_1a$. Then $KA = -iKe_1 . Kea = -i(e_1 - \alpha_3 e_2 + \alpha_2 e_3) ea$.

Also $(A \mid A) = -(\alpha_2^2 + \alpha_3^2) = (KA \mid KA)$.

Hence $AKA = \{e_1(e_1 - \alpha_3 e_2 + \alpha_2 e_3) ea\} ea = i(\alpha_2^2 + \alpha_3^2) ea = -i(A \mid A) ea$.

Thus $\alpha = \sin\alpha = \sqrt{\dfrac{-(AKA \mid AKA)}{(A \mid A)(KA \mid KA)}} = \sqrt{(ea \mid ea)} = \sqrt{(\alpha_1^2 + \alpha_2^2 + \alpha_3^2)}$.

(3) By properly choosing $\delta_1, \delta_2, \delta_3$ the line ed, which is the axis of the translation, can be made to be any line through e, while at the same time the parameter $\sqrt{\{\delta_1^2 + \delta_2^2 + \delta_3^2\}}$ can be made to assume any small value. Similarly the axis of rotation ea can be made to be any line through e by properly choosing $\alpha_1, \alpha_2, \alpha_3$, while at the same time the parameter $\sqrt{\{\alpha_1^2 + \alpha_2^2 + \alpha_3^2\}}$ can be made to assume any small value. Hence it follows from § 273 (2) that the combination of the three translations along ee_1, ee_2, ee_3 and the three rotations round the same axes may be made equivalent to any small congruent transformation whatever.

277. Associated System of Forces. (1) Let S denote the system of forces

$$\left\{\frac{\delta_1}{\gamma} e_2 e_3 + \frac{\delta_2}{\gamma} e_3 e_1 + \frac{\delta_3}{\gamma} e_1 e_2 - \alpha_1 ee_1 - \alpha_2 ee_2 - \alpha_3 ee_3\right\}.$$

Then it is immediately evident by performing the operations indicated [cf. §§ 99 (7) and 240 (5)] that*

$$Kx = x + i \mid xS.$$

(2) Similarly if P be any plane, then $KP = P + i \mid . P \mid S$.

For let $P = \mid p$. Then $Kp = p + i \mid pS$.

By taking the supplement of both sides of this equation

$$\mid Kp = K \mid p = KP = \mid p + i \mid\mid . pS = P + i \mid (\mid p \mid S) = P + i \mid . P \mid S.$$

Thus the system $\mid S$ bears the same relation to the transformation of planes that the system S bears to the transformation of points.

(3) It follows that, corresponding to every theorem referring to systems of forces, there exists a theorem referring to small congruent transformations. The system S will be called the associated system of the transformation. Also since S completely defines the transformation, it will be adopted as its name. Thus we shall speak of the transformation S.

* This formula has not been given before, as far as I am aware.

278. Properties deduced from the Associated System. (1) If the associated system be (with the previous notation) $\frac{\delta_1}{\gamma} e_2e_3 + \frac{\delta_2}{\gamma} e_3e_1 + \frac{\delta_3}{\gamma} e_1e_2$, the transformation is a translation along the line $e(\delta_1 e_1 + \delta_2 e_2 + \delta_3 e_3)$ of parameter $\sqrt{\{\delta_1^2 + \delta_2^2 + \delta_3^2\}}$. Hence if the associated system be the single force F, where F is anti-spatial, the transformation is a translation of axis $|F$ and of parameter $\sqrt{\{F|F\}}$.

If the associated system be the single force $(\alpha_1 ee_1 + \alpha_2 ee_2 + \alpha_3 ee_3)$, the transformation is a rotation round this line of parameter $\sqrt{\{\alpha_1^2 + \alpha_2^2 + \alpha_3^2\}}$. Hence if the associated system be the single force F, where F is spatial, the transformation is a rotation round F of parameter $\sqrt{\{-F|F\}}$.

(2) Hence the condition that the transformation S be a translation or a rotation is $(SS) = 0$. The additional condition, that it be a translation, is that $(S|S)$ be positive, and the additional condition, that it be a rotation, is that $(S|S)$ be negative.

(3) The system S can be reduced [cf. §§ 160 and 162] to two forces $F + F'$, in such a way that either one force is in a given line, or one force passes through a given point and the other force lies in a given plane. If both F and F' are spatial, the transformation has been reduced to two rotations round the two lines. If one (or both) of the forces be anti-spatial, instead of a rotation round the line of that force a translation along its polar line must be substituted.

Since the given point may be assumed to be spatial, and the pole of the given plane may be assumed to be a given spatial point, it follows that it is always possible to reduce a small congruent transformation to a rotation round an axis through one given spatial point and a translation along an axis through another given spatial point. The two given points may be chosen to coincide.

(4) The axis of the transformation is the axis of the system S.

Let F be a force on this axis, then

$$S = \theta F + i\phi |F.$$

If $F = \lambda ee_1 + \mu ee_2 + \nu ee_3 + \varpi e_2e_3 + \rho e_3e_1 + \sigma e_1e_2$; then the condition,

$$\lambda\varpi + \mu\rho + \nu\sigma = 0,$$

must be fulfilled.

Also
$$S = (\theta\lambda + \phi\varpi)\, ee_1 + (\theta\mu + \phi\rho)\, ee_2 + (\theta\nu + \phi\sigma)\, ee_3 + (\theta\varpi - \phi\lambda)\, e_2e_3 + (\theta\rho - \phi\mu)\, e_3e_1 + (\theta\sigma - \phi\nu)\, e_2e_3.$$

Hence with the previous notation,

$$\theta\lambda + \phi\varpi = -\alpha_1, \quad \theta\mu + \phi\rho = -\alpha_2, \quad \theta\nu + \phi\sigma = -\alpha_3,$$

$$\theta\varpi - \phi\lambda = \frac{\delta_1}{\gamma}, \quad \theta\rho - \phi\mu = \frac{\delta_2}{\gamma}, \quad \theta\sigma - \phi\nu = \frac{\delta_3}{\gamma}.$$

Therefore $\lambda = \dfrac{-\theta\alpha_1 - \phi\dfrac{\delta_1}{\gamma}}{\theta^2+\phi^2}$, $\varpi = \dfrac{\theta\dfrac{\delta_1}{\gamma} - \phi\alpha_1}{\theta^2+\phi^2}$, with four similar equations;

where the ratio of θ to ϕ is given by

$$\theta\phi\left(\alpha^2 - \frac{\delta^2}{\gamma^2}\right) + (\phi^2 - \theta^2)\frac{(\alpha_1\delta_1 + \alpha_2\delta_2 + \alpha_3\delta_3)}{\gamma} = 0,$$

and α and δ are put for $\sqrt{\{\alpha_1^2 + \alpha_2^2 + \alpha_3^2\}}$ and $\sqrt{\{\delta_1^2 + \delta_2^2 + \delta_3^2\}}$.

The equation of condition gives two ratios for $\theta : \phi$, and hence two lines are indicated as the axis. One of these is F and the other is $|F$; the spatial line is the axis.

The equation of condition for $\theta : \phi$ can be given in another form. For

$$(S|S) = -\alpha^2 + \frac{\delta^2}{\gamma^2}, \text{ and } (SS) = -2i\left\{\frac{\alpha_1\delta_1 + \alpha_2\delta_2 + \alpha_3\delta_3}{\gamma}\right\}.$$

Therefore $2\theta\phi\,(S|S) + (\phi^2 - \theta^2)\,i\,(SS) = 0$.

(5) The character of a small congruent transformation will be conceived as completely determined by its axis and the ratio of θ to ϕ; the remaining constant simply determines its intensity. In other words, the two transformations, of which the associated systems [cf. § 160 (1)] are S and λS, are of the same character with their intensities in the ratio $1 : \lambda$.

The performance of a given transformation λ times increases the intensity in the ratio $\lambda : 1$. For let S be the associated system.

Then $Kx = x + i\,|xS$, $K^2x = Kx + i\,|KxS = Kx + i\,|xS = x + 2i\,|xS$, neglecting squares of small quantities.

Hence $$K^\lambda x = x + \lambda i\,|xS.$$

279. Work. (1) Let any two spatial points a and b on a line ab be transformed into Ka and Kb, let the angle between aKa and ab be ψ and that between bKb and ab be χ, then

$$\sinh\frac{\overline{aKa}}{\gamma}\cos\psi = \sinh\frac{\overline{bKb}}{\gamma}\cos\chi$$

or $\overline{aKa}\cos\psi = \overline{bKb}\cos\chi$, since the transformation is small.

To prove this proposition, notice that

$$\sinh\frac{\overline{aKa}}{\gamma} = \sqrt{\frac{-(aKa\,|aKa)}{(a\,|a)(Ka\,|Ka)}} = \frac{\sqrt{\{-(aKa\,|aKa)\}}}{(a\,|a)},$$

and $$\cos\theta = \frac{-(aKa\,|ab)}{\sqrt{\{(aKa\,|aKa)(ab\,|ab)\}}}.$$

Hence $$\sinh\frac{\overline{aKa}}{\gamma}\cos\theta = \frac{-(aKa\,|ab)}{(a\,|a)\sqrt{\{-(ab\,|ab)\}}}.$$

Therefore we have to prove that

$$\frac{-(aKa\,|\,ab)}{(a\,|\,a)} = \frac{-(bKb\,|\,ab)}{(b\,|\,b)}.$$

Let the associated system of the transformation be S, where $S = \lambda(ab + cd)$. Then

$$Ka = a + i\,|\,aS = a + \lambda i\,|\,acd,$$

and

$$aKa = i\lambda a\,|\,acd.$$

Hence

$$(aKa\,|\,ab) = (ab\,|\,aKa) = i\lambda\,(ab\,.\,acd\,|\,a)$$
$$= i\lambda\,[ab\,\{(a\,|\,a)\,cd + (d\,|\,a)\,ac + (c\,|\,a)\,da\}]$$
$$= i\lambda\,(a\,|\,a)\,(abcd) = i\,(a\,|\,a)\,(abS).$$

Therefore

$$\frac{(aKa\,|\,ab)}{(a\,|\,a)} = i\,(abS) = \frac{(bKb\,|\,ab)}{(b\,|\,b)}.$$

We notice that if S be the associated system of the transformation

$$\overline{aKa}\cos\psi = \overline{bKb}\cos\chi = \frac{-i\,(abS)}{\sqrt{\{-(ab\,|\,ab)\}}}.$$

(2) *Definition of Work.* Let any point a on the line of a force of intensity ρ be transformed by a small congruent transformation to Ka, so that aKa makes an angle ψ with the force, then $\rho\,.\,\overline{aKa}\,.\cos\psi$ is said to be the work done by the force during the transformation.

It follows from the previous proposition that the work done by the force is the same for all points on its line.

(3) Let b be another point on the force (F) so that $F = ab$. Then

$$\rho = \sqrt{\{-(ab\,|\,ab)\}}.$$

Hence by (1) the work done by F during the transformation S is

$$-i\,(FS).$$

(4) Let the work done by any system of forces S' be defined to be the sum of the works done by the separate forces of the system. Thus let

$$S' = F + F' + F'' + \text{etc.}$$

Then the work done by S' during the transformation S is

$$-i\,(FS) - i\,(F'S) - i\,(F''S) - \text{etc.},$$

but this is

$$-i\,(SS').$$

We notice (α) that the work done by a system of forces during a small congruent transformation is the same however the system be resolved into component forces; and (β) that the work done by the system S' during the transformation λS is equal to the work done by the system S during the transformation $\lambda S'$; where λ is small, but the intensities of S and S' are not necessarily small.

If two systems be reciprocal, that is if $(SS') = 0$, then no work is done by either one during the transformation symbolized by the other.

280. Characteristic Lines. (1) Let the line joining any point with its transformed position, after a small congruent transformation, be called a characteristic line of the point: and let the line of intersection of a plane with its transformed position be called a characteristic line of the plane.

Thus if x be changed to Kx, the line xKx is the characteristic line of the point x; and if the plane P be changed to KP, the line PKP is the characteristic line of the plane P.

(2) Let L be any line, and KL its transformed position, and let L intersect KL, then L is the characteristic line of some point and also of some plane. For let L and KL intersect in Ka, and consider the points a and Ka. Since Ka lies on KL, then a lies on L, hence $L \equiv aKa$. Thus L is the characteristic line of the point a. Also consider the plane $P = K^{-1}a \,.\, a \,.\, Ka$, then $KP = aKaK^2a$. Hence aKa (i.e. L) is the characteristic line of the plane P.

(3) If S be the associated system, the characteristic line of any point x is $x \,|\, xS$, and the characteristic line of the plane P is $P \,| \,.\, P \,|S$.

(4) The locus of points x on the characteristic lines, which pass through a given point a, is given by $(axKax) = 0$. This is a quadric cone through the point a.

The equation can also be written $(aKa \,.\, xKx) = 0$.

Hence the characteristic lines of the points x are those characteristic lines, which intersect the characteristic line aKa.

The equation can also be written $(a \,|aS \,.\, x \,|xS) = 0$; that is [cf. § 167 (3)],

$$(axS)(ax \,|S) - \tfrac{1}{2}(SS)(ax \,|\, ax) = 0.$$

(5) Similarly if AP be a characteristic line lying in the plane A, then the planes P envelope the conic $(APKAP) = 0$, which lies in the plane A; that is to say, the characteristic lines in the plane A envelope a conic.

The equation can also be written $(AKA \,.\, PKP) = 0$. Hence the planes P are such that their characteristic lines intersect the characteristic line AKA.

The plane-equation of the conic can also be written

$$(APS)(AP \,|S) - \tfrac{1}{2}(SS)(AP \,|AP) = 0.$$

281. Elliptic Space. (1) The Kinematics of Elliptic Space can be developed in almost identically the same manner as that of Hyperbolic Space [cf. § 268], only with a greater simplicity.

The absolute quadric being now imaginary, the four corners of the self-corresponding tetrahedron in any congruent transformation must also be imaginary.

Let a_1, a_2, a_3, a_4 be these four corners. Let the lines a_1a_2 and a_3a_4 be real conjugate polar lines. Then a_1 and a_2 are conjugate imaginaries, and so are a_3 and a_4.

Thus let e_1 and e_2 be two real quadrantal points on a_1a_2, and let e_3 and e_4 be two real quadrantal points on a_3a_4. Then

$$a_1 = e^{-\frac{i\pi}{4}} e_1 - e^{\frac{i\pi}{4}} e_2, \quad a_2 = e^{\frac{i\pi}{4}} e_1 - e^{-\frac{i\pi}{4}} e_2, \quad a_3 = e^{-\frac{i\pi}{4}} e_3 - e^{\frac{i\pi}{4}} e_4, \quad a_4 = e^{\frac{i\pi}{4}} e_3 - e^{-\frac{i\pi}{4}} e_4.$$

The transformation changes a_1 into $e^{-i\alpha}a_1$, a_2 into $e^{i\alpha}a_2$, a_3 into $e^{-i\beta}a_3$, a_4 into $e^{i\beta}a_4$. Hence any point $x = \eta_1 e_1 + \eta_2 e_2 + \eta_3 e_3 + \eta_4 e_4$ is changed into

$$Kx = (\eta_1 \cos\alpha + \eta_2 \sin\alpha) e_1 + (\eta_2 \cos\alpha - \eta_1 \sin\alpha) e_2$$
$$+ (\eta_3 \cos\beta + \eta_4 \sin\beta) e_3 + (\eta_4 \cos\beta - \eta_3 \sin\beta) e_4.$$

Thus $$(Kx \mid Kx) = \eta_1^2 + \eta_2^2 + \eta_3^2 + \eta_4^2 = (x \mid x).$$

And $$(x \mid Kx) = (\eta_1^2 + \eta_2^2) \cos\alpha + (\eta_3^2 + \eta_4^2) \cos\beta.$$

(2) Thus any point $\eta_1 e_1 + \eta_2 e_2$ on e_1e_2 is transferred through a distance $\gamma\alpha$, any point on e_3e_4 through a distance $\gamma\beta$. Similarly any plane through e_1e_2 is transferred through an angle β, any plane through e_3e_4 is transferred through an angle α.

(3) The distance of Kx from e_1e_2 is δ, where

$$\sin\frac{\delta}{\gamma} = \sqrt{\frac{(e_1e_2Kx \mid e_1e_2Kx)}{(Kx \mid Kx)(e_1e_2 \mid e_1e_2)}}.$$

But

$$(e_1e_2Kx \mid e_1e_2Kx) = (\eta_3 \cos\beta + \eta_4 \sin\beta)^2 + (\eta_4 \cos\beta - \eta_3 \sin\beta)^2 = \eta_3^2 + \eta_4^2.$$

Hence $$\sin\frac{\delta}{\gamma} = \sqrt{\frac{\eta_3^2 + \eta_4^2}{(x \mid x)}}.$$

Thus the distance of Kx from e_1e_2 is the same as that of x from e_1e_2. Similarly for the distances from e_3e_4.

(4) Let α and β be called the parameters of the transformation. The transformation will be described as the transformation, axis e_1e_2, parameters α, β, or as the transformation, axis e_3e_4, parameters β, α.

The transformation, axis e_1e_2, parameters α, 0, will be called the translation, axis e_1e_2, parameter α; or else, the rotation, axis e_3e_4, parameter α.

(5) Any congruent transformation may be conceived as the combination of a rotation round and a translation along the same axis; or as the combination of two rotations round two reciprocally polar lines; or as the combination of two translations with two reciprocally polar lines as axes.

The distinction between translations and rotations, which exists in Hyperbolic Space, does not exist in Elliptic Space.

282. Surfaces of Equal Displacement. (1) The locus of points, for which the distance of displacement in the transformation axis e_1e_2, parameters α, β, is $\gamma\sigma$, is the quadric surface

$$\frac{(x|Kx)}{\sqrt{\{(x|x)(Kx|Kx)\}}} = \cos\sigma.$$

Hence employing the notation of the previous articles,

$$(\eta_1^2 + \eta_2^2)\cos\alpha + (\eta_3^2 + \eta_4^2)\cos\beta = \cos\sigma\,(\eta_1^2 + \eta_2^2 + \eta_3^2 + \eta_4^2).$$

Therefore $(\eta_1^2 + \eta_2^2)(\cos\alpha - \cos\sigma) + (\eta_3^2 + \eta_4^2)(\cos\beta - \cos\sigma) = 0.$

(2) The system of quadric surfaces found by varying σ is also easily shown to be the system which remains coincident with itself during the transformation. Its sections by planes through e_1e_2 or e_3e_4 (*i.e.* by planes perpendicular to e_3e_4 or e_1e_2) are circles.

283. Vector Transformations. (1) An interesting special case discovered by Clifford* arises in Elliptic Geometry which does not occur in Hyperbolic Geometry.

Let a congruent transformation, of which the parameters are numerically equal, be called a vector transformation. Thus, with the above notation, $\alpha = \pm\beta$.

Hence [cf. § 282 (1)] any point x is transferred through a distance $\gamma\alpha$. Accordingly in a vector transformation all points are transferred through the same distance, and similarly all planes are rotated round the same angle.

(2) Again, the line xKx is parallel to the axis e_1e_2.

For, taking $\alpha = \beta$,

$$\begin{aligned} xKx = &-(\eta_1^2 + \eta_2^2)\sin\alpha\,.\,e_1e_2 - (\eta_3^2 + \eta_4^2)\sin\alpha\,.\,e_3e_4 \\ &+ (\eta_1\eta_4 - \eta_3\eta_2)\sin\alpha\,.\,e_1e_3 - (\eta_1\eta_3 + \eta_2\eta_4)\sin\alpha\,.\,e_1e_4 \\ &+ (\eta_2\eta_4 + \eta_1\eta_3)\sin\alpha\,.\,e_2e_3 + (\eta_1\eta_4 - \eta_2\eta_3)\sin\alpha\,.\,e_2e_4 \\ = &-(\eta_1^2 + \eta_2^2)\sin\alpha\,.\,e_1e_2 - (\eta_3^2 + \eta_4^2)\sin\alpha\,.\,|e_1e_2 \\ &+ (\eta_1\eta_4 - \eta_3\eta_2)\sin\alpha\,(e_1e_3 - |e_1e_3) + (\eta_2\eta_4 + \eta_1\eta_3)\sin\alpha\,(e_2e_3 - |e_2e_3). \end{aligned}$$

Hence $xKx + |xKx = -(\eta_1^2 + \eta_2^2 + \eta_3^2 + \eta_4^2)\sin\alpha\,(e_1e_2 + |e_1e_2).$

Similarly if $\alpha = -\beta$,

$$xKx - |xKx = -(x|x)\sin\alpha\,(e_1e_2 - |e_1e_2).$$

Thus [cf. §§ 234 (2)] if α, α be the parameters of the vector transformation, all points are transferred along right-parallels to e_1e_2; and, if α, $-\alpha$ be the parameters, all points are transferred along left-parallels.

Let these transformations be called right-vector transformations and left-vector transformations respectively.

It follows therefore that any one of the lines parallel to the axis of a vector transformation may with equal right be itself conceived as the axis.

* Cf. Clifford, *Collected Papers, Preliminary Sketch of Biquaternions, loc. cit.* p. 370.

284. Associated Vector Systems of Forces. Let R be the unit right-vector system of forces $e_1e_2 + |e_1e_2$ [cf. § 236 (1)], then the right-vector transformation, axis e_1e_2 and parameter α, can be represented by

$$Kx = x \cos\alpha - \sin\alpha\,|xR.$$

This representation, unlike the preceding formula of § 277 and the subsequent formula of § 286, is not confined to the case when α is small. Let R be called the associated unit system of the right-vector transformation, axis e_1e_2.

Similarly if L be the unit left-vector system $e_1e_2 - |e_1e_2$, then the left-vector transformation, axis e_1e_2 and parameter β, can be represented by

$$K'x = x \cos\beta + \sin\beta\,|xL.$$

Let L be called the associated unit system of the left-vector transformation, axis e_1e_2.

285. Successive Vector Transformations. (1) If a right-vector transformation and a left-vector transformation be successively applied, the result is independent of the order of application*.

For let K and K' be the matrices denoting the vector transformations of the last article only with different axes, namely e_1e_2 and $e_1'e_2'$.

Then $$KK'x = K'x \cos\alpha - \sin\alpha\,|(K'x\,.\,R)$$
$$= x\cos\alpha\cos\beta + \cos\alpha\sin\beta\,|xL - \sin\alpha\cos\beta\,|xR$$
$$- \sin\alpha\sin\beta\,|(|xL\,.\,R).$$

Also $$K'Kx = Kx\cos\beta + \sin\beta\,|(Kx\,.\,L)$$
$$= x\cos\alpha\cos\beta - \sin\alpha\cos\beta\,|xR + \cos\alpha\sin\beta\,|xL - \sin\alpha\sin\beta\,|(|xR\,.\,L).$$

But $|(|xL\,.\,R) = ||xL\,.\,|R = -xL\,.\,R$; and similarly $|(|xR\,.\,L) = xR\,.\,L$. But, by § 167 equation (25), since $(RL) = 0$ [cf. § 235 (4)], it follows that

$$(xL\,.\,R) + (xR\,.\,L) = 0.$$

Hence finally, $$KK'x = K'Kx.$$

(2) Again, let R and R' be both of the same name, say both unit right-vector systems of the forms $e_1e_2 + |e_1e_2$, $e_1'e_2' + |e_1'e_2'$; and let K and K' be the corresponding matrices.

Then

$$K'Kx = x\cos\alpha\cos\beta - \cos\alpha\sin\beta\,|xR' - \cos\beta\sin\alpha\,|xR + \sin\alpha\sin\beta\,|(|xR'\,.\,R).$$

Now $$|(|xR'\,.\,R) = -xR'\,.\,|R = -xR'\,.\,R = -(RR')x + xR\,.\,R'.$$

Hence $$K'Kx = x\cos\alpha\cos\beta - \cos\alpha\sin\beta\,|xR' - \cos\beta\sin\alpha\,|xR$$
$$-(RR')\sin\alpha\sin\beta\,.\,x + \sin\alpha\sin\beta\,.\,xR\,.\,R'.$$

* Cf. Sir R. S. Ball, *loc. cit.* p. 370.

Similarly

$$KK'x = x\cos\alpha\cos\beta - \cos\alpha\sin\beta\,|xR' - \sin\alpha\cos\beta\,|xR - \sin\alpha\sin\beta\,.\,xR\,.\,R'.$$

Hence $K'Kx$ is not equal to $KK'x$.

Thus two vector transformations of the same name (left or right) applied successively produce different results according to their order*.

(3) The resultant transformation, which is equivalent to two successive vector transformations of the same name, is itself a vector transformation† of that name.

Let R and R' be the two unit associated right-vector systems with any axes, and let α and β be the respective parameters. Then it is proved in the preceding subsection that

$$KK'x = x\cos\alpha\cos\beta - \cos\alpha\sin\beta\,|xR' - \sin\alpha\cos\beta\,|xR - \sin\alpha\sin\beta\,.\,xR\,.\,R'.$$

Let K_1x be written for $KK'x$.

Also let x be a point at unit intensity, so that $(x\,|x) = 1$.

Then $\qquad (x\,|K_1x) = (x\,|x)\cos\alpha\cos\beta - \sin\alpha\sin\beta\,\{x\,|(xR\,.\,R')\}.$

Now $\qquad x\,|(xR\,.\,R') = x\,.\,(|xR\,.\,|R') = x\,.\,(R\,|x\,.\,R') = x\,.\,R\,|x\,.\,R'.$

Again $R = e_1e_2 + |e_1e_2 = xp + |xp$, where p is the unit point on the right-parallel to e_1e_2 through x and normal to x, so that $(x\,|p) = 0$.

Similarly $R' = e_1'e_2' + |e_1'e_2' = xp' + |xp'$, where p' is a similar point such that xp' is a right-parallel to $e_1'e_2'$.

Then $$R\,|x = xp\,|x = (x\,|x)\,p - (x\,|p)\,x = p.$$

Hence $$x\,|(xR\,.\,R') = xp\,(xp' + |xp') = (xp\,|xp') = \tfrac{1}{2}(RR').$$

Thus $$(x\,|K_1x) = \cos\alpha\cos\beta - \tfrac{1}{2}(RR')\sin\alpha\sin\beta.$$

Accordingly, if $\gamma\sigma$ be the distance through which x is transferred, remembering that $(K_1x\,|K_1x) = (x\,|x) = 1$, we find

$$\cos\sigma = \cos\alpha\cos\beta - \tfrac{1}{2}(RR')\sin\alpha\sin\beta.$$

Therefore the resultant distance of displacement of x to K_1x is the same for all positions of x. Therefore [cf. § 283 (1)] the transformation is a vector transformation of which the parameter is σ.

The proof is exactly similar if the two component transformations are left-vector transformations.

(4) It now remains to be proved that the resultant transformation is of the same name (left or right) as the component transformations. The method of proof will in fact prove the first part of the proposition also.

It is easy to prove that, with the notation of the previous sub-section,

$$KK'x = \cos\sigma\,.\,x - \cos\alpha\sin\beta\,.\,p' - \sin\alpha\cos\beta\,.\,p + \sin\alpha\sin\beta\,|(xp'p),$$

and $$K'Kx = \cos\sigma\,.\,x - \cos\alpha\sin\beta\,.\,p' - \sin\alpha\cos\beta\,.\,p + \sin\alpha\sin\beta\,|(xpp').$$

* Sir R. S. Ball, *loc. cit.*

† Cf. Sir R. S. Ball, *loc. cit.*

Then, if K_1 stand for KK', we find

$$xK_1x = -\sin\alpha\cos\beta\,.\,xp - \sin\beta\cos\alpha\,.\,xp' + \sin\alpha\sin\beta\,.\,|p'p,$$

since $$x\,|(xp'p) = (x\,|p)\,|xp' + (x\,|p')\,|px + (x\,|x)\,|p'p = |p'p.$$

Now let y be any other point, and let q and q' stand in the same relation to y as do p and p' respectively to x. Then

$$yK_1y = -\sin\alpha\cos\beta\,.\,yq - \sin\beta\cos\alpha\,.\,yq' + \sin\alpha\sin\beta\,.\,|q'q.$$

But $$xp + |xp = yq + |yq,\quad xp' + |xp' = yq' + |yq',$$

and by § 236 (2) $$pp' + |pp' = qq' + |qq'.$$

Hence $$xK_1x + |xK_1x = yK_1y + |yK_1y.$$

Therefore yK_1y and xK_1x are right-parallels, which was to be proved.

(5) These theorems, due to Sir Robert Ball, have been proved analytically in order to illustrate the algebraic transformations. They can however be more easily proved geometrically.

For consider two vector transformations of opposite names applied successively. Let the right-vector transformation transfer a to b, and the left-vector transformation transfer b to c [cf. fig. 1]. Complete the parallelogram $abcd$ [cf. § 237 (3)]. Then the left-vector transformation transfers a to d, and the right-vector transformation transfers d to c.

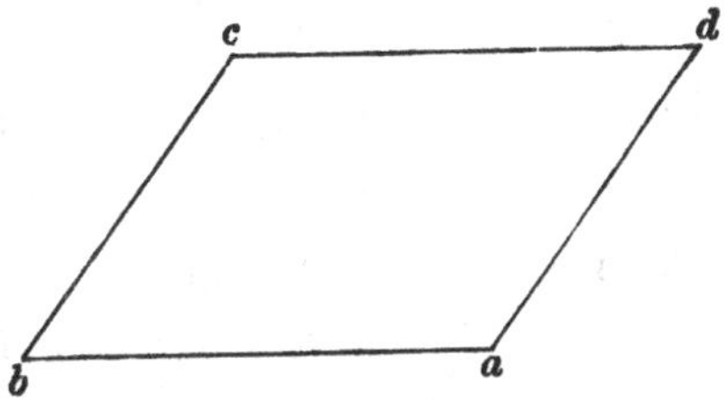

FIG. 1.

Hence in whatever order they are applied the same ultimate result is reached.

But [cf. fig. 2] if any other point a' is transferred by the same

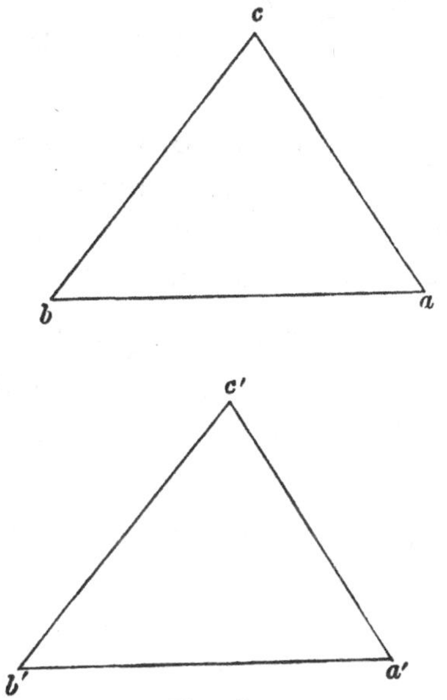

FIG. 2.

combination of transformations successively to b' and to c', $a'c'$ is not parallel to ac, since ab and $a'b'$ are not parallels of the same name as bc and $b'c'$ [cf. 237 (4)]. Hence the combination is not itself a vector transformation.

Consider again two vector transformations of the same name, say both right-vector transformations, call them T_1 and T_2.

Let T_1 transfer a to b, a' to b'; let T_2 transfer b to c and b' to c' [cf. fig. 2]. Then $\overline{ab} = \overline{a'b'}$ and $\overline{bc} = \overline{b'c'}$, and the pairs of lines are parallels of the same name. Hence by § 237 (4), $a'c'$ and ac are equal and parallels of that name.

Hence the resultant of the combination is itself a vector transformation of the same name.

But a parallelogram cannot be formed in which the opposite sides are parallels of the same name. Hence the combination of T_1 first, T_2 second, gives a different result from that of T_2 first, T_1 second.

286. Small Displacements. (1) The theory of small displacements in Elliptic Space is the same as that in Hyperbolic Space [cf. §§ 275 *et seqq.*].

Let $$S = \alpha_{34}e_1e_2 + \alpha_{42}e_1e_3 + \alpha_{23}e_1e_4 + \alpha_{14}e_2e_3 + \alpha_{13}e_4e_2 + \alpha_{12}e_3e_4,$$
where the coefficients α_{12}, etc., are small.

Then $Kx = x + |xS$, gives
$$Kx = (\xi_1 - \alpha_{12}\xi_2 - \alpha_{13}\xi_3 - \alpha_{14}\xi_4)\, e_1 + (\xi_2 + \alpha_{12}\xi_1 - \alpha_{23}\xi_3 - \alpha_{42}\xi_4)\, e_2$$
$$+ (\xi_3 + \alpha_{13}\xi_1 + \alpha_{23}\xi_2 + \alpha_{34}\xi_4)\, e_3 + (\xi_4 + \alpha_{14}\xi_1 + \alpha_{42}\xi_2 + \alpha_{34}\xi_3)\, e_4.$$
This is the most general type of small congruent transformation.

(2) Also $K|x = |Kx$. Let P be any plane, then
$$|KP = K|P = |P - |(|P \,.\, S).$$
Hence $$KP = -|K|P = P + ||(|P \,.\, S) = P - |(P|S).$$
Thus $|S$ bears the same formal relation to the transformation of a plane P, as S does to that of a point p.

(3) It follows, as in Hyperbolic Space, that corresponding to every theorem referring to a system of forces there exists a theorem referring to small congruent transformations. The system S will be called the associated system, and S will be used as the name of the transformation.

(4) Thus if $S = F + F'$, the transformation S is equivalent to two rotations round F and F', of parameters $\sqrt{(F|F)}$ and $\sqrt{(F'|F')}$ respectively. Every transformation possesses two axes, which are found as in Hyperbolic Space.

(5) Also [cf. § 160 (2)] if b be any point, $S = pb + P|b$. Thus the transformation is equivalent to a rotation, parameter $\sqrt{(pb|pb)}$, round the axis pb, and a rotation, parameter $\sqrt{(P|b|.P|b)}$, round the axis $P|b$: this last rotation can also be described as a translation, parameter $\sqrt{(P|b|.P|b)}$, along the axis $b|P$.

(6) Let any two points a and b on a line ab be transferred to Ka and Kb; let the angle between aKa and ab be θ, and that between bKb and ab be ϕ, then

$$\sin \frac{\overline{aKa}}{\gamma} . \cos \theta = \sin \frac{\overline{bKb}}{\gamma} . \cos \phi,$$

or, since the transformation is small,

$$\overline{aKa} . \cos \theta = \overline{bKb} . \cos \phi.$$

This theorem is proved (as in Hyperbolic Space) if we prove that

$$\frac{(aKa \mid ab)}{(a \mid a)} = \frac{(bKb \mid ab)}{(b \mid b)}.$$

Now let the associated system S be written in the form $\lambda (ab + cd)$.

Then $$Ka = a + \mid aS = a + \lambda \mid acd,$$

and $$aKa = \lambda a \mid acd = \lambda (a \mid d) \mid ac + \lambda (a \mid c) \mid da + \lambda (a \mid a) \mid cd.$$

Hence $$(aKa \mid ab) = (ab \mid aKa) = \lambda (a \mid a) (abcd).$$

Therefore $$\frac{(aKa \mid ab)}{(a \mid a)} = \lambda (abcd) = \frac{(bKb \mid ab)}{(b \mid b)}.$$

(7) The definition of work is the same as in Hyperbolic Space.

The work done by a force is (by the last proposition) the same for all points on its line.

If λS be the associated system of a transformation, where λ is small and S is not necessarily small, and S' be any system of forces, then the work done by S' during the transformation λS is $\lambda (SS')$. This is equal to the work done by S during the transformation $\lambda S'$. The proof of this theorem is exactly as in the case of Hyperbolic Space.

If two systems S, S' are reciprocal, so that $(SS') = 0$, then the work done by S (or S') during the transformation $\lambda S'$ (or λS) is zero.

CHAPTER VII.

CURVES AND SURFACES*.

287. CURVE LINES. (1) Let the Space be Elliptic (polar) and of three dimensions, though the resulting formulæ will in general hold for Hyperbolic Space. Let any point x be represented, as usual, by $\xi_1 e_1 + \xi_2 e_2 + \xi_3 e_3 + \xi_4 e_4$, where the co-ordinate points e_1, e_2, e_3, e_4 form a unit normal system.

Now let the co-ordinates of x, namely ξ_1, ξ_2, ξ_3, ξ_4, be functions of some variable τ. Then, as τ varies, x traces out a curve line.

When τ becomes $\tau + \delta\tau$, where $\delta\tau$ is indefinitely small, let x become

$$x + \dot{x}\delta\tau.$$

Then obviously $\dot{x} = \dot{\xi}_1 e_1 + \dot{\xi}_2 e_2 + \dot{\xi}_3 e_3 + \dot{\xi}_4 e_4$. Let x_1 stand for $x + \dot{x}\delta\tau$. Let $\ddot{x}$, $\dddot{x}$ etc. be derived in regular sequence by the same process as $\dot{x}$ is derived from x.

(2) Now as x changes its position to x_1, it might change its intensity as expressed in the above notation. Let it be assumed that the intensity of x remains always at unit intensity.

Hence $(x|x) = 1$, $(x|\dot{x}) = 0$, $(\dot{x}|\dot{x}) + (x|\ddot{x}) = 0$, and so on, by successive differentiations. Hence $\dot{x}$ is a point on the polar plane of x.

Also the same variation of τ to $\tau + \delta\tau$, which changes x to x_1, will change x_1 to x_2, where $x_2 = x_1 + \dot{x}_1\delta\tau = x + 2\dot{x}\delta\tau + \ddot{x}(\delta\tau)^2$, and will change x_2 to x_3, where $x_3 = x + 3\dot{x}\delta\tau + 3\ddot{x}(\delta\tau)^2 + \dddot{x}(\delta\tau)^3$, and so on.

(3) Let $\delta\sigma$ denote the length of the arc $\overline{xx_1}$, then

$$\sin\frac{\delta\sigma}{\gamma} = \frac{\delta\sigma}{\gamma} = \sqrt{\frac{(xx_1|xx_1)}{(x|x)(x_1|x_1)}} = \sqrt{(x\dot{x}|x\dot{x})}\,.\,\delta\tau$$
$$= \sqrt{\{(\dot{x}|\dot{x}) - (x|\dot{x})^2\}}\,\delta\tau = \sqrt{(\dot{x}|\dot{x})}\,\delta\tau.$$

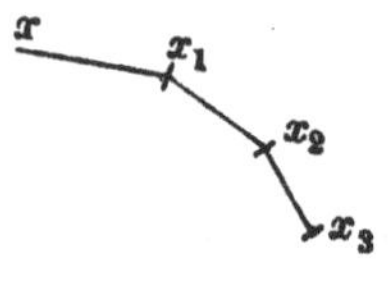

FIG. 1.

Therefore $$\frac{d\sigma}{d\tau} = \gamma\sqrt{(\dot{x}|\dot{x})}.$$

* This application of the Calculus of Extension has not been made before, as far as I am aware.

(4) Again, let $\delta\epsilon$ denote the angle of contingence of the curve at x corresponding to the small arc $\delta\sigma$. Then $\delta\epsilon$ denotes the angle between xx_1 and x_1x_2. Hence by considering the triangle xx_1x_2, we find [cf. § 216 (2)]

$$\delta\epsilon = \sin\delta\epsilon = \sqrt{\frac{(xx_1x_2 \mid xx_1x_2)}{(xx_1 \mid xx_1)(x_1x_2 \mid x_1x_2)}}.$$

Now $$xx_1x_2 = x\dot{x}\ddot{x}\,(\delta\tau)^3;$$

and $$(x_1x_2 \mid x_1x_2) = (xx_1 \mid xx_1) = (x\dot{x} \mid x\dot{x})(\delta\tau)^2, \text{ ultimately.}$$

Hence $$\delta\epsilon = \frac{\sqrt{(x\dot{x}\ddot{x} \mid x\dot{x}\ddot{x})}}{(x\dot{x} \mid x\dot{x})}\,\delta\tau = \frac{\sqrt{(x\dot{x}\ddot{x} \mid x\dot{x}\ddot{x})}}{(\dot{x} \mid \dot{x})}\,\delta\tau.$$

Also we may notice [cf. subsection (2)] that

$$(x\dot{x}\ddot{x} \mid x\dot{x}\ddot{x}) = \begin{vmatrix} 1, & 0, & -(\dot{x} \mid \dot{x}) \\ 0, & (\dot{x} \mid \dot{x}), & (\dot{x} \mid \ddot{x}) \\ -(\dot{x} \mid \dot{x}), & (\dot{x} \mid \ddot{x}), & (\ddot{x} \mid \ddot{x}) \end{vmatrix}$$
$$= (\dot{x} \mid \dot{x})(\ddot{x} \mid \ddot{x}) - (\dot{x} \mid \ddot{x})^2 - (\dot{x} \mid \dot{x})^3$$
$$= (\dot{x}\ddot{x} \mid \dot{x}\ddot{x}) - (\dot{x} \mid \dot{x})^3.$$

Hence $$\delta\epsilon = \frac{\sqrt{\{(\dot{x}\ddot{x} \mid \dot{x}\ddot{x}) - (\dot{x} \mid \dot{x})^3\}}}{(\dot{x} \mid \dot{x})}\,\delta\tau.$$

(5) The tangent line at x is the line xx_1, that is, the line $x\dot{x}$.

The normal plane at x is the plane through x perpendicular to xx_1. This plane is the plane $x \mid xx_1$, that is, $x \mid x\dot{x}$.

Now $$x \mid x\dot{x} = (x \mid \dot{x}) \mid x - (x \mid x) \mid \dot{x} = - \mid \dot{x}.$$

Hence $|\dot{x}$ is the normal plane of the curve at x.

The normal plane at x_1 is therefore $|\dot{x} + |\ddot{x}\delta\tau$.

Monge's 'polar line' of the curve at x is the line of intersection of the two planes, that is, the line $|\dot{x}\ddot{x}$.

(6) The osculating plane of the curve at x is xx_1x_2, that is, the plane $x\dot{x}\ddot{x}$.

The neighbouring osculating plane at x_1 is $x\dot{x}\ddot{x} + x\dot{x}\dddot{x}\delta\tau$.

The angle between these two planes, namely, $\delta\theta$, is the angle of torsion corresponding to the arc $\delta\sigma$. If the first plane be P and the second plane be Q, then

$$PQ = x\dot{x}\ddot{x} \,.\, x\dot{x}\dddot{x}\delta\tau = (x\dot{x}\ddot{x}\dddot{x})\, x\dot{x}\delta\tau.$$

Hence

$$\delta\theta = \sin\theta = \sqrt{\frac{(PQ \mid PQ)}{(P \mid P)(Q \mid Q)}} = \frac{(x\dot{x}\ddot{x}\dddot{x})\sqrt{(x\dot{x} \mid x\dot{x})}}{(x\dot{x}\ddot{x} \mid x\dot{x}\ddot{x})}\,\delta\tau = \frac{(x\dot{x}\ddot{x}\dddot{x})\sqrt{(\dot{x} \mid \dot{x})}}{(x\dot{x}\ddot{x} \mid x\dot{x}\ddot{x})}\,\delta\tau.$$

288. CURVATURE AND TORSION. (1) Let the 'measure of curvature' or the 'curvature' of the curve be defined to be the rate of increase of ϵ per unit length of σ, and let it be denoted by $\frac{1}{\rho}$.

Let the 'torsion' of the curve be defined to be the rate of increase of θ per unit length of σ, and let it be denoted by $\frac{1}{\kappa}$.

Then
$$\frac{1}{\rho} = \frac{d\epsilon}{d\sigma} = \frac{1}{\gamma}\frac{\sqrt{(x\dot{x}\ddot{x}\,|x\dot{x}\ddot{x})}}{(\dot{x}\,|\dot{x})^{\frac{3}{2}}} = \frac{1}{\gamma}\frac{\sqrt{\{(\dot{x}\ddot{x}\,|\dot{x}\ddot{x}) - (\dot{x}\,|\dot{x})^3\}}}{(\dot{x}\,|\dot{x})^{\frac{3}{2}}},$$
$$\frac{1}{\kappa} = \frac{d\theta}{d\sigma} = \frac{1}{\gamma}\frac{(x\dot{x}\ddot{x}\dddot{x})}{(x\dot{x}\ddot{x}\,|x\dot{x}\ddot{x})}.$$

Hence
$$\frac{1}{\rho^2\kappa} = \frac{1}{\gamma^3}\frac{(x\dot{x}\ddot{x}\dddot{x})}{(\dot{x}\,|\dot{x})^3}.$$

The condition, that a curve is plane, is $\frac{1}{\kappa} = 0$; that is, $(x\dot{x}\ddot{x}\dddot{x}) = 0$.

(2) Newton's geometrical formula for the curvature still holds. For let $\delta\sigma$ be the distance between the points x and $x + \delta x$, and let $\delta\nu$ be the perpendicular from $x + \delta x$ on to the tangent $x\dot{x}$.

Then [cf. § 226 (1)]
$$\frac{\delta\nu}{\gamma} = \sqrt{\frac{(x\dot{x}\delta x\,|x\dot{x}\delta x)}{(x\dot{x}\,|x\dot{x})}}.$$

But
$$\delta x = x + \dot{x}\delta\tau + \tfrac{1}{2}\ddot{x}\,(\delta\tau)^2.$$

Therefore
$$\delta\nu = \tfrac{1}{2}\gamma\sqrt{\frac{(x\dot{x}\ddot{x}\,|x\dot{x}\ddot{x})}{(x\dot{x}\,|x\dot{x})}}\,.\,(\delta\tau)^2 = \tfrac{1}{2}\gamma\sqrt{\frac{(x\dot{x}\ddot{x}\,|x\dot{x}\ddot{x})}{(\dot{x}\,|\dot{x})}}\,.\,(\delta\tau)^2.$$

Hence
$$\frac{\delta\nu}{(\delta\sigma)^2} = \frac{1}{2}\frac{1}{\gamma}\sqrt{\frac{(x\dot{x}\ddot{x}\,|x\dot{x}\ddot{x})}{(\dot{x}\,|\dot{x})^3}} = \frac{1}{2\rho}.$$

(3) The normal planes envelope the 'polar developable,' of which the edge of regression is the locus of intersection of three neighbouring normal planes.

The polar line $|\dot{x}\ddot{x}$ is a tangent to this edge of regression at the point corresponding to x.

The three normal planes at x, x_1, x_2 are
$$|\dot{x}, \quad |\dot{x} + |\ddot{x}\,.\,\delta\tau, \quad |\dot{x} + 2\,|\ddot{x}\,.\,\delta\tau + |\dddot{x}\,.\,(\delta\tau)^2.$$

Hence the point on the edge of regression of the polar developable which corresponds to x is $|\dot{x}\ddot{x}\dddot{x}$. Let the distance, ρ_1, of this point from x be called the radius of spherical curvature.

Then
$$\cos\frac{\rho_1}{\gamma} = \frac{(x\dot{x}\ddot{x}\dddot{x})}{\sqrt{(\dot{x}\ddot{x}\dddot{x}\,|\dot{x}\ddot{x}\dddot{x})}}.$$

(4) The polar line $|\dot{x}\ddot{x}$ meets the osculating plane in a point called the 'centre of curvature.'

This point is $x\dot{x}\ddot{x}\,|\dot{x}\ddot{x}$.

The distance of this point from x is called the radius of curvature—which is not in general equal to the inverse of the measure of curvature except in parabolic geometry.

In order to find the radius of curvature, which will be called ρ_2, we notice that $|\dot{x}\ddot{x}$ is perpendicular to $x\dot{x}\ddot{x}$. For the line through the point $|\dot{x}\ddot{x}\dddot{x}$ perpendicular to $x\dot{x}\ddot{x}$ is the line $|\dot{x}\ddot{x}\dddot{x}\,.\,|x\dot{x}\ddot{x} = |(\dot{x}\ddot{x}\dddot{x}\,.\,x\dot{x}\ddot{x}) = -(x\dot{x}\ddot{x}\dddot{x})\,|\dot{x}\ddot{x}$.

Hence neglecting the numerical factor, this is the line $|\dot{x}\ddot{x}$.

Therefore ρ_2 is really the distance of x from the polar line $|\dot{x}\ddot{x}$.

But the distance of x from any line F is

$$\gamma \sin^{-1}\left\{\sqrt{\frac{(xF\,|xF)}{(x\,|x)\,(F\,|F)}}\right\}.$$

Now $$x\,|\dot{x}\ddot{x} = (x\,|\ddot{x})\,|\dot{x} - (x\,|\dot{x})\,|\ddot{x} = (x\,|\ddot{x})\,|\dot{x}.$$

And $$(|\dot{x}\ddot{x}\,.\,||\dot{x}\ddot{x}) = (\dot{x}\ddot{x}\,|\dot{x}\ddot{x}).$$

Hence $$\sin\frac{\rho_2}{\gamma} = \sqrt{\frac{(x\,|\ddot{x})^2\,(\dot{x}\,|\dot{x})}{(\dot{x}\ddot{x}\,|\dot{x}\ddot{x})}} = \frac{(\dot{x}\,|\dot{x})^{\frac{3}{2}}}{\sqrt{(\dot{x}\ddot{x}\,|\dot{x}\ddot{x})}}.$$

Therefore $$\frac{\gamma^2}{\rho^2} = \operatorname{cosec}^2\frac{\rho_2}{\gamma_2} - 1 = \cot^2\frac{\rho_2}{\gamma}.$$

Hence $$\rho = \gamma\tan\frac{\rho_2}{\gamma}.$$

(5) The principal normal is the normal line in the osculating plane, that is the line,

$$x\dot{x}\ddot{x}\,|\dot{x} = (x\,|\dot{x})\,\dot{x}\ddot{x} + (\ddot{x}\,|\dot{x})\,x\dot{x} + (\dot{x}\,|\dot{x})\,\ddot{x}x = (\dot{x}\,|\ddot{x})\,x\dot{x} + (\dot{x}\,|\dot{x})\,\ddot{x}x.$$

The binormal is the normal perpendicular to the osculating plane, that is the line $x\,|x\dot{x}\ddot{x}$.

It will be useful later to notice that the intensity of $x\dot{x}\ddot{x}\,|\dot{x}$ is

$$\sqrt{\{(\dot{x}\,|\dot{x})\,(x\dot{x}\ddot{x}\,|x\dot{x}\ddot{x})\}}.$$

For if P and Q be two planes, then $(PQ\,|PQ) = (P\,|P)(Q\,|Q) - (P\,|Q)^2$; and by writing $x\dot{x}\ddot{x}$ for P and $|\dot{x}$ for Q the result follows.

289. Planar Formulæ. (1) The complete duality both of Elliptic and of Hyperbolic Space allows formulæ similar to the above to be deduced from the plane equation of a curve.

Let P denote the plane $v_1e_2e_3e_4 - v_2e_1e_3e_4 + v_3e_1e_2e_4 - v_4e_1e_2e_3$. Let v_1, v_2, v_3, v_4 be functions of one variable τ. Then the plane P, as it moves, envelopes a developable surface, of which the curve under consideration is the edge of regression.

Let $P_1 = P + \dot{P}\delta\tau$, where $\dot{P} = \dot{v}_1e_2e_3e_4 + \text{etc.}$, and let $\ddot{P}$ be derived from $\dot{P}$ as $\dot{P}$ is from P, and so on.

Let $P_2 = P + 2\dot{P}\,.\,\delta\tau + \ddot{P}\,(\delta\tau)^2$, $P_3 = \text{etc.}$ Let $(P\,|P) = 1$, so that

$$(P\,|\dot{P}) = 0,\ (\dot{P}\,|\dot{P}) + (P\,|\ddot{P}) = 0.$$

(2) The point on the curve corresponding to P is the point PP_1P_2, that is the point $P\dot{P}\ddot{P}$. The tangent line in P is the line $P\dot{P}$.

The angle of torsion $\delta\theta$ is given by

$$\delta\theta = \sqrt{\frac{(PP_1|PP_1)}{(P|P)(P_1|P_1)}} = \sqrt{(P\dot{P}|P\dot{P})} \, . \, \delta\tau = \sqrt{(\dot{P}|\dot{P})} \, . \, \delta\tau.$$

(3) The angle of contingence $\delta\epsilon$ is the angle between the lines PP_1 and P_1P_2. Hence by considering the stereometrical triangle P, P_1, P_2 and deducing the formula from that for a point-triangle by the theory of duality, we find

$$\delta\epsilon = \sin\delta\epsilon = \sqrt{\frac{(PP_1P_2|PP_1P_2)}{(PP_1|PP_1)(P_1P_2|P_1P_2)}} = \frac{\sqrt{(P\dot{P}\ddot{P}|P\dot{P}\ddot{P})}}{(\dot{P}|\dot{P})}\,\delta\tau$$
$$= \frac{\sqrt{\{(\dot{P}\ddot{P}|\dot{P}\ddot{P}) - (\dot{P}|\dot{P})^3\}}}{(\dot{P}|\dot{P})}\,\delta\tau.$$

(4) The length of the arc $\delta\sigma$ is obviously found from the analogous formula to that which gives the angle of torsion in the point-equation.

Thus
$$\frac{\delta\sigma}{\gamma} = \frac{(P\dot{P}\ddot{P}\dddot{P})\sqrt{(\dot{P}|\dot{P})}}{(P\dot{P}\ddot{P}|P\dot{P}\ddot{P})}\,\delta\tau.$$

Then
$$\frac{d\epsilon}{d\theta} = \frac{\sqrt{(P\dot{P}\ddot{P}|P\dot{P}\ddot{P})}}{(\dot{P}|\dot{P})^{\frac{3}{2}}} = \frac{\sqrt{\{(\dot{P}\ddot{P}|\dot{P}\ddot{P}) - (\dot{P}|\dot{P})^3\}}}{(\dot{P}|\dot{P})^{\frac{3}{2}}};$$

and
$$\frac{d\sigma}{d\theta} = \frac{\gamma(P\dot{P}\ddot{P}\dddot{P})}{(P\dot{P}\ddot{P}|P\dot{P}\ddot{P})}.$$

Hence
$$\rho = \frac{d\sigma}{d\epsilon} = \frac{\gamma(P\dot{P}\ddot{P}\dddot{P})(\dot{P}|\dot{P})^{\frac{3}{2}}}{(P\dot{P}\ddot{P}|P\dot{P}\ddot{P})^{\frac{3}{2}}},$$

$$\kappa = \frac{d\sigma}{d\theta} = \frac{\gamma(P\dot{P}\ddot{P}\dddot{P})}{(P\dot{P}\ddot{P}|P\dot{P}\ddot{P})}.$$

(5) The point $|\dot{P}$ is a point in the plane P, such that any line in P through $|\dot{P}$ is perpendicular to the tangent line $P\dot{P}$. For the point, where $|P\dot{P}$ meets P, is the point $P|P\dot{P} = (P|\dot{P})|P - (P|P)|\dot{P} = -(P|P)|\dot{P}$, and this is the point $|\dot{P}$, neglecting the numerical factor.

Thus $|\dot{P}$ is the point on the principal normal distant $\frac{1}{2}\pi\gamma$ from the point $P\dot{P}$.

The normal plane is the plane $P\dot{P}\ddot{P}|P\dot{P}$, this can easily be shown to be the plane $(\dot{P}|\dot{P})\ddot{P} + (\dot{P}|\dot{P})^2 P - (\dot{P}|\ddot{P})\dot{P}$.

290. Velocity and Acceleration. (1) Some of the main propositions respecting the Kinematics of a point in Elliptic or in Hyperbolic Space can now be easily deduced. It must be remembered, that in such spaces the idea of direction, as abstracted from the idea of the fully determined position of a line, does not exist. In Euclidean Space all parallel lines are said to have the same direction. Again, in considering the small displacements which a moving point has received in a small time $\delta\tau$, we must remember that the propositions of Euclidean Space are applicable to infinitely small figures in Elliptic or Hyperbolic Space.

As in the case of curved lines the reasoning will explicitly be confined to Elliptic Space.

(2) Let the variable τ of the preceding sections be now considered to be the time. Then the point x in the time $\delta\tau$ has moved to the position $x + \dot{x}\delta\tau$. The line joining these points is the line of the velocity, and the length of the arc traversed is its magnitude. Hence the linear element $x\dot{x}$ represents the line of the velocity, and its intensity represents the magnitude of the velocity. Hence $x\dot{x}$ may be said completely to represent the velocity. Hence at a time $\tau + \delta\tau$ the linear element $(x + \dot{x}\delta\tau)(\dot{x} + \ddot{x}\delta\tau)$, that is $x\dot{x} + x\ddot{x}\,.\,\delta\tau$, represents the velocity. Therefore, remembering that the propositions of Euclidean Space can be applied to the infinitely small figures which are being considered, the linear element $x\ddot{x}$ completely represents the acceleration.

(3) It is also obvious, from the applicability of the ideas of Euclidean Space to small figures, that two component velocities v_1 and v_2 along lines at an angle ω are equivalent to one resultant velocity of magnitude $\sqrt{(v_1^2 + v_2^2 + 2v_1v_2 \cos \omega)}$, making an angle θ with the line of v_1; where

$$\frac{v_1}{\sin(\omega - \theta)} = \frac{v_2}{\sin \theta} = \frac{v}{\sin \omega}.$$

The same theorem holds for accelerations.

Thus if $x\dot{x}$ and $x\dot{x}'$ denote two component velocities of the point x, then $x\dot{x} + x\dot{x}'$ represents the resultant velocity [cf. § 265 (4)]; and if $x\ddot{x}$ and $x\ddot{x}'$ represent two component accelerations, then $x\ddot{x} + x\ddot{x}'$ represents the resultant acceleration.

(4) The magnitude of the velocity is $\dot{\sigma}$, or $\gamma\sqrt{(\dot{x}\,|\,\dot{x})}$. It will now be shown that the acceleration is equivalent to two components, one along the tangent of magnitude $\ddot{\sigma}$, or $\gamma \dfrac{(\dot{x}\,|\,\ddot{x})}{\sqrt{(\dot{x}\,|\,\dot{x})}}$, and the other along the principal normal of magnitude $\dfrac{\dot{\sigma}^2}{\rho}$.

For a unit linear element along the tangent is $\dfrac{x\dot{x}}{\sqrt{(\dot{x}\,|\,\dot{x})}}$; also [cf. § 288 (5)] a unit linear element along the principal normal is $\pm \dfrac{x\dot{x}\ddot{x}\,|\,\dot{x}}{\sqrt{\{(\dot{x}\,|\,\dot{x})\,(x\dot{x}\ddot{x}\,|\,x\dot{x}\ddot{x})\}}}$, which can also be written $\pm \dfrac{(\dot{x}\,|\,\ddot{x})\,x\dot{x} - (\dot{x}\,|\,\dot{x})\,x\dot{x}}{\sqrt{\{(\dot{x}\,|\,\dot{x})\,(x\dot{x}\ddot{x}\,|\,x\dot{x}\ddot{x})\}}}$.

$$\text{Also } \gamma x\ddot{x} = \gamma\frac{(\dot{x}\,|\,\ddot{x})}{\sqrt{(\dot{x}\,|\,\dot{x})}} \cdot \frac{x\dot{x}}{\sqrt{(\dot{x}\,|\,\dot{x})}} + \gamma\sqrt{\left\{\frac{(x\dot{x}\ddot{x}\,|\,x\dot{x}\ddot{x})}{(\dot{x}\,|\,\dot{x})}\right\}}\,\frac{(\dot{x}\,|\,\dot{x})\,x\ddot{x} - (\dot{x}\,|\,\ddot{x})\,x\dot{x}}{\sqrt{\{(\dot{x}\,|\,\dot{x})\,(x\dot{x}\ddot{x}\,|\,x\dot{x}\ddot{x})\}}}$$

$$= \ddot{\sigma}\,\frac{x\dot{x}}{\sqrt{(\dot{x}\,|\,\dot{x})}} - \frac{\dot{\sigma}^2}{\rho}\,\frac{x\dot{x}\ddot{x}\,|\,\dot{x}}{\sqrt{\{(\dot{x}\,|\,\dot{x})\,(x\dot{x}\ddot{x}\,|\,x\dot{x}\ddot{x})\}}}.$$

Hence the acceleration is equivalent to two components as specified above.

(5) Let the point $\dot{x}$ be called the velocity-point of x, and $\ddot{x}$ the acceleration-point of x. The velocity of x is directed along the line from x towards $\dot{x}$ and the acceleration towards $\ddot{x}$. Furthermore, if v be the magnitude of the velocity and α of the acceleration, then,

$$v = \gamma \sqrt{(\dot{x} \,|\, \dot{x})};$$

and $$\alpha = \gamma \sqrt{(x\ddot{x} \,|\, x\ddot{x})} = \gamma \sqrt{\{(\ddot{x} \,|\, \ddot{x}) - (x \,|\, \ddot{x})^2\}} = \gamma \sqrt{\{(\ddot{x} \,|\, \ddot{x}) - (\dot{x} \,|\, \dot{x})^2\}}.$$

Hence $$\gamma^2\alpha^2 + v^4 = \gamma^4 (\ddot{x} \,|\, \ddot{x}).$$

291. THE CIRCLE. (1) To illustrate the formulæ of §§ 287, 288 consider the circle, radius α, in Hyperbolic Space [cf. §§ 216 (6) and 259].

Firstly let the centre be the spatial point e. Let e, e_1, e_2, e_3 be a system of normal points at unit intensities, spatial and anti-spatial respectively, and let ee_1e_2 be the plane of the circle.

Then if x be any point on the circle [cf. § 216 (6)], and ϕ be the parameter τ of § 287 (1),

$$x = e \cosh \frac{\alpha}{\gamma} + e_1 \sinh \frac{\alpha}{\gamma} \cos \phi + e_2 \sinh \frac{\alpha}{\gamma} \sin \phi,$$

$$\dot{x} = \sinh \frac{\alpha}{\gamma} \{- e_1 \sin \phi + e_2 \cos \phi\},$$

$$\ddot{x} = \sinh \frac{\alpha}{\gamma} \{- e_1 \cos \phi - e_2 \sin \phi\} = e \cosh \frac{\alpha}{\gamma} - x.$$

Therefore $$\frac{d\sigma}{d\phi} = \gamma \sqrt{-(\dot{x} \,|\, \dot{x})} = \gamma \sinh \frac{\alpha}{\gamma};$$

and $\sigma = \phi\gamma \sinh \frac{\alpha}{\gamma}$; and the length of the whole circumference is $2\pi\gamma \sinh \frac{\alpha}{\gamma}$.

Also

$$x\dot{x}\ddot{x} = \sinh \frac{\alpha}{\gamma} (e_1 \cos \phi + e_2 \sin \phi) . \sinh \frac{\alpha}{\gamma} (- e_1 \sin \phi + e_2 \cos \phi) . e \cosh \frac{\alpha}{\gamma}$$
$$= ee_1e_2 . \sinh^2 \frac{\alpha}{\gamma} \cosh \frac{\alpha}{\gamma}.$$

Hence $$\frac{d\epsilon}{d\phi} = \frac{\sqrt{\{x\dot{x}\ddot{x} \,|\, x\dot{x}\ddot{x}\}}}{-(\dot{x} \,|\, \dot{x})} = \cosh \frac{\alpha}{\gamma}.$$

The normal plane is $|\dot{x} = i \sinh \frac{\alpha}{\gamma} \{ee_2e_3 \sin \phi + ee_1e_3 \cos \phi\}$.

All the normal planes pass through the line ee_3, drawn through the centre perpendicular to the plane of the circle.

The measure of curvature is given by

$$\frac{1}{\rho} = \frac{1}{\gamma} \frac{\sqrt{(x\dot{x}\ddot{x} \,|\, x\dot{x}\ddot{x})}}{(-\dot{x} \,|\, \dot{x})^{\frac{3}{2}}} = \frac{1}{\gamma} \coth \frac{\alpha}{\gamma}.$$

Hence $$\rho = \gamma \tanh \frac{\alpha}{\gamma}.$$

When the radius is infinite, the measure of the curvature becomes $\frac{1}{\gamma}$, and this is its least possible value; when α is zero, the measure becomes infinite.

The binormals obviously all intersect in e_3, and form a cone with an anti-spatial vertex.

(2) Secondly, let the circle have an anti-spatial centre, e_2, and let it lie in the plane ee_1e_2; where e, e_1, e_2, e_3 form an unit normal system of points with e as spatial origin. Then [cf. § 259 (5)] the circle is a line of equal distance from the line ee_1.

Let β be the distance of a point x on the circle from this line. Then

$$\sinh \frac{\beta}{\gamma} = \sqrt{\frac{(xee_1 \mid xee_1)}{-(x \mid x)(ee_1 \mid ee_1)}} = \sqrt{(xee_1 \mid xee_1)},$$

when x, e, e_1 are at unit intensities, spatial and anti-spatial.

Hence if $$x = \xi e + \xi_1 e_1 + \xi_2 e_2,$$

we have $$\xi^2 - \xi_1^2 - \xi_2^2 = 1, \text{ and } \xi_2 = \sinh \frac{\beta}{\gamma}.$$

Hence we may put

$$\xi = \cosh \frac{\beta}{\gamma} \cosh \phi, \quad \xi_1 = \cosh \frac{\beta}{\gamma} \sinh \phi.$$

Thus $$x = e \cosh \frac{\beta}{\gamma} \cosh \phi + e_1 \cosh \frac{\beta}{\gamma} \sinh \phi + e_2 \sinh \frac{\beta}{\gamma},$$

$$\dot{x} = \cosh \frac{\beta}{\gamma} \{e \sinh \phi + e_1 \cosh \phi\},$$

$$\ddot{x} = \cosh \frac{\beta}{\gamma} \{e \cosh \phi + e_1 \sinh \phi\}$$

$$= x - e_2 \sinh \frac{\beta}{\gamma}.$$

The normal plane is given by

$$|\dot{x} = i \cosh \frac{\beta}{\gamma} \{-e_1e_2e_3 \sinh \phi - ee_2e_3 \cosh \phi\}.$$

The principal normal to the curve is the intersection of this plane with ee_1e_2; and it is perpendicular to the axis, since $|\dot{x}$ contains the polar line of the axis, viz. e_2e_3.

This normal meets the axis ee_1 in the point $ee_1|\dot{x}$, that is in the point $e_1 \sinh \phi + e \cosh \phi$. Call this the point y, then $\dot{y} = e \sinh \phi + e_1 \cosh \phi$.

Let $\delta\sigma'$ denote the distance between y and $y + \dot{y}\delta\phi$.

Then $$\frac{d\sigma}{d\phi} = \gamma \sqrt{-(\dot{x}|\dot{x})} = \gamma \cosh \frac{\beta}{\gamma},$$

and $$\frac{d\sigma'}{d\phi} = \gamma.$$

Hence [cf. § 262 (3)] $$\frac{d\sigma}{d\sigma'} = \cosh \frac{\beta}{\gamma}, \quad \sigma = \sigma' \cosh \frac{\beta}{\gamma}.$$

Again $$x\dot{x}\ddot{x} = - ee_1e_2 \cosh^2 \frac{\beta}{\gamma} \sinh \frac{\beta}{\gamma}.$$

Hence $$\frac{d\epsilon}{d\phi} = \frac{\cosh^3 \frac{\beta}{\gamma} \sinh \frac{\beta}{\gamma}}{\cosh^2 \frac{\beta}{\gamma}} = \sinh \frac{\beta}{\gamma},$$

and $$\frac{1}{\rho} = \frac{1}{\gamma} \tanh \frac{\beta}{\gamma}.$$

When β is infinite, $\rho = \gamma$; and when β is zero, the curvature is zero. In this latter case the curve is identical with the straight line which is the axis.

(3) Thirdly, let the centre of the circle be on the absolute, so that the curve is a limit-line. Let e be a spatial point on the curve, and let ee_1 be the normal at e, and ee_1e_2 be the plane of the curve; also let e, e_1, e_2, e_3 be a normal system of unit reference elements. Then from § 261 (3), the limit-line becomes the section of the limit-surface

$$(x|x) = \{x|(e+e_1)\}^2$$

by the plane ee_1e_2.

Thus, if x be of unit intensity, it can be exhibited in the form

$$x = (1 + \tfrac{1}{2}\theta^2) e + \tfrac{1}{2}\theta^2 e_1 + \theta e_2.$$

Then, if θ be the parameter τ of § 287 (1),

$$\dot{x} = \theta e + \theta e_1 + e_2, \quad \ddot{x} = e + e_1.$$

Hence $$\frac{d\sigma}{d\theta} = \gamma \sqrt{-(\dot{x}|\dot{x})} = \gamma, \quad \sigma = \theta\gamma.$$

Thus we can write

$$x = \left(1 + \frac{1}{2}\frac{\sigma^2}{\gamma^2}\right) e + \frac{1}{2}\frac{\sigma^2}{\gamma^2} e_1 + \frac{\sigma}{\gamma} e_2,$$

where σ is the length of the arc from e to x. Then writing x' and x'' for $\dot{x}$ and $\ddot{x}$ with this new variable σ,

$$x' = \frac{\sigma}{\gamma^2} e + \frac{\sigma}{\gamma^2} e_1 + \frac{1}{\gamma} e_2, \quad x'' = \frac{1}{\gamma^2}(e + e_1), \text{ and } xx'x'' = -\frac{1}{\gamma^3} ee_1e_2.$$

Hence $$\frac{d\epsilon}{d\sigma} = \frac{1}{\rho} = \frac{1}{\gamma}.$$

This agrees with the deduction above, that a circle of infinite radius is of curvature $\frac{1}{\gamma}$.

Again let y be a point on a second limit-line with the same centre as the first one, and at a constant normal distance δ from it; so that, if x and y be corresponding points on the first and second limit-lines respectively, the line xy is normal to both curves, and $\overline{xy}$ is equal to δ. Then if y be of unit intensity, it is easily proved that

$$y = \exp\left(-\frac{\delta}{\gamma}\right) x + \sinh \frac{\delta}{\gamma}(e + e_1).$$

Hence if σ be an arc of the x curve and σ' an arc of the y curve, and if σ be the independent variable,

$$\dot{y} = \exp\left(-\frac{\delta}{\gamma}\right) x'.$$

Hence $$\frac{d\sigma'}{d\sigma} = \gamma \sqrt{-(\dot{y}\,|\dot{y})} = \gamma \exp\left(-\frac{\delta}{\gamma}\right) \sqrt{-(x'\,|x')} = \exp\left(-\frac{\delta}{\gamma}\right).$$

Hence $$\sigma' = \exp\left(-\frac{\delta}{\gamma}\right) . \sigma.$$

This result is proved by both Lobatschewsky and J. Bolyai.

292. Motion of a Rigid Body. (1) Let $\dot{S}$ stand for the system

$$\dot{\alpha}_{34} e_1 e_2 + \dot{\alpha}_{24} e_1 e_3 + \dot{\alpha}_{23} e_1 e_4 + \dot{\alpha}_{14} e_2 e_3 + \dot{\alpha}_{13} e_4 e_2 + \dot{\alpha}_{12} e_3 e_4;$$

and let $\dot{S}\delta\tau$ be the associated system of the transformation which would displace the body from its position at time τ to its position at time $\tau + \delta\tau$. Let any point of the rigid body be x at time τ and $x + \dot{x}\delta\tau$ at time $\tau + \delta\tau$.

Then $$x + \dot{x}\delta\tau = x + |x\dot{S} . \delta\tau.$$

Hence $$\dot{x} = + |x\dot{S}.$$

Let $\dot{S}$ be called the associated system of the motion of the body.

(2) The theorem of § 286 (6) can be stated in the form: the resolved parts of the velocities of two points of a rigid body along the line joining these points are equal.

Thus if x, y be the two points, and $x\dot{x}$ and $y\dot{y}$ make angles θ and ϕ with xy, then

$$\overline{x\dot{x}} \cos\theta = \overline{y\dot{y}} \cos\phi.$$

(3) The velocity of any point x of the moving body is

$$x\dot{x} = x\,|x\dot{S}.$$

Hence the velocity of each point is perpendicular to its null plane with respect to $\dot{S}$.

(4) Again if $\dot{S}$ change to $\dot{S}+\ddot{S}\,.\,\delta\tau$ at the time $\tau+\delta\tau$, then

$$\ddot{x} = |x\ddot{S} + |\dot{x}\dot{S} = |x\ddot{S} + |\dot{x}\,|\dot{S} = |x\ddot{S} + x\dot{S}\,|\dot{S}.$$

(5) It is obvious that all the theorems which have been enunciated with respect to small congruent transformations hold with verbal alterations for the continuous motion of a rigid body.

Thus if $\dot{S}$ be a vector system of the form $a_1a_2 \pm |a_1a_2$, then $|\dot{S} = \pm \dot{S}$. Also suppose that $\dot{S}$ is constant with respect to the time. Then

$$\ddot{x} = \mp x\dot{S}\,.\,\dot{S} = \mp \tfrac{1}{2}(\dot{S}\dot{S})\,x.$$

Thus $x\ddot{x}$ is always zero and no point of the moving body has any acceleration. Thus each point of the body is moving uniformly in a straight line. This is a vector motion of the body.

293. Gauss' Curvilinear Co-ordinates. (1) Let x be any unit point on a surface in Elliptic Space. Then the co-ordinates of x, referred to any four reference elements, may be conceived as definite functions of two independent variables, θ and ϕ. And the two equations, $\theta=$ constant, $\phi=$ constant, represent two families of curves traced on the surface.

(2) Suppose that the unit point $x+\delta x$ corresponds to the values $\theta+\delta\theta$ and $\phi+\delta\phi$ of the variables. Then we may write δx in the form

$$\delta x = x_1\delta\theta + x_2\delta\phi + \tfrac{1}{2}\{x_{11}(\delta\theta)^2 + 2x_{12}\delta\theta\delta\phi + x_{22}(\delta\phi)^2\} + \text{etc.}$$

Since the point remains a unit point, we see by making $\delta\theta$ and $\delta\phi$ infinitely small, and by remembering that the ratio of $\delta\theta$ to $\delta\phi$ is arbitrary,

$$(x\,|x_1) = 0 = (x\,|x_2).$$

Hence x_1 and x_2 are in the polar plane of x.

In order to exhibit the meanings of x_1, x_2, x_{11}, etc., let $e_1e_2e_3e_4$ be a set of four unit quadrantal points; and let $x=\Sigma\xi e$. Then $x_1 = \Sigma\frac{\partial\xi}{\partial\theta}e$, $x_2 = \Sigma\frac{\partial\xi}{\partial\phi}e$, $x_{11} = \Sigma\frac{\partial^2\xi}{\partial\theta^2}e$, and so on; where the condition, $\xi_1^2+\xi_2^2+\xi_3^2+\xi_4^2=1$ is fulfilled.

It will be an obviously convenient notation to write $\frac{\partial x}{\partial\theta}$ for x_1, and so on, when occasion requires it.

(3) By differentiating the equations, $(x\,|x_1)=0=(x\,|x_2)$, with respect to θ and ϕ successively, we obtain

$$(x_1\,|x_1) = -(x\,|x_{11}),\ (x_2\,|x_2) = -(x\,|x_{22}),\ (x_1\,|x_2) = -(x\,|x_{12}).$$

(4) The distance $\delta\sigma$ between x and $x+\delta x$ is given by

$$\frac{(\delta\sigma)^2}{\gamma^2} = \sin^2\frac{\delta\sigma}{\gamma} = (x\delta x\,|x\delta x) = (xx_1\,|xx_1)(\delta\theta)^2 + 2(xx_1\,|xx_2)\,\delta\theta\delta\phi + (xx_2\,|xx_2)(\delta\phi)^2$$
$$= (x_1\,|x_1)(\delta\theta)^2 + 2(x_1\,|x_2)\,\delta\theta\delta\phi + (x_2\,|x_2)(\delta\phi)^2.$$

(5) The tangent line to the curve joining the points x and $x+\delta x$ is $x\delta x$, that is $x(x_1\delta\theta + x_2\delta\phi)$. Hence xx_1 and xx_2 are two tangent lines to the surface at x, and therefore the plane xx_1x_2 is the tangent plane at x.

The normal at x is the line $x|xx_1x_2$.

But $$x|xx_1x_2 = (x|x_2)|xx_1 + (x|x_1)|x_2x + (x|x)|x_1x_2 = |x_1x_2.$$

Hence the line $|x_1x_2$ is the normal at x.

294. Curvature of Surfaces. (1) Let $\delta\nu$ be the perpendicular from the point $x+\delta x$ on to the tangent plane at x.

Then [cf. § 224 (4)]

$$\frac{\delta\nu}{\gamma} = \frac{(xx_1x_2\delta x)}{\sqrt{\{(xx_1x_2|xx_1x_2)\}}}$$

$$= \frac{1}{2}\frac{(xx_1x_2x_{11})(\delta\theta)^2 + 2(xx_1x_2x_{12})\,\delta\theta\,\delta\phi + (xx_1x_2x_{22})(\delta\phi)^2}{\sqrt{\{x_1x_2|x_1x_2\}}}.$$

Let $\frac{1}{\rho}$ be the measure of the curvature of the normal section at x through the tangent line $x\delta x$. Then

$$\rho = \frac{1}{2}\frac{(\delta\sigma)^2}{\delta\nu} = \gamma\,\frac{\sqrt{\{x_1x_2|x_1x_2\}}\,.\,\{(x_1|x_1)(\delta\theta)^2 + 2(x_1|x_2)\,\delta\theta\,\delta\phi + (x_2|x_2)(\delta\phi)^2\}}{(xx_1x_2x_{11})(\delta\theta)^2 + 2(xx_1x_2x_{12})\,\delta\theta\,\delta\phi + (xx_1x_2x_{22})(\delta\phi)^2}.$$

(2) Now seek for the maximum and minimum values of ρ, when the ratio of $\delta\theta$ to $\delta\phi$ is varied. Let ρ_1 and ρ_2 be the maximum and minimum values found, and let $\delta\theta_1/\delta\phi_1$ and $\delta\theta_2/\delta\phi_2$ be the corresponding ratios of $\delta\theta/\delta\phi$.

Then $\rho_1/\gamma\sqrt{\{x_1x_2|x_1x_2\}}$ and $\rho_2/\gamma\sqrt{\{x_1x_2|x_1x_2\}}$ are the roots of the quadratic for ζ:

$$\{(xx_1x_2x_{11})\zeta - (x_1|x_1)\}\{(xx_1x_2x_{22})\zeta - (x_2|x_2)\} - \{(xx_1x_2x_{12})\zeta - (x_1|x_2)\}^2 = 0.$$

Hence $$\frac{1}{\rho_1\rho_2} = \frac{(xx_1x_2x_{11})(xx_1x_2x_{22}) - (xx_1x_2x_{12})^2}{\gamma^2\{(x_1|x_1)(x_2|x_2) - (x_1|x_2)^2\}^2};$$

$$\frac{1}{\rho_1} + \frac{1}{\rho_2} = \frac{(x_1|x_1)(xx_1x_2x_{22}) + (x_2|x_2)(xx_1x_2x_{11}) - 2(x_1|x_2)(xx_1x_2x_{12})}{\gamma\{(x_1|x_1)(x_2|x_2) - (x_1|x_2)^2\}^{\frac{3}{2}}}.$$

(3) The expression for $\frac{1}{\rho_1\rho_2}$ can be put in terms of $(x_1|x_1)$, $(x_2|x_2)$, $(x_1|x_2)$, and of their differential coefficients with respect to θ and ϕ.

For $(xx_1x_2x_{11})(xx_1x_2x_{22}) = (xx_1x_2x_{11}|xx_1x_2x_{22})$

$$= \begin{vmatrix} 1, & (x|x_1), & (x|x_2), & (x|x_{22}) \\ (x_1|x), & (x_1|x_1), & (x_1|x_2), & (x_1|x_{22}) \\ (x_2|x), & (x_2|x_1), & (x_2|x_2), & (x_2|x_{22}) \\ (x_{11}|x), & (x_{11}|x_1), & (x_{11}|x_2), & (x_{11}|x_{22}) \end{vmatrix}.$$

Hence, since $(x|x_1)=0=(x|x_2)$,

$$(xx_1x_2x_{11})(xx_1x_2x_{22})=\begin{vmatrix}(x_1|x_1), & (x_1|x_2), & (x_1|x_{22})\\ (x_1|x_2), & (x_2|x_2), & (x_2|x_{22})\\ (x_1|x_{11}), & (x_2|x_{11}), & 0\end{vmatrix}$$
$$+\{(x_1|x_1)(x_2|x_2)-(x_1|x_2)^2\}\{(x_{11}|x_{22})-(x|x_{11})(x|x_{22})\}.$$

Similarly

$$(xx_1x_2x_{12})^2=\begin{vmatrix}(x_1|x_1), & (x_1|x_2), & (x_1|x_{12})\\ (x_1|x_2), & (x_2|x_2), & (x_2|x_{12})\\ (x_1|x_{12}), & (x_2|x_{12}), & 0\end{vmatrix}$$
$$+\{(x_1|x_1)(x_2|x_2)-(x_1|x_2)^2\}\{(x_{12}|x_{12})-(x|x_{12})^2\}.$$

Now write $\frac{\partial}{\partial\theta}(x_1|x_1)$ in the form $(x_1|x_1)_1$, and $\frac{\partial}{\partial\phi}(x_1|x_1)$ in the form $(x_1|x_1)_2$, with a similar notation for the differential coefficients of similar quantities.

Then by successive differentiations of the equations $(x|x_1)=0$ and $(x|x_2)=0$, we find

$$(x|x_{11})=-(x_1|x_1),\ (x|x_{22})=-(x_2|x_2),\ (x|x_{12})=-(x_1|x_2);$$
$$(x_1|x_{11})=\tfrac{1}{2}(x_1|x_1)_1,\ (x_2|x_{22})=\tfrac{1}{2}(x_2|x_2)_2,\ (x_1|x_{12})=\tfrac{1}{2}(x_1|x_1)_2,\ (x_2|x_{12})=\tfrac{1}{2}(x_2|x_2)_1;$$
$$(x_2|x_{11})=(x_1|x_2)_1-\tfrac{1}{2}(x_1|x_1)_2,\ (x_1|x_{22})=(x_1|x_2)_2-\tfrac{1}{2}(x_2|x_2)_1.$$

Also $$(x_{11}|x_{22})-(x_{12}|x_{12})=(x_1|x_2)_{12}-\tfrac{1}{2}(x_2|x_2)_{11}-\tfrac{1}{2}(x_1|x_1)_{22}.$$

Hence $$\frac{\gamma^2\{(x_1|x_1)(x_2|x_2)-(x_1|x_2)^2\}^2}{\rho_1\rho_2}$$

$$=\begin{vmatrix}(x_1|x_1), & (x_1|x_2), & (x_1|x_2)_2-\tfrac{1}{2}(x_2|x_2)_1\\ (x_1|x_2), & (x_2|x_2), & \tfrac{1}{2}(x_2|x_2)_2\\ \tfrac{1}{2}(x_1|x_1)_1, & (x_1|x_2)_1-\tfrac{1}{2}(x_1|x_1)_2, & 0\end{vmatrix}$$
$$-\begin{vmatrix}(x_1|x_1), & (x_1|x_2), & \tfrac{1}{2}(x_1|x_1)_2\\ (x_1|x_2), & (x_2|x_2), & \tfrac{1}{2}(x_2|x_2)_1\\ \tfrac{1}{2}(x_1|x_1)_2, & \tfrac{1}{2}(x_2|x_2)_1, & 0\end{vmatrix}$$
$$+\{(x_1|x_1)(x_2|x_2)-(x_1|x_2)^2\}\{(x_1|x_2)^2-(x_1|x_1)(x_2|x_2)$$
$$+(x_1|x_2)_{12}-\tfrac{1}{2}(x_2|x_2)_{11}-\tfrac{1}{2}(x_1|x_1)_{22}\}.$$

Thus $\frac{1}{\rho_1\rho_2}$ is expressed in the required form.

(4) Hence follows the extension to Elliptic and Hyperbolic Space of Gauss' theorem with respect to the applicability to each other of two small elements of surface. It is evidently a necessary condition, that $\frac{1}{\rho_1\rho_2}$ should be the same for each element.

295. Lines of Curvature. (1) By the usual methods of the elementary Differential Calculus it is easily shown that the ratios $\delta\theta_1/\delta\phi_1$ and $\delta\theta_2/\delta\phi_2$, which give the directions of the lines of curvature (defined as lines of maximum or of minimum curvature) through x, are the roots of the following quadratic for $\delta\theta/\delta\phi$:

$$\{(x_1|x_2)(xx_1x_2x_{11})-(x_1|x_1)(xx_1x_2x_{12})\}(\delta\theta)^2+\{(x_2|x_2)(xx_1x_2x_{11})-(x_1|x_1)(xx_1x_2x_{22})\}\delta\theta\delta\phi$$
$$+\{(x_2|x_2)(xx_1x_2x_{12})-(x_1|x_2)(xx_1x_2x_{22})\}(\delta\phi)^2=0 \ldots\ldots \text{(i)}.$$

This equation can be put into another form.

For [cf. § 293 (3)]

$$(x_1|x_2)(xx_1x_2x_{11})-(x_1|x_1)(xx_1x_2x_{12})=(x|x_{11})(xx_1x_2x_{12})-(x|x_{12})(xx_1x_2x_{11})$$
$$=[\{(xx_1x_2x_{12})x_{11}-(xx_1x_2x_{11})x_{12}\}|x]=\{(xx_1x_2 \,.\, x_{11}x_{12})|x\},$$

by § 103 (3), equation (4).

But the product $\{(xx_1x_2 \,.\, x_{11}x_{12})|x\}$ is pure [cf. § 101], and therefore associative.

Hence $\{(xx_1x_2 \,.\, x_{11}x_{12})|x\}=(xx_1x_2|x \,.\, x_{11}x_{12})=(x_1x_2x_{11}x_{12})$,

since $xx_1x_2|x=(x|x)x_1x_2+(x_1|x)x_2x_1+(x_2|x)xx_1=x_1x_2$, by § 293 (2).

Similarly $(x_2|x_2)(xx_1x_2x_{11})-(x_1|x_1)(xx_1x_2x_{22})=(x_1x_2x_{11}x_{22})$,

and $(x_2|x_2)(xx_1x_2x_{12})-(x_1|x_2)(xx_1x_2x_{22})=(x_1x_2x_{12}x_{22})$.

Accordingly equation (i) takes the form

$$(x_1x_2x_{11}x_{12})\,\delta\theta^2+(x_1x_2x_{11}x_{22})\,\delta\theta\delta\phi+(x_1x_2x_{12}x_{22})\,\delta\phi^2=0 \ldots\ldots\ldots\ldots \text{(ii)}.$$

It easily follows from equation (i) that

$$(x_1|x_1)\,\delta\theta_1\delta\theta_2+(x_1|x_2)\{\delta\theta_1\delta\phi_2+\delta\theta_2\delta\phi_1\}+(x_2|x_2)\,\delta\phi_1\delta\phi_2=0.$$

(2) Let $x+\delta x$ and $x+\delta' x$ be any two neighbouring points to x on the surface, where

$$\delta x=x_1\delta\theta+x_2\delta\phi, \quad \delta' x=x_1\delta'\theta+x_2\delta'\phi.$$

Then the angle ψ between the two tangent lines $x\delta x$ and $x\delta' x$ is given by

$$\cos\psi=\frac{(x\delta x|x\delta' x)}{\sqrt{\{(x\delta x|x\delta x)(x\delta' x|x\delta' x)\}}}$$
$$=\gamma^2\frac{(x_1|x_1)\,\delta\theta\delta'\theta+(x_1|x_2)(\delta\theta\delta'\phi+\delta'\theta\delta\phi)+(x_2|x_2)\,\delta\phi\delta'\phi}{\delta\sigma\delta'\sigma},$$

where $\delta\sigma$ and $\delta'\sigma$ are the arcs between x and $x+\delta x$, x and $x+\delta' x$.

Corollary. The lines of curvature cut each other at right angles.

(3) Since $(x|\delta x)=0=(x|\delta' x)$, where δx and $\delta' x$ are infinitely small,

$$\cos\psi=\frac{(\delta x|\delta' x)}{\sqrt{\{(\delta x|\delta x)(\delta' x|\delta' x)\}}}.$$

Hence $$\sin\psi=\sqrt{\frac{(\delta x\delta' x|\delta x\delta' x)}{\{(\delta x|\delta x)(\delta' x|\delta' x)\}}}.$$

Therefore $\delta\sigma\delta\sigma'\sin\psi=\gamma^2\sqrt{\{\delta x\delta' x|\delta x\delta' x\}}$

$$=\gamma^2(\delta\theta\delta'\phi-\delta'\theta\delta\phi)\sqrt{\{(x_1|x_1)(x_2|x_2)\}}\,.\,\sin\omega;$$

where ω is the angle at x between the curves $\theta=$ constant, $\phi=$ constant.

(4) If the curves, $\theta=$ constant, $\phi=$ constant, be lines of curvature at all points, then the equation for the lines of curvature must reduce to

$$\delta\theta\delta\phi = 0.$$

Hence from subsection (1), equation (i),

$$(x_1 \mid x_2) = 0 = (xx_1x_2x_{12});$$

and these equations must hold for all values of θ and ϕ.

(5) Let $\frac{1}{\rho_1}$ be the measure of curvature of the normal section through xx_1, and $\frac{1}{\rho_2}$ of that through xx_2; where the θ and ϕ curves are lines of curvature.

The radius of curvature of any normal section [cf. § 294 (1)] is given by

$$\frac{1}{\rho}\{(x_1 \mid x_1)(\delta\theta)^2 + (x_2 \mid x_2)(\delta\phi)^2\} = \frac{(x_1 \mid x_1)}{\rho_1}(\delta\theta)^2 + \frac{(x_2 \mid x_2)}{\rho_2}(\delta\phi)^2.$$

The angle ψ, which the tangent line $x\delta x$ makes with the tangent line xx_1, is given by

$$\cos\psi = \frac{\sqrt{(x_1 \mid x_1)} \,.\, \delta\theta}{\sqrt{\{(x_1 \mid x_1)(\delta\theta)^2 + (x_2 \mid x_2)(\delta\phi)^2\}}},$$

$$\sin\psi = \frac{\sqrt{(x_2 \mid x_2)} \,.\, \delta\phi}{\sqrt{\{(x_1 \mid x_1)(\delta\theta)^2 + (x_2 \mid x_2)(\delta\phi)^2\}}}.$$

Hence
$$\frac{1}{\rho} = \frac{\cos^2\psi}{\rho_1} + \frac{\sin^2\psi}{\rho_2}.$$

This is Euler's Theorem.

(6) The condition for the θ and ϕ curves being lines of curvature may be put into a simpler form than that in subsection (4).

For we have $(x \mid x_1) = 0 = (x \mid x_2) = (x_1 \mid x_2) = (xx_1x_2x_{12})$.

Hence [cf. § 293 (3)] $(x \mid x_{12}) = 0$.

Hence since $(x \mid x_1) = 0 = (x \mid x_2) = (x \mid x_{12})$, either the three equations are not independent and x_{12} can be written in the form

$$\lambda x_1 + \mu x_2,$$

or x is of the form
$$\nu \mid x_1x_2x_{12}.$$

Taking the latter alternative, and substituting in the equation

$$(xx_1x_2x_{12}) = 0,$$

we find
$$(x_1x_2x_{12} \mid x_1x_2x_{12}) = 0.$$

But the condition $(P \mid P) = 0$ cannot be satisfied in Elliptic Space by a real plane area. Hence it implies, if the plane area is known to be real, $P = 0$.

Thus we are brought back to the first alternative, namely $x_1x_2x_{12} = 0$.

If the space be Hyperbolic, the condition, $(P|P)=0$, implies that all the points on the plane P be anti-spatial, except its point of contact with the absolute. Hence if x be a spatial point, (xP) cannot vanish. So again the first alternative is the only one satisfying the conditions.

296. Meunier's Theorem. The measure of curvature of the curve, $\phi=$ constant, is found from the formula of § 288 (1). Writing $\frac{1}{{}_1\rho}$ for it, ${}_1\rho$ is given by

$$ {}_1\rho = \gamma \frac{\{(x_1|x_1)\}^{\frac{3}{2}}}{\sqrt{\{xx_1x_{11}|xx_1x_{11}\}}}. $$

The measure of curvature of the normal section through xx_1 is given by

$$ \rho = \gamma \frac{(x_1|x_1)\sqrt{\{x_1x_2|x_1x_2\}}}{(xx_1x_2x_{11})}. $$

Hence
$$ \frac{{}_1\rho}{\rho} = \frac{(x_1|x_1)^{\frac{1}{2}}(xx_1x_2x_{11})}{\sqrt{\{(x_1x_{11}|x_1x_{11})(x_1x_2|x_1x_2)\}}}. $$

The osculating plane of the curve is xx_1x_{11}. Let χ be the angle between this plane and the normal section, which is the plane $x_1|x_1x_2$.

Then
$$ \cos\chi = \pm \frac{(x_1|x_1x_2 \,.\, |xx_1x_{11})}{\sqrt{\{(xx_1x_{11}|xx_1x_{11})(x_1|x_1x_2| \,.\, x_1|x_1x_2)\}}}. $$

Now
$$ x_1|x_1x_2 = (x_1|x_2)|x_1 - (x_1|x_1)|x_2; $$

hence
$$ (x_1|x_1x_2| \,.\, x_1|x_1x_2) = (x_1|x_1)(x_1x_2|x_1x_2). $$

And
$$ (x_1|x_1x_2 \,.\, |x_1x_2x_{11}) = -(xx_1x_2x_{11})(x_1|x_1). $$

Thus
$$ \cos\chi = \frac{(x_1|x_1)^{\frac{1}{2}}(xx_1x_2x_{11})}{\sqrt{\{(xx_1x_{11}|xx_1x_{11})(x_1x_2|x_1x_2)\}}} $$
$$ = \frac{{}_1\rho}{\rho}. $$

Therefore
$$ \rho\cos\chi = {}_1\rho. $$

297. Normals. (1) The normal at the point x is $N=|x_1x_2$, the normal at the point $x+\delta x$ is

$$ N' = |x_1x_2 + |\delta x_1x_2 + |x_1\delta x_2 + |\delta x_1\delta x_2. $$

Hence
$$ (NN') = |(x_1x_2\delta x_1\delta x_2) = (x_1x_2\delta x_1\delta x_2). $$

Now
$$ \delta x_1 = x_{11}\delta\theta + x_{12}\delta\phi, \quad \delta x_2 = x_{12}\delta\theta + x_{22}\delta\phi. $$

Therefore
$$ (NN') = (x_1x_2x_{11}x_{12})(\delta\theta)^2 + (x_1x_2x_{11}x_{22})\,\delta\theta\delta\phi + (x_1x_2x_{12}x_{22})(\delta\phi)^2. $$

Therefore in general normals at neighbouring points of the surface do not intersect. But [cf. § 295 (1) equation (ii)] normals at neighbouring points on a line of curvature do intersect,

(2) If the θ and ϕ curves are lines of curvature, then by § 295 (6) $x_1 x_2 x_{12} = 0$.

Hence $$(x_1 x_2 x_{11} x_{12}) = 0 = (x_1 x_2 x_{22} x_{12}).$$

Thus $$(NN') = (x_1 x_2 x_{11} x_{22})\, \delta\theta\delta\phi.$$

Hence neighbouring normals on the curve θ = constant, or on the curve ϕ = constant, intersect; that is to say, neighbouring normals on a line of curvature intersect.

298. Curvilinear Co-ordinates. (1) Let x be conceived as a function of three variables, θ, ϕ, ψ. Then the equations θ = constant, ϕ = constant, and ψ = constant, determine three families of surfaces. On the surface, θ = constant, x is a function of the variables ϕ and ψ; on the surface, ϕ=constant, it is a function of ψ and θ; on the surface, ψ = constant, a function of θ and ϕ.

Let $\dfrac{\partial x}{\partial \theta} = x_1$, $\dfrac{\partial x}{\partial \phi} = x_2$, $\dfrac{\partial x}{\partial \psi} = x_3$, with a corresponding notation for the higher differential coefficients.

(2) Now suppose that the three families of surfaces intersect orthogonally wherever they meet.

Then $$(x_1 \mid x_2) = 0 = (x_2 \mid x_3) = (x_3 \mid x_1).$$

Hence $(x_{13} \mid x_2) + (x_1 \mid x_{23}) = 0$, $(x_{12} \mid x_3) + (x_2 \mid x_{31}) = 0$, $(x_{23} \mid x_1) + (x_3 \mid x_{12}) = 0$.

Therefore $$(x_1 \mid x_{23}) = 0 = (x_2 \mid x_{31}) = (x_3 \mid x_{12}).$$

Also [cf. § 293 (2)] $(x \mid x_1) = (x \mid x_2) = (x \mid x_3) = 0$.

Hence since $(x_1 \mid x) = 0 = (x_1 \mid x_2) = (x_1 \mid x_3) = (x_1 \mid x_{23})$, it follows that
$$(x x_2 x_3 x_{23}) = 0.$$

But the equations $(x_2 \mid x_3) = 0 = (x x_2 x_3 x_{23})$ are the conditions [cf. § 295 (4)] that the ϕ and ψ curves should be lines of curvature on the surface, θ = constant. Thus the lines of intersection are lines of curvature on each surface. This is Dupin's Theorem.

299. Limit-Surfaces. As a simple illustration of some of the above formulæ, adapted to Hyperbolic Space, consider the limit-surface
$$\epsilon^2 (x \mid x) = (x \mid b)^2.$$

It has been proved [cf. § 261 (3)] that if the spatial origin e be taken on the surface, and if the line ee_1 be taken to be through the point b on the absolute, then the equation takes the form
$$(x \mid x) = \{x \mid (e + e_1)\}^2.$$

Now if x be at unit intensity, we may write
$$x = \left(1 + \frac{1}{2}\frac{\sigma_2^2 + \sigma_3^2}{\gamma^2}\right) e + \frac{1}{2}\frac{\sigma_2^2 + \sigma_3^2}{\gamma^2} e_1 + \frac{\sigma_2}{\gamma} e_2 + \frac{\sigma_3}{\gamma} e_3.$$

Then
$$x_1 = \frac{\partial x}{\partial \sigma_2} = \frac{\sigma_2}{\gamma^2} e + \frac{\sigma_2}{\gamma^2} e_1 + \frac{1}{\gamma} e_2,$$
$$x_2 = \frac{\partial x}{\partial \sigma_3} = \frac{\sigma_3}{\gamma^2} e + \frac{\sigma_3}{\gamma^2} e_1 + \frac{1}{\gamma} e_3.$$

Let $\delta\sigma$ be the element of arc between the points x and $x + \delta x$, then
$$\frac{(\delta\sigma)^2}{\gamma^2} = -(x_1 | x_1)\, \delta\sigma_2^2 - 2\,(x_1 | x_2)\, \delta\sigma_2 \delta\sigma_3 - (x_2 | x_2)\, \delta\sigma_3^2$$
$$= \frac{\delta\sigma_2^2 + \delta\sigma_3^2}{\gamma^2}.$$

Hence
$$\delta\sigma^2 = \delta\sigma_2^2 + \delta\sigma_3^2.$$

Accordingly the metrical properties of the surface must be the same as those of a Euclidean plane. The same result had been arrived at before [cf. § 262(6)] when it was proved, that the sum of the angles of a triangle formed by great circles on a sphere of infinite radius is equal to two right-angles.

The curvature $\left(\frac{1}{\rho}\right)$ of any normal section [cf. § 294(1)] is given by
$$\rho = \gamma \frac{\sqrt{\{-x_1 x_2 | x_1 x_2\}} \,.\, \{-(x_1 | x_1)\, \delta\sigma_2^2 - 2\,(x_1 | x_2)\, \delta\sigma_2 \delta\sigma_3 - (x_2 | x_2)\, \delta\sigma_3^2\}}{(x x_1 x_2 x_{11})\, \delta\sigma_2^2 + 2\,(x x_1 x_2 x_{12})\, \delta\sigma_2 \delta\sigma_3 + (x x_1 x_2 x_{22})\, \delta\sigma_3^2} = \gamma.$$

Hence every normal section is a limit-line, a result otherwise evident.

CHAPTER VIII.

TRANSITION TO PARABOLIC GEOMETRY.

300. PARABOLIC GEOMETRY. (1) The interest of Parabolic Geometry centres in the fact that it includes the three dimensional space of ordinary experience. Any generalization of our space conceptions, which does not at the same time generalize them into the more perfect forms of Hyperbolic or Elliptic Geometry, is of comparatively slight interest. We will therefore confine our investigations of Parabolic Geometry to space of three dimensions, in other words, to ordinary Euclidean space.

(2) The absolute quadric as represented by the point-equation has degenerated into the two coincident planes [cf. § 212]

$$(\alpha_1\xi_1 + \alpha_2\xi_2 + \alpha_3\xi_3 + \alpha_4\xi_4)^2 = 0.$$

The intensity of any point $x\,(= \Sigma\xi e)$ must therefore [cf. § 213] be conceived to be the square root of the left-hand side of this equation, that is, $\alpha_1\xi_1 + \dots + \alpha_4\xi_4$. The absolute plane itself being the locus of zero intensity.

(3) It is proved in § 87 that, if the intensities of the unit reference points be properly chosen, the equation of the absolute plane becomes

$$\xi_1 + \xi_2 + \xi_3 + \xi_4 = 0,$$

and the intensity of any point $\Sigma\xi e$ is $\Sigma\xi$.

The intensities of all points in this plane are zero. Hence, if a and b be any two points at unit intensity, the point $a - b$, which is at zero intensity, lies in the absolute plane.

(4) If three of the reference points, namely, u_1, u_2, u_3, be taken to be in the absolute plane, and e be any other reference point, then any point x is denoted by $\xi e + \xi_1 u_1 + \xi_2 u_2 + \xi_3 u_3$; and its intensity is ξ. Thus the expression $e + \Sigma\xi u$ is the typical form for all points at unit intensity.

301. PLANE EQUATION OF THE ABSOLUTE. (1) In order to discuss completely the formulæ for the measurement of distances and angles, it is requisite to write down the most general plane-equation of the absolute, which is consistent with the point-equation reducing to two coincident planes. This question was discussed in § 84 (4).

(2) Let any planar element be denoted by

$$P = \lambda u_1 u_2 u_3 - \lambda_1 e u_2 u_3 + \lambda_2 e u_1 u_3 - \lambda_3 e u_1 u_2.$$

Then it has been proved in § 84 (4) that the plane-equation

$$\alpha_{11}\lambda_1^2 + \ldots + 2\alpha_{12}\lambda_1\lambda_2 + \ldots = 0,$$

where the terms involving λ are omitted, necessarily implies the point equation,

$$\xi^2 = 0,$$

where any point x is written $\xi e + \xi_1 u_1 + \xi_2 u_2 + \xi_3 u_3$.

The fully determined absolute quadric may therefore be considered as a conic section lying in the absolute plane. The points on the absolute are the points of the plane, the planes enveloping the absolute are the planes touching the conic section. The absolute plane is also called the plane at infinity; and the conic section denoted by the plane-equation of the absolute may be called the absolute conic lying in the plane at infinity.

(3) Let this conic section be assumed to be imaginary, so that the elliptic measure of separation holds for planes [cf. § 211 (2)].

(4) It may be as well at this point to note that the operation of taking the supplement with respect to the absolute becomes entirely nugatory. The operation therefore symbolized by | will in Parabolic Geometry represent as at its first introduction in § 99 the fact that the reference points (whatever four points they may be) are replaced according to the following scheme,

$$e_1 \text{ by } e_2e_3e_4, \quad e_2 \text{ by } -e_1e_3e_4, \quad e_3 \text{ by } e_1e_2e_4, \quad e_4 \text{ by } -e_1e_2e_3.$$

This operation of taking the supplement, as thus defined, will (as previously) be useful in exhibiting the duality of the formulæ, when it exists. Its utility for metrical relations will be considered later [cf. Book VII., Chapter II.]

(5) It has been proved in § 212 that in either Elliptic or Hyperbolic space if we start with an absolute of the form,

$$\alpha\xi^2 + \alpha_{11}\xi_1^2 + \alpha_{22}\xi_2^2 + \alpha_{33}\xi_3^2 = 0,$$

and make it gradually degenerate to $\xi^2 = 0$, at the same time increasing the space-constant, then the distance between any two points x and y, where x is $\xi e + \xi_1 u_1 + \xi_2 u_2 + \xi_3 u_3$ and y is $\eta e + \eta_1 u_1 + \eta_2 u_2 + \eta_3 u_3$ takes the form

$$\frac{\sqrt{\{K_1(\xi_1\eta - \eta_1\xi)^2 + K_2(\xi_2\eta - \eta_2\xi)^2 + K_3(\xi_3\eta - \eta_3\xi)^2\}}}{\xi\eta}.$$

It will be observed that the assumption of the initial form of the absolute, from which the degeneration takes place, is equivalent to the assumption that eu_1, eu_2, eu_3 are mutually at right-angles.

The most general assumption for the plane-equation of the absolute is then [cf. § 84 (4)]

$$(\beta^2)(\lambda)^2 \equiv \beta_1^2\lambda_1^2 + \beta_2^2\lambda_2^2 + \beta_2^2\lambda_2^2 = 0.$$

And if θ be the angle between the two planes

$$\lambda u_1u_2u_3 - \lambda_1 eu_2u_3 + \lambda_2 eu_1u_3 - \lambda_3 eu_1u_2,$$

and

$$\lambda' u_1u_2u_3 - \lambda_1' eu_2u_3 + \lambda_2' eu_1u_3 - \lambda_3' eu_1u_2,$$

then [cf. § 211]

$$\cos\theta = \frac{(\beta^2)(\lambda)(\lambda')}{\sqrt{\{(\beta^2)(\lambda)^2 (\beta^2)(\lambda')^2\}}}.$$

(6) But the K's are not independent of the β's as these two detached forms of statement may suggest. In order to perceive the connection it is better to conduct the gradual degeneration of the absolute as follows.

Let the plane-equation of the absolute be

$$\beta^2\lambda^2 + \beta_1^2\lambda_1^2 + \beta_2^2\lambda_2^2 + \beta_3^2\lambda_3^2 = 0,$$

where β will ultimately be made to vanish.

Then the point-equation of the absolute is

$$\beta_1^2\beta_2^2\beta_3^2\xi^2 + \beta^2\beta_2^2\beta_3^2\xi_1^2 + \beta^2\beta_1^2\beta_3^2\xi_2^2 + \beta^2\beta_1^2\beta_2^2\xi_3^2 = 0.$$

Hence by reference to § 212 we see that

$$\frac{K_1}{\beta^2\beta_2^2\beta_3^2} = \frac{K_2}{\beta^2\beta_1^2\beta_3^2} = \frac{K_3}{\beta^2\beta_1^2\beta_2^2}.$$

Therefore if K be some finite constant,

$$K_1 = K\beta_2^2\beta_3^2,\ K_2 = K\beta_1^2\beta_3^2,\ K_3 = K\beta_1^2\beta_2^2.$$

Accordingly, when β is made to vanish, the distance between two points x and y takes the form

$$\frac{K\sqrt{\{\beta_2^2\beta_3^2(\xi_1\eta - \eta_1\xi)^2 + \beta_1^2\beta_3^2(\xi_2\eta - \eta_2\xi)^2 + \beta_1^2\beta_2^2(\xi_3\eta - \xi\eta_3)^2\}}}{\xi\eta}.$$

If x and y be two points of unit intensity, they are of the form $e + \Sigma\xi u$ and $e + \Sigma\eta u$, and their distance is

$$K\sqrt{\{\beta_2^2\beta_3^2(\xi_1 - \eta_1)^2 + \beta_1^2\beta_3^2(\xi_2 - \eta_2)^2 + \beta_1^2\beta_2^2(\xi_3 - \eta_3)^2\}}.$$

302. Intensities. (1) The intensities of the points which lie on the plane at infinity, which is the degenerate form represented by the point-equation of the absolute, are all zero according to the general law of intensity. It was explained in the chapter on Intensity [cf. § 86 (2)] that some special law of intensity, applying to these points on the locus of zero intensity, must be introduced.

(2) Consider two points x and x' on the line eu_1. Let $x = e + \xi_1 u_1$, $x' = e + \lambda\xi_1 u_1$, so that x and x' are at unit intensity. The distance $\overline{ex}$ is $K\beta_2\beta_3\xi_1$, the distance $\overline{ex'}$ is $K\beta_2\beta_3\lambda\xi_1$. Hence $\overline{ex'} = \lambda\overline{ex}$.

The difference of the two points x and e each at unit intensity is a point at infinity, in fact $x - e = \xi_1 u_1$; similarly $x' - e = \lambda \xi_1 u_1$. Hence $x - e$ and $x' - e$ denote the same point at infinity, but at intensities (according to some new law) which are proportional to the distances $\overline{ex}$ and $\overline{ex'}$.

(3) Let the intensity of a point at infinity be so defined that, if a and b be any two unit points at unit distance, the point $a - b$ is at unit intensity, positive or negative. Also let the three points u_1, u_2, u_3, used above, be at unit intensity.

Then any point $a = e + u_1$ is a unit point on eu_1 at unit distance from e. But its distance from e is $K\beta_2\beta_3$. Hence $K\beta_2\beta_3 = 1$. Similarly for points on eu_2 and eu_3.

Thus $\beta_1 = \beta_2 = \beta_3 = \beta$, say; and $K\beta^2 = 1$.

(4) Hence with these definitions, the plane-equation of the absolute is,

$$\lambda_1^2 + \lambda_2^2 + \lambda_3^2 = 0.$$

The angle θ between the two planes

$$\lambda\, u_1u_2u_3 - \lambda_1\, eu_2u_3 + \lambda_2\, eu_1u_3 - \lambda_3\, eu_1u_2,$$
$$\lambda' u_1u_2u_3 - \lambda_1' eu_2u_3 + \lambda_2' eu_1u_3 - \lambda_3' eu_1u_2,$$

is given by

$$\cos\theta = \frac{\lambda_1\lambda_1' + \lambda_2\lambda_2' + \lambda_3\lambda_3'}{\sqrt{\{(\lambda_1^2 + \lambda_2^2 + \lambda_3^2)(\lambda_1'^2 + \lambda_2'^2 + \lambda_3'^2)\}}}.$$

The distance between any two unit points $e + \Sigma\xi u$ and $e + \Sigma\eta u$ is

$$\sqrt{\{(\xi_1 - \eta_1)^2 + (\xi_2 - \eta_2)^2 + (\xi_3 - \eta_3)^2\}}.$$

The intensity of the first of the planar elements given above is

$$\sqrt{\{\lambda_1^2 + \lambda_2^2 + \lambda_3^2\}}.$$

The intensity of the point on the absolute plane, $\lambda_1 u_1 + \lambda_2 u_2 + \lambda_3 u_3$, is the distance between the points e and $e + \Sigma\lambda u$, that is, $\sqrt{\{\lambda_1^2 + \lambda_2^2 + \lambda_3^2\}}$.

(5) The transition* from Hyperbolic or from Elliptic Geometry to that of ordinary Space has now been fully investigated. The logical results of the definitions, which have finally been attained, will be investigated in the next book.

* Since Euclidean space is the limit both of Elliptic and of Hyperbolic space with infinitely large space-constants, it follows that the properties of figures in Elliptic or Hyperbolic Space, contained within a sphere of radius small compared to the space-constant, become ultimately those of figures in Euclidean space. Hence the experience of our senses, which can never attain to measurements of absolute accuracy, although competent to determine that the space-constant of the space of ordinary experience is greater than some large value, yet cannot, from the nature of the case, prove that this space is absolutely Euclidean.

303. Congruent Transformations. (1) It will however be instructive to work out the properties of Congruent Transformations for Parabolic Geometry in the same way as that in which they were discussed in the preceding chapter for Elliptic and Hyperbolic Geometry.

(2) The special properties of a congruent transformation are, as stated in § 268 (1), (α) the internal measure relations of any figure are unaltered by the transformation: (β) the transformation can be conceived as the result of another congruent transformation ρ times repeated, where ρ is any integer: (γ) real points are transformed into real points: (δ) the intensities of points are unaltered by transformation.

(3) It follows from (α) firstly that the plane at infinity is unaltered by the transformation; and secondly, that the degenerate quadric represented by the plane-equation of the absolute, which is a conic in the plane at infinity, is transformed into itself.

(4) Thus the plane at infinity is one semi-latent plane of a congruent transformation. It is proved in the next subsection that there must be at least three distinct latent points on this plane. Now, by reference to § 190, it can be verified that semi-latent planes, with at least three distinct latent points on them, only exist in the cases enumerated in § 190 (1), in § 190 (2), in § 190 (3) Cases I. and II., in § 190 (4) Cases I. and II., and in § 190 (5) Cases I. and II. But in each of these cases a semi-latent (or latent) line exists, which does not lie in the semi-latent plane containing the three distinct latent points. Now by Klein's Theorem [cf. § 200] the points on the absolute on this line are the latent points of the congruent transformation. But these points on the absolute are the two coincident points in which the line meets the plane at infinity. Hence the line is in general a semi-latent line with only one latent point on it, namely, the point at infinity.

(5) Now consider three unit points (u_1, u_2, u_3) on the plane at infinity, so that the three lines drawn to them from a unit point e, not on this plane, are at right-angles to each other. Then any point on the absolute is $\xi_1 u_1 + \xi_2 u_2 + \xi_3 u_3$, and any plane is $\lambda u_1 u_2 u_3 - \lambda_1 e u_2 u_3 + \lambda_2 e u_1 u_3 - \lambda_3 e u_1 u_2$. The plane-equation of the degenerate absolute conic is $\lambda_1^2 + \lambda_2^2 + \lambda_3^2 = 0$.

Hence, confining attention to points and lines on the absolute, any line on the absolute is $\lambda_1 u_2 u_3 + \lambda_2 u_3 u_1 + \lambda_3 u_1 u_2$: the line-equation of the absolute conic is $\lambda_1^2 + \lambda_2^2 + \lambda_3^2 = 0$; and its point-equation is $\xi_1^2 + \xi_2^2 + \xi_3^2 = 0$.

Now it is easily proved* that a linear transformation in two dimensions, which transforms a conic into itself, must be such that two of its latent

* Cf. Klein, *loc. cit.* p. 369.

points are on the conic, and a third is the pole of the line joining the other two.

Assume u_1 to be the latent point not on the absolute conic: then the polar of u_1 is the line u_2u_3. Let this line cut the absolute in the points v and v'. Then u_1, v, v' are the latent points of the transformation. Since the conic is imaginary, the points v and v' are conjugate imaginary points; and hence it is easily proved, that the three points v, v' and u_1 are necessarily distinct. The equation of the line u_2u_3 is $\xi_1 = 0$; hence v and v' are given by this equation and by $\xi_2^2 + \xi_3^2 = 0$. Thus we may write

$$v = e^{\frac{i\pi}{4}} u_2 + e^{-\frac{i\pi}{4}} u_3, \quad v' = e^{-\frac{i\pi}{4}} u_2 + e^{\frac{i\pi}{4}} u_3.$$

(6) Let the latent roots of the matrix be α, β, β' corresponding to u_1, v, v'. Then β and β' must be conjugate imaginaries; accordingly put

$$\beta = \beta_0 e^{i\delta}, \quad \beta' = \beta_0 e^{-i\delta}.$$

Again, considering the complete three dimensional transformation, the semi-latent line, not lying in the plane at infinity, corresponds to two equal roots. This repeated root must be real: hence the line also must be real, and cut the plane at infinity in a real latent point. Thus u_1 is the point in which the semi-latent line cuts the plane at infinity. Now if p be any point on this semi-latent line, and ϕ be the matrix representing the complete three dimensional transformation,

$$\phi p = \alpha p + \lambda u_1.$$

But from assumption (δ) of subsection (2), $\alpha = 1$.

Again*, in order that the conic may be transformed into itself, $\alpha^2 = \beta\beta'$. Hence $\beta_0^2 = 1$, and therefore $\beta_0 = 1$.

Thus finally the latent roots of the transformation are 1, 1, $e^{i\delta}$ and $e^{-i\delta}$.

(7) Now let u_2 and u_3 be transformed into u_2' and u_3'.

Then $$e^{\frac{i\pi}{4}} u_2' + e^{-\frac{i\pi}{4}} u_3' = e^{i\delta} v = e^{i\left(\delta+\frac{\pi}{4}\right)} u_2 + e^{-i\left(\delta-\frac{\pi}{4}\right)} u_3,$$

$$e^{-\frac{i\pi}{4}} u_2' + e^{\frac{i\pi}{4}} u_3' = e^{-i\delta} v' = e^{-i\left(\delta+\frac{\pi}{4}\right)} u_2 + e^{-i\left(\delta-\frac{\pi}{4}\right)} u_3.$$

Hence $$u_2' = u_2 \cos\delta + u_3 \sin\delta, \quad u_3' = u_3 \cos\delta - u_2 \sin\delta.$$

Also let e be any unit point on the semi-latent line cutting the absolute in u_1. Then [cf. § 200 (2)] any point $e + \xi u_1$ on this line is transformed to $e + (\xi + \gamma) u_1$. Thus all points on this line are displaced through the same distance γ. Let this line be called the axis of the transformation.

Any point $e + \Sigma \xi u$ becomes

$$e + (\xi_1 + \gamma) u_1 + (\xi_2 \cos\delta - \xi_3 \sin\delta) u_2 + (\xi_3 \cos\delta + \xi_2 \sin\delta) u_3.$$

* Cf. Klein, *loc. cit.* p. 369.

(8) If $\delta = 0$, the transformation is called a translation. The axis of a translation is indeterminate, since any line parallel to eu_1 possesses the same properties with regard to it as eu_1.

If $\gamma = 0$, the transformation is a rotation. Every point on the axis eu_1 of the rotation is a latent point.

If any point at a finite distance is unchanged by a congruent transformation, then the axis must pass through that point, and $\gamma = 0$. Hence the transformation is a rotation.

BOOK VII.

APPLICATION OF THE CALCULUS OF EXTENSION TO GEOMETRY.

CHAPTER I.

VECTORS.

304. INTRODUCTORY. (1) The analytical formulæ applicable to Euclidean space relations were arrived at, under the name of Parabolic Geometry, as a special limiting case of a generalized theory of distance. We will now start afresh, and, apart from any generalized theory, will consider the applicability of the Calculus of Extension to the investigation of Euclidean Geometry of three dimensions. Neither will it be endeavoured to assume a minimum of axioms and definitions in Geometry, and thence to build up the whole science by the aid of the Calculus. Such a scientific point of view was adopted in the investigation of the generalized metrical theory of the previous book. At present the propositions of elementary analytical Geometry will be assumed as known, and the suitability of the Calculus for geometrical investigation demonstrated by their aid. It may be further noticed that the propositions, which fall under the head of what is ordinarily called Projective Geometry, have been sufficiently exemplified in Book III., so that now metrical propositions will be chiefly attended to.

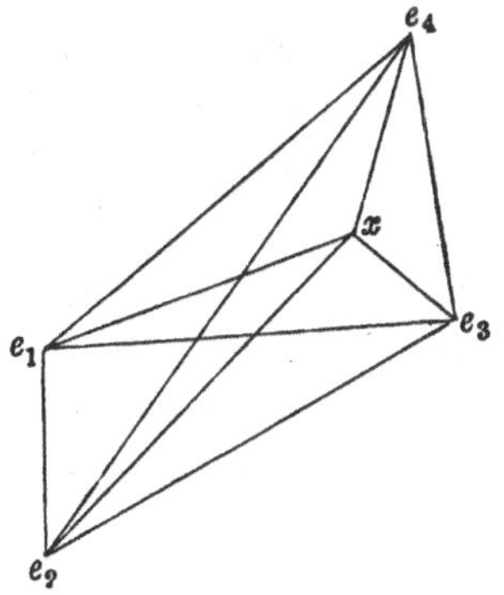

FIG. 1.

(2) Let the points e_1, e_2, e_3, e_4 form a tetrahedron; and let x be any other point. Let the co-ordinates of x in tetrahedral co-ordinates be ξ_1, ξ_2, ξ_3, ξ_4 referred to the fundamental tetrahedron $e_1e_2e_3e_4$; so that, for

instance, ξ_1 is the ratio of the volume of the tetrahedron $xe_2e_3e_4$ to that of the fundamental tetrahedron. Similarly for ξ_2, ξ_3, and ξ_4. Also ξ_1 is positive when x is on the same side of the plane $e_2e_3e_4$ as the point e_1, and ξ_1 is negative when x is on the other side, with similar conventions for the signs of the other co-ordinates. With these conventions the co-ordinates of x always satisfy the equation,

$$\xi_1 + \xi_2 + \xi_3 + \xi_4 = 1.$$

(3) Now let e_1, e_2, e_3, e_4 also stand for four reference elements of the first order [cf. §§ 20 and 94] in the calculus, and let x denote the element $\xi_1 e_1 + \xi_2 e_2 + \xi_3 e_3 + \xi_4 e_4$. And let x be at unit intensity [cf. § 87], when $\xi_1 + \xi_2 + \xi_3 + \xi_4 = 1$; and be at intensity λ, when $\xi_1 + \xi_2 + \xi_3 + \xi_4 = \lambda$.

(4) Then, when x is at unit intensity, the co-ordinates ξ_1, ξ_2, ξ_3, ξ_4 of subsections (2) and (3) can be identified. For [cf. §§ 64 and 65] if x and y be any two points with tetrahedral co-ordinates ξ_1, ξ_2, ξ_3, ξ_4 and η_1, η_2, η_3, η_4 respectively, then the point z which divides the line xy in the ratio $\lambda : \mu$, so that $\overline{zx}$ is to $\overline{zy}$ as λ is to μ, has as its co-ordinates

$$(\mu\xi_1 + \lambda\eta_1)/(\lambda + \mu),\ (\mu\xi_2 + \lambda\eta_2)/(\lambda + \mu),\ (\mu\xi_3 + \lambda\eta_3)/(\lambda + \mu),\ (\mu\xi_4 + \lambda\eta_4)/(\lambda + \mu).$$

Thus if x and y also stand for unit elements in the calculus, the point z stands for the element $(\mu x + \lambda y)/(\lambda + \mu)$, and is also at unit intensity as thus represented.

Thus conversely $(\mu x + \lambda y)$ can be made to represent any point on the straight line xy, by a proper choice of λ/μ.

(5) For instance let x, y, z denote the three angular points of a triangle at unit intensity. The middle points of the sides, also at unit intensity, are $\frac{1}{2}(y + z)$, $\frac{1}{2}(z + x)$, $\frac{1}{2}(x + y)$. Any points, not necessarily at unit intensity, on the three medians respectively are

$$\tfrac{1}{2}\lambda(y + z) + \mu x,\ \tfrac{1}{2}\lambda'(z + x) + \mu' y,\ \tfrac{1}{2}\lambda''(x + y) + \mu'' z.$$

It is obvious therefore that the three medians meet in the point $(x + y + z)$, which, as thus represented, is at intensity 3.

305. Points at Infinity. (1) The point $\mu x - \lambda y$, assuming λ and μ to be positive, divides the line xy externally in the ratio λ to μ. In particular, the point $x - y$ divides xy externally in a ratio of equality, and is therefore the point on xy at an infinite distance. It is to be noticed that the intensity of $x - y$ is necessarily zero, and therefore that the intensity of $\lambda(x - y)$ is also zero. Thus the plane at infinity is the locus of points at zero intensity [cf. § 86 (1)].

(2) A special law of intensity must therefore be assumed to hold for the points on the plane at infinity [cf. § 86 (2)]. Thus if x, y, z be three collinear points at unit intensity, $y - x$ and $z - x$ both denote the same point at infinity, but not at the same intensity according to this special law.

Suppose for instance that z divides the distance between x and y in the ratio λ to μ, so that

$$z = (\mu x + \lambda y)/(\lambda + \mu). \quad \text{Then } z - x = \frac{\lambda}{\lambda + \mu}(y - x).$$

Hence the intensity of $z - x$ is to that of $y - x$ in the ratio of the distance $\overline{xz}$ to that of the distance $\overline{xy}$ [cf. § 302 (2)].

(3) Any law of intensity may be assumed to hold in the plane at infinity, which preserves this property [cf. § 85 (2)]. But great simplicity is gained by defining the distance $\overline{xy}$ as the intensity of the element $y - x$.

(4) Let the line ab be parallel to the line cd and of the same length.

Fig. 2.

Let the points a, b, c, d be at unit intensity. Then the elements $b - a$ and $d - c$ are the same point at infinity at the same intensity.

Hence $b - a = d - c$. Therefore $a - c = b - d$, and the symbols express the fact, that ac and bd are equal and parallel. Also $a + d = b + c$; which expresses the fact, that ad and bc bisect each other.

306. Vectors. (1) Let a point at infinity be called a vector line, or shortly, a vector. A vector may be conceived as a directed length associated with any one of the series of parallel lines in its direction.

Thus if u denote the vector parallel to ab and cd, and of length equal to ab or cd reckoned from a to b or from c to d, then $u = b - a = d - c$.

(2) The conception of vectors is rendered clearer by the introduction of the idea of steps, which is explained in § 18. Thus the addition of u to a is the step by which we pass from a to b, for $a + u = b$; and the intensity of u measures the length of the step. Since also $c + u = d$, we must reckon, in accordance with this definition, all parallel steps in the same sense and of the same length as equivalent [cf. § 3].

Again if v denote the vector, or step, from d to b, then $v = b - d$. So $u + v = d - c + b - d = b - c$. Thus the sum of two steps is found by the parallelogram law.

(3) The fundamental tetrahedron may be chosen to have for its corners any unit point e and three independent vectors u_1, u_2, u_3, each of unit length. Any point x is then symbolized by $\xi e + \xi_1 u_1 + \xi_2 u_2 + \xi_3 u_3$, and the intensity of x is ξ [cf. § 87 (4)]. Thus if x be at unit intensity, it is written $e + \xi_1 u_1 + \xi_2 u_2 + \xi_3 u_3$. Thus the lines eu_1, eu_2, eu_3 are three Cartesian axes, and ξ_1, ξ_2, ξ_3 are the Cartesian co-ordinates of the point. For let e_1, e_2, e_3 be three unit points on the lines eu_1, eu_2, eu_3 respectively, and each at unit distance from e. Then $u_1 = e_1 - e$, $u_2 = e_2 - e$, $u_3 = e_3 - e$. Also let $x = \xi e + \xi_1 e_1 + \xi_2 e_2 + \xi_3 e_3$, where ξ, ξ_1, ξ_2, ξ_3 are tetrahedral co-ordinates of x. Then

$$x = (\xi + \xi_1 + \xi_2 + \xi_3) e + \xi_1 (e_1 - e) + \xi_2 (e_2 - e) + \xi_3 (e_3 - e) = e + \xi_1 u_1 + \xi_2 u_2 + \xi_3 u_3.$$

But ξ_1 is the ratio of the tetrahedron exe_2e_3 to the tetrahedron $ee_1e_2e_3$, that is, the ratio of that Cartesian co-ordinate of x, measured on eu_1, to a unit length. Similarly for ξ_2 and ξ_3. Hence ξ_1, ξ_2, ξ_3 may be considered as the Cartesian co-ordinates of x, referred to the axes eu_1, eu_2, eu_3.

(4) Any vector can be written in the form

$$\xi_1 u_1 + \xi_2 u_2 + \xi_3 u_3.$$

A vector of the form $\lambda u + \mu v$ must denote a vector parallel to the series of planes which are parallel to the pair of vectors u and v.

307. LINEAR ELEMENTS. (1) A linear element, or the product of two points, must be conceived as a magnitude associated with a definite line. Thus, if a and b be two points, the linear element ab is a magnitude associated with the definite line ab.

Suppose that c is another point on ab such that the length from a to c is λ times the length from a to b. Then $\overline{ac} = \lambda \overline{ab}$, $\overline{cb} = (1 - \lambda)\, \overline{ab}$. Therefore $c = (1 - \lambda) a + \lambda b$, and c is at unit intensity, if a and b are also supposed to be at unit intensity. But $ac = \lambda ab$. Hence the intensity of ac is λ times that of ab, when the length $\overline{ac}$ is λ times the length $\overline{ab}$.

(2) We may therefore define the intensity of the product of two unit points as the length of the line joining them.

If the two points a and b are at intensities α and β, the intensity of ab is $\alpha\beta$ times the length $\overline{ab}$.

(3) The vector $b - a$ (a and b being at unit intensities) and the product ab should be carefully compared.

The intensity of each is defined as the length $\overline{ab}$, but they are magnitudes of different kinds. For $b - a$ is a directed length associated with any one of the infinite set of straight lines parallel to ab, and is an extensive magnitude of the first order, being really a point at infinity. While ab

must be conceived as a directed length associated with the one definite line ab, and is an extensive magnitude of the second order.

(4) Also ab can be written in the form $a(b-a)$, since $aa=0$. Hence the linear element ab may be conceived as the vector $b-a$, fixed down or anchored to a particular line; and the unit point a, as a factor, may be conceived as not affecting the intensity, but as representing the operation of fixing the vector.

(5) Also if c be any other unit point on the line ab, then

$$(a-c)(b-a)=0;$$

since $a-c$ and $b-a$ represent the same point at infinity at different intensities. Hence $a(b-a)=\{c+(a-c)\}(b-a)=c(b-a)$. Thus any other unit point in the line ab may be substituted as a factor in place of a.

(6) Hence if a, b, c be three collinear unit points,

$$ab+bc=ac.$$

For by (5) $b(c-b)=a(c-b)$; and hence

$$ab+bc=a(b-a)+b(c-b)=a(b-a)+a(c-b)=a(c-a)=ac.$$

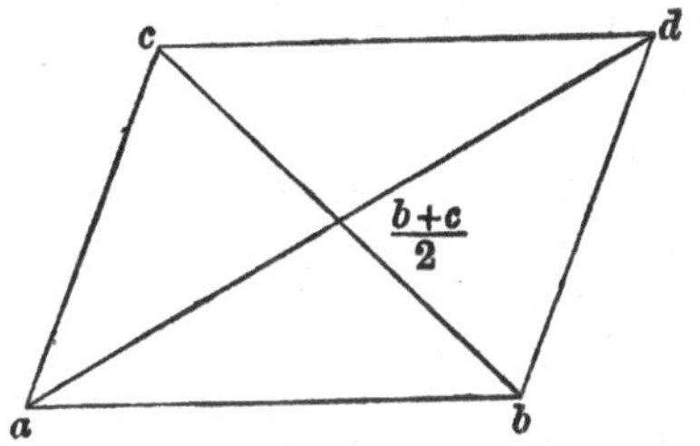

FIG. 3.

If a, b, c be not collinear, then

$$ab+ac=2\,.\,a\,\frac{b+c}{2}=ad,$$

where d is the opposite corner of the parallelogram found by completing the parallelogram ab, ac.

308. VECTOR AREAS. (1) A product of two vectors will be called a vector area.

If uv be any vector area, then only the intensity is altered when any two vectors parallel to the system of parallel planes defined by u and v are substituted for u and v.

For let $$u_1=\lambda_1 u+\mu_1 v, \quad u_2=\lambda_2 u+\mu_2 v;$$

then $$u_1u_2=(\lambda_1\mu_2-\lambda_2\mu_1)\,uv.$$

Hence u_1u_2 denotes the same vector area as uv only at different intensity.

(2) From any point e draw two lines ep and eq representing in magnitude and direction the vectors u and v respectively, and complete the parallelogram $eprq$. Also draw ep_1 and ep_2 to represent the vectors u_1 and u_2 respectively, and complete the parallelogram $ep_1p_3p_2$.

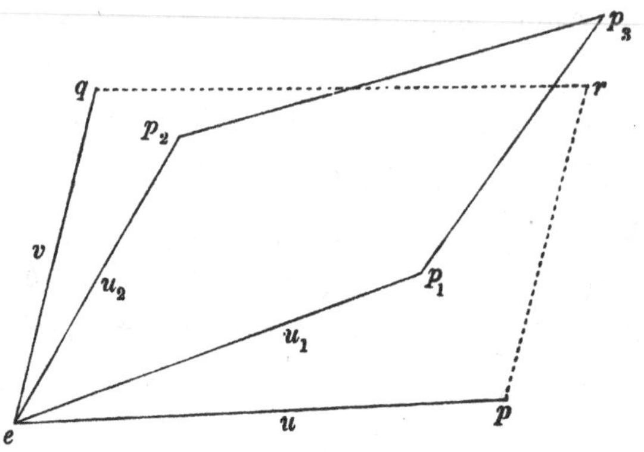

FIG. 4.

Then, conceiving eu and ev as two Cartesian axes and assuming that the vectors u and v are of equal length, the co-ordinates of p_1 and p_2 are λ_1, μ_1 and λ_2, μ_2 respectively.

Hence the area of the parallelogram $ep_1p_3p_2$ is to that of the parallelogram $eprq$ as $(\lambda_1\mu_2 - \lambda_2\mu_1)$ is to unity. Therefore the intensities of uv and u_1u_2 are in the ratio of the areas of the parallelograms formed by uv and u_1u_2.

(3) But the intensities of vector areas are necessarily zero according to the general definition [cf. § 307 (2)] of the intensity of a linear element. For if au be a linear element where u is a vector of length δ and a is a point at intensity α, then the intensity of au is $\alpha\delta$. But when a becomes a vector, α is zero. Therefore the intensity of a product of two vectors is necessarily zero.

Accordingly a special definition must be adopted for the intensity of vector areas; and the above investigation shews that we may consistently adopt the definition that the intensity is the area of the parallelogram formed by completing the parallelogram eu, ev.

The intensity of uv will be considered by convention as positive when by traversing the perimeter of the parallelogram so as first to move in the direction of u and then of v, the direction of motion is clockwise relatively to the enclosed area.

Then in the above figure for uv, we start from e and traverse ep which represents u and then pr which represents v, and the motion is anti-clockwise so that the area is negative.

(4) A vector area will be conceived as possessing an aspect or direction, namely the aspect of the system of parallel planes which are parallel to the two vectors. A line parallel to this system of planes will be called parallel to the plane of the area, or parallel to its aspect.

309. Vector Areas as Carriers. (1) The addition of a vector area to any linear element, which is parallel to its aspect, simply transfers the linear element to a parallel line without altering its intensity.

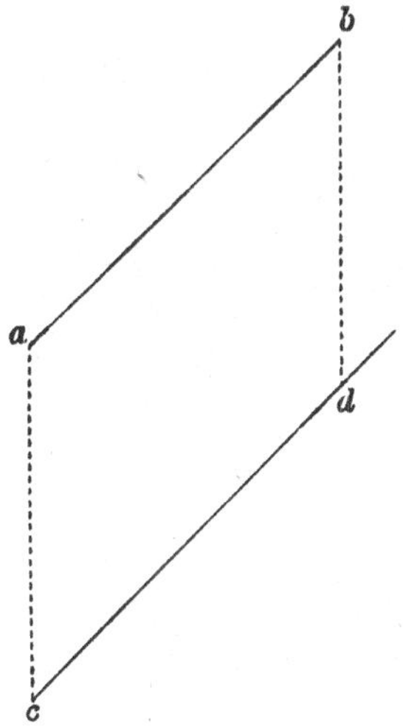

Fig. 5.

For $ab = a(b-a) = \{c + (a-c)\}(b-a) = c(b-a) + (a-c)(b-a)$.

Now let $$d - c = b - a.$$

Then $$c(d-c) + (a-c)(d-c) = a(b-a).$$

Thus the addition to cd of the vector area $(a-c)(d-c)$ transfers it to ab, which is an equal and parallel linear element.

(2) It is also to be noticed that, if cd is conceived as continuously moved into its new position by being kept parallel to itself with its ends on ca and db, then it sweeps through the area $abdc$, which is the area of the parallelogram representing the intensity of the vector area.

(3) Let x and y denote any two unit points

$$e + \xi_1 u_1 + \xi_2 u_2 + \xi_3 u_3, \quad \text{and} \quad e + \eta_1 u_1 + \eta_2 u_2 + \eta_3 u_3.$$

Then by multiplication

$$xy = e.\{(\eta_1 - \xi_1) u_1 + (\eta_2 - \xi_2) u_2 + (\eta_3 - \xi_3) u_3\}$$
$$+ (\xi_2\eta_3 - \xi_3\eta_2) u_2u_3 + (\xi_3\eta_1 - \xi_1\eta_3) u_3u_1 + (\xi_1\eta_2 - \xi_2\eta_1) u_1u_2.$$

This is the form which any linear element must assume. Any vector area takes the less general form

$$\lambda_1 u_2u_3 + \lambda_2 u_3u_1 + \lambda_3 u_1u_2.$$

310. PLANAR ELEMENTS. (1) A planar element, or the product of three points, must be conceived as a magnitude associated with a definite plane.

Thus if abc be a planar element formed by the product of the three points a, b, and c, then abc is a magnitude associated with the definite plane abc.

(2) Let u_1 and u_2 be two unit vectors parallel to this plane but not parallel to one another, and let e be any other point in it. Then we may write

$$a = e + \alpha_1 u_1 + \alpha_2 u_2, \quad b = e + \beta_1 u_1 + \beta_2 u_2, \quad c = e + \gamma_1 u_1 + \gamma_2 u_2.$$

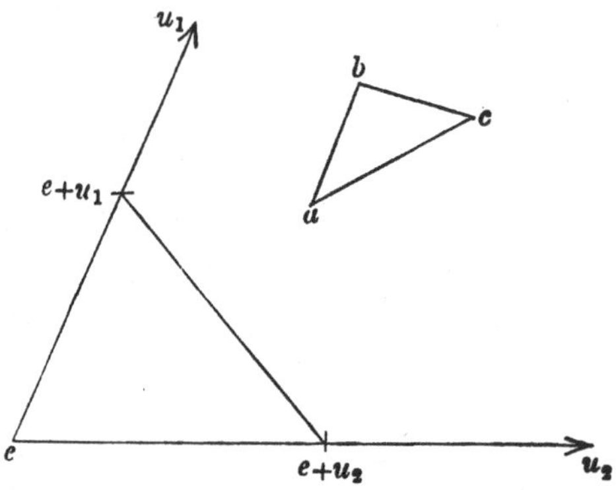

FIG. 6.

Hence
$$abc = \begin{vmatrix} 1 & 1 & 1 \\ \alpha_1 & \beta_1 & \gamma_1 \\ \alpha_2 & \beta_2 & \gamma_2 \end{vmatrix} eu_1u_2.$$

Also
$$eu_1u_2 = e(e+u_1)(e+u_2).$$

Therefore the intensity of the planar element abc is to that of the planar element $e(e+u_1)(e+u_2)$ in the ratio of the area of the triangle formed by a, b, c to that of the triangle formed by e, $e+u_1$, $e+u_2$.

(3) We may therefore consistently define the intensity of the planar element abc as twice the area of the triangle abc. Also the convention will be made that the intensity is positive when the order of letters in abc directs that the perimeter of the triangle be traversed in a clockwise direction.

If a be at intensity α, b at intensity β, c at intensity γ, then the intensity abc is $2\alpha\beta\gamma$ times the area of the triangle abc.

(4) In comparing a vector area with a planar element it must be noticed that a vector area is conceived as an area associated with any one of a series of parallel planes, while a planar element is conceived as an area associated with a definite plane.

The planar element abc, where a, b, c are unit points, can be written in the form $a(b-a)(c-a)$. Then $(b-a)(c-a)$ is a vector area of which the area representing the intensity is the same in magnitude and sign as the area representing the intensity of the planar element abc. Accordingly if U represent a vector area, and a be any unit point, then the planar element aU may be conceived as the tying down of the vector area to the particular plane of the parallel system which passes through a. Also this operation of fixing the vector area makes no change in the intensity.

311. VECTOR VOLUMES. (1) A product of three vectors will be called a vector volume.

The intensity of a vector volume is necessarily zero according to the general definition of the intensity of a planar element. For if U be any vector area of area δ and a any point of intensity α, then the intensity of aU is $\alpha\delta$. Accordingly, when a becomes a vector and α is therefore zero, the intensity of the planar element vanishes.

A special definition of the intensity of a vector volume must therefore be adopted.

(2) We may first notice that all vector volumes are simply numerical multiples of any assigned vector volume. For let u_1, u_2, u_3 be any three non-coplanar vectors. Then since there can only be three independent vectors, any other vectors u, v, w can be written respectively in the forms

$$\xi_1 u_1 + \xi_2 u_2 + \xi_3 u_3, \quad \eta_1 u_1 + \eta_2 u_2 + \eta_3 u_3, \quad \zeta_1 u_1 + \zeta_2 u_2 + \zeta_3 u_3.$$

Then by multiplication

$$uvw = \begin{vmatrix} \xi_1, & \xi_2, & \xi_3 \\ \eta_1, & \eta_2, & \eta_3 \\ \zeta_1, & \zeta_2, & \zeta_3 \end{vmatrix} u_1 u_2 u_3.$$

Thus any vector volume is a numerical multiple of $u_1 u_2 u_3$.

(3) Also let two parallelepipeds be formed with lines representing respectively u_1, u_2, u_3 and u, v, w as conterminous edges. Then the intensities of $u_1 u_2 u_3$ and uvw are in the ratio of the volumes of these parallelepipeds. Thus we may consistently define the intensity of a vector volume as the volume of the corresponding parallelepiped.

312. VECTOR VOLUMES AS CARRIERS. (1) The addition of any vector volume to a planar element transfers the planar element to a parallel plane without altering its intensity.

For consider any planar element abc and any vector volume V. Then we may write

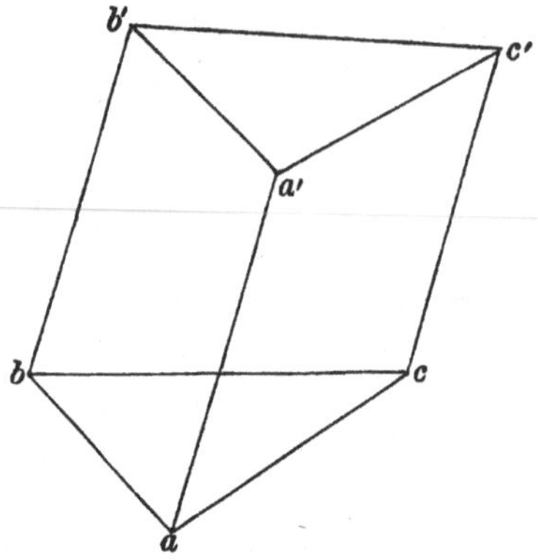

FIG. 7.

$$abc = a(b-a)(c-a),$$

and
$$V = u(b-a)(c-a),$$

where u is some vector.

Hence
$$abc + V = (a+u)(b-a)(c-a) = a'(b'-a')(c'-a'),$$

where $a'-a$ is the vector u, and $a'b'$, $a'c'$ are equal and parallel to ab and ac respectively.

(2) Also it is obvious that if abc moves continuously into its new position remaining parallel to itself with its corners on aa', bb', cc' respectively, it sweeps out a volume equal to half the volume of V.

313. PRODUCT OF FOUR POINTS. (1) Since the complete region is of three dimensions, the product of four points is a mere numerical quantity. Let e_1, e_2, e_3, e_4 be any four unit points forming a tetrahedron, and let a, b, c, d be any four other unit points, also expressible in the forms $\Sigma\alpha e$, $\Sigma\beta e$, $\Sigma\gamma e$, $\Sigma\delta e$.

Then
$$(abcd) = \begin{vmatrix} \alpha_1, & \alpha_2, & \alpha_3, & \alpha_4 \\ \beta_1, & \beta_2, & \beta_3, & \beta_4 \\ \gamma_1, & \gamma_2, & \gamma_3, & \gamma_4 \\ \delta_1, & \delta_2, & \delta_3, & \delta_4 \end{vmatrix} (e_1e_2e_3e_4).$$

Accordingly (from a well-known proposition respecting tetrahedral coordinates) the numbers expressed by the two products $(abcd)$ and $(e_1e_2e_3e_4)$ are in the ratios of the tetrahedrons $abcd$ and $e_1e_2e_3e_4$.

(2) Let the product of four points, such as $abcd$, be defined to be equal to the volume of the parallelepiped, which has the three lines ab, ac, ad as conterminous edges.

Also
$$\{abcd\} = \{a(b-a)(c-a)(d-a)\}.$$

Hence $\{abcd\} = (aV)$, where V is a vector volume of volume equal to the volume $(abcd)$.

314. POINT AND VECTOR FACTORS. (1) It has now been proved that every non-vector product of an order higher than the first may be conceived as consisting of two parts, the point factor, which will be conceived as of unit intensity, and the vector factor. Also the intensity of the product, which is either a length, or an area, or a volume, is also the intensity of the vector factor.

(2) Thus any linear element can be written in the form au, where u is a vector line and a is a unit point; any planar element in the form aM, where M is a vector area; any numerical product of four points in the form (aV), where V is a vector volume.

(3) Also since a is a unit point and not a vector, it follows that $au = 0$ involves $u = 0$, and $aM = 0$ involves $M = 0$, and $aV = 0$ involves $V = 0$.

Thus $au = au'$, involves $u = u'$;

and $aM = aM'$, involves $M = M'$;

and $aV = aV'$, involves $V = V'$.

(4) Again, if a and b be two unit points in the line au, then $au = bu$.

If a and b be two unit points in the plane aM, then $aM = bM$.

If a and b be any two unit points whatever, then $(aV) = (bV)$.

315. Interpretation of Formulæ. (1) It will serve as an illustration of the above discussion to observe the geometrical meanings of the leading formulæ of the Calculus of Extension in this application of it.

In the first place, let the complete region be a plane so that the multiplication of two lines is regressive [cf. § 100]. Let p, q, r, s be four points, and let t be the point of intersection of the two lines pq and rs.

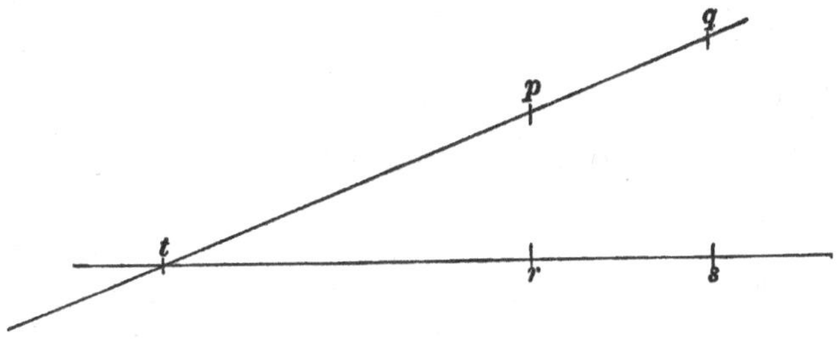

Fig. 8.

Then $$t \equiv pq \, . \, rs = (pqs)\, r - (pqr)\, s = (prs)\, q - (qrs)\, p.$$

Hence t divides rs in the ratio of the area of the triangle rpq to that of the triangle spq; and the section is external, if the order of the letters in rpq and spq makes the circuit of the triangles in the same direction; and it is internal, if the circuits are made in opposite directions.

Similarly t divides pq in the ratio of the area prs to qrs.

(2) In the second place, let three dimensional space form the complete region. Then the products of a line and a plane, and also of two planes, are regressive.

Let p, q, r, s, t be any five points, and let st meet the plane pqr in x.

Then $$x \equiv pqr \, . \, st = (pqrt)\, s - (pqrs)\, t = (pqst)\, r + (rpst)\, q + (qrst)\, p.$$

Hence x divides st in the ratio of the volumes of the tetrahedrons $pqrs$ and $pqrt$; and the section is external, if the products $(pqrs)$ and $(pqrt)$ are of the same sign, that is, if s and t are on the same side of the plane pqr: otherwise the section is internal.

Also, the last form for x states that the areal co-ordinates of x referred to the triangle pqr are in the ratio of the volumes of the tetrahedrons $qrst$, $rpst$, $pqst$.

(3) We may also notice here that according to these formulæ any five points in space are connected by the equation

$$(qrst)\,p - (prst)\,q + (pqst)\,r - (pqrt)\,s + (pqrs)\,t = 0.$$

The formulæ for the line of intersection of two planes abc and def are

$$abc \,.\, def = (abcf)\,de + (abce)\,fd + (abcd)\,ef$$
$$= (adef)\,bc + (bdef)\,ca + (cdef)\,ab.$$

The geometrical meanings of these formulæ are obvious, though they would be rather lengthy to describe.

316. VECTOR FORMULÆ. (1) Some of these formulæ take a special form, if four vectors u_1, u_2, v_1, v_2 be substituted for four of the points. The special peculiarities arise from the fact that the product of four vectors is necessarily zero; and that if V be any vector volume, and a and b any two unit points, then $(aV) = (bV)$.

(2) The formula for five points becomes

$$(au_2v_1v_2)\,u_1 - (au_1v_1v_2)\,u_2 + (au_1u_2v_2)\,v_1 - (au_1u_2v_1)\,v_2 = 0.$$

(3) Again, $au_1u_2 \,.\, v_1v_2 = (au_1u_2v_2)\,v_1 - (au_1u_2v_1)\,v_2 = (au_1v_1v_2)\,u_2 - (au_2v_1v_2)\,u_1$.

Also $\quad av_1v_2 \,.\, u_1u_2 = (av_1v_2u_2)\,u_1 - (av_1v_2u_1)\,u_2 = (av_1u_1u_2)\,v_2 - (av_2u_1u_2)\,v_1$.

Hence $\quad au_1u_2 \,.\, v_1v_2 + av_1v_2 \,.\, u_1u_2 = 0.$

Or, if M and M' be any two vector areas,

$$aM \,.\, M' + aM' \,.\, M = 0.$$

Again, it is obvious that, if a and b be any two unit points,

$$aM \,.\, M' = bM \,.\, M'.$$

317. OPERATION OF TAKING THE VECTOR. (1) Let a unit vector volume be denoted by the symbol $\mathfrak{U}$, and let the sense of $\mathfrak{U}$ be such that $(a\mathfrak{U}) = 1$, where a is any point of unit intensity.

Also if u be any vector volume, and if M be any vector area, then $(u\mathfrak{U}) = 0$, and $(M\mathfrak{U}) = 0$.

(2) Now, if F be any linear element, it can be written in the form au and $F\mathfrak{U} = au \,.\, \mathfrak{U} = (a\mathfrak{U})\,u = u$.

Similarly if P denote any planar element, it can be written in the form aM, and $P \,.\, \mathfrak{U} = aM \,.\, \mathfrak{U} = (a\mathfrak{U}) \,.\, M = M$.

Hence the operation of multiplying $\mathfrak{U}$ on to any non-vector element of any order yields the vector factor of that element. This operation will therefore be called the operation of taking the vector.

(3) We may notice that, if this operation be applied to a vector, the result is zero; and if to a point, the result is the intensity of the point with its proper sign.

(4) Thus if any force be ab where a and b are unit points, by taking the vector we have by the ordinary rule of multiplication

$$ab \,.\, \mathfrak{U} = (a\mathfrak{U})\, b - (b\mathfrak{U})\, a = b - a.$$

Again, if any plane area be abc where a, b, and c are unit points, taking the vector we have

$$abc \,.\, \mathfrak{U} = (a\mathfrak{U})\, bc + (b\mathfrak{U})\, ca + (c\mathfrak{U})\, ab = bc + ca + ab,$$

which is therefore the required vector factor.

(5) In considering the effect of this operation on regressive products, it is well to notice that, if p be any point, the product $(p\mathfrak{U})$ can be conceived both as progressive and as regressive. Therefore the multiplication of $\mathfrak{U}$ on to any pure regressive product still leaves a pure regressive product, which is therefore associative.

(6) The regressive product $au \,.\, bM$ is a point, so taking its vector must yield the intensity of the point. Also the product $au \,.\, bM \,.\, \mathfrak{U}$ is a pure regressive product and is therefore associative.

Hence $$au \,.\, bM \,.\, \mathfrak{U} = au\, (bM \,.\, \mathfrak{U}) = (auM).$$

Therefore (auM), which also equals (buM), is the intensity of the point.

(7) Again, $aM \,.\, bM'$ is a linear element, and its vector factor is given by

$$aM \,.\, bM' \,.\, \mathfrak{U} = aM \,.\, (bM' \,.\, \mathfrak{U}) = aM \,.\, M' = -bM' \,.\, M.$$

Also, since the result is a vector, it is evident that any unit point c can be substituted for a or b in these two formulæ for the vector factor. These results should be compared with the last formulæ in § 316 (3).

(8) Finally $aM \,.\, bM' \,.\, cM''$ is a point. To find its intensity take the vector, then

$$aM \,.\, bM' \,.\, cM'' \,.\, \mathfrak{U} = aM \,.\, bM' \,.\, (cM'' \,.\, \mathfrak{U}) = aM \,.\, bM' \,.\, M''.$$

(9) As an illustration of these formulæ, let us find the vector factor of $abc \,.\, def$. Then by subsection (4) of this article

$$abc \,.\, def \,.\, \mathfrak{U} = abc \,.\, (def \,.\, \mathfrak{U}) = abc \,.\, (ef + fd + de)$$
$$= -def \,.\, (bc + ca + ab).$$

Also

$$abc\,.\,(ef+fd+de)=(abcd)(f-e)+(abce)(d-f)+(abcf)(e-d).$$

(10) Again, let a be any unit point, and F be any linear element. Then F can be written bc, where b and c are unit points.

Hence $\quad aF\,.\,\mathfrak{U}=abc\,.\,\mathfrak{U}=(bc+ca+ab)$

$$=F+a(b-c).$$

But $\quad c-b=F\,.\,\mathfrak{U}.$

Therefore $\quad aF\,.\,\mathfrak{U}=F-a\,.\,F\mathfrak{U}.$

(11) Let F be any linear element. Then the linear element through any point d parallel to F is $d\,.\,F\mathfrak{U}$. Thus if F be in the form ab, where a and b are unit points, the parallel line through d is $d(b-a)$.

Let P be any planar element. Then the plane through any point d parallel to P is $d\,.\,P\mathfrak{U}$. Thus if P be in the form abc, where a, b, and c are unit points, the parallel plane through d is $d(bc+ca+ab)$.

318. THEORY OF FORCES. (1) The theory of forces or linear elements, as discussed in Book V., holds in the Euclidean Space now under discussion. But some further propositions involving vectors must be added.

(2) In § 160 (2) it is proved that, if a be any given point, and A any given planar element, then any system of forces S can be written in the form

$$S=ap+AP,$$

where p and P are respectively a point and planar element depending on the system S.

Now let A denote a vector volume; then AP denotes some vector area, call it M. Also ap can be written in the form au, where u is a vector.

Thus $\quad S=au+M.$

(3) The vector u is independent of the point a. For taking the vector of both sides

$$S\mathfrak{U}=au\,.\,\mathfrak{U}+M\mathfrak{U}=au\,.\,\mathfrak{U}=u.$$

Hence, since u can be written in the form $S\mathfrak{U}$, it is independent of any special method of writing S.

(4) Let $S\mathfrak{U}$ be called the 'principal vector' of S. It is the sum of the vector parts of those separate forces which can be conceived as forming S.

Let the vector area M be called the vector moment of the system round the point a, or the couple of S with respect to a.

Let a be called the base point to which the system is reduced.

(5) Also M depends on the position of a. For $aS=aM$.

Hence $M=aS\,.\,\mathfrak{U}$, and therefore M is the vector factor of the planar element aS, which is the planar-element representing the null plane with respect to a.

The same results respecting M and u follow directly from § 317 (10). For by adding the results of applying the formula of that subsection to each component force of S, we at once obtain

$$S = a \,.\, S\mathfrak{u} + aS \,.\, \mathfrak{u}.$$

Let a' be any other unit point, and let M' be the vector moment of S with respect to it. Then

$$S = au + M = \{a' + (a - a')\}\, u + M = a'u + \{(a - a')\, u + M\}.$$

Hence $$M' = (a - a')\, u + M.$$

(6) Also $(SS) = 2\,(auM)$; and since uM is a vector volume,

$$(auM) = (a'uM) = (a'uM').$$

And since $aM = aS$, $(SS) = 2\,(auS)$, where u is the principal vector of S.

(7) Again evidently

$$(aa'M) = (aa'M') = (aa'S).$$

And $$(auS) = \tfrac{1}{2}\,(SS) = (a'uS).$$

Therefore $\{(a - a')\, uS\} = 0$, where a and a' are any two unit points. This is only an expression of the fact that uS is a vector volume, where u is the principal vector of S. In fact from § 167 (2) we have

$$uS = S \,.\, S\mathfrak{u} = \tfrac{1}{2}\,(SS)\, \mathfrak{u}.$$

Thus the plane at infinity is the null plane of the principal vector.

(8) To find the locus of base points with the same vector moment M.

Let a be one such point and x any other such point. Then by hypothesis

$$S = x \,.\, S\mathfrak{u} + M = a \,.\, S\mathfrak{u} + M.$$

Hence $x\,(S - M) = 0$. But $S - M$ is the linear element $a \,.\, S\mathfrak{u}$. Therefore the equation, $x\,(S - M) = 0$, denotes that x lies on a straight line parallel to $S\mathfrak{u}$.

(9) Let M_0 be any given vector area, then if θM_0 be the vector moment of S (round an appropriate base point) which is parallel to M_0, $uS = \theta u M_0$, where u is the principal vector of S. Hence if a be any unit point

$$(auS) = \tfrac{1}{2}\,(SS) = \theta\,(auM_0).$$

Therefore $$\theta = \frac{1}{2}\, \frac{(SS)}{(auM_0)}.$$

Thus the locus of a point x such that the vector moment of S with respect to it is parallel to M_0—or in other words, the point of which the null plane with respect to S is parallel to M_0—is given by

$$x\left(S - \frac{1}{2}\, \frac{(SS)}{(auM_0)}\, M_0\right) = 0.$$

But it was proved in § 162 (2) that the conjugate of any line ab is $S - \frac{1}{2}\frac{(SS)}{(abS)}ab$. Hence the conjugate of any vector area M_0 is a straight line parallel to the principal vector, and this line is also the locus of points corresponding to which the vector moments are parallel to M_0.

319. Graphic Statics. (1) It will illustrate the methods of the Calculus of Extension as applied to Euclidean Space, if we investigate at this point the ordinary graphic construction for finding the resultant of any number of forces lying in one plane.

(2) Let the given system, S, of coplanar forces be also denoted by $a_1u_1 + a_2u_2 + \ldots + a_\nu u_\nu$; where a_1u_1, a_2u_2, etc. are given forces (cf. fig. 9). We require to construct their resultant.

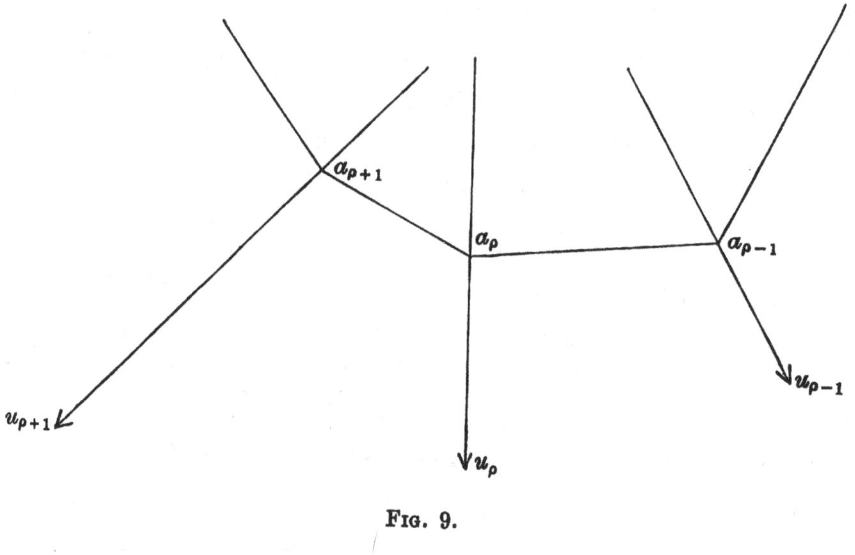

Fig. 9.

Let v be any arbitrarily assumed vector in the plane. Then

$$\left.\begin{aligned}
u_1 &= v + (u_1 - v),\\
u_2 &= (v - u_1) + (u_1 + u_2 - v),\\
u_3 &= (v - u_1 - u_2) + (u_1 + u_2 + u_3 - v),\\
&\ldots\ldots\ldots\ldots\ldots\ldots\ldots\ldots\\
u_\rho &= (v - u_1 \ldots - u_{\rho-1}) + (u_1 + u_2 + \ldots + u_\rho - v),\\
&\ldots\ldots\ldots\ldots\ldots\ldots\ldots\ldots
\end{aligned}\right\}\ldots\ldots\ldots\ldots(1).$$

Thus

$$\begin{aligned}
S = a_1v &+ (a_2 - a_1)(v - u_1) + \ldots\\
&+ (a_\rho - a_{\rho-1})(v - u_1 \ldots - u_{\rho-1}) + \ldots\\
&+ (a_\nu - a_{\nu-1})(v - u_1 \ldots - u_{\nu-1})\\
&+ a_\nu(u_1 + u_2 + \ldots + u_\nu - v) \quad\ldots\ldots\ldots\ldots(2).
\end{aligned}$$

(3) The equations (1) giving the vector parts of the forces are equivalent to starting from any point b_0 (cf. fig. 10) and drawing b_0c to represent v and b_0b_1 to represent u_1. Then b_1c represents $v - u_1$. Also from b_1 draw b_1b_2

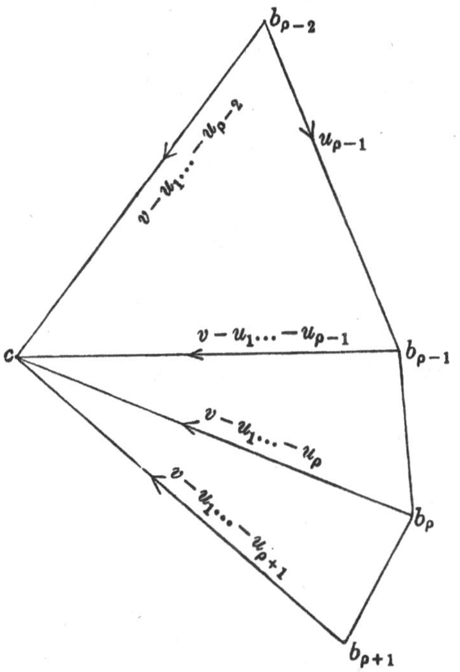

FIG. 10.

to represent u_2, then b_2c represents $v - u_1 - u_2$, and so on. This is the ordinary Graphic construction for the force polygon with any pole c.

(4) To simplify the expression for S notice that $a_1, a_2, \dots a_\nu$ may be any points on the lines of the forces; hence we may assume (cf. figs. 9 and 10) that $a_2 - a_1$ is drawn parallel to $v - u_1$, $a_3 - a_2$ parallel to $v - u_1 - u_2$, and so on.

Then $$S = a_1v + a_\nu(u_1 + u_2 + \dots + u_\nu - v).$$

Thus the resultant force passes through the point d which is the point of intersection of a_1v and $a_\nu(u_1 + u_2 + \dots + u_\nu - v)$, and is the force

$$d(u_1 + u_2 + \dots + u_\nu).$$

(5) This gives the ordinary construction for a funicular polygon, thus: start from any point a_0, draw a_0a_1 parallel to v, a_1a_2 parallel to $v - u_1$, and so on, finally $a_\nu a_{\nu+1}$ parallel to $v - u_1 - u_2 \dots - u_\nu$. Then the resultant passes through the point of intersection of a_0a_1 and $a_\nu a_{\nu+1}$.

(6) Suppose that two different funicular polygons are drawn, namely $a_0a_1 \dots a_\nu$ and $a_0'a_1' \dots a_\nu'$, corresponding to the arbitrary assumptions of two

different vectors v and v' respectively with which to commence the constructions (cf. fig. 11). We will prove the well-known theorem that the points of intersection of corresponding sides are collinear.

For $a_\rho a_\rho'$ is parallel to u_ρ, hence $a_\rho u_\rho = a_\rho' u_\rho$.

Again, $a_\rho a_{\rho-1}$ is parallel to $v - u_1 - \ldots - u_{\rho-1}$, hence

$$a_\rho (v - u_1 - \ldots - u_{\rho-1}) = a_{\rho-1} (v - u_1 - \ldots - u_{\rho-1}).$$

Similarly, $a_\rho' (v' - u_1 - \ldots - u_{\rho-1}) = a'_{\rho-1} (v' - u_1 - \ldots - u_{\rho-1})$.

Therefore $a_1 v - a_1' v' = a_1 (v - u_1) - a_1' (v' - u_1)$

$$= a_2 (v - u_1) - a_2' (v' - u_1) = a_2 (v - u_1 - u_2) - a_2' (v - u_1 - u_2)$$

$$= a_3 (v - u_1 - u_2) - a_3' (v - u_1 - u_2) = \text{etc.}$$

Let $a_{\rho-1} a_\rho$ and $a'_{\rho-1} a_\rho'$ intersect in $d_{\rho-1}$, then

$$a_1 v - a_1' v' = d_0 (v - v'),\ a_2 (v - u_1) - a_2' (v' - u_1) = d_1 (v - v'),$$

and so on.

Hence $$d_0 (v - v') = d_1 (v - v') = \ldots = d_r (v - v').$$

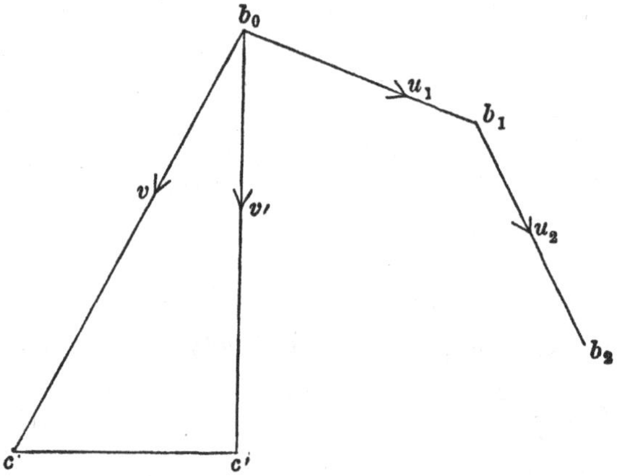

FIG. 11.

Thus the points d_0, d_1, etc. all lie on a straight line parallel to $v - v'$.

Also if the force polygon $b_0 b_1 b_2$ etc. be identical in the two cases (cf. fig. 11), and if $b_0 c$ represents v and $b_0 c'$ represents v', then cc' is parallel to $v - v'$.

NOTE. Grassmann considers vectors in the *Ausdehnungslehre von* 1862; but not in connection with points. The formulæ and ideas of the present and the next chapter are, I believe, in this respect new. The two operations of 'Taking the Vector' and of 'Taking the Flux' [cf. § 325] are, I believe, new operations which have never been defined before. Since this note was in print I have seen the work of M. Burali-Forti, mentioned in the *Note on Grassmann* at the end of this volume.

CHAPTER II.

VECTORS (*continued*).

320. SUPPLEMENTS. (1) The theory of supplements and of inner multiplication has important relations to vector properties.

Take any self-normal quadric [cf. §§ 110, 111], real or imaginary, and let e_1, e_2, e_3, e_4 be four real unit points forming a self-conjugate tetrahedron with respect to this quadric. Let $\epsilon_1, \epsilon_2, \epsilon_3, \epsilon_4$ be respectively the normal intensities [cf. § 109 (3)] of these reference points, where $\epsilon_1, \epsilon_2, \epsilon_3, \epsilon_4$ are any real or pure imaginary quantities. Then

$$(e_1|e_1) = 1/\epsilon_1^2, \quad (e_2|e_2) = 1/\epsilon_2^2, \quad (e_3|e_3) = 1/\epsilon_3^2, \quad (e_4|e_4) = 1/\epsilon_4^2.$$

Also if x be any point $\Sigma\xi e$, then

$$(x|x) = \frac{\xi_1^2}{\epsilon_1^2} + \frac{\xi_2^2}{\epsilon_2^2} + \frac{\xi_3^2}{\epsilon_3^2} + \frac{\xi_4^2}{\epsilon_4^2}.$$

Suppose that y, which is $\Sigma\eta e$, is on the polar plane of x with respect to the quadric, then

$$(x|y) = \frac{\xi_1\eta_1}{\epsilon_1^2} + \frac{\xi_2\eta_2}{\epsilon_2^2} + \frac{\xi_3\eta_3}{\epsilon_3^2} + \frac{\xi_4\eta_4}{\epsilon_4^2} = 0.$$

Also all the points normal to x with respect to the quadric must lie on this plane.

(2) Hence the pole of the plane at infinity is the only point which has three vectors, not coplanar, normal to it. This is the point

$$\frac{\xi_1}{\epsilon_1^2} = \frac{\xi_2}{\epsilon_2^2} = \frac{\xi_3}{\epsilon_3^2} = \frac{\xi_4}{\epsilon_4^2} = \frac{\xi_1 + \xi_2 + \xi_3 + \xi_4}{\epsilon_1^2 + \epsilon_2^2 + \epsilon_3^2 + \epsilon_4^2}.$$

This point is the centre of the quadric. Let it be denoted by e, where e is at unit intensity, and let the normal intensity of e be ϵ. Then we may write

$$(\epsilon_1^2 + \epsilon_2^2 + \epsilon_3^2 + \epsilon_4^2)\, e = \epsilon_1^2 e_1 + \epsilon_2^2 e_2 + \epsilon_3^2 e_3 + \epsilon_4^2 e_4.$$

Hence $$(e|e) = \frac{1}{\epsilon^2} = \frac{1}{\epsilon_1^2 + \epsilon_2^2 + \epsilon_3^2 + \epsilon_4^2}.$$

Therefore $$\epsilon^2 = \epsilon_1^2 + \epsilon_2^2 + \epsilon_3^2 + \epsilon_4^2.$$

(3) But e_1, e_2, e_3, e_4 are any four points forming a normal system with respect to the quadric. Hence the last equation proves that, when the quadric has not its centre at infinity (in which case $\epsilon_1^2 + \epsilon_2^2 + \epsilon_3^2 + \epsilon_4^2 = 0$), the sum of the squares of the normal intensities of any normal system of points is constant and is equal to the square of the normal intensity of the centre.

321. RECTANGULAR NORMAL SYSTEMS. (1) Now let e be the centre of the self-normal quadric, and let u_1, u_2, u_3 be three unit vectors forming with e a normal system with respect to the quadric. We may assume without loss of generality that the normal intensity of e is unity. Since u_1, u_2, u_3 lie in the locus of zero intensity their normal intensities according to the general definition of intensity holding for all points, are zero. But [cf. §§ 109 (3) and 110 (1)] let the normal lengths (or intensities) of u_1, u_2, u_3 be α_1, α_2, α_3, according to the special definition of intensity for vectors [cf. § 305 (3)].

Then $\quad (e\,|\,e) = 1, \quad (u_1\,|\,u_1) = \dfrac{1}{\alpha_1^2}, \quad (u_2\,|\,u_2) = \dfrac{1}{\alpha_2^2}, \quad (u_3\,|\,u_3) = \dfrac{1}{\alpha_3^2}.$

(2) Also any point x at unit intensity is of the form $e + \Sigma\xi u$. Hence

$$(x\,|\,x) = 1 + \frac{\xi_1^2}{\alpha_1^2} + \frac{\xi_2^2}{\alpha_2^2} + \frac{\xi_3^2}{\alpha_3^2}.$$

Thus the self-normal quadric is,

$$\frac{\xi_1^2}{\alpha_1^2} + \frac{\xi_2^2}{\alpha_2^2} + \frac{\xi_3^2}{\alpha_3^2} + 1 = 0;$$

and is accordingly purely imaginary, unless one or more of α_1, α_2, α_3 are pure imaginaries.

(3) It is obvious from the equation of the self-normal quadric that, if u_1, u_2, u_3 be any three mutually normal vectors, then eu_1, eu_2, eu_3 are three conjugate diameters of the quadric.

In general one set and only one set of such conjugate diameters are mutually at right angles. But if the quadric be a sphere with a real or imaginary radius, then all such sets are at right angles. In such a case let the normal systems be called rectangular normal systems. The centre (e) of the self-normal sphere will be called the Origin.

322. IMAGINARY SELF-NORMAL SPHERE. (1) Firstly, let the sphere be imaginary with radius $\sqrt{(-1)}$. Then $\alpha_1 = \alpha_2 = \alpha_3 = 1$.

Hence if u_1, u_2, u_3 be any set of unit vectors at right angles, then

$$(u_1\,|\,u_1) = 1 = (u_2\,|\,u_2) = (u_3\,|\,u_3);$$

and

$$(u_2\,|\,u_3) = 0 = (u_3\,|\,u_1) = (u_1\,|\,u_2).$$

Also if e be the centre of the sphere,

$$(e\,|\,u_1) = 0 = (e\,|\,u_2) = (e\,|\,u_3),$$

and

$$(e\,|\,e) = 1.$$

Again, $|e = u_1u_2u_3$, $|u_1 = -eu_2u_3$, $|u_2 = -eu_3u_1$, $|u_3 = -eu_1u_2$.

And $$|eu_1 = u_2u_3, \quad |u_2u_3 = eu_1,$$

and $$|eu_2 = u_3u_1, \quad |u_3u_1 = eu_2,$$

and $$|eu_3 = u_1u_2, \quad |u_1u_2 = eu_3.$$

Also $$|u_1u_2u_3 = -e, \; |eu_2u_3 = u_1, \; |eu_3u_1 = u_2, \; |eu_1u_2 = u_3.$$

(2) Let v be any vector $\xi_1u_1 + \xi_2u_2 + \xi_3u_3$, then $(v|v) = \xi_1^2 + \xi_2^2 + \xi_3^2$. Hence $(v|v)$ is the square of the length of v.

Again, let v and v' be two vectors $\xi_1u_1 + \xi_2u_2 + \xi_3u_3$ and $\xi_1'u_1 + \xi_2'u_2 + \xi_3'u_3$, of lengths ρ and ρ' respectively. Then

$$(v|v') = \xi_1\xi_1' + \xi_2\xi_2' + \xi_3\xi_3' = \rho\rho' \cos\theta,$$

where θ is the angle between the two vectors.

Thus $$\cos\theta = \frac{(v\ v')}{\sqrt{\{(v|v)(v'|v')\}}}, \text{ and } \sin\theta = \sqrt{\frac{(vv'|vv')}{(v|v)(v'|v')}}.$$

(3) Again, let M be any vector area $\xi_1u_2u_3 + \xi_2u_3u_1 + \xi_3u_1u_2$, and let μ be its area.

Then $$(M|M) = \xi_1^2 + \xi_2^2 + \xi_3^2 = \mu^2.$$

Also let M' be another vector area in a plane making an angle θ with that of M; let M' be written $\xi_1'u_2u_3 + \xi_2'u_3u_1 + \xi_3'u_1u_2$, and let its intensity be μ'.

Then $$(M|M') = \xi_1\xi_1' + \xi_2\xi_2' + \xi_3\xi_3' = \mu\mu' \cos\theta.$$

(4) The inner squares and products of points and linear elements in general have no important signification.

It will be observed that these formulæ for the product of two vectors or of two vector areas are entirely independent of the centre of the self-normal quadric.

The expressions $(M|M)$ and $(u|u)$ will be often shortened into M^2 and u^2, on the understanding that the normal system is rectangular and the radius of the self-normal sphere is $\sqrt{(-1)}$.

323. Real Self-Normal Sphere. (1) Secondly, let the self-normal sphere be real with radius unity. Then with the notation of § 321

$$\alpha_1 = \alpha_2 = \alpha_3 = \sqrt{(-1)}.$$

(2) Hence if u_1, u_2, u_3 be any set of unit vectors mutually at right-angles, then

$$(u_1|u_1) = -1 = (u_2|u_2) = (u_3|u_3);$$

and $$(u_2|u_3) = 0 = (u_3|u_1) = (u_1|u_2).$$

Also, if e be the centre of the sphere,

$$(e|u_1) = 0 = (e|u_2) = (e|u_3);$$

and $$(e|e) = 1.$$

Thus if v and v' be any two vectors $\Sigma\xi u$ and $\Sigma\xi' u$, of lengths ρ and ρ', and making an angle θ with each other,

$$(v|v') = -(\xi_1\xi_1' + \xi_2\xi_2' + \xi_3\xi_3') = -\rho\rho'\cos\theta.$$

And in fact a set of formulæ can be deduced analogous to those which obtain in the first case, when the sphere is imaginary with radius $\sqrt{(-1)}$.

(3) But the constant introduction of the negative sign is very inconvenient, we will therefore for the future, unless it is otherwise expressly stated, assume that rectangular normal systems are employed and that the self-normal sphere is imaginary as in § 322.

324. GEOMETRICAL FORMULÆ. (1) If u and v are vectors the square of the length of $u + v$ is given by

$$(u+v)^2 = (u+v)|(u+v) = u^2 + v^2 + 2(u|v)$$
$$= u^2 + v^2 + 2\sqrt{(u^2v^2)}\cos\theta,$$

where θ is the angle between u and v.

(2) To express the area of the triangle abc. Assume that a, b and c are at unit intensities. Then taking the vector

$$abc\,.\,\mathfrak{U} = bc + ca + ab = (a-b)(a-c).$$

Hence, if Δ be the required area,

$$\Delta^2 = \tfrac{1}{4}(abc\,.\,\mathfrak{U})^2 = \tfrac{1}{4}\{(a-b)(a-c)|(a-b)(a-c)\}$$
$$= \tfrac{1}{4}[(a-b)^2(a-c)^2 - \{(a-b)|(a-c)\}^2].$$

If α, β, γ be the lengths of the sides, and $\boldsymbol{\alpha}$, $\boldsymbol{\beta}$, $\boldsymbol{\gamma}$ the corresponding angles, $(a-b)^2 = \gamma^2$, $(a-c)^2 = \beta^2$, $(a-b)|(a-c) = \beta\gamma\cos\boldsymbol{\alpha}$.

Thus $$\Delta^2 = \tfrac{1}{4}\beta^2\gamma^2\sin^2\boldsymbol{\alpha}.$$

(3) The angle, θ, between two linear elements C_1 and C_2 is given [cf. § 322 (2)] by

$$\cos\theta = \frac{(C_1\mathfrak{U}|C_2\mathfrak{U})}{\sqrt{\{(C_1\mathfrak{U})^2(C_2\mathfrak{U})^2\}}}.$$

The angle, θ, between two planar elements P and Q is given [cf. § 322 (3)] by

$$\cos\theta = \frac{(P\mathfrak{U}|Q\mathfrak{U})}{\sqrt{\{(P\mathfrak{U})^2(Q\mathfrak{U})^2\}}}.$$

(4) The length of the perpendicular from any point a on to the plane of the planar element P is obviously [cf. §§ 311 (3) and 322 (3)]

$$\frac{(aP)}{\sqrt{(P\mathfrak{U})^2}}.$$

(5) To find the shortest distance between two lines C_1 and C_2.

Let a_1 and a_2 be any two unit points on C_1 and C_2 respectively. Then $C_1 = a_1\,.\,C_1\mathfrak{U}$, and $C_2 = a_2\,.\,C_2\mathfrak{U}$. Also the required perpendicular (ϖ) is equal to the perpendicular from a_2 on to the plane $a_1\,.\,C_1\mathfrak{U}\,.\,C_2\mathfrak{U}$.

Hence $$\varpi = \frac{(a_2 a_1 \,.\, C_1 \mathfrak{U} \,.\, C_2 \mathfrak{U})}{\sqrt{(C_1 \mathfrak{U} \,.\, C_2 \mathfrak{U})^2}} = \frac{(C_1 C_2)}{\sqrt{(C_1 \mathfrak{U} \,.\, C_2 \mathfrak{U})^2}}.$$

(6) If γ_1 and γ_2 be the lengths of the linear elements C_1 and C_2, θ the angle between them, and ϖ the length of the common perpendicular, then [cf. § 322 (2)]

$$(C_1 \mathfrak{U} \,.\, C_2 \mathfrak{U})^2 = \gamma_1^2 \gamma_2^2 \sin^2 \theta.$$

Hence $$(C_1 C_2) = \gamma_1 \gamma_2 \varpi \sin \theta.$$

We may notice here that

$$\sin \theta = \sqrt{\{1 - \cos^2 \theta\}} = \sqrt{\frac{(C_1 \mathfrak{U} \,.\, C_2 \mathfrak{U} \,|.\, C_1 \mathfrak{U} \,.\, C_2 \mathfrak{U})}{(C_1 \mathfrak{U})^2 (C_2 \mathfrak{U})^2}}.$$

325. Taking the Flux. (1) It is often necessary to express the vector-line normal to a vector-area, or the vector-area perpendicular to a vector-line. This is accomplished by a combination of operations, which we will call the operation of taking the 'Flux.' Let v be any vector

$$\xi_1 u_1 + \xi_2 u_2 + \xi_3 u_3,$$

and let e be the origin. Then

$$|ev = \xi_1 \,|eu_1 + \xi_2 \,|eu_2 + \xi_3 \,|eu_3 = \xi_1 u_2 u_3 + \xi_2 u_3 u_1 + \xi_3 u_1 u_2.$$

Accordingly $|ev$ is the vector-area perpendicular to the vector v. Also the length of v is $\sqrt{\{\xi_1^2 + \xi_2^2 + \xi_3^2\}}$ times the unit of length, and the area of $|ev$ is $\sqrt{\{\xi_1^2 + \xi_2^2 + \xi_3^2\}}$ times the unit of area. The vector-area $|ev$ will be called the flux of the vector v.

Again, let M be the vector-area

$$\xi_1 u_2 u_3 + \xi_2 u_3 u_1 + \xi_3 u_1 u_2.$$

Then $$|eM = \xi_1 \,|eu_2 u_3 + \xi_2 \,|eu_3 u_1 + \xi_3 \,|eu_1 u_2 = \xi_1 u_1 + \xi_2 u_2 + \xi_3 u_3.$$

Hence the vector-line $|eM$, which has been defined as the flux of M, is perpendicular to the plane of M.

The operation of taking the flux can of course be applied to non-vector elements, but the results are of no interest as permanent formulæ and are easily worked out afresh when required.

(2) It will be noticed that the result of the operation on vector elements is really entirely independent of the position of e, the centre of the self-normal sphere.

Also furthermore the operation is capable of an alternative form. For

$$|ev = |e\,|v = \mathfrak{U}\,|v = -\,|v \,.\, \mathfrak{U}\,; \text{ and } |eM = |e\,|M = \mathfrak{U}\,|M = |M \,.\, \mathfrak{U}.$$

Hence, except in respect to sign, the operation is identical with that of taking the vector of the supplement of the vector-line or of the vector-area.

(3) For these two reasons it is desirable to adopt a single symbol for the combination of operations denoted by $|e$. Let $|ev$ and $|eM$ be written $\mathfrak{F}v$ and $\mathfrak{F}M$ respectively.

We notice that $\mathfrak{F}\mathfrak{F}v = v$ and $\mathfrak{F}\mathfrak{F}M = M$.

Also the operation is distributive as regards addition; but it is not distributive as regards multiplication.

326. Flux Multiplication. (1) The operation of multiplying the flux of a vector-line or vector-area into a vector-line or vector-area will be called Flux Multiplication. Its formulæ are almost identical with those of Inner Multiplication [cf. § 119]. They are all independent of the position of the origin: this fact will be obvious, if it be remembered that $(e\mathfrak{U}) = (e'\mathfrak{U})$, where e and e' are any two unit points.

(2) Firstly, it is obvious that

$$(v\mathfrak{F}v) = (\xi_1^2 + \xi_2^2 + \xi_3^2)\,\mathfrak{U} = (v\,|v)\,\mathfrak{U},$$
$$(M\mathfrak{F}M) = (\xi_1^2 + \xi_2^2 + \xi_3^2)\,\mathfrak{U} = (M\,|\,M)\,\mathfrak{U}.$$

Similarly $$(v\mathfrak{F}v') = (v\,|\,v')\,\mathfrak{U} = (v'\mathfrak{F}v),$$
and $$(M\mathfrak{F}M') = (M\,|\,M')\,\mathfrak{U} = (M'\mathfrak{F}M).$$

(3) Again, $v_1\mathfrak{F}v_2v_3 = v_1\,(|v_2\,.\,|v_3\,.\,\mathfrak{U}) = (v_1\,|v_3)\,\mathfrak{U}\,|v_2 + (v_1\,|v_2)\,|v_3\,.\,\mathfrak{U}$

$$= (v_1\,|\,v_3)\,\mathfrak{F}v_2 - (v_1\,|v_2)\,\mathfrak{F}v_3 \quad\text{.........(i).}$$

And $$v_2v_3\mathfrak{F}v_1 = 0 = M\mathfrak{F}v_1 \quad\text{.........(ii).}$$

Equation (i) may be compared with

$$v_1\,|v_2v_3 = (v_1\,|v_3)\,|v_2 - (v_1\,|v_2)\,|v_3.$$

From equation (ii), it follows as a particular case that

$$\mathfrak{F}v_1\,.\,\mathfrak{F}v_2 = 0.$$

(4) From equation (i), replacing v_1 by $\mathfrak{F}uu'$, we have

$$\mathfrak{F}uu'\,.\,\mathfrak{F}vv' = (\mathfrak{F}uu'\,.\,|v')\,\mathfrak{F}v - (\mathfrak{F}uu'\,.\,|v)\,\mathfrak{F}v'.$$

But [cf. § 99 (4)] $$\mathfrak{F}M\,|v = |(eMv) = (eMv).$$

Hence $$\mathfrak{F}uu'\,.\,\mathfrak{F}vv' = (euu'v')\,\mathfrak{F}v - (euu'v)\,\mathfrak{F}v' \quad\text{.........(iii).}$$

As a particular case of equation (iii) we deduce

$$\mathfrak{F}vv'\,.\,\mathfrak{F}vv'' = (evv'v'')\,\mathfrak{F}v \quad\text{.........(iv).}$$

(5) Let a be any unit point, then

$$a\mathfrak{F}u = -a\,(|u\,.\,\mathfrak{U}) = -(a\mathfrak{U})\,|u + (a\,|u)\,\mathfrak{U} = -|u + (a\,|u)\,\mathfrak{U}.$$

Hence

$$a\mathfrak{F}u\,.\,\mathfrak{F}v = -\{|u - (a\,|u)\,\mathfrak{U}\}\,\mathfrak{F}v = -|u\,.\,\mathfrak{F}v = |u\,.\,|v\,.\,\mathfrak{U} = |uv\,.\,\mathfrak{U} = \mathfrak{F}uv \quad\text{......(v).}$$

Also if b be another unit point, and M and M' be vector-areas,

$$aM\,.\,b\mathfrak{F}M' = (aM\mathfrak{F}M')\,b - (aMb)\,\mathfrak{F}M' = (M\,|M')\,b - (aMb)\,\mathfrak{F}M' \quad\text{......(vi).}$$

And as particular cases

$$\left.\begin{aligned} aM\,.\,b\mathfrak{F}M &= M^2b - (aMb)\,\mathfrak{F}M \\ aM\,.\,a\mathfrak{F}M' &= (M\,|M')\,a \end{aligned}\right\}\quad\text{.........(vii).}$$

327. Geometrical Formulæ. (1) To find the foot of the perpendicular from the vertex d of the tetrahedron $abcd$ on to the opposite face.

The required point [cf. §§ 317 (4) and 325 (1)] is $abc \,.\, d\mathfrak{F}(bc + ca + ab)$, and by the transformation of equation (vii) this can be written

$$(bc + ca + ab)^2 d - (abcd)\,\mathfrak{F}(bc + ca + ab).$$

(2) To find the line perpendicular to two given lines.

Let C_1 and C_2 be linear elements on the two given lines. Then $C_1\mathfrak{U} \,.\, C_2\mathfrak{U}$ is a vector-area perpendicular to the required line. Hence $\mathfrak{F}(C_1\mathfrak{U} \,.\, C_2\mathfrak{U})$ is a vector parallel to the required line. But the required line intersects both the lines C_1 and C_2. Hence it lies in both the planes $C_1\mathfrak{F}(C_1\mathfrak{U} \,.\, C_2\mathfrak{U})$ and $C_2\mathfrak{F}(C_1\mathfrak{U} \,.\, C_2\mathfrak{U})$. The required line is therefore

$$C_1\mathfrak{F}(C_1\mathfrak{U} \,.\, C_2\mathfrak{U}) \,.\, C_2\mathfrak{F}(C_1\mathfrak{U} \,.\, C_2\mathfrak{U}).$$

(3) Now if the two lines C_1 and C_2 are given respectively in the forms a_1u_1 and a_2u_2, this expression for the line of the common perpendicular can be transformed by equation (i).

For
$$C_1\mathfrak{F}(C_1\mathfrak{U} \,.\, C_2\mathfrak{U}) = a_1u_1\mathfrak{F}u_1u_2 = a_1 \,.\, u_1\mathfrak{F}u_1u_2$$
$$= (u_1 \mid u_2)\, a_1\mathfrak{F}u_1 - (u_1 \mid u_1)\, a_1\mathfrak{F}u_2,$$

and similarly
$$C_2\mathfrak{F}(C_2\mathfrak{U} \,.\, C_1\mathfrak{U}) = (u_1 \mid u_2)\, a_2\mathfrak{F}u_2 - (u_2 \mid u_2)\, a_2\mathfrak{F}u_1.$$

Hence the required line is now [cf. § 102 (7)] in the form

$$(u_1 \mid u_2)^2\, a_1\mathfrak{F}u_1 \,.\, a_2\mathfrak{F}u_2 + (u_1 \mid u_1)(u_2 \mid u_2)\, a_1\mathfrak{F}u_2 \,.\, a_2\mathfrak{F}u_1$$
$$- (u_1 \mid u_2)(u_2 \mid u_2)(a_1a_2\mathfrak{F}u_1)\,\mathfrak{F}u_1 - (u_1 \mid u_2)(u_1 \mid u_1)(a_1a_2\mathfrak{F}u_2)\,\mathfrak{F}u_2.$$

328. The Central Axis. (1) It follows as a particular case of § 318 (9) that any system S can be written in the form $au + \varpi\mathfrak{F}u$; where a is any point on a certain line parallel to u, called the Central Axis, and ϖ is called the pitch. Also [cf. § 318 (9)]

$$\varpi = \tfrac{1}{2}\frac{(SS)}{(au\mathfrak{F}u)} = \tfrac{1}{2}\frac{(SS)}{u^2} = \tfrac{1}{2}\frac{(SS)}{(S\mathfrak{U})^2}.$$

(2) It is obvious that S can be written, without the use of u, in the form

$$a \,.\, S\mathfrak{U} + \varpi\mathfrak{F}(S\mathfrak{U}).$$

Also other expressions for ϖ are found from

$$(S \mid S) = u^2(1 + \varpi^2) = (S\mathfrak{U})^2(1 + \varpi^2).$$

Hence
$$1 + \varpi^2 = \frac{(S \mid S)}{(S\mathfrak{U})^2}, \text{ and } \frac{2\varpi}{1 + \varpi^2} = \frac{(SS)}{(S \mid S)}.$$

(3) We are now prepared to discuss the signification of the addition of any vector-area M to any force system S.

It has been proved [cf. § 309] that the addition of M to any linear element C_1 in its plane is equivalent to the transference of C_1, kept parallel to itself, to a new position, so that in the transference C_1 sweeps out an area $\sqrt{(M)^2}$ with the aspect of M.

Let S be written in the form $au + \varpi \mathfrak{F}u$. Then if M and $\mathfrak{F}u$ have the vector w in common, M can be written in the form $\lambda\mathfrak{F}u + wu$.

Thus
$$S + M = au + \varpi\mathfrak{F}u + \lambda\mathfrak{F}u + wu$$
$$= (a + w)\, u + (\varpi + \lambda)\, \mathfrak{F}u.$$

Hence the component of M which is parallel to the central axis (namely, wu) transfers the central axis according to the ordinary rule for transferring a linear element parallel to the plane of the vector; and the component of M which is perpendicular to the central axis (namely, $\lambda\mathfrak{F}u$) simply alters the pitch. It will be noticed that the principal vector of S is unaltered.

329. PLANES CONTAINING THE CENTRAL AXIS. (1) The plane through the central axis of S and parallel to any vector v is
$$Sv - \varpi\,(v\,|\,.\,S\mathfrak{U})\,\mathfrak{U},$$
where ϖ is the pitch of S.

For, writing S in the form $au + \varpi\mathfrak{F}u$, the required plane is
$$auv = (S - \varpi\mathfrak{F}u)\, v = Sv - \varpi v\mathfrak{F}u = Sv - \varpi\,(v\,|u)\,\mathfrak{U}.$$

If v be perpendicular to $S\mathfrak{U}$, the plane becomes Sv.

(2) Hence the plane containing the central axis of S, and perpendicular to the given vector area M, is
$$S\mathfrak{F}M - \varpi\,(SM)\,\mathfrak{U}.$$

For by subsection (1) the plane is $S\mathfrak{F}M - \varpi\,(\mathfrak{F}M\,.\,|S\mathfrak{U})\,\mathfrak{U}$.

Now $\quad (\mathfrak{F}M\,|S\mathfrak{U}) = (|eM\,.\,|S\mathfrak{U}) = (eM\,.\,S\mathfrak{U}) = (eM\mathfrak{U}\,.\,S) = (MS)$;

since the three terms eM, S and $\mathfrak{U}$, form a pure regressive product [cf. 101 (3)].

330. DUAL GROUPS OF SYSTEMS OF FORCES. (1) The metrical properties of dual groups—and therefore also of quadruple groups [cf. §§ 169 and 170]—can now be discussed.

Let S_1 and S_2 be any two systems defining a dual group. Let S_1 be written in the form $a_1v_1 + \varpi_1\mathfrak{F}v_1$, and S_2 in the form $a_2v_2 + \varpi_2\mathfrak{F}v_2$.

Then
$$(S_1S_2) = (a_1v_1a_2v_2) + \varpi_1\,(a_1v_1\mathfrak{F}v_2) + \varpi_2\,(a_2v_2\mathfrak{F}v_1)$$
$$= (a_1v_1a_2v_2) + (\varpi_1 + \varpi_2)\,(v_1\,|v_2).$$

Accordingly [cf. § 324 (6)] if θ be the angle between a_1v_1 and a_2v_2, and δ the shortest distance between them,
$$(S_1S_2) = \sqrt{\{v_1^2v_2^2\}}\,.\,\{d\sin\theta + (\varpi_1 + \varpi_2)\cos\theta\}.$$

(2) Let $\frac{(S_1S_2)}{\sqrt{\{(S_1\mathfrak{U})^2(S_2\mathfrak{U})^2\}}}$ be called the virtual coefficient of the two systems S_1 and S_2. This virtual coefficient can accordingly also be written in the form $\{d\sin\theta+(\varpi_1+\varpi_2)\cos\theta\}$. The idea of the virtual coefficient, and this latter form of it are due to Sir R. S. Ball in his *Theory of Screws*.

(3) Since the condition that the two systems be reciprocal can be written

$$d\sin\theta+(\varpi_1+\varpi_2)\cos\theta=0,$$

it follows that if the central axes of two reciprocal systems intersect, either they are at right angles, or the sum of the pitches is zero.

331. Invariants of a Dual Group. (1) A vector-area parallel to the two central axes is $S_1\mathfrak{U}\,.\,S_2\mathfrak{U}$. This vector-area is an invariant [cf. § 177] of the system. For let

$$S=\lambda_1S_1+\lambda_2S_2,\quad S'=\mu_1S_1+\mu_2S_2.$$

Then $$S\mathfrak{U}\,.\,S'\mathfrak{U}=(\lambda_1\mu_2-\lambda_2\mu_1)\,S_1\mathfrak{U}\,.\,S_2\mathfrak{U}.$$

Hence all the central axes of the dual group are perpendicular to the same vector $\mathfrak{F}(S_1\mathfrak{U}\,.\,S_2\mathfrak{U})$.

(2) The plane through the central axis of S_1, and parallel to the normal to the vector-area $S_1\mathfrak{U}\,.\,S_2\mathfrak{U}$ is, by § 329 (2), $S_1\mathfrak{F}(S_1\mathfrak{U}\,.\,S_2\mathfrak{U})$.

Hence the line of the shortest distance between the central axes of S_1 and S_2 is

$$S_1\mathfrak{F}(S_1\mathfrak{U}\,.\,S_2\mathfrak{U})\,.\,S_2\mathfrak{F}(S_1\mathfrak{U}\,.\,S_2\mathfrak{U}).$$

But this linear element is an invariant of the group. For substituting S and S' for S_1 and S_2, we find by the use of the previous subsection

$$\begin{aligned}&S\,.\,\mathfrak{F}(S\mathfrak{U}\,.\,S'\mathfrak{U})\,.\,S'\mathfrak{F}(S\mathfrak{U}\,.\,S'\mathfrak{U})\\ &=(\lambda_1\mu_2-\lambda_2\mu_1)^2(\lambda_1S_1+\lambda_2S_2)\,\mathfrak{F}(S_1\mathfrak{U}\,.\,S_2\mathfrak{U})\,.\,(\mu_1S_1+\mu_2S_2)\,\mathfrak{F}(S_1\mathfrak{U}\,.\,S_2\mathfrak{U})\\ &=(\lambda_1\mu_2-\lambda_2\mu_1)^3\,S_1\mathfrak{F}(S_1\mathfrak{U}\,.\,S_2\mathfrak{U})\,.\,S_2\mathfrak{F}(S_1\mathfrak{U}\,.\,S_2\mathfrak{U}).\end{aligned}$$

Hence all the central axes of the systems of a dual group intersect the same line at right angles. Let this line be called the axis of the group.

332. Secondary Axes of a Dual Group. (1) In any dual group there are in general two, and only two, reciprocal systems with their central axes at right angles.

For let S and S' be two such systems. Then $(SS')=0$, gives

$$\lambda_1\mu_1(S_1S_1)+\lambda_2\mu_2(S_2S_2)+(\lambda_1\mu_2+\lambda_2\mu_1)(S_1S_2)=0;$$

and $(S\mathfrak{U}\,|\,S'\mathfrak{U})=0$, gives

$$\lambda_1\mu_1(S_1\mathfrak{U})^2+\lambda_2\mu_2(S_2\mathfrak{U})^2+(\lambda_1\mu_2+\lambda_2\mu_1)(S_1\mathfrak{U}\,|\,S_2\mathfrak{U})=0.$$

From these two equations by eliminating μ_1/μ_2, we deduce

$$\lambda_1^2\{(S_1S_1)(S_1\mathfrak{U}\,|\,S_2\mathfrak{U})-(S_1S_2)(S_1\mathfrak{U})^2\}-\lambda_2^2\{(S_2S_2)(S_1\mathfrak{U}\,|\,S_2\mathfrak{U})-(S_1S_2)(S_2\mathfrak{U})^2\}$$
$$+\lambda_1\lambda_2\{(S_1S_1)(S_2\mathfrak{U})^2-(S_2S_2)(S_1\mathfrak{U})^2\}=0.$$

Let α/β and α'/β' be the two roots of this quadratic for λ_1/λ_2. Then

$$\alpha S_1+\beta S_2 \quad\text{and}\quad \alpha' S_1+\beta' S_2$$

are two reciprocal systems with their central axes at right angles, and there are only two such systems belonging to the group S_1S_2.

(2) It follows from § 330 (3) that the central axes of these two systems must intersect. Hence they intersect at a point on the axis of the group. Let this point be called the 'Centre of the Group.'

Also let the two central axes of these systems be called the 'Secondary Axes of the Group'; and let the systems S_1 and S_2 be called the 'Principal Systems' of the group.

Let the plane through the centre perpendicular to the axis be called the 'Diametral Plane of the Group.'

333. THE CYLINDROID. (1) Let the assemblage of central axes of a dual group, each axis being conceived as associated with its pitch, be termed a Cylindroid. (Cf. Sir R. S. Ball's *Theory of Screws.*)

Take the centre of the group as the origin e, let eu_1 and eu_2 be the two secondary axes of the group and eu_3 the axis of the group. Then eu_1, eu_2, eu_3 are a system of rectangular axes. Assume that u_1, u_2, u_3 are unit vectors.

Let ϖ_1 and ϖ_2 be the pitches associated respectively with eu_1 and eu_2, then we may write

$$S_1=eu_1+\varpi_1\mathfrak{F}u_1=eu_1+\varpi_1u_2u_3,$$
$$S_2=eu_2+\varpi_2\mathfrak{F}u_2=eu_2+\varpi_1u_3u_1.$$

(2) Then any other system S of the group, with its principal vector of unit length and making an angle θ with u_1, is $S_1\cos\theta+S_2\sin\theta$.

Hence

$$S=e(u_1\cos\theta+u_2\sin\theta)+\varpi_1\cos\theta\,.\,u_2u_3+\varpi_2\sin\theta\,.\,u_3u_1$$
$$=(e+\xi_3u_3)(u_1\cos\theta+u_2\sin\theta)$$
$$+(\varpi_1\cos\theta+\xi_3\sin\theta)\,u_2u_3+(\varpi_2\sin\theta-\xi_3\cos\theta)\,u_3u_1;$$

where ξ_3 is a quantity which can be expressed in terms of ϖ_1, ϖ_2, and θ.

Now assume that the central axis of S cuts eu_3 in the point $e+\xi_3u_3$, and that ϖ is the pitch of S.

Then $(\varpi_1\cos\theta+\xi_3\sin\theta)\,u_2u_3+(\varpi_2\sin\theta-\xi_3\cos\theta)\,u_3u_1$

$$=\varpi\mathfrak{F}(u_1\cos\theta+u_2\sin\theta)=\varpi\cos\theta\,.\,u_2u_3+\varpi\sin\theta\,.\,u_3u_1.$$

Hence

$$\varpi_1\cos\theta+\xi_3\sin\theta=\varpi\cos\theta,$$
$$\varpi_2\sin\theta-\xi_3\cos\theta=\varpi\sin\theta.$$

Thus $$\varpi = \varpi_1 \cos^2\theta + \varpi_2 \sin^2\theta,$$
and $$\xi_3 = (\varpi_2 - \varpi_1) \sin\theta \cos\theta.$$
These equations completely define the cylindroid.

(3) If $x = e + \xi_1 u_1 + \xi_2 u_2 + \xi_3 u_3$ be any point on one of the central axes, the equation of the surface, on which it lies, is obviously
$$\xi_3 (\xi_1^2 + \xi_2^2) = (\varpi_2 - \varpi_1) \xi_1 \xi_2.$$

(4) The director forces of the group are the two systems of zero pitch. Hence the angles their lines make with eu_1 are given by
$$\varpi_1 \cos^2\theta + \varpi_2 \sin^2\theta = 0.$$

It follows that the directrices of the group are real if $\varpi_1\varpi_2$ be negative, and are imaginary if $\varpi_1\varpi_2$ be positive.

The distances from the centre of the points of intersection of the directrices with the axis are easily seen to be $\pm\sqrt{(-\varpi_1\varpi_2)}$.

334. THE HARMONIC INVARIANTS. (1) The harmonic invariant $H(x)$ [cf. § 179 (3)] of any point x can be written $2xS_1 . S_2$; where S_1 and S_2 are the principal systems of the group.

Hence, with the notation of § 333 (1), by an easy reduction
$$H(x) = 2(xeu_1u_2)\,e + 2\varpi_2 (xeu_3u_1)\,u_1 - 2\varpi_1 (xeu_2u_3)\,u_2 + 2\varpi_1\varpi_2 (xu_1u_2u_3)\,u_3.$$

(2) Thus if x lie on the plane eu_1u_2, $H(x)$ is a vector.

If x be the centre of the group, the harmonic invariant is the vector u_3 parallel to the axis of the group.

Also the harmonic invariant of u_3 is the centre of the group.

(3) Also by a similar proof the harmonic plane-invariant of eu_1u_2 is the plane at infinity; and therefore conversely the harmonic plane-invariant of the plane at infinity is the diametral plane of the group.

Hence $$H(\mathfrak{U}) = \lambda eu_1u_2.$$
This can be easily verified by direct transformation.

For $$H(\mathfrak{U}) = 2\mathfrak{U}S_1 . S_2 = 2u_1 (eu_2 + \varpi_2 u_3 u_1) = -2eu_1u_2.$$

335. TRIPLE GROUPS. (1) Any triple group defines a quadric surface, called the director quadric. It was proved in § 186 (4) that if p, p_1, p_2, p_3 be the vertices of a self-conjugate tetrahedron of this quadric, then a set of three reciprocal systems are given by
$$\left.\begin{aligned} S_1 &= pp_1 + \mu_1 p_2 p_3, \\ S_2 &= pp_2 + \mu_2 p_3 p_1, \\ S_3 &= pp_3 + \mu_3 p_1 p_2; \end{aligned}\right\} \dots\dots\dots\dots\text{(i)},$$
where μ_1, μ_2, μ_3 are three numbers, real or imaginary, depending on the given group.

Also one of the vertices of the tetrahedron may be arbitrarily chosen.

Let the point p be taken to be the centre of the quadric, then p_1, p_2, p_3 are on the plane at infinity, that is to say, they are vectors. Hence if the centre of the quadric be called c, and v_1, v_2, v_3 be unit vectors in the directions of three conjugate diameters, a set of reciprocal systems can be written,

$$\left.\begin{aligned} S_1 &= cv_1 + \mu_1 v_2 v_3, \\ S_2 &= cv_2 + \mu_2 v_3 v_1, \\ S_3 &= cv_3 + \mu_3 v_1 v_2. \end{aligned}\right\} \quad \ldots\ldots\ldots\ldots \text{(ii)}.$$

Thus if S_1, S_2, S_3 be any three reciprocal systems of the group, $S_1\mathfrak{U}$, $S_2\mathfrak{U}$, $S_3\mathfrak{U}$ are the directions of a set of three conjugate diameters of the quadric.

(2) If u_1, u_2, u_3 be three unit vectors in the directions of the principal axes of the quadric, then the three corresponding systems;

$$\left.\begin{aligned} \Sigma_1 &= cu_1 + \varpi_1 u_2 u_3 = cu_1 + \varpi_1 \mathfrak{F} u_1, \\ \Sigma_2 &= cu_2 + \varpi_2 u_3 u_1 = cu_2 + \varpi_2 \mathfrak{F} u_2, \\ \Sigma_3 &= cu_3 + \varpi_3 u_1 u_2 = cu_3 + \varpi_3 \mathfrak{F} u_3, \end{aligned}\right\} \quad \ldots\ldots\ldots\ldots \text{(iii)},$$

will be called the principal systems of the group.

336. THE POLE AND POLAR INVARIANTS. (1) The polar invariant of the point x is denoted in § 183 (3) by $P(x)$ and is the polar plane of x with respect to the quadric [cf. § 185 (1)]. Similarly the pole invariant of the plane X is $P(X)$ and is the pole of the plane X. Thus the centre of the quadric is the point $P(\mathfrak{U})$.

(2) Now let S_1, S_2, S_3 denote a set of reciprocal systems.

Then $$P(\mathfrak{U}) = 2\mathfrak{U}S_1 \,.\, S_2 \,.\, S_3 = 2S_1\mathfrak{U} \,.\, S_2 \,.\, S_3.$$

But $2S_1\mathfrak{U} \,.\, S_2$ is the diametral plane of the dual group S_1S_2 [cf. 334 (3)]. Hence the centre is the null point of this plane with respect to S_3.

Accordingly the diametral plane of any dual subgroup contained in the triple group passes through the centre of the quadric. Also the centre of the quadric is the null point of such a diametral plane with respect to the system of the triple group reciprocal to the corresponding dual subgroup.

(3) It may be noticed that, if $H_1(x)$, $H_2(x)$, $H_3(x)$ denote the harmonic invariants of x with respect to the dual groups S_2S_3, S_3S_1, S_1S_2 respectively, then [cf. § 184 (2)] $\lambda c = P(\mathfrak{U}) = H_1(S_1\mathfrak{U}) = H_2(S_2\mathfrak{U}) = H_3(S_3\mathfrak{U})$.

These are four alternative methods of expressing the centre of the quadric.

(4) Since the diametral planes of the three dual groups are

$$c \,.\, S_2\mathfrak{U} \,.\, S_3\mathfrak{U}, \quad c \,.\, S_3\mathfrak{U} \,.\, S_1\mathfrak{U}, \quad c \,.\, S_1\mathfrak{U} \,.\, S_2\mathfrak{U},$$

it follows that they intersect in pairs in the edges $c \,.\, S_1\mathfrak{U}$, $c \,.\, S_2\mathfrak{U}$, $c \,.\, S_3\mathfrak{U}$.

(5) The diametral plane, which bisects chords parallel to any vector v, is obviously $P(v)$. Thus the diametral plane, of which the conjugate chords are parallel to any line L, is $P(L\mathfrak{U})$.

The diametral plane parallel to any vector area M is $P(\mathfrak{U})M$; the vector parallel to its conjugate chords is $P\{P(\mathfrak{U})M\}$.

337. Equation of the Associated Quadric. (1) The condition, that any point p is on the quadric, can by § 187 (6) be written in the form

$$P(p)p = 0.$$

Now let ρ_1 be the length of the semi-diameter $c \,.\, S_1\mathfrak{U}$. Then the point $p = c + \rho_1 v_1$ is on the quadric.

Hence by § 335, equations (ii),

$$pS_1 \,.\, S_2 = (cv_1v_2v_3)\{\mu_1\mu_2v_3 + \mu_1\rho_1v_2\},$$

$$pS_1 \,.\, S_2 \,.\, S_3 = (cv_1v_2v_3)\{\mu_1\mu_2\mu_3v_1v_2v_3 - \mu_1\rho_1cv_2v_3\}.$$

And $$pS_1 \,.\, S_2 \,.\, S_3 \,.\, p = -(cv_1v_2v_3)^2\{\mu_1\mu_2\mu_3 + \mu_1\rho_1^2\}.$$

Hence $$\rho_1^2 = -\mu_2\mu_3 = -\frac{1}{4}\frac{(S_2S_2)(S_3S_3)(S_1\mathfrak{U})^2}{(c \,.\, S_1\mathfrak{U} \,.\, S_2\mathfrak{U} \,.\, S_3\mathfrak{U})^2};$$

with similar expressions for ρ_2 and ρ_3.

(2) Let $\Sigma_1, \Sigma_2, \Sigma_3$ be the principal systems of the group, and u_1, u_2, u_3 unit vectors parallel to the principal vectors of $\Sigma_1, \Sigma_2, \Sigma_3$.

Let it be assumed that

$$\Sigma_1 = cu_1 + \varpi_1 \mathfrak{F} u_1 = cu_1 + \varpi_1 u_2 u_3,$$

$$\Sigma_2 = cu_2 + \varpi_2 \mathfrak{F} u_2 = cu_2 + \varpi_2 u_3 u_1,$$

$$\Sigma_3 = cu_3 + \varpi_3 \mathfrak{F} u_3 = cu_3 + \varpi_3 u_1 u_2.$$

Let x be the point $c + \xi_1 u_1 + \xi_2 u_2 + \xi_3 u_3$; then ξ_1, ξ_2, ξ_3 are the rectangular co-ordinates of x referred to the three axes cu_1, cu_2, cu_3.

Also $$P(x) = 2x\Sigma_1 \,.\, \Sigma_2 \,.\, \Sigma_3$$
$$= -2\{\varpi_1\xi_1cu_2u_3 + \varpi_3\xi_2cu_3u_1 + \varpi_3\xi_3cu_1u_2 - \varpi_1\varpi_2\varpi_3u_1u_2u_3\}$$
$$= 2|\{\varpi_1\xi_1u_1 + \varpi_2\xi_2u_2 + \varpi_3\xi_3u_3 + \varpi_1\varpi_2\varpi_3c\};$$

where c is the centre of the rectangular normal system.

The equation of the quadric is $\{xP(x)\} = 0$, that is

$$\frac{\xi_1^2}{\varpi_2\varpi_3} + \frac{\xi_2^2}{\varpi_3\varpi_1} + \frac{\xi_3^2}{\varpi_1\varpi_2} = -1.$$

338. Normals. (1) The vector parallel to the normal at a point x on the quadric is $\mathfrak{F}\{P(x)\mathfrak{U}\}$; and this can be transformed into $|P(x-c)$, where x is a unit point, and c is the centre of the quadric.

The first expression requires no proof, if it be remembered that $P(x)$ is a planar element in the tangent plane at x.

(2) To prove the second expression, we have

$$\mathfrak{F}\{P(x)\,\mathfrak{U}\} = |\{c\,.\,P(x)\,\mathfrak{U}\}.$$

Now, $c\,.\,P(x)\,\mathfrak{U} = (c\mathfrak{U})\,P(x) - \{cP(x)\}\,\mathfrak{U} = P(x) - \{x\,P(c)\}\,\mathfrak{U}.$

But $P(c) = \lambda\mathfrak{U}$. Hence $\{xP(c)\} = \lambda$.

Hence $\mathfrak{F}\{P(x)\,\mathfrak{U}\} = |P(x) - \lambda\,|\mathfrak{U} = |\{P(x) - P(c)\} = |P(x-c)$.

Thus the plane $P(x-c)$ is the diametral plane perpendicular to the normal.

339. Small Displacements of a Rigid Body. (1) If any rigid body be successively displaced according to the specifications of two small congruent transformations [cf. § 268 (1) and § 303], it is obviously immaterial which of the two transformations is applied first, so long as small quantities of the second order are neglected.

Now let e be any origin and eu_1, eu_2, eu_3 any set of rectangular axes, u_1, u_2, u_3 being of unit length. Assume the three small translations defined by the vectors $\lambda\alpha_1u_1$, $\lambda\alpha_2u_2$, $\lambda\alpha_3u_3$, and the three small rotations with axes eu_1, eu_2, eu_3 through angles $\lambda\delta_1$, $\lambda\delta_2$, $\lambda\delta_3$ successively applied in any order; where λ is a small fraction whose square may be neglected, and α_1, α_2, α_3, δ_1, δ_2, δ_3 are not necessarily small.

(2) Then any point $x = e + \Sigma\xi u$ becomes Kx, where

$$Kx = e + (\xi_1 + \lambda\alpha_1 + \lambda\delta_2\xi_3 - \lambda\delta_3\xi_2)\,u_1 + (\xi_2 + \lambda\alpha_2 + \lambda\delta_3\xi_1 - \lambda\delta_1\xi_3)\,u_2 + (\xi_3 + \lambda\alpha_3 + \lambda\delta_1\xi_2 - \lambda\delta_2\xi_1)\,u_3.$$

Hence the combined effect is equivalent to the combination of the small translation $\lambda\,(\alpha_1u_1 + \alpha_2u_2 + \alpha_3u_3)$, and of the small rotation round an axis $e\,(\delta_1u_1 + \delta_2u_2 + \delta_3u_3)$ through an angle $\lambda\sqrt{(\delta_1^2 + \delta_2^2 + \delta_3^2)}$. But by properly choosing α_1, α_2, α_3 and δ_1, δ_2, δ_3 these can be made to be any small translation and any small rotation with its axis through e.

Accordingly the above linear transformation is equivalent to the most general form of small congruent transformation.

(3) Let S denote the system of linear elements

$$\alpha_1u_2u_3 + \alpha_2u_3u_1 + \alpha_3u_1u_2 + \delta_1eu_1 + \delta_2eu_2 + \delta_3eu_3.$$

Then $xS = (\alpha_1 + \delta_2\xi_3 - \delta_3\xi_2)\,eu_2u_3 + (\alpha_2 + \delta_3\xi_1 - \delta_1\xi_3)\,eu_3u_1 + (\alpha_3 + \delta_1\xi_2 - \delta_2\xi_1)\,eu_1u_2 + (\alpha_1 + \alpha_2 + \alpha_3)\,u_1u_2u_3.$

Hence
$$Kx = x + \lambda\mathfrak{F}\,(Sx\,.\,\mathfrak{U}).$$

Let S be termed the system of linear elements associated with the transformation, or more shortly the associated system. And conversely λS will be used as the name of the transformation.

(4) If two small congruent transformations λ_1S_1 and λ_2S_2 be successively applied, then neglecting small quantities of the second order the combined effect is that of the single transformation $\lambda_1S_1 + \lambda_2S_2$.

For
$$K_1x = x + \lambda_1\mathfrak{F}(S_1x \,.\, \mathfrak{U}),$$
$$K_2K_1x = x + \lambda_1\mathfrak{F}(S_1x \,.\, \mathfrak{U}) + \lambda_2\mathfrak{F}(S_2x \,.\, \mathfrak{U}),$$
neglecting $\lambda_1\lambda_2$.

It is now obvious that every theorem respecting systems of linear elements possesses its analogue respecting small congruent transformations.

(5) When S is a single linear element through any point e, the transformation λS is a rotation with its axis through e.

When S is a vector area, the transformation λS is a translation in a direction perpendicular to the vector area.

The transformation λS can be decomposed into two rotations round any two conjugate lines of S.

(6) Let S be written in the form $av + \varpi\mathfrak{F}v$, where v is a unit vector, then
$$Kx = x + \lambda\mathfrak{F}(avx \,.\, \mathfrak{U}) + \lambda\varpi\mathfrak{F}\{\mathfrak{F}v \,.\, x \,.\, \mathfrak{U}\}.$$
Now
$$\mathfrak{F}\{\mathfrak{F}v \,.\, x \,.\, \mathfrak{U}\} = \mathfrak{F}\{\mathfrak{F}v\} = v.$$
Hence
$$Kx = x + \lambda\mathfrak{F}(avx \,.\, \mathfrak{U}) + \lambda\varpi v.$$

Now λav denotes a rotation through an angle λ round the axis av; and $\lambda\varpi\mathfrak{F}v$ denotes a translation parallel to v through a distance $\lambda\varpi$.

Thus the axis of the transformation is the central axis of the associated force system, and the pitch of the force system is the ratio of the distance of the translation to the angle of the rotation. Let this pitch also be called the pitch of the transformation.

(7) Whatever point x may be, we may write S in the form $xv + M$. Hence $Kx = x + \lambda\mathfrak{F}M$. In other words, $\lambda Sx \,.\, \mathfrak{U}$ defines the translation required to bring x into its final position.

340. Work. (1) If the force F pass through the point x, and x be displaced to $x + \lambda v$, where v is a vector and λ is small, then the work done by F is $\lambda(F\mathfrak{F}v)$.

This is obviously in accordance with the common definition of work.

The work can also be written in the equivalent forms, $\lambda(F\mathfrak{U} \mid v)$ and $\lambda(v \mid F\mathfrak{U})$.

(2) Let $F = xu$, and let λv be the displacement of x produced by the congruent transformation λS.

Then
$$v = \mathfrak{F}(Sx \,.\, \mathfrak{U}).$$
Hence
$$\mathfrak{F}v = Sx \,.\, \mathfrak{U}.$$
And the work done by F is
$$\lambda\{xu \,.\, (Sx \,.\, \mathfrak{U})\} = -\lambda\{u \,.\, x(Sx \,.\, \mathfrak{U})\} = -\lambda\{u \,.\, xS\} = \lambda(FS).$$

It then follows, that the work done by a force F during the small congruent transformation λS is the same at whatever point of its line of action F be supposed to be applied.

(3) Let F_1, F_2, etc. be any number of forces acting on the rigid body during its transformation. Then the sum of the work done by them is

$$\lambda (F_1 S) + \lambda (F_2 S) + \ldots = \lambda \{(F_1 + F_2 + \ldots) S\} = \lambda (S'S);$$

where S' is the force system $F_1 + F_2 + \ldots$.

Hence the work done by the force system S' during the small congruent transformation λS is equal to that done by the force system S during the small congruent transformation $\lambda S'$.

Also if S and S' be reciprocal, the work done is in both cases zero.

CHAPTER III.

Curves and Surfaces.

341. Curves. (1) Let the point $x\ (= e + \Sigma\xi u)$ be conceived to be in motion, so that ξ_1, ξ_2, ξ_3 are continuous functions of the single variable τ, which is the time.

Then $x + \dot{x}\delta\tau = e + \Sigma\xi u + \delta\tau\Sigma\dot{\xi}u$, is the position of the point at the time $\tau + \delta\tau$. Hence the vector $\dot{x}$, which is $\Sigma\dot{\xi}u$, represents the velocity of x in magnitude and direction.

Similarly the vector $\ddot{x}$, or $\Sigma\ddot{\xi}u$, represents the acceleration of x in magnitude and direction. Let $\dddot{x}$, $\ddddot{x}$, etc. be formed according to the same law.

(2) Let σ be the length of the arc traversed during the time τ, and $\sigma + \delta\sigma$ that traversed during the time $\tau + \delta\tau$.

Then
$$\frac{d\sigma}{d\tau} = \sqrt{(\dot{x})^2}.$$

(3) The tangent line to the path of x at the time τ is $x\dot{x}$, that at the time $\tau + \delta\tau$ is $x\dot{x} + x\ddot{x}\delta\tau$.

Let $\delta\epsilon$ be the angle of contingence between these tangent lines, then by § 322 (2)

$$\sin\delta\epsilon = \delta\epsilon = \sqrt{\frac{(\dot{x}\ddot{x}\,|\,\dot{x}\ddot{x})\,(\delta\tau)^2}{(\dot{x})^4}}$$
$$= \frac{\sqrt{(\dot{x}\ddot{x}\,|\,\dot{x}\ddot{x})}}{(\dot{x})^2}\,\delta\tau.$$

(4) Hence, if ρ be the principal radius of curvature of the path,

$$\frac{1}{\rho} = \frac{d\epsilon}{d\sigma} = \frac{\sqrt{(\dot{x}\ddot{x}\,|\,\dot{x}\ddot{x})}}{\{(\dot{x})^2\}^{\frac{3}{2}}}.$$

Therefore
$$\frac{1}{\rho^2} = \frac{(\dot{x}\ddot{x}\,|\,\dot{x}\ddot{x})}{\{(\dot{x})^2\}^3} = \frac{(\ddot{x}\,|\,\ddot{x})}{(\dot{x})^4} - \frac{(\dot{x}\,|\,\ddot{x})^2}{(\dot{x})^6}.$$

342. Osculating Plane and Normals. (1) At the end of a second interval $\delta\tau$, x has moved to $x + 2\dot{x}\delta\tau + 2\ddot{x}(\delta\tau)^2$.

Therefore the osculating plane is $x\dot{x}\ddot{x}$.

The vector factor of this product is $\dot{x}\ddot{x}$. Hence the direction of the binormal is that of the vector $\mathfrak{F}\dot{x}\ddot{x}$. The binormal is $x\mathfrak{F}\dot{x}\ddot{x}$.

(2) The neighbouring osculating plane is $x(\dot{x}\ddot{x} + \dot{x}\dddot{x}\delta\tau)$. The vector factor of the neighbouring binormal is $\mathfrak{F}(\dot{x}\ddot{x} + \dot{x}\dddot{x}\delta\tau)$. Let $\delta\lambda$ be the angle of torsion.

Then [cf. 322 (2)] $$\frac{d\lambda}{d\tau} = \sqrt{\frac{(\mathfrak{F}\dot{x}\ddot{x} \,.\, \mathfrak{F}\dot{x}\dddot{x} \,|\, . \mathfrak{F}\dot{x}\ddot{x} \,.\, \mathfrak{F}\dot{x}\dddot{x})}{(\dot{x}\ddot{x})^4}};$$

since $$\mathfrak{F}\dot{x}\ddot{x} \,.\, \mathfrak{F}(\dot{x}\ddot{x} + \dot{x}\dddot{x}\delta\tau) = \mathfrak{F}\dot{x}\ddot{x} \,.\, \mathfrak{F}\dot{x}\dddot{x}\delta\tau.$$

Now by § 326 (4) equation (iv), $\mathfrak{F}\dot{x}\ddot{x} \,.\, \mathfrak{F}\dot{x}\dddot{x} = (e\dot{x}\ddot{x}\dddot{x})\mathfrak{F}\dot{x}$.

Hence $$\frac{d\lambda}{d\tau} = \sqrt{\frac{(e\dot{x}\ddot{x}\dddot{x})^2(\dot{x})^2}{(\dot{x}\ddot{x})^4}} = \frac{(e\dot{x}\ddot{x}\dddot{x})\sqrt{(\dot{x})^2}}{(\dot{x}\ddot{x})^2}.$$

Now let $\frac{1}{\kappa}$ be the measure of the torsion.

Then $$\frac{1}{\kappa} = \frac{(e\dot{x}\ddot{x}\dddot{x})}{(\dot{x}\ddot{x} \,|\, \dot{x}\ddot{x})}.$$

Thus $$\frac{1}{\rho^2\kappa} = \frac{(e\dot{x}\ddot{x}\dddot{x})}{(\dot{x})^6}.$$

(3) The normal plane at x is $x\mathfrak{F}\dot{x}$.

The principal normal at x is the line N, where
$$N = x\dot{x}\ddot{x} \,.\, x\mathfrak{F}\dot{x} = (x\dot{x}\mathfrak{F}\dot{x})\,x\ddot{x} - (x\ddot{x}\mathfrak{F}\dot{x})\,x\dot{x} = (\dot{x})^2 x\ddot{x} - (\dot{x}\,|\,\ddot{x})\,x\dot{x}.$$

The vector parallel to the principal normal is therefore
$$(\dot{x})^2\ddot{x} - (\dot{x}\,|\,\ddot{x})\,\dot{x}.$$

343. Acceleration. In order to resolve the acceleration along the tangent and the principal normal, notice that
$$\ddot{x} = \frac{(\dot{x}\,|\,\ddot{x})}{(\dot{x})^2}\dot{x} + \frac{(\dot{x})^2\ddot{x} - (\dot{x}\,|\,\ddot{x})\,\dot{x}}{(\dot{x})^2}.$$

Now let t and n be unit vectors along the tangent and principal normal respectively.

Then $$\dot{x} = (\dot{x}^2)^{\frac{1}{2}}\,t,\quad (\dot{x})^2\ddot{x} - (\dot{x}\,|\,\ddot{x})\,\dot{x} = \frac{(\dot{x})^4}{\rho}\,n.$$

Hence $$\ddot{x} = \frac{(\dot{x}\,|\,\ddot{x})}{\{(\dot{x})^2\}^{\frac{1}{2}}}\,t + \frac{(\dot{x})^2}{\rho}\,n.$$

Therefore the acceleration is equivalent to a component $\frac{d\{(\dot{x})^2\}^{\frac{1}{2}}}{d\tau}$ along the tangent, and $\frac{(\dot{x})^2}{\rho}$ along the normal.

344. Simplified Formulæ. (1) In order to develope more fully the theory of curves, let us make use of the simplification introduced by the supposition that the curve is traversed with uniform velocity by the moving point. We may then take σ as the independent variable, and use dashes to denote differential coefficients with respect to σ.

(2) Collecting our formulæ, we have in this case

$$(x')^2 = 1, \text{ and therefore } (x'|x'') = 0,\ (x'')^2 + (x'|x''') = 0.$$

$$\frac{1}{\rho^2} = (x''|x''),\quad \frac{1}{\rho^2\kappa} = (ex'x''x''').$$

The tangent line is xx'; the normal plane is $x\mathfrak{F}x'$; the osculating plane is $xx'x''$; the binormal is $x\mathfrak{F}x'x''$; the principal normal is xx''.

(3) To find Monge's polar line of the point x of the curve, notice that the normal plane at the point $x + x'\delta\sigma$ is $(x + x'\delta\sigma)\{\mathfrak{F}x' + \mathfrak{F}x''.\delta\sigma\}$, that is, $x\mathfrak{F}x' + (x\mathfrak{F}x'' + \mathfrak{U})\delta\sigma$, since $(x'\mathfrak{F}x') = \mathfrak{U}$.

Hence the polar line is $x\mathfrak{F}x'.(x\mathfrak{F}x'' + \mathfrak{U})$, that is, $x\mathfrak{F}x'.x\mathfrak{F}x'' + \mathfrak{F}x'$.

It is obvious that $x\mathfrak{F}x'.x\mathfrak{F}x''$ is some line through x. Assume that it is xv.

Then $$v = xv.\mathfrak{U} = x\mathfrak{F}x'.x\mathfrak{F}x''.\mathfrak{U} = x\mathfrak{F}x'.\mathfrak{F}x'' = \mathfrak{F}x'x'',$$
by § 326 (5). Hence the polar line of x is $x\mathfrak{F}x'x'' + \mathfrak{F}x'$.

(4) The centre of curvature is the point where this line meets the osculating plane, that is the point

$$(x\mathfrak{F}x'x'' + \mathfrak{F}x').xx'x'' = (x'x'')^2 x + \mathfrak{F}x'.xx'x''.$$

Now $\mathfrak{F}x'.xx'x'' = -|x'.\mathfrak{U}.xx'x'' = |x'.x'x'' = -(x''|x')x' + (x'|x')x'' = x''$.

Hence the centre of principal curvature is the point $x + \dfrac{x''}{(x'x'')^2}$, that is the point $x + \rho^2 x''$.

345. Spherical Curvature. (1) The centre of spherical curvature is the point where three neighbouring normal planes intersect, that is the point

$$x\mathfrak{F}x'.(x\mathfrak{F}x'' + \mathfrak{U}).x\mathfrak{F}x''' = (x\mathfrak{F}x'x'' + \mathfrak{F}x').x\mathfrak{F}x''',$$

by use of the transformation of § 344 (3).

Now by the rule of the middle factor [cf. § 102 (7)]

$$x\mathfrak{F}x'x''.x\mathfrak{F}x''' = (x\mathfrak{F}x'x''.\mathfrak{F}x''')x = (x\mathfrak{F}x'''\mathfrak{F}x'x'')x.$$

Also $$\mathfrak{F}x'''\mathfrak{F}x'x'' = |(ex'''.ex'x'') = (ex'x''x''')|e = (ex'x''x''')\mathfrak{U}.$$

Hence $$x\mathfrak{F}x'x''.x\mathfrak{F}x''' = (ex'x''x''')x.$$

Also by § 326 (5) $\mathfrak{F}x'.x\mathfrak{F}x''' = \mathfrak{F}x'''x' = -\mathfrak{F}x'x'''$.

Hence the centre of spherical curvature is

$$x - \frac{\mathfrak{F}x'x'''}{(ex'x''x''')},\ \text{that is } x - \rho^2\kappa\mathfrak{F}x'x'''.$$

(2) Therefore, if ρ_1 be the radius of spherical curvature,

$$\rho_1^2 = \rho^4\kappa^2 (x'x''')^2 = \rho^4\kappa^2 \{(x''')^2 - (x' \mid x''')^2\}$$
$$= \rho^4\kappa^2 \{(x''')^2 - (x'' \mid x'')^2\} = \rho^4\kappa^2 (x''')^2 - \kappa^2.$$

(3) Now $(x''')^2$ can be found by squaring the determinant $(ex'x''x''')$.

Thus [cf. § 342 (2)] $\dfrac{1}{\rho^4\kappa^2} = (ex'x''x''' \mid ex'x''x''')$

$$= \begin{vmatrix} 1, & 0, & (x' \mid x''') \\ 0, & (x'' \mid x''), & (x'' \mid x''') \\ (x' \mid x'''), & (x'' \mid x'''), & (x''' \mid x''') \end{vmatrix}$$

$$= \frac{(x''')^2}{\rho^2} - (x'' \mid x''')^2 - \frac{(x' \mid x''')^2}{\rho^2}.$$

Now $\dfrac{1}{\rho^2} = (x'' \mid x'') = -(x' \mid x''')$. And $-\dfrac{2}{\rho^3}\dfrac{d\rho}{d\sigma} = 2(x'' \mid x''')$.

Hence $$\frac{1}{\rho^4\kappa^2} = \frac{(x''')^2}{\rho^2} - \frac{1}{\rho^6}\left(\frac{d\rho}{d\sigma}\right)^2 - \frac{1}{\rho^6}.$$

Therefore $$\rho_1^2 = \rho^2 + \kappa^2\left(\frac{d\rho}{d\sigma}\right)^2 + \kappa^2 - \kappa^2 = \rho^2 + \kappa^2\left(\frac{d\rho}{d\sigma}\right)^2.$$

This is the well-known formula for ρ_1.

346. LOCUS OF CENTRE OF CURVATURE. (1) It is easy to see that the inner products of the various differential coefficients of x, such as x', x'', etc., can be expressed in terms of ρ and κ and their differential coefficients with respect to σ.

Let ρ_1 be written for $\sqrt{\left\{\rho^2 + \kappa^2\left(\dfrac{d\rho}{d\sigma}\right)^2\right\}}$.

Then we have by successive differentiation

$$(x')^2 = 1,\ (x' \mid x'') = 0,\ (x'')^2 + (x' \mid x''') = 0,$$
$$3(x'' \mid x''') + (x' \mid x^{\text{iv}}) = 0,\ 3(x''')^2 + 4(x'' \mid x^{\text{iv}}) + (x' \mid x^{\text{v}}) = 0,$$

and so on.

Also $(x'')^2 = \dfrac{1}{\rho^2} = -(x' \mid x''')$. Hence $(x'' \mid x''') = -\dfrac{1}{\rho^3}\dfrac{d\rho}{d\sigma} = -\dfrac{1}{3}(x' \mid x^{\text{iv}})$.

Again $(x''')^2 = \dfrac{\rho_1^2}{\rho^4\kappa^2} + \dfrac{1}{\rho^4} = \lambda^2$, say. And $(x''' \mid x^{\text{iv}}) = \lambda\dfrac{d\lambda}{d\sigma}$.

Also $$\frac{d}{d\sigma}(x'' \mid x''') = (x''')^2 + (x'' \mid x^{\text{iv}}) = -\frac{d}{d\sigma}\left(\frac{1}{\rho^3}\frac{d\rho}{d\sigma}\right).$$

Thus $(x'' \mid x^{\text{iv}})$ is expressed in the required form, and so on,

(2) Let y be the centre of principal curvature of the curve at the point x. Let σ' be the arc of the locus of y measured from some fixed point. Then when σ is the independent variable, and $\dot{y}$ is used for $\frac{dy}{d\sigma}$,

$$y = x + \rho^2 x'', \; \dot{y} = x' + 2\rho \frac{d\rho}{d\sigma} x'' + \rho^2 x'''.$$

Hence
$$\left(\frac{d\sigma'}{d\sigma}\right)^2 = (\dot{y})^2 = 1 + 4\rho^2 \left(\frac{d\rho}{d\sigma}\right)^2 (x'')^2 + 2\rho^2 (x' | x''')$$
$$+ 4\rho^3 \frac{d\rho}{d\sigma} (x'' | x''') + \rho^4 (x''')^2$$
$$= 1 + 4\left(\frac{d\rho}{d\sigma}\right)^2 - 2 - 4\left(\frac{d\rho}{d\sigma}\right)^2 + \frac{\rho_1^2}{\kappa^2} + 1 = \frac{\rho_1^2}{\kappa^2}.$$

Therefore
$$\frac{d\sigma'}{d\sigma} = \frac{\rho_1}{\kappa}.$$

(3) Also if $\delta\epsilon'$ be the angle of contingence corresponding to $\delta\sigma'$, then

$$\frac{d\epsilon'}{d\sigma} = \frac{\sqrt{(\dot{y}\ddot{y} | \dot{y}\ddot{y})}}{(\dot{y})^2}.$$

But
$$\ddot{y} = \left(1 + 2\left(\frac{d\rho}{d\sigma}\right)^2 + 2\rho \frac{d^2\rho}{d\sigma^2}\right) x'' + 4\rho \frac{d\rho}{d\sigma} x''' + \rho^2 x^{\text{iv}}.$$

Thus by mere multiplication $\frac{d\epsilon'}{d\sigma}$, and hence $\frac{d\epsilon'}{d\sigma'}$, is expressed in terms of ρ and κ and of their differential coefficients with respect to σ.

347. Gauss' Curvilinear Co-ordinates. (1) Let x be any unit point on a given surface. Then the co-ordinates of x referred to any four reference elements may be conceived as definite functions of two independent variables θ and ϕ. Then the two equations, θ = constant, and ϕ = constant, represent two families of curves traced on the surface.

(2) Suppose that the unit point $x + \delta x$ corresponds to the values $\theta + \delta\theta$ and $\phi + \delta\phi$ of the variables. Then δx is the vector representing the line joining the point (θ, ϕ) with the point $(\theta + \delta\theta, \phi + \delta\phi)$ of the surface.

Also δx can be written in the form

$$\delta x = (x_1 \delta\theta + x_2 \delta\phi) + \tfrac{1}{2}(x_{11} \overline{\delta\theta}|^2 + 2x_{12}\delta\theta\delta\phi + x_{22}\overline{\delta\phi}|^2) + \dots,$$

where x_1, x_2, x_{11}, x_{12}, x_{22} are vectors. Hence, if e be the origin,

$$(e | x_1) = 0 = (e | x_2) = (e | x_{11}) = (e | x_{12}) = (e | x_{22}).$$

(3) In order to exhibit the meanings of these vectors, let e be any origin and eu_1, eu_2, eu_3 rectangular unit axes.

Then x is $e + \Sigma \xi u$, and x_1 is $\Sigma \frac{\partial \xi}{\partial \theta} u$, x_2 is $\Sigma \frac{\partial \xi}{\partial \phi} u$, x_{11} is $\Sigma \frac{\partial^2 \xi}{\partial \theta^2} u$, and so on.

It will at times be an obviously convenient notation to write $\frac{\partial x}{\partial\theta}$ for x_1, $\frac{\partial x}{\partial\phi}$ for x_2, $\frac{\partial^2 x}{\partial\theta^2}$ for x_{11}, and so on.

(4) The distance $\delta\sigma$ between x and $x+\delta x$ is given by

$$(\delta\sigma)^2 = (\delta x)^2 = (x_1|x_1)(\delta\theta)^2 + 2(x_1|x_2)\delta\theta\delta\phi + (x_2|x_2)(\delta\phi)^2.$$

(5) The tangent line of the curve joining the points is

$$x(x_1\delta\theta + x_2\delta\phi).$$

Now let $x+\delta_1 x$ be a neighbouring point on the curve $\phi=$ constant, and $x+\delta_2 x$ on the curve $\theta=$ constant.

Then $x+\delta_1 x = x + x_1\delta\theta + \frac{1}{2}x_{11}(\delta\theta)^2$, $x+\delta_2 x = x + x_2\delta\phi + \frac{1}{2}x_{22}\overline{\delta\phi}|^2$.

Hence two tangent lines to the curves ϕ and θ respectively are xx_1 and xx_2. Accordingly the tangent plane at x is xx_1x_2.

The normal at x is $x\mathfrak{F}x_1x_2$.

(6) The angle ω between the tangents at x to the θ curve and the ϕ curve is given by

$$\cos\omega = \frac{(x_1|x_2)}{\sqrt{\{(x_1|x_1)(x_2|x_2)\}}}, \quad \sin\omega = \sqrt{\frac{(x_1x_2|x_1x_2)}{(x_1|x_1)(x_2|x_2)}}.$$

(7) Let $\delta\nu$ be the perpendicular from $x+\delta x$ on to the tangent plane.

Then [cf. § 324 (4)] $\delta\nu = \frac{(xx_1x_2\delta x)}{\sqrt{(x_1x_2|x_1x_2)}} = \frac{(ex_1x_2\delta x)}{\sqrt{\{(x_1|x_1)(x_2|x_2)-(x_1|x_2)^2\}}}$

$$= \frac{1}{2}\,\frac{(ex_1x_2x_{11})(\delta\theta)^2 + 2(ex_1x_2x_{12})\delta\theta\delta\phi + (ex_1x_2x_{22})(\delta\phi)^2}{\sqrt{(x_1x_2|x_1x_2)}}.$$

348. Curvature. (1) Let ρ be the radius of curvature of the normal section through x and $x+\delta x$. Then $\rho = \frac{1}{2}\frac{(\delta\sigma)^2}{\delta\nu}$.

Hence $\rho = \frac{\sqrt{(x_1x_2|x_1x_2)}\,.\,\{(x_1|x_1)(\delta\theta)^2 + 2(x_1|x_2)\delta\theta\delta\phi + (x_2|x_2)(\delta\phi)^2\}}{(ex_1x_2x_{11})(\delta\theta)^2 + 2(ex_1x_2x_{12})\delta\theta\delta\phi + (ex_1x_2x_{22})(\delta\phi)^2}$.

(2) Now seek for the maximum and minimum values of ρ when the ratio of $\delta\theta$ to $\delta\phi$ is varied. Let ρ_1 and ρ_2 be the maximum and minimum values found, and let $\delta\theta_1$ to $\delta\phi_1$ and $\delta\theta_2$ to $\delta\phi_2$ be the corresponding ratios of $\delta\theta$ to $\delta\phi$. Then $\rho_1/\sqrt{(x_1x_2|x_1x_2)}$ and $\rho_2/\sqrt{(x_1x_2|x_1x_2)}$ are the roots of the following quadratic for ζ:

$$\{(ex_1x_2x_{11})\zeta - (x_1|x_1)\}\{(ex_1x_2x_{22})\zeta - (x_2|x_2)\} - \{(ex_1x_2x_{12})\zeta - (x_1|x_2)\}^2 = 0.$$

Hence $$\frac{1}{\rho_1\rho_2} = \frac{(ex_1x_2x_{11})(ex_1x_2x_{22}) - (ex_1x_2x_{12})^2}{\{(x_1|x_1)(x_2|x_2) - (x_1|x_2)^2\}^2},$$

$$\frac{1}{\rho_1} + \frac{1}{\rho_2} = \frac{(x_1|x_1)(ex_1x_2x_{22}) + (x_2|x_2)(ex_1x_2x_{11}) - 2(x_1|x_2)(ex_1x_2x_{12})}{\{(x_1|x_1)(x_2|x_2) - (x_1|x_2)^2\}^{\frac{3}{2}}}.$$

(3) The expression for $\frac{1}{\rho_1\rho_2}$ can be put in terms of $(x_1|x_1)$, $(x_2|x_2)$, $(x_1|x_2)$, and of their differential coefficients with respect to θ and ϕ.

For, since by § 347 (2) $(e|x_1) = 0 = \ldots = (e|x_{22})$,

$$(ex_1x_2x_{11})(ex_1x_2x_{22}) = (ex_1x_2x_{11}|ex_1x_2x_{22}) = \begin{vmatrix} (x_1|x_1), & (x_1|x_2), & (x_1|x_{22}) \\ (x_2|x_1), & (x_2|x_2), & (x_2|x_{22}) \\ (x_{11}|x_1), & (x_{11}|x_2), & (x_{11}|x_{22}) \end{vmatrix},$$

and

$$(ex_1x_2x_{12})^2 = \begin{vmatrix} (x_1|x_1), & (x_1|x_2), & (x_1|x_{12}) \\ (x_1|x_2), & (x_2|x_2), & (x_2|x_{12}) \\ (x_1|x_{12}), & (x_2|x_{12}), & (x_{12}|x_{12}) \end{vmatrix}$$

Hence $(ex_1x_2x_{11})(ex_1x_2x_{22}) - (ex_1x_2x_{12})^2$

$$= \begin{vmatrix} (x_1|x_1), & (x_1|x_2), & (x_1|x_{22}) \\ (x_1|x_2), & (x_2|x_2), & (x_2|x_{22}) \\ (x_1|x_{11}), & (x_2|x_{11}), & 0 \end{vmatrix} - \begin{vmatrix} (x_1|x_1), & (x_1|x_2), & (x_1|x_{12}) \\ (x_1|x_2), & (x_2|x_2), & (x_2|x_{12}) \\ (x_1|x_{12}), & (x_2|x_{12}), & 0 \end{vmatrix}$$
$$+ \{(x_1|x_1)(x_2|x_2) - (x_1|x_2)^2\}\{(x_{11}|x_{22}) - (x_{12}|x_{12})\}.$$

Now let $(x_1|x_1)_1$, $(x_1|x_1)_2$, etc. stand for $\frac{\partial}{\partial\theta}(x_1|x_1)$, $\frac{\partial(x_1|x_1)}{\partial\phi}$, etc.

Then
$$(x_1|x_1)_1 = 2(x_1|x_{11}),\quad (x_1|x_1)_2 = 2(x_1|x_{12}),$$
$$(x_2|x_2)_1 = 2(x_2|x_{12}),\quad (x_2|x_2)_2 = 2(x_2|x_{22}),$$
$$(x_1|x_2)_1 = (x_{11}|x_2) + (x_1|x_{12}),\quad (x_1|x_2)_2 = (x_{12}|x_2) + (x_1|x_{22}),$$
$$(x_1|x_1)_{22} = 2(x_{12}|x_{12}) + 2(x_1|x_{122}),\quad (x_2|x_2)_{11} = 2(x_{12}|x_{12}) + 2(x_2|x_{112}),$$
$$(x_1|x_2)_{12} = (x_{11}|x_{22}) + (x_{12}|x_{12}) + (x_{112}|x_2) + (x_1|x_{122}).$$

Hence $(x_{11}|x_{22}) - (x_{12}|x_{12}) = (x_1|x_2)_{12} - \frac{1}{2}(x_1|x_1)_{22} - \frac{1}{2}(x_2|x_2)_{11}$.

Thus $\frac{1}{\rho_1\rho_2}$ can be expressed in the required manner.

349. Lines of Curvature. (1) Also if ζ_1 and ζ_2 stand for
$$\rho_1/\sqrt{(x_1x_2|x_1x_2)} \text{ and } \rho_2/\sqrt{(x_1x_2|x_1x_2)} \text{ respectively.}$$

Then the ratios $\delta\theta_1$ to $\delta\phi_1$ and $\delta\theta_2$ to $\delta\phi_2$ are given by

$$\frac{\delta\theta_1}{(ex_1x_2x_{12})\zeta_1 - (x_1|x_2)} = \frac{-\delta\phi_1}{(ex_1x_2x_{11})\zeta_1 - (x_1|x_1)},$$

and

$$\frac{\delta\theta_2}{(ex_1x_2x_{12})\zeta_2 - (x_1|x_2)} = \frac{-\delta\phi_2}{(ex_1x_2x_{11})\zeta_2 - (x_1|x_1)}.$$

(2) By the aid of the quadratic for ζ_1 and ζ_2 which have been found, it can easily be seen that the lines of curvature are given by

$$\{(x_1|x_2)(ex_1x_2x_{11}) - (x_1|x_1)(ex_1x_2x_{12})\}(\delta\theta)^2$$
$$+ \{(x_2|x_2)(ex_1x_2x_{11}) - (x_1|x_1)(ex_1x_2x_{22})\}\,\delta\theta\,\delta\phi$$
$$+ \{(x_2|x_2)(ex_1x_2x_{12}) - (x_1|x_2)(ex_1x_2x_{22})\}(\delta\phi)^2 = 0.$$

And therefore it follows that

$$(x_1 | x_1)\, \delta\theta_1 \delta\theta_2 + (x_1 | x_2) \{\delta\theta_1 \delta\phi_2 + \delta\theta_2 \delta\phi_1\} + (x_2 | x_2)\, \delta\phi_1 \delta\phi_2 = 0.$$

(3) Let $x + \delta x$ and $x + \delta' x$ be two neighbouring points to x, where

$$\delta x = x_1 \delta\theta + x_2 \delta\phi + \ldots, \text{ and } \delta' x = x_1 \delta'\theta + x_2 \delta'\phi + \ldots.$$

Then the angle ψ between the two tangent lines $x\delta x$ and $x\delta' x$ is given by

$$\cos\psi = \frac{(\delta x | \delta' x)}{\sqrt{\{(\delta x | \delta x)(\delta' x | \delta' x)\}}}$$

$$= \frac{(x_1 | x_1)\, \delta\theta\, \delta'\theta + (x_1 | x_2)(\delta\theta\, \delta'\phi + \delta\phi\, \delta'\theta) + (x_2 | x_2)\, \delta\phi\, \delta'\phi}{\delta\sigma\, \delta'\sigma},$$

$$\sin\psi = \frac{(\delta\theta\, \delta'\phi - \delta'\theta\, \delta\phi) \sqrt{(x_1 x_2 | x_1 x_2)}}{\delta\sigma\, \delta'\sigma}.$$

Hence

$$\delta\sigma\, \delta'\sigma \cos\psi = (x_1 | x_1)\, \delta\theta\, \delta'\theta + (x_1 | x_2)(\delta\theta\, \delta'\phi + \delta'\theta\, \delta\phi) + (x_2 | x_2)\, \delta\phi\, \delta'\phi,$$

and [cf. § 347 (6)]

$$\delta\sigma\, \delta'\sigma \sin\psi = (\delta\theta\, \delta'\phi - \delta'\theta\, \delta\phi) \sin\omega \sqrt{\{(x_1 | x_1)(x_2 | x_2)\}}.$$

Corollary. The tangents to the lines of curvature at x are at right angles.

(4) The conditions that the θ curves and the ϕ curves should be lines of curvature at each point are, from subsection (2) and the corollary of subsection (3), that the equations,

$$(x_1 | x_2) = 0 = (e x_1 x_2 x_{12}),$$

should obtain at each point.

350. DUPIN'S THEOREM. (1) Let x be conceived as a function of three variables θ, ϕ, ψ. Then the equations $\theta = \text{constant}$, $\phi = \text{constant}$, and $\psi = \text{constant}$, determine three families of surfaces. On the surface, $\theta = \text{constant}$, x is a function of the variables ϕ and ψ; on the surface, $\phi = \text{constant}$, a function of ψ and θ; on the surface, $\psi = \text{constant}$, of θ and ϕ.

Let $\frac{\partial x}{\partial\theta} = x_1$, $\frac{\partial x}{\partial\phi} = x_2$, $\frac{\partial x}{\partial\psi} = x_3$, with a corresponding notation for the higher differential coefficients.

(2) Now suppose that the three families of surfaces intersect orthogonally. Then

$$(x_1 | x_2) = 0, \quad (x_2 | x_3) = 0, \quad (x_3 | x_1) = 0 \quad \ldots\ldots\ldots\ldots(1).$$

Hence by differentiation

$$(x_{13} | x_2) + (x_1 | x_{23}) = 0, \quad (x_{12} | x_3) + (x_2 | x_{13}) = 0, \quad (x_{23} | x_1) + (x_3 | x_{12}) = 0$$

Hence

$$(x_1 | x_{23}) = 0 = (x_2 | x_{31}) = (x_3 | x_{12}) \quad \ldots\ldots\ldots\ldots(2).$$

The condition that the lines of intersection of θ and ϕ with the surface ψ should be lines of curvature is $(e x_1 x_2 x_{12}) = 0$.

But from equations (1) $x_1 x_2 \equiv \mathfrak{F} x_3$, and hence we may write $x_1 x_2 = \lambda \mathfrak{F} x_3$. And therefore from equations (2)

$$(e x_1 x_2 x_{12}) = \lambda\, (e \mathfrak{F} x_3 . x_{12}) = \lambda\, (x_3 | x_{12}) = 0.$$

Hence the lines of intersection are lines of curvature of the surfaces on which they lie.

351. Euler's Theorem. (1) Let the curves $\theta = \text{constant}$ and $\phi = \text{constant}$ be lines of curvature, so that

$$(x_1 | x_2) = 0 = (ex_1x_2x_{12}).$$

Let ρ_1 be the radius of curvature of the normal section through xx_1 and ρ_2 of that through xx_2.

The radius of curvature of any normal section is given by

$$\frac{1}{\rho}\{(x_1 | x_1)(\delta\theta)^2 + (x_2 | x_2)(\delta\phi)^2\} = \frac{(x_1 | x_1)}{\rho_1}(\delta\theta)^2 + \frac{(x_2 | x_2)}{\rho_2}(\delta\phi)^2.$$

(2) The angle ψ, which the tangent line $x\delta x$ makes with the tangent line xx_1, is given by

$$\cos\psi = \frac{\sqrt{(x_1 | x_1)} \,.\, \delta\theta}{\sqrt{\{(x_1 | x_1)(\delta\theta)^2 + (x_2 | x_2)(\delta\phi)^2\}}},$$

$$\sin\psi = \frac{\sqrt{(x_2 | x_2)} \,.\, \delta\phi}{\sqrt{\{(x_1 | x_1)(\delta\theta)^2 + (x_2 | x_2)(\delta\phi)^2\}}}.$$

Hence $$\frac{1}{\rho} = \frac{\cos^2\psi}{\rho_1} + \frac{\sin^2\psi}{\rho_2}.$$

352. Meunier's Theorem. (1) The principal radius of curvature of the curve, $\phi = \text{constant}$, is

$$\frac{\{(x_1)^2\}^{\frac{3}{2}}}{\sqrt{\{x_1x_{11} | x_1x_{11}\}}} = {}_1\rho, \text{ say}.$$

The radius of curvature of the normal section through xx_1 is

$$\rho = \frac{(x_1)^2 \{(x_1x_2)^2\}^{\frac{1}{2}}}{(ex_1x_2x_{11})}.$$

Hence $$\frac{{}_1\rho}{\rho} = \frac{(ex_1x_2x_{11})\{(x_1)^2\}^{\frac{1}{2}}}{\sqrt{\{(x_1x_2)^2(x_1x_{11})^2\}}}.$$

(2) The osculating plane of the curve ϕ is xx_1x_{11}, the normal section is $xx_1\mathfrak{F}x_1x_2$. If χ be the angle between these planes, it is given by

$$\cos\chi = \frac{(x_1\mathfrak{F}x_1x_2 | x_1x_{11})}{\sqrt{\{(x_1\mathfrak{F}x_1x_2)^2(x_1x_{11})^2\}}}.$$

But $$x_1\mathfrak{F}x_1x_2 = (x_1 | x_2)\mathfrak{F}x_1 - (x_1 | x_1)\mathfrak{F}x_2.$$

Therefore $$(x_1\mathfrak{F}x_1x_2 | x_1x_{11}) = (x_1 | x_1)(ex_1x_2x_{11}).$$

And $$(x_1\mathfrak{F}x_1x_2)^2 = (x_1)^2(x_1x_2 | x_1x_2).$$

Hence $$\cos\chi = \frac{\{(x_1)^2\}^{\frac{1}{2}}(ex_1x_2x_{11})}{\sqrt{\{(x_1x_2)^2(x_1x_{11})^2\}}} = \frac{{}_1\rho}{\rho}.$$

Note. I do not think that any of the formulæ or proofs of the present chapter have been given before in terms of the Calculus of Extension.

CHAPTER IV.

Pure Vector Formulæ.

353. Introductory. (1) A simple and useful form of the Calculus of Extension for application to physical problems is arrived at by dropping altogether the representation of the point as the primary element, and only retaining vectors. The relations between vectors of unit length will give the expressions in terms of the Calculus of Extension for the formulæ of Spherical Trigonometry. Also many formulæ of Mathematical Physics can be immediately translated into this notation. These vectors may [cf. § 210 (3)] also be conceived as the elements of a two-dimensional region, and their metrical relations are those of two-dimensional Elliptic Geometry.

(2) Let i, j, k represent any three unit vectors at right angles. Then any other vector x takes the form $\xi i + \eta j + \zeta k$.

(3) We will recapitulate the forms which the formulæ assume in this case. It will be obvious that, as stated above, they form a special case of Elliptic Geometry.

$$|i = jk, \quad |j = ki, \quad |k = ij.$$

$$|jk = i = ||i, \quad |ki = j = ||j, \quad |ij = k = ||k.$$

$$(j|k) = 0 = (k|i) = (i|j).$$

$$(i|i) = (j|j) = (k|k) = (ijk) = 1.$$

(4) The multiplication formulæ are the ordinary formulæ for a two-dimensional region: we mention them all for the sake of convenience of reference.

Let x, y be any two vectors; and X, Y any two vector areas.

Then
$$xy = -yx, \quad XY = -YX.$$

Also xy represents a vector area, and XY a vector: the vector area xy is parallel to both vectors x and y, and the vector XY is parallel to the intersection of the vector areas X and Y.

Again, $(x \mid y) = (y \mid x)$, $(X \mid Y) = (Y \mid X)$;
and the result is in each case a purely numerical quantity.

354. Lengths and Areas. (1) The length of the vector x is $\sqrt{(x \mid x)}$. The angle θ between two vectors x and x' is given by

$$\cos\theta = \frac{(x \mid x')}{\sqrt{\{(x \mid x)(x' \mid x')\}}}, \quad \sin\theta = \sqrt{\frac{(xx' \mid xx')}{\{(x \mid x)(x' \mid x')\}}}.$$

(2) Any vector area X takes the form $\xi jk + \eta ki + \zeta ij$.

The magnitude of the area is $\sqrt{(X \mid X)}$.

The angle θ between two vector areas X and X' is given by

$$\cos\theta = \frac{(X \mid X')}{\sqrt{\{(X \mid X)(X' \mid X')\}}}, \quad \sin\theta = \sqrt{\frac{(XX' \mid XX')}{\{(X \mid X)(X' \mid X')\}}}.$$

(3) Also $\mid X$ denotes a vector line of length $\sqrt{(X \mid X)}$, and $\mid x$ denotes a vector area of magnitude $\sqrt{(x \mid x)}$.

It will be useful at times to employ the term 'flux' to denote a vector area.

(4) Let ξ and η be the lengths of the vectors x and y, and let θ be the angle between them. Then the magnitude of the vector area xy is

$$\sqrt{\{xy \mid xy\}} = \sqrt{\{(x \mid x)(y \mid y) - (x \mid y)^2\}} = \xi\eta \sin\theta.$$

Again, let ξ and η be the magnitudes of the vector areas X and Y, and θ the angle between their planes. Then the length of the vector XY is

$$\sqrt{\{XY \mid XY\}} = \sqrt{\{(X \mid X)(Y \mid Y) - (X \mid Y)^2\}} = \xi\eta \sin\theta.$$

355. Formulæ. (1) The extended rule of the middle factor [cf. § 103] gives the following formulæ:

$$X \,.\, xy = (Xy)\,x - (Xx)\,y \quad\text{.........(i)};$$

$$x \,.\, XY = (xY)\,X - (xX)\,Y \quad\text{.........(ii)}.$$

The second can also be deduced from the first by taking supplements.

(2) The same rule also gives the following formulæ for inner multiplication:

$$xy \mid z = (x \mid z)\,y - (y \mid z)\,x \quad\text{.........(iii)};$$

$$z \mid xy = z \,.\, \mid x \mid y = (z \mid y) \mid x - (z \mid x) \mid y \quad\text{.........(iv)}.$$

(3) Also from § 105,

$$xy \,.\, XY = (xX)(yY) - (xY)(yX) \quad\text{.........(v)}.$$

And writing $\mid u$ for X and $\mid v$ for Y, we deduce

$$(xy \mid uv) = (x \mid u)(y \mid v) - (x \mid v)(y \mid u) \quad\text{.........(vi)}.$$

Similarly, the supplemental formula

$$(XY \mid UV) = (X \mid U)(Y \mid V) - (X \mid V)(Y \mid U) \quad\text{.........(vii)}.$$

(4) Two particular cases of the formulæ (vi) and (vii) have been already used above, namely,

$$(xy \mid xy) = (x \mid x)(y \mid y) - (x \mid y)^2 \qquad \text{(viii)};$$

$$(XY \mid XY) = (X \mid X)(Y \mid Y) - (X \mid Y)^2 \qquad \text{(ix)}.$$

It will be convenient to write any expression of the form $(x \mid x)$ in the form $(x)^2$ or x^2, and $(X \mid X)$ in the form $(X)^2$ or X^2. Thus $(xy)^2$ stands for $(xy \mid xy)$.

356. THE ORIGIN. By conceiving the vectors drawn from any arbitrary origin O, any vector x may be considered as representing a point. Thus it is the point P such that the line from O to P can be taken to represent the vector x in magnitude and direction. This origin however is not symbolized in the present application of the Calculus.

357. NEW CONVENTION. (1) Before proceeding with the development of this Calculus it will be advisable explicitly to abandon, for this chapter only, the convention [cf. § 61 (1)] which has hitherto been rigorously adhered to, that letters of the Italic alphabet represent algebraic extraordinaries and letters of the Greek alphabet numerical quantities of ordinary algebra.

As a matter of practical use and not merely of theoretical capabilities it would be found so necessary by any investigator in mathematical physics to continually form the Cartesian equivalents of his equations—if only for comparison with other investigations—that the capabilities of the Greek alphabet for the representation of numerical quantities would not be found sufficient. The German alphabet is found by most people difficult to write and to read. But let the following convention, which is a modified form of one adopted by Oliver* Heaviside, be adopted.

(2) Let all letters of the Greek alphabet denote numerical quantities.

Let all letters, capital and small, of the Latin alphabet *without subscripts* denote respectively vector areas and vectors; except that in formulæ concerned with Kinematics or with Mathematical Physics t always denotes the time.

Let i, j, k denote invariably three rectangular unit vectors.

If x denote any vector, let x_1, x_2, x_3 be *numerical quantities* denoting the magnitudes of the resolved parts of x in the directions i, j, k respectively: so that $x = x_1 i + x_2 j + x_3 k$.

Let x_0 be a numerical quantity denoting the magnitude of x. Thus

$$x_0 = \sqrt{(x)^2} = \sqrt{(x_1^2 + x_2^2 + x_3^2)}.$$

* Cf. 'On the Forces, Stresses, and Fluxes of Energy in the Electromagnetic Field,' *Phil. Trans.* 1892.

If X denote any flux, let X_1, X_2, X_3 be *numerical quantities* denoting the magnitudes of the resolved parts of X along the unit fluxes jk, ki, ij: so that $X = X_1 jk + X_2 ki + X_3 ij$. Let X_0 denote the magnitude of the flux, so that

$$X_0 = +\sqrt{(X)^2} = \{X_1^2 + X_2^2 + X_3^2\}^{\frac{1}{2}}.$$

(3) This notation avoids a too rapid consumption of the letters of the alphabet, and shews at a glance the relationships of the various symbols employed.

We note as obvious truths;

if $$X = |x, \text{ then } X_0 = x_0,\ X_1 = x_1,\ X_2 = x_2,\ X_3 = x_3.$$

Also we note that if $x = |X$, then $|x = ||X = X$, and the same results follow.

358. SYSTEM OF FORCES. (1) Let forces represented by the vectors f, f', ..., act at points denoted by the vectors x, x', ..., drawn from any assigned origin.

Then any force f at x is equivalent to f at the origin and a vector area xf representing the moment of f at x about the origin.

(2) Hence the system is equivalent to Σf at origin and the vector area Σxf, representing a couple; which may be called the vector moment of the system about the origin.

If L be this 'vector area,' L_1, L_2, L_3 are the three moments of the system about axes through the origin parallel to i, j, k.

359. KINEMATICS. (1) Let any point in space be determined [cf. § 350 (1)] by the three generalized co-ordinates (θ, ϕ, ψ). It will be called the point (θ, ϕ, ψ). If the point be referred to three rectangular axes, the rectangular co-ordinates will be written x_1, x_2, x_3, and the point will be represented by the vector x.

If θ, ϕ, ψ be conceived as the co-ordinates of a moving particle, they are functions of the time. Let u be the vector which denotes the velocity of the particle at each instant; then corresponding to each position (θ, ϕ, ψ) there is a definite velocity u. Hence u must be conceived as a function of θ, ϕ, ψ: that is to say, if i, j, k be any three fixed rectangular vectors, and $u = u_1 i + u_2 j + u_3 k$, then u_1, u_2, u_3 are functions of θ, ϕ, ψ.

Since θ, ϕ, ψ are functions of the time, u can also be conceived as a function of the time.

(2) Let u be the velocity of the point at the time t, and $u + \dot{u}\delta t$ at the time $t + \delta t$. Then when δt is made infinitely small, $\dot{u}$ is the acceleration.

Also evidently,

$$\dot{u} = \dot{u}_1 i + \dot{u}_2 j + \dot{u}_3 k = \frac{du}{dt}.$$

(3) The aspect of the osculating plane of the curve traced by the point is represented by the vector-area $u\dot{u}$.

The binormal is represented by the vector $|u\dot{u}$.

The normal plane is represented by the vector-area $|u$.

The principal normal is represented by the unit vector n, where

$$n = \frac{u\dot{u}\,|u}{\sqrt{(u\dot{u}\,|u)^2}} = \frac{(u\,|u)\,\dot{u} - (u\,|\dot{u})\,u}{\sqrt{\{(u\,|u)\,(u\dot{u}\,|u\dot{u})\}}}.$$

The distance traversed in the short time δt is $\sqrt{(u)^2} \,.\, \delta t$.

(4) The angle $\delta\epsilon$ between the directions of motion at the times t and $t + \delta t$ is given by

$$\delta\epsilon = \sin \delta\epsilon = \sqrt{\frac{(u\dot{u}\,|u\dot{u})}{(u)^4}} \,.\, \delta t.$$

The radius of curvature is $\dfrac{\{(u)^2\}^{\frac{3}{2}}}{\sqrt{(u\dot{u}\,|u\dot{u})}}$.

(5) Thus

$$\dot{u} = \frac{\sqrt{(u\dot{u}\,|u\dot{u})}}{\sqrt{(u\,|u)}} \,.\, n + \frac{(u\,|\dot{u})}{(u\,|u)}\,u$$

$$= \frac{(u)^2}{\rho}\,n + \frac{d\sqrt{(u)^2}}{dt} \,.\, \frac{u}{\sqrt{(u)^2}}.$$

This represents the ordinary normal and tangential resolution of the acceleration.

360. A CONTINUOUSLY DISTRIBUTED SUBSTANCE. (1) Many branches of Mathematical Physics depend upon the investigation of the kinematical properties of substances (ordinary matter or some other medium) distributed continuously throughout all, or some portion of, space. The continuously distributed substance will possess various properties dependent on its motion and on other intrinsic properties. Let any quantity associated with a particle of matter, which does not require a direction for its specification, be termed scalar, according to Hamilton's nomenclature.

(2) Then scalar quantities, such as the density, and vector quantities, such as the acceleration, are associated at each point with the existence of the continuously distributed matter.

These quantities, scalar or vector, may be associated either with the varying elements of matter occupying given points of space, or with the given elements of matter occupying varying points in space.

(3) If the quantities be thus associated in the first way with the given points of space, then the co-ordinates—say θ, ϕ, ψ—of any point are not to

be considered as functions of the time. Let χ be any scalar function of the matter at the point (θ, ϕ, ψ) at any time t, then at the subsequent instant $t+\delta t$ a fresh element of matter occupies the position (θ, ϕ, ψ). Let its corresponding scalar function at the time $t+\delta t$ be $\chi+\frac{\partial \chi}{\partial t}\delta t$. Thus χ is conceived as expressed in the form $\chi(\theta, \phi, \psi, t)$, where θ, ϕ, ψ are not functions of t. The operators

$$\frac{\partial}{\partial \theta}, \frac{\partial}{\partial \phi}, \frac{\partial}{\partial \psi}, \frac{\partial}{\partial t}$$

applied to χ have therefore the relative properties of operators denoting partial differentiation.

Call $\frac{\partial}{\partial t}$ the stationary differential operator with respect to the time.

(4) Similarly if u be any vector function of the matter at the given point (θ, ϕ, ψ) at any time t, then at the subsequent instant $t+\delta t$ the corresponding vector function of the new element which occupies the position (θ, ϕ, ψ) can be written $u+\frac{\partial u}{\partial t}\delta t$. Also it is obvious that

$$\frac{\partial u}{\partial t} = \frac{\partial u_1}{\partial t} i + \frac{\partial u_2}{\partial t} j + \frac{\partial u_3}{\partial t} k.$$

Let $\frac{\partial u}{\partial t}$ and $\frac{\partial \chi}{\partial t}$ be abbreviated into u' and χ', or into u_t, χ_t.

(5) We shall assume, except where the limitation is expressly stated, that the scalar and vector functions spoken of are continuous functions of the variables: and that if χ be any scalar and u any vector, χ, u_1, u_2, u_3 have finite and continuous partial and stationary differential coefficients with respect to θ, ϕ, ψ, t.

(6) If the quantities be associated with the given particles of matter, let the co-ordinates θ, ϕ, ψ mark the position of any given particle at the time t. Then at the subsequent time $t+\delta t$, the co-ordinates of that particle have become $\theta+\dot{\theta}\delta t$, $\phi+\dot{\phi}\delta t$, $\psi+\dot{\psi}\delta t$. Also if χ be any scalar function of that particle at the time t, the same function of the same particle at the time $t+\delta t$ will be denoted by $\chi+\dot{\chi}\delta t$ or by $\chi+\frac{d\chi}{dt}\delta t$. The function χ can be conceived now as a function of θ, ϕ, ψ, t and written $\chi(\theta, \phi, \psi, t)$, where θ, ϕ, ψ are functions of the time. Thus

$$\frac{d\chi}{dt} = \frac{\partial \chi}{\partial t} + \frac{\partial \chi}{\partial \theta}\dot{\theta} + \frac{\partial \chi}{\partial \phi}\dot{\phi} + \frac{\partial \chi}{\partial \psi}\dot{\psi}.$$

Similarly if u be any vector associated with the particle, at the time $t+\delta t$ the same vector function associated with the same particle is $u+\dot{u}\delta t$ or $u+\frac{du}{dt}\delta t$.

Thus obviously
$$\frac{du}{dt} = \frac{du_1}{dt} i + \frac{du_2}{dt} j + \frac{du_3}{dt} k;$$

where
$$\frac{du_1}{dt} = \frac{\partial u_1}{\partial t} + \dot{\theta} \frac{\partial u_1}{\partial \theta} + \dot{\phi} \frac{\partial u_1}{\partial \phi} + \dot{\psi} \frac{\partial u_1}{\partial \psi},$$

with two similar equations. Hence
$$\frac{du}{dt} = \frac{\partial u}{\partial t} + \dot{\theta} \frac{\partial u}{\partial \theta} + \dot{\phi} \frac{\partial u}{\partial \phi} + \dot{\psi} \frac{\partial u}{\partial \psi}.$$

Call $\frac{d}{dt}$ the mobile differential operator with respect to the time.

361. Hamilton's Differential Operator. (1) Let the position of any point be denoted by the vector x which is represented by the line drawn to it from any arbitrarily assumed origin. Then x_1, x_2, x_3 are the rectangular co-ordinates of the point referred to axes parallel to i, j, k; and x_1, x_2, x_3 may be conceived as taking the place of the unspecified co-ordinates θ, ϕ, ψ of the previous investigations.

(2) Let χ be any scalar function of position at a given instant. Then $i\frac{\partial \chi}{\partial x_1} + j\frac{\partial \chi}{\partial x_2} + k\frac{\partial \chi}{\partial x_3}$ obviously represents the rate of change of χ at the point x in the direction of the normal to the surface χ = constant, which passes through x. It follows that the function $i\frac{\partial \chi}{\partial x_1} + j\frac{\partial \chi}{\partial x_2} + k\frac{\partial \chi}{\partial x_3}$ is independent of the directions of the vectors i, j, k so long as they are a rectangular set.

Let the symbol operator, $i\frac{\partial}{\partial x_1} + j\frac{\partial}{\partial x_2} + k\frac{\partial}{\partial x_3}$, be written ∇, and called Hamilton's* Differential Operator, or more shortly, the Hamiltonian. Its properties were first fully investigated by Prof. Tait** for the very similar case of quaternions.

(3) The Hamiltonian may accordingly be conceived as operating on a vector by means of the conventions
$$\nabla u = \nabla u_1 . i + \nabla u_2 . j + \nabla u_3 . k,$$
$$(\nabla \mid u) = (\nabla u_1 \mid i) + (\nabla u_2 \mid j) + (\nabla u_3 \mid k).$$

(4) Hence $\nabla u = jk\left(\frac{\partial u_3}{\partial x_2} - \frac{\partial u_2}{\partial x_3}\right) + ki\left(\frac{\partial u_1}{\partial x_3} - \frac{\partial u_3}{\partial x_1}\right) + ij\left(\frac{\partial u_2}{\partial x_1} - \frac{\partial u_1}{\partial x_2}\right)$.

$|\nabla u$ is called the Curl of the vector u, ∇u is the Curl-flux of the vector u.

(5) Also
$$(\nabla \mid u) = \frac{\partial u_1}{\partial x_1} + \frac{\partial u_2}{\partial x_2} + \frac{\partial u_3}{\partial x_3}.$$

$(\nabla \mid u)$ is called the Divergence of the vector u and is a scalar quantity.

* Cf. Hamilton's *Lectures on Quaternions*, Lecture VII. § 620.

** Cf. his *Elementary Treatise on Quaternions*, 1st Edition, 1867, 3rd Edition, 1890.

(6) It is obvious with this symbolic use of ∇ that it can be treated as a vector as far as formal algebraical transformations are concerned, so long as in the product it is kept to the left of the quantity which it operates on, and so long as those quantities to its right on which it does not operate are noted.

(7) Thus in accordance with the rest of our algebraical notation we may write $\nabla|\nabla\chi$ in the form $\nabla^2\chi$, where [cf. § 357 (2)]

$$\nabla^2\chi = \frac{\partial^2\chi}{\partial x_1^2} + \frac{\partial^2\chi}{\partial x_2^2} + \frac{\partial^2\chi}{\partial x_3^2}.$$

It is obvious that $\nabla\nabla\chi = 0$; $\nabla\nabla u = 0$. Again, $\nabla|\nabla . u$ becomes $\nabla^2 u$, which is $\nabla^2 u_1 . i + \nabla^2 u_2 . j + \nabla^2 u_3 . k$.

(8) An important example of the possibility of formal algebraic transformations of expressions involving ∇ is as follows:

If a, b, c be any vectors,

$$ab|c = (a|c)\,b - (b|c)\,a = -|(c|ab) = |(c|ba).$$

Hence $$(b|c)\,a = b\,(c|a) - |(c|ba).$$

Now putting ∇ for both b and c and u for a, we find

$$\nabla^2 u = \nabla(\nabla|u) - |(\nabla|\nabla u).$$

362. Conventions and Formulæ. (1) The symbol ∇ is to be assumed as operating on all the subsequent vectors in a product in which it stands, in the absence of some special mark attached to a vector. If a vector such as v is not operated on by a preceding ∇, let it be written with a bar on the top, thus $\bar{v}$. For instance, ∇uv implies that ∇ operates both on u and v; but $\nabla u\bar{v}$ implies that ∇ operates on u and not on v, and $\nabla\bar{u}v$ implies that ∇ operates on v and not on u. Similarly $\nabla(\bar{u}|v)$ implies that ∇ operates on v and not on u.

(2) The advantage of affixing a sign to a vector not operated on by ∇ is that, as far as the vanishing of a product is concerned owing to the formal laws of multiplication [cf. § 93 (4)] a vector behaves differently according as it is or is not operated on by ∇.

For instance if u, v, are any two vectors, $uvu = vuu = 0$. This is true by reason of the formal laws of multiplication. Now substitute the symbolic vector ∇ for v, then $u\nabla u = \nabla u\bar{u}$, and this is not zero. Thus it is convenient, as far as formal multiplication is concerned, to reckon u and $\bar{u}$ as different vectors. It sometimes conduces to clearness in tracing the algebraic transformations to preserve the bar over a vector even when it is placed in front of ∇; thus $\nabla u\bar{u} = \bar{u}\nabla u$. In such cases the bar may obviously be placed or dropped without express mention.

(3) The following are important examples:

$$\left.\begin{array}{l}\nabla(u|v)=\nabla(\bar{u}|v)+\nabla(u|\bar{v}),\\ \nabla(u|\bar{u})=\tfrac{1}{2}\nabla u^2\end{array}\right\}\quad\text{hence}\quad\text{(i).}$$

$$\nabla uv=\nabla u\bar{v}+\nabla\bar{u}v=v\nabla u-u\nabla v \quad\text{(ii).}$$

Also [cf. 361 (8)]

$$u|\nabla u=-|(\nabla u|\bar{u})=|\nabla(\bar{u}|u)-(\bar{u}|\nabla)|u=\tfrac{1}{2}|\nabla u^2-(u|\nabla)|u.$$

But $$|(u|\nabla u)=|u\,.\,\nabla u=-\nabla u|\bar{u}.$$

Therefore $$\left.\begin{array}{r}\nabla u|\bar{u}=(u|\nabla)u-\tfrac{1}{2}\nabla u^2,\\ (u|\nabla)u=\nabla u|\bar{u}+\tfrac{1}{2}\nabla u^2\end{array}\right\}\quad\text{(iii).}$$

or

(4) If the operation ∇ is repeated in a product, a little care must be exercised so that the use of the bar may be unambiguous. For instance the following transformation exemplifies this remark.

We wish to operate on $\nabla u|\bar{u}$ with ∇. The new operation of course operates both on the u and the $\bar{u}$ of the existing expression; since the bar merely refers to the existing operation ∇. Write $\nabla u|\bar{u}=-|u\,.\,\nabla u$.

Then $$\nabla(|u\,.\,\nabla u)=\nabla(|\bar{u}\,.\,\nabla u)+\nabla(|u\,.\,\overline{\nabla u}),$$
where obviously the newly placed bars refer to the ∇ outside the bracket.

(5) But the introduction of a new symbol, such as $|v=\nabla u$ is often the simplest solution of the difficulty. For instance $\nabla u|\bar{u}$ becomes $|vu$.

An important example is arrived at by operating with $\nabla|$ on $\nabla u|\bar{u}$. Write $|v$ for ∇u.

Then $\nabla|\,.\,|vu=\nabla vu=u\nabla v-v\nabla u$, by equation (ii).

But $$v\nabla u=(v|v)=v^2=(\nabla u)^2,$$

and $$u\nabla v=u\,.\,\nabla|\nabla u=(u|\nabla)(\nabla|u)-u|\nabla^2 u.$$

Hence $$\nabla vu=(u|\nabla)(\nabla|u)-u|\nabla^2 u-(\nabla u)^2 \quad\text{(iv),}$$

where $$v=|\nabla u.$$

(6) It will be convenient to adhere to the further convention that ∇ immediately preceding a scalar such as ϕ does not operate on a subsequent vector unless some stop is placed between the ∇ and the scalar.

Thus $\nabla\phi u$ has the same meaning as $\nabla\phi\bar{u}$, but $\nabla\,.\,\phi u$ implies that ∇ operates on ϕu. This convention is useful in dealing with such expressions as $\nabla\phi\nabla\psi$: it avoids the clumsy form $\nabla\phi\overline{\nabla\psi}$.

It is however often better to place bars where there is a risk of mistake, so as not unduly to burden the memory with conventions.

(7) The preceding transformations have brought into prominence the symbolic operator $(u|\nabla)$. It is a scalar operator, and in the Cartesian notation [cf. § 357 (2)] is $u_1\dfrac{\partial}{\partial x_1}+u_2\dfrac{\partial}{\partial x_2}+u_3\dfrac{\partial}{\partial x_3}=u_0\dfrac{d}{d\sigma}$, where $d\sigma$ is an element of arc at the point x in the direction of u.

It follows that, if u be the velocity of the matter at the point x,

$$\frac{d}{dt}=\frac{\partial}{\partial t}+u_1\frac{\partial}{\partial x_1}+u_2\frac{\partial}{\partial x_2}+u_3\frac{\partial}{\partial x_3}=\frac{\partial}{\partial t}+(u\,|\nabla).$$

363. POLAR CO-ORDINATES. (1) The analytical transformations of ∇ into polar and cylindrical co-ordinates can be easily established.

Let P be the point x, and O the origin: let the position of P be defined

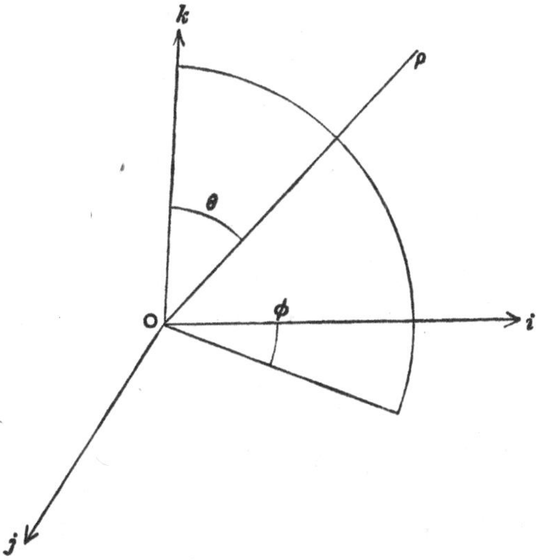

FIG. 1.

by the length ρ of OP, the angle θ between OP and the direction of k, the angle ϕ between the plane through OP and k and the plane ki.

It may be noted that by the convention of § 357 (2) ρ has also been denoted by x_0.

(2) Let r be the *unit* vector in the direction of OP, thus $r=\frac{x}{\rho}=\frac{x}{x_0}$.

Let v be the unit vector perpendicular to the plane through OP and k, positive in the direction of ϕ increasing.

Let u be the unit vector perpendicular to OP in the above plane and positive in the direction of θ increasing.

Thus $(r\,|r)=1=(u\,|u)=(v\,|v)$; and $(r\,|u)=0=(r\,|v)=(u\,|v)$;

and $$u=|\,vr,\quad v=|\,ru,\quad r=|\,uv.$$

(3) Also $$r=i\cos\phi\sin\theta+j\sin\phi\sin\theta+k\cos\theta.$$

Hence $$v\equiv rk\equiv(i\sin\phi-j\cos\phi)\sin\theta;$$

therefore $$v=j\cos\phi-i\sin\phi.$$

And $$u=|vr=i\cos\phi\cos\theta+j\sin\phi\cos\theta-k\sin\theta.$$

(4) Again,
$$\nabla = r\frac{\partial}{\partial\rho} + u\frac{\partial}{\rho\partial\theta} + v\frac{\partial}{\rho\sin\theta\partial\phi}.$$

Therefore
$$(\nabla\,|\,r) = \left(u\,\Big|\,\frac{\partial r}{\rho\partial\theta}\right) + \left(v\,\Big|\,\frac{\partial r}{\rho\sin\theta\partial\phi}\right) = \left(u\,\Big|\,\frac{u}{\rho}\right) + \left(v\,\Big|\,\frac{v}{\rho}\right) = \frac{2}{\rho};$$
$$(\nabla\,|\,v) = \left(v\,\Big|\,\frac{\partial v}{\rho\sin\theta\partial\phi}\right) = \left\{v\,\Big|\,\frac{-(j\sin\phi + i\cos\phi)}{\rho\sin\theta}\right\} = 0.$$
$$(\nabla\,|\,u) = \left(u\,\Big|\,\frac{\partial u}{\rho\partial\theta}\right) + \left(v\,\Big|\,\frac{\partial u}{\rho\sin\theta\partial\phi}\right) = -\left(\frac{u\,|\,r}{\rho}\right) + \left(v\,\Big|\,\frac{v\cos\theta}{\rho\sin\theta}\right) = \frac{\cot\theta}{\rho}.$$

Hence
$$\nabla^2 = \frac{\partial^2}{\partial\rho^2} + \frac{\partial^2}{\rho^2\partial\theta^2} + \frac{\partial^2}{\rho^2\sin^2\theta\partial\phi^2} + \frac{2}{\rho}\frac{\partial}{\partial\rho} + \frac{\cot\theta}{\rho^2}\frac{\partial}{\partial\theta}.$$

(5) Again, let p be any vector function of the position of P, and let
$$p = \varpi_1 r + \varpi_2 u + \varpi_3 v,$$
where r, v, u are the three rectangular unit vectors as defined above which correspond to the position of P, and ϖ_1, ϖ_2, ϖ_3, are scalar functions of ρ, θ, ϕ.

Then
$$(\nabla\,|\,p) = (r\,|\,\nabla\varpi_1) + (u\,|\,\nabla\varpi_2) + (v\,|\,\nabla\varpi_3) + \varpi_1(\nabla\,|\,r) + \varpi_2(\nabla\,|\,u) + \varpi_3(\nabla\,|\,v)$$
$$= \frac{\partial\varpi_1}{\partial\rho} + \frac{\partial\varpi_2}{\rho\partial\theta} + \frac{\partial\varpi_3}{\rho\sin\theta\partial\phi} + \frac{2\varpi_1}{\rho} + \frac{\varpi_2\cot\theta}{\rho}.$$

(6) Again, let q be the curl of p, and let $q = \kappa_1 r + \kappa_2 u + \kappa_3 v$.

Then
$$|q = \nabla p = \nabla\varpi_1\bar{r} + \nabla\varpi_2\bar{u} + \nabla\varpi_3\bar{v} + \varpi_1\nabla r + \varpi_2\nabla u + \varpi_3\nabla v.$$

Now $\nabla r = u\dfrac{u}{\rho} + v\dfrac{v}{\rho} = 0;$

$$\nabla u = -u\frac{r}{\rho} + v\frac{v\cos\theta}{\rho\sin\theta} = \frac{ru}{\rho};$$
$$\nabla v = -v\frac{(j\sin\phi + i\cos\phi)}{\rho\sin\theta} = -v\frac{(r\sin\theta + u\cos\theta)}{\rho\sin\theta} = \frac{\cot\theta}{\rho}uv - \frac{1}{\rho}vr.$$

Hence
$$|q = uv\left[-\frac{1}{\rho\sin\theta}\frac{\partial\varpi_2}{\partial\phi} + \frac{1}{\rho}\frac{\partial\varpi_3}{\partial\theta} + \frac{\varpi_3\cot\theta}{\rho}\right]$$
$$+ vr\left[\frac{1}{\rho\sin\theta}\frac{\partial\varpi_1}{\partial\phi} - \frac{\partial\varpi_3}{\partial\rho} - \frac{\varpi_3}{\rho}\right] + ru\left[-\frac{1}{\rho}\frac{\partial\varpi_1}{\partial\theta} + \frac{\partial\varpi_2}{\partial\rho} + \frac{\varpi_2}{\rho}\right].$$

Thus
$$\kappa_1 = \frac{1}{\rho}\frac{\partial\varpi_3}{\partial\theta} - \frac{1}{\rho\sin\theta}\frac{\partial\varpi_2}{\partial\phi} + \frac{\varpi_3\cot\theta}{\rho},$$
$$\kappa_2 = \frac{1}{\rho\sin\theta}\frac{\partial\varpi_1}{\partial\phi} - \frac{\partial\varpi_3}{\partial\rho} - \frac{\varpi_3}{\rho},$$
$$\kappa_3 = \frac{\partial\varpi_2}{\partial\rho} - \frac{1}{\rho}\frac{\partial\varpi_1}{\partial\theta} + \frac{\varpi_2}{\rho}.$$

364. CYLINDRICAL CO-ORDINATES. (1) Employing cylindrical co-ordinates, let x_3 denote the length ON in the annexed figure, and σ denote the length NP. Also let v denote the same vector as in § 363, and w denote a unit vector parallel to NP.

(2) Then $v = j\cos\phi - i\sin\phi,\ w = i\cos\phi + j\sin\phi;$

and $w = |\,vk,\ v = |\,kw,\ k = |\,wv.$

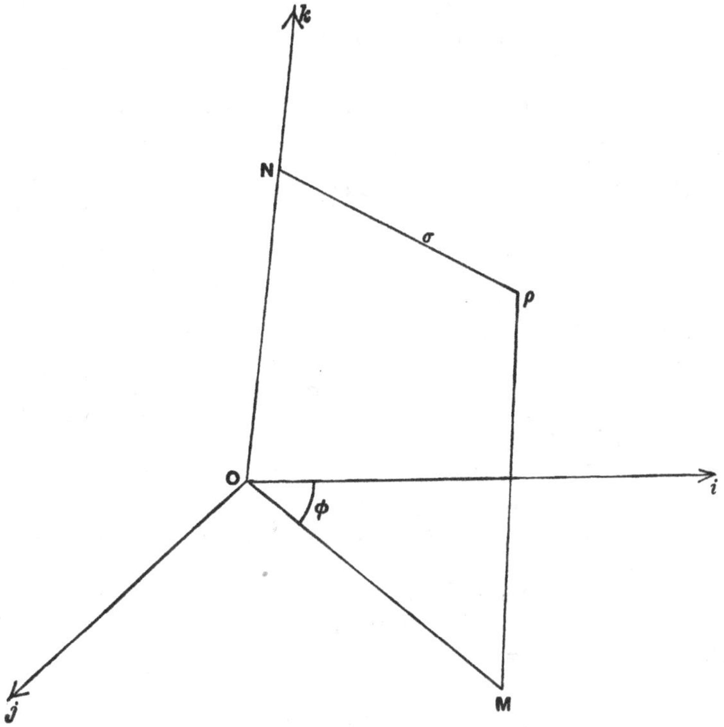

FIG. 2.

Also $$\nabla = w\frac{\partial}{\partial\sigma} + v\frac{\partial}{\sigma\partial\phi} + k\frac{\partial}{\partial x_3}.$$

And $$\nabla\,|v = -v\left|\frac{w}{\sigma} = 0,\ \nabla\,|w = v\right|\frac{v}{\sigma} = \frac{1}{\sigma}.$$

Hence $$\nabla^2 = \frac{\partial^2}{\partial\sigma^2} + \frac{\partial^2}{\sigma^2\partial\phi^2} + \frac{\partial^2}{\partial x_3^2} + \frac{1}{\sigma}\,\frac{\partial}{\partial\sigma}.$$

Again, $$\nabla v = -v\frac{w}{\sigma} = \frac{wv}{\sigma},\ \nabla w = v\,\frac{v}{\sigma} = 0.$$

(3) Let any vector p be written $\varpi_1 w + \varpi_2 v + \varpi_3 k$, and let its curl q be written $\kappa_1 w + \kappa_2 v + \kappa_3 k$.

Then $$(\nabla\,|p) = (\nabla\varpi_1|\overline{w}) + (\nabla\varpi_2|\overline{v}) + (\nabla\varpi_3|k) + \varpi_1(\nabla\,|w) + \varpi_2(\nabla\,|v)$$

$$= \frac{\partial\varpi_1}{\partial\sigma} + \frac{1}{\sigma}\,\frac{\partial\varpi_2}{\partial\phi} + \frac{\partial\varpi_3}{\partial x_3} + \frac{\varpi_1}{\sigma}.$$

$$|q = \nabla\varpi_1\,.\,\overline{w} + \nabla\varpi_2\,.\,\overline{v} + \nabla\varpi_3\,.\,k + \varpi_1\nabla w + \varpi_2\nabla v$$

$$= wv\left[-\frac{1}{\sigma}\,\frac{\partial\varpi_1}{\partial\phi} + \frac{\partial\varpi_2}{\partial\sigma} + \frac{\varpi_2}{\sigma}\right] + vk\left[-\frac{\partial\varpi_2}{\partial x_3} + \frac{1}{\sigma}\,\frac{\partial\varpi_3}{\partial\phi}\right] + kw\left[\frac{\partial\varpi_1}{\partial x_3} - \frac{\partial\varpi_3}{\partial\sigma}\right].$$

Hence
$$\kappa_3 = \frac{\partial \varpi_2}{\partial \sigma} - \frac{1}{\sigma}\frac{\partial \varpi_1}{\partial \phi} + \frac{\varpi_2}{\sigma},$$
$$\kappa_1 = \frac{1}{\sigma}\frac{\partial \varpi_3}{\partial \phi} - \frac{\partial \varpi_2}{\partial x_3},$$
$$\kappa_2 = \frac{\partial \varpi_1}{\partial x_3} - \frac{\partial \varpi_3}{\partial \sigma}.$$

365. ORTHOGONAL CURVILINEAR CO-ORDINATES*. (1) The formulæ may be generalized thus: let l, m, n be three unit rectangular vectors associated with any point P, such that the system of vectors suffers a continuous change in direction as the point P moves in a continuous line to any other point P'. Let $\delta\sigma_1$, $\delta\sigma_2$, $\delta\sigma_3$ be elements of arc traversed by the point as it moves through small distances from P in the directions l, m, n respectively.

(2) Let P be determined by three curvilinear co-ordinates θ_1, θ_2, θ_3, such that during the small motion $\delta\sigma_1$, θ_1 becomes $\theta_1 + \delta\theta_1$, and θ_2 and θ_3 are unaltered; with two other similar specifications. Also assume that
$$\delta\sigma_1 = \frac{\delta\theta_1}{h_1}, \quad \delta\sigma_2 = \frac{\delta\theta_2}{h_2}, \quad \delta\sigma_3 = \frac{\delta\theta_3}{h_3};$$
where h_1, h_2, h_3 are functions of θ_1, θ_2, θ_3.

Then [cf. § 361 (2)] $\nabla = l\frac{h_1\partial}{\partial\theta_1} + m\frac{h_2\partial}{\partial\theta_2} + n\frac{h_3\partial}{\partial\theta_3}$.

(3) Thus if ϕ be any scalar function of θ_1, θ_2, θ_3,
$$\nabla\phi = l\frac{h_1\partial\phi}{\partial\theta_1} + m\frac{h_2\partial\phi}{\partial\theta_2} + n\frac{h_3\partial\phi}{\partial\theta_3}.$$

Also
$$\nabla^2\phi = \frac{h_1\partial}{\partial\theta_1}\left(\frac{h_1\partial\phi}{\partial\theta_1}\right) + \dots + \frac{h_1\partial\phi}{\partial\theta_1}(\nabla\,|\,l) + \dots\dots\dots\dots\text{(i)}.$$

Again, let p be any vector, and let $p = \varpi_1 l + \varpi_2 m + \varpi_3 n$.

Then
$$(\nabla\,|\,p) = \frac{h_1\partial\varpi_1}{\partial\theta_1} + \frac{h_2\partial\varpi_2}{\partial\theta_2} + \frac{h_3\partial\varpi_3}{\partial\theta_3} + \varpi_1(\nabla\,|\,l) + \varpi_2(\nabla\,|\,m) + \varpi_3(\nabla\,|\,n)\ \dots\text{(ii)}.$$

Similarly
$$\nabla p = \nabla\varpi_1\bar{l} + \nabla\varpi_2\bar{m} + \nabla\varpi_3\bar{n} + \varpi_1\nabla l + \varpi_2\nabla m + \varpi_3\nabla n$$
$$= mn\left[\frac{h_2\partial\varpi_3}{\partial\theta_2} - \frac{h_3\partial\varpi_2}{\partial\theta_3}\right] + nl\left[\frac{h_3\partial\varpi_1}{\partial\theta_3} - \frac{h_1\partial\varpi_3}{\partial\theta_1}\right] + lm\left[\frac{h_1\partial\varpi_2}{\partial\theta_1} - \frac{h_2\partial\varpi_1}{\partial\theta_2}\right]$$
$$+ \varpi_1\nabla l + \varpi_2\nabla m + \varpi_3\nabla n\dots\dots\dots\text{(iii)}.$$

* As far as I am aware, the methods of transformation of the present and the two preceding articles have not been employed before. The methods are the analogue in this Algebra of Webb's method of Vector-Differentiation, published in the *Messenger of Mathematics*, 1882, and fully explained and applied to this case in Love's *Treatise on the Mathematical Theory of Elasticity*, § 119.

Thus when $(\nabla|l)$, $(\nabla|m)$, $(\nabla|n)$, ∇l, ∇m, ∇n have been obtained, the formulæ for transformation are complete.

(4). Now $\theta_1 =$ constant, $\theta_2 =$ constant, $\theta_3 =$ constant, form three sets of mutually orthogonal surfaces.

Hence $$l = |mn = \frac{1}{h_1}\nabla\theta_1 = \frac{1}{h_2h_3}|\nabla\theta_2\nabla\theta_3.$$

Hence $$(\nabla|l) = \frac{1}{h_2h_3}\nabla . \nabla\theta_2\nabla\theta_3 + \nabla\theta_2\nabla\theta_3\nabla\frac{1}{h_2h_3}.$$

Now $$\nabla . \nabla\theta_2\nabla\theta_3 = \nabla\nabla\theta_2 . \nabla\theta_3 - \nabla\theta_2 . \nabla\nabla\theta_3 = 0.$$

Therefore
$$(\nabla|l) = \nabla\theta_2\nabla\theta_3\nabla\frac{1}{h_2h_3} = h_2h_3\left\{|l . \nabla\frac{1}{h_2h_3}\right\} = h_2h_3(l|\nabla)\frac{1}{h_2h_3} = h_1h_2h_3\frac{\partial}{\partial\theta_1}\frac{1}{h_2h_3}.$$

Similarly
$$(\nabla|m) = h_1h_2h_3\frac{\partial}{\partial\theta_2}\frac{1}{h_3h_1}, \text{ and } (\nabla|n) = h_1h_2h_3\frac{\partial}{\partial\theta_3}\frac{1}{h_1h_2}.$$

Hence from equation (ii) $$(\nabla|p) = h_1h_2h_3\left\{\frac{\partial}{\partial\theta_1}\frac{\varpi_1}{h_2h_3} + \frac{\partial}{\partial\theta_2}\frac{\varpi_2}{h_3h_1} + \frac{\partial}{\partial\theta_3}\frac{\varpi_3}{h_1h_2}\right\}.$$

(5) Again, $$\nabla l = \nabla . \frac{1}{h_1}\nabla\theta_1 = \nabla\frac{1}{h_1} . \nabla\theta_1 = \frac{1}{h_1^2}\nabla\theta_1\nabla h_1 = \frac{1}{h_1}l\nabla h_1$$
$$= \frac{h_2}{h_1}lm\frac{\partial h_1}{\partial\theta_2} - \frac{h_3}{h_1}nl\frac{\partial h_1}{\partial\theta_3}.$$

Similarly $$\nabla m = \frac{h_3}{h_2}\frac{\partial h_2}{\partial\theta_3}mn - \frac{h_1}{h_2}\frac{\partial h_2}{\partial\theta_1}lm, \quad \nabla n = \frac{h_1}{h_3}\frac{\partial h_3}{\partial\theta_1}nl - \frac{h_2}{h_3}\frac{\partial h_3}{\partial\theta_2}mn.$$

Hence from equation (iii)
$$|\nabla p = \left\{\frac{h_2\partial\varpi_3}{\partial\theta_2} - \frac{h_3\partial\varpi_2}{\partial\theta_3} + \varpi_2\frac{h_3}{h_2}\frac{\partial h_2}{\partial\theta_3} - \varpi_3\frac{h_2}{h_3}\frac{\partial h_3}{\partial\theta_2}\right\}l + \text{etc.}$$
$$= h_2h_3\left(\frac{\partial}{\partial\theta_2}\frac{\varpi_3}{h_3} - \frac{\partial}{\partial\theta_3}\frac{\varpi_2}{h_2}\right)l + h_3h_1\left(\frac{\partial}{\partial\theta_3}\frac{\varpi_1}{h_1} - \frac{\partial}{\partial\theta_1}\frac{\varpi_3}{h_3}\right)m$$
$$+ h_1h_2\left(\frac{\partial}{\partial\theta_1}\frac{\varpi_2}{h_2} - \frac{\partial}{\partial\theta_2}\frac{\varpi_1}{h_1}\right)n.$$

Accordingly if q be the curl of p, that is $|\nabla p$; and if q be written
$$\kappa_1 l + \kappa_2 m + \kappa_3 n,$$
then
$$\kappa_1 = h_2h_3\left(\frac{\partial}{\partial\theta_2}\frac{\varpi_3}{h_3} - \frac{\partial}{\partial\theta_3}\frac{\varpi_2}{h_2}\right),$$
$$\kappa_2 = h_3h_1\left(\frac{\partial}{\partial\theta_3}\frac{\varpi_1}{h_1} - \frac{\partial}{\partial\theta_1}\frac{\varpi_3}{h_3}\right),$$
$$\kappa_3 = h_1h_2\left(\frac{\partial}{\partial\theta_1}\frac{\varpi_2}{h_2} - \frac{\partial}{\partial\theta_2}\frac{\varpi_1}{h_1}\right).$$

These formulæ of course include as special cases the preceding formulæ for polar and cylindrical co-ordinates [cf. §§ 363 and 364].

366. VOLUME, SURFACE, AND LINE INTEGRALS. (1) Let $d\tau$ stand for an element of volume at the point x. Let dS be a vector-area representing in magnitude and direction an element of surface at the point x. Then

$$dS = jkdS_1 + kidS_2 + ijdS_3;$$

where $$dS_1 = dx_2dx_3, \quad dS_2 = dx_3dx_1, \quad dS_3 = dx_1dx_2.$$

Let $|dS$ denote the normal, positive when drawn outwards.

Let dx be a vector line denoting in magnitude and direction an element of arc at x. Then $dx = idx_1 + jdx_2 + kdx_3$, and dx_0 is often denoted by $d\sigma$.

(2) Then the well-known theorem connecting the volume and surface integrals of any continuous function of position within a closed surface is

$$\iiint(\nabla\,|u)\,d\tau = \iint(udS).$$

(3) Green's Theorem can be written

$$\iiint(\nabla\phi\,|\overline{\nabla\psi})\,d\tau = \iint(\phi\nabla\psi dS) - \iiint\phi\nabla^2\psi d\tau$$
$$= \iint(\psi\nabla\phi dS) - \iiint\psi\nabla^2\phi\,d\tau.$$

This can obviously be deduced from subsection (2) by writing $\phi\nabla\psi$ for u, then

$$(\nabla\,|u) = (\nabla\,|\phi\nabla\psi) = (\nabla\phi\,|\overline{\nabla\psi}) + (\phi\nabla^2\psi).$$

(4) Stokes' Theorem connecting line integrals and surface integrals is expressed by

$$\iint(\nabla u\,|dS) = \int(u\,|dx),$$

where the line integral is taken completely round any closed circuit, and the surface integral is taken over any surface with its edge coincident with the surface.

367. THE EQUATIONS OF HYDRODYNAMICS. (1) Let the vector u denote the velocity of a frictionless fluid at any point represented by the vector x drawn from an arbitrarily chosen origin. Let ρ be the density at that point, and ϖ the pressure. Let the vector f denote the external force per unit mass at x. Also let the vector q denote the curl of u, so that $q = |\nabla u$. The vector q defines the vortex motion at each point of the fluid: portions of the fluid, for which $q = 0$ at each point, are moving irrotationally. Then the fundamental equation of motion is

$$\frac{d\rho u}{dt} = -\nabla\varpi + \rho f \quad \text{..............................(i)}.$$

(2) Assume in the first place that the fluid is homogeneous and incompressible: also that f is derivable from a force potential ψ, so that

$$f = -\nabla\psi.$$

Equation (i) becomes

$$\frac{du}{dt} = -\nabla\left(\frac{\varpi}{\rho} + \psi\right).$$

This can be transformed [cf. § 362 (7)] into

$$\frac{\partial u}{\partial t} + (u \mid \nabla)\, u = -\nabla\left(\frac{\varpi}{\rho} + \psi\right) \quad\text{.......................(ii).}$$

But by equation (iii) of § 362 (3)

$$(u \mid \nabla)\, u = \mid qu + \tfrac{1}{2}\nabla u^2.$$

Hence $$\frac{\partial u}{\partial t} + \mid qu = -\nabla\left(\frac{\varpi}{\rho} + \psi + \tfrac{1}{2}u^2\right) \quad\text{.....................(iii).}$$

(3) The equation of continuity becomes

$$(\nabla \mid u) = 0 \quad\text{..................................(iv).}$$

(4) These equations are independent of any special co-ordinate system. Thus let θ_1, θ_2, θ_3 denote any set of orthogonal curvilinear co-ordinates, so that $\theta_1 =$ constant, $\theta_2 =$ constant, $\theta_3 =$ constant, denote three systems of mutually orthogonal surfaces. Let l, m, n and h_1, h_2, h_3 have the meanings assigned to them in § 365.

Let $$u = v_1 l + v_2 m + v_3 n, \quad q = \kappa_1 l + \kappa_2 m + \kappa_3 n.$$

Then $\kappa_1 = h_2 h_3 \left(\dfrac{\partial}{\partial\theta_2}\dfrac{v_3}{h_3} - \dfrac{\partial}{\partial\theta_3}\dfrac{v_2}{h_2}\right)$, with two similar equations for κ_2 and κ_3.

Now l, m, n are independent of t. Hence equation (iii) splits up into three equations of the type

$$\frac{\partial v_1}{\partial t} + \kappa_2 v_3 - \kappa_3 v_2 = -\frac{h_1\partial}{\partial\theta_1}\left(\frac{\varpi}{\rho} + \psi + \tfrac{1}{2}u^2\right) \quad\text{..................(v).}$$

And equation (iv) becomes

$$\frac{\partial}{\partial\theta_1}\frac{v_1}{h_2 h_3} + \frac{\partial}{\partial\theta_2}\frac{v_2}{h_3 h_1} + \frac{\partial}{\partial\theta_3}\frac{v_3}{h_1 h_2} = 0 \quad\text{......................(vi).}$$

These are the general equations of motion of a homogeneous incompressible fluid referred to any orthogonal curvilinear co-ordinates. They include as special cases the equations referred to polar or to cylindrical co-ordinates.

368. Moving Origin. (1) Equation (iii) of the preceding article may be extended to the case of a moving origin. Suppose that the origin moves with velocity v, then v may be a function of t, but of course is not a function of position.

The point, which at time t is defined by the vector x, at the time $t+\delta t$ is defined by the vector $x - v\delta t$.

Let $\frac{\delta}{\delta t}$ denote the stationary differential operator with respect to the time *relatively to the moving origin*, so that $\frac{\delta}{\delta t}$ gives the rate of change at a *moving* point which is defined by the *constant* vector x drawn from the moving origin.

Then $$\frac{\delta}{\delta t} = \frac{\partial}{\partial t} + (v \mid \nabla).$$

Hence equation (iii) of § 367 becomes

$$\frac{\delta u}{\delta t} - (v \mid \nabla) u + \mid qu = -\nabla \left(\frac{\varpi}{\rho} + \psi + \tfrac{1}{2} u^2 \right) \quad \text{..............(vii)}.$$

(2) Equation (ii) of § 367 becomes

$$\frac{\delta u}{\delta t} - (v \mid \nabla) u + (u \mid \nabla) u = -\nabla \left(\frac{\varpi}{\rho} + \psi \right).$$

Now let $u' = u - v$. Then u' is the velocity of the fluid at any point relatively to the origin.

Substitute $u' + v$ for u and remember that $(u \mid \nabla) v = 0 = (v \mid \nabla) v$, since v is not a function of x.

Also if $\dot{v}$ be the acceleration of the origin,

$$\frac{\delta u}{\delta t} = \frac{\delta u'}{\delta t} + \dot{v}.$$

Hence the equation of motion becomes

$$\frac{\delta u'}{\delta t} + \dot{v} + (u' \mid \nabla) u' = -\nabla \left(\frac{\varpi}{\rho} + \psi \right) \quad \text{..................(viii)}.$$

The equation of continuity is $(\nabla \mid u') = 0$.

(3) The curl of u' is the same as that of u, since

$$\mid \nabla u = \mid \nabla (u' + v) = \mid \nabla u'.$$

Hence equation (viii) can be transformed into

$$\frac{\delta u'}{\delta t} + \dot{v} + \mid qu' = -\nabla \left(\frac{\varpi}{\rho} + \psi + \tfrac{1}{2} u'^2 \right).$$

Furthermore, since $\dot{v}$ is not a function of position, $\dot{v} = \nabla (\dot{v} \mid x)$.

Hence finally

$$\frac{\delta u'}{\delta t} + \mid qu' = -\nabla \left\{ \frac{\varpi}{\rho} + \psi + (\dot{v} \mid x) + \tfrac{1}{2} u'^2 \right\} \quad \text{...............(ix)}.$$

(4) Therefore a uniform motion of the origin does not affect the form of the hydrodynamical equation, when the velocity is reckoned relatively to the origin.

An acceleration of the origin adds a term to the force potential.

The vortices are the same whatever motion be assigned to the origin.

Therefore by suitable modifications of ψ, equations (ii) and (iii) of § 367 may be looked on as the typical hydrodynamical equations, whether the origin be at rest or be moving in any way.

369. Transformations of Hydrodynamical Equations. (1) Operating on (iii) of § 367 with ∇

$$\nabla \frac{\partial u}{\partial t} + \nabla \,|qu = 0.$$

But
$$\nabla \frac{\partial u}{\partial t} = \left| \frac{\partial q}{\partial t} \right. .$$

Also $\nabla \,|qu = (\nabla \,|u)\,|\bar{q} + (\nabla \,|\bar{u})\,|q - (\nabla \,|q)\,|\bar{u} - (\nabla \,|\bar{q})\,|u = (u \,|\nabla)\,|q - (q \,|\nabla)\,|u;$

since
$$(\nabla \,|u) = 0 = (\nabla \,|q).$$

Hence by taking the supplement

$$\frac{\partial q}{\partial t} + (u \,|\nabla)\, q = (q \,|\nabla)\, u.$$

This can also be written

$$\frac{dq}{dt} = (q \,|\nabla)\, u \qquad \text{(x).}$$

(2) Again, operate on (iii) with $\nabla|$.

Now
$$\nabla| \frac{\partial u}{\partial t} = \frac{\partial}{\partial t}(\nabla \,|u) = 0.$$

Hence
$$\nabla qu = -\nabla^2 \left(\frac{\varpi}{\rho} + \psi + \tfrac{1}{2}u^2 \right) \qquad \text{(xi).}$$

Also by § 362 equation (iv)

$$\nabla qu = (u \,|\nabla)(\nabla \,|u) - u\nabla^2 u - (\nabla u)^2 = -u\nabla^2 u - (\nabla u)^2.$$

370. Vector Potential of Velocity. (1) Assume that there are no boundaries to the fluid which extends to infinity in all directions; also that the vortices [cf. § 367 (1)] only extend to a finite distance from the origin. Now $q = |\nabla u$.

Hence [cf. § 355 (2) equation (iv)]

$$\nabla q = \nabla \,|\nabla u = |\nabla\, (\nabla \,|u) - |\nabla^2 u = -|\nabla^2 u.$$

Therefore by the ordinary theory of the potential, since by assumption $q = 0$ at all points beyond a certain finite distance from the origin,

$$|u = \frac{1}{4\pi} \iiint \frac{\nabla' q'}{\sqrt{(x - x')^2}}\, d\tau';$$

where q' represents the curl of the velocity at the point x', and ∇' stands for $i\frac{\partial}{\partial x_1'} + j\frac{\partial}{\partial x_2'} + k\frac{\partial}{\partial x_3'}$, and $d\tau'$ is an element of volume at x'.

(2) The integration may be assumed to be confined within any surface large enough to contain all the vortices and such that none of them lie on the surface.

Integrate by parts, and remember that by the assumption q' is zero at all points of the surface.

Then [cf. § 362 (1)] $$|u = \frac{-1}{4\pi}\iiint \nabla' \frac{q'}{\sqrt{(x-x')^2}} d\tau'.$$

Now $$\nabla' \frac{1}{\sqrt{(x-x')^2}} = -\nabla \frac{1}{\sqrt{(x-x')^2}}.$$

Hence $$|u = \frac{1}{4\pi}\iiint \nabla \frac{q'}{\sqrt{(x-x')^2}} d\tau' = \frac{1}{4\pi}\nabla \iiint \frac{q'}{\sqrt{(x-x')^2}} d\tau' \quad\text{......(xii)}.$$

Hence $\frac{1}{4\pi}\iiint \frac{q'}{\sqrt{(x-x')^2}} d\tau'$ is a vector such that u is its curl. Let this vector be denoted by p, then $u = |\nabla p$.

(3) Also by integrating by parts, it is easily seen that

$$(\nabla | p) = \frac{1}{4\pi}\iiint \nabla \left| \frac{q'}{\sqrt{(x-x')^2}} d\tau' = -\frac{1}{4\pi}\iiint \nabla' \right| \frac{q'}{\sqrt{(x-x')^2}} d\tau'$$
$$= \frac{1}{4\pi}\iiint \frac{(\nabla' | q')}{\sqrt{(x-x')^2}} d\tau' = 0;$$

since $$(\nabla | q) = 0.$$

The vector p is called the vector potential of the velocity.

(4) The same suppositions as to the absence of boundaries and as to vorticity enable us similarly to solve equation (xi).

For by the ordinary theory of the potential equation (xi) can be transformed into

$$\frac{\varpi}{\rho} + \psi + \tfrac{1}{2}u^2 + \text{constant} = \frac{1}{4\pi}\iiint \frac{\nabla' q'u'}{\sqrt{(x-x')^2}} d\tau'.$$

Now integrating by parts exactly as above,

$$\frac{\varpi}{\rho} + \psi + \tfrac{1}{2}u^2 + \gamma = \frac{1}{4\pi}\nabla \iiint \frac{q'u'}{\sqrt{(x-x')^2}} d\tau' \quad\text{............(xiii)}.$$

Hence if L denote the flux $\frac{1}{4\pi}\iiint \frac{q'u'}{\sqrt{(x-x')^2}} d\tau$, then

$$\frac{\varpi}{\rho} + \psi + \tfrac{1}{2}u^2 + \gamma = \nabla L = \frac{\partial L_1}{\partial x_1} + \frac{\partial L_2}{\partial x_2} + \frac{\partial L_3}{\partial x_3}.$$

371. Curl Filaments of Constant Strength. (1) Let v be any vector which at each point is definitely associated with the fluid at that point: the magnitude and direction of v may depend, wholly or in part, on the velocity of the fluid and on its differential coefficients, and it may depend partly on other properties of the fluid not here specified. Let it be assumed that the components of v, namely v_1, v_2, v_3, and their differential coefficients are single-valued and continuous functions. Let r denote the curl of v, so that $r = |\nabla v$.

Lines formed by continually moving in the direction of the r of the point are called the curl lines of the vector v. Since r fulfils the solenoidal condition, namely $(\nabla | r) = 0$, such lines must either be closed or must begin and end on a boundary.

(2) A curl filament is formed by drawing the curl lines through every point of any small circuit in the fluid. If dS be the vector area at any point of a curl filament, then rdS is called the strength of the filament. It follows from the solenoidal condition by a well-known proposition that the strength of a given curl filament is the same at all points of it.

Let any finite circuit be filled in with any surface, then by § 366 (4),

$$\iint (rdS) = \int (v | dx);$$

where the line integral is taken round the circuit.

(3) Let us now find the condition that the sum of the strengths of the filaments, which pass through any circuit consisting of given particles of the fluid, may be independent of the time. Also for the sake of brevity assume that the region of space considered is not multiply connected.

The condition is $$\frac{d}{dt}\int (v | dx) = 0.$$

This becomes $$\int \left(\frac{dv}{dt} | dx\right) + \int (v | du) = 0.$$

Now $$du = (dx | \nabla) u.$$

Therefore $$(v | du) = (dx | \nabla)(\bar{v} | u) = \nabla (\bar{v} | u) | dx.$$

Hence $$\int \left(\frac{dv}{dt} | dx\right) + \int (v | du) = 0,$$

becomes $$\int \left[\left\{\frac{dv}{dt} + \nabla (\bar{v} | u)\right\} | dx\right] = 0.$$

Now if ψ be any scalar function of x and t which together with its differential coefficient is continuous and single-valued,

$$\int (\nabla \psi | dx) = 0,$$

where the integration is completely round the circuit. Hence if ψ be some such scalar function, we deduce

$$\frac{dv}{dt} + \nabla(\bar{v} \,|\, u) = -\nabla\psi.$$

(4) Also
$$\frac{dv}{dt} = \frac{\partial v}{\partial t} + (u \,|\, \nabla)\, v.$$

Now

$$(u \,|\, \nabla)\, v = \nabla(v \,|\, \bar{u}) + \nabla v \,|\, \bar{u} = \nabla(v \,|\, \bar{u}) - |\bar{u} \,.\, \nabla v = \nabla(v \,|\, \bar{u}) - |.\, u \,|\, \nabla v = \nabla(v \,|\, \bar{u}) - |\bar{u}r.$$

Hence
$$\frac{\partial v}{\partial t} - |ur + \nabla(v \,|\, u) = -\nabla\psi.$$

Now put
$$\chi = \psi + (v \,|\, u).$$

Then
$$\frac{\partial v}{\partial t} + |ru = -\nabla\chi \quad \text{......(xiv)}.$$

(5) To eliminate χ we operate with ∇, then

$$\left|\frac{\partial r}{\partial t} + \nabla \,|\, ru = 0.\right.$$

Now
$$\nabla \,|\, ur = (\nabla \,|\, r) \,|\, u - (\nabla \,|\, u) \,|\, r.$$

But
$$(\nabla \,|\, r) = 0 = (\nabla \,|\, u) \,|.$$

Hence
$$\nabla \,|\, ur = (r \,|\, \nabla) \,|\, u - (u \,|\, \nabla) \,|\, r.$$

Therefore the equation becomes after taking supplements,

$$\frac{\partial r}{\partial t} + (u \,|\, \nabla)\, r = (r \,|\, \nabla)\, u.$$

This is
$$\frac{dr}{dt} = (r \,|\, \nabla)\, u \quad \text{......(xv)}.$$

(6) This condition should be compared with equation (x). It follows from the comparison that the strengths of all vortex filaments are constant. In other words, that if equation (xiv) be conceived as an equation to find the unknown vector r, then q is one solution for r. But q is not necessarily the most general solution. Thus there are other curl filaments in the fluid with the same property of constancy.

(7) But equations (xiv) and (xv) are more general than these enunciations would suggest. For in the derivation of (xiv) neither the equation of continuity for an incompressible substance nor the kinetic equation of fluid motion were used. It follows that if the motion of any continuous substance be assumed given, so that u is a given function of x and t, then any vector v, as defined in subsection (1), with its curl r which satisfies equation (xiv) is such that the curl filaments are of constant strength.

(8) Equation (xv) involves the equation $(\nabla \,|\, u) = 0$. Hence this equation holds for any incompressible substance moving in any continuous manner.

An extended form of (xv) can be deduced by writing $(\nabla|u)=\theta$, where θ is a known function of x and t, since u is such a function.

Hence $$\nabla|ur=(r|\nabla)|u-(u|\nabla)|r-\theta|r.$$

Therefore $$\frac{\partial r}{\partial t}+(u|\nabla)r=(r|\nabla)u-\theta r;$$

that is, $$\frac{dr}{dt}=(r|\nabla)u-\theta r \qquad \text{(xvi)}.$$

372. Carried Functions. (1) Let ϕ be a scalar function of x and t such that for all values of t any surface $\phi=\gamma$, where γ is any particular constant, represents the same sheet of particles of the substance. Then the function ϕ will be called a carried function*.

(2) The analytical condition which ϕ must satisfy is

$$\frac{d\phi}{dt}=0,$$

or as it may be written,

$$\frac{\partial\phi}{\partial t}+(u|\nabla\phi)=0.$$

Also $$\frac{d}{dt}\nabla\phi=\frac{\partial}{\partial t}\nabla\phi+(u|\nabla)\nabla\phi=\nabla\frac{\partial\phi}{\partial t}+\nabla(\bar{u}|\nabla)\phi$$

$$=\nabla\frac{d\phi}{dt}-\nabla(u|\nabla)\phi+\nabla(\bar{u}|\nabla)\phi=-\nabla(u|\nabla)\phi+\nabla(\bar{u}|\nabla)\phi$$

$$=-\nabla(\overline{\nabla\phi}|u) \qquad \text{(xvii)}.$$

(3) Now let ϕ and ψ be any two carried functions. Then by equation (xvii)

$$\frac{d}{dt}(\nabla\phi\nabla\psi)=-\nabla(\overline{\nabla\phi}|u)\overline{\nabla\psi}-\overline{\nabla\phi}\nabla(\overline{\nabla\psi}|u)$$

$$=-\nabla\{(\overline{\nabla\phi}|u)\overline{\nabla\psi}-(\overline{\nabla\psi}|u)\overline{\nabla\phi}\}=-\nabla\{\overline{\nabla\phi}\,\overline{\nabla\psi}|u\} \qquad \text{(xviii)};$$

by § 355 (2) equation (iii).

(4) Also if ϕ, ψ, χ be any three carried functions

$$\frac{d}{dt}(\nabla\phi\nabla\psi\nabla\chi)=-\nabla\{(\overline{\nabla\phi}|u)\overline{\nabla\psi}\,\overline{\nabla\chi}+(\overline{\nabla\psi}|u)\overline{\nabla\chi}\,\overline{\nabla\phi}+(\overline{\nabla\chi}|u)\overline{\nabla\phi}\,\overline{\nabla\chi}\}$$

$$=-\nabla\{(\overline{\nabla\phi}\,\overline{\nabla\psi}\,\overline{\nabla\chi})|u\},$$

by the extended rule of the middle factor, where the product of three vectors is treated as an extensive magnitude formed by progressive multiplication.

Hence

$$\frac{d}{dt}(\nabla\phi\nabla\psi\nabla\chi)=-(\nabla\phi\nabla\psi\nabla\chi)(\nabla|u)=-\theta(\nabla\phi\nabla\psi\nabla\chi) \qquad \text{(xix)}.$$

* These functions for a perfect fluid have been investigated by Clebsch in *Crelle*, Bd. LVI. 1860, and by M. J. M. Hill, in the *Transactions of the Cambridge Philosophical Society*, Vol. XIV. 1883.

This result is obtained by Hill, without the use of the Calculus of Extension, in the paper cited. The brevity of the necessary analysis by this method is to be noted.

(5) Putting δ for the determinant $(\nabla\phi\nabla\psi\nabla\chi)$, equation (xix) can be written

$$\frac{d\delta}{dt} = -\theta\delta,$$

and it follows at once that

$$\frac{d^2\delta}{dt^2} = -(\dot\theta - \theta^2)\,\delta,\ \frac{d^3\delta}{dt^3} = -(\ddot\theta - 3\theta\dot\theta + \theta^3)\,\delta, \text{ and so on.}$$

Hence if none of the series θ, $\dot\theta$, $\ddot\theta$, and so on, are infinite, then all the successive mobile differential coefficients of δ with regard to the time are zero when δ is zero.

Hence if δ is zero at each point at any one instant, it remains zero at all subsequent times.

373. Clebsch's Transformations. (1) The curl filaments, defined by § 371, equation (xvi), move with the substance with unaltered strength. Let two systems of surfaces be drawn at any instant on which the curl lines lie. Then if these surfaces be defined at any instant by the carried functions ϕ and ψ, the intersections of the two systems at any subsequent instant will define the curl lines.

Therefore, remembering that $\nabla\phi$ and $\nabla\psi$ are vectors at each point respectively perpendicular to the surfaces $\phi = \text{constant}$, and $\psi = \text{constant}$, passing through that point, we may write $r \equiv |\nabla\phi\nabla\psi = \varpi\,|\nabla\phi\nabla\psi$, where ϖ is some function of x and t.

(2) But from equation (xvi), $\dfrac{dr}{dt} = (r\,|\nabla)\,u - \theta r.$

Now $$\frac{dr}{dt} = \dot\varpi\,|\nabla\phi\nabla\psi + \varpi\left|\frac{d}{dt}\nabla\phi\nabla\psi\right. = \dot\varpi\,|\nabla\phi\nabla\psi - \varpi\,|\nabla\,\{\overline{\nabla\phi}\,\overline{\nabla\psi}\,|u\}.$$

Also
$$\begin{aligned}(r\,|\nabla)\,u - \theta r &= \varpi\,\{(\nabla\phi\nabla\psi\nabla)\,u - |\nabla\phi\nabla\psi\,(\nabla\,|u)\}\\ &= \varpi\,|\{(\nabla\phi\nabla\psi\nabla)\,|u - \nabla\phi\nabla\psi\,(\nabla\,|u)\}\\ &= \varpi\,|\{\overline{\nabla\psi}\nabla\,(\overline{\nabla\phi}\,|u) + \nabla\overline{\nabla\phi}\,(\overline{\nabla\psi}\,|u)\}\\ &= \varpi\,|\nabla\,\{-\overline{\nabla\psi}\,(\overline{\nabla\phi}\,|u) + \overline{\nabla}\phi\,(\overline{\nabla\psi}\,|u)\}\\ &= -\varpi\,|\nabla\,\{\overline{\nabla\phi}\,\overline{\nabla\psi}\,|u\}.\end{aligned}$$

By equating these results we obtain $\dot\varpi\,|\nabla\phi\nabla\psi = 0$.

But by hypothesis the vector $|\nabla\phi\nabla\psi$ is not null. Therefore $\dot\varpi = 0$.

Hence ϖ is a carried function of the substance.

Let ϖ be replaced by the carried function χ. Thus

$$r = \chi \,|\nabla\phi\nabla\psi, \text{ and } |r = \chi\nabla\phi\nabla\psi.$$

(3) Now the solenoidal condition $(\nabla\,|r) = 0$, gives

$$\nabla\,.\,\chi\nabla\phi\nabla\psi = 0, \text{ that is } \nabla\chi\nabla\phi\nabla\psi = 0,$$

since

$$\nabla\,.\,\nabla\phi\nabla\psi = \nabla\nabla\phi\,.\,\nabla\psi - \nabla\phi\,.\,\nabla\nabla\psi = 0.$$

But $\nabla\chi\nabla\phi\nabla\psi$ is the well-known Jacobian whose vanishing is the condition that χ is a function of ϕ and ψ, where t is regarded as a constant.

Hence $$\chi = f(\phi,\ \psi,\ t).$$

But since ϕ, ψ, χ are carried functions, $\frac{d\chi}{dt} = \frac{\delta f}{\delta t} = 0$; where $\frac{\delta f}{\delta t}$ means that ϕ and ψ are regarded as constant. Hence χ is a function of ϕ and ψ only, where t is regarded as a variable. Thus $\chi = f(\phi,\ \psi)$.

(4) It is now easy to prove that the most general form for these curl filaments, which satisfy equation (xvi), is

$$r = |\nabla\phi\nabla\psi \qquad\text{(xx)}.$$

For let ϖ be a carried function of the form $f(\phi,\ \psi)$. Then $\frac{\partial\varpi}{\partial\phi}$ and $\frac{\partial\varpi}{\partial\psi}$ are carried functions of the same form. Then we have proved that the most general form for r is $\frac{\partial\varpi}{\partial\phi}\,|\nabla\phi\nabla\psi$.

But $$\nabla\varpi = \frac{\partial\varpi}{\partial\phi}\nabla\phi + \frac{\partial\varpi}{\partial\psi}\nabla\psi.$$

Hence $$\nabla\varpi\nabla\psi = \frac{\partial\varpi}{\partial\phi}\nabla\phi\nabla\psi.$$

Thus the most general form can be converted into $|\nabla\varpi\nabla\psi$, which is the form stated in equation (xx).

(5) Now $$\nabla\phi\nabla\psi = \nabla\,.\,\phi\nabla\psi = -\nabla\,.\,\psi\nabla\phi.$$

Hence from the preceding subsections of this article the most general form of the solution of equation (xvi) for the vector v, of which r is the curl, is given by

$$v = \phi\nabla\psi + \nabla\varpi \qquad\text{(xxi)},$$

where ϕ and ψ are carried functions, and ϖ is any continuous function of x and t.

(6) We can also solve for the function χ which appears in equation (xiv) in terms of ϕ, ψ and ϖ. It is to be noted that the χ of equation (xiv) is not a carried function.

Now by equation (xx)

$$|ru = \nabla\phi\nabla\psi\,|u = (u\,|\nabla\phi)\,\nabla\psi - (u\,|\nabla\psi)\,\nabla\phi = -\frac{\partial\phi}{\partial t}\nabla\psi + \frac{\partial\psi}{\partial t}\nabla\phi;$$

since [cf. § 272 (2)]

$$(u\,|\nabla\phi) = \frac{d\phi}{dt} - \frac{\partial\phi}{\partial t} = -\frac{\partial\phi}{\partial t}, \quad (u\,|\nabla\psi) = -\frac{\partial\psi}{\partial t}.$$

Also $$\frac{\partial v}{\partial t} = \frac{\partial}{\partial t}\{(\phi\nabla\psi) + \nabla\varpi\} = \phi_t\nabla\psi + \phi\nabla\psi_t + \nabla\varpi_t,$$

where ϕ_t is written for $\frac{\partial\phi}{\partial t}$, and so on.

Hence equation (xiv) can be written

$$\phi\nabla\psi_t + \nabla\varpi_t + \psi_t\nabla\phi = -\nabla\chi.$$

Therefore $$\nabla\{\chi + \phi\psi_t + \varpi_t\} = 0.$$

Now only the differential coefficients of χ appear in equation (xiv), so we may with perfect generality write

$$\chi = -\phi\psi_t - \varpi_t \text{..........................(xxii).}$$

Equations (xxi) and (xxii) are the extension of Clebsch's transformations for the velocity of a perfect fluid.

374. Flow of a Vector. (1) The flow of a vector v along any unclosed curve will be defined to be the integral

$$\int(v\,|dx),$$

where the lower limit is the starting point of the line, curved or straight, and the upper limit is the end point.

(2) If the vector v be such that its curl filaments are of constant strength, then its flow between any two points P and Q along a defined line takes by equation (xxi) the form

$$\int_P^Q \phi d\psi + \varpi_Q - \varpi_P.$$

(3) The part $\int_P^Q \phi d\psi$ is such that it is independent of the time if the same line of particles be always considered. But it does in general depend on the special line of particles chosen, and is not completely defined by the terminal particles.

The part $\varpi_Q - \varpi_P$ is completely defined by the terminal particles, but varies with the time.

(4) Suppose that the mobile differential coefficient of the flow of any vector v along any line of particles in the substance is always equal to the

flow of some vector p along the same line of particles, then p will be called the *motive vector* of the flow of v.

The definition of p is therefore $\frac{d}{dt}\int(v\,|\,dx) = \int(p\,|\,dx)$.

By attending to the derivation of equation (xiv) it is easy to see by the use of the same analysis as there employed that

$$p = \frac{\partial v}{\partial t} + |ru + \nabla \varpi \quad\text{.......................(xxiii)},$$

where ϖ is some single-valued scalar function of x and t, r is the curl of v, u is the velocity of the substance at the point x.

This equation should be compared with the equations of Electromotive Force in Clerk Maxwell's *Electricity and Magnetism*, Vol. II., Article 598.

NOTE. The present chapter is written to shew that formulæ and methods which have been developed by Hamilton and Tait for Quaternions are equally applicable to the Calculus of Extension. The pure vector formulæ have some affinity to those of the very interesting algebra developed by Prof. J. W. Gibbs, of Yale, U.S.A., and called by him Vector Analysis. Unfortunately the pamphlet called, 'Elements of Vector Analysis,' New Haven, 1881—4, in which he developed the algebra, is not published, and therefore is not generally accessible to students. The algebra is explained and used by Oliver Heaviside, *loc. cit.* p. 550; it will be noticed in its place among the Linear Algebras.

NOTE ON GRASSMANN.

H. Grassmann's *Ausdehnungslehre von* 1844 was republished by him in 1878 (Otto Wigand, Leipzig).

A note by the publisher in this edition states that the author died while the work was passing through the press. A complete edition of Grassmann's Mathematical and Physical Works (he also wrote important papers on Comparative Philology) with admirable notes is now being published under the auspices of the Royal Saxon Academy of Sciences, edited by F. Engel (Leipzig, Teubner) Band I. Theil I. 1894, Band I. Theil II. 1896; the remaining parts are not yet published (December 1897). I have not been able to make any substantial use of this admirable edition: the present work has been many years in composition and already nearly two years in the press; and the parts most closely connected with Grassmann's own work were, for the most part, the first written.

It must be distinctly understood that the present work does not pretend to exhaust the suggestions in Grassmann's two versions of the 'Ausdehnungslehre': I only deal with those parts, which I have been able to develope and to bring under one dominant idea. Thus Grassmann's important contribution to the theory of Pfaff's Equation by the use of the Calculus of Extension, given in the *Ausdehnungslehre* of 1862, is not touched upon here. It is explained in Forsyth's work, *Theory of Differential Equations*, Part I. Chapter V.

The following list of the mathematical papers of Grassmann is taken from the Royal Society Catalogue of Scientific Papers.

Theorie der Centralen, *Crelle* XXIV. 1842; and XXV. 1843.

Ueber die Wissenschaft der extensiven Grösse oder die Ausdehnungslehre, *Grunert, Archiv* VI. 1845.

Neue Theorie der Electrodynamik, *Poggend. Annal.* LXIV. 1845.

Grundzüge zu einer rein geometrischen Theorie der Curven, mit Anwendung einer rein geometrischen Analyse, *Crelle* XXXI. 1846.

Geometrische Analyse geknüpft an die von Leibnitz erfundene geometrische Characteristik, *Leipzig, Jablon. Preisschr.* (No. 1) 1847.

Ueber die Erzeugung der Curven dritter Ordnung durch gerade Linien, und über geometrische Definitionen dieser Curven, *Crelle* XXXVI. 1848.

Der allgemeine Satz über die Erzeugung aller algebraischer Curven durch Bewegung gerader Linien, *Crelle* XLII. 1851.

Die höhere Projectivität und Perspectivität in der Ebene; dargestellt durch geometrische Analyse, *Crelle* XLII. 1851.

Die höhere Projectivität in der Ebene, dargestellt durch Functionsverknüpfungen, *Crelle* XLII. 1851.

Erzeugung der Curven vierter Ordnung durch Bewegung gerader Linien, *Crelle* XLIV. 1852.

Zur Theorie der Farbenmischung, *Poggend. Annal.* LXXXIX. 1853; and *Phil. Mag.* XII. 1854.

Allgemeiner Satz über die lineale Erzeugung aller algebraischer Oberflächen, *Crelle* XLIX. 1855.

Grundsätze der stereometrischen Multiplication, *Crelle* XLIX. 1855.

Ueber die verschiedenen Arten der linealen Erzeugung algebraischer Oberflächen, *Crelle* XLIX. 1855.

Die stereometrische Gleichung zweiten Grades, und die dadurch dargestellten Oberflächen, *Crelle* XLIX. 1855.

Die stereometrischen Gleichungen dritten Grades, und die dadurch erzeugten Oberflächen, *Crelle* XLIX. 1855.

Sur les différents genres de multiplication, *Crelle* XLIX. 1855.

Die lineale Erzeugung von Curven dritter Ordnung, *Crelle* LII. 1856.

Ueber eine neue Eigenschaft der Steiner'schen Gegenpunkte des Pascal'schen Sechsecks, *Crelle* LVIII. 1861.

Bildung rationaler Dreiecke. Angenäherte Construction von π, *Archiv Math. Phys.* XLIX. 1869.

Lösung der Gleichung $x^3 + y^3 + z^3 + u^3 = 0$ in ganzen Zahlen, *Archiv Math. Phys.* XLIX. 1869.

Elementare Auflösung der allgemeinen Gleichung vierten Grades, *Archiv Math. Phys.* LI. 1870.

Zur Theorie der Curven dritter Ordnung, *Göttingen Nachrichten*, 1872.

Ueber zusammengehörige Pole und ihre Darstellung durch Producte, *Göttingen Nachrichten*, 1872.

Die neuere Algebra und die Ausdehnungslehre, *Math. Annal.* VII. 1874.

Zur Elektrodynamik, *Crelle* LXXXIII. 1877.

Die Mechanik nach den Principien der Ausdehnungslehre, *Math. Annal.* XII. 1877.

Der Ort der Hamilton'schen Quaternionen in der Ausdehnungslehre, *Math. Annal.* XII. 1877.

Verwendung der Ausdehnungslehre für die allgemeine Theorie der Polaren und den Zusammenhang algebraischer Gebilde (posthumous), *Crelle* LXXXIV. 1878.

An obituary notice will be found in the *Zeitschrift Math. Phys.* Vol. XXIII. 1878, by Prof. F. Junghans of Stettin.

The works on the Calculus of Extension by other authors deal chiefly with the application of the Calculus to Euclidean Space of three dimensions, to the Theory of Determinants, and to the Theory of Invariants and Covariants in ordinary Algebra. Thus they hardly cover the same ground as the parts of the present work, dealing with Grassmann's Calculus, except so far as all are immediately, or almost immediately, derived from Grassmann's own work. Some important and interesting works have been written, among them are:

Abriss des geometrischen Kalküls, by F. Kraft, Leipzig 1893 (Teubner).

Die Ausdehnungslehre oder die Wissenschaft von den extensiven Grössen in strenger Formel-Entwicklung, by Robert Grassmann, Stettin 1891.

System der Raumlehre, by V. Schlegel, Part I. 1872, Part II. 1875, Leipzig (Teubner).

Calcolo Geometrico, by G. Peano, 1888, Turin (Fratelli Bocca).

Introduction à la Géométrie Différentielle, suivant la Méthode de H. Grassmann, by C. Burali-Forti, 1897, Paris (Gauthier-Villars et fils).

The Directional Calculus, by E. W. Hyde, 1890, Boston (Ginn and Co.).

I did not see the above-mentioned work by C. Burali-Forti till the whole of the present volume was in print. It deals with the theory of Vectors and of Curves and Surfaces in Euclidean Space, in a similar way to that in which they are here dealt with in Chapters I., III., and IV. § 359 of Book VII. The operation of taking the Vector is explained and defined. The formulæ of multiplication in so far as they involve supplements are however pure vector formulæ: some interesting investigations are given which I should like to have included: the application to Gauss' method of curvilinear co-ordinates is also pointed out.

Buchheim's and Homersham Cox's important papers have already been mentioned [cf. notes pp. 248, 370]. I find that Buchheim has already proved [cf. Proc. Lond. Math. Soc. Vol. XVIII.] the properties of skew matrices of § 155: also the extension of the idea of Supplements in Chapter III. Book IV. is to some extent the same as his idea of taking the polar [cf. Proc. of Lond. Math. Soc. Vol. XVI.]. I had not noticed this, when writing the above chapter. He does not use the idea of 'normal intensity'; accordingly his point of view is rather different. He does not bring out the fundamental identity of his process of taking the polar with Grassmann's process of taking the supplement.

Homersham Cox has also written a paper*, *Application of Grassmann's Ausdehnungslehre to Properties of Circles*, Quarterly Journal of Mathematics, October, 1890.

There are two papers by E. Lasker, *An Essay on the Geometrical Calculus*, Proc. of the London Math. Soc. Vol. XXVIII. 1896 and 1897. The paper applies the Calculus to Euclidean space of n dimensions and to point-groups in such a space. It contains results which I should like to have used, if I had seen it in time.

Helmholtz uses Grassmann's Calculus, as far as concerns addition, in his *Handbuch der physiologischen Optik*, § 20, pp. 327 to 330 (2nd Edition).

* In this paper by a slip of the pen the words 'Outer' and 'Inner' as applied to multiplication are interchanged.

INDEX.

The references are to pages.

W.